Botany for NEET
and Other
Entrance Examinations

Related Books

Titles	Authors
— 9 Year Questions: Sorted and Solved CSIR UGC NET Life Science	Kailash Chaudhary
— Agricultural Extension A Glimpse	S. Jananai
— Agricultural Science At Your Fingertips	Jagan Singh Gora
— Basic Concepts of Plant Biotechnology With MCQ`s	Vijay Prakash
— Biology for NEET (English & Hindi) (NCERT Based Objective) (Set)	Sumant Verma
— Competitive Nematology	N.G. Ravichandran
— Comprehensive Forestry (for RFO and GPSC Examinations)	R.L. Meena
— CSIR NET: Part A	Christy Varghese
— CSIR-NET General Aptitude: A New Outlook	Christy Varghese
— Dr. Singhs NEET PG MCQ`s for Medical Students	Brijendra Singh
— Fundamentals of Agriculture *2nd Ed.* (Vol. 1-2) (Set)	R.L. Arya
— Glimpse on General Agriculture	P. Laxman Rao
— Indiras Objective Agricultural Extension	R.L. Arya
— Indiras Objective Agriculture *2nd Ed.*	R.L. Arya
— Indiras Objective Agronomy *3rd Ed*	R.L. Arya
— कृषि जगत : जेट JET एवं कृषि पर्यवेक्षक परीक्षाउपयोगी (द्वितीय संस्करण)	एम.एल. जाट
— MCQ`s in Agricultural Microbiology	D. G.Panpatte
— MCQ`s on Soil and Water Conservation Engineering	Ramesh Verma
— MCQ`s in Plant Breeding Biotechnology and Seed Science	K. Vanangamudi
— MCQ`s in Microbiology: Advanced	Balaram Mohapatra
— Model Test Paper Agriculture	R.L. Arya
— Model Test Paper of Seed Science and Technology	K.Vanangamudi
— Multiple Choice Questions (MCQ) in Food Technology	P.M. Kotecha
— Objective Agribusiness Management *3rd Ed*	S.R. Panigrahy
— Objective Animal Physiology for JRF, SRF, ARS, NET, Civil Services & other Competitive Examinations	Arpita Mohapatra
— Objective Environmental Pollution and Climate Change	M. Prasanthrajan
— Objective Food Science & Technology *3rd Ed*	Deepak Mudgil
— Objective Forestry: for All Competitive Examinations *3rd Ed*	K.T. Parthiban
— Objective Life Science MCQ`s for Life Science Examination *4th Ed*	Kailash Choudhary
— Objective Plant Pathology *2nd Ed*	Susanta Banik
— Objective Plant Physiology *3rd Ed*	P. Dwivedi
— Objective Seed Science and Technology *2nd Ed*	K. Vanangamudi
— Practice Paper and Mock Test for NEET : Botany	Khwaja Salahuddin
— Universal Forestry *2nd Ed.* (UPSC, PCS, ARS/SRF/JRF/AFO, State PG & Ph.D. Entrance Examinations and Interviews of All Forest Services)	Mohit Husain
— Universal Objective Forestry *2nd Ed.* for IFS, JRF, SRF, NET, ACF, RFO, AFO, IBPS and Other Allied Exams	Mohit Husain

Botany for NEET
and Other
Entrance Examinations

Khwaja Salahuddin
Assistant Professor of Botany
L.N. Mithila University,
Darbhanga, Bihar

Published by:

SCIENTIFIC PUBLISHERS (INDIA)
5A, New Pali Road, P.O. Box 91
Jodhpur 342 001, India
E-mail: info@scientificpub.com
Website: www.scientificpub.com

ISBN: 978-93-90749232
eISBN: 978-93-90749249

Printed in India

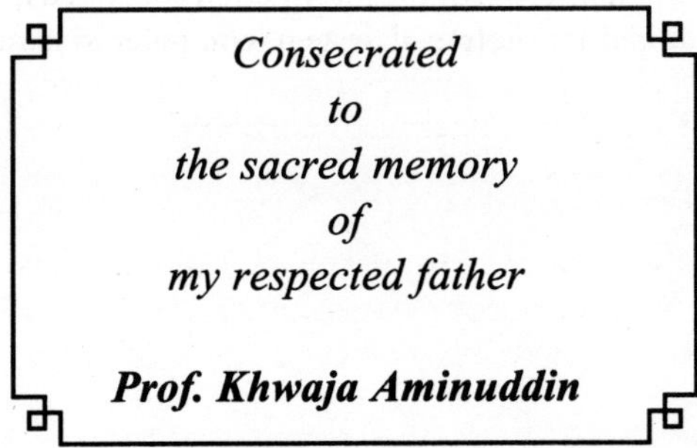

Consecrated
to
the sacred memory
of
my respected father

Prof. Khwaja Aminuddin

PREFACE

It gives me immense pleasure to write this book in response to requests from a large number of students and friends keeping the requirements of medical aspirants in mind. I have been teaching Botany for the last several years and realised the problems of students in Botany. It has taken more than three years in bringing out this book and I am sure that the book will be liked by the teachers and students. It helps students to understand the subject more precisely. In fact, the subject Biology is packed with plethora of information and there are several books on Botany in the market and each book has its own shortcomings so students are unable to purchase all. Even if students are accessible to all the books, usually may not have the time to consult them all. I have incorporated all the relevant materials from the competition point of view by avoiding unnecessary verbosity and repetitions.

I have singled out the entire content of Biology into Botany and Zoology. I have divided the entire content of this book into 24 chapters. I have designed the book according to the latest syllabus of NCERT and further I have incorporated three chapters viz. 'Biotechnology: Principles and processes', 'Biotechnology and its applications' and 'Microbes in Human Welfare' in this book. Special focus has been given on those topics which are usually tougher. The thorough and comprehensive information on the said topics followed by the limited number of most probable questions are given so students are advised to go through theories first and then after solve the multiple choice questions. Each chapter contains short answer type questions and long answer type questions in the end for plus two students.

I have included all the materials containing relevant information systematically in lucid language in readable and interesting manner without increasing the volume of book. I am convinced that the students will be able to perform well with greater confidence in the competitive medical examinations such as NEET, AIIMS and JIPMER. This book will also cater to the needs of all such students who are associated with Botany.

I have taken utmost care to make this book flawless and indispensable for the students but chance of some minor errors cannot be ruled out. I will welcome your suggestion for improvement.

I am sure that this book is my maiden venture and shall prove more useful to the students who aspire to be a doctor. I wish the medical aspirants all the best in their endeavour.

Dr. Khwaja Salahuddin

ACKNOWLEDGEMENTS

I want to extend my heartfelt thanks to all my teachers for their constant inspiration and encouragements. I am grateful to the authors of various articles and books which form the basis of this book.

I feel highly indebted to Prof. S.M. Zufair, Col. Ishrat Nabi, Hashmat Nabi, Mr. Shakeb Nabi, Mr. Azmat Nabi, Shakir Jamal, Md. Moizuddin, Sofia Nabi, Shafqat Nabi, Dr. Ahmad Ali, Dr. Ram Prasad, Ms. Sabiha Yasmin, Dr. Rajesh Bhatt, Dr. Manish Jani, Dr. Manish Visavadiya, Dr. Kamlesh Shah, Dr. Bharat Katariya, Dr. Suresh Gor, Dr. Virendra Kumar Soni, Mr. Reyaz Alam, Dr. P.K. Patel, Ms. Mohsena Ahmad,Dr. Nizamuddin Ansari, Mr. Abdus Sattar, Dr. Ashok Babriya, Md.Moizur Rahman, Mr. Amar Thakur, Dr.Dipen Ghimiray, Dr. Ram Prasad Tyagi, Mr. Firoze Ahmad, Mr. Pawan Kumar, Mr. Trilochan Mishra, Dr. Shafiqur Rahman, Dr. Khalid Anwar, Mr. Seraj Alam, Mr. Azmat Ali, Mr. Muzaffar Jamal,Dr. Dileep Kumar, Dr. Dheeraj Gautam, and Dr. Fauzia Farheen for showering extraordinary kindness and evincing keen interest in this book.

I am thankful to Dr. Anshuman Agarwal, Mr. Pankaj Mathur, Ms. Sonali Bhattacharya, Ms. Mridula Minaxi and Ms. Pratima Tiwari of Resonance eduventures Ltd. Kota; Mr. Manoj Kumar, Mr. Bharat Lal, Dr. Manish Sharma and Javed Rishi of Allen career institute and Dr. Pankaj Kumar, Mr. Gholam Rasool, Mr. Sandeep Rana,Mr. Shushant Gupta and Ms. Sushmita Sorout of Aakash institute for their sympathy and peptalk.

I can't forget the galaxy of my intellectual colleagues of erstwhile Resonance eduventures Ltd. viz. Mr. Ameer Basha, Mr. Sudhanshu Kumar Jha, Mr. Harishankar Nath, Dr.Abhishek Priyaranjan, Mr. Ankush Kumar Singhal, Mr.Anurag Kohli, Mr. Aakash Kumar, Mr. Shivshant Kumar, Mr. Rajesh Kumar, Mr. Deepak Soner, Mr.Karnkabeer Gupta, Mr. Sunil Kumar, Mr. Navin Ram, Mr. Mayank Agrawal, Mr. Sanjay Yadav, Mr. Rahul Khalkar, Mr. Neeraj Jain, Mr. Brijesh Roy, Ms. Rupa Kundu, and Mr. Anand Kumar for making several useful suggestions and further keeping pressure on me to complete the write-up on priority basis.

I submit myself to my mother and offer if at all I have everything at my disposal at her feet with most supplication and reverence. I have the privilege of her blessings day in and day out. This book is nothing but fruition of her blessings and blessings alone. I pray, may she live long.

I dedicate it to my father the late Prof. Khwaja Aminuddin who did all that he could in his life time for the well-being, prosperous living and all-round betterment of kith and kin including myself. I pray to God, may his soul rest in eternity.

I appreciate my spouse, Ms. Shabana Nabi and daughters, Sara and Simra who always supported my venture even when they were greatly inconvenienced due to my preoccupations with the manuscript.

Last but not least I would like to express my appreciation to my Publishers, Mr. Tanay Sharma and his entire team for bringing out the book in nice form.

Dr. Khwaja Salahuddin
Assistant Professor of Botany
MRM College (LNMU)
Darbhanga

CONTENTS

SYLLABUS OF NEET (NCERT)

Unit I: Diversity in Living world

What is living?; Biodiversity, Need for classification, Three domains of life, Taxonomy and systematic, Concept of species and taxonomical hierarchy, Binomial nomenclature, Tools of study of taxonomy-museums, zoos, herbaria, botanical gardens. Five kingdoms classification- Salient features and classification of Monera, Protista and Fungi into major groups. Lichens, Viruses and viroids. Salient features and classification of plants into major groups-Algae, Bryophytes, Pteridophytes, Gymnosperms and Angiosperms (three to five salient and distinguishing features and at least two examples of category), Angiosperms-classification up to class, characteristic features and examples).

Salient features and classification of animals - Nonchordates up to phyla level and chordates up to classes level (three to five salient features and at least two examples).

Unit II: Structural organisation in animals and plants

Morphology and modifications, Tissues and anatomy and functions of different parts of flowering plants- root, stem, leaf, inflorescence- cymose and racemose, flower, fruit and seed.

Animal tissues: Morphology, anatomy and functions of different systems (digestive, circulatory, respiratory, nervous and reproductive) of an insect (cockroach).

Unit III: Cell structure and function

Cell theory and cell has the basic unit of life: Structure of prokaryotic and eukaryotic cell, plant cell and animal cell, cell envelope, cell membrane, cell wall, cell organelles- structure and function, Endomembrane system- Endoplasmic reticulum, Golgi bodies, Lysosomes, Vacuoles, Mitochondria, Ribosomes, Plastids, Microbodies, Cytoskeleton, cilia, flagella and centrioles (Ultrastructure and function), nucleus-nuclear membrane, chromatin, nucleolus.

Chemical constituents of living cells

Biomolecules- structure and function of proteins, carbohydrates, lipids, nucleic acids, enzymes-types, properties, enzyme actions.

Cell Division: Cell cycle, mitosis, meiosis and their significance

Unit IV: Plant physiology

Transport in plants: Movement of water, gases and nutrients, Cell to cell transport- diffusion, facilitated diffusion, active transport, Plant-water relations-Imbibition, water potential, osmosis and plasmolysis, long distance transport of water- Absorption, apoplast, symplast, Transpiration pull, root pressure and guttation, Transpiration-opening and closing of stomata, uptake and translocation of mineral nutrients-transport of food, phloem transport, mass flow hypothesis, diffusion of gases.

Mineral nutrition: Essential minerals, macro and micronutrients and their role, deficiency symptoms, mineral toxicity, elementary idea of hydroponics as a method to study mineral nutrition, Nitrogen metabolism- Nitrogen cycle, biological nitrogen fiuxation.

Photosynthesis: Photosynthesis as a means of autotrophic nutrition, site of photosynthesis takes palce, pigments involved in photosynthesis, Photochemical and biosynthetic phases of photosynthesis, Cyclic and non-cyclic photophosphorylation, Chemiosmotic hypothesis, photorespiration, C_3 and C_4 pathways, Factors affecting photosynthesis.

Respiration: Exchange of gases, cellular respiration-Glycolysis, fermentation (anaerobic), TCA cycle and electron transport system (aerobic), Energy relations, Number of ATP molecules generated, amphibolic pathways, respiratory quotient.

Plant growth and development: seed germination, Phases of plant growth and growth rate, conditions of growth, differentiation and redifferentiation, sequence of developmental process in a plant cell, growth regulators-auxin, gibberellin, cytokinin, ethylene, ABA, seed dormancy, vernalisation, photoperiodism

Unit V: Human Physiology

Digestion and absorption: Alimentary canal and digestive glands, Role of digestive enzymes and gastrointestinal hormones, peristalsis, digestion, absorption and assimilation of proteins, carbohydrates and fats, caloric value of proteins, carbohydtrates and fats. Egestion, nutritional and digestive disorders - PEM, indigestion, constipation, vomiting, jaundice and diarrhoea.

Breathing and respiration: Respiratory organs in animals, respiratory systems in humans, mechanism of breathing and its regulation in humans-exchange of gases, transport of gases and regulation of respiration, respiratory volumes, disorders related to respiration-asthma, emphysema, occupational respiratory disorders.

Body fluids and circulation: Composition of blood , blood groups, coagulation of blood, Composition of lymph and its function, human circulatory system- structure of human heart and blood vessels, cardiac cycle, cardiac output, ECG, double circulation, regulation of cardiac activity, disorders of circulatory system- hypertension, coronary artery disease, angina pectoris, heart failure.

Excretory products and their elimination: modes of excretion- ammonotelism, ureotelism, uricotelism, Human excretory system- structure and function, urine formation, osmoregulation, regulation of kidney function- rennin-angiotensin, atrial natriuretic factor, ADH and Diabetes insipidus, Role of other organs in excretion, Disorders, Uraemia, Renal failure, Renal calculi, Nephritis, Dialysis and artificial kidney.

Neural control and coordination: Neuron and nerves, nervous system in humans- central nervous system, peripheral nervous system and visceral nervous system, generation and conduction of nerve impulse, Reflex action, Sense organs, Elementary structure and function of eye and ear.

Chemical coordination and regulation: Endocrine glands and hormones, Human endocrine system- hypothalamus, pituitary, pPineal, Thyroid, parathyroid, Adrenal, Pancreas, Gonads, mechanism of hormone action (elementary idea), Role of hormones as messengers and regulators, Hypo-and hyperactivity and related disorders (Common disorders e.g., Dwarfism, Acromegaly, Cretinism, Goiter, Exopthalmic goiter, Diabetes and disorders.

Unit VI: Reproduction

Reproduction in organisms: Reproduction, a characteristic feature of all organisms for continuation of species, modes of reproduction-asexual and sexual, Asexual reproduction, modes- Binary fission, sporulation, budding, gemmule, fragmentation, vegetative propagation in plants.

Sexual reproduction in flowering plants: Flower structure, Development of male and female gametophytes, Pollination-types, agencies and examples, Outbreeding devices, Pollen-pistil interaction, Double fertilisation, Post fertilization events, Development of endosperm and embryo, Development of seed and formation of fruit, special modes-apomixis, parthenocarpy, polyembryony, Significance of seed and fruit formation.

Human reproduction: Male and female reproductive systems, Microscopic anatomy of testis and ovary, Gametogenesis-spermatogenesis and oogenesis, Menstrual cycle, Fertilisation, Embryo development up to blastocyst formation, implantation, Pregnancy and placenta formation(elementary idea), Lactation (Elementary idea).

Reproductive Health: Need for reproductive health and prevention of sexually transmitted diseases (STD), Birth control-Need and methods, Contraception and Medical termination of pregnancy (MTP), Amniocentesis, Infertility and assisted reproductive technologies- IVF, ZIFT, GIFT (Elementary idea for general awareness).

Unit VII: Genetics and Evolution

Heredity and variation: Mendelian inheritance, deviations from Mendelism-Incomplete dominance, codominance, multiple alleles and inheritance of blood groups, Pleiotropy, Elementary idea of polygenic inheritance, Chromosomal theory of inheritance, Chromosomes and genes, Sex determination in humans, birds, honey bee; Linkage and crossing over, Sex linked inheritance- Haemophilia, colour blindness, Mendelian disorders in humans- Thalassaemia, Chromosomal disorders in humans- Down's syndrome, Turner's syndrome and Klinefelter's syndrome

Molecular basis of inheritance: Search for genetic material and DNA as genetic material, Structure of DNA and RNA, DNA packaging, DNA replication, central dogma, Transcription, genetic code, translation; Gene expression and regulation- Lac operon, Genome and human genome project, DNA finger printing

Evolution: origin of life, biological evolution and evidences for biological evolution from palaeontology, comparative anatomy, embryology and molecular evidence; Darwin's contribution, Modern synthetic theory of evolution, Mechanism of evolution- variation (Mutation and recombination) and Natural selection with examples; types of natural selection, Gene flow and genetic drift, Hardy-Weinberg's principle, adaptive radiation, Human evolution

Unit VIII: Biology and Human Welfare

Health and disease: Pathogens, parasites causing human diseases (malaria, filariasis, ascariasis, typhoid, pneumonia, common cold, amoebiasis, ring worm), Basic concepts of immunology- vaccines, cancer, HIV and AIDS, Adolescence, drug and alcohol abuse

Improvement in food production: Plant breeding, tissue culture, single cell protein, Biofortification, Apiculture and animal husbandry

Microbes in human welfare: Role of microbes in household food, food processing and industrial production, sewage treatment, energy generation, Biocontrol agents and biofertilisers

Unit IX: Biotechnology and its application:

Principles and process of biotechnology: Genetic engineering (Recombinant DNA technogy), Applications of biotechnology in health and agriculture, Human insulin and vaccine production, Gene therapy, Genetically modified organisms- Bt crops, transgenic animals, Biosafety issues- biopiracy and bioptaents

Unit X: Ecology and Environment

Organisms and environment: Habitat and niche, Population and ecological adaptations, Population interactions-mutualism, competition, predation, parasitism, Population attributes-growth, birth rate and death rate, age distribution

Ecosystem: Patterns, components, productivity and decomposition, Energy flow, Pyramids of number, biomass and energy, nutrient cycling (Carbon and phosphorus), Ecology of succession, Ecological services- carbon fixation, pollination, oxygen release.

Biodiversity and its conservation: Concepts of biodiversity, patterns of biodiversity, Importance of biodiversity, Law of biodiversity, Biodiversity conservation, Hotspots, endangered organisms, extinction, Red data book, Biosphere reserves, National parks and sanctuaries

Environmental issues: Air pollution and its control, Water pollution and its control, Agrochemicals and their effects, Solid waste management, Radioactive waste management, Greenhouse effect and global warming, Ozone depletion, Depletion, Any three case studies as success stories addressing environmental issues

1

The Living World

Biology (Gr. *Bios*=life, logos=study) is a branch of science dealing with the study of living organisms.

The term 'Biology' was coined by Jean Baptiste de Lamarck (1801).

Aristotle (384-322 B.C) is regarded as father of Biology.

Botany (Gr. *Botane*=herb) is study of plants.

Theophrastus (370-287 B.C) is regarded as Father of Botany.

Living organism is highly organised, structurally complicated, self replicating, evolving and self regulating interactive system. It is considered as an open system which has a constant flow of free energy. Even very simple single-celled organisms are remarkably complex. Inside each cell, atoms make up molecules which in turn form cell organelles. In multicellular organisms, similar cells form tissues which in turn form organ. The organs work together to form organ system.

Properties of living organisms

1. **Growth:** It is irreversible quantitative increase in number (mass) and size (weight) of cells or tissues or organism or its parts. In nonliving objects, growth is extrinsic (increase in mass from outside). But in living organisms, growth is intrinsic (increase in mass of body from inside). Growth occurs slowly as well as rapidly.

   In unicellular organisms, growth is due to increase in number of cells but in multicellular organisms, it is due to increase in number and size of cells.

   In plants, growth is indefinite (continuous growth throughout life span) but in animals, it is definite (growth reaches a certain size and then stops).

   Growth requires cell division, cell elongation and secretion of extracellular matrix or apoplasmic substances. Thus growth occurs due to synthesis of protoplasmic substances (living matter like cytoplasm and nucleus) and apoplasmic substances (nonliving materials formed by cells).

   It is not a defining property because nonliving things present in living organisms show growth.

2. **Reproduction:** It is production of new offspring (progeny) showing features more or less similar to their parents. The unicellular organisms divide to form two new cells by first duplicating their DNA but multicellular organisms produce specialised reproductive cells that will form new individuals.

   In unicellular organisms, both growth and reproduction are mutually inclusive but in multicellular organisms, both are mutually exclusive (if growth occurs then it cannot be said that reproduction will take place). It occurs by following means:

   (a) Vegetative: It produces organisms by fragmentation of vegetative parts or budding.

   (b) Asexual: It produces organisms without the use of gametes by binary fission, spores, tissue culture, grafting etc.

(c) Sexual: It produces organisms with the joining of gametes by isogamy, anisogamy and oogamy. It cannot be defining property because worker of honey bee; mule (hybrid of male ass and female horse), hinny (hybrid of female ass and male horse) and infertile human couple are sterile because they are unable to reproduce.

3. **Metabolism:** It is sum total of all chemical reactions taking place in the cells of a living organism. It is an organised process by which all living organisms release energy and utilise it for various functions such as growth, development, reproduction, movement etc. Metabolism can be demonstrated outside the body cell or in cell free system. An isolated metabolic reaction in a test tube (outside body) is neither living nor nonliving. The *in vitro* reactions are not living things but surely living reactions. It is a defining property because it occurs in all living organisms. It shows two types of mechanisms such as

   (a) Anabolism: It is synthesis of complex substance from simpler ones. It stores energy e.g., Photosynthesis.

   (b) Catabolism: It is breakdown of complex substance into simpler ones. It releases energy e.g., Respiration.

   There are three cases such as

   (a) If anabolism is equal catabolism, then there will be no growth.

   (b) If anabolism is greater than catabolism, then there will be growth.

   (c) If anabolism is lesser than catabolism, there will be degrowth.

4. **Consciousness:** It is response to external physical, chemical or biological stimuli. It is either positive if movement is towards a stimulus or negative if it is away from stimulus. All organisms from prokaryotes to eukaryotes sense their surroundings and show response. The plants respond to external factors like light, temperature, water, pollutants etc and animals perceive stimuli by their well developed sense organs (eyes, ears. taste buds).

   It is most obvious and technically complicated feature.

5. **Homeostasis:** It is the maintenance (constancy) of favourable dynamic internal environment in spite of changes in external environment. It must be stable for smooth functioning of metabolic processes. The term 'Homeostasis' was coined by Cannon (1932).It occurs at all levels from cell level to the level of ecosystem.

   The organisms that live in cold climates, have body structures (fur, feather, blubber, fat retaining energy) that help them withstand low temperatures and conserve body heat. The organisms of hot climate have methods (perspiration in humans or panting in dogs) that help them to shed excess body heat. It occurs through

   (a) Self regulatory mechanisms using genetic clock.

   (b) Feedback system in which rate of formation of product is regulated by gathering information about the amount left out at anytime. It exhibits the process of thermoregulation by two modes such as **Poikilotherm** where the energy produced during metabolic reactions is lost to environment hence their body temperature varies with that of environment and **Homoiotherm** where the energy produced during reactions is not lost but retained by hairs, feathers, subcutaneous fat hence body temperature is kept constant.

6. **Protoplasm:** It is a distinctive and fundamental material of which all living organisms are composed. It is physical basis of life. It contains water, proteins, lipids, carbohydrates and inorganic materials. It is crystallo-colloidal, slightly alkaline and highly organised material which produces energy of life. It shows Brownian movement and Tyndall effect.

7. **Cellular organisation:** All living organisms possess cell which is the basic structural and functional unit of life. The cell will form tissue which develops organs. The living organisms may be unicellular and multicellular on the basis of number of the cell. The complexity of living organism begins with the cell. It is a defining feature.
8. **Genetic continuity:** All living organisms possess genetic material (DNA) for the expression of trait from generation to generation. It is the basis of inheritance and performs so many vital functions.
9. **Adaptation:** It is the condition in which living organisms cope with environmental stresses and pressure to lead successful life in given environment. It can be morphological (structural), physiological and behavioural. The living organisms show remarkable ability to adjust to their environment. These may be (i) short term in which changes are developed in response to particular condition for short time and later disappear and (ii) long term in which changes are inheritable and occur to adjust the environment permanently. If organism fails to adjust then it will perish or become extinct.
10. **Movement:** It is natural change of position. It is general tendency of all living organisms to change their location. The animals exhibit clear movement but sessile animals and plants in general show movement of parts.
11. **Death:** It is ultimate reality of all living organisms to die. It is necessary for the population to be checked and recycling of materials. Some organisms such as Bacteria and *Amoeba* are said to be immortal because they divide by binary fission hence they do not undergo natural death.

Diversity: The planet earth has a large number of living organisms such as plants, animals, fungi etc. Around 1.7 million species (1.2 million animals and 0.5 million plants) are found at different geographical areas. Due to diversity, it is very difficult to study each and every plants and animals so these are required to be classified for further study.

Taxonomy: It is the arrangement of different types of organisms on the basis of shared characteristics. It is study of characterisation (description), identification, nomenclature and classification of living organisms. The term 'Taxonomy' was coined by AP de Candolle (1813). He composed '*Theorie elementare de la botanique*'.

Father of Taxonomy is Carolus Linnaeus.

Father of Indian Taxonomy is Henry Santapau.

Linnaeus composed (i) Philosophia Botanica in 1751 in which principles of binomial nomenclature were mentioned. (ii) Species Plantarum in 1753 in which nomenclature of 5900 plants was given and (iii) Systema Naturae (10th ed) in 1758 in which description of 4326 animals was made.

Systematics: It was coined by Linnaeus. It is process of organising taxonomic information about organisms into a logical classification. It deals with kinds and diversity of all organisms and their existing relationship. It is taxonomy plus phylogeny (evolutionary history of organisms). It includes those areas of biology concerned primarily with similarities and differences between organisms.

Biosystematics

It provides information regarding habit, habitat, characters, affinities, and diversity of organisms.

(a) It provides information of evolution.

(b) It helps in identification of newly discovered organisms and fossils.

Types of Taxonomy: Turril divided taxonomy into 3 types:

1. **Alpha taxonomy (Classical taxonomy):** It is based on morphological characters, compilation of flora and monograph. Classical taxonomists are Theophrastus, Aristotle.
2. **Beta Taxonomy (Modern or Explorative taxonomy):** It is based on morphological, anatomical, embryological, palynological, physiological, cytological and biochemical characters. It deals with identification of plants.

   It is also known as Biosystematics.
3. **Omega taxonomy (Encyclopaedic taxonomy):** It is based on all available known information about plants. It deals with evolutionary patterns.

   It is also known as Neosystematics (term coined by Julian Huxley).

New Branches of Taxonomy

(a) **Chemotaxonomy:** It is based on studies of chemical constituents such as polysaccharides, amino acids, proteins, fatty acids, nucleic acids (DNA sequences), alkaloids, pigments (betacyanin in beet root), aromatic compounds, tannins, crystals in the form of cystolith (Calcium carbonates) and raphides (Calcium oxalates).

(b) **Cytotaxonomy:** It is based on cytological studies such as number, structure and meiotic behaviour of chromosomes.

(c) **Karyotaxonomy:** It is based on studies of chromosomal bands (light and dark band).

(d) **Numerical Taxonomy (Phenetics/Taximetrics or computer aided taxonomy):** It was proposed by Adanson. It is based on numerical or statistical methods in which large number of characters are observed and recorded. It determines the similarities and dissimilarities or primitiveness and advancement of several characters at the same time. It has the following features:

(i) If content of information in taxa is increased then the classification will be better.

(ii) Every character is of equal weight in creating natural taxa.

(iii) Classification is based on percentage similarity coefficient for each operational taxonomic units (OTU).

Dendrogram: It is family tree of organisms based on numerical taxonomy.

Cladogram: It is family tree of organisms based on phylogeny (showing ancestry).

Taxonomic category: It is the position or status or level in a hierarchal classification of organisms.

Taxon: It was coined by Adolph Mayr. It is the scientific term of category. It is taxonomic group or unit of organisms of any rank in any category. It is group of genetically related individuals having distinct characters from those of other groups. It was introduced for the first time in 1956 by ICBN. All fishes or birds form taxon. Cats, mammals and animals represent taxa at different level or category.

Category	Taxon	Category	Taxon
Kingdom	Plantae	Kingdom	Animalia
Division	Angiospermae	Phylum	Chordata
Class	Dicotyledonae	Class	Mammalia
Order	Polemoniales	Order	Primata
Family	Solanaceae	Family	Hominoidea
Genus	*Solanum*	Genus	*Homo*
Species	*tuberosum*	Species	*sapiens*

Taxonomic hierarchy: It is classification of organisms in a definite but descending sequence of categories. Hence all categories constitute hierarchy.

Linnaeus firstly proposed hierarchy (Ranking system) in which 5 categories were mentioned. But he did not incorporate Division and Family.

Class-------- Order------- Genus------ Species------ Variety

Later Taxonomic Hierarchy mentions 7 obligate categories.

Kingdom---Division/Phylum---Class----Order---Family---Genus---Species

(Phylum is used in place of Division in animals.)

1. **Kingdom:** It is the highest taxonomic category. It is a group of related divisions. All plants are placed in kingdom-Plantae.
2. **Division:** It is a group of related classes. It was coined by Eichler. It mentions suffix-phyta. In animals, the Phylum is equivalent to the Division. The term 'Phylum' was coined by Cuvier.
3. **Class:** It is a group of related orders. It was coined by Linnaeus. It mentions suffix-phyceae (algae) or -opsida (plants such as Bryophytes, Pteridophytes and Gymnosperms). If subclass exists then it will mention suffix -idea for plants. The orders such as Ranales, Polemoniales and Personales have been placed in class, Dicotyledonae.
4. **Order:** It is a group of related families having few similar features. It was coined by Linnaeus. It mentions suffix-ales. If suborder exists then it mentions suffix-ineae. The families such as Solanaceae and Convolvulceae have been placed in the order, Polemoniales.
5. **Family:** It is a group of related genera having some similar or common features. It was coined by John Ray. It mentions suffix -aceae (plants) or –idea (animals). If there is sub-family then suffix will be –oideae. The three different genera such as *Solanum, Atropa* and *Nicotiana* have been placed in the family, Solanaceae.

   Tribe: It is between family and genus. It mentions suffix -eae (plants) and -ini (animals)
6. **Genus:** It is a group of closely related species having many similar or common features. It was coined by Tournefort. It occupies a special position in classification hence species cannot be named without assigning it to a genus. The plants such as Potato (*Solanum tuberosum*) and Brinjal (*Solanum melongena*) are two species but both belong to same genus-*Solanum*. Genus is of two types:
   (a) Monotypic genus: Some genera have only one species. e.g., *Ginkgo biloba*, *Homo sapiens*.
   (b) Polytypic genus: Some genera have more than one species. e.g., *Solanum* has *tuberosum* and *melongena*; *Panthera* has *leo* and *tigris*.
7. **Species:** It is basic (smallest) unit of biological classification. It was coined by John Ray. It is group of naturally interbreeding population with the ability to produce fertile offspring but reproductively isolated. It is group of closely related individuals with similar morphological, anatomical, biochemical, and cytological characters. It is of two types:
   (a) Monotypic species (Microspecies): The species without differentiation of sub-species or variety. It is the species in which variations are less but it reproduces asexually. e.g., Bacteria (*Escherichia coli*), Cyanobacteria (*Anabaena azollae*).
   (b) Polytypic species: (Macrospecies): The species in which more than one varieties or subspecies are found. It is of three types:
   (i) Biotype: The members of same species inhabit similar environment and they show some permanent genetic variations. The members cannot breed among themselves.e.g., *Brassica*

oleracea var. *botrytis* (Cauliflower), *Brassica oleracea* var. *capitata* (Cabbage) and *Brassica oleracea* var. *pristis* (Knol kohl).

(iii) Ecotype: The members of same species inhabit different environment and they show some permanent genetic variations. The members can breed themselves but due to geographical barrier they cannot interbreed or it is genetically different population. e.g., *Corvus splendens splendens* (Indian crow), *Corvus splendens insolence* (Myanmarese crow) and *Corvus splendens perotegatus* (Srilankan crow).

(iii) Ecades (Ecophenes): The members of same species show some temporary non-genetic variations due to environment or It is a form of a plant modified by its habitat but not heritable. e.g., Money plant is dark green (in shade) and yellow (in sunlight).

Features of categories

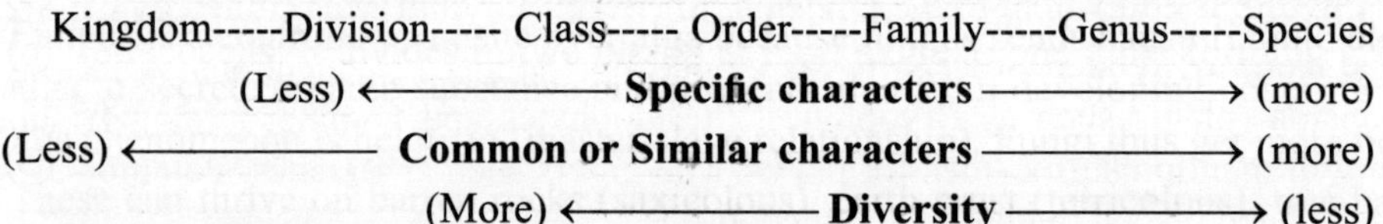

Types of species:

1. Morphospecies: The species which are based on morphological characters only.
2. Taxonomic species: The species which have definite binomial name.
3. Sympatric species: The species which live in same geographical area.
4. Allopatric species: The species which live in different geographical areas.
5. Parapatric species: The species which live in narrow overlap of two different areas.
6. Synchronic species: The species which belong to same period of time. e.g., *Archaeopteryx and Dinosaur*
7. Allochronic species: The species which belong to different period of time. e.g., Dinosaur and Man.
8. Sibling species (Cryptic species): True species which do not breed (reproductively isolated).
9. Palaeospecies: These species which are known from fossils only. e.g., Dinosaurs.
10. Neontological species: The species which are living.

Institutions for Nomenclature

1. International Code of Botanical Nomenclature (**ICBN**)
2. International Code of Zoological Nomenclature (ICZN)
3. International Code of Viral Nomenclature (ICVN)
4. International Code for Nomenclature of Bacteria (ICNB)
5. International Code of Nomenclature for Cultivated plants (ICNCP)

ICBN: It was established at 12th International Botanical congress held at Leningrad in 1975 to provide uniformity and to avoid confusion regarding nomenclature of plants. It came into existence in 1978. The latest 19th International Botanical Congress was held in Shenzen, China in July 2017. It has three types of articles- Principles, Rules and Recommendations. The most important rule is principle of priority. The principles include

(a) Botanical nomenclature is independent of zoological nomenclature.

(b) The nomenclature of taxonomic group is based on priority of publication.

Now ICBN has been changed to International code of Nomenclature (ICN) in July, 2011.

Plant names became effective from 1st May, 1753 and animal names from 1st August, 1758.

Binomial Nomenclature: It was firstly introduced by Gaspard Bauhin in his book 'Pinax' (1623) but it was finally proposed by Linnaeus (1753). There are several **rules** which are as follows:

1. The name of organism should consist of genus and species.
2. The name of genus should be written first followed by species name.
3. The first letter of name of genus should be in capital and that of species in small.
4. The name of genus and species should consist of 3 to 12 letters.
5. The botanical name should be taken from Latin and Greek languages.
6. The botanical name (genus and species) should be printed in italics. If it is handwritten then it should be underlined separately.
7. The name of discoverer of plant should be written after species in last and in abbreviation. e.g., *Mangifera indica* Linn
8. The name of family and other categories should be based on type genus.
9. The name of family and other categories should not be printed in italics.
10. If a species name is transferred or revised then the name of original worker should be written but in parenthesis. e.g., *Syzygium cumini* (L) Skeels
11. If the name of species is compound (having two words) then a hyphen should be mentioned between these words.
12. In case of two or more names, the name given first is recognised as valid name and all other names will be called synonyms.

Trinomial Nomenclature: It was proposed by Lamarck. In case of plants, it includes name of genus, species and variety. e.g., *Brassica oleracea botrytis* (cauliflower), *Brassica oleracea capitata* (cabbage) but in case of animals, it includes name of genus, species and sub-species. e.g., *Homo sapiens sapiens* (Human being), *Corvus splendens splendens* (Indian crow)

Nomenclature Types: It acts as source for naming a certain taxonomic group. The ICBN recognised following several types:

1. Holotype: It is original description of type specimen by author.
2. Isotype: It is duplicate of holotype so this is other specimen collected at same place, same time by same author.
3. Neotype: It is new type specimen when original description (holotype) is missing.
4. Paratype: It is actual description of holotype with some additional information.
5. Syntype: It is two or more type specimen when original description (holotype) is missing.
6. Lectotype: It is used as substitute when holotype is lost.

Nomenclature terms

1. Autonym: It is name with same species and subspecies. e.g., *Corvus splendens splendens*
2. Synonym: It is more than one name of taxon. e.g., *Catharanthus roseus* (=*Vinca rosea*) for periwinkle and *Malus pumila* (=*Pyrus malus*) for apple.

3. Homonym: It is same name for two different plants. e.g., *Prunus dulcis* for almond and plum.
4. Basionym: It is name on which new combination is based.
5. Tautonym: It is name with same genus and species. It is only for animals. e.g., *Bison bison, Rattus rattus, Naja naja, Gorilla gorilla, Catla catla.*

Taxonomic aids

These are meant for storing and preserving the specimens and their information.

1. **Herbarium:** It is storehouse of preserved plant specimens on sheet. It is arranged in the sequence based on accepted system of classification. Caesalpino prepared first herbarium of 768 plants. The international size of sheet is 11 ½" × 16 ½" or 29 × 41 cm. It mentions (a) Date of collection, (b) Place of collection. (c) Local (vernacular) name, (d) English name, (e) Botanical name, (f) Name of family and (g) Name of collector.

   Procedure:

   (a) Collection: Specimen should contain roots, stem, leaf, flower and fruits.
   (b) Pressing: It is done by blotting paper.
   (c) Drying: It is done to remove moisture.
   (d) Poisoning: Dried specimen is sprayed with fungicides (0.1% $HgCl_2$) to check the growth of fungus.
   (e) Mounting: Specimen is mounted on sheet except succulents.
   (f) Stitching: It is properly fixed on sheet.
   (g) Labelling: The parts of specimen were properly named.
   (h) Deposition: The specimen is now submitted.

   Examples:

   1. Herbarium of Royal Botanical Garden, Kew (England): It was established in 1841. It is the largest containing 60 lakh specimens.
   2. Herbarium of Indian Botanical Garden, Kolkata (India): It was established in 1787 containing 10 lakh specimens.
   3. Herbarium of National Botanical Research institute (NBRI), Lucknow (India): It was established in 1948 containing 40,000 specimens.

   Importance:

   (a) It helps in identification of unknown plants.
   (b) It is used to preserve specimens.
   (c) It provides information regarding flora of different places.
   (d) It is used as quick reference for alpha taxonomic studies and research.

2. **Botanical Garden:** It is large area of land where plants of different types and areas are grown.

   Examples:

   1. Royal Botanical Garden, Kew, (England)
   2. Indian Botanical Garden, Kolkata
   3. National Botanical Garden, Lucknow
   4. Lloyd Botanical Garden, Darjeeling

Importance:

(a) It helps in *ex situ* conservation of plants.

(b) It helps in identification of local flora.

(c) It helps in knowing the taxonomic affinities.

3. **Zoological Garden or Zoological Park or Zoo:** It is protected area or enclosed space where live wild animals are kept under human care.

   Examples:

   1. Kruger Zoo (South Africa): It is the largest zoo in the world.
   2. National Zoological Park, New Delhi (India).

   Importance:

   (a) It helps in *ex situ* conservation of animals.

   (b) It is used in breeding of rare animal (fauna).

   (c) It serves as recreation and education.

4. **Museum:** It is collection of preserved plants and animals in the container or jar in preservative solution. It is established to preserve algae, fungi, bryophytes (mosses), pteridophytes (ferns) and parts of gymnosperms since they cannot be kept in herbaria. Likewise different animals esp. lower ones are also preserved.

   Importance:

   (a) It is used for study and reference.

Identification kits

1. **Flora:** It is a book which provides information esp. habitat and distribution of plants of a particular area.

   Examples:

   1. Flora Indica: It was written by William Roxburgh (Father of Indian Botany).
   2. Flora of British India: It was written by Bentham and Hooker.
   3. Flora of Delhi: It was written by Maheshwari.

2. **Manual:** It is a book containing complete listing and description of plants of a particular area for identification process.

   Examples:

   1. Manual of Mangroves
   2. Manual of cultivated plants

3. **Monograph:** It is a printed material which has information of only one taxon.
4. **Catalogue:** It is alphabetical arrangement of species of a particular area describing their features.
5. **Key:** It is artificial and analytical device containing set of alternate (similar or dissimilar) characters of various plants and animals for easy and quick identification. It was developed by John Ray.

   It is based on set of contrasting characters generally in a pair called couplet and each character of couplet or statement is called lead. It is used in preparing flora and monograph. It is of two types:

   (a) Indented (Yoked) key: It is a set of choice between two or more statements or characters of species. It is generally less followed.

(b) **Bracketed (Parallel) key:** It is a set of contrasting characters in pairs. It is most popular hence more followed.

SHORT ANSWER TYPE QUESTIONS

1. Define metabolism.
2. What do you mean by homeostasis?
3. Why growth and reproduction are not defining features?
4. Distinguish between taxonomy and systematic.
5. How alpha taxonomy differs from beta taxonomy?
6. What do you mean by numerical taxonomy?
7. What is difference between dendrogram and cladogram?
8. Explain species with examples.
9. Distinguish between sympatric and allopatric species.
10. Distinguish between taxon and category.
11. Explain Linnean hierarchy.
12. Explain monotypic genus and polytypic genus.
13. Distinguish between binomial and trinomial nomenclature.
14. Define botanical garden and mention its importance.
15. Discuss the types of key.

LONG ANSWER TYPE QUESTIONS

1. Describe the different properties of living organisms.
2. What is ICBN? Discuss the different rules of binomial nomenclature.
3. What is herbarium? Describe its procedure with examples.
4. Describe the all categories of taxonomic hierarchy.
5. Explain the different identification kits.

MULTIPLE CHOICE QUESTIONS

1. The growth and reproduction are mutually inclusive in
   (a) *E. coli* and *Fucus* (b) *E. coli* and *Chlamydomonas*
   (c) *Chlamydomonas* and Moss (d) Moss and Fern
2. The most obvious and technically complicated feature of living organism is
   (a) Growth (b) Metabolism (c) Consciousness (d) Homeostasis
3. Reproduction is not exhibited by
   (a) Worker bee (b) Mule (c) Hinny (d) All of the above
4. Systematics involve
   (a) Taxonomy and ontogeny (b) Taxonomy and phylogeny
   (c) Numerical taxonomy and karyotaxonomy (d) Cytotaxonomy and karyotaxonomy
5. The term 'Neosystematics' was coined by
   (a) Linnaeus (b) de Candolle (c) Julian Huxley (d) Adanson
6. At which level of hierarchy homeostasis operate?
   (a) Cell level (b) Population level (c) Ecosystem level (d) All of above
7. The family tree of organisms on the basis of phylogeny is called
   (a) Dendrogram (b) Histogram (c) Karyogram (d) Cladogram
8. Linnean hierarchy does not include
   (a) Class (b) Order (c) Family (d) Variety
9. Which of the following categories is used in place of 'Division' in animals?
   (a) Domain (b) Phylum (c) Class (d) Variety
10. Which of the following categories is used in place of 'Subspecies' of plants?
    (a) Form (b) Section (c) Series (d) Variety
11. The suffix of 'Family' is
    (a) -ae (b) -eae (c) -ceae (d) -aceae
12. The suffix of 'Tribe' is
    (a) -ae (b) -eae (c) -ceae (d) -aceae
13. The 'Tribe' is placed between
    (a) Order and family (b) Family and Genus
    (c) Genus and Species (d) Species and Variety
14. Which is not correct with respect to taxonomy?
    (a) Numerical taxonomy is also known as phenetics
    (b) Beta taxonomy is also known as biosystematics
    (c) All birds form a taxon
    (d) Taxonomy includes phylogeny
15. The exceptions to 'Biological species concept' in natural condition are
    (a) Mule and Tigon (b) Tigon and Liger (c) Mule and Hinny (d) Hinny and Tigon

16. The specific characters are more in
   (a) Class (b) Order (c) Family (d) Genus
17. The diversity is less in
   (a) Class (b) Order (c) Family (d) Genus
18. The polytypic species is shown by the genus
   (a) Solanum (b) Brassica (c) Lilium (d) All of the above
19. Dinosaur and *Archeopteryx* belong to
   (a) Allochronic species (b) Synchronic species
   (c) Allopatric species (d) Sympatric species
20. The name of plants became effective from
   (a) 1-5-1751 (b) 1-5-1753 (c) 1-5-1758 (d) 1-5-1759
21. *Prunus dulcis* is homonym for
   (a) Plum and Almond (b) Plum and Walnut
   (c) Almond and Black berry (d) Cashew and Walnut
22. If a particular specimen was originally described by author, it is called
   (a) Isotype (b) Holotype (c) Paratype (d) Lectotype
23. The botanical name of cauliflower is
   (a) *Brassica oleracea* var. *botrytis* (b) *Brassica oleracea* var. *capitata*
   (c) *Brassica oleracea* var. *pristis* (d) *Brassica oleracea* var. *glomerata*
24. Which of the following nomenclature terms is used especially for animals?
   (a) Autonym (b) Basionym (c) Homonym (d) Tautonym
25. Which of the following processes is not used in the preparation of herbarium?
   (a) Pressing (b) Cutting (c) Drying (d) Poisoning
26. The specimen for herbaria are treated to prevent any fungal attack by
   (a) 0.1% mercuric chloride (b) 0.1% mercurous chloride
   (c) Acetic acid (d) Phenol
27. Herbarium does not mention
   (a) Local name of plant (b) Name of family of plant
   (c) Name of collector (d) Name of order of plant
28. Which of the following taxonomic kits is used for actual identification of plants?
   (a) Flora (b) Manual (c) Monograph (d) All of the above
29. Key is an artificial device coined by
   (a) Linnaeus (b) John Ray (c) Eichler (d) Julian Huxley
30. The largest herbarium is found in
   (a) USA (b) England (c) India (d) Russia
31. In India, the largest Botanical Garden is situated at
   (a) Lucknow (b) Delhi (c) Kolkata (d) Mumbai

32. The *ex situ* conservation of organisms can be done at
    (a) Herbarium and Botanical garden (b) Botanical garden and Zoological park
    (c) Museum and Zoological park (d) Botanical garden and Museum
33. Which of the following taxonomic aids serves as a quick referral system in taxonomic studies?
    (a) Botanical garden (b) Zoological park (c) Museum (d) Herbarium
34. Dinosaur and human being belong to
    (a) Sympatric species (b) Allopatric species
    (c) Synchronic species (d) Allochronic species
35. Biological epithet includes
    (a) Generic name and specific name
    (b) Specific name and variety name
    (c) Generic name, specific name and author citation
    (d) Specific name, variety name and author citation
36. Which of the following is incorrect regarding metabolism?
    (a) It involves catabolism and anabolism
    (b) It cannot be demonstrated outside the body cell
    (c) It is an organised process
    (d) It is a defining property.
37. The ICBN came into existence in the year
    (a) 1975 (b) 1978 (c) 1981 (d) 1986
38. Potato and brinjal belong to
    (a) Same genus but different species (b) Same species but different varieties
    (c) Same families but different genus (d) Different families but same genus
39. Which of the following is incorrect regarding herbarium?
    (a) It was first prepared by Caesalpino
    (b) The size of sheet is 29 × 41 cm
    (c) It mentions the local name
    (d) It cannot be reference of alpha taxonomic studies
40. The species which live in same geographical are known as
    (a) Allopatric species (b) Sympatric species
    (c) Parapatric species (d) Synchronic species
41. The phylogenetic relationship can be established by
    (a) Isoenzymes (b) Alkaloids
    (c) Alcoholic dehydrogenase (d) Morphology
42. The phylogenetic relationship can beprecisely established by
    (a) Amino acid sequences (b) DNA
    (c) rRNA (d) mRNA

43. Molecular clock of evolution could be traced on the basis of
    (a) Comparison of short arm of 16 S rRNA
    (b) DNA finger printing
    (c) Fossil study
    (d) Substitution in amino acids of polypeptides due to mutation
44. When two species are morphologically almost identical but reproductively isolated, are termed as
    (a) Taxonomic species (b) Ecotypes (c) Sibling species (d) Morphospecies
45. A group of species which are phylogenetically closer but they are lacking common ancestor. Such group is regarded as
    (a) Monophyletic (b) Polyphyletic (c) Paraphyletic (d) Sympatric
46. Different hybrid forms of same parentage are termed as
    (a) Apomicts (b) Nothomorphs (c) Race (d) Variety
47. Which of the following techniques is used among plants for genetic relationship?
    (a) Amino acid sequence (b) Serological investigation
    (c) Isoenzyme profiling (d) Chromosome morphology
48. ICBN is not applicable for
    (a) Algae (b) Bryophytes (c) Angiosperms (d) Cultivated plants
49. Besides nomenclature of plants in wild, ICBN also gives binomial names for
    (a) Fungi (b) Cultivated plants
    (c) Bacteria and fungi (d) Fungi and cultivated plants
50. ICBN was established at
    (a) 10th International Botanical congress
    (b) 11th International Botanical congress
    (c) 12th International Botanical congress
    (d) 13th International Botanical congress
51. Which of the following is not a characteristic of a living organism?
    (a) Reproduction (b) Complex chemical organisation
    (c) Adaptation (d) Differentiation from cells to tissues
52. Which of the following is capitalised in printed scientific name?
    (a) Species (b) Genus (c) Family (d) Order
53. Each category of taxonomic hierarchy refers to as a unit of
    (a) Systematics (b) Nomenclature (c) Identification (d) Classification
54. The common characteristic between tomato and potato will be maximum at the level of
    (a) Genus (b) Family (c) Order (d) Division
55. Each statement in the key is called
    (a) Lead (b) Manual (c) Catalogue (d) Monograph
56. The true species are
    (a) Interbreeding (b) Sharing the same niche
    (c) Feeding on the same food (d) Reproductively isolated

57. Which of the following is 'Domain' taxon?
    (a) Bacteria (b) Eubacteria (c) Archaebacteria (d) Protista
58. Which of the following types of honey bee is an exception to reproduction?
    (a) Worker (b) Queen (c) Drone (d) All of the above
59. Which of the following is incorrect about protoplasm?
    (a) It is crystallo-colloid (d) It is slightly alkaline
    (c) It shows Brownian movement (d) It includes cell wall
60. The diversity is maximum in
    (a) Class (b) Order (c) Family (d) Genus

ANSWERS

1. b	2. c	3. d	4. b	5. c	6. d	7. d	8. c	9. b	10. d
11. d	12. b	13. b	14. d	15. c	16. d	17. d	18. b	19 .b	20. b
21. a	22. b	23. a	24. d	25. b	26. a	27. d	28. b	29. b	30. b
31. c	32. b	33. d	34. d	35. c	36. b	37. b	38. a	39. d	40. b
41. a	42. b	43. d	44. c	45. b	46. b	47. a	48. a	49. c	50. c
51. d	52. b	53. d	54. b	55. a	56. d	57. a	58. a	59. d	60. a

2

Biological Classification

Biological classification

Biological classification is study of arranging organisms into groups on the basis of their similarities and dissimilarities. It is very important because

(a) It helps in easy identification of organisms.

(b) It gives an idea of diversity of living organisms.

(c) It gives an idea of origin of organism and their interrelationship

(d) It gives an order of evolution of different types of organisms.

Several classifications have been proposed which are as follows:

1. **Aristotle** proposed first scientific classification based on a few external characters. He classified plants into Herbs, Shrubs and Trees on the basis of morphological characters (growing habit). He also classified animals on the basis of availability of RBCs into **Anaima** (animals without red blood. e.g., Invertebrates) and **Enaima** (animals having red blood. e.g., Vertebrates).

2. **Two kingdom system**: It was proposed by Linnaeus (1758). He divided living organisms into two kingdoms on the basis of (a) Presence or absence of cell wall, (b) Mode of nutrition, (c) Locomotion and (d) Response to external stimuli

Plantae	Animalia

Demerits:

(a) The organisms are very much inconsistent.

(b) It failed to place a few dual organisms such as *Chlamydomonas, Euglena*, Slime moulds, *Rhizopus*, and Mushroom which showed their disputed status.

(c) The position of fungi is not clear.

(d) It failed to distinguish between unicellular and multicellular organisms.

(e) It failed to distinguish between prokaryotic and eukaryotic organisms.

(f) It failed to distinguish between algae (photosynthetic) and fungi (non-photosynthetic).

3. **Three Kingdom system**: It was proposed by Haeckel (1866). He included Protista for the first time. He divided living organisms into three kingdoms:

Protista	Plantae	Animalia

Demerits:

(a) Bacteria, Cyanobacteria, Slime moulds, Protozoans, Algae and Fungi etc were placed in Protista in spite of dissimilarities between these organisms.

4. **Four Kingdom System**: It was proposed by Copeland (1956). He included Monera for the first time. He divided living organisms into four kingdoms:

Monera	Protista	Plantae	Animalia

Demerits: (a) Slime moulds, protozoans, algae and fungi were included in Protista.

5. **Five Kingdom System**: It was proposed by Whittaker (1969). He included Fungi for the first time. He divided living organisms into five kingdoms on the basis of following features:

Features	Monera	Protista	Fungi	Plantae	Animalia
Body organisation	Unicellular	unicellular	Multicellular	Multicellular	Multicellular
Cell structure	Prokaryotic	Eukaryotic	Eukaryotic	Eukaryotic	Eukaryotic
Presence or absence of cell wall	Present	Mostly present	Present	Present	Absent
Composition of cell wall	Peptidoglycan	Cellulose	Chitin	Cellulose	Absent
Mode of nutrition	Mostly heterotrophic	Autotrophic and Heterotrophic	Heterotrophic (Absorptive)	Autotrophic	Heterotrophic (Ingestive)
Ecological Life style	Mostly decomposer	Producers, Consumers Decomposers	Decomposers	Producers	Consumers

Merits:

(a) It is natural and more scientific

(b) It follows phylogenetic system.

(c) It gives more emphasis on cellular organisation and mode of nutrition.

(d) Fungi have been placed separately due to their heterotrophic mode of nutrition.

(e) The disputed organisms such as *Euglena* and slime moulds have been placed in Protista.

(f) Plantae have been more homogeneous with the exclusion of fungi and some unicellular algae.

(g) Animalia has become more homogeneous with the exclusion of Protozoans.

Demerits:

(a) Viruses have not been placed properly.

(b) Archaebacteria and bacteria have not been separated.

(c) Archaebacteria and *Mycoplasma* have been placed in kingdom Monera.

(d) Monera and Protista do not show monophyletic origin (developing from single ancestor).

(e) Slime moulds showing fungal features have been placed in Protista.

(f) Protista is an artificial group and diverse.

(g) The unicellular algae such as *Volvox* and *Chlamydomonas* have been placed in kingdom Plantae.

(h) Algae have been placed in three kingdoms such as Monera, Protista and Plantae.

6. **Six kingdom system**: It was proposed by Carl Woese (1990). He divided all organisms into three Domains on the basis of 16S ribosomal RNA.

Domains		**Kingdoms**
1. Archaea	———	1. Archaebacteria
2. Bacteria	———	2. Eubacteria

3. Eukarya ——— 3. Protista
——— 4. Fungi
——— 5. Plantae
——— 6. Animalia

He proposed that Archaea, Bacteria and Eukarya each developed separately from an ancestor with less developed genetic machinery called progenote. The ribosomal RNA unlike genome, has changed more slowly hence there was slow evolution of rRNA. So it is used for comparing the ancestry of different strains of bacteria. The sequencing of 23S rRNA provided more information for classification because the evolutionary changes in rRNA occurred in different segments at different times. The r-RNA has been referred to as a molecular clock because of independent rate of changes in sequences.

Thus he separated Archaebacteria from Eubacteria on the basis of absence of peptidoglycan in cell wall and occurrence of branched chain lipid in the membrane of archaebacteria.

Archaebacteria

These are ancient bacteria present in extreme or hostile condition of environment which prevailed on primitive earth. These appeared to be the oldest living fossils. These are free living, non-motile, and Gram negative. These are heterogeneous group of prokaryotes which are different from bacteria. These are of mainly three types:

1. **Methanogens:** e.g., *Methanococcus, Methanobacterium, Methanosarcina.*
   (a) These are present in marshy areas, water logged field, wetland, gobar gas plant, lake mud, marine sediment, anaerobic sludge digester, intestinal tract of animals and human, and rumen of cattle.
   (b) These are rounded in shape (cocci form).
   (c) These are chemoautotrophs and obligate anaerobes.
   (d) These convert CO_2, formic acid and acetic acid (fermentation products of other anaerobes) to methane.
   (e) These help in digestion of cellulose in ruminants.
2. **Halophiles:** e.g., *Halococcus, Halobacterium.*
   (a) These are present in sea, salted lake, salted fish, salted hides etc.
   (b) These are rounded in shape (cocci form).
   (c) These are chemoheterotrophs and facultative anaerobes.
   (d) These tolerate high salt concentration due to the presence of special lipids in membrane and mucilage covering.
   (e) They form purplish pigmented membrane (made up of bacteriorhodopsin) which protects from strong sunlight.
   (f) They utilise light energy to run their metabolism and produce ATP but they cannot utilise this ATP for the synthesis of food.
3. **Thermoacidophiles: e.g.,** *Pyrococcus, Thermoplasma, Sulpholobus*
   (a) These are present in hot sulphur spring, thermal vent.
   (b) These grow at high temperature (80^0C) and acidity (pH=2).
   (c) These are chemoautotrophs facultative anaerobes.

(d) These tolerate high temperature due to the presence of homopolar bonds in their protein.
(e) These convert sulphur to sulphur dioxide under aerobic condition and further oxidise to form sulphuric acid which makes environment acidic.
(f) These convert sulphur to hydrogen sulphide under anaerobic condition.

Archaebacteria	Bacteria
These are extremophiles.	These are mesophiles
Their cell wall is not made up of peptidoglycan. Instead, protein and non-cellulosic polysaccharides are present.	Their cell wall is made up of peptidoglycan.
Their membrane lipids have branched chain hydrocarbons linked to glycerol by ether linkage.	Their membrane lipid has straight chain hydrocarbons linked to glycerol by ester linkage.
They have introns.	They do not have introns.
They have several RNA polymerase enzymes	They have only one RNA polymerase enzyme.
They have an initial amino acid, methionine, for protein synthesis.	They have formylated methionine for protein synthesis.
They are not sensitive to streptomycin and chloramphenicol.	They are sensitive to streptomycin and chloramphenicol.

Bacteria

Bacteria are the simplest, most primitive and microscopic organisms.

Bacterium was discovered by Antonie Van Leeuwenhoek (1675) from rain water using compound microscope of his own design and called it 'Wild animalcule'.

The term 'Bacterium' was coined by Ehrenberg (1838).

Smallest bacterium	*Mycoplasma gallisepticum* (0.2μm). It is equal to the largest virus.
Largest bacterium	*Thiomargarita namibiensis* (100 μm). It is visible to unaided eye

Father of Microbiology (study of microbes) = Louis Pasteur

Father of Bacteriology (study of bacteria) = Robert Koch

Contributions of Pasteur:

(a) He proposed germ theory of disease.
(b) He discovered that microbes can cause fermentation
(c) He developed vaccine for Rabies (Immunisation technique).
(d) He disproved the theory of spontaneous generation.
(e) He discovered the existence of life in the absence of oxygen.
(f) He developed Pasteurisation technique.

Contributions of Robert Koch:

(a) He proved germ theory of disease.
(b) He isolated *Bacillus anthracis, Mycobacterium tuberculosis* and *Vibrio cholerae.*
(c) He cultured bacteria for the first time.
(d) He developed staining method for bacteria.
(e) He proposed **Koch's postulates** to establish a causal relationship between a microbe and a disease. His postulates are as follows:

(a) The suspected microorganism must always be found in diseased individual but never in healthy one.
(b) The microorganism must be isolated in pure culture on a nutrient medium.
(c) The same disease must develop when the isolated microorganism is inoculated into a healthy individual.
(d) The same organism must be reisolated from the experimentally infected host.

Features of Bacteria:

(a) These are found in every habitat (ubiquitous) but mostly in soil (one gram of soil contains 40 million bacterial cells and 1 ml. of water contains one million bacterial cells).
(b) These show simple structure but their behaviour is complex.
(c) These are prokaryotic and unicellular but some show colonial or filamentous structure. It is one tenth size of eukaryotic cell.
(d) The cell wall is made up of peptidoglycan or murein or mucopeptide.
(e) The cell has single envelope type of organisation.
(f) The flagella are single stranded and made up of flagellin protein.
(g) The cell lacks membrane bound cell organelles such as mitochondria, chloroplast, endoplasmic reticulum, Golgi bodies etc.
(h) The respiratory enzymes are present along the infolding of plasma membrane called mesosome.
(i) The ribosome is 70S and found freely in cytoplasm.
(j) It does not show cyclosis or cytoplasmic streaming.
(k) Its DNA is single (haploid), circular, double stranded, and naked (without histone proteins).
(l) It shows most extensive metabolic diversity.
(m) Its mode of nutrition is autotrophic as well as heterotrophic.
(n) Its mode of respiration is aerobic as well as anaerobic.
(o) It undergoes asexual reproduction by transverse binary fission, or endospore or conidia.
(p) It does not have mitosis and meiosis.
(q) Its cell division is amitotic type.
(r) It lacks sex organs hence there is no formation of gametes.
(s) It does not undergo sexual reproduction.

Shapes of Bacteria:

1. **Bacillus (Pl. Bacilli):** It is rod-shaped. It is most common.
   (a) Monobacillus: It is single rod like cell. e.g., *Rhodospirillum*
   (b) Diplobacillus: It divides in one plane and remains paired. e.g., *E. coli*
   (c) Streptobacillus: It divides several times in one plane forming several bacteria attached in chain. e.g., *Bacillus anthracis*
   (d) Trichome: Cells of chains have large area of contact with each other. e.g., *Beggiatoa*
   (e) Palisade like: Cells are lined side by side and at angles to one another. e.g., *Corynebacterium*
   (f) Hyphae: Cells are arranged in long branched filaments. e.g., *Streptomyces*

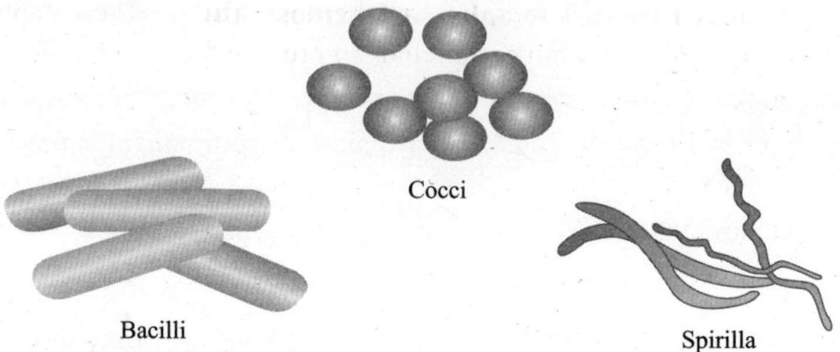

2. **Coccus (Pl. Cocci):** It is spherical.
   (a) Micrococcus (Monococcus): It is single. e.g., *Micrococcus flavus*
   (b) Diplococcus: It divides in one plane and remains paired. e.g., *Diplococcus pneumoniae*
   (c) Tetracoccus: It divides in two planes at right angle to one other forming group of four cells. e.g., *Tetracoccus cechii*
   (d) Streptococcus: It divides several times in one plane forming several bacteria attached in chain. e.g., *Streptococcus lactis*
   (e) Staphylococcus: It divides several times irregularly and all are attached together just like bunch of grapes. e.g., *Staphylococcus aureus*
   (f) Sarcinae: It divides in three planes at right angle to one another forming cubical packet of 8, 16, 32. e, g., *Sarcina*
3. **Spirillum (Pl. Spirilla)**: It is thick, rigid spiral shaped. It has several twists with several turns. e.g., *Heliobacter pylori, Spirillum* .
4. **Vibrion (Pl. Vibrio):** It is comma shaped or slightly curved. It shows less than one complete twist or turn. e.g., *Vibrio cholerae*
5. **Spirochaetes**: It is thin, highly flexible and tightly curved e.g., *Treponema pallidum.*
6. **Pleomorphic**: It has different shape and form. e.g., *Rhizobium, Arthrobacter, Azotobacter*

Mode of nutrition:

1. **Autotrophs**: They utilise CO_2 as carbon source for synthesis of food.
   (a) Photoautotrophs (Photolithotrophs) or Photosynthetic: These absorb and utilise light energy to prepare organic food from CO_2.
   Carbon source=CO_2
   Energy source=light
   Electron donor=$H_2S/Na_2S_2O_3$

   $$6CO_2 + 12H_2S \xrightarrow{\text{Light, bacterioviridin}} C_6H_{12}O_6 + 6H_2O + 12S$$
   e.g., *Chlorobium* (Green sulphur bacteria)

   $$6CO_2 + 3Na_2S_2O_3 + 15H_2O \xrightarrow{\text{Light, bacteriochlorophyll}} C_6H_{12}O_6 + 6NaHSO_4 + 6H_2O$$
   e.g., *Chromatium* (Purple sulphur bacteria)

(b) **Chemoautotrophs (Chemolithotrophs) or Chemosynthetic:** These capture energy released by oxidation of inorganic compounds to prepare food from CO_2. These bacteria help in recycling of nutrients.

Carbon source: CO_2

Energy source: Oxidation of inorganic compounds (NH_3, NO_2^-, NO_3^-, Fe^{+3}, H_2, H_2S)

Electron donor: Inorganic compound

$2NH_3 + 3O_2 \longrightarrow 2NO_2^- + 2H_2O + 2H^+ + \text{Energy (78kcal)}$

e.g., *Nitrosococcus, Nitrosomonas*

$2NO_2^- + O_2 \longrightarrow 2NO_3^- + \text{Energy (17 kcal)}$

e.g., *Nitrococcus, Nitrobacter*

$4FeCO_3 + 6H_2O + O_2 \longrightarrow 4Fe(OH)_3 + 4CO_2 + \text{Energy}$

e.g., *Ferrobacillus* (Iron bacteria)

$2H_2 + O_2 \longrightarrow 2H_2O + \text{Energy.}$

e.g., *Hydrogenomonas* (Hydrogen bacteria)

$2H_2S + O_2 \longrightarrow 2H_2O + 2S + \text{Energy}$

e.g., *Beggiatoa* (Sulphur bacteria)

2. **Heterotrophs:** These require organic compound as carbon source.

(a) **Photoheterotrophs (Photoorganotrophs)**:

Carbon source: Organic compound

Energy source: Light

Electron donor: Organic compound (alcohol/ acid)

e.g., *Chloronema* , *Chloroflexus* (Green non-sulphur bacteria)

Rhodopseudomonas, Rhodospirillum (Purple non-sulphur bacteria)

(b) **Chemoheterotrophs (Chemoorganotrophs):**

Carbon source: Organic compound

Energy source: Oxidation-reduction of organic compound

Electron donor: Organic compound

These are of three types:

(a) **Parasites:** These obtain nutrition for growth from living plants and animals. These may be **Obligate parasite** (These organisms survive and grow only in living cells) and **Facultative parasites** (These are basically saprotrophs but become parasitic under certain condition. e.g., *Bacillus, E. coli, Salmonella, Pseudomonas, Agrobacterium, Xanthomonas*

(b) **Saprotrophs:** These obtain nutrition from dead organic matter by breaking down complex organic compound into simpler forms with the secretion of extracellular enzymes. e.g., *Bacillus ramosus, Bacillus mycoides*. Majority of bacteria are aerobic saprotrophs.

(c) **Symbionts:** These obtain nutrition from host plants and provide nitrates to plants by establishing symbiotic relationship. These produce humus from animals and plant remains and cause decay of materials and useful to man. e.g., *Rhizobium*

Terms:

1. **Fermentation:** It is anaerobic breakdown of carbohydrate
2. **Putrefaction:** It is anaerobic breakdown of protein
3. **Decay:** It is aerobic breakdown of organic compound

Bacterial photosynthesis	Green plant photosynthesis
There is one photochemical reaction which involves only one pigment system.	There are two photochemical reactions which involve two pigment systems.
It occurs at wavelength longer than 700 nm.	It occurs at wavelength in a range of 400-700 nm.
The reaction centre is B-890.	The reaction centres are P-680 and P-700.
It involves only cyclic photophosphorylation.	It involves both non-cyclic and cyclic photophosphorylation.
The electron donor is H_2S.	The electron donor is H_2O.
Evolution of O_2 does not occur.	Evolution of O_2 occurs.
The reductant is NADH + H^+	The reductant is NADPH + H^+
Emerson's enhancement effect is not seen.	Emerson's enhancement effect is seen

Mode of respiration: Bacteria show the following types of respiration

1. **Obligate aerobe:** These grow only in the presence of oxygen. e.g., *Bacillus subtilis*, *Azotobacter*, *Vibrio*
2. **Obligate anaerobes**: These grow only in the absence of oxygen. e.g., *Clostridium*
3. **Facultative aerobes**: These are basically anaerobes but change to aerobic mode if oxygen is available. e.g., *Chlorobium, Chloroflexa*
4. **Facultative anaerobes**: These are basically aerobes but change to anaerobic mode if oxygen is unavailable. e.g., *Escherichia coli*

Mode of reproduction:

1. **Asexual reproduction**: It takes place by following methods:
   (a) **Transverse binary fission**: It is most common method which occurs during favourable condition of temperature, moisture and nutrients. It involves the following steps:
      (i) **Replication of DNA:** The genetic material of bacteria (nucleoid) replicates (divides) amitotically forming two circular nucleoids after undergoing theta mode in bidirectional manner.
      (ii) **Partitioning or separation of DNA**: The nucleoid is attached to mesosome which starts dividing because of the synthesis of membrane between nucleoid-mesosome attachment sites. Then each mesosome is pushed to the middle of daughter cell. Finally two nucleoids get properly partitioned into two daughter cells.
      (iii) **Cross-wall formation**: The transverse septum is formed due to invagination of plasma membrane and simultaneously cross wall is formed and grows centripetally forming two equal sized organisms.
   (b) **Budding:** It occurs in *Rhodopseudomonas*
   (c) **Conidia:** It occurs on *Streptomyces*
   (d) **Endospore:** It occurs in *Bacillus, Clostridium, Thermaoactinomyces* (mostly in Gram positive bacteria).It occurs in unfavourable condition (exhaustion of nutrients). It is thick-walled,

highly resistant body (tolerates very high temperature) formed inside the cell (endogenous). It is single formed per cell. It contains a central core (prespore) made up of DNA and spore cytoplasm. Its formation involves the following steps:

(i) Transverse septum (but no transverse wall) is formed which separates central core (prespore) from rest of cell content.

(ii) Forespore is formed when prespore develops membrane outside it.

(iii) Cortex is formed and deposited between two membranes surrounding forespore.

(iv) Spore coat is formed over cortex. Spore now becomes refractile.

(v) Ca^{+2} and dipicolinic acid are synthesised to provide heat resistance to endospore.

(vi) Mature endospore is released by autolysis of parent cell.

(vii) Endospore germinates to form vegetative cell on the onset of favourable condition when spore coat bursts by imbibition of water.

Since only one endospore is formed per cell, then endospore is a means to ensure survival (perennation).

2. **Sexual reproduction (Gene recombination):** It does not occur in bacteria because there is no gamete formation and fertilisation but exchange of genetic material occurs hence there is gene recombination which occurs in the following three process:

   (a) Transformation: It is the transfer of DNA from donor cell to recipient cell when both cells are not in contact. It was discovered by Griffith (1928) in *Diplococcus pneumoniae.*

   (b) Conjugation: It is transfer of DNA from donor cell to recipient cell during a period of contact. It was discovered by Joshua Lederberg and Edward Tatum (1946) in K_{12} strain of *E.coli.*

   (c) Transduction: It is the transfer of DNA from donor cell to recipient cell by means of a temperate λ-bacteriophage. It was discovered by Zinder and Lederberg (1952) in *Salmonella typhimurium.*

Mycoplasma

It comes under class Mollicutes. It was discovered by Nocard and Roux (1898) from pleural fluid of cattle. It is the smallest and simplest microorganism. It was earlier known as pleura Pneumonia like Organism (PPLO).

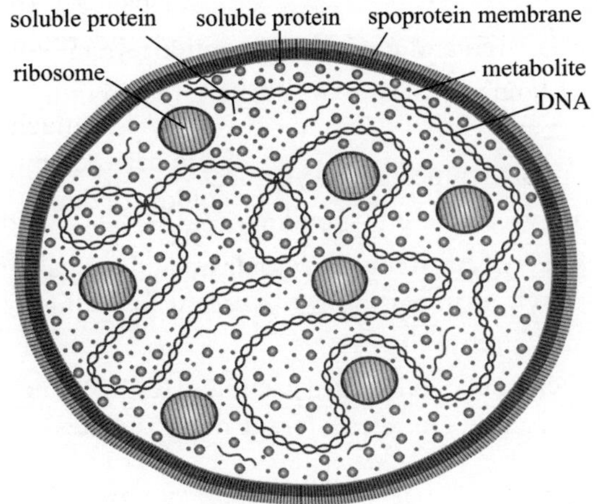

Fig: ***Mycoplasma***

Features:

(a) It is spherical, non-motile, Gram negative, facultative anaerobe, chemoheterotroph (parasitic) and highly pleomorphic (of different shape hence called Joker of Plant kingdom).

(b) It does not have cell wall.

(c) It is filterable through bacterial filter(of size 450 nm)

(d) It grows artificially in culture and forms fried egg shaped colonies

(e) It is resistant to penicillin because it attacks on cell wall which is absent in *Mycoplasma.*

(f) It is sensitive to Erythromycin, Streptomycin and Tetracycline.

(g) It has linear, double-stranded, and coiled DNA where G+C content is low.

(h) It causes serious disease in plants such as little leaf of Brinjal, Sandal spike and Witch's broom of potato, groundnut, strawberry etc.

(i) It causes Pleuropneumonia in animals (by *Mycoplasma pneumoniae*), Urethritis in human beings (by *Mycoplasma agentalum*) and sterility in human beings (by *Mycoplasma hominis, Spiroplasma*).

Cyanobacteria:

The term 'Cyanobacteria' was coined by ICBN (1978). It is also known as Blue Green algae or Cyanophyceae. It was originated in Precambrian period.

Features:

(a) These are ubiquitous (mostly fresh water and some are terrestrial and marine).

(b) These may be unicellular (e.g., *Chrococcus, Chamaesiphon*), multicellular colonial (e.g., *Microcystis*), filamentous (e.g., *Anabaena, Nostoc, Oscillatoria, Spirulina*) and heterotrichous (e.g., *Scytonema, Tolypothrix*).

(c) It is the largest photosynthetic bacterium.

(d) Cell is extremely simple, larger than bacteria and Gram negative.

(e) It is enclosed in a layer of thick, dense and undulating gelatinous sheath which is made up of pectin. It protects cell from desiccation and holds the cells in colony together. Thus it is also known as myxophyceae.

(f) Its cell wall is rigid, four layered and made up of peptidoglycan.

(g) Its protoplast is simple without any plastid. It is differentiated into peripheral dense chromoplasm and central clear centroplasm.

(h) It is photoautotroph and photosynthesis is oxygenic.

(i) Its reserve food is cyanophycean starch (showing negative test to iodine).

(j) Its photosynthetic thylakoids occur freely in cytoplasm and are called chromatophore. It contains pigments such as chlorophyll-a (imparting green colour to thallus), β-carotene, xanthophyll and phycocyanin (giving blue colour to thallus).

(k) It contains cytoplasmic inclusions such as gas vacuoles (for buoyancy), α-granules (reserve food), β-granules (lipid droplet), volutin granules (storage of phosphates), polyhedral bodies (containing DNA and RNA).

(l) It shows gliding movement.

(m) Cyanobacteria show Gaidukov phenomenon or chromatic adaptation where colour changes depending upon wavelength of light received.

(n) These show Nitrogen fixing ability.

(o) These form water blooms in polluted water bodies causing release of neurotoxins which make water undrinkable.

(p) These play a role in recycling of nutrients like N, P, S, Fe etc.

(q) These undergo vegetative reproduction by fragmentation of hormogonia (part of filament).

(r) These undergo asexual reproduction by

(i) Akinete: It is thick-walled, rounded, large sized, pale yellow. It lacks pigments. It tolerates desiccation and high temperature hence it is perennating structure. It germinates to form new individual after a period of rest under unfavourable condition.

(ii) Heterocyst: It is clear, thick-walled, two layered, barrel-shaped and pale yellow structure. It is terminal as in *Gloeotricha* or intercalary as in *Nostoc, Scytonema* and lateral as in *Nostochopsis*. It is absent in *Oscillatoria*. It obtains nutrition from adjacent vegetative cell. It retains PS-I activity but lacks PS-II and RuBP carboxylase hence it neither fixes CO_2 nor releases O_2 in light. It takes part in N_2-fixation and contributes to fixed N_2 in the form of amino acid (Glutamine) because heterocyst provides anaerobic condition to cyanobacteria.

(s) These do not show sexual reproduction due to lack of sex organs.

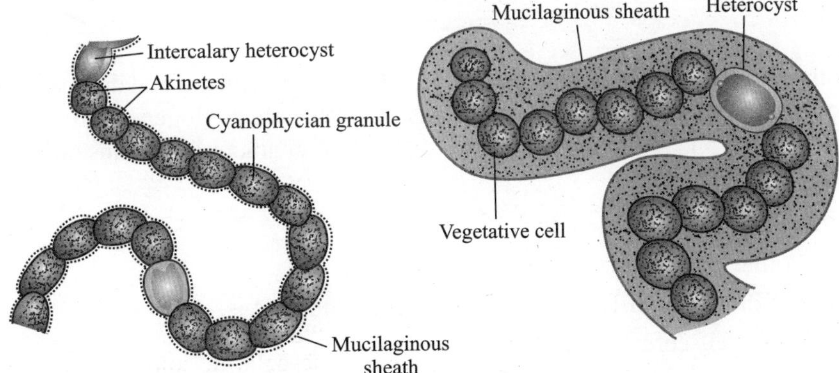

Fig: ***Akinete and Heterocyst***

Classification:

Tribe I: Coccogoneae: These are unicellular and non filamentous. These undergo reproduction by fission.

Order 1. Chroococcales

Order 2. Chamaesiphonales

Tribe II: Hormogoneae: These are filamentous. These undergo reproduction by hormogonia.

Order 3. Oscillatoriales

Order 4. Nostocales

Order 5. Scytonemales

Order 6. Stigonematales

Order 7. Rivulariales

Bacteria	Cyanobacteria
Cell is smaller in size.	Cell is larger in size.
Cell wall is one to two layered.	Cell wall is four layered.
Flagella are present.	Flagella are absent.
These are mostly anaerobes.	These are aerobes.
These are mostly heterotrophs (non-photosynthetic).	These are autotrophs (photosynthetic).
These have glycogen as reserve food.	These have cyanophycean starch as reserve food.
These take up H_2S as electron donor and release Sulphur.	These take up H_2O as electron donor and release O_2.
These do not contain heterocyst and akinetes.	These contain heterocyst and akinetes.

Economic importance:

(a) *Spirulina* is rich in protein (70%) and is eaten.
(b) *Nostoc commune* is eaten (edible).
(c) *Lyngbia* is source of antibiotic.
(d) It increases soil fertility by fixing nitrogen (biofertiliser) but in rice field, *Aulosira* is used.
(e) It helps in reclamation of alkaline (usar) soil into fertile land. (It is grown in alkaline soil, causing lowering of pH and enhancing N, P and organic matter).

Protista

The term 'Protista' was coined by Ernst Haeckel (1866). It is an artificial and diverse group. The protists are microscopic, unicellular and eukaryotic organisms. These are phylogenetically or evolutionary link between fungi, Plantae and Animalia. These are difficult to characterise because of great diversity.

General features:

(a) These are mostly aquatic (plankton).
(b) These are solitary or colonial unicellular organisms.
(c) These do not show any differentiation into tissues and organs.
(d) These have no well-defined boundaries.
(e) These have membrane bound cell organelles.
(f) These show 9+2 arrangement of cilia and flagella; if present.
(g) These show cytoplasmic streaming or cyclosis.
(h) These are autotrophic as well as heterotrophic.
(i) These are aerobic.
(j) These show asexual reproduction by binary fission (cell division).
(k) These show sexual reproduction by isogamy, anisogamy and oogamy.
(l) These do not show the presence of embryo.

1. **Diatoms (Desmids):** Division: Chrysophyta
   Class: Bacillariophyceae (Golden Brown Algae)
   (a) These are known as jewels of plant kingdom.
   (b) These are freshwater as well as marine.

(c) These are passively floating hence non-flagellate.

(d) These are of Pennate type (bilaterally symmetrical or rectangular) or Centric type (radially symmetrical or circular).

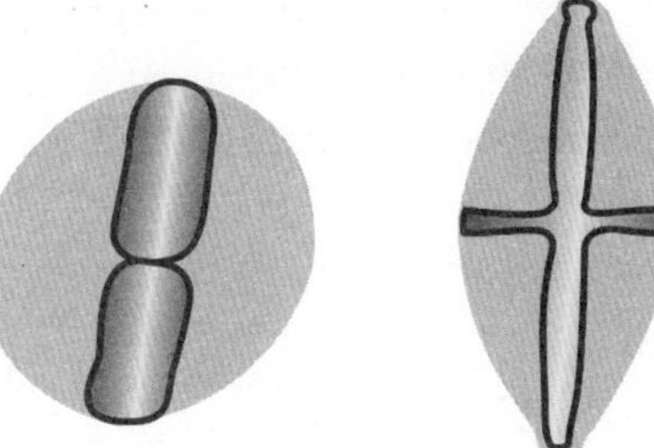

Fig: ***Diatoms***

(e) Body is made up of two thin overlapping shells (halves)-outer and larger epitheca and younger and smaller hypotheca. These shell look like soapbox.

(f) Cell wall is made up of cellulose and silica.

(g) Cell has several discoid or two plate like chromatophore.

(h) These are photosynthetic due to presence of chlorophyll a and -c.

(i) The reserve food is leucosin or chrysolaminarin (carbohydrate) and oil.

(j) These are chief producers of ocean.

(k) Asexual reproduction takes place by binary fission (cell division).

(l) Sexual reproduction takes place by isogamy, anisogamy and oogamy. Zygote forms diploid auxospore that develops new set of vegetative cell.

(m) These form diatomite or diatomaceous earth or kiesulghur when these settle at the bottom of sea after death.

(n) These are indicator of sewage pollution.

e.g., *Naviculla, Pinnularia, Milosira*

2. Dinoflagellates: Division: Pyrrophyta

Class: Dinophyceae (Dark yellow algae)

(a) These are mostly marine.

(b) Body has longitudinal groove (sulcus) and transverse groove (annulus/cingulus). These may be of different colours.

(c) These are biflagellate-heterokont and oriented at right angle to each other thus they are also called whirling whip.

(d) Cell wall is made up of cellulose and pectin.

(e) These show mesokaryotic cellular organisation.

(f) These are photosynthetic due to presence of chlorophyll-a and -c.

(g) Reserve food is starch and oil.

(h) These contain non-contractile vacuole called pusule which helps in excretion and osmoregulation.

(i) Asexual reproduction takes place by binary fission (cell division).

(j) Sexual reproduction takes place by isogamy, anisogamy and oogamy.

(k) Life cycle is haplontic (involving zygotic meiosis) as well as diplontic (involving gametic meiosis).

(l) These cause red tide (sea looks red) due to rapid multiplication of *Gonyaulax.*

(m) Some dinoflagellates such as *Noctiluca, Pyrodinium* (Fire algae) show bioluminescence (emit light at night which makes sea surface illuminated).

(n) These produce powerful toxins such as saxitonin and brevitoxin that cause respiratory failure in vertebrates esp. fishes.

e.g., *Gymnodinium, Gonyaulax, Noctiluca, Pyrodinium*

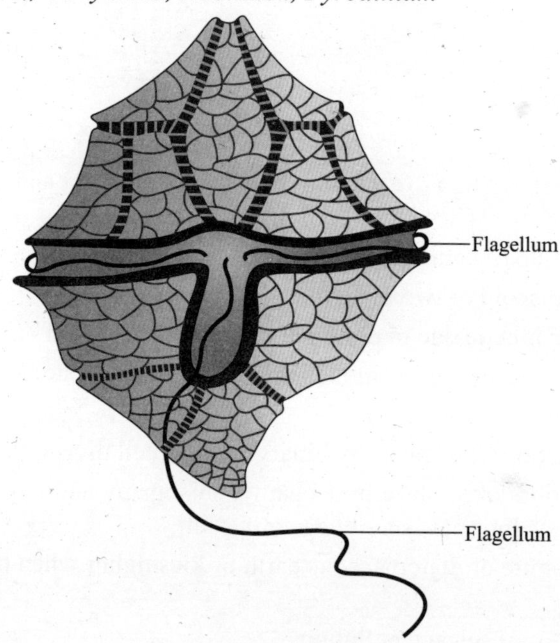

Fig: *Dinoflagellate*

3. **Euglenoids:** Division: Euglenophyta
Class: Euglenophyceae

(a) These are freshwater esp. in stagnant water.

(b) Body is spindle shaped having blunt anterior end and pointed posterior end. The anterior end has invagination containing cytostome (mouth), cytopharynx (canal) and reservoir. Eye spot (stigma) is situated at anterior end that is involved in perception of light stimulus.

(c) Body is covered with protein rich layer called pellicle which makes body elastic or flexible. Pellicle has oblique but parallel stripes called myonemes.

(d) These are biflagellate and heterokont (of different length).

(e) Cell wall is absent (animal feature).

(f) Chloroplast is present which contains pyrenoid.

(g) These are photosynthetic (autotrophic) in the presence of light and chlorophyll a and - b (just like higher plants). These also become heterotrophic (holozoic) in the absence of light. Hence mode of nutrition is mixotrophic.

(h) Reserve food is paramylon starch (β-1,3 glucan).

(i) These contain contractile vacuole (at anterior end) which helps in osmoregulation.

(j) These contain food vacuole.

(k) Asexual reproduction takes place by longitudinal binary fission.

(l) Sexual reproduction is absent.

(m) These develop encystment to tide over unfavourable condition.

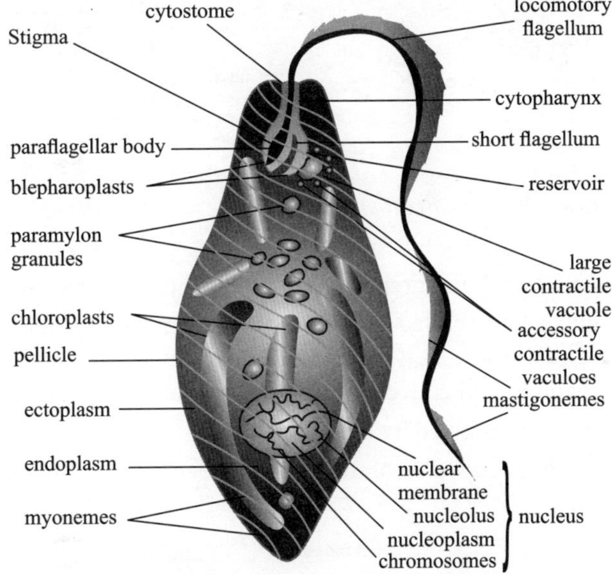

Fig: *Euglena*

4. **Slime moulds**: Division: Myxomycota
Class: Myxomycetes

These are also known as consumer-decomposer protists or fungus animals or saprophytic protists. These are on the transition line between fungi and protista. These are of two types:

(i) Acellular or True slime moulds: e.g., *Dictyostellium, Fuligo, Physarum*

(a) These are aquatic.

(b) Somatic body is made up of active, free-living, naked, amoeboid, diploid and multinucleate mass of protoplasm called plasmodium. It is covered with slime and formed during favourable condition. It moves over the surface of substratum by pseudopodia (amoeboid movement).

(c) These are heterotrophic (holozoic or ingestive) due to absence of chlorophyll (animal like feature).

(d) These form a fruiting body called sporocarp by differentiation under unfavourable condition. Sporocarp contains sporangia which form highly resistant spores by meiosis under unfavourable condition. Sporangium contains an intricate network of cytoplasmic threads called capillitium. Spore has cellulosic cell wall (plant feature). Spore is dispersed by air currents and germinates to form one, biflagellate and spindle shaped swarm cell.

(e) Asexual reproduction takes place by spores.

(f) Sexual reproduction takes place by isogamy.

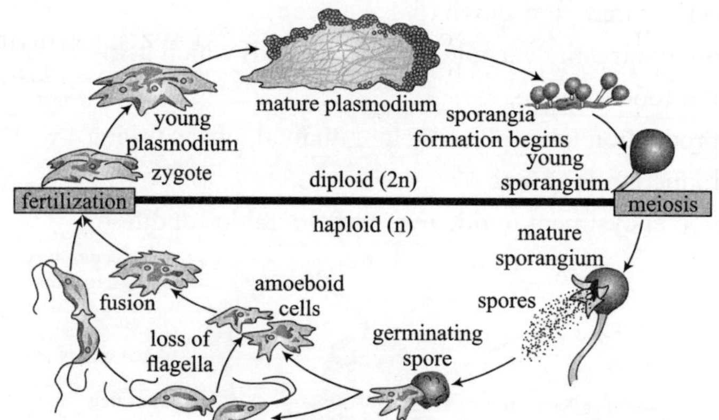

(ii) Cellular slime moulds: e.g., *Protostelium*

(a) These are terrestrial.

(b) Somatic body is made up of aggregated amoeba like organisms called pseudoplasmodium.

(c) These form a fruiting body called sporocarp which contains sporangia. Sporangium develops spores. Spore has cellulosic cell wall. Spore germinates to form non-flagellate, wall-less, uninucleate and haploid cell called myxamoebae which act as gamete.

(d) Sexual reproduction takes place by anisogamy.

(e) Zygote directly forms pseudoplasmodium which becomes multinucleate by repeated division.

(f) Life cycle is diplontic.

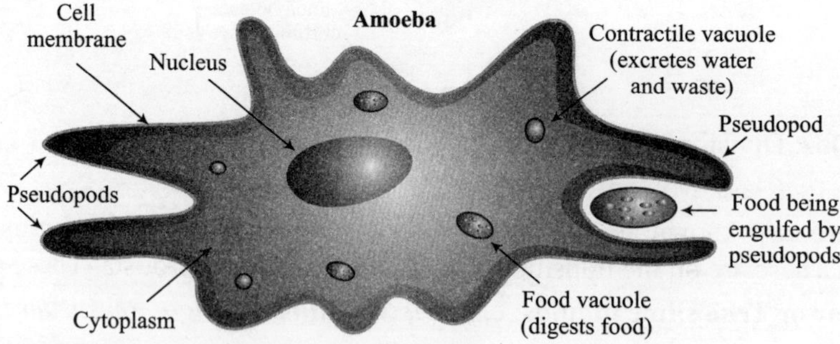

5. **Protozoans:** These are considered as 'first animals' or predator protists.

Class 1: Rhizopoda (Sarcodina): e.g., *Amoeba, Entamaoeba*

Class 2: Ciliata: e.g., *Paramoecium, Balantidium*

Class 3: Flagellata (Mastigophora): e.g., *Leishmania, Trypanosoma*

Class 4: Sporozoa: e.g., *Monocystis, Plasmodium*

(a) These are freshwater as well as marine.

(b) These are solitary as well as colonial.

(c) These are free-living as well as parasites.

(d) These are animal like microorganisms that lack cell wall.

(e) These move through locomotary organelles such as pseudopodia (false feet), cilia and flagella which help in locomotion and food capture.

(f) These are heterotrophic (holozoic or ingestive and saprozoic involving diffusion or active transport of soluble nutrients such as glucose, amino acids across plasma membrane.

(g) These undergo intracellular digestion through general body surface.

(h) These undergo respiration and excretion through general body surface by diffusion.

(i) Asexual reproduction takes place through transverse binary fission, multiple fission.

(j) Sexual reproduction takes place through conjugation, hemixis, automixis.

(k) These form cysts (encystment) during unfavourable condition.

Fungi

Fungi are multicellular, haploid, nonvascular, achlorophyllous, spore bearing and heterotrophic organisms. Fungi are major decomposers of wastes (along with bacteria) and help in recycling of inorganic and organic materials present on earth.

- Study of fungi is called Mycology (*Mykes* = Mushroom).
- Father of Mycology: Pier Antonio Micheli (He wrote Nova Genera Plantarum).
- Father of modern mycology/Plant pathology: de Bary
- Father of Systemic Mycology: E.M. Fries (Linnaeus of Mycology)
- Father of Indian Mycology: E.J. Butler

Smallest fungus	*Sacchromyces cerevisiae* (Yeast)
Largest fungus	*Armillaria* (Honey mushroom)

Features:

(a) These are ubiquitous and cosmopolitan. These show great diversity in morphology and habitat.

(b) These grow well at 20-30^{0}C and pH=6.

(c) Body is thalloid. Thallus is unicellular (Yeast) or made up of fine, tubular and thread like filaments called hyphae.

- Hyphae become thick and root-like called rhizomorph. e.g., *Agaricus*.
- Hyphae become terminally swollen to penetrate host and cause infection. e.g., *Erysiphe*.
- Hyphae become closely packed and compact as in *Claviceps*.
- Hyphae develop appendages to called haustoria which penetrate host cell for nutrition. These occur commonly.
- Hyphae are usually interwoven into a network called mycelium.
- Mycelia also become multinucleate and non-septate hence called coenocytes. e.g., Oomycetes and Zygomycetes.
- Mycelia form chain of cells by budding forming pseudomycelium. e.g., yeast.
- Mycelia often aggregate into tissue like structure called plectenchyma which is of two types:

  (i) Prosenchyma: Several hyphae come close together and run parallel to each other.

  (ii) Pseudoparenchyma: Several hyphae are closely packed.

(d) Cell wall is made up of chitin (a polymer of N-acetyl glucosamine). It is absent in Oomycetes where cellulose is present. Chitin is also absent in yeast where mannan (polymer of mannose) and glucan (polymer of glucose) are found.

(e) Lomasome (invagination of plasma membrane) is present between cell wall and plasma membrane.

(f) Chloroplast is absent.

(g) Golgi body is unicisternal.

(h) These are heterotrophs (saprophytes, parasites and symbionts) due to absence of chloroplast and chlorophyll molecules. These may be

- **Obligate saprophytes:** These totally depend on dead organic matter.
- **Facultative saprophytes:** These are normally parasites but can also be saprophytes.
- **Obligate parasites:** These totally depend on host for nutrition.
- **Facultative parasites:** These are normally saprophytes but can also be parasites.

(i) Reserve food is glycogen.

(j) Sex organs are unicellular and non-jacketed.

(k) Vegetative reproduction takes place by fragmentation of hyphae (common), budding, fission.

(l) Asexual reproduction takes place by different spores such as

- **Zoospores** (oval, unstalked, thin, endogenous and flagellate)
- **Sporangiospore** (rounded, unstalked, thin, endogenous)
- **Conidia** (rounded, stalked, thin, exogenous and non-flagellate)
- **Chlamydospore** (rounded, unstalked, thick).

(m) Fungi are monoecious (homothallic) and dioecious (heterothallic).

(n) Fungi are mostly eucarpic (a part of thallus is converted into reproductive structure). Some are holocarpic (entire thallus is converted into reproductive structure).

(o) Sexual reproduction involves plasmogamy (fusion of two protoplasm of either two motile or non-motile gametes takes place bringing two compatible nuclei close together), karyogamy (fusion of two nuclei to form diploid nucleus) and meiosis (formation of haploid spores from zygote).

(p) Sexual reproduction takes place by following methods:

(a) Planogametic copulation: It is fusion of naked gametes. It involves

Isogamy: It is fusion of morphologically similar gametes.

Anisogamy: It is fusion of morphologically and physiologically dissimilar gametes.

Oogamy: It is fusion of smaller, active and motile sperm (formed due to repeated division of male gametangium) with larger, passive and non-motile egg (formed directly from female gametangium without division).

(b) Gametangial contact: Both male and female gametangia lie in close contact by fertilisation tube. The gametes are never released. Then one or several male nuclei shift to female gametangia where fusion takes place to form oospore. The gmetangia never fuse.

(c) Gametangial copulation: The two isogametangia come in close contact and their contents are mixed after dissolving their walls at the point of contact. Both fuse to form thick-walled zygospore.

(d) Somatogamy: The two vegetative cells or two vegetative hyphae bearing nuclei of different parental strain fuse after dissolution of wall to form karyon formation. In this sex organs and gametes are not formed. It occurs in *Agaricus, Morchella, Peziza.*

(e) **Spermatisation:** It occurs when non-motile and spore like spermatium (male gamete) releases its contents into trichogyne (female receptive hyphae or female gamete). In this sex organs are absent.

(q) Sexual reproduction shows gradual and progressive simplification from lower to higher forms.
Oogamy→Anisogamy→Isogamy→Gametangial copulation→Spermatisation→Somatogamy

(r) Embryo is not formed.

Classification: Fungi have been divided into five classes based on following criteria.

(a) Nature of somatic hyphae: either unicellular or myceloid; if latter septate or non-septate.
(b) Types of asexual spore-zoospore, sporangiospore, conidia.
(c) Presence or absence of sexual stage
(d) Nature of life cycle: haplontic, diplontic, haplo-diplontic, dilobiontic, haploid dikaryotic

1. Oomycetes (Egg fungi) or Phycomycetes (Algal fungi):

(a) These are aquatic.
(b) Mycelia are branched and coenocytic (multinucleate and non-septate).
(c) Cell wall is made up of cellulose.
(d) Flagella (whiplash or tinsel type) are found.
(e) Asexual reproduction takes place by biflagellate and heterokont zoospore.
(f) Sexual reproduction takes place by gametangial contact to form oospore.

Examples:

- *Phytophthora infestans*: It causes late blight of potato (Irish famine in1845).
- *Synchytrium endobioticum*: It causes wart disease of potato
- *Albugo candia*: It causes white rust of crucifers.
- *Pythium debarynum*: It causes damping off of seedlings in tobacco, chilly
- *Sclerospora*: It causes green ear of pearl millet.
- *Saprolegnia* (Water mould): It causes salmon disease on fish gill.
- *Pilobolus*: It is coprophilous fungus.

2. Zygomycetes (Conjugating fungi):

(a) Mycelia are branched and coenocytic.
(b) Zoospore is absent.
(c) Asexual reproduction takes place by unflagellate sporangiospore.
(d) Sexual reproduction takes place by gametangial copulation forming zygospore.

Examples:

- *Mucor* (Pin mould/Dung mould): It produces trisporic acid.
- *Rhizopus* (Bread mould): It shows heterothallism as discovered by Blakeslee where conjugation occurs between + and – strains of hyphae.

3. Ascomycetes (Sac fungi): It is the largest class.

(a) These are mostly coprophilous (growing on cow dung).
(b) Mycelia are branched and septate.
(c) Motile structure is absent.

(d) Asexual reproduction takes place by conidia.
(e) Ascocarp (fruiting body) containing asci is formed.
(f) Crozier formation occurs in formation of ascus.
(g) Dikaryophase is small between plasmogamy and karyogamy.
(h) Sexual reproduction takes place by usually 8 (sometimes 4), linear and endogenous ascospore.

Examples:

- *Penicillium* (Blue green mould): Fruiting body is cleistothecium (spherical).
- *Peziza* (Cup fungus): Fruiting body is apothecium (cup-shaped).
- *Claviceps* (Ergot fungus): Fruiting body is perithecium (flask-shaped).
- *Neurospora* (Red or pink mould): *Drosophila* of Plant kingdom
- *Aspergillus* (Black mould): Guinea pig of plant kingdom or weed of laboratory.
- *Saccharomyces* (True yeast or sugar fungus): Asci are formed.
- *Candida* (False yeast): Asci are not formed.
- *Morchella* (Morels): It is edible and proteinaceous.
- *Tuberaestivum* (Truffle): It is edible.
- *Erysiphe*: It causes powdery mildew of pea.

4. **Basidiomycetes (Club fungi):**

(a) Mycelia are branched and septate.
(b) Dolipore septum (centrally located and barrel shaped pore) is present.
(c) Motile structure is absent.
(d) Mycelia may be primary (short-lived containing monokaryotic cells) and secondary (long lived containing dikaryotic cells).
(e) Clamp connection is seen in dikaryotic hyphae. It forms dikaryon.
(f) Vegetative reproduction takes place by fragmentation.
(g) Asexual spore is absent.
(h) Dikaryophase is large and present between plasmogamy and karyogamy.
(i) Basidiocarp (fruiting body) containing basidia is formed.
(j) Sexual reproduction takes place by 4, exogenous basidiospores.

Examples:

- *Agaricus* (Mushroom/Gill fungus): It is edible.
- *Puccinia graminis tritici* (rust fungus): It causes black stem rust of wheat
- *Ustilagotritici* (Smut fungus): It causes loose smut of wheat.
- *Tilletia*: It causes bunt of wheat.
- *Polyporus* (Bracket fungus): It causes heart rot disease.
- *Amanita* (Toad stool): It contains hallucinogens.
- *Lycoperdon* (Puff ball): It is fleshy edible.

5. **Deuteromycetes** (Fungi Imperfecti):

(a) These are mostly decomposers. Some are parasitic or saprophytic.
(b) Mycelia are branched and septate.

(c) Asexual reproduction takes place by conidia.
(d) Asexual fruiting bodies such as synnema, sporodochium, pycnidium, acervulus are formed.
(e) Sexual reproduction is absent hence perfect stage is not known. If sexual stage was discovered then it was moved to either ascomycetes or basidiomycetes.
(f) These help in mineral cycling.

Examples:

- *Alternaria solanii*: It causes early blight of potato
- *Colletotrichum falcatum*: It causes red rot of sugarcane
- *Fusarium oxysporum*: It causes wilt of red gram
- *Cercospora personata*: It causes tikka disease of groundnut.
- *Helminthosporium oryzae*: It causes brown leaf spot disease of rice (Bengal famine in 1942).
- *Pyrecularia oryzae*: It causes blast disease of rice.
- *Trichoderma* : It is used in biocontrol.

Mycorrhiza:

It is symbiotic association of fungi with roots of higher plants. It was discovered by Frank (1885).

In this association, fungus gets nutrition (carbohydrate) from roots of higher plants and in return, higher plants uptake water and minerals due to spread of fungal hyphae in soil. It occurs in most of the plants which live in nutrient (organic matter) poor soil. The association is not very specific. Mycorrhiza does not form root hair and root cap.

Types: There are two types of mycorrhiza.

Ectomycorrhiza	Endomycorrhiza
Fungal hyphae form sheath or mantle around roots (Hartig net) because these do not penetrate the root.	Fungal hyphae do not form sheath or mantle because these penetrate root intercellularly and intracellularly.
The cells of root secrete sugar into intercellular space for fungal hyphae.	The cells of root do not secrete sugar.
It does not form rounded structures (vesicles) and branch like structures (arbuscles).	It forms arbuscles and vesicles hence called Vesicular Arbuscular Mycorrhiza (VAM).
It is less common.	It is more common.
It is shown by *Boletus, Lacterius* (Ascomycetes) and *Amanita* (Basidiomycetes).	It is shown by *Glomus, Rhizoctonia* (Zygomycetes).
It occurs in roots of Birch, *Eucalyptus*, *Pinus*, *Quercus*.	It occurs in roots of Tomato, Pepper, Coffee, Cardamom (Crop plants), Grasses, Orchids.

Economic importance:

(a) Mycorrhiza helps in absorption of water and minerals (esp. N, P, S, Ca) because fungus increases the surface absorbing area of roots.
(b) It protects the roots and provides resistance in plants because fungus secretes antimicrobial substances.
(c) It increases the solubilisation of humus because fungus secretes several growth promoting substances.
(d) It increases the fertility of soil and is used as biofertiliser.

(e) It minimises the requirement of chemical fertilisers since it does not cause pollution.

(f) It increases the tolerance of plants to salinity and drought condition in soil.

Lichen

It is composite or dual organism formed as a result of symbiotic association of algae and fungi. The term 'Lichen' was coined by Theophrastus. Father of Lichenology is Acharius.

Features:

(a) Lichens are cosmopolitan in distribution but in India these are abundantly found in Eastern Himalaya.

(b) These are perennial, variously coloured and show very slow growth.

(c) These show permanent symbiotic association (consortium) of algae and fungi. The alga provides food to fungus and in return, fungus provides protection, water and nitrogenous substance to alga. Fungus is a controlled parasite over alga because fungus sends haustoria into algal cells, induces alga to secrete organic substance and prevents alga from developing covering of pectin. Hence this phenomenon is helotism (master-slave relationship). Fungi thus get more benefit than algae.

(d) These can thrive on barren rocks (saxicolous), earth crust (terricolous), tree bark (corticolous), wood (lignocolous).

(e) These have algal component (**phycobiont**) which is shown mostly by chlorophyceae (esp. *Trebouxia, Trentepohlia, Pleurococcus*) and some Cyanophyceae (esp. *Anabaena, Calothrix, Gloeocapsa*) as well as fungal component (**mycobiont**) which is shown by ascomycetes. The algal component is minor (5%) and fungal component is major (95%). Both components do not change within a species even under varying habitat.

(f) Thallus shows following structures:

- **Cyphellae:** These are small, circular, scattered depression on underside. These are aerating structures which help in exchange of gases (analogous to stomata of higher plants). These are found in *Parmelia*.
- **Cephalodia:** These are gall like outgrowth on upper side. These retain moisture. These are found in *Peltigera*.

(g) Vegetative reproduction takes place by fragmentation of thallus.

(h) Asexual reproduction takes place by soredia (small and rounded outgrowth of thallus), isidia (elongated outgrowth of thallus) and conidia.

(i) Sexual reproduction takes place by fungal partner (Ascomycetes) through ascospore.

Types:

1. **Crustose Lichen:** This forms a thin crust on substratum. e.g., *Dermatocarpon* (Stone mushroom), *Rhizocarpon, Graphis, Lecanora, Strigula.*
2. **Foliose lichen:** This forms a flat, branched and leaf like thallus attached to substratum by rhizoid like structure (rhizine). e.g., *Cetraria, Loboria, Parmelia, Peltigera* (Dog lichen), *Physcia, Chowdhuria*
3. **Fruticose lichen:** This forms much branched, bushy and pendant to upright. e.g., *Cladonia* (Reindeer moss), *Usnea, Ramalina*

Economic Importance:

(a) Lichens are indicators of SO_2 pollution.

(b) Crustose lichens are pioneer colonisers (communities) in xeric succession (xerosere).

(c) Lichens such as *Cetraria icelandica* (Iceland moss), *Cladonia* and *Dermatocarpon* are edible.

(d) *Rocella tinctoria* is source of litmus solution and orcein dye (Orchill).

(e) *Peltigera* is used to cure hydrophobia.

Virus:

It is simple, ultramicroscopic, non-cellular infectious agent. Its unit is called virion which is fully assembled mature particle capable of causing infection. The term 'Virus' was coined by Louis Pasteur. Father of Virology is Beijerinck. It is neither prokaryotic nor eukaryotic because it has no cell structure. It is considered as connecting link between non living organism and living organism.

Non-living properties:

(a) Virus does not have protoplasm and cell organelles.

(b) It does not show any type of metabolism.

(c) It can be crystallised (as done by Stanley (1935) on TMV and Polio virus)

(d) It does not secrete enzymes.

(e) It does not grow on culture medium.

(f) It can be precipitated, stored and dissolved in water.

(g) It is inert chemical or molecule outside host cell and exists as virion.

(h) It is autocatalytic.

(i) It remains unaffected by antibiotics.

Living properties:

(a) It grows and multiplies (reproduces) inside host cell by redirecting host metabolism. Its life cycle includes
- Attachment (adsorption) of virion to susceptible host
- Penetration (injection)
- Synthesis of nucleic acid and proteins
- Assembly of proteins and packaging of nucleic acid
- Virion (progeny) release from host cell

(b) It has either DNA or RNA as genetic material.

(c) It shows mutation due to nucleic acid (as proved by Luria, Delbruck and Lwoff). HIV is most mutable.

(d) It is sensitive to temperature, chemicals and UV radiations.

(e) It has genetic continuity and has definite race.

(f) It shows antigenic properties.

(g) It can direct protein synthesis by using ribosomes of host cell.

(h) It can be killed by autoclaving

(i) It causes diseases in plants and animals hence it is obligate intracellular parasite.

Symmetry of virus: Viruses show the following three types of symmetry:

1. Helical (Cylindrical): TMV, Influenza virus, Rabies virus

2. Polyhedral (Icosahedral): Herpes simplex virus, Polio virus, Turnip mosaic virus
3. Complex (Mixed symmetrical): Bacteriophage, Polio virus.

Disease Symptoms of viruses:

1. Stunting (reduced growth)
2. Mosaic (dark green, light green and yellow angular areas)
3. Mottling (development of different colours)
4. Necrosis (death of tissues)
5. Leaf rolling
6. Curling
7. Yellowing
8. Vein clearing

Classification of Viruses: Lowff, Home and Tournier proposed LHT system which was adopted.

- Phylum Vira: It was divided into two subphyla:
- Deoxyvira: Deoxyhelica, Deoxycubica and Deoxybinala
- Ribovira: Ribohelica and Ribocubica

Virus family	Nature of genome	Examples
Parvoviruses	Single stranded DNA	Coliphage Ψ × 174, Canine parvovirus
Adenoviruses, Pox viruses	Double stranded DNA	Small pox virus, Cytomegalo virus.
Picornaviruses	Single stranded RNA	Polio virus, Influenza virus.
Retroviruses	Double stranded RNA	Reovirus, Rice dwarf virus

Types of viruses:

1. **Plant virus:** Genetic material in plant viruses is usually single stranded RNA.
   - **Exceptions:** The ds RNA is found in Rice dwarf virus, Wound tumour virus (Reo virus). The ds DNA is found in Cauliflower mosaic virus.

Smallest plant virus	Tobacco necrosis virus & Alfalfa mosaic virus (17nm)
Largest plant virus	Citrus tristeza virus (2000 × 12 nm)

   - **Tobacco mosaic Virus**: It was the first virus to be discovered by Dmitri Ivanovsky (1893) and found it smaller than bacteria because it passed through bacterial proof filter. Beijerinck (1898) proposed that the filtrate of tobacco leaves caused infection or tobacco mosaic disease (dark green, light green, yellow angular areas on leaves) in healthy plants and called it '*Contagium Vivum Fluidum*' (Infectious living fluid). Adolph Mayer discovered infective nature of tobacco mosaic disease which was developed by applying sap of infected leaves onto leaves of healthy plants. Pirie and Bawden (1937) discovered nucleo-proteinaceous nature of virus.

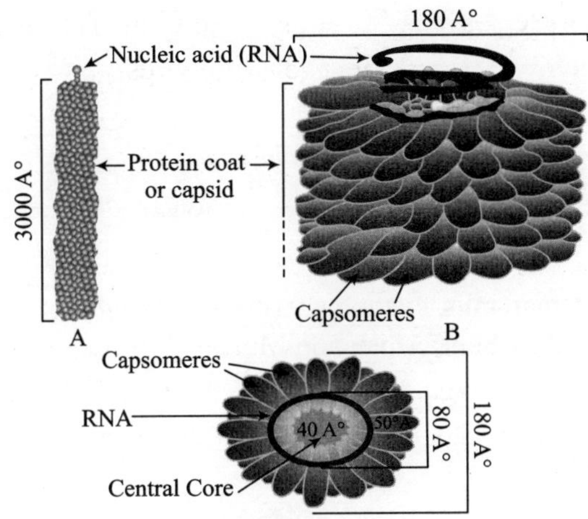

Fig: ***Tobacco mosaic virus***

Features:

(a) It is most thoroughly studied virus.

(b) It was first virus to be seen under electron microscope.

(c) It was first virus to be purified.

(d) It was first virus from which infectious nucleic acid was obtained.

(e) It is rod shaped and total length of each rod has 130 helicals.

(f) Its size is 3000 A^0 x180 A^0.

(g) It is made up of RNA (5%) and protein (95%).

(h) RNA is single stranded, linear, helical, grooved and intertwined with protein. It has 6400 nucleotides. RNA performs self replication and synthesis of specific proteins.

(i) Protein coat is capsid which protects RNA and has 2130 capsomeres (subunits of protein). Each capsomere contains 158 amino acids.

(j) The ratio of nucleotides and capsomeres is 3:1.

2. **Animal Virus:** Genetic material is mostly double stranded DNA.

   Exceptions: The ss-DNA is found in Parvo virus

   The ss-RNA is found in Retro virus (Rous sarcoma virus, HIV), Polio virus and Influenza virus.

   The ds-RNA is found in Reo virus (Wound tumour virus).

Smallest animal virus	Foot and mouth disease of virus (20 nm)
Largest animal virus	Small pox virus (350 × 250 × 100 nm)

3. **Bacterial virus:** Genetic material is mostly double stranded DNA. Genetic material is mostly double stranded DNA

   Exceptions: The ss-DNA is found in ϕ x174, fd, M-13,

   The ss-RNA is found in F_2, QB, R_{17},MS_2

   The ds-RNA is found in ϕ_6

Smallest bacterial virus	Coliphage P_2 (2 nm)
Largest bacterial virus	Pandora virus

Types of viruses:

(a) Cyanophage: It is virus that kills cyanobacteria esp. *Lyngbya*, *Phormidium* and *Plectonema* (LPP-1). It was discovered by Schfferman and Morris (1962).

(b) Mycophage: It is virus that kills *Agaricus bisporus*. It was discovered by Sinden (1957).

(c) Bacteriophage: It is virus that kills bacteria. It was discovered by Twort and Felix D' Herelle (1957).

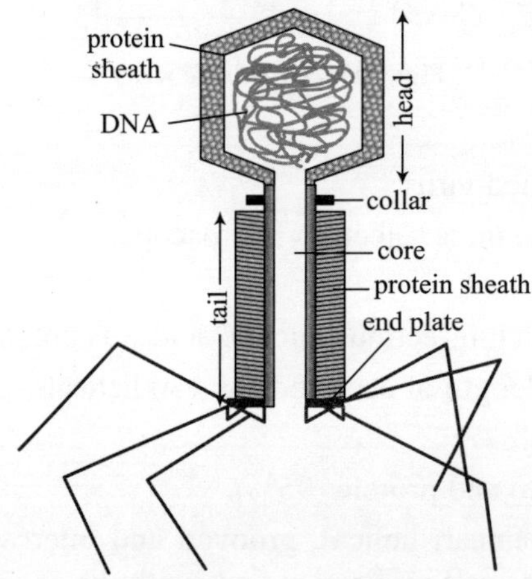

Fig: *Bacteriophage*

Features of bacteriophage:

(a) It is tadpole shaped.

(b) It has polyhedral (hexagonal) head (size = 950 × 650A°) and binal (helical) tail.

(c) Head has double stranded DNA surrounded by protein coat called capsid. Amount of DNA is 40% and that of protein is 60%. Capsid has 2130 protein subunits called capsomeres. Each capsomere has 158 amino acids. DNA of T even phage (T_2, T_4) has hydroxymethyl cytosine instead of cytosine of normal DNA.

(d) Tail is attached with Head by collar. It is proteinaceous and surrounded by sheath having 24 annular rings (striations). At lower end, hexagonal base plate contains 6 small tail pins and 6 large tail fibres. Tail fibres help in attachment of phage to host cell.

(e) It is obligate parasite.

Types of Bacteriophage:

1. **Temperate phage:** It is non-infective phage. In this lysis does not occur. Viral genome gets integrated with host genome forming prophage (lysogeny). e.g., λ-phage
2. **Virulent phage:** It is infective phage. In this, lysis occurs. e.g., T_2, T_4- bacteriophage.

Economic Importance:

1. T_2 bacteriophage was used by Hershey and chase to prove that DNA was genetic material.
2. Bacteriophage was used by Zinder and Lederberg to bring about transduction in bacteria.
3. Bacteriophage was used to destroy bacteria to purify water body
4. Cyanophages were employed to control algal bloom.

Sub-viral particles:

These are smaller particles than viruses. These are of three types:

1. **Virusoid:** It was discovered by Randle (1981). It is small, circular and single stranded RNA. It lives in protein coat of large virus. It replicates within host and does not cause infection.
2. **Viroid:** It was discovered by Diener and Raymer (1967). It is extremely simple and one of the smallest known infectious agent. It contains RNA only but protein is absent. RNA is single stranded, circular, covalently closed and tightly folded. It contains 260-380 nucleotides having low molecular weight. RNA replicates in nucleus of host cell unlike virus which replicates in host cytoplasm. It causes diseases only in higher plants. It is transmitted mechanically or through wind. It causes the following diseases:
   (a) Cadang cadang of coconut
   (b) Potato spindle tuber disease (PSTD)
   (c) Citrus exocortis (stunt)
   (d) Chrysanthemum stunt
3. **Prion:** It was discovered by Stanley Prusiner (1972). It is the smallest known infectious agent (100 times smaller than the smallest virus). It is rod shaped. It contains proteins (PrP) only but does not have nucleic acids (DNA/RNA). It undergoes autonomous replication in spite of absence of nucleic acid. It is latent and resistant to heat (135^{0}C for 18 min), UV radiations, ionising radiation and protease. It is sensitive to urea, phenol, sodium dodecyl sulphate (SDS). Prion protein (PrP) is not sedimentable and cannot be separated by centrifugation. PrP occurs in infectious and non-infectious cells. It converts folded protein into PrP. It is called slow virus because disease occurs very slowly. It causes diseases only in animals. It causes the following diseases:
   (a) Scrapie disease: It occurs in sheep and goat in which there is loss of wool due to scrapping of body on rough surface.
   (b) Kuru disease: It is a degenerative disease of CNS in human being of New Guinea.
   (c) Mad Cow disease: It is also known as Bovine Spongiform Encephalopathy in which soft brain and cavities are developed in brain of cattle.
   (d) Creutzfeldt-Jakob disease: It occurs in human being in which there is loss of memory and coordination.

SHORT ANSWER TYPE QUESTIONS

1. Discuss the importance of biological classification.
2. Mention the demerits of two kingdom system of classification.
3. Disting uish between methanogens and halophiles.
4. Distinguish between archaebacteria and bacteria.
5. What are the different shapes of bacteria?

6. Write short notes on *Mycoplasma*.
7. Distinguish between bacteria and cyanobacteria.
8. Draw a well labelled diagram of *Euglena*.
9. Discuss the five characteristic features of fungi.
10. What are the main features which put Fungi in different kingdom?
11. Discuss the importance of mycorhiza.
12. Discuss the economic importance of lichen
13. Discuss the salient features of tobacco mosaic virus.
14. Draw a well labelled diagram of bacteriophage.
15. Distinguish between viroid and prion.

LONG ANSWER TYPE QUESTIONS

1. Describe the different criteria of five kingdom system of classification.
2. Discuss the different modes of nutrition in bacteria.
3. Describe the important features of each group of protista.
4. Describe the characteristic features of different classes of fungi.
5. What are the living and non-living properties of virus?

MULTIPLE CHOICE QUESTIONS

1. Human immune deficiency virus (HIV) has a protein coat and a genetic material which is
   (a) Single stranded RNA (b) Double stranded RNA
   (c) Single stranded DNA (d) Double stranded DNA
2. Which of the following plant viruses has DNA?
   (a) Tobacco mosaic virus (b) Tomato mosaic virus
   (c) Potato mosaic virus (d) Cauliflower mosaic virus
3. Phages that show lysogenic cycle are called
   (a) Virulent phages (b) Temperate phages
   (c) Avirulent phages (d) Lytic phages
4. The most complex animal virus is
   (a) Polio virus (b) Pox virus (c) Parrot fever virus (d) Influenza virus
5. Which of the following viruses was first to be cultured *in vitro* in human cell?
   (a) Polio virus (b) Pox virus (c) Influenza virus (d) HIV
6. Which of the following statements is incorrect about *Rhizopus*?
   (a) Coenogametes are found.
   (b) Trisporic acid is a hormone involved in sexual reproduction.
   (c) The common bread mould is *Rhizopus stolonifer.*
   (d) Coenocytic condition is not found.
7. The disease 'Aspergillosis' caused by *Aspergillus* which infected
   (a) Liver (b) Kidney (c) Lung (d) Ear
8. Koch's postulates are not applicable to
   (a) *Mycobacterium leprae* (b) *Mycobacterium tuberculosis*
   (c) *Diplococcus pneumonia* (d) *Vibrio cholera*
9. Bacteria responsible for nitrification come under
   (a) *Bacillus* (b) *Coccus* (c) *Spirilla* (d) *Vibrio*
10. The kingdom 'Fungi' was included for the first time in
    (a) Two kingdom system (b) Three kingdom system
    (c) Four kingdom system (d) Five kingdom system
11. In Five kingdom system classification, greater emphasis was given on
    (a) Cellular organisation and mode of nutrition
    (b) Cellular organisation and Life style
    (c) Composition of cell wall and Cellular organisation
    (d) Composition of cell wall and Life style
12. Carl Woese divided all organisms into three domains on the basis of
    (a) 23 S rRNA (b) 5S rRNA (c) 28S rRNA (d) 16 S rRNA
13. Which of the following archaebacteria are found in hot spring?
    (a) Methanogens (b) Halophiles (c) Thermoacidophiles (d) All of the above

14. Which of the following is incorrect regarding archaebacteria?
    (a) Methanogens are found in rumen of cattle.
    (b) Halophiles have purplish pigmented membrane
    (c) Thermo-acidophiles grow at high temperature and acidity.
    (d) All archaebacteria are motile and Gram negative.
15. Halophiles tolerate high salt concentration due to the presence of
    (a) Special lipids in membrane
    (b) Mucilage covering
    (c) Special protein and mucilage covering
    (d) Special lipid and mucilage covering
16. Which of the following is incorrect regarding bacteria?
    (a) They have simple structure and complex behaviour
    (b) They do not show cyclosis
    (c) They show most extensive metabolic diversity
    (d) They form gametes
17. Most of the bacteria are
    (a) Photoautotrophs
    (b) Chemoautotrophs
    (c) Photoheterotrophs
    (d) Chemoheterotrophs
18. *Chromatium* is also known as
    (a) Green sulphur bacteria
    (b) Purple sulphur bacteria
    (c) Green non-sulphur bacteria
    (d) Purple non-sulphur bacteria
19. Which of the following is incorrect regarding endospore of bacteria?
    (a) It is thick walled and highly resistant
    (b) It is endogenous
    (c) It occurs during unfavourable period
    (d) It is several formed per cell
20. Which of the following is incorrect regarding cyanobacteria?
    (a) They are Gram negative
    (b) Their photosynthesis is anoxygenic
    (c) They form algal blooms
    (d) They contain gas vacuoles.
21. Which of the following is incorrect regarding heterocyst of cyanobacteria?
    (a) It is thick walled structure
    (b) It takes part in asexual reproduction
    (c) It plays a role in nitrogen fixation
    (d) It undergoes vegetative reproduction.
22. Which of the following is incorrect regarding protista?
    (a) These are unicellular and eukaryotic
    (b) These have no well defined boundaries
    (c) These are anaerobes
    (d) These show 9+2 arrangement of cilia and flagella
23. Diatoms are photosynthetic due to the presence of
    (a) Chlorophyll-a and b
    (b) Chlorophyll-a and c
    (c) Chlorophyll-a and d
    (d) Chlorophyll-a only
24. Auxospores are found in
    (a) Diatoms
    (b) Dinoflagellates
    (c) Euglenoids
    (d) Slime moulds

25. Which of the following dinoflagellates causes red tide in sea?
(a) *Gymnodinium* (b) *Gonyaulax* (c) *Pyrodinium* (d) *Noctiluca*
26. Euglenoids resemble higher plants in having
(a) Reserve food (b) Composition of cell wall
(c) Food vacuole (d) Chlorophyll pigments
27. Which of the following is not an animal feature in Euglenoids?
(a) Presence of eye spot (b) Absence of cell wall
(c) Heterotrophic mode of nutrition in dark (d) Longitudinal binary fission
28. The spore of slime moulds is made up of
(a) Starch (b) Protein (c) Cellulose (d) Glycogen
29. The somatic body of slime moulds is made up of
(a) Thallus (b) Sporocarp (c) Pseudoplasmodium (d) Plasmodium
30. The locomotary organelles are absent in
(a) Rhizopoda (b) Ciliata (c) Flagellata (d) Sporozoa
31. The mycelium in lower fungi is coenocytic which is
(a) Binucleate and septate (b) Uninucleate and septate
(c) Multinucleate and aseptate (d) Multinucleate and aseptate
32. Cell wall in fungi is made up of chitin which is a polymer of
(a) NAG (b) NAM (c) NAG and NAM (d) Amino acids
33. Sex organs in fungi are
(a) Unicellular and jacketed (b) Multicellular and non-jacketed
(c) Multicellular and jacketed (d) Unicellular and non-jacketed
34. Which of the following spores is not involved in asexual reproduction?
(a) Oospore (b) Conidia (c) Sporangiospore (d) Zoospore
35. Irish famine is caused by
(a) *Alternaria solanii* (b) *Synchytrium endobioticum*
(c) *Helminthosporium oryzae* (d) *Phytophthora infestans*
36. Which is incorrectly matched?
(a) *Mucor*-Pin mould (b) *Rhizopus*-Bread mould
(c) *Aspergillus*-Black mould (d) *Penicillium*-Red mould
37. Which of the following fungi is involved in biocontrol?
(a) *Alternaria* (b) *Mucor* (c) *Colletotrichum* (d) *Trichoderma*
38. Sexual reproduction is absent in
(a) Oomycetes (b) Zygomycetes (c) Ascomycetes (d) Deuteromycetes
39. Dikaryophase is longer and present between plasmogamy and karyogamy in
(a) Phycomycetes (b) Ascomycetes (c) Basidiomycetes (d) Deuteromycetes
40. Crozier formation occurs in
(a) Phycomycetes (b) Ascomycetes (c) Basidiomycetes (d) Deuteromycetes

41. Clamp connection is seen in
(a) Phycomycetes (b) Ascomycetes (c) Basidiomycetes (d) Deuteromycetes
42. Which of the following classes of fungi is involved in mineral cycling?
(a) Phycomycetes (b) Ascomycetes (c) Basidiomycetes (d) Deuteromycetes
43. Gametangial copulation in fungus results in the formation of
(a) Oospore (b) Zygospore (c) Sporangiospore (d) Zoospore
44. Endomycorrhiza is shown by
(a) *Lacterias* (b) *Boletus* (c) *Amanita* (d) *Glomus*
45. The main algal and fungal components in lichens are, respectively,
(a) Cyanophyceae and Oomycetes (b) Chlorophyceae and Ascomycetes
(c) Chlorophyceae and Basidiomycetes (d) Phaeophyceae and Ascomycetes
46. Virus is an inert chemical or molecule outside host cell and exists as
(a) Viroid (b) Prion (c) Virusoid (d) Virion
47. In bacteriophage, the nature of genetic material is
(a) ss RNA (b) ss DNA (c) ds RNA (d) ds DNA
48. The protein subunit in TMV is known as
(a) Capsule (b) Capsid (c) Capsomere (d) Cap
49. Which of the following is incorrect with respect to sub-viral particles?
(a) Viroid does not contain protein (b) Prion does not contain nucleic acid
(c) Viroid causes PSTD (d) Prion is the largest known infectious agent
50. Mad cow disease is caused by
(a) Viroid (b) Virion (c) Prion (d) A bacterium
51. First organisms evolved on earth were
(a) Aerobic heterotrophs (b) Anaerobic heterotrophs
(c) Aerobic autotrophs (d) Anaerobic autotrophs
52. Mycorhiza works as
(a) Modified root (b) A root hair in adverse condition
(c) Vegetative propagule (d) Mechanical support
53. The electron donor in photoautotrophic bacteria is
(a) H_2O (b) H_2S
(c) Both a and b (d) Organic compound
54. Fungi imperfecti is also known as
(a) Phycomycetes (b) Ascomycetes (c) Basidiomycetes (d) Deuteromycetes
55. The unit of virus is known as
(a) Viroid (b) Virion (c) Prion (d) Virusoid
56. Endomycorrhiza is seen in
(a) *Pinus* (b) *Betula* (c) *Eucalyptus* (d) Orchids

57. The dominant pigment in cyanobacteria is
(a) Phycocyanin (b) Fucoxanthin (c) Phycoerythrin (d) Beta carotene
58. Which of the archebacteria belongs to hot sulphur springs?
(a) Methanogens (b) Halophiles (c) Thermoacidophiles (d) All of the above
59. The locomotary organelles in *Leischmania* and *Trypanosoma* are
(a) Cilia (b) Flagella (c) Pseudopodia (d) None
60. Virus is also known as
(a) Obligate intracellular parasite (b) Facultative intracellular parasite
(c) Obligate intracellular saprophyte (d) Facultative intracellular saprophyte
61. Who classified animals into two groups on the basis of presence or absence of RBCs?
(a) Aristotle (b) Linnaeus (c) Copeland (d) Whittaker
62. How many bacteria are produced in 3 hours if a bacterium divides once in half a n hour?
(a) 16 (b) 32 (c) 64 (d) 128
63. Which of the following processes are involved in the reproduction of protists?
(a) Binary fission and budding (b) Cell fusion and zygote formation
(c) Spore formation and cyst formation (d) All of the above
64. Which of the following fungi is used in biochemical and genetic research?
(a) Mucor (b) Neurospora (c) Aspergillus (d) Tichoderma
65. Fungi show sexual reproduction by all the following spores except
(a) Conidia (b) Zygospore (c) Oospore (d) Ascospore
66. In basidiomycetes, the vegetative reproduction takes place by
(a) Endospore (b) Conidia (c) Akinetes (d) Fragmentation
67. The term '*Contagium vivum fluidum*' was proposed by
(a) Louis Pasteur (b) Beijerinck (c) Stanley Prusiner (d) Leewenhoek
68. The algal component of lichen mostly belongs to the class
(a) Chlorophyceae (b) Phaeophyceae (c) Rhodophyceae (d) Cyanophyceae
69. Which of the following is incorrect statement?
(a) Diatoms do not contain locomotary organelles
(b) Euglena shows mixotrophic nutrition
(c) Slime moulds belong to class myxomycetes
(d) Sporozoa class of Protozoa contains locomotary organelles.
70. Potato spindle tuber disease is caused by
(a) Prion (b) Viroid (c) Bacteria (d) Protozoa

ANSWERS

1.a	2.d	3.b	4.c	5.a	6.d	7.c	8.a	9.a	10.d
11.a	12.d	13.c	14.d	15.d	16.d	17.d	18.b	19.d	20.b

21.d	22.c	23.b	24.a	25.b	26.d	27.d	28.c	29.d	30.d
31.c	32.a	33.c	34.a	35.d	36.d	37.d	38.d	39.c	40.b
41.c	42.d	43.b	44.d	45.b	46.d	47.d	48.c	49.d	50.c
51.c	52.a	53.b	54.d	55.b	56.d	57.a	58.c	59.b	60.b
61.a	62.c	63.d	64.b	65.a	66.d	67.b	68.a	69.d	70.b

3

Plant Kingdom

Plant kingdom comprises five groups which are organised as such:

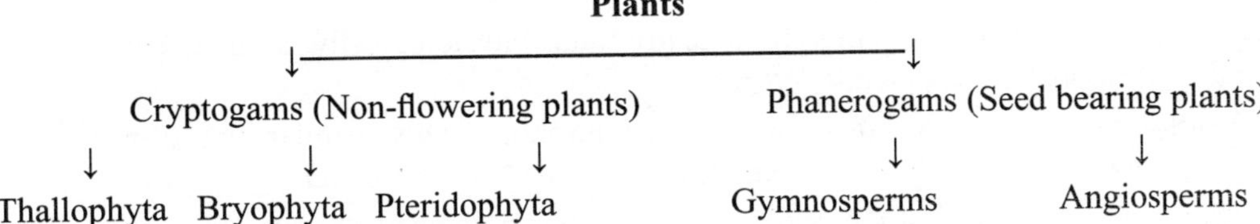

General characters of Plants:

1. These are multicellular and eukaryotic.
2. These have cell wall which is made up of cellulose.
3. These are autotrophic (photosynthetic) due to the presence of pigments.
4. These store food in the form of starch.
5. These possess rhizoids or roots for fixation and absorption of water.
6. These complete the life cycle in two alternating phases- a diploid or sporophyte and the haploid or gametophyte.

Algae

Algae come under thallophyta (young shoot) which was coined by Endlicher. Algae are simple, non-flowering, chlorophyllous, non-vascular and aquatic group of plants. These were originated from cambrian period. Study of algae is known as phycology (Gr. *Phykos*=sea weed). Algae are known as children of sea.

- Father of phycology is Fritsch.
- Father of Indian phycology is M.O.P. Iyenger

Smallest alga	*Chlamydomonas* (a green alga)
Largest alga	*Macrocystis* (a brown alga)

Characteristic features:

1. Algae are ubiquitous (found everywhere). These are mostly aquatic (freshwater forms are called pond scum and marine forms are called sea weed).
2. Plant body is gametophyte. It is made up of thallus which does not contain organised tissues hence roots, stem and leaves are not differentiated. Thallus shows great variation.
3. Mechanical and vascular tissues are absent
4. Cell wall is made up of cellulose, galactans (polymer of galactose), mannans (polymer of mannose) and small amount of minerals ($CaCO_3$).

5. These are photoautotrophs due to the presence of different chlorophyll molecules.
6. Reserve food is mostly starch.
7. Vegetative reproduction takes place by fragmentation (as in filamentous algae).
8. Asexual reproduction takes place by binary fission, (as in unicellulor algae) budding, palmella stage and different spores such as zoospore (thin walled and motile), aplanospore (thin walled and non motile) and hypnospore (thick walled and non motile).
9. Sex organs are unicellular and non-jacketed; if multicellular, then all cells are fertile.
10. Sexual reproduction involves plasmogamy (fusion of cytoplasm of gametes), karyogamy (fusion of nuclei of gametes) and meiosis.
11. Sexual reproduction occurs in three ways:
    (a) Isogamy: It is fusion of morphologically and physiologically similar gametes. e.g., *Chlamydomonas debaryanum, Ulothrix.*
    (b) Physiological anisogamy: It is fusion of morphologically similar but physiologically (behavioural) dissimilar (+ and -) gametes. e.g., *Spirogyra, Zygnema*
    (c) Anisogamy: It is fusion of morphologically and physiologically dissimilar gametes. In this, male gamete is smaller and more active while female gamete is larger and less active. e.g., *Chlamydomonas braunii*
    (d) Oogamous: It is most advanced type in which the protoplast of male gametangium undergoes repeated division to form smaller, motile and smaller gamete and the protoplast of female gametangium does not divide but directly converted into larger, non-motile and passive egg. e.g., *Chlamydomonas coccifera, Volvox, Oedogonium, Fucus.*
12. Embryo is not formed after fertilisation.
13. The process of reproduction increases in complexity from lower to higher forms.
14. Sporophyte is not dependent on gametophyte.
15. Life cycle is haplontic in most of algae. Vegetative phase is haploid and zygotic meiosis occurs. Some other life cycles are as such
    (a) Diplontic: Vegetative phase is diploid and gametic meiosis occurs. e.g., *Fucus, Sargassum* etc.
    (b) Haplo-Diplontic: There is alternation of gametophytic and sporophytic phase and it shows sporic meiosis. It may be Isomorphic (morphologically similar): *Cladophora, Ectocarpus, Ulva* Heteromorphic (morphologically different): *Laminaria*
    (c) Haplobiontic (diphasic): There are two haploid phases-(i) sexual individual and (ii) carpospores bearing stage alternating with diploid phase (zygote).e.g., *Nemalion*
    (d) Haplobiontic (triphasic): There is one gametophytic phase and two sporophytic phases such as carposporophyte and tetrasporophyte.
    (e) Diplobiontic: (triphasic): There are two diploid phases (i) diploid carposporophyte, and (ii) tetrasporic individual alternating with single haploid phase (sexual individual). e.g., *Polysiphonia.*

Classification: Fritsch divided algae into 11 classes (now 7 classes) in his book 'Structure and Reproduction of Algae' on the basis of

(a) Composition of cell wall
(b) Types of pigment

(c) Flagellation
(d) Composition of reserve food
(e) Mode of reproduction.

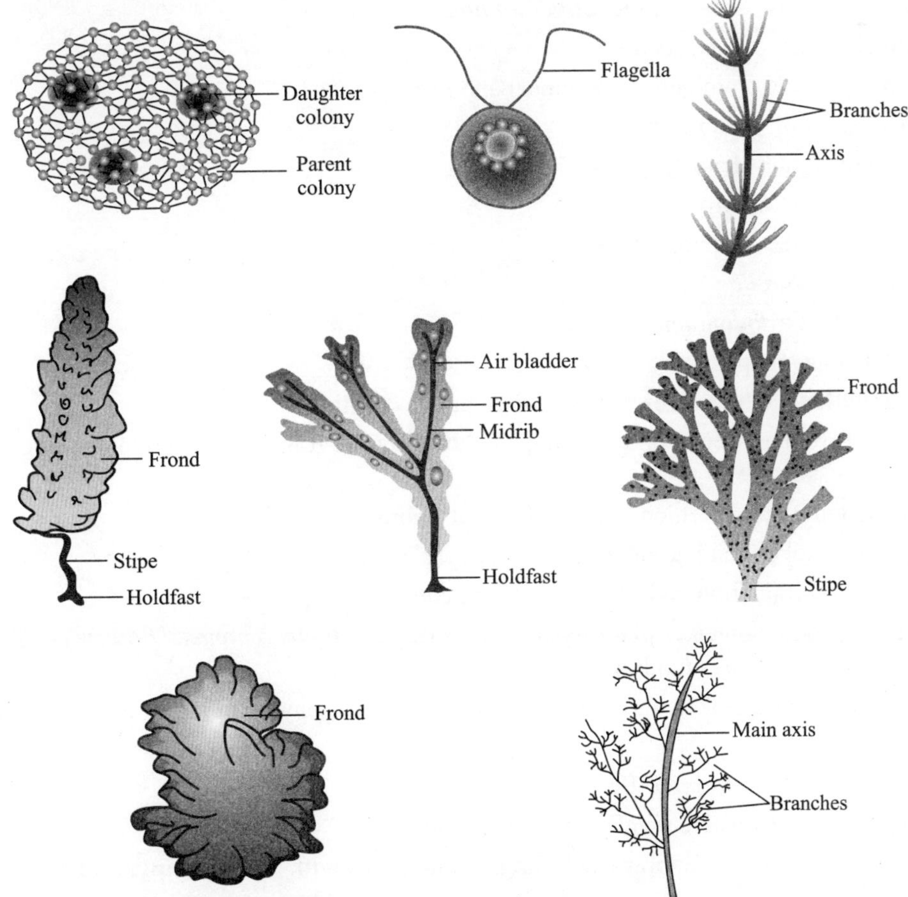

Fig. *(a) Green algae (i) Volvox (ii) Chlamydomonas (iii) Chara*

(b) Brown algae (i) Laminaria (ii) Fucus (iii) Dictyota

(c) Red algae (i) Porphyra (ii) Polysiphonia

Class 1: Chlorophyceae:

(a) These are mostly fresh water (90%).

(b) Thallus may be

- Unicellular and motile- *Chlamydomonas* (the smallest gametophytic plant)
- Unicellular and non-motile- *Chlorella*
- Filamentous, unbranched and attached- *Ulothrix, Oedogonium*
- Filamentous, unbranched and free-living- *Spirogyra*.
- Filamentous and branched- *Cladophora*.

- Colonial and motile- *Volvox, Gonium*
- Colonial and non-motile- *Scenedesmus, Hydrodictyon*
- Coenocytic- *Caulerpa*
- Heterotrichous- *Coleochaete, Stigoclonium*
- Palmella stage- *Tetraspora*.

(c) Cell wall is made up of cellulose (inner) and pectin (outer).

(d) Chloroplast is of variable shape

- Spiral or ribbon shaped: *Spirogyra*
- Cup shaped: *Chlamydomonas*
- Girdle or collar shaped: *Ulothrix*
- Discoid: *Chlorella*
- Reticulate: *Oedogonium*
- Stellate (star shaped): *Zygnema*

(e) Chlorophyll-a and b are present and β-carotene is dominant.

(f) Pyrenoid is proteinaceous body having sheath of starch. It is found in envelope of chloroplast. It synthesises starch.

(g) Reserve food is starch which is in the form of grains.

(h) Flagella are isokont and acronematic (anteriorly inserted).

(i) Vegetative reproduction takes place by fragmentation.

(j) Asexual reproduction takes place by zoospores (e.g., *Chlamydomonas*, *Ulothrix*) and aplanospore (e.g., *Chlamydomonas*; and hypnospore).

(k) Sexual reproduction takes place by isogamy, anisogamy, physiological anisogamy and oogamy.

Orders:

(i) Volvocales: Thallus with motile cell.

(ii) Chlorococcales: Cells are single and non-motile

(iii) Ulotrichales: Thallus is unbranched. Single chloroplast with 1 or more pyrenoid.

(iv) Cladophorales: Cells are cylindrical and multinucleate.

(v) Chaetophorales: Thallus is heterotrichous.

(vi) Oedogoniales: Cell division results in cap formation; Reticulate chloroplast.

(vii) Conjugales: Thallus is unbranched.

(viii) Siphonales: Thallus is coenocytic; Chloroplast many and discoid.

(ix) Charales: Thallus is differentiated into nodes and internode. Sex organs are nucule (female) and globule (male).

(x) Ulvales: Thallus is tubular.

2. Phaeophyceae (Brown algae):

(a) These are mostly marine (found in cooler sea) and lithophytes (growing attached to rocks).

(b) Thallus is differentiated into hold fast (for attachment), stipe (stalk) and frond (leaf like). Air vesicles are present on frond for providing buoyancy.

(c) Thallus is heterotrichous (different types of branches):

- Prostrate and erect system: *Fritschiella*

- Reduced prostrate and developed erect system: *Ectocarpus*
- Developed prostrate and reduced erect system: *Coleochaete.*

(d) Cell wall is made up of cellulose and algin.
(e) Chlorophyll-a and-c are present in Chromatophore.
(f) Dominant pigment is fucoxanthin.
(g) Reserve food is laminarin (sugar) and mannitol (alcohol) in pyrenoid.
(h) Flagella are heterokont (of different length) and stigonematic (laterally inserted).
(i) Pyrenoid is naked (without starch sheath) and stalked.
(j) Conducting or trumpet hyphae are present in large forms for transport of food.
(k) Vegetative reproduction takes place by fragmentation.
(l) Asexual reproduction takes place by pear shaped biflagellate zoospores.
(m) Sexual reproduction takes place by isogamy (e.g., *Ectocarpus*). anisogamy (e.g., *Cutleria*) and oogamy (e.g., *Fucus, Dictyota , Laminaria*).

Orders:

(i) Ectocarpales: Thallus is filamentous; Growth is trichothallic; Reproductive organs-Unilocular or plurilocular sporangia; Isomorphic alternation of generation.
(ii) Laminariales: Gametophyte is microscopic and dioecious; Sporophyte is large and parenchymatous; Sporangia (sori) on stipe and blade.
(iii) Fucales: Thallus is parenchymatous; sex organs are in conceptacles; It does not show alternation of generation.
(iv) Dictyotales: Thallus is parenchymatous; Growth by apical cell; Asexual reproduction by teraspores; Isomorphic alternation of generation.

3.Rhodophyceae (Red algae):

(a) These are predominantly marine (found in warmer sea) and lithophytes. Only freshwater red alga is *Batrachospermum.*
(b) Thallus may be
- Unicellular (e.g., *Porphyridium*);
- Pseudoparenchymatous (e.g., *Gelidium*);
- Filamentous -uniaxial filament (e.g., *Batrachospermum*)
- Multiaxial filament (e.g., *Polysiphonia*).

(c) Cell wall is made up of cellulose and pectin.
(d) Flagella are absent.
(e) Chlorophyll a and d are present in chromatophore (photosynthetic lamellae).
(f) Dominant pigment is phycoerythrin.
(g) Pyrenoid is absent.
(h) The reserve food is Floridian starch (it stains red with iodine test and resembles amylopectin (branched).
(i) Vegetative reproduction is not very common.
(j) Asexual reproduction takes place by unflagellate spores such as tetraspores, carpospores, monospore etc.

(k) Sexual reproduction takes place by oogamy. The gametes are non-motile.

(l) Male sex organ is spermatangium and female sex organ is carpogonium.

(m) Post fertilisation is very complex and occurs in carpogonium (flask-shaped).

4. Xanthophyceae (Yellow green algae):

5. Chrysophyceae (Yellow green algae):

6. Cryptophyceae:

7. Chloromonadinea (Bright green algae):

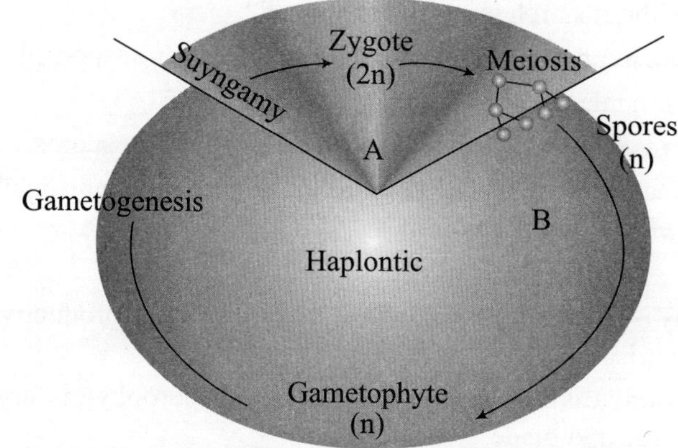

Fig: *Haplontic life cycle*

Evolution of Plants:

Tetrasporeine line: It is most accepted.

Chlamydomonas (motile and unicellular)→T*etraspora*(non-motile colony)→*Ulothrix* (filamentous)→ *Draparnaldiopsis* (heterotrichous)→ *Ulva* (parenchymatous)→*Chara* (branched multilayered)

Economic Importance:

(a) Algae produce organic matter as a result of photosynthesis and provide oxygen to aquatic organisms.

(b) Some algae such as *Chlorella, Scenedesmus , Ulva, Chondrus, Porphyra* etc. are source of proteins.

(c) Some brown algae such as *Fucus, Sargassum, Laminaria* etc. are used as fodder for animals.

(d) A few red algae such as *Gelidium, Gracilaria, Gigartina, Pterocladia* produce agar-agar which is dried hydrophilic and phycocolloidal polysaccharide (galactose monomer) obtained from the cell wall of algae. It is used in food, pharma and cosmetic industries. It is also used in media preparation as solidifying agent.

(e) A few brown algae such as *Laminaria* and *Fucus* which produce algin which phycocolloidal salt of alginic acid obtained from the cell wall of algae. It is used as thickener for preparing polish, cream, soup, sauce etc.

(f) A red alga, *Chondrus* produces carrageenan which is phycocolloidal sulphate polysaccharide present in its cell wall. It is used as stabiliser in cosmetic and paint industries.

(g) A green alga, *Chlorella* produces an antibiotic, chlorellin.

(h) Brown algae (kelps) are source of iodine and soda.

(i) A few green algae such as *Chlamydomonas*, *Chlorella, Scenedesmus*, *Pandorina* etc. remove organic matter and coliform bacteria by aerobic oxidation (pumping of O_2 as a result of photosynthesis).

(j) *Chlorella* has been used in space by astronauts because it has rapid rate of photosynthesis hence it produces oxygen and food to space traveller.

(k) *Chlorella* has also been used as experimental material in photosynthesis for studying path of carbon.

(l) *Acetabularia* has been used in cytology to prove nucleus as storehouse of genetic information.

Bryophyta

The term 'Bryophyta' (Gr. *Bryon*=moss, phyton=plant) was coined by Robert Brown (1864). Bryophytes were originated in devonian period.

Father of bryology: Hedwig

Father of Indian bryology: SR. Kashyap

Smallest bryophyte	*Zoopsis argentea* (a liver wort)
Largest bryophyte (gametophyte)	*Dawsonia* (a moss) =70 cm
Fossil bryophytes	*Ricciopsis, Hepaticites, Naiadetia*
Aquatic bryophytes	*Riccia fluitans, Ricciocarpus natans, Riella*
Saprophytic Bryophytes	*Bauxbaumia, Cryptothallus*

Features:

1. These grow on moist and shady places for leading vegetative phase as well as live in water for leading reproductive phase hence called amphibious plants. Western Himalaya is known as gold mine of bryophytes.
2. Plant body is differentiated into root like rhizoids for attachment with substratum, stem-like and leaf like structure but there is no true root, stem and leaf.
3. Vascular tissues are absent.
4. Main plant body is gametophyte which is dominant, free living, green (photosynthetic), independent and develops from spore. It is attached to soil so it can absorb water and minerals.
5. Sex organs are multicellular and jacketed.
6. Male sex organ is antheridium which encloses mass of sperms mother cell (androcyte) producing biflagellate and spirally coiled sperms that release on water.
7. Female sex organ is archegonium which is flask shaped containing tubular neck and swollen venter. Neck encloses 6-10 rows of neck canal cells while venter consists of a smaller and sterile ventral canal cell and a larger and fertile egg.
8. Sperm is attracted chemotactically (sugar) to egg.
9. Water is required for fertilisation.
10. Fertilisation occurs inside archegonium.
11. Zygote does not undergo resting period. It is first cell of sporophyte. It divides mitotically to form embryo.
12. Embryo is exoscopic (apex of embryo develops from outer wall) and later forms sporophyte.
13. Sporophyte is well-developed and differentiated into foot, seta and capsule.

14. Sporophyte is nutritionally (parasitically) dependent on gametophyte because it is aerial and has no soil contact and thus it is non-photosynthetic.
15. Sporogenous tissue (archesporium) shows progressive sterilisation from lower to higher forms.
16. Spore mother cell undergoes sporic meiosis inside capsule to form haploid spores. Bryophytes are homosporous (only one type of spores).
17. Spores are disseminated by wind and germinate immediately.
18. Vegetative reproduction takes place by progressive death and decay of thallus, fragmentation of gemmae, tubers, buds, adventitious branches and protonema.
19. Asexual reproduction by spores is absent.
20. Life cycle is haplo-diplontic and heteromorphic alternation of generation.
21. Bryophytes do not attain great height because (i) there is no root, (ii) there is no vascular tissue and mechanical tissue, (iii) there is no cuticle on the body, and (iv) there is requirement of water.

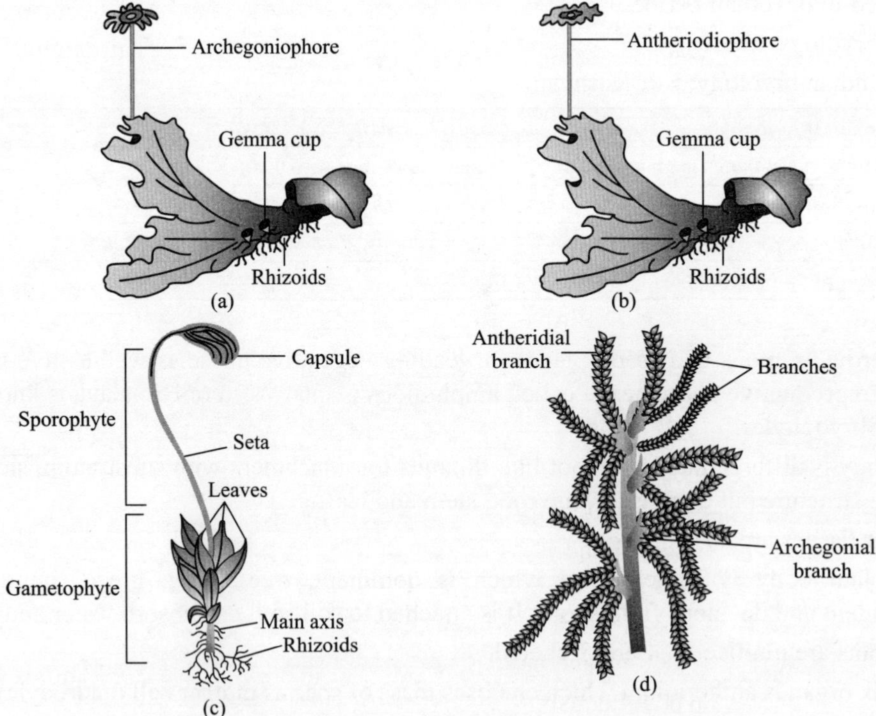

Fig. ***Bryophytes: A liverwoert-Marchantia (a) Female thallus (b) Male thallus Mosses (c) Funarta, gametophyte and sporophyte (d) Sphagrium gametophyte***

Classification:

1. **Hepaticopsida** (Liverworts):
   (a) Main plant body is thalloid, liver-like, dorso-ventrally flattened, and dichotomously branched.
   (b) Rhizoids are unicellular, non-septate, unbranched, smooth walled and tuberculate and found on ventral surface of thallus.
   (c) Thallus anatomically consists of upper (dorsal) assimilatory region having air chambers and air pores and the lower (ventral) storage region.

(d) Chloroplast is without pyrenoid.
(e) Scales are multicellular and found at ventral surface of thallus.
(f) Gemmae are green, multicellular, stalked bud and found on dorsal surface of thallus.
(g) Sex organs are dioecious (e.g., *Marchantia*) and monoecious (e.g., *Riccia*).
(h) Sex organs are embedded on dorsal surface of thallus (homothallic-*Riccia*) and projected from thallus in the form of stalked antheridiophore and archegoniophore (heterothallic-*Marchantia*).
(i) Sporophyte is simple, primitive and found at dorsal surface of thallus.
(j) Sporophyte is completely dependent on gametophyte.
(k) Seta is delicate and short-lived.
(l) Capsule does not contain stomata and chlorophyll hence non-photosynthetic. It does not show external differentiation.
(m) Sporogenous tissue (archesporium) develops from endothecium. It forms elaters and spores. Elaters are dead, elongated and hygroscopic having spiral band of thickening. These help in spore dispersal.
(n) Capsule splits irregularly by death and decay of spore wall.
(o) Vegetative reproduction takes place by death and decay of thallus and tubers.
(p) Asexual reproduction takes place due to germination of gemma. Each gemma has two thalli because two notches are present on opposite lateral side.

2. **Anthocerotopsida** (Hornworts):
(a) Plant body is thalloid. Thallus is simple in form, dorso-ventrally flattened variously lobed without midrib, air-chambers and air-pores.
(b) Rhizoids are unicellular and smooth walled. Tuberculate rhizoids are absent.
(c) Scales are absent.
(d) Sex organs are embedded in thallus on dorsal side.
(e) Sperms are biflagellate and whiplash type.
(f) Sporophyte is elongated and is differentiated into bulbous foot, buried in thallus and long, upright, and emerged cylindrical capsule.
(g) Seta is absent.
(h) Capsule is photosynthetic and has simple functional stomata.
(i) Capsule grows up continuously with the help of basal intercalary meristem.
(j) Capsule contains a central and sterile columella which develops from entire endothecium.
(k) Sporogenous tissue (archesporium) develops from inner layer of amphithecium.
(l) Archesporium gives rise to spores and elaters (no thickening band hence called pseudoelaters).
(m) Archesporium is dome shaped arching over columella at top.

3. **Bryopsida (Musci):**
(a) Main plant body is laterally branched. It shows

Primary protonema: It is green, thalloid, filamentous and branched and juvenile stage. It develops from germination of spore. It resembles green alga. It disappears after the establishment of leafy gametophore.

Gametophore: It is leafy (foliose), erect, spiral and mature stage. It is differentiated into axis, rhizoid and leaves but without midrib. It is monoecious.

(b) Rhizoids are multicellular, branched and oblique septate.

(c) Air-chambers and air-pores are absent.

(d) Scales are absent.

(e) Gemmae are absent.

(f) Sex organs are always monoecious. These develop in the axil of leaf. Antheridium is club shaped and stalked and it is surrounded by perigonial leaves. Archegonium is flask-shaped and surrounded by perichaetial leaves. It arises from apical cell of stem thereby terminating its growth (acrogynous condition). Both are intermingled with paraphysis. Largest archegonium is found in *Funaria*.

(g) Sporophyte is complex and advanced. It is partially dependent on gametophyte. It is differentiated into foot, seta and capsule.

(h) Seta is strong and long lived. It performs conduction.

(i) Capsule is differentiated into operculum, theca and paraphysis. It contains stomata and chlorophyll hence photosynthetic.

(j) Sporogenous tissue (archesporium) develops from outer layer of endothecium as in *Funaria* and from inner layer as in *Sphagnum*. The entire archesporium forms spores.

(k) Columella is sterile structure develops from inner layer of endothecium.

(l) Elaters are absent.

(m) Spore sac has many small spores.

(n) Split of capsule for spore dispersal is regular. There is elaborate mechanism of spore dispersal.

(o) Vegetative reproduction takes place by bulbil which is rounded bud like structure on rhizoid. It germinates to form protonema which forms thallus. It also occurs by secondary protonema which is formed from the structure other than spores such as rhizoids, stem and leaves. It breaks into a number of segments containing bud and each gives rise to leafy gametophores. e.g., *Funaria* (Cord moss), *Polytrichum* (Hair cap moss), *Sphagnum* (bog/peat moss), *Pogonatum* (Maiden hair moss), *Andrea* (Granite moss), *Grimmia* (Black moss), *Tortula* (Twisted moss).

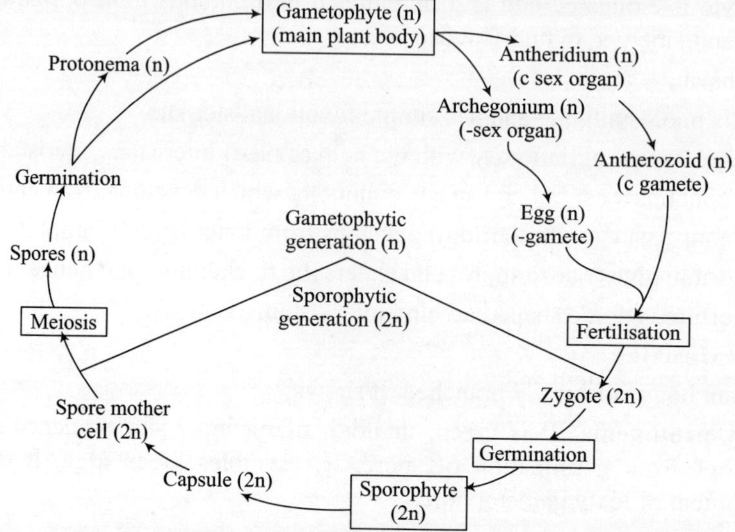

False moss: These plants belong to the groups other than bryophyte.

- Algae: *Chondrus crispus* (Irish moss/Sea moss), *Gracilaria* (Ceylon moss)
- Lichen: *Cladonia rangiferina* (Reindeer moss), *Cetraria icelandica* (Iceland moss), *Evernia* (Oak moss).
- Pteridophytes: *Lycopodium* (Club moss), *Selaginella* (Little club moss/spike moss), *Selaginella rupestris* (Bird's nest moss).
- Angiosperms: *Silene* (Champion moss), *Tillandsia* (Spanish moss/ Florida moss).

Economic importance:

(a) Bryophytes such as *Grimmia, Tortula Polytrichum* are link or transitional communities of primary succession in xerosere.

(b) *Marchantia polymorpha* is used in treatment of liver and lung disease.

(c) *Sphagnum* is source of peat which is dried, compressed, spongy and accumulation of partially decomposed plant due to anaerobic and highly acidic condition in wetland bog. It is also used as fuel and in surgical dressing due to its antiseptic property.

(d) *Funaria* is an indicator plant for copper.

(e) *Atrichium* is used as pollution indicator (bioindicator).

Pteridophyta

The term 'pteridophyta' (Gr. *Pteron*=feather, phyton=plant) was coined by Haeckel. Tracheophyta includes pteridophyta, gymnosperms and angiosperms. Pteridophytes were originated in silularian period of paleozoic era. Thus age of pteridophytes is carboniferous period. These are primitive vascular cryptogams (first land plants). These are called botanical snakes due to circinate coiling in young leaves.

Father of pteridology: Bowers

Father of Indian pteridology: S. S. Bir

Smallest pteridophyte	*Azolla*
Largest pteridophyte	*Cyathea*
Oldest fossil pteridophytes	*Rhynia, Cooksonia*

Features:

1. These are mostly herbaceous (annual and perennial).
2. These grow on moist cool and shady places (mostly terrestrial).
3. Main plant body is sporophyte which is dominant, free-living, independent, photosynthetic and develops from zygote.
4. Plant is divided into true roots (rhizome), stem and leaves.
5. Vascular tissues are present and form stele which is medullated (with pith) and non-medullated (without pith). Xylem lacks vessel and phloem lacks companion cell. Cambium is absent hence there is no secondary growth.
6. Ramenta is present in stem and leaves.
7. Leaves are microphyllous (small, simple and unveined) and megaphyllous (large, pinnately compound and branched veined). Young leaves show circinnate vernation.
8. Sporophyll is leaf-like structure which bears sporangia (spore bearing organ) on its abaxial

side. Sporophyll may be uniform as in ferns; compact (cone like) as in *Selaginella* or forms specialised structure called sporocarp as in *Marsilea*.

9. Sporangia may be single or in group (sori) protected by indusium.
10. The development of sporangia is of two types:
    - **Leptosporangiate:** It develops from single superficial cell. It divides transversely into outer and inner cell. The outer cell forms entire sporangium. The inner cell remains inactive and forms stalk of sporangium.
    - **Eusporangiate:** It develops from a group of superficial cells which divides periclinally differentiating into an outer layer of jacket initials and inner layer of primary sporangium cells. The jacket cells eventually produce sporangium; the primary sporogenous cell undergoes nuclear division.
11. Sporangia show the following types of arrangement:
    - Simple: All sporangia grow and mature together. e.g., *Ophioglossum*.
    - Gradate: Mature sporangia at centre and immature at periphery. e.g., *Marsilea*.
    - Mixed: Mature and immature sporangia are arranged irregularly. e.g., *Pteris*.
12. These are mostly homosporous (same type of spores) but heterosporous (different types of spores) condition is seen in *Selaginella, Isoetes, Azolla, Marsilea, Salvinia*.
13. Spore germinates from to form small, multicellular, short-lived, independent (free-living), photosynthetic (green), thalloid and heart-shaped gametophyte called prothallus. It is monoecious and protandrous.
14. The development of gametophyte is either exosporic (outside spore wall) as in homosporous forms which are monoecious or endosporic (inside spore wall) as in heterosporous forms which are dioecous.
15. Sex organs are multicellular and jacketed and these develop from ventral surface of prothallus. Sex organs may be embedded or projected.
16. Antheridium (male sex organ) is round and sessile which develops multi-flagellate and spiral sperms as in ferns but it forms numerous and biflagellate sperms as in *Selaginella*.
17. Archegonium (female sex organ) contains venter and neck. Venter divides to form smaller and sterile ventral canal cell and larger and fertile egg. There is single binucleate neck canal cell.
18. Fertilisation is internal and zooidogamous in which male gametes swim in a film of water to the female gametes (archegonium). Pteridophytes are found in narrow geographical region due to dependence on water for fertilisation.
19. Sperm is chemotactically (malic acid/fumaric acid) attracted to archegonia. (Smallest sperm is seen in *Isoetes*).
20. Zygote produces a small multicellular embryo for its nutrition.
21. Embryo is formed but no ovule. Embryo is endoscopic (apex is towards base of archegonia or pushed inside interior of female gametophyte.
22. Embryo develops into a mature diploid sporophyte.
23. Life cycle is diplo-haplontic or heteromorphic or heterologous alternation between independent sporophyte and gametophyte generation,
24. Vegetative reproduction takes place.
25. Asexual reproduction occurs by apogamy.

Classification:

1. **Psilopsida**: It is primitive and the simplest.
   (a) Sporophyte is rootless and leafless.
   (b) It is differentiated into underground prostrate stem and aerial and erect short system.
   (c) It shows dichotomous branching.
   (d) Sporangia are found in axil of small leaf like appendages.
   e.g., Whisk fern (*Psilotum*)
2. **Lycopsida:**
   (a) Leaves are microphyllous and spirally arranged.
   (b) Sporangia are found singly at axial surface of sporophyll.
   (c) Homosporous as well as heterosporous conditions are found.
   e.g., Little club moss/Spike moss (*Selaginella*), Club moss (*Lycopodium*), Quill wort (*Isoetes*).
3. **Sphenopsida:**
   (a) Sporophyte has joint axis differentiated into roots and internodes.
   (b) Leaves are microphyllous and show whorled phyllotaxy.
   (c) Sporangia are found in groups at the end of branches on sporangiophore.
   (d) These are homosporous.
   e.g., Horse tail (*Equisetum*), Adder's tongue fern (*Ophioglossum*)

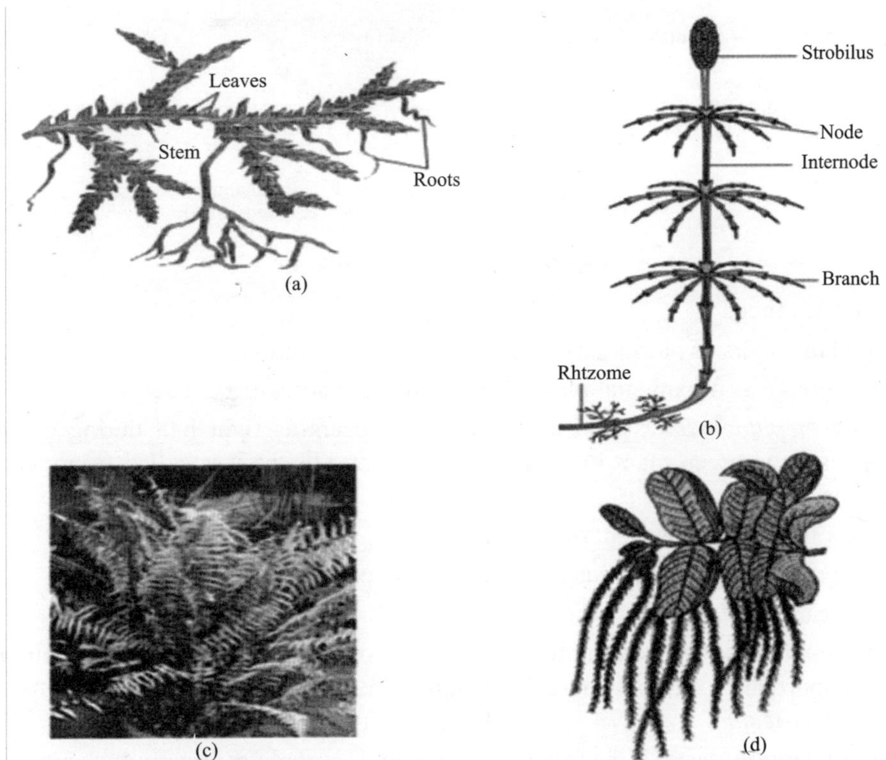

Fig: *Selaginella* (a), *Equisetum* (b), Fern (c) and *Salvinia* (d)

4. **Pteropsida**: It is the largest and highly advanced.
   (a) Sporophyte consists of underground stem called rhizome.
   (b) Leaves are megaphyllous and pinnately compound.
   (c) Sporangia are aggregated to form sori covered by indusium and at ventral surface of sporophyll. e.g., Male shield fern (*Dryopteris*), Sun fern (*Pteris*), Maiden hair fern or Walking fern (*Adiantum*), Tree fern (*Cyathea*) Filmy fern (*Hymenophyllum*), Water ferns (*Azolla, Marsilea, Salvinia*), Bracken fern (*Pteridium*).

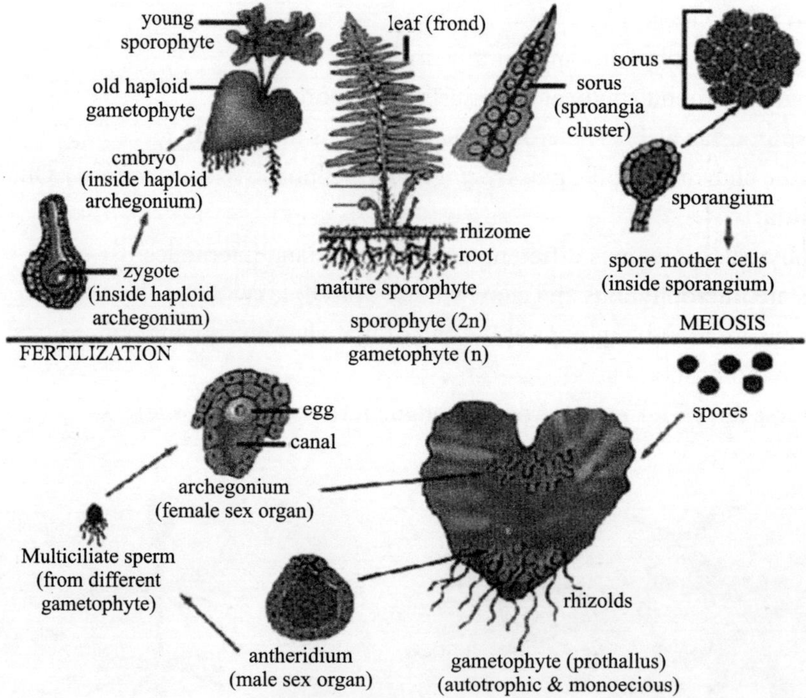

Fig: *Diplo-haplontic life cycle*

Economic importance:

(a) The fossil members of pteridophytes constitute coal formation.
(b) Ferns are grown as ornamental plants because of their attractive foliage.
(c) *Selaginella lepidophylla* is sold as novelty. It appears as tight ball during dry period and later opens up in green plant in the presence of water hence it is called resurrection plant or Sanjeevani.
(d) *Equisetum* is an indicator plant of gold in soil.
(e) *Azolla* is grown commercially as biofertiliser due to its nitrogen fixing ability.
(f) The sporocarp of *Marsilea* is edible.

Apospory in pteridophytes: It is the formation of diploid gametophyte directly from one or more vegetative cell of sporophyte without spore formation or meiosis. It was reported by Druery (1884) in a fern, *Arthyrium filix-femina*.

Apogamy in pteridophytes: It is the formation of haploid sporophyte directly one or more vegetative cell of gametophyte except egg without fertilisation. It was reported by Farlow (1874) in *Pteris cretica*.

Pre-requisites for seed habit: For the first time, seed habit was originated in pteridophytes. Following features are required to form seed in any plants:

(a) Production of two types of spores (heterospory).

(b) Formation of one megaspore (female gametophyte) within megasporangium.

(c) Presence of additional protective coat called integument around megasporangium.

(d) Permanent retention of megaspore inside megasporangium.

(e) *In situ* fertilisation and development of embryo.

Gymnosperms

The term 'Gymnosperm' (Gr. *Gymnos*=naked, sperma=seed) was coined by Theophrastus. These are seed bearing plants but seeds are not enclosed. These are smallest group of plants having around 900 and 70 living genera. These were originated in Carboniferous period of Paleozoic era.

Smallest gymnosperm	*Zamia pygmaea*
Tallest gymnosperm	Giant Redwood (*Sequoia sempervirens*)
Gymnosperm of highest life span	*Macrozamia* (10,000 years)

Features:

1. The gymnosperms are mostly evergreen trees or shrubs (no herbs). These grow in temperate region where they maintain their dominance over angiosperms.
2. These are xerophytic in nature.
3. Main plant body is sporophyte which is divided into roots, stem and leaves.
4. Root is well developed and is of tap root. Root shows diarch to polyarch vascular bundle with exarch condition. The root of *Cycas* (coralloid root) is negatively geotropic and does not bear root hairs and root cap. It shows symbiotic association with *Anabaena* for nitrogen fixation. But the root of *Pinus* shows symbiotic association with fungi to form ectomycorrhiza.
5. Stem is woody and unbranched, soft and manoxylic (containing parenchyma) in *Cycas* but branched, hard and pycnoxylic (without parenchyma) in *Pinus*. Xylem contains tracheids and parenchyma with bordered pits but lacks vessels. *Ephedra* and *Gnetum* exceptionally show the presence of vessel. Phloem contains fibres and parenchyma but lacks sieve tube and companion cell. Instead it contains sieve cell. *Cycas* has polyxylic vascular bundle (many cambial rings and vascular bundle) but *Pinus* has monoxylic vascular bundle (one cambial ring and vascular bundle). Secondary growth occurs and annual rings are formed.
6. Leaf is dimorphic and spirally arranged. The foliage leaf is green and pinnately compound as in *Cycas* and simple and needle-like (acicular) as in *Pinus*. The scaly leaves are minute and deciduous. There is no venation of leaf. Leaf has thick cuticle, thick epidermis having sunken stomata and sclerenchymatous hypodermis. Transfusion tissue is found in endodermis around vascular bundle. It is composed of tracheid and parenchyma cells. It helps in lateral conduction of water.
7. Sporophyll is leaf bearing spores. It may be microsporophyll and megasporophyll which aggregate to form male and female cone, respectively. *Cycas* has the largest male cone.

Microsporophyll	Equivalent to stamen of angiosperms
Microsporangium	Equivalent to pollen sac of angiosperms
Microspore	Equivalent to pollen grain of angiosperms

Megasporophyll	Equivalent to carpel of angiosperms
Megasporangium	Equivalent to nucellus of ovule of angiosperms
Megaspore	Equivalent to embryo sac of angiosperms

8. Eusporangiate development of sporangia occurs.
9. Heterospory condition is seen. Microspore germinates to form male gametophyte and megaspore germinates to form female gametophyte.
10. Male and female gametophytes are highly reduced, embedded in sporophyte and not free-living (independent). These develop from sporophytic structure.
11. Gymnosperms are mostly dioecious (e.g., *Cycas*) and few are monoecious (e.g., *Pinus*).
12. Antheridium is male sex organ. It bears non-flagellate sperms but *Cycas* shows multiciliate and top-shaped sperm.
13. Archegonium is female sex organ. It contains ventral canal cell and egg. *Cycas* has no neck canal cells but shows the largest egg.
14. Pollination is anemophilous (through wind).
15. Pollen grains shed at 3-celled stage as in *Cycas* and 4-5 celled stage as in *Pinus*.
16. Pollens are deposited in pollen chamber in nucellus of ovule.
17. Fertilisation does not require water. It is siphonogamous (with the formation of pollen tube) and zooiodogamous (sperm reaches the archegonium by swimming in the fluid present in archegonial chamber).
18. Zygote forms embryo.
19. Endosperm is haploid (develops before fertilisation).
20. Double fertilisation does not occur. It occurs in *Ephedra* only.
21. Embryo develops from a part or lower half of zygote (meroblastic development). It shows endoscopic condition.
22. Polyembryony occurs but usually one embryo survives.
23. Embryo shows radicle, plumule, haustorium and cotyledon.
24. Ovule is orthoropous and unitegmic (one integument has outer fleshy, middle stony and inner fleshy). *Cycas thourasii* has the largest ovule (7cm in length).
25. Ovule is transformed into seed which is not covered but exposed on the surface of megasporophyll.
26. Seed has two cotyledons as in *Cycas* and three to fifteen (polycotyledonous) as in *Pinus*. Seed undergoes a resting period and during the period seeds are dispersed over vast areas. Germination of seed is hypogeal.
27. Seed represents three generations:
    (a) Parent (old) sporophytic generation is represented by nucellus and integument.
    (b) Gametophytic generation is represented by endosperm.
    (c) Future (new) sporophytic generation is represented by embryo.
28. Life cycle is diplontic.

Classification:

1. **Pteridospermopsida:**
    (a) Leaf is large, pinnately compound and frond-like having large leaf trace.

(b) Spores are aggregated to form synangia.

(c) Sporangia are borne on leaves which never form cones.

e.g., *Caytonia, Glossopteris*

2. Cycadopsida:

(a) Stem is unbranched and manoxylic.

(b) Leaf is pinnately compound having direct and girdle like trace.

e.g., *Cycas* (Sago palm), *Cycadeoidea, Nilssonia, Pentoxylon.*

3. Coniferopsida: (largest class)

(a) Stem is branched and pycnoxylic.

(b) Leaf is simple.

e.g., *Pinus, Taxus, Ginkgo* (Maiden hair tree or Pagoda tree), *Cedrus deodara* (the strongest soft wood).

4. Chlamydospermopsida:

(a) Xylem contains vessels.

(b) Leaves are dimorphic.

(c) Strobilus is compound.

e.g., *Ephedra* (Jointed fir), *Gnetum, Welwitchia.*

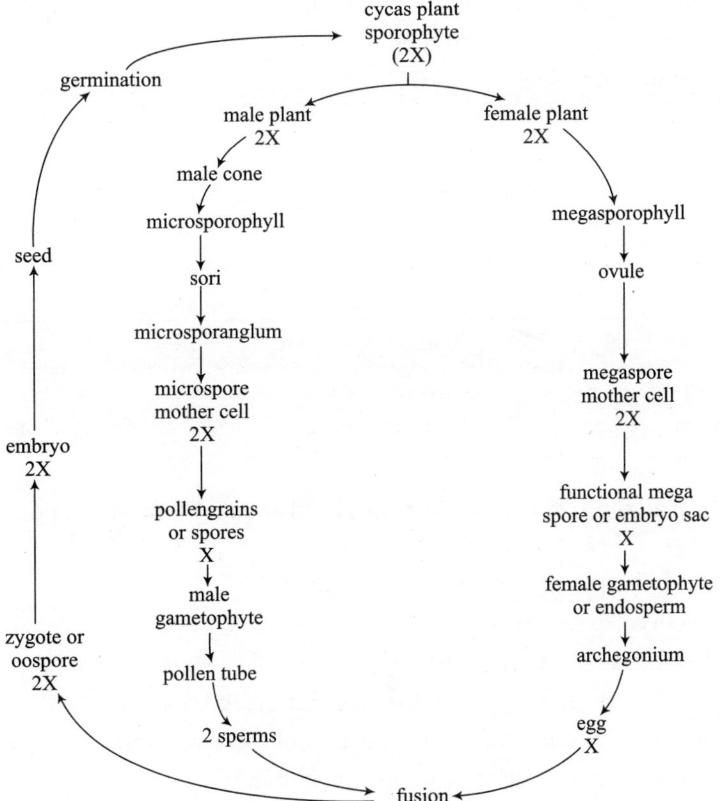

Fig: *Diplontic life cycle of Cycas*

Economic importance:

(a) *Cycas revoluta, Araucaria excelsa* (Christmas tree), *Araucaria imbricata* (Monkey puzzle), *Pinus, Cupressus* are grown as ornamental plants.

(b) Stem and seed of *Cycas circinnalis* yield a starch called sago.

(c) Seed of *Pinus gerardiana* is source of chilgoza which is edible.

(d) Stem of *Pinus roxburghii* (Chir), *Pinus wallichiana* (Kail) and *Pseudotsuga menzeiss* (Douglas fir) are source of timber.

(e) Stem of *Pinus roxburghii* (Chir) is source of resin and turpentine oil.

(f) Bark of *Taxus* produces taxol which is anticancerous compound.

(g) Stem of *Juniperus virginniana* (Red cedar) produces red cedar wood oil used in microscopy.

(h) Stem of *Abies balsamia* produces a mounting agent, canada balsam used in preparation of permanent slide.

(i) Stem or whole plant of *Ephedra* produces ephedrine used in respiratory trouble.

(j) Wood pulp of *Picea* (Spruce) is source of paper.

Adaptations to land plants:

(a) Gametophytic generation is highly reduced and is protected inside sporophytic tissue.

(b) Pollination occurs through wind.

(c) Fertilisation does not require water.

(d) Fertilisation occurs through pollen tube.

(e) Ovule is developed into seed.

Angiosperms

The term 'Angiosperm' (Gr. *Angeion* = naked; sperma = seed) was coined by Oswald Tippo. Angiosperms are flowering plants in which seeds are enclosed inside fruit. These are the largest group of plants and most advanced. Angiosperms were originated in Cretaceous period of Mesozoic era and appeared about 130 million years ago.

Smallest angiosperm	*Wolffia microscopica* (0.1 mm)
Tallest angiosperm	*Eucalyptus regnans* (114 m)
Largest spreading tree	Banyan tree (*Ficus benghalensis*)-Father of forest

Features:

1. These occur in all types of habitat (terrestrial and aquatic). Marine angiosperms are *Thalassia* and *Zostera*.
2. Main plant body is sporophyte which is dominant.
3. Sporophyte is divided into root, stem and leaves
4. Mechanical and Vascular tissues are well developed in vegetative parts.
5. Vessels are found in xylem for long distance and quick transport
6. Sieve tube is found in phloem for translocation of food material.
7. Secondary growth occurs due to presence of cambium.
8. Bark is present which is meant for protection.
9. Sporophylls are aggregated to form flower.

10. These are mostly monoecious and some are dioecious.
11. Male sex organ is stamen and female sex organ is carpel.
12. Dispersal of fruits and seeds takes place.
13. Heterospory condition is seen.
14. Megaspore is single and functional and permanently retained in nucellus.
15. Embryo sac is well developed.
16. Gametophyte is dependent on sporophyte.
17. Archegonia are absent.
18. Pollination occurs by several agencies.
19. Fertilisation does not require water.
20. Fertilisation is siphonogamous (presence of pollen tube).
21. Double fertilisation is seen.
22. Endosperm is triploid which occurs after fertilisation due to triple fusion.
23. Fertilised ovule ripens into seed.
24. Fertilised ovary ripens into fruit.

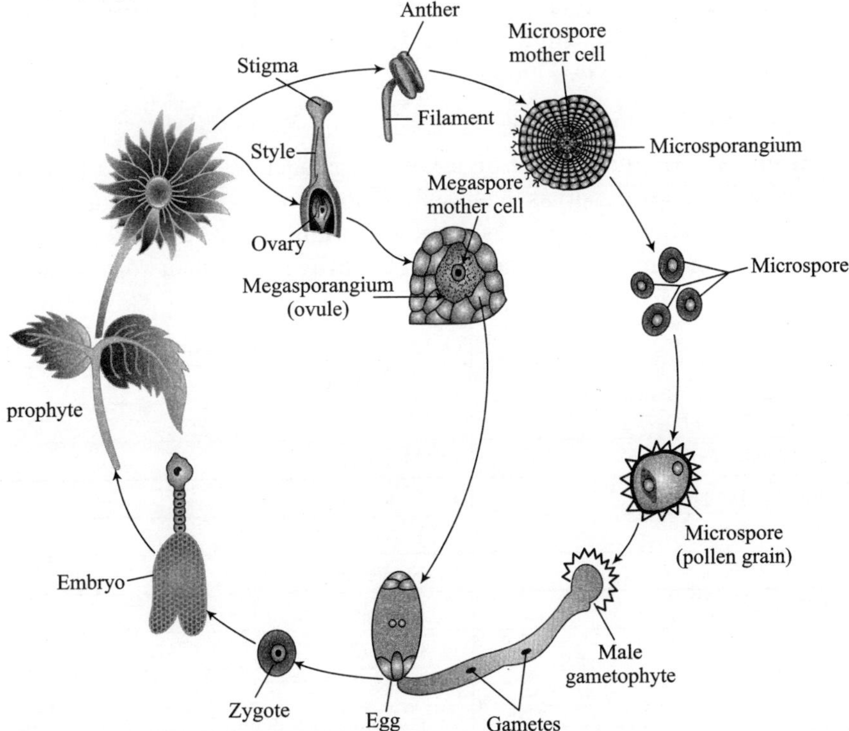

Fig. ***Life cycle of an angiospher***

Classification of Angiosperms:

There are several terms for angiosperms which are as follows:

1. Anthophyta (a division) by Bessey (1915) who composed 'Phylogenetic Taxonomy of Flowering Plants'.

2. Magnoliophyta (a division) by Takhtajan who wrote 'A system of phylogeny of flowering plants' and Cronquist who wrote 'Evolution and classification of flowering plants'.
3. Annonopsida (a class) by Dahlgren.

There are three approaches to classify the angiosperms:

1. **Artificial system** (Utilatarian): It is based on a few superficial vegetative characters chosen arbitrarily. Several taxonomists proposed different classifications which are as follows:

Theophrastus (320 BC): He composed De Historia Plantarum (the oldest book) and enquiry of plants. He described 480 plants. He pointed out fundamental difference between dicots and monocots. He classified plants into four groups on the basis of habitat.

Herbs	Under shrubs	Shrubs	Trees

He further classified herbs into Annuals, Biennials and Perennials.

Caesalpino: He proposed methodical classification of plants based on definite morphological characters. He first suggested the use of embryological characters in Taxonomy. He classified plants into two groups:

Woody	Herbaceous

John Ray: He composed 'Historia Generalis plantarum'. He considered all parts of plant for classification which is now basis of natural system. He was first to divide seed plants into Dicots and Monocots. He classified Herbs into two groups:

Imperfectae (Non-flowering plants)	Perfectae (flowering plants).

Carolus Linnaeus (1753): He is father of Taxonomy/Modern Botany. He described 7800 plants. He considered stamen (male sex organ) as the basis and noted constancy in the number of stamens that is why his classification is known as sexual system and arithmetical system. He proposed 24 classes based on number and arrangement of stamens (23 classes were of Phanerogams and 1 class (24th) belonged to Cryptogams.

1. Monandria (1 stamen)	13. Polyandria (20 or more attached to receptacle)
2. Diandria (2 stamens)	14. Didynamia
3. Triandria (3 stamens)	15. Tetradynamia
4. Tetrandria (4 stamens)	16. Monadelphia
5. Pentandria (5 stamens)	17. Diadelphia
6. Hexandria (6 stamens)	18. Polyaldelphia
7. Heptandria (7 stamens)	19. Syngenesia
8. Octandria (8 stamens)	20. Gynandria
9. Enneandria (9 stamens)	21. Monoecia
10. Decandria (10 stamens)	22. Diocea
11. Dodecandria (11-19 stamens)	23. Polygamia (both unisexual and bisexual flowers on same plant).
12. Icosandria (20 or more attached to calyx)	24. Cryptogamia (from algae to pteridophytes)

Demerits:

(a) Only a few vegetative characters are considered.

(b) Vegetative characters are changed according to environment.

(c) Natural affinities and phylogenetic relationships were not considered.

(d) It is based on dissimilarities rather similarities because two plants which are similar in many characters are separated on the basis of single character.

2. **Natural system (Horizontal/2-D system):** It is based on form relationships. The taxonomists such as de Candolle, Jussieu, Robert Brown, Bentham & Hooker proposed different classifications under this system.

Bentham and Hooker: They wrote 'Genera Plantarum' in 3 volumes (1st in 1862, 2nd in 1872 and 3rd in 1883). They proposed classification considering large number of characters based on similarities and dissimilarities among organisms. They gave an idea about natural relationships amongst organisms. It is based on de Candolle's system. It is widely accepted classification for practical utility. They recognised 202 families, 7,569 genera and 97,205 species.

Phanerogams:

	↓	↓	↓
Class	1. Dicotyledonae	2. Gymnosperms	3. Monocotyledonae

Dicotyledonae has been divided into 3 sub-classes: It contains 165 families.

Sub-class:

1. **Polypetalae** (Petals are free): It is divided into 3 series.
   (a) Thalamiflorae: Ovary is hypogynous. It is divided into 6 orders.
   Orders:Ranales,Parietales, Polygalinae, Caryophyllinae, Guttiferales,Malvales
   (b) Disciflorae: Ovary is hypogynous; Disc is present below ovary.
   Orders: Geraniales, Olacales, Celastrales, Sapindales
   (c) Calyciflorae: Ovary is epigynous.
   Orders: Rosales, Myrtales, Passiflorales, Ficoidales, Umbellales.
2. **Gamopetalae** (Petals are fused). It is divided into 3 series.
   (a) Inferae: Ovary is inferior; Stamen is equal to petal lobes.
   Orders: Rubiales, Asterales, Campanales
   (b) Heteromerae: Ovary is superior; Carpels are more than 2.
   Orders: Ericales, Primulales, Ebenales
   (c) Bicarpellatae: ovary is superior; Carpels are two.
   Orders: Gentianales, Polemoniales, Personales, Lamiales.
3. **Monochlamydae**: It has one whorl of perianth. It is divided into 8 series:
   (a) Curvembryae: Embryo is curved.
   (b) Multiovulate aquaticae: Aquatic plants with numerous ovules
   (c) Multiovulate terrestris: Terrestrial plants with numerous ovules
   (d) Microembryae: embryo is very small.
   (e) Daphnales: Monocarpellary ovary with single ovule.
   (f) Achlamydosporae: Inferior ovary
   (g) Unisexuales: Flowers are unisexual.
   (h) Ordines anomali: Uncertain or abnormal relationship in families.

Gymnospermae: It contains 3 families. It is divided into 3 orders:

(a) Cycadales: e.g., *Cycas*
(b) Coniferales: e.g., *Pinus*
(c) Gnetales: e.g., *Gnetum*

Monocotyledonae: It contains 34 families. It is divided into 7 series:

(a) Microspermae: Seeds are small; Ovary is inferior
(b) Epigynae: Seeds are large; Ovary is inferior.
(c) Coronariae: Perianth is coloured; Ovary is superior.
(d) Calycinae: Perianth is green; ovary is superior
(e) Nudiflorae: Perianth is absent.
(f) Apocarpae: Carpels are free.
(g) Glumaceae: Perianth is reduced.

Merits:

(a) It is simple in nature because it is based on actual observation.
(b) It focused on floral characters.
(c) It provides a key for the identification of plants.
(d) Polypetalae is considered as most primitive and Monochlamydae as most advanced.
(e) Ranales order was placed first in dicots and it is quite reasonable.
(f) Glumaceae is considered most advanced among monocots.

Demerits:

(a) Evolutionary criteria or phylogenetic relationship were not followed.
(b) Gymnospermae has been placed between dicots and monocots. It is only for convenience rather than affinities. It is extremely anomalous.
(c) In polypetalae, plants with superior ovary have been placed first than plants with inferior ovary while in Gamopetalae, plants with inferior ovary have been placed first.
(d) Cucurbitaceae has been placed in polypetalae in spite of having fused petals.
(e) The sub-class, Monochlamydae, is an entirely artificial group.
(f) Monocots have been directly divided into series.
(g) Orchidaceae, being an advanced family under series Microspermae, has been considered primitive.

4. **Phylogenetic system (3-D system)**: Engler & Prantl, Eichler, Hutchinson, Bessey, Cronquist, Takhtajan. It is based on evolutionary relationship or tendencies of organisms. The organisms belonging to same taxa are believed to have a common ancestor and may be represented in the form of family tree called cladogram.

Engler & Prantl: They composed 'die Naturlichen Pflanzenfamilien' in 23 volumes. They proposed classification based on increasing complexity of flowers. They recognised 280 families.

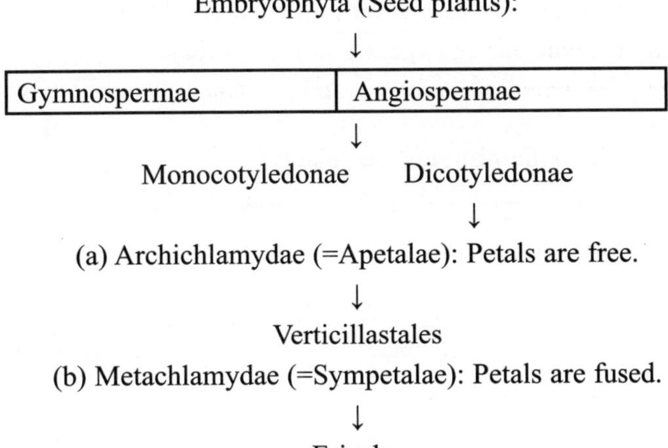

Demerits:

(a) Monocots were considered primitive and dicots as advanced.

(b) Monophyletic origin of angiosperms.

Hutchinson: He composed 'Families of Flowering plants'(2 volumes). He proposed classification and recognised 411 families based on 24 primitive or advanced characters which are as follows:

(a) Evolution is both upwards and downwards. The former leads to preservation and latter leads to suppression (reduction) of characters.

(b) Evolution does not involve all organs of plants at same time.

(c) Evolution is generally consistent.

(d) Shrubs and trees are primitive whereas herbs and climbers are advanced.

(e) Terrestrial is primitive whereas aquatic is advanced.

(f) Perennial is primitive whereas biennial and annual are advanced.

(g) Normal habit is primitive whereas parasitic and saprophytic habits are advanced.

(h) Dicots are primitive whereas monocots are advanced.

(i) Simple leaf is primitive whereas compound leaf is advanced.

(j) Spiral leaf is primitive whereas cyclic leaf is advanced.

(k) Monoecious condition is primitive whereas dioecious is advanced.

(l) Solitary flower is primitive whereas inflorescence is advanced.

(m) Bisexual flower is primitive whereas unisexual flower is advanced.

(n) Actinomorphic condition is primitive whereas zygomorphic is advanced.

(o) Petaliferous condition is primitive whereas apetalous condition is advanced.

(p) Polypetalous condition is primitive whereas gamopetalous condition is advanced.

(q) Imbricate aestivation is primitive whereas valvate aestivation is advanced.

(r) Free anther is primitive whereas fused anther is advanced.

(s) Many stamens in a flower are primitive whereas few stamens are advanced.

(t) Polycarpy is primitive whereas oligocarpy (a few carpels) is advanced.

(u) Apocarpous condition is primitive whereas syncarpous is advanced.

(v) Hypogynous condition is primitive whereas epigynous condition is advanced.

(w) Single fruit is primitive whereas aggregate fruit is advanced.

(x) Endospermic seed with small embryo is primitive whereas non-endospermic seed with large embryo is advanced.

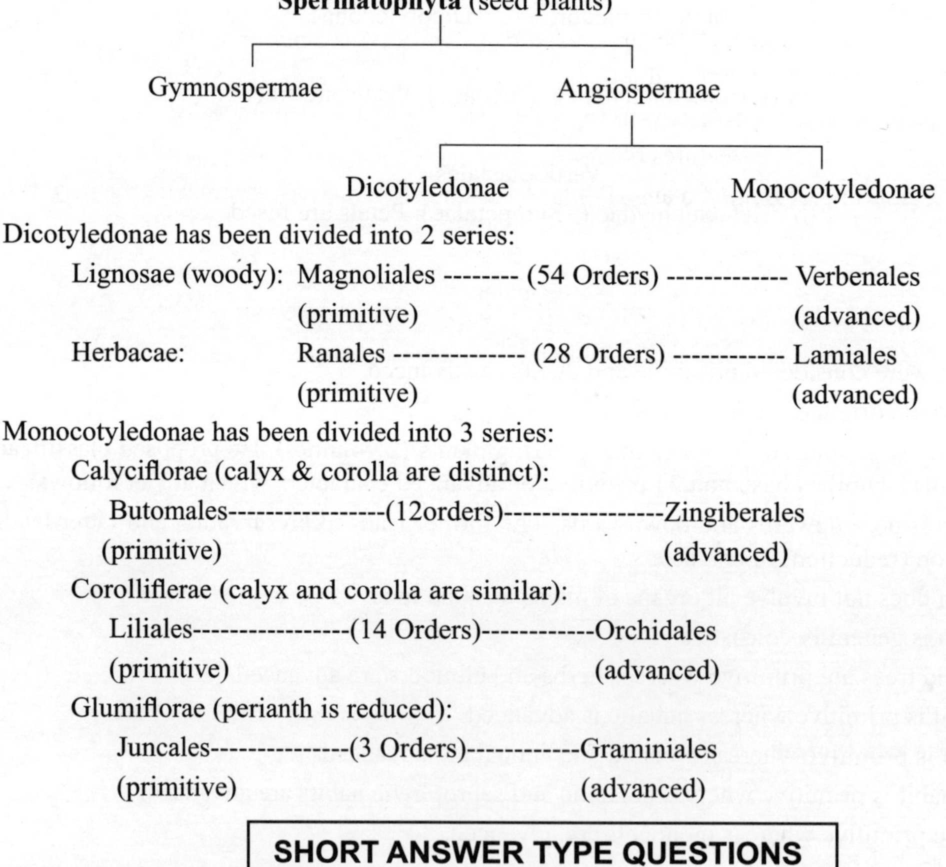

Dicotyledonae has been divided into 2 series:

Lignosae (woody): Magnoliales -------- (54 Orders) ------------- Verbenales
(primitive) (advanced)

Herbacae: Ranales -------------- (28 Orders) ------------ Lamiales
(primitive) (advanced)

Monocotyledonae has been divided into 3 series:

Calyciflorae (calyx & corolla are distinct):

Butomales-----------------(12orders)-----------------Zingiberales
(primitive) (advanced)

Corolliflerae (calyx and corolla are similar):

Liliales-----------------(14 Orders)------------Orchidales
(primitive) (advanced)

Glumiflorae (perianth is reduced):

Juncales---------------(3 Orders)------------Graminiales
(primitive) (advanced)

SHORT ANSWER TYPE QUESTIONS

1. Discuss the different modes of sexual reproduction in algae.
2. What are the different criteria of classification in algae?
3. Discuss the economic importance of algae.
4. Discuss the range of thallus structure in green algae.
5. Explain the haplo-diplontic life cycle in bryophytes.
6. Draw a well-labeled diagram of dioecious condition of *Marchantia*.
7. What are the prerequisites for seed habits in plants?
8. Distinguish between bryophytes and pteridophytes.
9. How the life cycle of pteridophytes differs from bryophytes?
10. Distinguish between leptosporangiate and eusporangiate development.

11. What are the adaptations to land plants?
12. Distinguish between *Cycas* and *Pinus*.
13. What are the demerits of artificial classification?
14. What do you mean by phylogenetic system of classification and give its examples.
15. Distinguish between gymnosperms and angiosperms.

LONG ANSWER TYPE QUESTIONS

1. Describe the characteristic features of green algae, brown algae and red alage.
2. Distinguish between liverworts and mosses.
3. Describe the characteristic features of pteridophytes and explain its life cycle.
4. Describe the characteristic features of gymnosperms and explain its life cycle.
5. Describe the outline of Bentham and Hooker's classification and mention their demerits.

MULTIPLE CHOICE QUESTIONS

1. Agar-agar is obtained mostly from *Gelidium* which belongs to
   (a) Chlorophyceae (b) Phaeophyceae (c) Rhodophyceae (d) Cyanophyceae
2. The position of flagella in the members of brown algae is
   (a) Apical (b) Lateral (c) Subapical (d) Basal
3. Motile stages are absent in
   (a) Green algae (b) Brown algae (c) Red algae (d) Mosses
4. Haplontic life cycle is common in algae but diplontic life cycle is seen in
   (a) *Chlamydomonas* (b) *Spirogyra* (c) *Fucus* (d) *Ulothrix*
5. Heterotrichous habit is seen in
   (a) *Fucus* (b) *Ectocarpus* (c) *Chara* (d) *Chlamydomonas*
6. Which of the statements is true for red algae?
   (a) Pyrenoid is present. (b) Chlorophyll a and -c are present
   (c) Reserve food material is floridean starch (d) Dominant pigment is fucoxanthin.
7. Which of the statements is true for brown algae?
   (a) Flagella are not seen (b) Algin is not found in cell wall
   (c) Frond is seen (d) Pyrenoid is absent
8. Which of the following types of meiosis occurs in algae?
   (a) Sporic (b) Zygotic (c) Gametic (d) All of the above.
9. The type of sex organs in algae is
   (a) Unicellular and non-jacketed (b) Unicellular and jacketed
   (c) Multicellular and non-jacketed (d) Multicellular and jacketed.
10. Embryphyta does not include
    (a) Algae, Bryophytes (b) Brypohytes, pteridophytes
    (c) Gymnosperms and angiosperms (d) Algae only
11. Archegoniatae does not include
    (a) Algae and Bryophytes (b) Bryophytes and Pteridophytes
    (c) Gymnosperms and angiosperms (d) Algae and angiosperms
12. The capsule of sporophyte is photosynthetic in
    (a) *Riccia* (b) *Marchantia* (c) *Anthoceros* (d) *Sphagnum*
13. In which plant of bryophytes does archesporium arise from endothecium?
    (a) *Marchantia* (b) *Sphagnum* (c) *Funaria* (d) *Anthoceros*
14. Heterospory and ligulate leaves are the characteristic features of
    (a) *Lycopodium* (b) *Selaginella* (c) *Equisetum* (d) *Adiantum*
15. Prothallus in pteridophytes is
    (a) Monoecious and protogynous (b) Monoecious and protandrous
    (c) Dioecious and protogynous (d) Dioecious and protandrous

16. Spores of *Funaria* on germination gives rise to
    (a) Primary protonema (b) Secondary protonema
    (c) Gametophores (d) All of the above
17. Sporophyte is partially dependent on gametophyte in
    (a) Liverworts (b) Hornworts (c) Mosses (d) All of the above
18. Sporophylls are compact and cone like in
    (a) Ferns (b) *Selaginella* (c) *Azolla* (d) *Marsilea*
19. Which of the following is wrong with respect to pteridophytes?
    (a) Young leaves show circinate vernation (b) Phloem lacks companion cells
    (c) Sex organs are multicellular and jacketed (d) Life cycle is diplontic
20. Circinate vernation in ferns refers to
    (a) Uncoiling of new leaves from base to apex (b) Coiling of new leaves from base to apex
    (c) Shedding of leaves from base to apex (d) Arrangement of leaves
21. *Cycas* is vegetatively propagated by
    (a) Corms (b) Bulbs (c) Bulbils (d) Suckers
22. Which of the following is wrong with respect to gymnosperms?
    (a) Phloem contains sieve cells (b) Leaf has sunken stomata
    (c) Polyembryony is seen (d) Ovule is anatropous
23. Which of the following is wrong with respect to *Cycas*?
    (a) The root of *Cycas* shows Nitrogen fixing ability
    (b) Stem is unbranched
    (c) Leaf is pinnately compound
    (d) It is monoecious
24. The sperm of *Cycas* is
    (a) Top shaped and multiciliate (b) Pear shaped and biflagellate
    (c) Pear shaped and unflagellate (d) Top shaped and biflagellate
25. In Gymnosperms, endosperm is formed by the
    (a) Fusion of egg with one of polar nucleus (b) Fusion of egg with two polar nuclei
    (c) Fusion of egg with synergid cells (d) Germination of megaspore
26. The wing in seed of *Pinus* develops from
    (a) Nucellus (b) Ovuliferous scale (c) Endosperm (d) Integument
27. The winged pollen grains of *Pinus* are produced in
    (a) Pollen chamber (b) Anther
    (c) Tapetum (d) Microsporangium
28. The edible part of *Pinus* seed is
    (a) Pericarp (b) Female gametophyte
    (c) Perisperm (d) Endosperm

29. The seed of *Pinus* is
    (a) Adaxial and endospermic (b) Adaxial and non-endospermic
    (c) Abaxial and endospermic (d) Abaxial and non-endospermic
30. If the number of chromosome in pollen grain of *Pinus* is 12, then the number of chromosome in endosperm cell will be
    (a) 6 (b) 12 (c) 18 (d) 36
31. Which of the following sets of gymnosperms is considered as living fossils?
    (a) *Cycas* and *Gnetum* (b) *Pinus* and *Ginkgo*
    (c) *Pinus* and *Ephedra* (d) *Cycas* and *Ginkgo*
32. Which group of plants showed dominant sporophyte for the first time?
    (a) Bryophyta (b) Pteridophyta (c) Gymnosperms (d) Angiosperms
33. Two distinct generations present in a single life history are found in
    (a) Bryophytes only (b) Pteridophytes only
    (c) Bryophytes and Angiosperms (d) Thallophytes and Bryophytes
34. Largest archegonium is seen in
    (a) *Selaginella* (b) *Pinus* (c) *Funaria* (d) *Cycas*
35. Which of the following statements is wrong with respect to angiosperms?
    (a) Bark is formed (b) Archegonia are absent
    (c) Double fertilisation occurs (d) Life cycle is haplo-diplontic
36. Isomorphic alternation of generations is found in
    (a) *Chara* (b) *Vaucheria* (c) *Oedogonium* (d) *Ectocarpus*
37. Sexual reproduction in algae results in the formation of
    (a) Zygospore (b) Zoospore (c) Oospore (d) Zygote
38. Agranal chloroplast is seen in
    (a) Algae (b) Bryophytes (c) Pteridophytes (d) Angiosperms
39. The largest gametophyte is found in
    (a) *Cycas* (b) *Nephrolepis* (c) *Polytrichum* (d) *Pinus*
40. Gymnosperms are called soft wood spermatophytes because they lack
    (a) Cambium (b) Tracheids (c) Phloem fibres (d) Xylem fibres
41. Megasporophyll of *Cycas* is homologous to
    (a) Sepal (b) Petal (c) Stamen (d) Carpel
42. An ovule of *Cycas*
    (a) One archegonium (b) Two archegonia (c) 2-4 archegonia (d) 2-8 archegonia
43. The gametophyte of *Funaria* is
    (a) Monoecious and autoecious (b) Dioecious and heteroecious
    (c) Monoecious and heteroecious (d) Dioecious and autoecious
44. Meiosis does not take place in the formation of gamete from
    (a) Prothallus (b) Protonema (c) Promycelium (d) Sporangium

45. Primitive type of stomata is found in
    (a) Axis of moss (b) Leaf of moss
    (c) Apophysis of moss capsule (d) All of the above
46. Which of the following features is incorrect regarding bryophytes?
    (a) Dominant stage is gametophyte (b) Cone is not formed
    (c) Antherozoids are biflagellate (d) Prothallus is seen
47. Which of the following features is incorrect regarding pteridophytes?
    (a) They are mostly heterosporous (b) Seed habit is shown by *Selaginella*
    (c) Circinnate vernation is seen (d) Antherozoids are mostly multiflagellate
48. Elater mechanism for spore dispersal is shown by
    (a) *Riccia* (b) *Marchantia* (c) *Funaria* (d) *Sphagnum*
49. Transfusion tissue found in the leaves of gymnosperms helps in
    (a) Lateral conduction of water (b) Downward conduction of water
    (c) Upward conduction of water (d) All of the above
50. Life cycle in angiosperms is
    (a) Haplontic (b) Diplontic (c) Haplo-diplontic (d) All of the above
51. Sporophyte is very reduced in
    (a) Algae (b) Bryophytes (c) Pteridophytes (d) Angiosperms
52. Multicellular and oblique septa are found in
    (a) Liverworts (b) Hornworts (c) Mosses (d) All of the above
53. Exoscopic embryo is found in
    (a) Bryophytes only (b) Pteridophytes only
    (c) Both of the above (d) Angiosperms only
54. Zooiodogamous fertilisation occurs in
    (a) Bryophytes and pteridophytes (b) Pteridophytes and gymnosperms
    (c) Gymnosperms and angiosperms (d) All of the above
55. Sexual system of classification was given by
    (a) Benthem and Hooker (b) Hutchinson
    (c) Engler and Prantl (d) Linnaeus
56. The ovules in gymnosperms are
    (a) Orthotropous and bitegmic (b) Orthotropous and unitegmic
    (c) Anatropous and unitegmic (d) Anatropous and bitegmic
57. Sex organs are projected in
    (a) *Riccia* (b) *Marchantia* (c) *Funaria* (d) *Sphagnum*
58. Sperms are biflagellate in
    (a) *Selaginella* (b) *Cycas* (c) *Pinus* (d) All of the above
59. Pyrenoid in algae synthesises
    (a) Cellulose (b) Protein (c) Glycogen (d) Starch

60. Secondary protonema in mosses is not developed from
(a) Spore (b) Rhizoid (c) Stem (d) Leaf
61. Land plants arose during
(a) Jurassic (b) Cambrian (c) Carboniferous (d) Ordovician
62. A system of classification in which a large number of traits are considered, is known as
(a) Artificial (b) Natural (c) Phylogenetic (d) Synthetic
63. Multicellular and oblique septate rhizoids are seen in
(a) *Riccia* (b) *Marchantia* (c) Both a and b (d) *Funaria*
64. Simple sporophyte is found in
(a) *Riccia* (b) *Marchantia* (c) *Funaria* (d) *Sphagnum*
65. Unicellular and non-jacketed sex organs are seen in
(a) Algae (b) Btryophytes (c) Pteridophytes (d) Gyumnosperms
66. Iodine and soda are produced from
(a) Green algae (b) Brown algae (c) Red algae (d) All of the above
67. Cup shaped chloroplast is seen in
(a) *Spirpgyra* (b) *Ulothrix* (c) *Chlorella* (d) *Chlamydomonas*
68. The roots of *Pinus* show
(a) Symbiotic nitrogen fixation (b) Formation of mycorhiza
(c) Presence of xylem vessel (d) All of the above
69. Pollination in gymnosperms takes place through
(a) Water (b) Wind (c) Insects (d) Birds
70. Siphonogamous fertilisation occurs in
(a) Algae (b) Bryophytes (c) Pteridophytes (d) Angiosperms
71. Aplanospores in algae are
(a) Thin walled and motile (b) Thin walled and non-motile
(c) Thick walled and motile (d) Thick walled and non-motile
72. Which of the following algae is filamentous, unbranched and attached?
(a) *Ulothrix* (b) *Spirogyra* (c) *Chlorella* (d) *Volvox*
73. Globule and nocule are male sex organ and female sex organs, respectively are found in
(a) *Chara* (b) *Volvox* (c) *Chlamydomonas* (d) *Fucus*
74. Conducting or trumpet hyphae are seen in
(a) Green algae (b) Brown algae
(c) Red algae (d) Dark yellow algae
75. Floridean starch is reserve food material in
(a) Green algae (b) Brown algae
(c) Red algae (d) Dark yellow algae
76. Which of the following bryophytes shows projected sex organs
(a) *Riccia* (b) *Marchantia* (c) *Sphagnum* (d) *Funaria*

77. Fucoxanthin is dominant pigment in
    (a) Green algae (b) Brown algae
    (c) Red algae (d) Dark yellow algae
78. Sporangia in some pteridophytes are in groups called sori which are protected by
    (a) Ramenta (b) Sporophyll (c) Rhizophore (d) Indusium
79. Ferns are the plants to appear first after a forest fire and they survive due to
    (a) Spores (b) Leaves (c) Rhizomes (d) Sporophylls
80. Fruits are not formed in gymnosperms due to absence of
    (a) Ovary (b) Pollination (c) Fertilisation (d) seed

ANSWERS

1. c	2. b	3.c	4.c	5.b	6.c	7.c	8.b	9.a	10.d
11.d	12.d	13.c	14.b	15.b	16.a	17.c	18.b	19.d	20.a
21.c	22.d	23.d	24.a	25.d	26.b	27.d	28.d	29.a	30.b
31.d	32.b	33.c	34.c	35.d	36.d	37.a	38.a	39.c	40.d
41.d	42.d	43.a	44.a	45.c	46.d	47.a	48.b	49.a	50.b
51.a	52.c	53.a	54.b	55.d	56.b	57.b	58.a	59.d	60.a
61.d	62.b	63.d	64.a	65.a	66.b	67.d	68.b	69.b	70.d
71.b	72.a	73.a	74.b	75.c	76.b	77.b	78.d	79.c	80.a

4

Morphology of Flowering Plants

Morphology of Flowering Plants

Morphology is study of form, size, colour, structure and relative position of various parts of organisms.

Forms of Plants

Theophrastus (Father of Botany) classified plants into three categories on the basis of their shape and size.

1. **Herbs:** These are small sized plants. There is no single main stem and all branches emerging out of soil are not woody in nature. The herbs may be
   (a) Annuals: These plants complete their life cycle (from seed germination to seed formation) in one growing season of a year lasting from a few weeks to a few months. Such plants are monocarpic (bearing flowers and fruits only once). e.g., cereals, *Ranunculus*
   (b) Biennials: These plants complete their life cycle in two growing seasons in two years (vegetative phase in one year and reproductive phase in next year). Such plants are monocarpic (producing flowers and fruits once). e.g., Radish, Turnip, Carrot, Henbane etc.
   (c) Perennials: These plants complete their life cycle in many years. Such plants are polycarpic (bearing flowers and fruits every year). e.g., Banana, *Cynodon*, ginger, turmeric etc.
2. **Shrubs:** These are medium sized plants. There is no single main stem but several woody branches are developed. These are Perennial e.g., Henna, rose, china rose, *Capparis* etc.
3. **Trees:** These are large sized plant. There is single main stem and all branches are woody arising from a certain height. These are perennial and polycarpic. The trees have the following variations:
   (a) Caudex: Unbranched stem or trunk having crown of leaves at top. e.g., coconut palm, date palm etc.
   (b) Excurrent: Main stem or trunk tapers from base to apex but lateral branches give a cone-like appearance. e.g., *Casuarina, Eucalyptus, Pinus, Polyalthia* etc.
   (c) Deliquescent: Main stem or trunk is distinguishable only for some distance after which it is replaced by a number of large branches which form a dome-shaped crown. e.g., *Dalbergia*, mango etc.

Parts of a flowering plant

A flowering plant consists of root, stem, leaf, flower, fruit and seed.

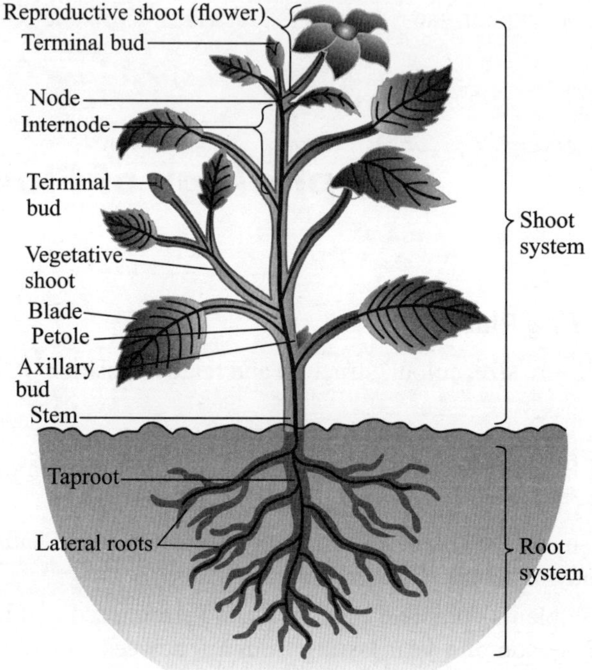

Root

It is underground and cylindrical part of plant. The longest root is found in *Adansonia digitata* but rootless angiosperms are *Aldrovanda, Wolffia, Balanophora, Utricularia, Ceratophyllum* etc.

Features:

(a) It is positively geotropic, negatively phototropic and positively hydrotropic
(b) It is usually non-green part.
(c) It does not bear nodes and internodes.
(d) It usually does not form buds.
(e) It bears unicellular hairs.
(f) It develops endogenous branches.
(g) It is protected by root cap at its tip. .

Parts of a typical root:

1. **Zone of root cap:** It is present at apical region which is cap like and conical. Simple root cap is found in most of the plants. Multiple root cap is found in screwpine (*Pandanus*). The root cap is regenerated if damaged. It protects young apical meristematic cells against soil particles. It secretes mucilage which lubricates the passage of root while penetrating the soil particles. Root pockets are found instead of root cap in *Eichhornia, Pistia, Lemna* (hydrophytes) in which these function as balancers. These root pockets cannot be regenerated if damaged.
2. **Zone of Cell division**: It is found just behind root cap hence it is subterminal. The cells are in active state of division where cells are thin walled having dense cytoplasm, large nucleus and

reduced vacuoles or without vacuoles. This is meristematic zone which produces root cap and other regions.

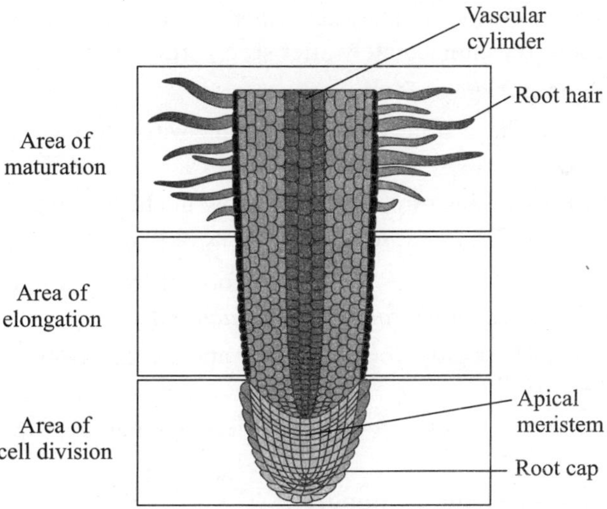

Fig: ***Different zones of root***

3. **Zone of cell elongation:** It is a region below meristematic zone. In this rapid growth in length occurs. The cells are elongated and show large central vacuole filled with sap due to fusion of large number of small vacuoles (vacuolation).
4. **Zone of cell differentiation or maturation**: The cells are differentiated into permanent tissues of root such as cortex, endodermis and stele depending upon the functions. It bears unicellular root hairs. In hydrophytes, root hairs are absent because they absorb water through general body surface.

Types of Root: It is of two types.

1. **Tap root:** It develops from continuous growth of radicle. Basically it is primary root. It gives rise to secondary roots at certain angles and later forms tertiary roots. Tap root together with secondary roots and tertiary roots constitute tap root system. If tap root grows indefinitely it is called racemose type of root system and if it stops growing after sometime then it is known as cymose type of root system.

It is usually found in dicots.

Modifications of Tap root: For storage of food

(a) **Fusiform root:** It is thick in middle and tapers towards top and bottom. It bears secondary and tertiary roots. e.g., Radish

(b) **Napiform root**: It is broad at top and tapers abruptly downwards. e.g., Beet, Turnip etc.

(c) **Conical root:** It is wide at top and tapers gradually downwards. e.g., Carrot

(d) **Tuberous root:** It stores food irregular way. e.g., Four o' clock plant (*Mirabilis*), *Cassava, Trichosanthes* etc.

Special type:

(e) **Respiratory roots (Pneumatophores):** These come out of the soil to obtain oxygen with the help of several aerating pores called pneumathodes. These are negatively geotropic and grow vertically upwards in marshy or salty areas. e.g., *Rhizophora, Heriteria, Avicennia, Sonneratia, Penniseria, Ceriops* (Mangrove plants) etc.

2. **Adventitious root**: It grows from any part other than radicle such as hypocotyl, base or node of stem because radicle dies immediately after seed germination. It is usually found in monocots.

Modifications: For storage of food

(a) **Tuberous roots:** These store food in irregular way and have no regular shape. e.g., Sweet potato (*Ipomoea batatas*)

(b) **Fasciculated roots:** These are fleshy and grow in clusters. e.g., *Asparagus,* Dahlia

(c) **Palmate roots:** These are fleshy and fuse to form finger like. e.g., *Orchis*, *Aconitum*

(d) **Moniliform or beaded roots:** These store food at frequent intervals looking like bead. e.g., Bitter gourd (*Momordica charantia), Portulaca, Vitis*.

(e) **Nodulose roots:** These store food at or near apex. e.g., Arrow root (*Maranta*), Mango ginger (*Curcuma amada*).

(f) **Annulated root:** These store food in ring like manner. e.g., Ipecac (*Cephaelis*), *Psychotria* etc.

Modifications: For mechanical support

(a) **Prop roots:** These arise from aerial branches of stem and hang downwards in air. These become large, woody and penetrate the soil to provide additional support. e.g., Banyan (*Ficus benghalensis*)

(b) **Stilt roots (Brace roots):** These arise obliquely from the lower node of stem. These move downwards and grow in clusters. e.g., Maize, Sugarcane, *Sorghum*. In Screwpine (*Pandanus*), stilt roots arise from upper node.

(c) **Climbing roots:** These arise from node of stem and help in climbing by fixing to support. e.g., Money plant (*Pothos*), Betel (*Piper betle*).

(d) **Clinging roots:** These arise from the node of stem and axil of leaves and pierce the substratum for fixation. e.g., Orchid, Ivy

(e) **Buttress root:** These arise from base of stem, become irregular and spread in different direction. e.g., *Ficus religiosa*, Red silk cotton (*Bombax*), *Salmalia*, *Terminalia.*

(f) **Contractile roots:** These are fleshy and thick and their apical part contract and fix the plant in soil in vertical direction. e.g., *Canna*, Onion, *Crocus, Lilium*.

Modifications: For vital functions

(a) **Assimilatory roots (Photosynthetic roots):** These are long, slender, hanging and arise from branches. These develop chlorophyll and perform photosynthesis. These have finger like attaching organs called haptera. e.g., *Trapa, Tinospora, Taeniophyllum, Podostemon.*

(b) **Floating roots:** There are two types of roots which arise from node. The floating root is white, soft, spongy, aerial and store air to help in floating but the normal root remains in water. e.g., *Jussiaea (=Ludwigia*).

(c) **Hygroscopic roots (Aerial roots):** These are aerial, hanging and develop porous and multilayered epidermal tissue called velamen which absorbs moisture from air (atmosphere). e.g., Orchid (Vanda).

(d) **Sucking roots (Parasitic roots):** These arise from stem and develop haustoria which obtain water, minerals and food from xylem and phloem of stem and leaf of host plant. e.g., Dodder (*Cuscuta*), Broom rape (*Orobanche*), Watch weed (*Striga*).

Special types:

(a) Fibrous roots: These are thin, underground, thread like and equal in length arising from the node of horizontal stem in groups. e.g., Grasses, wheat, barley (Monocot plants).

(b) Foliar root: These arise from the leaf margin. e.g., *Bryophyllum*

(c) Seminal root: These arise from coleorhiza or from base of radicle. These behave like primary root.

(d) Lateral roots: These arise from pericycle region of other root. It is initially horizontal and later downwards.

Main Functions:

(a) It absorbs water and minerals and transports the different parts of plant.

(b) It fixes the plant body in the soil (Anchorage) and provides support to the plant.

Special functions:

(a) Storage of food (b) Assimilation (d) Vegetative reproduction

Stem

Stem is ascending and aerial part of plant that bears leaf, flower and fruit.

Features:

(a) It develops from plumule.

(b) It bears nodes, internodes and apical bud. If stem having solid node and hollow internode then it is called culm e.g., Bamboo, wheat

(c) It bears multicellular hairs.

(d) It is negatively geotropic and positively phototropic.

(e) It bears exogenous branches.

Types of Stem:

1. Erect stems: These are upright and ascending. e.g., Majority of Angiosperms.
2. Weak stems: These are thin and delicate which cannot be upright without any support. These are of several types:
   (a) Trailers: The plants which creep on ground without rooting at interval. e.g., *Eichhornia, Euphorbia prostrate*
   (b) Creepers: The plants which creep on ground rooting at intervals. e.g., Grass, Strawberry.
   (c) Twiners: The plants of weak stem in which stem coils or twines around upright support. e.g., Bean, *Clitoria*, Railway creeper (*Ipomoea carica*), Rangoon creeper (*Quisqualis*) etc.
   (d) Climbers: The plants which climb on upright support by special clinging structures like tendril, roots, hooks. e.g., Grapevine (*Vitis*)
   (e) Lianas: These are woody twiners and climbers. e.g., *Bauhinia vahlii, Aristolachia, Hyptage* etc.

Branching of stem

(a) Unbranched: Maize, sugarcane, coconut

(b) Branched: Most of the plants

Modifications of stem:

1. **Underground modifications of stem**: The modified part remains inside soil. Its function is storage of food and also helps in perennation.

   (a) **Rhizome:** Stem is modified into thick, fleshy and branched rhizome. It grows horizontally below soil. It bears distinct nodes and internodes. It bears adventitious roots on lower side of node. It bears scale leaves at node.

   e.g., Ginger, turmeric, *Canna*, fern

   (b) **Corm:** Stem is modified into short, thick, unbranched and massively laden with food and arises from its base. It grows vertically below soil. It bears horizontal circular nodes (lines). It bears adventitious roots on lower side of node. It bears scale leaves bearing axillary bud on node.

   e.g., Zaminkand (*Amorphophallus), Colocasia, Crocus, Gladiolus, Colchicum*

   (c) **Tuber**: Stem is modified into oval or spherical and unbranched tuber arising from the axil of lower leaf (branch). It has no fixed orientation. It bears node (depression or eye) and internode. The eye is modified axillary bud (1-3) in the axil of small scale leaf. It does not bear adventitious roots.

   e.g., Potato, Jerusalem artichoke (*Helianthus tuberosus*)

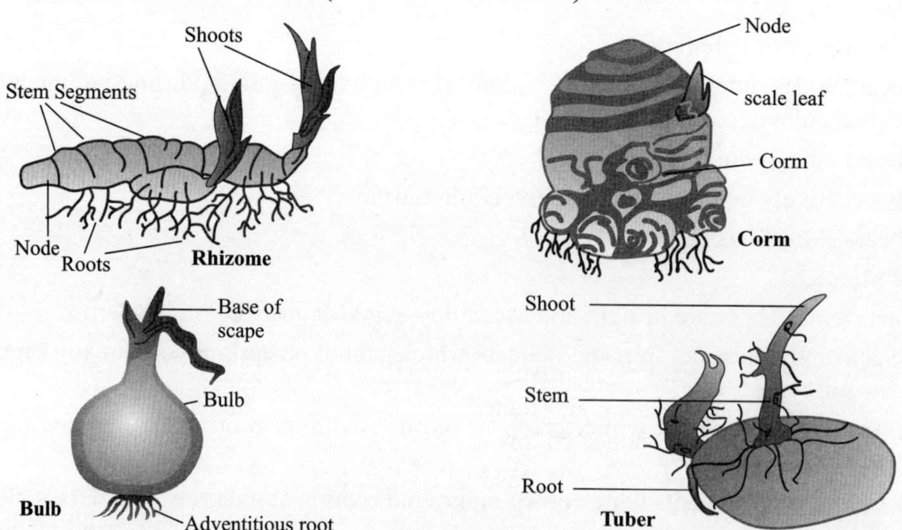

   (d) **Bulb:** Stem is modified into short, condensed and discoid bulb. It bears short internodes and fleshy scale leaves in concentric manner. It bears numerous adventitious roots from the lower side. It bears apical bud which gives rise to aerial flowering shoot (scape). It is of three types:

   (i) **Simple Tunicated bulb**: It is covered with dry scaly leaves (tunic). e.g., Onion, tulip,, daffodil etc. (The characteristic smell of Onion is due to sulphurous compound (allyl sulphide) present in fleshy leaves)

   (ii) **Compound Tunicated bulb**: It contains several bulbets which are covered with tunic. e.g., Garlic (Garlic contains Sulphur containing amino acid, S-allyl-cystine sulphoxide).

   (iii) **Scaly (Imbricate) bulb**: It is not compact body and not covered with tunic. Fleshy scale leaves lie loose without forming concentric circles. e.g., Lily.

2. **Subaerial modifications**: Some part of stem lives underground or on water and remaining part of stem is aerial. The branch breaks forming a new plant (vegetative reproduction). These are of the following types:

   (a) Runner: It is a long branch (internode) running horizontally on soil surface in different directions. It bears adventitious roots at nodes on lowerside. It bears axillary bud. It is called sobole (underground runner/ creeping stem) if internode is thick as in *Agropyron*.

   e.g., Lawn grass (*Cynodon*), wood sorrel (*Oxalis*), *Centella,*

   (b) Stolon: It is slender lateral branch arising from lower node of stem. It grows aerially for some time and bends downwards to touch the ground. It bears adventitious roots at nodes on lower side and new shoots from its terminal bud.

   e.g., Colocasia, jasmine, *Mentha*, *Vallisneria*

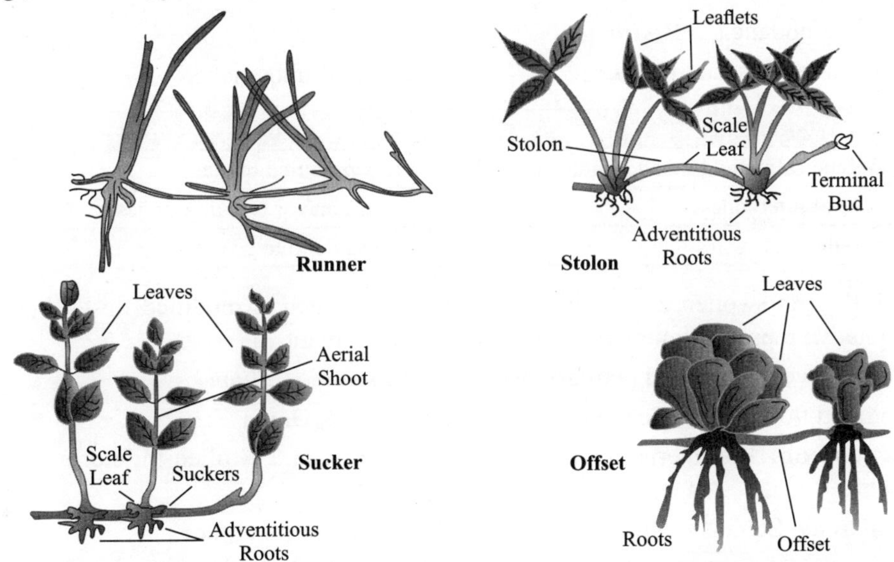

   (c) Sucker: It is lateral branch arising from lower node of stem. It grows horizontally below soil surface to some distance and then comes out obliquely to form aerial leafy shoot. It bears adventitious roots at nodes on lower side.

   e.g., Banana, *Chrysanthemum*, rose, pineapple.

   (d) Offset: It is short and thick horizontal branch (internode) arising from main stem. It bears cluster of roots below from node and rosette of leaves above.

   e.g., Water hyacinth (*Eichhornia*), water lettuce (*Pistia*).

3. **Aerial modification**: The entire part of stem remains above ground. It performs specific and different functions.

   (a) Stem tendril: It is thin and spirally coiled branch arising from node of branch. It helps plant to climb.

   Modified axillary bud: e.g., Cucumber, *Passiflora*

   Modified apical bud: e.g., Grapevine (*Vitis vinifera*)

   Modified floral bud or inflorescence: e.g., *Antigonon, Cardiospermum*

(b) Stem thorn: It is straight, hard and pointed structure which protects from animals (defensive) and also checks transpiration. It has vascular connection.

Modified axillary bud: e.g., Lemon, Orange, Duranta, *Bougainvillea*, Wood apple.

Modified apical (terminal) bud: e.g., *Carissa*

(c) Phylloclade: It is flattened, fleshy and green structure of unlimited growth bearing several nodes and internodes. It performs photosynthesis and also stores water and food. It helps plants to grow in dry habitat (xeric or xerophytic).

Flattened: e.g., Cactus (*Opuntia),* cocoloba (*Muhlebenckia*)

Cylindrical: e.g., *Casuarina, Euphorbia.*

(d) Cladode: It is green and scaly leaf like structure of limited growth bearing a few internodes. It performs photosynthesis.

One internode: e.g., *Asparagus*

Two internodes: e.g., *Ruscus*

Phylloclade	Cladode
It is modified stem and branch (whole shoot).	It is modified branch.
It has several internodes.	It has one or two internodes.
It has indefinite growth.	It has definite growth.

(e) Bulbil: It is swollen and multicellular structure which is modified fleshy bud. It helps in vegetative reproduction on being shed from parent plant.

Modified vegetative bud (axillary bud): e.g., *Dioscorea, Lilium*

Modified floral bud (apical bud): e.g., *Agave*, Garlic, Onion

(f) Pseudobulb: The internode of stem becomes fleshy and swollen for storing moisture. e.g., Orchid

(g) Thalamus: Stem is modified into floral axis bearing floral organs

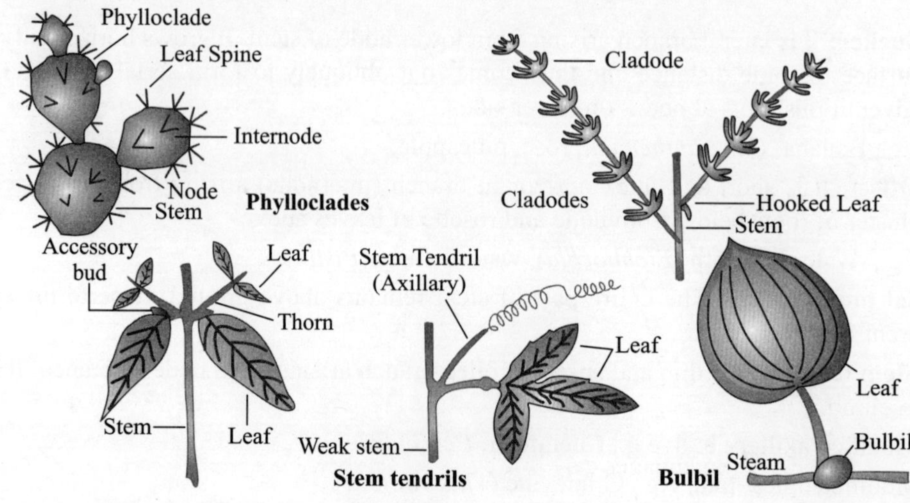

Bud

It is young, undeveloped and condensed shoot having a growing point surrounded by closely packed immature leaves.

Types of bud: On the basis of position on plant.

1. **Normal bud:** It is borne on stem. It may be
   (a) Apical (Terminal) bud: It is present at the apex of main stem and branches. It is responsible for elongation of plant. e.g., Cabbage (the largest bud)
   (b) Lateral (Axillary) bud: It is present in the axil of leaf.
2. **Adventitious bud**: It arises from part other than node.
   Radical bud: It grows from root. e.g., *Dalbergia, Ipomoea*
   Cauline bud: It grows from stem. e.g., *Artocarpus*, rose.
   Foliar bud: It grows from leaf. e.g., *Begonia, Bryophyllum*

Leaf

It is flat and lateral or flattened outgrowth of stem or branch. It arises from node and usually bears a bud in its axil. Its fresh weight is minimum at afternoon and maximum at morning. It grows definitely to a specific pattern and shape and then stops.

Smallest leaf	*Wolffia microscopica*
Largest leaf	*Victoria regia* (diameter=1.5m)
Longest leaf	*Raphia ruffia* (length=10m)

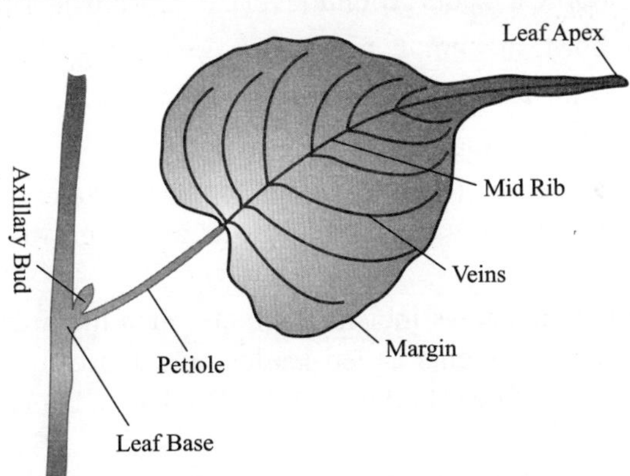

Fig. ***Leaf of Ficus religiosa (pipal) showing various parts of the leaf***

Parts of leaf

1. **Leaf base (Hypopodium):** It is lowermost part of leaf attached with stem on one side and with petiole on other side.
   It expands into sheath encircling stem in monocots.
   It bears two lateral appendages called stipules in dicots except Solanaceae and Liliaceae.
   Leaf base is swollen called pulvinus. It is responsible for shock movement (siesmonasty) or drooping of leaves. e.g., Touch-me-not (*Mimosa pudica*), mango, pea, gram, bean, *Ficus* etc..
   It forms outgrowth called ligule between leaf base and leaf lamina. e.g., Grasses.

2. **Petiole (Mesopodium):** It is long stalk which holds leaf at a certain height. It mechanically links the leaf to the stem for transport of materials.
   It is hollow (fistular) in Papaya.
   It is modified into wing in Citrus plants.
   Petiole is absent in leaf hence called sessile as in Rice, wheat, *Gloriosa, Calotropis.*
3. **Leaf lamina=Leaf blade (Epipodium)**: It is uppermost and expanded part of leaf. It produces prominent midrib at centre and a number of lateral veins. These veins form skeleton of leaf and function as channels for transport of water, minerals and food.

Surface of leaf: It may be glaucous (shinning and smooth) and glabrous (shinning and dull). There are two surface of leaf:

Abaxial= lower or dorsal or away from the stem of plant.

Adaxial= upper or ventral or facing towards the stem of plant.

Types of leaf: On the basis of origin and function

(a) Cotyledonary leaves: These are embryonic and nutritive leaves after seed germination.
(b) Bract leaves: These are protective leaves attached at the base of flower.
(c) Foliage leaves: These are green and photosynthetic leaves.
(d) Scale leaves: These are non-green (non-photosynthetic) and protective present on bud.
(e) Sporophylls: These are spore bearing leaves. These are reproductive in function and equivalent to stamen and carpel of angiosperms.

Types of leaf: On the basis of insertion

(a) Radical: Leaves are borne at root (ground level). e.g., Radish, Turnip.
(b) Cauline: Leaves are borne at stem. e. g., Maize
(c) Ramal: Leaves are borne at branch of stem. e.g., Plum, *Dalbergia, Ficus*

Types of leaf: On the basis of structure

1. **Simple leaf**: Leaf lamina is entire (there is no cut or incision) and if incised the incisions do not touch the midrib. It shows the presence of bud in its axil. e.g., Mango. Cucurbits, China rose, Guava.
2. **Compound leaf**: Leaf lamina shows incisions which touch the midrib or petiole. Hence leaf is broken up into a number of segments called leaflets which are free from one another. A leaflet appears as a leaf but it does not bear bud in its axil. The leaflet contains stipule called stipel as in bean. It is of two types:
   (a) Pinnately compound leaf: The leaflets are attached laterally on both sides at midrib known as rachis.
      (i) Unipinnate: The leaflets are attached directly at both sides of rachis (mid rib). It may be paripinnate (having even number of leaflets) in Cassia, Tamarind and imparipinnate (having odd number of leaflets) in Neem, Rose etc.
      (ii) Bipinnate: The leaflets are borne on secondary axis. e.g., *Acacia, Mimosa*
      (iii) Tripinnate: The leaflets borne on tertiary axis. e.g., *Ficus, Moringa*
      (iv) Decompound: Leaf lamina is divided into narrow segments because rachis divided repeatedly without any definite pattern. e.g., Carrot, Coriander, Fennel etc.

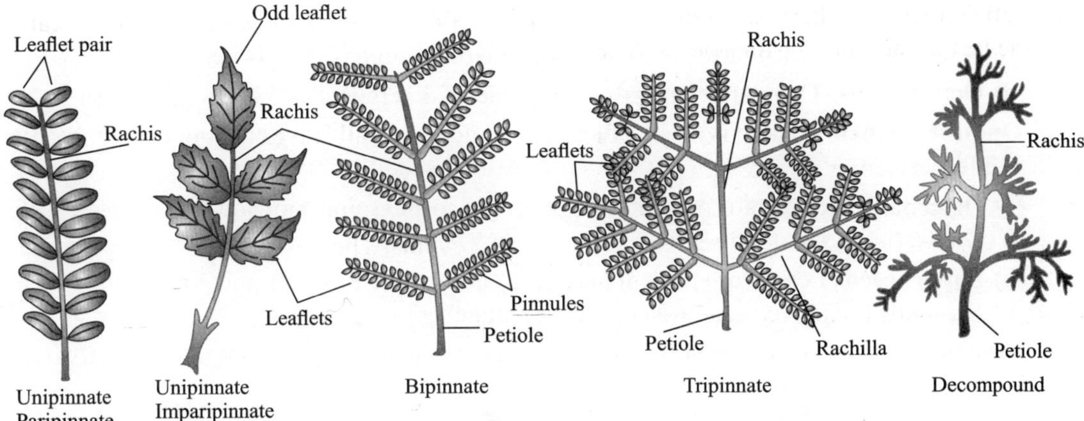

(b) Palmately compound leaf: The leaflets are joined at the tip of petiole. It may be

(i) Unifoliate: There is one leaflet. e.g., Citrus

(ii) Bifoliate: There are two leaflets which arise from petiole. e.g., Balanites, *Hardwickia.*

(iii) Trifoliate: There are three leaflets which arise from petiole. e.g., Oxalis, wood apple., *Trifolium*

(iv) Quadrifoliate: There are four leaflets which arise from petiole. e.g., *Marsilea.*

(v) Multifoliate (digitate): There are five or more leaflets which arise from petiole. e.g., *Bombax*, lupin.

Duration of leaf:

(a) Caducous: Leaves fall off just after their formation. e.g., *Opuntia*

(b) Deciduous: Leaves fall off at the end of growing season. e.g., Mulberry, Poplar.

(c) Persistent (Evergreen): Leaves remain on the plant for more than one year or leaves fall off individually at different times so that plants remain evergreen. e.g., Mango

Phyllotaxy: It is the arrangement of leaves on main stem or branches. It is three types:

1. **Alternate (Spiral):** It is most common. There is a single leaf at a node. Leaves arising at successive nodes alternate with each other at certain angle of divergence (angle between leaves of successive nodes).

   (a) Distichous: There are two rows and angle of divergence is 180°. e.g., *Annona*, Vanda

(b) Tristichous: There are three rows and angle of divergence is 120°. e.g., *Cyperus*

(c) Penatstichous: There are five rows and angle of divergence is 144°. e.g., *Ficus*

(d) Octastichous: There are eight rows and angle of divergence is 135°. e.g., Papaya

2. **Opposite (Spiral)**: There is a pair of leaves at a node and all leaves are opposite to each other. It is of types:

   (a) Decussate: Leaves are present at right angle to next upper or lower pair. e.g., *Calotropis*, Guava (in horizontal branches), *Mussaenda, Ocimum*.

   (b) Superposed: Leaves are present parallel to next upper or lower pair e.g., *Quisqualis, Syzygium* (=*Eugenia*), Guava (in vertical branches).

3. **Whorled/cyclic:** There are three or more leaves at a node. e.g., *Nerium,Hydrilla, Alstonia*

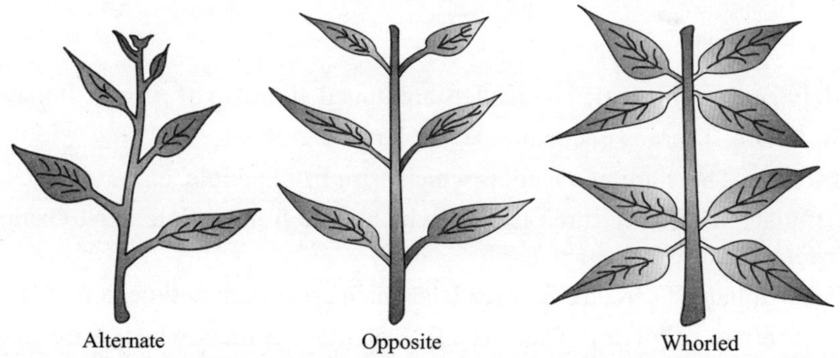

Heterophylly: It is occurrence of different types of leaf on same plant. It comes under plasticity of growth which is the ability of plants to follow different pathways in response to environment. Plasticity is irreversible stretching. It is of several types.

1. Environmental heterophylly: The submerged leaves are much dissected, linear and narrow but aerial or floating leaves are expanded and broad due to change in environment including soil, temperature, humidity and air current. e.g., Buttercup (*Ranunculus*), *Limnophila, Sagittaria*
2. Developmental heterophylly: The young or juvenile leaves are different in shape than those in mature leaves. e.g., Cotton, Coriander, Larkspur (*Delphinium*), Mustard, Pea, Eucalyptus.
3. Habitual heterophylly. The leaves of different incisions and shape occur at same time. e.g., Jack tree (*Artocarpus heterophyllus*).

Venation: It is the arrangement of veins on leaf lamina. Veins provide shape and mechanical strength to the lamina. Veins are hard structure which contains xylem and phloem. It is of two types

1. **Reticulate:** Veins are irregularly distributed forming a network. It occurs in dicots except *Calophyllum, Corymbium, Eryngium*.

   (a) Pinnate: One single main vein or mid rib gives rise to several veins on lateral side. e.g., *Ficus*, Guava, Mango etc.

   (b) Palmate: Veins radiate outward from the base of the leaf.

      (i) Convergent: Several veins start from base and merge at tip. e.g., Bay leaf, plum

      (ii) Divergent: Several veins start from base and go towards margin. e.g., Castor, cucurbits, papaya.

2. **Parallel venation**: Veinlets are arranged parallel to each other. It occurs in Monocots except *Alocasia, Colocasia, Dioscorea, Smilax* etc.

(a) Pinnate: One single main vein or midrib gives rise to lateral veins on both sides. e.g., Banana, *Canna*, ginger, turmeric etc.

(b) Palmate: Veins radiate outward from the base of the leaf.

(i) Convergent: Several main veins start from base and go towards tip. e.g., Bamboo, grass, rice, sugarcane, maize, wheat, *Eichhornia.*

(ii) Divergent: Several main veins start from base and go towards margins. e.g., Fan palm.

Modifications of Leaf:

1. **Leaf tendril**: Leaf or part of leaf is modified into slender and coiled structure which helps plant to climb.
   - Whole leaf modified: e.g., Wild pea (*Lathyrus aphaca*). In this stipule performs photosynthesis.
   - Terminal leaflet modified: e.g., Sweet Pea (*Lathyrus odoratus*), Garden Pea (*Pisum sativum)*
   - Leaf apex modified: e.g., Glory lily (*Gloriosa*)
   - Petiole modified: e.g., *Clematis*, *Nepenthes*, Garden nasturtium (*Tropaeolum)*
   - Stipule modified: e.g., *Smilax*
2. **Leaf spine:** Leaf or part of leaf is modified into sharp pointed structure called spine which is defensive in function.
   - Whole leaf modified into spine: e.g., Cactus (*Opuntia*), Barberry.
   - Partial leaf modified into spine: e.g., Prickly poppy (*Argemone*).
   - Leaf apex modified into spine: e.g., Date palm, *Yucca*
   - Stipule modified into spine: e.g., *Acacia*, Plum (*Zizyphus*)
3. **Scale leaf:** Leaf is modified into thin, dry, membranous, stalkless and non-photosynthetic scale leaf which protects axillary bud. e.g., *Asparagus, Casuarina, Ruscus.*
4. **Pitcher:** Leaf lamina is modified into a pitcher like structure (20-30 cm in height) which captures, kills and digests insects to obtain nitrogenous compounds esp. amino acid. In this leaf apex is modified to lid to attract insect, rim contains nectariferous glands and pitcher contains digestive fluid. e.g., Pitcher plant (*Nepenthes*), Venus fly trap (*Dionea), Dischidia, Darlingtonia, Sarracenia.*

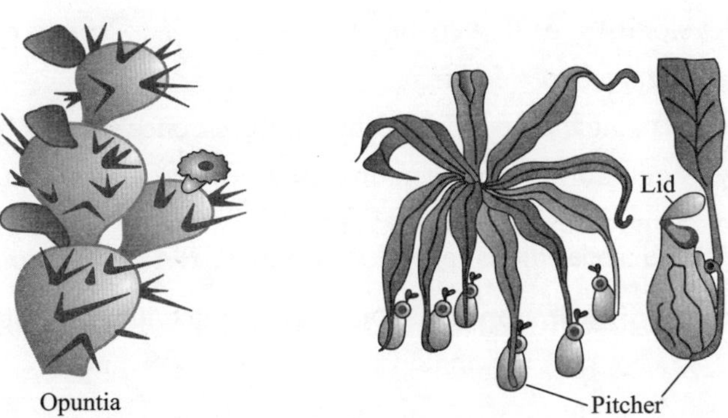

5. **Phyllode:** Petiole of leaf is modified into flattened, green and looks like leaf to perform photosynthesis when normal leaf of plant falls off at seedling stage. e.g., Australian acacia (*Acacia melanoxylon), Parkinsonia.*

6. **Bladder:** Leaf shows segmented structure and some of the segments are modified to form bladder showing sensitive hairs, branched bristle, trap valve and internal or external gland for trapping aquatic animals such as water fleas, Cyclops etc. e.g., Bladderwort (*Utricularia*).
7. **Leaf root:** One leaf of each node is modified into root for maintaining balance on water. e.g., *Salvinia*
8. **Leaf hook:** Three terminal leaflets of compound leaf are modified into hooks which help in climbing. e.g., *Bignonia.*
9. **Succulent leaf:** Leaf is modified into fleshy structure which store water or mucilage. e.g., *Aloe vera, Bryophyllum.*

Functions of leaf: (a) Photosynthesis **(b)** Transpiration and (c) Exchange of gases

Terms:

1. Phyllome: Presence of all the leaves on a plant.
2. Convulate: Folding of leaves lengthwise along midrib
3. Conduplicate: Bending of upper half of leaf upon the lower half
4. Ptyxis: Folding of individual leaf in bud condition.
5. Vernation: Arrangement of leaves in bud condition. It is of three types:
   (a) Circinnate vernation: Rolling (coiling) of leaves from apex to base. e.g., Fern, *Cycas*
   (b) Convolute vernation: Rolling of leaves from one margin to the other or with the lower surface outermost.
   (c) Involute vernation: Rolling of leaves from both margins on opposing sides of the leaf to form two tubes meeting at midrib.

Inflorescence

It is the arrangement and distribution of group of flowers on mother axis called peduncle. Morphologically it is the modified part of the shoot. Peduncle is axis of inflorescence. The enlarged peduncle is called cob found in maize. The flattened peduncle is called receptacle (torus).

The largest inflorescence is seen in *Agave* (12m).

Cauliflory: It is the development of flowers on old stem as in *Artocarpus.*

Tassel: It is staminate inflorescence of corn.

Solitary flower: Flower is single and does not form inflorescence. It may be

(a) Solitary axillary: Single flower is present in the axil of leaf. e.g., Bean, *Petunia*

(b) Solitary terminal: Single flower is present at the tip of main stem. e.g., *Crotalaria, Magnolia,* Poppy.

(c) Solitary intercalary: Single flower is present at intercalary position. e.g., Bottle brush (*Callistemon*).

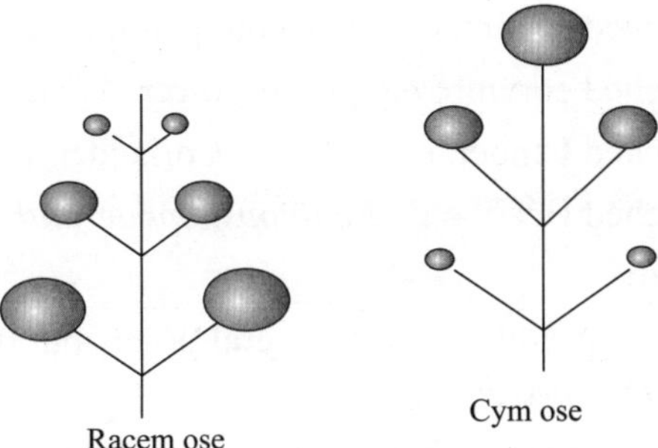

Types of inflorescence:

1. **Racemose inflorescence:** Growth of axis is continuous (indefinite/indeterminate or monopodial) as it does not end in a terminal flower. It keeps forming lateral flowers. Flowers are laterally and acropetally arranged (younger and inner flowers are at top and older and outer flowers are at base). Flowers open centripetally (peripheral flowers open first followed by central flowers).

Main axis is elongated:

(a) **Raceme:** Flowers are stalked (pedicellate). e.g., Radish, mustard, Tamarind, *Delphinium,* Snapdragon *(Antirrhinum).*

(b) **Spike:** Flowers are stalkless (sessile) and arise in axils of bracts. e.g., *Achyranthes, Adhatoda, Amaranthus*, barley etc.

(c) **Spikelet:** It is small spike and arises in axil of two fertile glumes or lemma (modified bract). At the base of flower, there is membranous structure called palea. e.g., Members of Poaceae (=Graminae) such as wheat, rice etc.

(d) **Catkin (Amentum):** It is pendulous and bears only unisexual and sessile flowers. e.g., *Casuarina,* Mulberry (*Morus*), *Betula, Quercus*, Poplar, Willow.

(e) **Spadix**: The axis is fleshy and flowers are covered with large coloured bract called spathe. e.g., Banana, Maize, *Colocasia,* Palm (Monocots).

Main axis is flattened or suppressed;

(f) **Head (Capitulum or Anthodium):** Sessile flowers are directly produced on thalamus called florets which are covered by involucres of bracts. There are two types of florets on receptacle:

- Disc florets: These have central, actinomorphic, bisexual and tubular flowers.
- Ray florets: These have peripheral, zygomorphic, unisexual, and strap shaped (ligulate) flowers.

Head is of two types:

- Homogamous head: It does not have both ray florets and disc florets. Either ray or disc florets are found. e.g., *Sonchus, Zinnia*
- Heterogamous head: It has both ray florets and disc florets. e.g., Sunflower.

It is advanced or most perfect type of inflorescence due to aggregation of florets which ensure maximum protection to flowers. e.g., Asteraceae (=Compositae) family

Compound racemose:

(a) Compound or branched raceme (Panicle): e.g., *Azadirachta, Cassia, Delonix*, Mango, *Yucca*

(b) Compound or branched spike or spike of spikelet: e.g., Members of Poaceae family

(c) Compound or branched spadix: e.g., Coconut palm, Date palm

(d) Compound or branched corymb: e.g., Cauliflower, Apple

(e) Compound or branched Umbel: e.g., Carrot, Coriander, Fennel

(f) Compound or branched Head: e.g., *Echinops, Sphaeranthes*

Main axis is shortened:

(g) Corymb: Flowers are produced from several points but they reach same level. e.g., *Cassia*, Candytuft (*Iberis*), *Delonix, Iberis*

(h) Umbel: Flowers are produced from one point and they reach same level. e.g., Members of Apiaceae (Umbelliferae) family such as Carrot, Coriander, Cumin, Fennel etc.

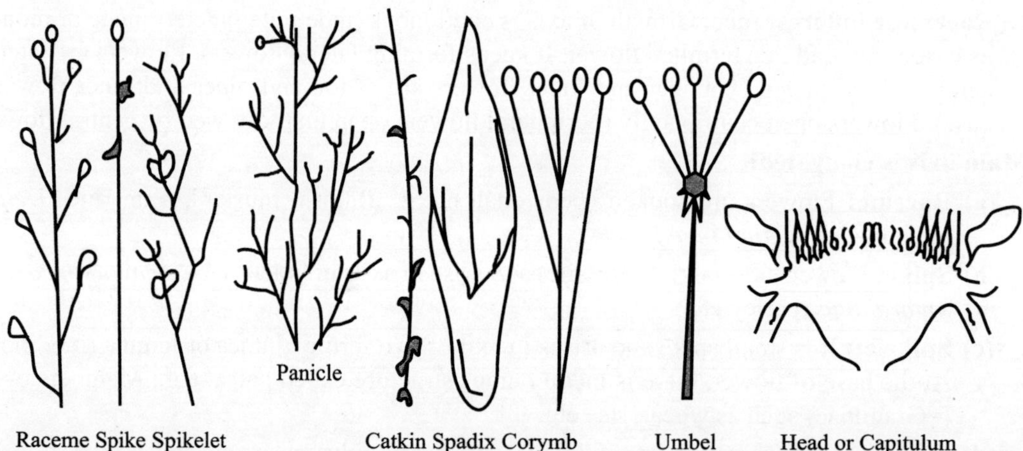

2. Cymose inflorescence:

Growth of main axis is definite/determinate or sympodial and ends in a terminal flower. Flowers are basipetally arranged (smaller flowers are at base and larger flowers are at top. Thus maturity starts from top. Flowers open centrifugally.

(a) Uniparous (Monochasial) cyme: Flowers are produced only on one lateral side. It shows following two conditions:

Helicoid cyme: The successive flowers are found in helical manner. e.g., *Begonia, Juncus, Saraca, Myositis* etc.

Scorpoid cyme: The successive flowers are found in zig-zag manner. e.g., Cotton, *Drosera*, Onion, *Ranunculus.*

(b) Biparous (Dichasial cyme): Flowers are produced on both sides from two lateral branches which in turn end in a flower. e.g., *Ixora*, Jasmine, *Nyctanthes,* Bougainvillea, Pink, teak etc.

(c) Multiparous (Multichasial) cyme: Numerous flowers are produced on lateral side. e.g., *Calotropis, Hamelia.*

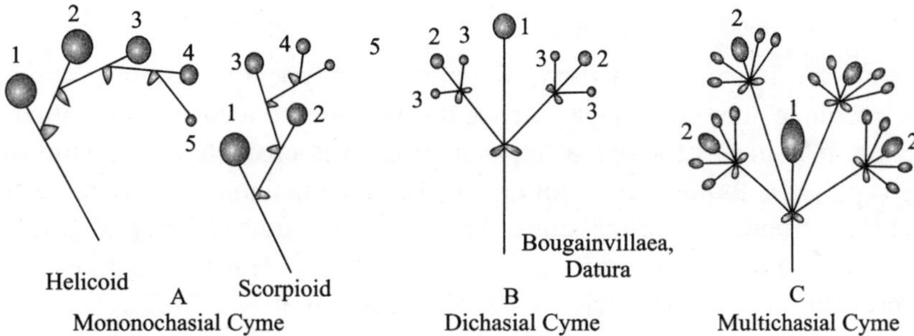

3. **Special type of Inflorescence:**
   (a) **Cyathium:** It is cup-shaped and looks like a single flower. All flowers are enclosed by involucres of bract. Nectariferous glands are present. At the centre there is a single female flower on long stalk which is surrounded by many male flowers. e.g., Poinsettia (*Euphorbia pulcherimma), Pedilanthus.*
   (b) **Hypanthodium (Syconium):** Peduncle and receptacle are flask-shaped and fleshy. Inside there is hollow central cavity where flowers are unisexual and not exposed. Male flowers are present towards apical opening, female flowers are present at the base and gall flowers (female sterile & short styled) are present below male flowers. e.g., Banyan tree (*Ficus benghalensis*), Fig (*Ficus carica*), Gular (*Ficus glomerata*) etc.
   (c) **Verticillaster:** It is condensed form of cymose (dichasial) inflorescence in which cluster of sessile flowers borne on dichasial cyme ends in monochasial cyme (two scorpoid cyme). These sessile flowers appear in pseudowhorls (verticil) and develop from each of two opposite axillary leaves. e.g., Sacred basil (*Ocimum), Salvia, Leucas, Mentha* etc.

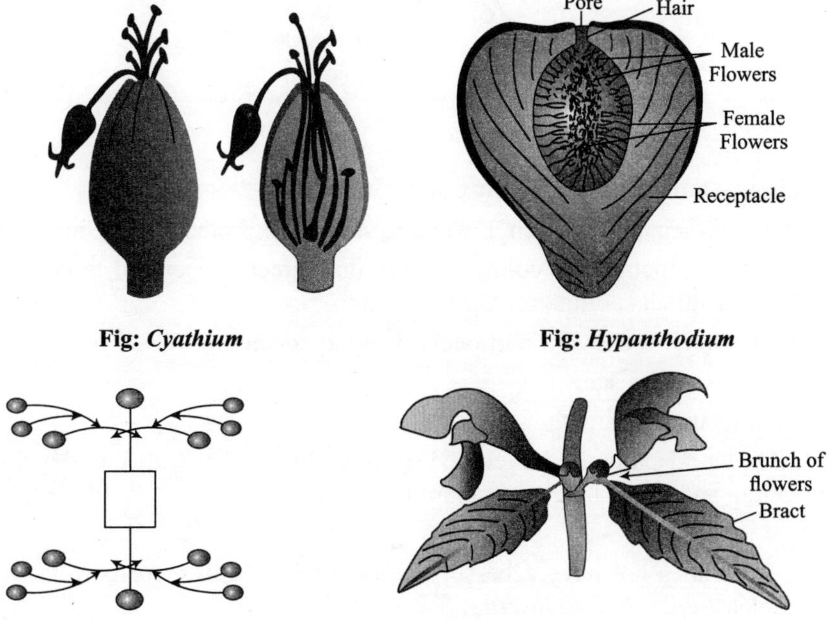

Fig: ***Cyathium***

Fig: ***Hypanthodium***

Fig: *Verticillaster*

Flower

It is specialised condensed or modified shoot in which shoot apical meristem is converted into floral meristem. In this internode does not elongate hence the axis gets condensed. It is called as reproductive unit in angiosperms. It is meant for sexual reproduction. It is characteristic feature of angiosperms. Study of flowers is called anthology. Its swollen part is called thalamus. There are usually four nodes and highly reduced internode in the thalamus. The stalk of a flower is called pedicel. Flowers are stalkless (without pedicel) in *Achyranthes, Adhatoda*, Mulberry (*Morus)* etc. Flowers are involved in pollination and fertilisation thus these help in sexual reproduction.

Mother axis: It is part of shoot on which flower is borne. It is always posterior in position. A part of flower is called posterior when it is towards mother axis or facing stem and a part of flower is anterior when it is opposite to mother axis or facing the subtending bract.

Smallest flower	*Wolffia* (0.1mm)
Largest flower	*Rafflesia arnoldi* (1m)
Largest flower in India	*Sapria himalayensis* (a root parasite)
National flower of India	Lotus (*Nelumbo nucifera*)

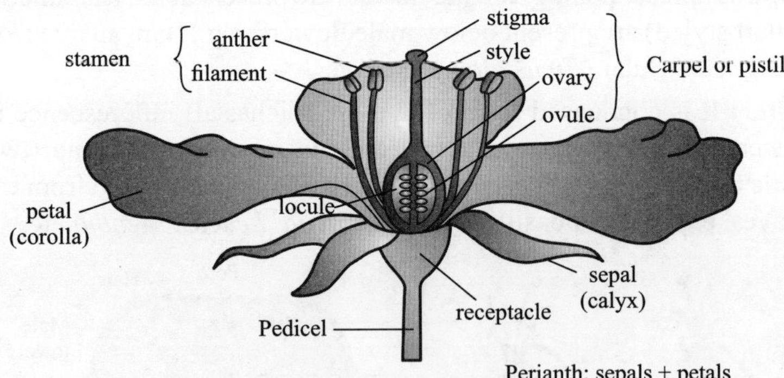

Parts of flower:

1. **Accessory outer whorl:** It contains calyx and corolla.
   (a) Calyx: Its unit is sepal. It is green, lower and smaller. It protects other inner floral parts.
   (b) Corolla: Its unit is petal. It is coloured other than green, upper and larger. It protects inner floral parts and attracts pollinators for pollination.
2. **Essential inner whorl:** It contains androecium and gynoecium.
   (a) Androecium: Its unit is stamen.
   (b) Gynoecium: Its unit is carpel.

Development: The floral parts are developed in ascending order. Calyx is developed first, then corolla and androecium and in the last, gynoecium is developed.

Types of flower:

1. Achlamydeous: It does not have accessory whorl (neither calyx nor corolla) hence flower is naked. e.g., Betel, *Euphorbia, Poinsettia.*
2. Monochlamydeous: It has only one accessory whorl (perianth). e.g., Castor

3. Dichlamydeous: It has both accessory whorls (calyx and corolla) due to differentiation of perianth. It is of two types:
   (a) Homochlamydeous: Both whorls (calyx and corolla) are similar in all respects. e.g., *Michelia*
   (b) Heterochlamydeous: Both whorls (calyx and corolla) are dissimilar in all respects. It is most common.

Terms:

1. Monoecious: Both male and female sex organs are present in same plant.
   (a) Bisexual (Hermaphrodite): Both male and female sex organs are present in same flower. It is common.
   (b) Unisexual: Male and female sex organs are present in different flowers. e.g., Castor, Coconut, Cucurbits, Maize, *Acalypha.*
2. Dioecious: Male and female sex organs are present in separate plants. These are always unisexual. e.g., Date palm, Mulberry, Papaya, *Cannabis*, *Coccinia*, Spinach, hollyhock.
3. Trioecious: Plant contains male, female and bisexual flowers. e.g., *Silene.*
4. Andromonoecious: Plant contains male and bisexual flower. e.g., Lily
5. Gynomonoecious: Plant contains female and bisexual flower. e.g., Sun flower
6. Polygamous: Plant contains male, female, bisexual and neuter flowers. e.g., Mango, Cashew, *Polygonum.*

Types of internode: In many flowers, internode is not seen clearly due to condensed thalamus but some flowers where internode is elongated and distinct. It may be

1. Anthophore: It is long internode between calyx and corolla. e.g., *Silene*
2. Androphore: It is long internode between corolla and androecium. e.g., *Passiflora*
3. Gynophore: It is long internode between androecium and gynoecium. e.g., *Capparis*
4. Gynandrophore: It shows both androphore and gynophore. e.g., *Gynandrophis*

Merosity: It is the number of floral parts in whorl. It is of two types:

1. **Isomerous**: All floral parts except carpel are similar in shape, size, colour, and origin. Each whorl of flower has equal number of parts. It may be
   (a) Dimerous (having two sets or two floral parts in each whorl): e.g., Poppy
   (b) Trimerous (having three sets): e.g., Members of Liliaceae family
   (c) Tetramerous (having four sets): e.g., Members of Brassicaceae (=Cruciferae)
   (d) Pentamerous (having five sets): e.g., Members of Fabaceae, Solanaceae
2. **Heteromerous:** All floral parts are dissimilar in shape, size, colour and origin. Each whorl of flower has unequal number of parts.

Anthotaxy: It is the arrangement of flowers.

1. Cyclic: Floral parts are arranged in circle. e.g., Brassicaceae, Malvaceae.
2. Spiral: Floral parts are arranged in spiral. e.g., Magnoliaceae
3. Spirocyclic (Hemicyclic): Some are in circle and some spiral. e.g., Ranunculaceae.

Symmetry:

1. Actinomorphic (Radially symmetrical): A flower can be divided into two equal parts by any vertical plane passing through centre. e.g., Solanaceae, Liliaceae, Brassicaceae (Cruciferae).

2. Zygomorphic (Bilaterally symmetrical): A flower can be divided into two equa parts by only one vertical plane passing through centre. e.g., Fabaceae, *Ocimum, Delphinium.*
3. Asymmetrical: A flower cannot be divided into two equal parts by any plane. e.g., *Canna, Opuntia.*

Bract (Hypsophyll): It is special leaf like structure where flower develops in its axil. If bract is small scale like structure, it is called bracteole.

Modifications of bract:

(a) Leafy (Foliaceous): It is green and leaf like. e.g., *Acalypha, Adhatoda*
(b) Glume: It is small and dry. e.g., Poaceae (=Graminae)
(c) Scaly bracteole: It is present at the base of each floret. e.g., Sunflower
(d) Epicalyx: It is whorl of bracteoles below calyx. e.g., Malvaceae
(e) Involucre: It is green and in one or more whorl below inflorescence. e.g., Asteraceae
(f) Petaloid bract: It is brightly coloured. e.g., *Bougainvillea*, Poinsettia
(g) Spathe: It is large and encloses inflorescence. e.g., Banana, Coconut, Maize

Parts of a flower:

1. **Calyx:** Its unit is sepal. It may be polysepalous when sepals are free and gamosepalous when sepals are united or fused. Sepal is usually green but in *Clematis and Salvia,* it is called petaloid where it is coloured other than green to attract pollinators. Odd sepal is usually posterior when it is in line with mother axis but anterior in Fabaceae family.

   Duration of sepals:

   (a) Caducous: Sepals fall off at the time of opening flower. e.g., Poppy
   (b) Deciduous: Sepals fall off at the time of maturity. e.g., Mustard
   (c) Persistent: Sepals remain attached to fruit for providing protection. e.g., Tomato, Brinjal.

   Modifications of sepal:

   (a) Pappus: Sepal is thread like and helps in dispersal. e.g., Asteraceae
   (b) Leafy: Sepal is large to attract pollinators. e.g., *Mussaenada*
   (c) Bilabiate: Sepal has two lips. e.g., *Salvia*
   (d) Campanulate: Sepal is bell shaped. e.g., China rose
   (e) Hooded: Sepal is hood like. e.g., *Aconitum*
   (f) Infundibuliform: Sepal is funnel like. e.g., *Atropa*
   (g) Spurred: Sepal is beak like to store nectar. e.g., *Delphinium*
   (h) Tubular: Sepal appears as tube like. e.g., *Datura*, Tobacco.

2. **Corolla:** Its unit is petal. It may be polypetalous when petals are free and gamopetalous when petals are fused or united. It forms corona which is additional whorl formed by transverse splitting of corolla. It may be coronary corona (lobed) in *Calotropis* and staminal corona in *Nerium*.

   Modifications of petal:

   (a) Cruciform: Petals are cross-like. e.g., Brassicaceae
   (b) Bilabiate: Petals have two separate lips. e.g., *Ocimum, Salvia*
   (c) Campanulate: Petals are bell shaped. e.g., *Cuscuta, Physalis*
   (d) Infundibular: Petals are funnel shaped. e.g., *Datura*

(e) Spurred: Petals are beak like. e.g., Larkspur (*Delphinium*)

(f) Tubular: Petals are tube-like. e.g., Disc floret of sun flower.

(g) Ligulate: Petals are strap-shaped. e.g., Ray floret of sunflower

(h) Caryophyllaceous: Petals are clawed. e.g., Pink, Carnation

(i) Rosaceous: Petals are rose like. e.g., Rose

Perianth: Its unit is tepal. It is a structure when there is no distinction between sepal and petal. It is found in Monocots. It may be polyphyllous when tepals are free and gamophyllous when tepals are united.

(a) Tepal is sepaloid when it is green. e.g., Date palm

(b) Tepal is petaloid when it is coloured other than green. e.g., *Asphodelus.*

Aestivation: It is the arrangement of sepals and petals with respect to one another in a flower. It is of several types.

(a) Valvate: The margins of adjacent sepal do not overlap each other. e.g., Solanaceae, Liliaceae.

(b) Twisted (Contorted): The one margin of adjacent sepal overlaps each other only in one direction regularly. e.g., Members of Asteraceae (sunflower), Malvaceae (Cotton, Lady's finger).

(c) Ascending imbricate: The posterior petal is completely inner; one is completely outer and rest are partly outer and partly inner. e.g., Members of Caesalpinaceae (*Cassia, Delonix*).

(d) Quincuncial: Two petals are completely outer; two are completely inner and rest one petal is partly outer and partly inner. e.g., *Ranunculus*, guava, *Ipomoea.*

(e) Vexillary (Descending imbricate): Posterior petal (vexillum or standard) is completely outer; two anterior fused petals (keel or carina) are inner and two lateral petals (alae or wing) are partly outer and partly inner. e.g., Members of Fabaceae

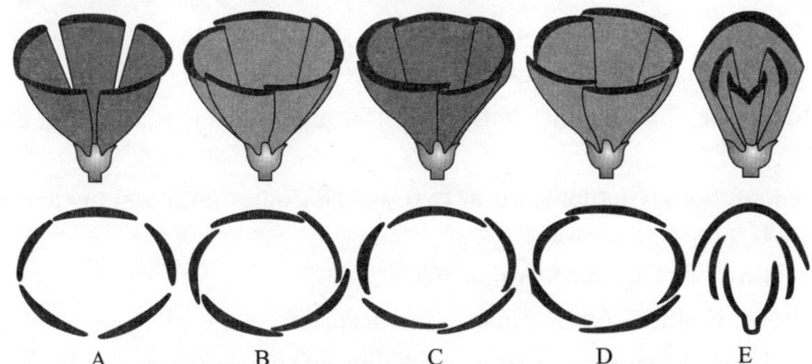

Different types of aestivation of calyx and corolla
A, Valvate, B, Twisted, C. Imbricate, D. Quincuncial, E. Vexillary

Positions of floral parts on thalamus:

(a) Hypogynous: The floral parts such as sepal, petal and stamen are situated below ovary hence ovary is superior. e.g., Members of Solanaceae, Malvaceae, Brassiccaceae, Liliaceae etc.

(b) Epigynous: The floral parts such as sepal, petal and stamen are situated above ovary hence ovary is inferior. e.g., Members of Cucurbitaceae, Ray florets of Asteraceae, Guava.

(c) Perigynous: The floral parts such as sepal, petal and stamen are in same line with ovary hence ovary is either half superior or half inferior. e.g., Rose, Plum, Peach.

3. **Androecium**: Its unit is stamen (microsporophyll). It is third whorl and male sex organ of flower. It consists of anther, connective and filament. Fertile stamen produces pollen grains. If stamen is sterile (does not produce pollen), then it is called staminode as seen in *Cassia.*

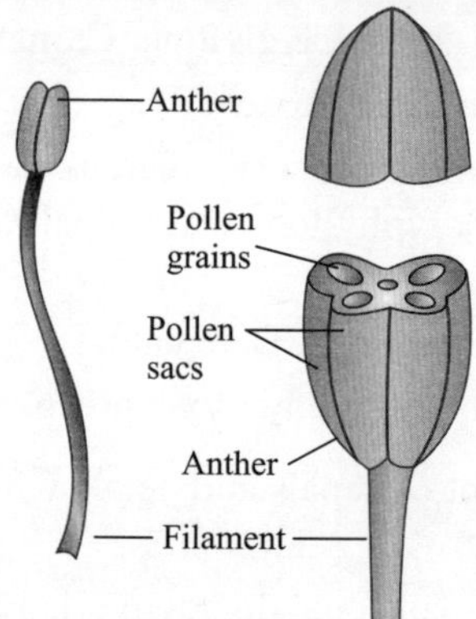

Fig: *Stamen*

Length of stamens:

(a) Isostemonous: Stamens are of same length. It is common

(b) Heterostemonous: Stamens are of different length. It is of two types

(i) Didynamous: There are 4 stamens in which two are outer and shorter and two are inner and longer. e.g., Members of Lamiaceae (=Labiatae), Verbenaceae.

(ii) Tetradynamous: There are 6 stamens in which two are outer and shorter and four are inner and longer. e.g., Members of Brassicaceae (=Cruciferae).

Arrangement of stamens:

(a) Alternipetalous: Stamens are alternate with petals.

(b) Antipetalous: Stamens are opposite to petals

(c) Diplostemonous: Stamens are in two whorls-outer alternipetalous and inner antipetalous. e.g., *Cassia*

(d) Obdiplastemonous: Stamens are in two whorls- outer antipetalous and inner alternipetalous. e.g., Pink (*Dianthus*), castor

Cohesion of stamen (Union of similar parts):

(a) Polyandrous: Stamens (anther and filament) are free. e.g., Papaya, Poppy, *Ranunculus*

(b) Monadelphous: Filaments are fused in one bundle but anthers are free. e.g., Members of Malvaceae (China rose, cotton, lady's finger)

(c) Diadelphous: Filaments are fused in two bundles but anthers are free. e.g., Members of Fabaceae

(d) Polyadelphous: Filaments are fused in several bundles but anthers are free. e.g., Citrus plants (Lemon, orange), castor, cotton, *Bombax* etc.

(e) Syngenesius: Filaments are free but anthers are fused. e.g., Members of Asteraceae (Sunflower, Marigold).

(f) Synandrous: Filament and anther are fused throughout their length. e.g., Cucurbits

Adhesion of stamens (Union of dissimilar parts):

(a) Epipetalous: Stamens are fused with petals. e.g., Solanaceae, Asteraceae

(b) Epiphyllous: Stamens are fused with tepals. e.g., Liliaceae

(c) Gynandrous: Stamens are fused with carpels. e.g., *Calotropis*.

Condition of anther

(a) Monothecous: Anther is one lobed having two microsporangia (bisporangiate). e.g., Members of Malvaceae (Cotton, lady's finger)

(b) Dithecous: Anther is two lobed having four microsporangia (tetrasporangiate). e.g., most common.

(c) Introrse: Face of anther is towards centre of flower. e.g., Tomato

(d) Extrorse: Face of anther is towards periphery. e.g., *Ranunculus*

Fixation of anther:

(a) Adnate: Filament is attached to the total length of anther on the back. e.g; *Magnolia, Ranunculus, Michelia, Nymphaea.*

(b) Basifixed: Filament is attached to the base of anther. e.g., *Datura*, Mustard

(c) Dorsifixed: Filament is attached to anther by its dorsal side. e.g., *Passiflora*

(d) Versatile: Filament is attached to back of anther which swings freely in all directions. e.g., Members of Poaceae family (Grasses, Wheat).

4. **Gynoecium**: Its unit is carpel. It consists of ovary, style and stigma. Carpel is centrally placed in thalamus. If it is sterile then it is called pistillode. It may be apocarpous (carpels are free) and syncarpous (carpels are fused or united). In some plants, the gynoecium shows the following types:

   (a) Monocarpellary: e.g., Members of Fabaceae and Poaceae

   (b) Bicarpellary and syncarpous: e.g., Members of Solanaceae

   (c) Tricarpellary and syncarpous: e.g., Members of Liliaceae

   (d) Pentacarpellary and syncarpous: e.g., China rose (*Hibiscus*)

   (e) Multicarpellary and apocarpous: e.g., *Michelia, Ranunculus*

   (f) Multicarpellary and Syncarpous: e.g., Poppy (*Papaver*)

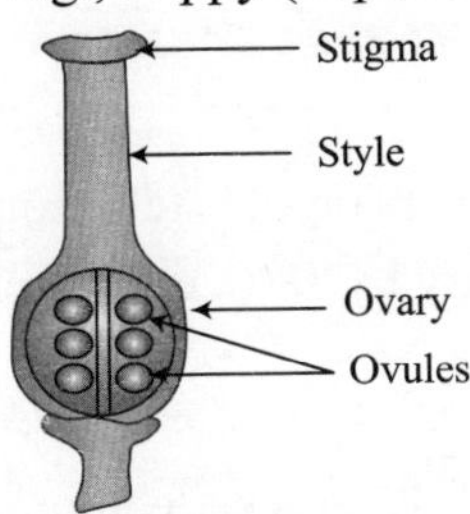

Fig: ***The different parts of carpel***

Placentation: It is the arrangement and position of placentae and ovules in ovary. Placenta is cushion-like structure to which ovules are attached. It is of the following types:

(a) Marginal: Ovary is unilocular and ovules arise from ventral suture (along the junction of both margins of ovary). e.g., Fabaceae.

(b) **Axile:** Ovary is multilocular and ovules arise from central axis. It is most common. e.g., Solanaceae, Malvaceae, Liliaceae.

(c) **Parietal:** Ovary is unilocular and ovules arise from inner wall of ovary. e.g., Brassicaceae, Cucurbitaceae.

(d) **Free-central:** Ovary is unilocular and ovules arise from central axis but septum is absent. e.g., Pink (*Dianthus*), *Primula*, *Silene* (Members of Caryophyllaceae).

(e) **Basal:** Ovary is unilocular and ovules arise from base of ovary. It is considered as advanced. e.g., Asteraceae, Poaceae

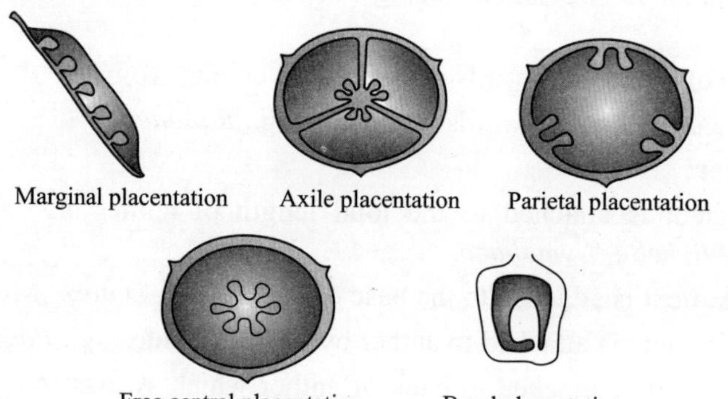

Types of style:

(a) Terminal style: Style arises from terminal part of ovary. e.g., *Datura*, *Solanum*

(b) Lateral style: Style arises from one side of ovary. e.g., Mango

(c) Gynobasic style: Style arises from the base of ovary. e.g., *Ocimum* and other members of Lamiaceae (=Labiatae).

(d) Stylopodium: Style is swollen at its base. e.g., Coriander and other members of Apiaceae (=Umbelliferae).

Types of Stigma:

(a) Capitate: It is rounded. e.g., China rose

(b) Forked: It is divided. e.g., *Tridax*

(c) Feathery: It is brush like. e.g., Grasses

Terms:

1. Cauliflory: It is production of flowers on old stem from dormant bud. e.g., *Artocarpus*, *Ficus*.
2. Resupination: It is rotation of ovary by 180°. e.g., Orchids

Fruit

It is ripened or fertilised ovary. Study of fruits is called Pomology. It consists of pericarp (formed from wall of ovary) and seed (formed from ovule). Pericarp is differentiated into epicarp, mesocarp and endocarp. Fruit may be False when it is developed from thalamus along with ovary. Such fruits are called pseudocarp. e.g., Apple, pear and all composite fruits.

Largest fruit	Double coconut (*Ladoicea maldivica*) = 18 kg
Smallest fruit	*Wolffia*

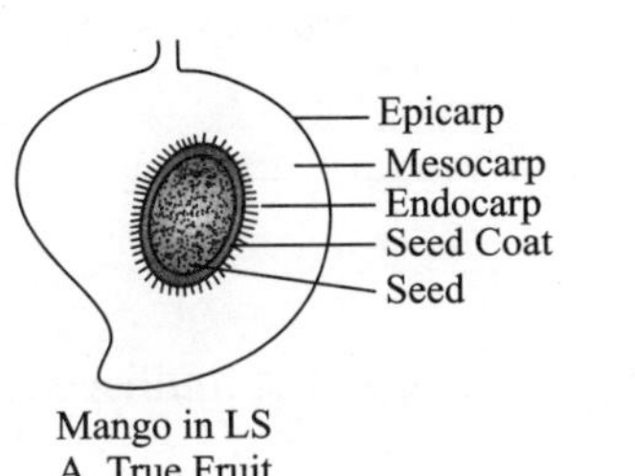

Mango in LS
A. True Fruit

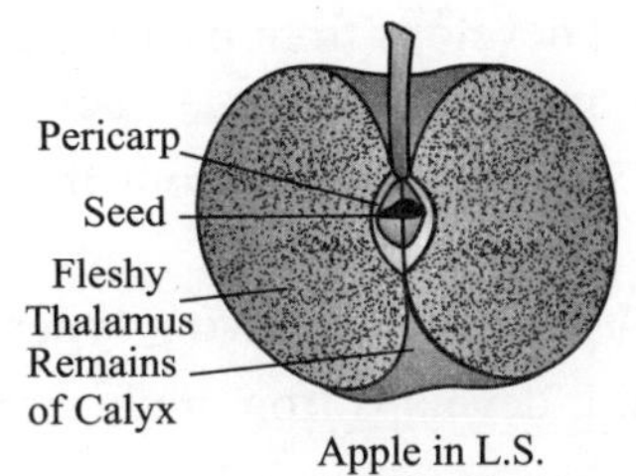

Apple in L.S.
B. False Fruit

Fruit may be parthenocarpic (seedless) when it is developed without pollination and fertilisation. e.g., Banana, Grapes, Orange, Pineapple etc.

Types of true fruit:

1. **Simple fruits**: These develop from syncarpous (either monocarpellary or bicarpellary or polycarpellary) ovary. These are categorised into four types:

(a) **Dry dehiscent fruits**: Fruits split open. These are of following types:

(i) **Legume (Pod):** It develops from monocarpellary ovary and dehiscence occurs by both halves. e.g., Beans (Fabaceae family)

(ii) **Follicle:** It develops from bicarpellary ovary and dehiscence occurs by one half. e.g., *Calotropis, Michelia*, Periwinkle (*Catharanthus*).

(iii) **Siliqua:** It develops from bicarpellary ovary and dehiscence occurs by both halves. e.g., Brassicaceae. Flat siliqua is known as silicula. e.g., *Capsella*, *Iberis*.

(iv) **Capsule:** It develops from multicarpellary ovary and dehiscence occurs by pores (Poricidal) as in *Argemone*, Poppy; by locules (Loculicidal) as in Cotton, Lady's finger; by septa (Septicidal) as in Linseed, Mustard and by fragments of septa (Septifragal) as in *Datura*.

(b) **Dry indehiscent fruits**: These are one seeded fruit which does not split inside. These are of following types:

(i) **Achene:** It develops from monocarpellary ovary and pericarp is not fused with seed coat. e.g., *Boerhaavia*, rose.

(ii) **Caryopsis**: It develops from monocarpellary ovary and pericarp is fused with seed coat. e.g., Members of Poaceae (=Graminae)

(iii) **Cypsela**: It develops from bicarpellary and inferior ovary. Persistent pappus is attached with fruit. e.g., Asteraceae.

(iv) **Nut:** It develops from multicarpellary ovary and epicarp is rough or hard. e.g., Cashew nut, litchi, hazel nut, *Trapa*.

(v) **Samara**: It is winged fruit and wing develops from pericarp. e.g., *Dioscorea, Shorea*.

(c) **Dry Schizocarpic fruits**: Fruits split into segments with one or several seeded structure called mericarp. These are of following types:

(i) **Lomentum:** It develops from monocarpellary ovary and fruit splits into one seeded structure at maturity. e.g., Tamarind, Groundnut

(ii) **Cremoacarp:** It develops from bicarpellary ovary and fruit splits into two seeded structure at maturity. e.g., Carrot, Coriander

(iii) **Regma:** It develops from tricarpellary ovary and fruit splits into three seeded structure at maturity. e.g., Castor

(iv) **Carcerulus:** It develops from bicarpellary ovary and fruit splits into four seeded structure at maturity. e.g., members of Lamiaceae

(v) **Double samara**: It develops from bicarpellary ovary and fruit splits into two samara at maturity. e.g., *Acer*

(d) **Fleshy fruits: These are of following types:**

(i) **Amphisacra**: It develops from multicarpellary ovary. It is multiseeded and epicarp is stony. e.g., Wood apple

(ii) **Balausta:** It develops from multicarpellary ovary. Testa is juicy. e.g., Pomegranate

(iii) **Berry:** It develops from bicarpellary ovary. It is multiseeded. e.g., Banana, Brinjal, Grapes, Guava, Tomato, gooseberry etc.

(iv) **Drupe:** It develops from multicarpellary ovary. It is one seeded. Endocarp is stony. e.g., Mango, Almond, Coconut, Plum, Walnut, Olive, cherry, apricot etc.

(v) **Hesperidium:** It develops from multicarpellary ovary. Endoacarp is segmented but epicarp and mesocarp are fused forming rind or skin. e.g., Lemon, Orange.

(vi) **Pepo:** It develops from tricarpellary inferior ovary. It is multiseeded. e.g., Cucurbits, Papaya.

(vii) **Pome:** It develops from multicarpellary ovary. e.g., Apple, Pear

2. **Aggregate Fruits**: These develop from multicarpellary and apocarpous ovary. It is a group of fruitlets (simple fruit) called etaerios. These are of following types:

(i) **Etaerio of achene**: Strawberry, lotus, *Ranunculus, Clematis*

(ii) **Etaerio of Berry**: Custard apple

(iii) **Etaerio of Drupe**: Raspberry

(iv) **Etaerio of follicle**: *Aconitum, Delphinium*

3. **Composite or Multiple fruits:** These develop from inflorescence. Each flower produces a fruit but these mature into a single mass. These are regarded as false fruits. These are of following types:

1. **Sorosis:** Perianth and bract are fused and fleshy. It develops from
   - Female spike: e.g., Pineapple
   - Female catkin: e.g., Mulberry
   - Female spadix: e.g., Jackfruit
2. **Syconus:** It develops from fleshy hypanthodium inflorescence. e.g.,Fig, Banyan tree, Peepal.

Types of fruits with edible parts:

Plant	Botanical name	Type	Edible parts
Mango	*Mangifera indica*	Drupe	Mesocarp
Grapes	*Vitis vinifera*	Berry	Pericarp and Placenta
Banana	*Musa sapientum*	Berry	Mesocarp and Endocarp
Tomato	*Lycopersicon esculentum*	Berry	Mesocarp and Placenta
Guava	*Psidium guajava*	Berry	Thalamus and pericarp
Papaya	*Carica papaya*	Berry	Mesocarp
Date palm	*Phoenix dactylifera*	Berry	Fleshy pericarp

Cucumber	*Cucumis sativus*	Pepo	Mesocarp, endocarp and Placenta
Lemon	*Citrulus aurantifolia*	Hesperidium	Juicy endocarp (hairs)
Orange	*Citrus reticulate*	Hesperidium	Juicy endocarp
Apple	*Malus pumila (=Pyrus malus)*	Pome	Fleshy thalamus
Pear	*Pyrus communis*	Pome	Fleshy thalamus
Pineapple	*Ananas comosus*	Sorosis	Receptacle, bract & perianth
Jack fruit	*Artocarpus heterophyllus*	Sorosis	Receptacle, bract and perianth
Cashew nut	*Anacardium accidentale*	Nut	Cotyledon and stalk
Coconut	*Cocos nucifera*	Drupe	Endosperm
Walnut	*Juglans regia*	Drupe	Lobed cotyledon
Water melon	*Cucumis melo*	Pepo	Mesocarp
Plum	*Prunus domestica (=Prunus dulcis)*	Drupe	Epicarp and mesocarp
Black berry	*Syzygium cumuni*	Drupe	Mesocarp
Mulberry	*Morus alba*	Sorosis	Fleshy receptacle
Strawberry	*Fragaria indica*	Etaerio of achene	Fleshy receptacle
Raspberry	*Rubus idaeus*	Etaerio of drupe	Fleshy receptacle
Fig	*Ficus carica*	Syconus	Fleshy inflorescence
Litchi	*Litchi chinensis*	Nut	Fleshy aril
Sapodilla	*Achras sapota*	Berry	Mesocarp
Tamarind	*Tamarindus indica*	Lomentum	Mesocarp
Wood apple	*Aegle marmelos*	Amphisacra	Endocarp and placentae
Custard apple	*Annona squamosa*	Etaerio of berry	Fleshy pericarp
Pomegranate	*Punica granatum*	Berry	Fleshy testa
Pumpkin	*Cucurbita moschata*	Pepo	Mesocarp
Lady's finger	*Abelmoschus esculentus*	Capsule	Whole fruit
Ground nut	*Arachis hypogea*	Lomentum	Seed
Almond	*Prunus dulcis*	Drupe	Seed
Pista	*Pistacia vera*	Drupe	Cotyledon

Seed

It is fertilised or ripened ovule which contains embryo. Study of seed is called Spermology.

Smallest or lightest seed	Orchis (0.004mg)
Largest or heaviest seed	Double coconut (*Ladoicea maldivica*)=10 kg

Parts of seed: A typical seed consists of

1. **Seed coat**: It is outer protective covering. Its outer and hard layer is called testa and inner layer is called tegmen (usually fused with testa).

   It has a fine pore called micropyle on surface

It has a scar called hilum where funicle (stalk of ovule/seed) is attached.

If seed coat is absent then such seed is called kernel.

2. **Embryo**: It has young or miniature plant enclosed within seed coat. It has two main parts:
   (a) **Cotyledon:** It is leaf of embryo. It is either one or two in number.
   It is absent in *Cuscuta*. The largest cotyledon is seen in *Capsella*. It stores food and becomes fleshy. If it does not store food then it becomes papery and thin.
   (b) **Tigellum:** It is main axis of embryo which is attached with cotyledon. Its micropylar end is called radicle and its opposite end is called plumule. The part of axis below point of attachment of cotyledon or between cotyledon and radicle is called hypocotyl and the part of axis above point of attachment of cotyledon or between cotyledon and plumule is called epicotyl. The hypocotyl terminates with radicle and epicotyl terminates with plumule.
3. **Endosperm:** It is nutritive tissue formed as a result of triple fusion. In monocots, its outer covering is called aleurone layer which separates endosperm from embryo. It stores food when cotyledon cannot store food. Seed may be:
   (a) **Endospermic** (albuminous): Seed contains endosperm. It is found in monocots and few dicots such as Castor, Custard apple, Cotton, Papaya, Sunflower, Rubber).
   (b) **Nonendospermic** (exalbuminous): Seed does not contain endosperm because it was fully consumed by developing embryo. It is found in dicots and few monocots such as *Alisma*, Orchid, *Pothos, Sagittaria, Vallisneria*).

Dicot seed	Monocot seed
Seed coat is not fused with pericarp (fruit wall).	Seed coat is fused with pericarp (fruit wall).
Seed has two cotyledon.	Seed has one shield shaped cotyledon known as scutellum.
Embryo is large and centrally placed.	Embryo is small and placed on one side.
Endosperm is small and food is stored in cotyledon.	Endosperm is bulky where food is stored.
Aleurone layer is absent.	Aleurone layer is found outside endosperm.
Radicle is not covered with coleorhiza.	Radicle is covered with coleorhiza.
Plumule is not covered with coleoptile.	Plumule is covered with coleoptiles
Plumule is situated between two cotyledons	Plumule is situated at one end near cotyledon.

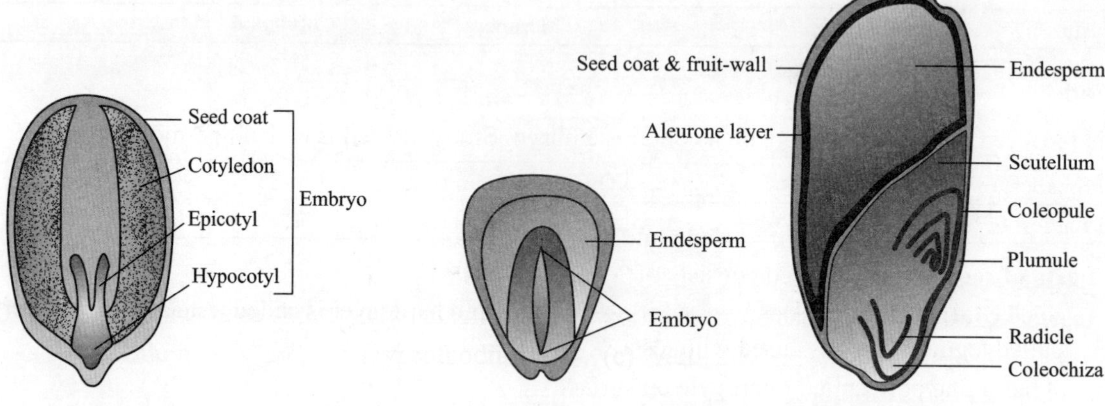

Fig: *Dicot seed*

Fig: *Monocot seed*

Special structures: There are several structures which develop from various parts of ovule during post fertilisation development.

1. **Perisperm:** It is remnant of nucellus in seed. Nucellus is usually consumed after fertilisation but sometimes it is left. It is seen in Beet, Black pepper, *Nymphaea.*
2. **Caruncle:** It is an outgrowth of outer integument either on the side of funicle or all around micropyle. It is whitish and fleshy structure. It helps in seed germination because it is hygroscopic by absorbing water from soil and passes to embryo. It is seen in Castor.
3. **Aril:** It is fleshy appendage or outgrowth which arises from funicle or testa or both. It surrounds the ovule. It is regarded as third integument. It is seen in *Litchi*, Nutmeg etc.

Types of seed:

1. **Orthodox seed:** It tolerates low moisture (dehydration), low temperature (freezing) and anaerobic condition. e.g., Cereals, Legumes.
2. **Recalcitrant seed:** It gets killed under low moisture and low temperature. It is regarded as non-biodegradable. e.g., Coconut, Jackfruit, Tea.

Advantages:

(a) Seed helps the plants to colonise in other areas as aresult of dispersal.
(b) Seed nourishes the seedling till photosynthesis because it has enough food
(c) It protects young embryo due to hard seed coat.
(d) It produces new genetic combination which leads to variation.

Seed dormancy: It is the condition in which seeds are unable to germinate even after getting all necessary conditions. During this period growth of seeds remain suspended which occurs due to internal factors and some specific environmental factors.

Causes of dormancy:

(a) Hard seed coat impermeable to gases: e.g., Apple
(b) Hard seed coat impermeable to chemicals: e.g., *Xanthium*
(c) Hard seed coat impermeable to water: e.g., Legumes, *Chenopodium, Trigonella*
(d) Mechanically resistant seed (embryo is unable to grow): e.g., *Alisma*, *Amaranthus, Capsella,* Mustard.
(e) Immaturity of embryo at the time of shedding: e.g., *Eranthis, Ginkgo, Ranunculus*
(f) Absence of light (Negatively photoblastic): e.g., Lettuce, Onion, Tobacco, Tomato, *Phlox* etc.
(g) Absence of after-ripening period (seeds undergo a period of rest even when condition is favourable): e.g., Iris, Peach, Poplar, Tulip etc.
(h) Presence of excessive salts at seed shedding: e.g., *Atriplex*
(i) Presence of inhibitors (ABA, NH_3, Coumarin, Phenolics) in seeds: e.g., Cucurbits, *Xanthium*, *Iris*

Methods to break seed dormancy (or to start seed germination)

1. Scarification: It is method of softening and weakening of seed coat by hot water, acids, alcohol (chemical method) and by thrashing with machine, rupturing with hammer and cutting from scalpel (mechanical method). It is done in *Cassia, Melilotus*
2. Stratification: It is treatment of seed in well aerated and moist condition at very low temperature (0-5°C) just after freezing point.

3. Alternating high and low temperature: This is done in seed of immature embryo and difference of temperature should be 10°C to 20°C.
4. Exposure to light: Red light (<730 nm) breaks dormancy (seed germination) and Far red light (>730nm) maintains dormancy (inhibits seed germination).
5. Soaking seeds in chemicals (KNO_3, H_2SO_4, thiourea, gibberellin, cytokinin, ethylene).
6. Destruction of Inhibitors:

Importance:

(a) Seeds can be stored during unfavourable period (drought, cold, high temperature)
(b) Seeds can remain alive in soil for several years.
(c) Seed dormancy is essential for seed disposal.

Seed germination: It is sprouting of seed and growth of embryo present inside seed into a seedling capable of independent existence.

Process:

(a) Seed absorbs water through micropyle by imbibition and osmosis.
(b) Hydration of cell organelles occurs
(c) Embryo is activated.
(d) Seed coat swells, softens and ruptures by cell elongation.
(e) Metabolic activity stars and respiratory enzymes become active.
(f) Seed respires initially anaerobically and later aerobically when O_2 enters into seed coat and releases heat energy.
(g) Seed gets requisite energy for the growth of embryo.
(h) Embryo secretes hydrolytic enzymes such as amylase, protease, lipase which convert insoluble reserve food (starch, protein, oil) into soluble substances.
(i) Embryo utilises these soluble substances
(j) Cell division starts in radicle and plumule.
(k) Radicle firstly comes out of seed coat and forms tap root.
(l) Plumule comes out rupturing seed coat and forms shoot and
(m) Finally seedling is formed.

Types of seed germination:

Epigeal germination	Hypogeal germination
Cotyledon comes out of soil surface by greater elongation of hypocotyl (part of axis below cotyledon). Cotyledon is flat, green and leaf like.	Cotyledon remains inside or just on surface but plumule comes out of soil by greater elongation of epicotyl. In this, hypocotyl does not grow much.
e.g., Bean, Castor, Cotton, Cucurbits, Mustard, Papaya, Poppy, Tamarind, Mustard, Onion	e.g., Cereals, Gram, Coconut, Date palm, Fan palm, Groundnut, Mango, Jack fruit, Pea.

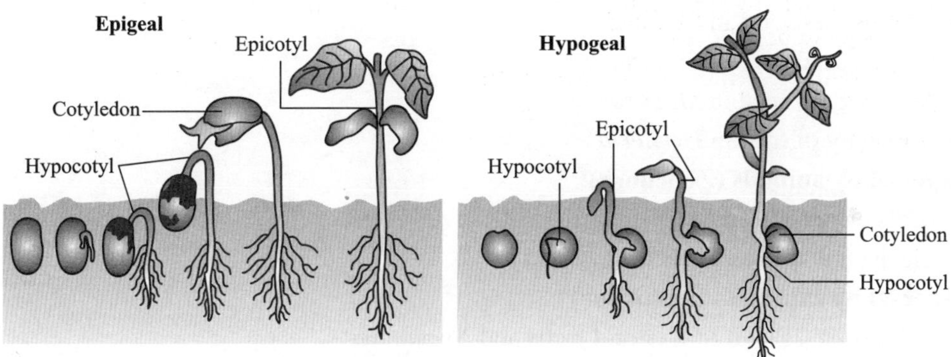

Factors:

(a) Water: It causes softening of seed coat due to seed coat.

(b) Light: Red light is stimulatory but far red light and blue light are inhibitory to seed germination.

(c) Temperature: The required optimum temperature is 25-30°C.

(d) Oxygen: It is essential for seed germination except Rice and *Typha*.

(e) Availability of food:

(f) Viability: It is ability to germinate. It is maximum in Lotus (*Nelumbo*).

(g) Inhibitors: These chemicals such as phenolics, tannins, alkaloids, lactone, mustard oil, scopoletin etc. inhibit seed germination.

Vivipary: It is germination of seed while it is still attached to parent plant or before detachment from fruit. In this seed does not undergo dormancy thus forms seedling. Due to its increasing weight the seedling separates from the parent plant and falls down into mud or soil hence plant is established. It is halophytic adaptation and found in *Rhizophora* and other mangrove plants. It is not desirable in crop plants because seed cannot be stored in this condition.

Dispersal of fruit and seed: Dispersal is must for colonisation of new areas, avoiding competition and production of mixed population. It is also a wasteful method as majority of seeds does not find suitable place for germination. It occurs by various agencies

1. Dispersal by wind (Anemochory):
   - Wing of seed in *Cinchona*, *Moringa, Tecoma* etc
   - Wing of fruit in *Dioscorea, Shorea, Acer, Hiptage, Terminalia*
   - Censor mechanism: The seeds of *Argemone, Aristolachia, Antirrhinum*, Poppy etc. are scattered after dehiscence of fruit when shaken by wind.
   - Parachute mechanism (Appendages keep the seeds in air for longer span of time and disperses at a good distance. It occurs through

     (a) Pappus (modified calyx in persistent hair like structure) in *Sonchus*

     (b) Persistent hairy style in fruit in *Clematis, Narvelia*

     (c) Balloon like appendage (modified calyx) in *Physalia*

     (d) Light seeds and fruits in Orchids
2. **Dispersal of water (Hydrochory):** Seeds and fruits have floating devices in the form of spongy outer coats:

(a) Fibrous mesocarp in Coconut
(b) Spongy seed coat in *Nymphaea*
(c) Air cavity of seed in *Mucuna*
(d) Air cavity of fruit in *Heriteria*

3. **Dispersal by animals (Zoochory):**
   - Birds disperse seeds of *Viscum*
   - Elephant disperses seeds of *Rafflesia*
4. **Dispersal by explosion (Autochory)**:
   (a) Jerk disperses fruit of Castor, *Oxalis* and *Nyctanthes*
   (b) Jaculator mechanism disperses seed of *Barleria, Ruellia, Andrographis*
   (c) Noise disperses seed of *Bauhinia vahilii*

Description of Families

Largest family	Asteraceae
Second largest family	Orchidaceae
Third largest family	Fabaceae
Fourth largest family	Poaceae
Fifth largest family	Rubiaceae
Smallest family	Nympheaceae

The various symbols and their full forms are as follows:

Symbol or abbreviation	Full form
Br	– bracteate condition
Ebr	– ebracteate condition
Brl	– bracteolate
Epil	– epicalyx
0 (zero)	– absence of a particular whorl
∝	– indefinite number of floral parts in a whorl
⊕	– actinomorphic (regular) condition
%	– zygomorphic (irregular) condition
κ	– calyx (sepals)
C	– corolla (petals)
P	– perianth (tepals)
A	– androecium (stamens)
G	– gynoecium (carpels)
♂	– unisexual, staminate (male) flower
♀	– unisexual, pistillate (femal) flower
⚥	– bisexual flower
( )	– cohesion of floral parts in a whorl

( )	– partial or slight cohesion of the parts
$\overset{\frown}{P\,A}$	– epiphyllous stamens
$\overset{\frown}{C\,A}$	– epipetalous stamens
$\underline{G}$	– hypogyny (ovary superior)
G–	– perigyny (ovary half-inferior)
$\overline{G}$	– epigyny (ovary inferior)
G(2)	– bicapellary syncarpous
$G\underline{(2)}$	– bicarpellary syncarpous ovary superior
G2	– bicarpellary free carpels
K_{2+2}	– presence of two whorls of calyx, each of 2 sepals (Brassicaceae or Cruciferae)

Old name of family	New name of family
Graminae	Poaceae
Leguminosae	Fabaceae
Cruciferae	Brassicaceae
Labiatae	Lamiaceae
Umbelliferae	Apiaceae
Palmae	Arecaceae
Compositae	Asteraceae
Guttiferae	Clusiaceae

Fabaceae:

It is also known as bean family. Its earlier name was Papilionaceae, a subfamily of family leguminosae.

Classification: (According to Bentham and Hooker's system)

Kingdom	Plantae
Division	Phanerogamia
Class	Dicotyledonae
Subclass	Polypetalae
Series	Calyciflorae
Order	Rosales
Family	Fabaceae

Vegetative features:

- Habit: Herbs, shrubs and trees
- Root: Tap root. Root nodule is mostly seen.
- Stem: Erect or climbing
- Leaf: Alternate, pinnately compound or simple, leaf base shows pulvinus, stipulate, reticulate venation.

Diagnostic features:

- **Inflorescence**: Raceme or solitary axillary
- **Flower:** Ebracteate, Pedicellate, Zygomorphic, Bisexual, Hypogynous, Complete, Pentamerous
- **Calyx:** Sepals- 5, Gamosepalous, Odd sepal anterior, Valvate aestivation
- **Corolla:** Petals-5 (unequal), Polypetalous, Largest petal upper and posterior, Descending imbricate (Vexillary) aestivation,
- **Androecoum:** Stamens-10, Diadelphous, Free stamen posterior and fused stamens anterior, Anther dithecous, dorsifixed and introrse
- **Gynoecium:** Monocarpellary, Ovary superior, Marginal placentation, Ovary unilocular, numerous ovules
- **Fruit:** Mostly legume or pod
- **Seed:** Non-endospermic

Floral formula: % ⚥ $K_{(5)}\ C_{1+2+(2)}\ A_{(9)+1}\ G_1$

Floral diagram

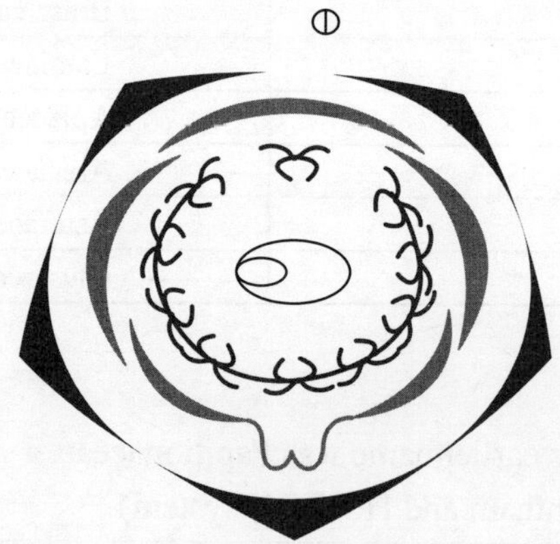

Economic importance:

1. **Food (Pulses):** Gram/Chickpea(*Cicer arietinum*), Red gram/Pigeon pea (*Cajanus cajan*), Green gram (*Phaseolus radiatas=Vigna radiata*), Black gram (*Phaseolus mungo*), Garden pea (*Pisum sativum*), Sweet pea/ Grass pea (*Lathyrus odoratus*), Cow pea (*Vigna unguiculata*), Soya bean (*Glycine max*), French bean/Kidney bean (*Phaseolus vulgaris*), Hyacinth bean (*Dolichos lablab*), Broad bean (*Vicia faba*), Lentil (*Lens culinaris*)
2. **Food (Spices):** Fenugreek (*Trigonella foenumgraecum*)
3. **Fodder/forage crops**: Alfalfa (*Medicago sativa*), Cluster bean (*Cyamopsis tetragonoloba*), Berseem/Egyptian clover (*Trifolium alexandrium*), Senji/Indian clover (*Melilotus alba*).
4. **Oil:** Groundnut/peanut (*Arachis hypogea*): Seed produces oil.
5. **Fibre:** Sunhemp (*Crotolaria juncea*): The secondary phloem of stem produces fibres.
6. **Medicine:** Liquorice (*Glycyrhiza glabra*): Root is useful for cough and cold

   Jeweller's weight/ratti (*Abrus precatorius*): Leaf is useful part.
7. **Dye:** Indigo (*Indigofera tinctoria*): Leaf produces dye.

8. **Timber:** Indian redwood/shisham (*Dalbergia sisso*)
9. **Ornamental plants:** Lupin (*Lupinus arboreus*), Flame of forest (*Butea monosperma*).

Solanaceae:

It is also called potato family.

Classification: (According to Bentham and Hooker's system)

Kingdom	Plantae
Division	Phanerogamia
Class	Dicotyledonae
Subclass	Gamopetalae
Series	Bicarpellatae
Order	Polemoniales
Family	Solanaceae

Vegetative characters:

- Habit: Mostly herbs
- Root: Tap root
- Stem: Mostly herbaceous
- Leaf: Alternate, simple, exstipulate, reticulate venation

Diagnostic features:

- **Inflorescence:** Helicoid cyme or solitary axillary
- **Flower:** Ebracteate, Pedicellate, Actinomorphic, Bisexual, Hypogynous, Heterochlamydeus, Complete, Pentamerous
- **Calyx:** Sepals-5, Gamosepalous, Valvate aestivation, odd sepal posterior, Persistent.
- **Corolla:** Petals-5, Gamopetalous, Valvate aestivation,
- **Androecium:** Stamens-5, Epipetalous, Alternipetalous, Anther dithecous, basifixed and introrse
- **Gynoecium:** Bicarpellary, Syncarpous, Ovary superior, Axile placentation, Ovary bilocular (multilocular due to false septum as in Tomato, Datura etc), Many ovules in each locule, Swollen placenta, Carpel (ovary) obliquely placed (Posterior carpel to the right about 45° from median and anterior to left).
- **Fruit:** Capsule or berry
- **Seed:** Endospermic

Floral formula: $\oplus$ ⚥ $K_{(5)}\ \overset{\frown}{C_{(5)}\ A_5}\ \underline{G_{(2)}}$

Floral diagram

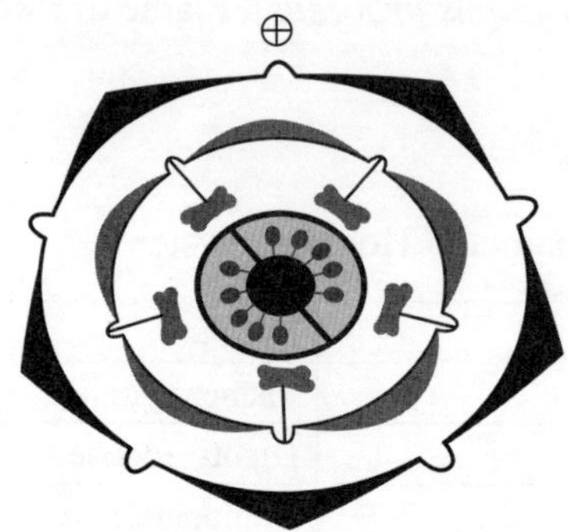

Economic importance:

- **Food:** Potato (*Solanum tuberosum*): Underground stem is used as vegetable.
  Brinjal/Egg plant (*Solanum melongena*): Fruit is used as vegetable.
  Tomato (*Lycopersicon esculentum*): Fruit is used as vegetable.
  Chilly (*Capsicum annum*): Fruit is used as condiment.
  Red pepper (*Capsicum frutescens*): Fruit is used as vegetable.
  Gooseberry (*Physalis peruviana):*
- **Medicine:** Deadly nightshade (*Atropa belladonna*: Root produces a sedative called belladonna, and its leaf produces an alkaloid, atropine.
  Black nightshade (*Solanum nigrum*): Fruit is given in fevers, diarrhoea and hydrophobia.
  Jimson weed/Datura (*Datura stramonium*): It is sedative.
  Ashwagandha (*Withania somnifera*): It is aphrodisiac but root is used in rheumatism.
  Henbane (*Hyoscyamus niger*): It produces alkaloids such as hyoscyamine and scopolamine.
- **Narcotics:** Tobacco (*Nicotiana tabacum*): Leaf produces nicotine and anabasin.
- **Ornamental plants:** Night jasmine/queen of night (*Cestrum nocturnum*), Henbane (*Hyoscyamus niger*):Yesterday-today-tomorrow (*Brunfelsia hopeana*), *Petunia*.

Liliaceae

It is also called Lily family. It comes under monocots.

Classification: (According to Bentham and Hooker's system)

Kingdom	Plantae
Division	Phanerogamia
Class	Monocotyledonae
Series	Coronarieae
Order	Liliales
Family	Liliaceae

Vegetative characters:

- Habit: Mostly herbs, perennial
- Root: Adventitious root (contractile root) and fibrous root
- Stem: It is modified into underground bulb, corm etc.
- Leaf: Alternate, exstipulate, parallel venation

Diagnostic features:

- **Inflorescence:** Variable; Solitary axillary, raceme or cyme; Inflorescence may develop on a special aerial stem called scape
- **Flower:** Mostly bracteate, Pedicellate, Actinomorphic, Bisexual, Incomplete, Hypogynous, Trimerous
- **Perianth:** Tepals-6, Gamophyllous, Arranged in two whorls of 3 each, Polyphyllous/ Polytepalous, Valvate aestivation, Petaloid tepal
- **Androecium:** Stamens-6, Arranged in two whorls of 3 each, Polyandrous, Epitepalous (Epiphyllous), Antiphyllous, Anther dithecous, introrse, versatile.
- **Gynoecium:** Tricarpellary, Syncarpous, Axile placentation, Ovary superior, Trilocular, Two ovules in each locule. Stigma trifid (trilobed).
- **Fruit:** Capsule or berry
- **Seed:** Endospermic

Floral formula: $\mathbf{Br} \oplus ⚥\ \widehat{\mathbf{P}_{(3+3)}\ \mathbf{A}_{3+3}}\ \mathbf{G}_{(3)}$

Floral diagram:

Economic importance:

- **Food:** Onion (*Allium cepa*): Edible part is bulb and fleshy leaves.
  Garlic (*Allium sativum*): Edible part is bulb.
  Garden Asparagus (*Asparagus officinalis*): Edible part is fasciculated root.
- **Medicine:** Aloe (*Aloe vera*) produces a drug, aloin which is laxative or purgative. Roots of *Smilax* produce 'Sarsaparilla' which is blood purifier.
- **Fibre:** Yucca (*Yucca filamentosa*), Sisal / century plant (*Agave Americana*): leaf fibre

- **Chemical**: Autumn Crocus *(Colchicum autumnale)*: Its corm and seed produce an alkaloid colchicine which causes chromosomal doubling (polyploidy).
- **Rodenticide**: *Urginea indica* produces rat poison from its bulb.
- **Ornamental plants:** Glory lily (*Gloriosa superba*), Lily (*Lilium bulbifera*), *Asparagus racemosus*, *Dracaena terniflora*, Dragger plant (*Yucca alolifolia*), Mother in law's tongue (*Sansivieria*), Tulip (*Tulipa gesneriana), Smilax, Ruscus, Fritillaria*

SHORT ANSWER TYPE QUESTIONS

1. Draw a well labelled diagram of different zones of root.
2. Define the respiratory root with examples.
3. How corm of underground modification of stem differs from tuber?
4. Distinguish between phylloclade and cladode.
5. Distinguish between thorn and spine.
6. Explain types of phyllotaxy with examples.
7. Distinguish between racemose inflorescence and cymose inflorescence.
8. Explain the different symmetries of flower.
9. Discuss the position of floral parts with respect to ovary.
10. Discuss the several conditions of cohesion of stamens in flower.
11. What do you mean by aggregate fruits. Give examples.
12. Distinguish between dicot seed and monocot seed.
13. Explain the causes of seed dormancy.
14. Draw a floral diagram of Solanaceae family.
15. Mention the botanical name of three plants of Fabaceae, Solanaceae and Liliaceae families.

LONG ANSWER TYPE QUESTIONS

1. Describe the various modifications of leaf with examples.
2. Discuss the special types of inflorescence.
3. What is aestivation? Describe its various types with diagrams.
4. What is placentation? Describe its several types with diagrams.
5. Mention the five distinguishing features of each Fabaceae, Solanaceae, and Liliaceae families with their floral formulae.

MULTIPLE CHOICE QUESTIONS

1. Roots are absent in
   (a) *Wolffia* (b) *Victoria* (c) *Nymphaea* (d) All of the above
2. A stem with distinct solid node and hollow internode is called
   (a) Sobole (b) Culm (c) Scape (d) Coleoptile
3. Most reduced stem occurs in
   (a) Rhizome (b) Corm (c) Bulb (d) Tuber
4. Prickles in rose are
   (a) Exogenous in origin (b) Endogenous in origin
   (c) Modified stipule (d) Modified leaf
5. The lateral roots arise from
   (a) Pericycle cells (b) Endodermal cells (c) Cortical cells (d) Hypodermal cells
6. Which of the following features would distinguish monocot from a dicot plant?
   (a) Aestivation (b) Vernation (c) Phyllotaxy (d) Venation
7. Fibrous roots develop in maize from
   (a) Lower nodes (b) Upper nodes (c) Upper internodes (d) Radicle
8. If a growing plant is decapitated
   (a) Leaves become yellow (b) Growth stops
   (c) Axillary buds are inactivated (d) Axillary buds are activated
9. Tunicated bulb is found in:
   (a) Lily (b) Onion (c) Garlic (d) *Oxalis*
10. Maximum growth in root occurs
    (a) Just behind the root apex (b) At the root apex
    (c) In the light (d) In the dark
11. Dicotyledonous leaf showing parallel venation is seen in:
    (a) *Smilax* (b) *Ranunculus* (c) *Calophyllum* (d) *Dioscorea*
12. Velamen tissue is found in:
    (a) Aerial roots of Orchids (b) Roots of Screwpine
    (c) Leaves of *Ficus* (d) Stem of Orchids
13. The plumule in monocot seeds is covered by
    (a) Coleorhiza (b) Coleoptile (c) Scutellum (d) Aleurone layer
14. A fibrous root system is adapted better than tap root system to perform the function of:
    (a) Absorption of water and minerals (b) Transport of water and organic food
    (c) Anchorage of plants into the soil (d) Storage of food
15. The presence of sheathing leaf base and a ligule are characteristic of:
    (a) Fern leaf (b) Banana leaf (c) *Cycas* leaf (d) Grass leaf
16. Sessile flowers of only one sex are found in the inflorescence are called:
    (a) Spadix (b) Catkin (c) Umbel (d) Panicle

17. Which one of the following fruits is not type of composite or multiple fruits?
(a) Pineapple (b) Mulberry (c) Jack fruit (d) Custard apple
18. Which of the following aestivation shows arrangement of sepals in one direction only?
(a) Valvate (b) Twisted (c) Vexillary (d) Quincuncial
19. Multiple root cap is present in:
(a) *Pandanus* (b) *Eichhornia* (c) *Pistia* (d) *Lemna*
20. Largest and edible bud is found in:
(a) Cabbage (b) Cauliflower (c) Onion (d) *Agave*
21. Imparipinnate leaf is the one where:
(a) Leaflets are borne in pairs (b) Leaflets are small
(c) Leaflets are large (d) Rachis is terminated by an odd leaflet
22. The main difference between thorns and prickles is in having:
(a) Endogenous origin (b) Vascular supply (c) Bark (d) All of the above
23. Circular internodes are present in
(a) Groundnut (b) Ginger (c) *Amorphophallus* (d) Turmeric
24. Entire leaf is modified into leaf tendrils in:
(a) *Lathyrus odoratus* (b) *Lathyrus aphaca* (c) *Gloriosa* (d) *Clematis*
25. Persistent calyx is feature of
(a) Fabaceae (b) Solanaceae (c) Liliaceae (d) Malvaceae
26. Which of the following condition belongs to adhesion of stamen?
(a) Syngenesius (b) Diadelphous (c) Synandrous (d) Epipetalous
27. Which of the following inflorescence shows false whorl
(a) Cyathium (b) Spadix (c) Hypanthodium (d) Verticillaster
28. The number of stamens in tetradynamous condition is
(a) 2 (b) 4 (c) 6 (d) 8
29. The number of leaf on each node of opposite phyllotaxy is
(a) One (b) Two (c) Three (d) Many
30. The root hairs are formed in
(a) Zone of maturation (b) Zone of elongation (c) Zone of division (d) Zone of cap
31. Which of the following inflorescence is highly evolved?
(a) Spikelet (b) Head (c) Cyathium (d) Hypanthodium
32. Leaf apex is modified into spine in
(a) *Gloriosa* (b) Pea (c) *Smilax* (d) Date palm
33. Which of the following is not correctly matched ?
(a) Stilt root-Maize (b) Clinging root-Orchid
(c) Photosynthetic root-*Tinospora* (d) Sucking root- *Jussiea*
34. Which of the following vegetative characters is highly conserved and used for classification of plants?
(a) Phyllotaxy (b) Venation of leaf
(c) Presence of bud (d) Node and internode

35. Fabaceae family belongs to order
    (a) Rosales (b) Personales (c) Passiflorales (d) Polemoniales
36. Monothecous anther is seen in:
    (a) Cotton and Tulip (b) *Calotropis* and lady's finger
    (c) *Calotropis* and *Ficus* (d) Lady's finger and cotton
37. The condition of didynamous stamen is
    (a) Two outer and longer stamen (b) Four outer and shorter stamen
    (c) Four outer and shorter stamen (d) Two inner and longer stamen
38. Ventral suture condition is seen in
    (a) Parietal placentation (b) Marginal placentation
    (c) Axilc placentation (d) Basal placentation
39. Which of the following placentations is multilocular?
    (a) Marginal (b) Axile (c) Parietal (d) Free central
40. *Withania sominifera* produces an important drug belongs to
    (a) Liliaccae (b) Malvaceae (c) Solanaceae (d) Fabaceae
41. The formula of petals in Fabaceae family is
    (a) 1+ (2) + (2) (b) (1) + (2) + 2 (c) 1 + 2 + (2) (d) 1 + (2) + 2
42. Ovary is obliquely placed in
    (a) Fabaccae (b) Solanaceae (c) Liliaceae (d) All of the above
43. Perigynous condition is seen in
    (a) Rose, Plum, *Ficus* (b) Plum, *Ficus*, *Morus*
    (c) *Morus* Pineapple, Radish (d) Plum, Peach, Rose
44. Unisexual flowers in dioecious condition are been in
    (a) Mulberry, Castor, Rice (b) Papaya, Date palm, Mulberry
    (c) Papaya Date palm, Maize (d) Maize, Mulberry, Papaya
45. The placentation which develops from inner wall of ovary is called
    (a) Axile (b) Basal (c) Parietal (d) Free central
46. Syngenesius condition exhibits
    (a) Filamens are free but anther are united (b) Filaments are united but anthers are free
    (c) Both filaments and anthers are free (d) Filaments anthers are united
47. Trimerous condition is found in
    (a) Fabaceae (b) Solanaceae (c) Liliaceae (d) Brassicaceae
48. The colour of sepaloid petal is:
    (a) Green (b) Coloured other than green
    (c) White (d) Mosaic coloured
49. The number of stamens in Fabaceae is
    (a) 5 (b) 10 (c) 8 (d) 5
50. In which aestivation, all sepals are partly outer and partly inner?
    (a) Valvate (b) Twisted
    (c) Descending imbricate (d) Quincuncial

51. Composite or multiple fruits are seen in
(a) Mulberry (b) Pineapple (c) Jackfruit (d) all of the above
52. Zygomorphic and diadelphous condition are seen in
(a) Pulses (b) Cereals (c) Cucurbits (d) Citrus plants
53. Perianth is found in
(a) Fabaceae (b) Solanaceae (c) Liliaceae (d) all of the above
54. Stilt roots are produced by
(a) Lower internodes (b) Upper internodes (c) Lower nodes (d) Upper nodes
55. Pneumatophores growing in waterlogged soils are helpful in
(a) Transpiration (b) Guttation (c) Food storage (d) Respiration
56. Leafless stem of onion which is produced to bear flowers is called
(a) Rachis (b) Peduncle (c) Scape (d) Torus
57. Pseudostem is found in
(a) Grass (b) Banana (c) Bamboo (d) Cauliflower
58. Axillary buds arise
(a) Endogenously from pericycle
(b) Endogenously from the inner cortex
(c) Exogenously from pericycle
(d) Exogenously from the inner cortex
59. Lateral branch growing obliquely from underground stem is called
(a) Bulbil (b) Sucker (c) Offset (d) Stolon
60. Stem is reduced to disc in
(a) Rhizome (b) Corm (c) Bulb (d) Tuber
61. Ginger is a stem, not a root because
(a) It stores food material
(b) It grows parallel to surface
(c) It has nodes and internodes
(d) It does not have chlorophyll
62. Finely dissected leaves are commonly found in
(a) Submerged plants
(b) Free floating plants
(c) Fixed floating plants
(d) Amphibious plants
63. The petiole is modified into leafy green structure known as
(a) Phylloclade (b) Cladode (c) Phyllode (d) Bract
64. In sweet pea, the tendrils are modified
(a) Stipule (b) Petiole (c) Leaflet (d) Leaf apex
65. Veins of leaf are useful in
(a) Transport of water
(b) Transport of minerals
(c) Transport of organic solutes
(d) All of the above
66. A compound leaf looks like a simple leaf in
(a) *Calotropis* (b) Citrus (c) Mango (d) Mustard
67. The time taken for the development of two adjacent leaves is called
(a) Leaf index (b) Plastochron index (c) Foliar time index (d) Stomatal index

68. Obdiplastemonous arrangement of stamens shows
    (a) Inner antipetalous and outer alternipetalous
    (b) Outer antipetalous and inner alternipetalous
    (c) Both inner and outer stamens are alternipetalous
    (d) Both inner and outer stamens are antipetalous
69. Swollen placenta is characteristic feature of
    (a) Fabaceae (b) Solanaceae (c) Liliaceae (d) Malvaceae
70. Eyes on potato are modified
    (a) Axillary bud (b) Apical bud (c) Internode (d) Scars
71. A monocarpic plant is that which
    (a) produces only one seed per fruit (b) produces only one fruit in its lifetime
    (c) has only one carpel in its flower (d) flowers only once in its life cycle
72. The term 'Epicalyx' represents
    (a) Involucre (b) An additional whorl of calyx
    (c) A whorl of bracteoles (d) A whorl of bracts
73. The coloured and attractive part in *Bougainvillea* is
    (a) Bract (b) Sepal (c) Petal (d) Leaf
74. The vexillum or standard in Pea flowers constitutes
    (a) Calyx (b) Corolla (c) Androecium (d) Gynoecium
75. Which of the following parts is considered equivalent to perianth?
    (a) Lodicules (b) Glumes (c) Inferior palea (d) Superior palea
76. For the study of parts of a flower, most suitable flower would be of
    (a) Mustard (b) Cucumber (c) Sunflower (d) Champak
77. When the inflorescence axis is branched, it is called
    (a) Head (b) Panicle (c) Umbel (d) Cyme
78. Three types of flowers are found in
    (a) Cyathium (b) Hypanthodium (c) Verticillaster (d) Capitulum
79. The most highly evolved type of inflorescence is
    (a) Hypanthodium (b) Cyathium (c) Capitulum (d) Spadix
80. Which of the following plant sets is monocarpic?
    (a) *Bambusa* and *Physalis* (b) *Agave* and *Pinus*
    (c) *Cycas* and *Pinus* (d) *Bambusa* and *Agave*
81. Nectiferous glands are found in
    (a) Cyathium (b) Hypanthodium (c) Verticillaster (d) Head
82. Trimerous and epiphyllous conditions are found in
    (a) Fabaceae (b) Solanaceae (c) Liliaceae (d) All of the above
83. Potato is included in Solanaceae family because
    (a) It has gamopetalous corolla (b) It has epipetalous stamen
    (c) It has pentamerous flowers (d) It has obliquely placed ovary

84. Tegmen develops from
    (a) Inner integument (b) Outer integument (c) Nucellus (d) Micropyle
85. Caruncle is a specialised growth develops from
    (a) Inner integument (b) Outer integument (c) Nucellus (d) Micropyle
86. Replum is a false septum found in
    (a) Lily (b) Pea (c) Mustard (d) All of the above
87. Geocarpic fruit is found in
    (a) Mustard (b) Potato (c) Sweet potato (d) Groundnut
88. Aleurone layer of maize grain is specially rich in
    (a) Carbohydrate (b) Fats (c) Oil (d) Proteins
89. Hypogeal germination occurs in
    (a) Cereals (b) Beans (c) Castor (d) Cucurbits
90. The plants such as *Lupin* and *Tulipa* belong to the families, respectively
    (a) Solanaceae and Liliaceae (b) Liliaceae and Fabaceae
    (c) Fabaceae and Liliaceae (d) Liliaceae and Solanaceae
91. Fibrous root system is found in
    (a) Monocots (b) Dicots (c) Gymnosperms (d) Bryophytes
92. The character of flower which is represented by floral formula but not by floral diagram is
    (a) Aestivation (b) Position of gynoecium
    (c) Placentation (d) Adhesion of stamens
93. Gall flowers are characteristic of inflorescence
    (a) Capitulum (b) Hypanthodium (c) Cyathium (d) Verticillaster
94. Colchicine is an alkaloid produced from *Colchicum autumnale* which belongs to
    (a) Fabaceae (b) Solanaceae (c) Liliaceae (d) Malvaceae
95. Censor mechanism works for dispersal of seeds in
    (a) Poppy (b) Bean (c) Cucurbits (d) Berry
96. Vivipary germination of seed is seen in
    (a) *Lilium* (b) *Artocarpus* (c) *Rhizophora* (d) *Eugenia*
97. Perisperm is remnant of nucellus seen in
    (a) Beet (b) Black pepper (c) *Nymphaea* (d) All of the above
98. Which of the following is incorrect about monocot seed?
    (a) Seed coat is fused with pericarp (b) Scutellum is seen
    (c) Aleurone layer is found (d) Radicle is not covered with coleorhiza
99. The plants such as *Lupinus, Physalis, Petunia* and *Tulipa* belong to
    (a) One family (b) Two families (c) Three families (d) Four families
100. Which inflorescence shows single female flower which is surrounded by several male flowers?
    (a) Cyathium (b) Hypanthodium (c) Verticillaster (d) Head

ANSWERS

1.a	2.b	3.c	4.d	5.a	6.d	7.a	8.d	9.b	10.a
11.d	12.a	13.a	14.c	15.d	16.b	17.d	18.b	19.a	20.a
21.d	22.d	23.d	24.c	25.b	26.d	27.d	28.c	29.b	30.a
31.b	32.d	33.d	34.b	35.a	36.d	37.d	38.b	39.b	40.c
41.c	42.b	43.d	44.b	45.c	46.a	47.c	48.a	49.b	50.b
51.d	52.a	53.c	54.c	55.d	56.c	57.b	58.d	59.b	60.c
61.c	62.a	63.c	64.c	65.d	66.b	67.b	68.b	69.b	70.a
71.d	72.d	73.a	74.b	75.a	76.c	77.b	78.b	79.c	80.d
81.c	82.c	83.d	84.a	85.b	86.c	87.d	88.d	89.a	90.c
91.a	92.b	93.b	94.c	95.a	96.c	97.d	98.d	99.c	100.a

5

Anatomy of Flowering Plants

Anatomy deals with the internal organisation of plants. The plants are made up of cells and these are organised into tissues. These tissues in turn develop into organs which vary in their internal structure.

Tissue: It is group of cells having same origin, structure (form) and functions. The term tissue was coined by N. Grew. Father of plant anatomy is Nehemiah Grew.

Plant tissue is divided into two types**: Meristematic tissue and Permanent tissue**.

Meristematic tissue (Meristem): The term 'Meristem' was coined by Nageli (1859).

Meristem is a group of young, immature and undifferentiated cells which have capacity to divide. It is localised and growing region of plant body and contributes to the growth of plants.

Features

(a) These cells are spherical, oval or polygonal in shape but isodiametric.

(b) These are thin-walled, elastic, compactly arranged and homogeneous.

(c) These are actively dividing and metabolising cells hence their surface area per unit volume and nuclear-cytoplasmic ratio are high.

(d) These lack intercellular space but interconnected by plasmodesmata.

(e) The cell wall is made up of cellulose. It is thin containing primary wall only.

(f) The cell has small vacuoles but sometimes vacuoles are absent.

(g) The cell has dense cytoplasm. Nucleus is large. Mitochondria are present. Plastids are absent but proplastids are present. Other cell organelles are either absent or present in non-functional state. The ergastic substances are absent.

(h) These cells have no reserve food (carbohydrate, protein and fats).

Types: On the basis of origin

1. **Promeristem (Primordial meristem):** These are embryonic meristem found at the apices of embryonic root and shoot. These give rise to primary meristem by differentiation.
2. **Primary Meristem:** These develop right from the beginning from promeristem and persist throughout life. These are present in apex of root and shoot. These are found in monocots and herbaceous dicots. These are responsible for primary growth or first growth (increase in length of root and stem). e.g., Fascicular (Intrafascicular) cambium.
3. **Secondary Meristem:** These develop at later stage of life by dedifferentiation of cells of primary permanent tissue. These are lateral in position. These are found in shrubs and trees of dicots. These give rise to secondary tissues. These are responsible for secondary growth (increase in girth or width or circumference of root and stem. e.g., Interfascicular cambium, cork cambium.

Types: On the basis of location

1. **Apical Meristem:** These are primary meristem. These are present in apex of roots, stem and reproductive shoots (buds). These are responsible for primary growth (increase in length of root and stem).
2. **Lateral Meristem:** These are secondary meristem. These are present in older parts of dicot root and stem and lie parallel to their long axis (sides). These divide periclinally (radially and tangentially). These produce secondary xylem and phloem and medullary rays. These are responsible for secondary growth (increase in width or girth or circumference). e.g., Vascular cambium (fascicular and interfascicular cambium), cork cambium.
3. **Intercalary cambium:** These are primary meristem. These are present between regions of permanent or mature tissues such as base of node as in Mint; base of internode as in monocots and base of leaf as in *Pinus*. These are short lived and soon converted into permanent tissues after completing active growth period. These are detached part of apical meristem that have been separated from apex by layers of permanent tissue. These are responsible for primary growth (increase in length of plant except at apex.

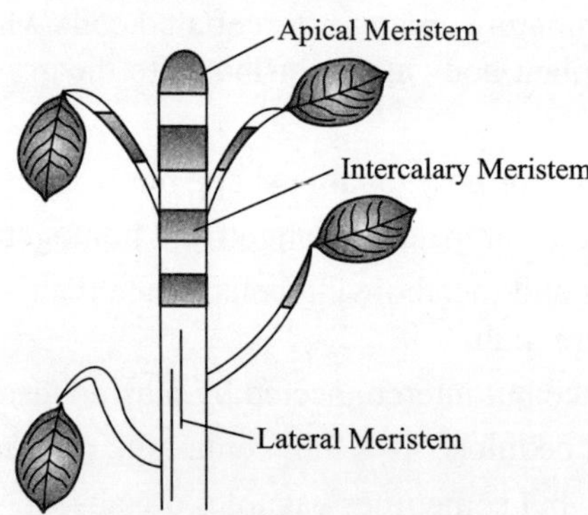

Fig. ***Types of meristem***

Types: On the basis of plane of division

1. **Rib (File) meristem:** Cell division occurs in one plane forming a row of cells. It increases in length. e.g., cortex and pith of young stem, most of filamentous algae.
2. **Plate meristem:** Cell division occurs in two plane (at right angle to each other) forming plate of cells. It increases in area. e.g., epidermis
3. **Mass meristem:** Cell division occurs in all planes forming a mass of cells. it increases in volume. e.g., endosperm, embryo, sporangia.

Types: On the basis of rate of cell division

1. **Summit:** Rate of mitotic cell division is slow.
2. **Flank:** Rate of mitotic cell division is fast. It gives rise to leaf primordia.

Functions of meristem

(a) These are responsible for growth in plant by adding new cells followed by differentiation.

(b) These form new organs such as bud, leaf and flower.

(c) These heal the injured parts of plant body.

Shoot apex organisation: Shoot apex contains meristematic cells. It is situated just above the youngest leaf primordial. It gives rise to lateral branches of stem and leaf.

Theories to explain shoot apex organisation

1. **Apical cell theory:** It was proposed by Hofmeister and supported by Nageli. It is applicable for shoot apex. It is single apical cell with three cutting faces in shoot apex. It is responsible for entire growth of plant. It is applicable for algae, bryophytes and pteridophytes.
2. **Histogen theory:** It was proposed by Hanstein. It is applicable for shoot apex. There are three groups of initials or zones or histogens or tissue builders such as
   (a) **Dermatogen:** It is outermost zone which forms epidermis.
   (b) **Periblem:** It is middle zone which forms cortex and endodermis.
   (c) **Plerome:** It is innermost zone which forms stele.
3. **Tunica-Corpus theory:** It was proposed by Schmidt. It is applicable for shoot apex. There are two zones which are as follows:
   (a) **Tunica:** It is single layered, smaller and divides anticlinally (at right angle to the surface of apex). It gives rise to epidermis hence there is surface growth.
   (b) **Corpus:** It is several layered, larger, and divides anticlinally and periclinally (tangentially and radially) in all planes. It gives rise to cortex, endodermis, and stele (pericycle, phloem, xylem and pith) hence there is growth in volume.

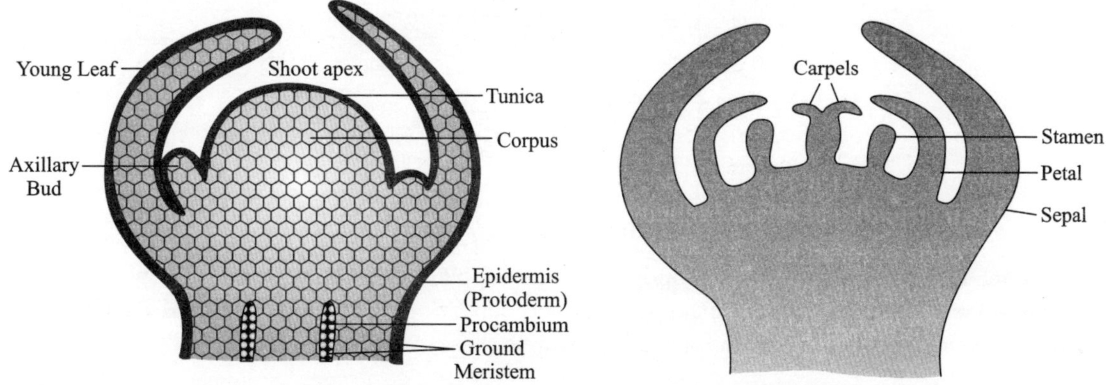

Fig. ***Vegetative shoot apex and reproductive shoot apex***

Vegetative shoot apex	Reproductive shoot apex
It is conical or dome shaped.	It is flat in shape.
It is protected by young leaves.	It is protected by sepals.
It shows indeterminate growth.	It shows determinate growth.
It gives rise to leaves and bud spirally.	It gives rise to sepals, petals, stamens and carpels in whorl.

Root apex organisation: Root apex contains meristematic cells. It is situated just below apex due to presence of cap hence it is sub-apical or sub-terminal.

Theories to explain root apex:

1. **Quiescent Centre Concept:** It was proposed by Clowe (1957) on maize. It is applicable for root apex. Quiescent centre is cup like region of cells lying between root cap and active meristematic region at centre. It is inactive or passive region of root where cells have very small nucleus having fewer mitochondria and endoplasmic reticulum. The cell has low concentration of DNA, RNA and protein. The cell undergoes very few divisions. It functions as reserve meristem when other meristematic cells are damaged. It is the site of auxin synthesis. There are four regions on the basis of function.
   - (a) **Protoderm:** It gives rise to epiblema (epidermis bearing root hairs).
   - (b) **Ground meristem:** It gives rise to cortex, endodermis, pith, and medullary rays.
   - (c) **Procambium:** It is narrow elongated parenchymatous cell which gives rise to vascular bundle (Primary xylem and phloem) and cambium.
   - (d) **Calyptrogen:** It gives rise to root cap (root apex of monocots).

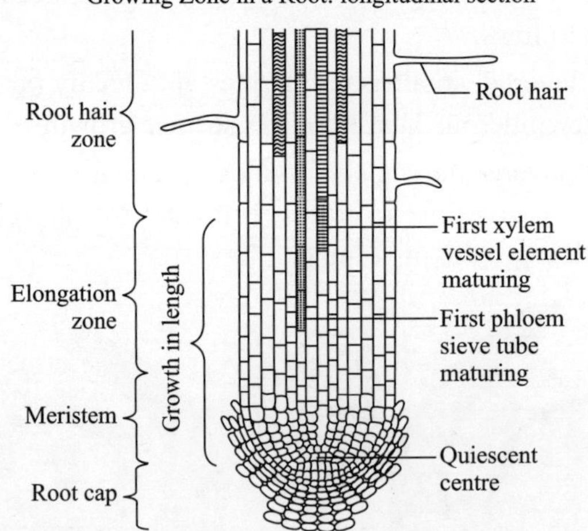

Fig. ***Growing region in root***

2. **Korper-Kappe theory:** It was proposed by Schwapp (1917). The cells of root apex divide first transversely and then one of the cell divides vertically. Some cells form inverted T called Korper (cap) and some cells form straight T called Kappe (base).

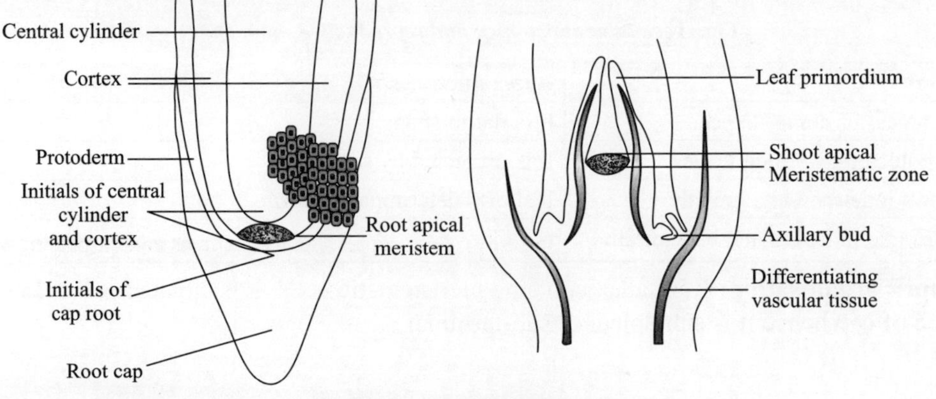

Fig. ***Root apex and shoot apex***

Root apex	Shoot apex
It is cup shaped and shorter.	It is dome shaped and longer.
It is formed at radicle end of embryo.	It is formed at plumule end of embryo.
It does not form lateral appendages.	It forms lateral appendages.
It is subterminal or subapical because it is covered by root cap.	It is terminal or apical.

Note:

- Epidermis is absent in both root tip and shoot tip.
- There are 5 shoot apical meristems likely to be present in a twig of a plant possessing 4 branches and 26 leaves.
- In virus infected plants, the meristematic tissues in both apical and axillary buds are free from viruses because cell division in meristems is faster than the rate of viral multiplication.

Types of Cell: On the basis of the ability to divide

1. Undifferentiated cells (**Undifferentiation**): These usually possess power of division hence these are unspecialised. e.g., root and shoot apex (Apical meristem), cambium, zygote., Stem cell (animal)
2. Differentiated cells (**Differentiation**): These do not undergo division hence these become specialised and undergo maturation and qualitative permanent change with respect to structure and function. e.g., Root and shoot apex (Apical meristem) form parenchyma, collenchyma, sclerenchyma, aerenchyma, chlorenchyma, xylem, phloem, stomata, mesophyll cell.
3. Dedifferentiated cells (**Dedifferentiation**): The cells of permanent or specialised tissues lose their specialisation and regain the power of division and form new cells. e.g., Interfascicular cambium, cork cambium.
4. Redifferentiated cells (**Redifferentiation**): The dedifferentiated cells again lose their ability to divide and become specialised and permanent cells. e.g., Secondary xylem, Secondary phloem, Phellem, phelloderm, Callus in suspension culture.

Permanent Tissues: It is tissue in which cells (living or dead) have lost their capacity to divide. It develops from meristematic tissue. It is of two types on the basis of composition.

Simple permanent tissue: It is structurally similar and performs same function.

1. Parenchyma: It was coined by Grew.

It is evolved first hence it is most primitive. It is known as packing tissue. It is most common or abundantly found in all parts of plant. It is morphologically and physiologically unspecialised. Cells are living, usually spherical, oval or elongated but isodiametric (equal in length and width), thin walled (having 14 facets or sides) and loosely arranged containing intercellular spaces. Cell wall is made up of cellulose. It is found in cortex and pith of root and stem; mesophyll of leaf and floral parts; pulp and rind of fruit; Cotyledon and endosperm of seed. It is metabolically active and retains the ability to divide at maturity (interfascicular cambium, cork cambium). It is known as prosenchyma when it is elongated or fibre like. It is also associated with xylem and phloem.

Types:

1. Angular parenchyma: *Ficus, Vitis*
2. Lamellar parenchyma: *Rheum*
3. Tubular parenchyma: *Malva, Salvia*

Functions:

(a) Parenchyma stores food materials (carbohydrate, proteins and fats).

(b) It performs photosynthesis when it contains chloroplast hence it is called chlorenchyma.

(c) It helps in floating when it contains air cavity due to separation of cells (schizogenous cavity) hence it is known as aerenchyma.

(d) It stores ergastic substances such as oils, tannins, minerals in the form of crystals hence it is known as idioblast.

(e) It secretes enzymes, nectars and resins.

(f) It helps in secondary growth because it may become meristematic.

(g) It provides rigidity to softer parts.

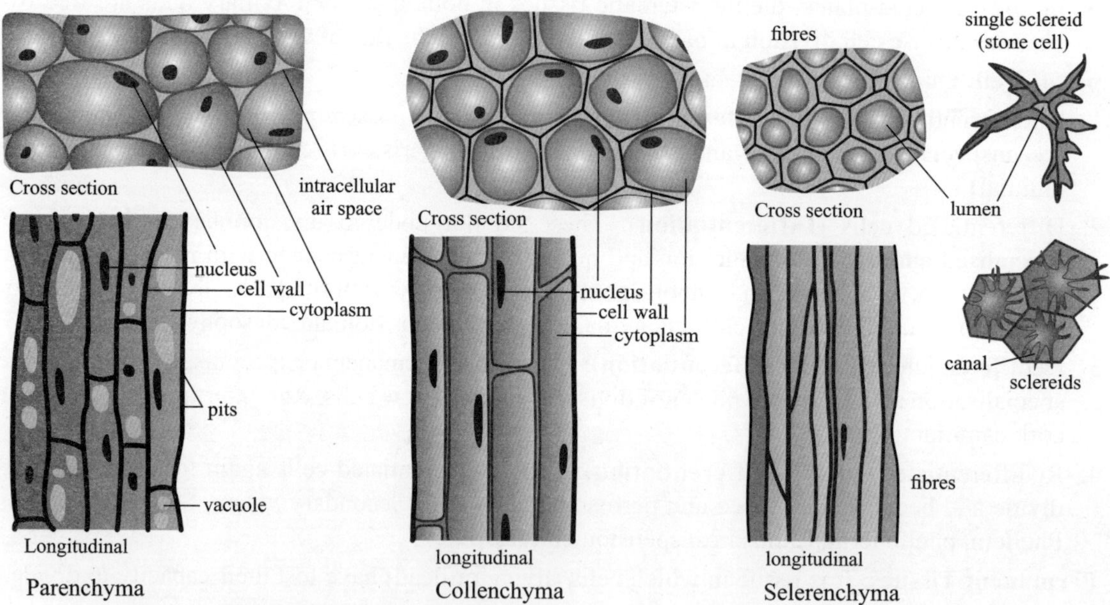

2. **Collenchyma**: It was coined by Schleiden.

   It is known as mechanical tissue. Cells are living, usually oval, elongated or polygonal, thick-walled, (unevenly at corners due to deposition of pectin, cellulose and hemicellulose), compactly arranged without intercellular space. Cell wall is made up of pecto-cellullose. Cell contains cytoplasm having prominent nucleus and functional cell organelles but vacuole is absent. It is hypodermal in position. It is found in angles of dicot stem; midrib and petiole of dicot leaves. It is absent in roots and monocots. It is not associated with xylem and phloem.

   Types:

   1. Angular collenchyma: It is most common. The thickenings occur at angles. It is found in hypodermis of stem of Solanaceae.
   2. Lacunate (Tubular) collenchyma: Thickenings occur on the wall bordering the intercellular spaces. It is found in hypodermis of stem of cucurbits; fruit wall.
   3. Lamellar collenchyma: Thickenings occur on tangential wall forming plate like structure. It is found in hypodermis of stem of sunflower; petiole.

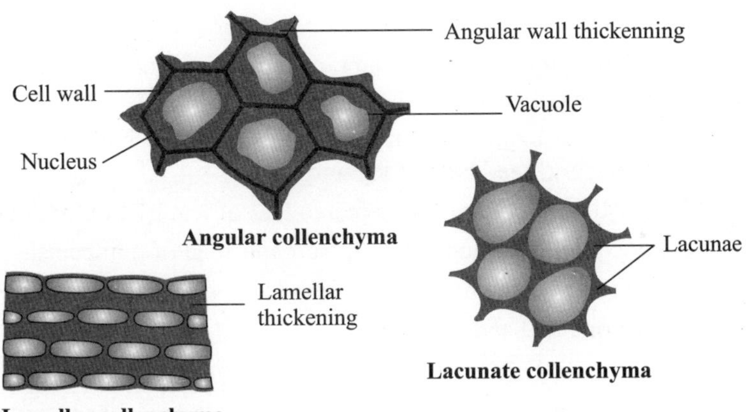

Types of collenchyma

Functions:

(a) It provides mechanical support to growing parts of stem and leaf.

(b) It provides elasticity and flexibility to plant parts while bending.

(b) It is sometimes photosynthetic due to presence of chlorophyll.

3. **Sclerenchyma**: It was coined by Mettenius.

It is also known as mechanical tissue. Cells are dead, elongated, narrow, thick walled (evenly or uniformly), and compactly arranged without intercellular space. Cell does not contain cytoplasm, nucleus, vacuole and cell organelles. Cell wall of cell is made up of lignocellulose. Cell wall has depression or cavities called simple pits where secondary wall material is not deposited. It is found in all parts of plants. It is associated with xylem and phloem. It is of two types:

(a) Fibre: It is long and unbranched with tapered (narrow) or pointed end wall. It shows large lumen. It is present in bundle and overlaps each other. It is found in cortex, pericycle, xylem and phloem of stem. It provides mechanical or tensile strength while bending, pulling or compression without breaking off.

Types of fibre:

1. Soft fibres: Pericycle of stem produces fibre of flax (*Linum*) and secondary phloem (bast) of stem produces fibres of Jute, Sun hemp, Ramie etc.
2. Hard (structural) fibres: Leaf produces fibres of Sisal (*Agave*) and Manila hemp (*Musa textilis*).
3. Surface fibres: Mesocarp of fruit produces fibres in Coconut; seed hairs produce fibres in Red silk cotton (*Bombax*) and seed coat (testa) of seed produces fibres in cotton.

(b) Sclereid (Stone cell/Sclerotic cell): It shows great variation in shape (rounded, elongated or bone like) and without tapering end wall. It shows small or reduced lumen due to thickening of cell wall. It is scattered singly. It is found in different regions of root, stem, leaf (of tea), flower, fruit wall (of nuts), fruit pulp (of guava, pear) and seed coat (legumes). It provides rigidity (stiffnesss) and grittiness.

Types of Sclereid

1. Brachysclereid: It is short, rounded or oval, unbranched. It is found in stem; fruits of guava, pear, apple, sapota, hard endocarp of coconut.
2. Macrosclereid: It is large, elongated, and rod like. It is found in bark of dicot stem; protective scales of onion and garlic; and seed coat of legumes.
3. Asterosclereid: It is star like. It is found in stem and leaf of tea, hydrophytes and xerophytes.
4. Osteosclereid: It is bone like. It is found in leaf, fruit and seed of monocots.
5. Trichosclereid: It is elongated and hair like. It is found in aerial root of *Monstera*; leaves of Olive.
6. Filiform sclereid: It is elongated and fibre like.

Meristematic tissues	Permanent tissues
These are actively and continuously dividing.	These do not divide and carries out specialised function.
Cells are undifferentiated.	Cells are differentiated.
Cells are smaller and isodiametric.	Cells are variable in shape.
Cell wall is thin.	Cell wall is thick.
Nuclei are large but vacuoles and intercellular spaces are absent.	Nuclei are small but vacuoles and intercellular spaces are present.
Metabolic rate is very high.	Metabolic rate is very low.
It does not attain definite form or size.	It attains definite size or form.

Complex Permanent tissue: It is structurally dissimilar but performs same function. It is of two types:

1. **Xylem** (Hadrome/wood): The term 'xylem' was coined by Nageli.

   It is dead cell, sclerified and durable. It conducts water and minerals. It is differentiated into tracheary elements (tracheids and vessels), xylem parenchyma and xylem fibres. It consists of the following parts:

   (a) **Tracheid:** It is primitive or firstly formed. It is unicellular, long (1mm), tubular, thick, imperforated. It has narrow lumen (its diameter is shorter than that of vessel). It has oblique and tapered end wall which contains pit by which all tracheids are connected. It is dead (without protoplasm) due to deposition of lignin hence it is lignified. It forms long row one above the other. It is arranged parallel to long axis of plant. It has several types of thickenings of lignin such as annular (ring like), spiral, scalariform (ladder like), reticulate (network like), pitted. Pitted thickening is advanced and shows maximum deposition of lignin. It is found in pteridophytes, gymnosperms and angiosperms. It helps in short distance transport of water and mineral hence it is less efficient.

   (b) **Vessel** (Tracheae): It is advanced or later formed. It is multicellular (formed by fusion of several cells in a row due to dissolution of septa or end wall), long (10 cm), tubular, thick and perforated. It is dead (without protoplasm) due to deposition of lignin hence it is lignified. It has wide lumen (greater than that of tracheid). It has rounded and flat end wall of uniform diameter which is perforated. It has several types of thickenings of lignin such as annular (ring-like), spiral, scalariform (ladder-like), reticulate (network like), pitted. Pitted thickening is advanced and shows maximum deposition of lignin. It is absent in pteridophytes but present in *Selaginella, Equisetum, Pteridium, Marsilea*. It is absent in Gymnosperms but present in *Gnetum, Ephedra, Welwitchia*. It is present in Angiosperms except *Dracaena, Yucca*, and members of Winteraceae, Tetracentraceae and Trochodendraceae families. It is the largest in *Eucalyptus* and Oak. It helps in long distance quick transport of water and minerals hence it is more efficient.

(c) Xylem Parenchyma: It is thin walled and abundant. It is only living part of xylem. It may be apotracheal in which parenchyma is diffused or not associated with vessel, and paratracheal in which parenchyma is associated with vessel. It forms outgrowth called tyloses when xylem gets injured or becomes inactive. Tyloses always pass through pit, if not passes then it is called tylosoid (as in resin canal). It stores food materials (starch, fat, tannin). It helps in short distance transport and radial conduction of water.

(d) Xylem fibre: It is dead, long, narrow, thick. It has narrow lumen and tapered end wall. It is more abundant in secondary xylem than primary xylem. It provides mechanical support but does not help in transport of water.

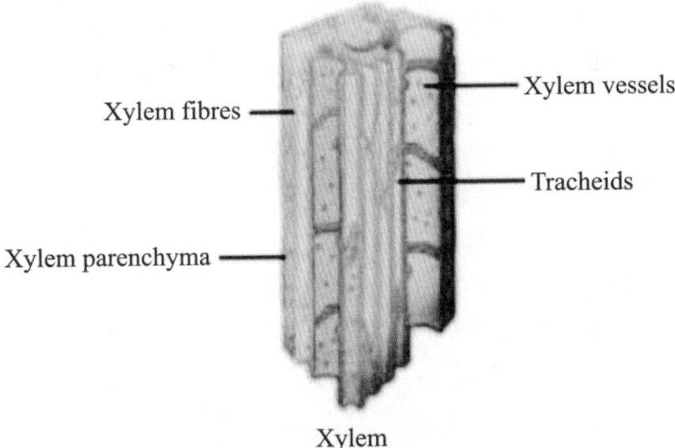

Xylem

Types of Xylem: On the basis of origin

1. **Primary Xylem:** It is developed from procambium during primary growth. It is of two types:

Protoxylem	Metaxylem
It is smaller in size	It is larger in size
It is firstly formed.	It is later formed.
It shows annular and spiral thickening.	It shows reticulate and pitted thickenings.

2. **Secondary Xylem:** It is developed from vascular cambium during secondary growth.

Pit: It is thin region or cavity in the secondary wall where lignin is not deposited. It contains pit chamber and pit membrane. It is found in tracheid of gymnosperms and angiosperms; and vessel of Angiosperms. It helps in transport of sap between tracheids. It is of two types:

(i) Simple pit: Pit chamber is of uniform width and primary wall is not covered by secondary layer.

(ii) Bordered pit: Pit chamber is flask shaped and primary wall is covered by secondary layer. It is found in *Pinus*.

2. **Phloem** (Leptome or bast): The term 'phloem' was coined by Nageli. It is living cell and transports food materials in the form of sucrose. It consists of the following parts:

(a) Sieve tube: It was discovered by Hartig. It is long, tubular, slender and thin walled. It is formed by end to end fusion of cells. It is arranged longitudinally. It is made up of pecto-cellulose. It has end wall (sieve-plate) having numerous sieve pores. Each sieve pore is lined by a layer of callose (a carbohydrate). It is only living plant cell without nucleus at maturity but since it is attached with living companion cell hence it is considered as living. Its cytoplasm lacks nucleus, vacuole, Golgi complex and ribosomes. It contains P-protein which helps in transport of material from leaf to storage organ in downward direction and later from storage organ to growing region in upward direction. It also seals pores after wounding. In winter sieve pores get plugged with callose hence transport of food is slowed down and in summer transport resumes when callose gets dissolved. It is present in angiosperms but it is absent in pteridophytes and gymnosperms where sieve cells are present. Sieve cells are uniform and further these lack sieve plates.

(b) Companion cell: It is long, narrow, thin walled, and specialised parenchymatous cell. It is living because it contains nucleus and dense cytoplasm. It does not have sieve plate. It is connected to sieve tube with simple pits. It resembles sieve tube because both develop from same cell (ontogenetically same) hence called sister cells. It is absent in pteridophytes and gymnosperms where albuminous cells are present instead of companion cell. It helps in loading food material from mesophyll cell to sieve tube. It helps in transport of food by maintaining pressure gradient in sieve tube.

(c) Phloem parenchyma: It is living, elongated, cylindrical and thin walled. It absent in Monocots. It stores food materials as well as mucilage, latex and resins etc. It also helps in radial transport of food material.

(d) Phloem fibre (Bast fibre): It is dead (due to lack of protoplasm), long, narrow, thick walled and tapered. It is absent in primary phloem but present in secondary phloem. It provides mechanical support.

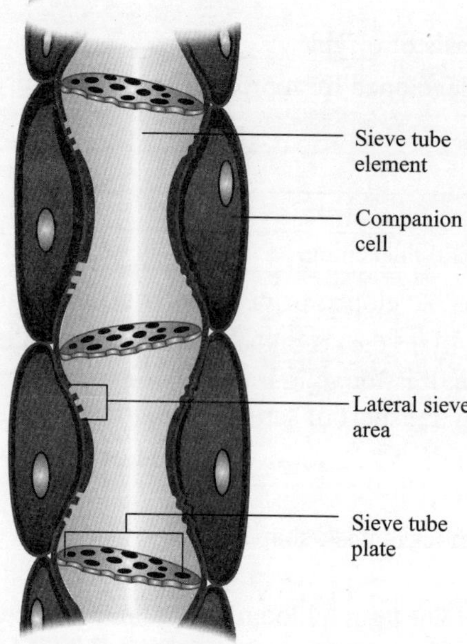

Fig. ***Components of Phloem***

Types of Phloem: On the basis of origin

1. Primary Phloem: It originates from procambium. It is of two types:

Protophloem	Metaphloem
It is firstly formed.	It is later formed.
It is smaller in size.	It is larger in size.
It has narrow sieve tube.	It has large sieve tube.

2. Secondary Phloem: It develops from vascular cambium.

Types of Phloem: It is of three types on the basis of position

1. External phloem: It is normal type and present outside xylem. It is quite common in gymnosperms and angiosperms.
2. Internal or Intraxylary phloem: It originates from procambium. It is primary phloem. It is present on innerside of primary xylem. e.g., members of Apocyanaceae, Asclepiadaceae, Convolvulaceae, Solanaceae families.
3. Included or Interxylary phloem: It originates from cambium and is present in groups within secondary xylem. It is anomalous feature found in *Salvadora*, *Leptidaenia*, *Amaranthus*, *Boerhaavia.Chenopodium* etc.

Sieve Cell	Sieve tube
It is longer and narrower channel with tapering end wall.	It is shorter and broader channel with broad end wall.
Sieve areas are present throughout lateral walls and end walls	Sieve areas are present only at end wall.
Companion cells are absent.	Companion cells are present.
Albuminous cell is present.	Albuminous cell is absent.
It is found in Pteridophytes and Gymnosperms.	It is found in angiosperms.

Tissue System: It is the group of different plant tissues irrespective of their position in the plant body performing a common function. Sachs (1875) divided tissue system into three systems:

Epidermal Tissue system: It develops from outermost layer of apical meristem. It consists of

1. **Epidermis:** It is outermost and single layered but multiple or multiseriate epidermis is seen in velamen of aerial root of Orchids and leaf of *Ficus, Nerium* and *Peperomia* (on upperside) and Maize (on upper side and lower side). It is living, thin walled, compactly arranged and parenchymatous. Its cells have small cytoplasm and large vacuole. It is not uniform in structure and function due to presence of many types of hairs and guard cells of stomata. It bears unicellular hairs in roots and multicellular hairs in stem. It secretes cutin, waxy substance which forms cuticle. Cuticle is absent in roots and hydrophytes but thick in xerophytes. It contains cystolith ($CaCO_3$) hence called lithocyst. It protects plants from desiccation (drying) and infection.
2. **Stomata:** The epidermis bears small opening called stomata. It is absent in root. It is present in herbaceous stem and leaf. Its number varies from 14-1038 per square mm. but average number is around 200. It is kidney or bean shaped in dicots and dumbbell shaped in monocots. It has a pore called stoma bounded by two guard cells and surrounded by several subsidiary or accessory cells collectively called stomatal apparatus. The inner wall of guard cell is thicker and non-elastic. Guard cell unlike subsidiary cell bears small chloroplast but performs no photosynthesis due to

absence of RuBisCO in spite of the presence of pigment systems. Guard cell is heterotrophic and import sugar from adjacent cells. Guard cell has large vacuole and nucleus and not linked by plasmodesmata to subsidiary cells. Stomata performs transpiration and exchange of gases (O_2 and CO_2).

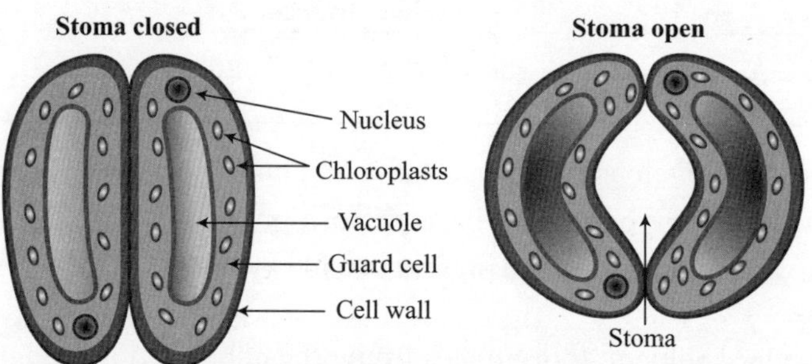

Fig. ***Internal structure of Stomata***

Types of Stomata: On the basis of origin

1. Haplochelic: Guard cell and subsidiary cell are not related and both originate separately. e.g., *Cycas, Pinus, Ginkgo*
2. Syndetochelic: Guard cell and subsidiary cell originate from same single mother cell. e.g., *Gnetum, Welwitchia.*

Types of Stomata: On the basis of number of subsidiary cells

1. Diacytic: There are two subsidiary cells which are right angle to guard cell. e.g., Caryophyllaceae, Lamiaceae.
2. Paracytic: There are two subsidiary cells which are parallel to guard cell. e.g., Magnoliaceae, Rubiaceae.
3. Anisocytic: There are three subsidiary cells. e.g., Brassicaceae, Solanaceae.
4. Actinocytic: There are four subsidiary cells. e.g., Commelinaceae, Musaceae.
5. Anomocytic: There are five subsidiary cells e.g., Malvaceae, Ranunculaceae.

Distribution of stomata:

1. Astomatic type (Potamogeton type): Stomata are absent. e.g., *Potamogeton, Vallisneria* (submerged hydrophytes), Algae
2. Amphistomatic (Potato type): Stomata are present on both surface but more on lower side as in mesophytes; but more on upper side. e.g., Sciophytes.
3. Amphistomatic (Oat type): Stomata are present equally on both surface of leaf. e.g., Monocots
4. Epistomatic (Water lily type): Stomata are present on upper side only. e.g., Water lily, Lotus, *Nymphaea*.
5. Hypostomatic (Apple type): Stomata are present on lower side. e.g., Apple, Banyan, Oak, Peach, Walnut.
6. Sunken stomata: Stomata is present deep inside. e.g., xerophytes.

3. **Appendages**: The epidermis contains the following two types of appendages:

(a) Hairs: These are unicellular in roots and multicellular in stem.

(b) Trichomes: These are multicellular structure on stem. These check transpiration. These are also protective and secretive.

Ground Tissue system: It is developed from ground meristem or partly from periblem and partly from plerome.

1. **Hypodermis**: It is 2-3 layered. It is absent in roots. It is collenchymatous in dicot stem and sclerenchymatous in monocot stem. It is protective and photosynthetic.
2. **Cortex:** It is several layered, thin walled and parenchymatous. It usually contains intercellular spaces. It contains several types of cells such as resin duct, oil gland, laticifers. It stores food materials.
3. **Endodermis:** It is single layered, elongated or barrel shaped and compactly arranged. It contains a thickened band called **Casparian strips** on radial and tangential wall of roots. It is formed by a waxy substance, suberin which makes water impermeable. The Casparian strips force water and minerals through the membrane of endodermis to protect from plasmolysis; prevents water entry into xylem through apoplast; and regulate water and minerals in plants. It stores starch hence called starch sheath. It controls the movement of water between cortex and xylem hence acts as water jacket between vascular and non-vascular region. It regulates root pressure.
4. **Pericycle:** It is outermost part of stele. It is usually one layered (but multilayered pericycle is seen in root of *Agave* and *Smilax*), thin walled and parenchymatous. It gives rise to lateral root in monocot root and gives rise to lateral root and vascular cambium in dicot root. It provides mechanical support and acts as storage organ.
5. **Pith:** It is thin walled, parenchymatous and present at centre. Intercellular spaces are present or absent. It stores food materials as well as excretory materials (tannin and phenol).

Vascular tissue system: It develops from procambium. It includes xylem and phloem. These are usually associated to form vascular bundle. Cambium is a strip of primary meristem between xylem and phloem. It is of two types:

1. **Vascular cambium:** It is formed inside stele. It is of two types;

   Fascicular cambium: It is developed between xylem and phloem of vascular bundle during secondary growth.

   Interfascicular cambium: It is developed between two vascular bundles during secondary growth.
2. **Cork Cambium:** It is formed inside cortex during secondary growth.

Types of vascular bundle:

1. Open: Cambium is present between xylem and phloem. e.g., Dicots
2. Closed: Cambium is absent between xylem and phloem. e.g., Monocots.

Conditions of Protoxylem:

- **Exarch:** Protoxylem is present at periphery and metaxylem at centre. Primary xylem is developed from periphery to centre (centripetal development). e.g., Roots
- **Endarch:** Protoxylem is at centre and metaxylem at periphery. Primary xylem is developed from centre to periphery (centrifugal). e.g., Stem
- **Mesarch:** Protoxylem is at centre and surrounded by metaxylem. e.g., Rhizome of Ferns, Roots of *Ophioglossum*.

Types of vascular bundle: It is of three types on the basis of relative position of xylem and phloem.

1. **Radial:** Xylem and phloem are present in separate bundles which are arranged alternately on different radii. e.g., roots.
2. **Conjoint**: Both xylem and phloem are present in same bundles which are situated on same radii. It may be of two types:
   (a) **Collateral:** Xylem forms inner part and phloem forms outer part. e.g., stem.
   (b) **Bicollateral:** Xylem is surrounded by two layers of phloem from both inner and outer side hence xylem is sandwiched. e.g., Stem of cucurbits and *Solanum*.
3. **Concentric:** One of two vascular tissues (xylem or phloem) occupies at centre and is completely surrounded by another (xylem or phloem). In this cambium is absent. It may be of two types:
   (a) **Amphivasal** (Leptocentric): Xylem surrounds phloem from peripheral side. e.g., Stem of *Dracaena* and *Yucca*.
   (b) **Amphicribal** (Hadrocentric): Phloem surrounds xylem from peripheral side. e.g., Stem of Pteridophytes.

Exceptions:

1. Cortical vascular bundle: It is seen in *Casuarina, Nyctanthes*
2. Medullary vascular bundle: It is seen in *Amaranthus, Achyranthus, Bougainvillea,Mirabilis*.

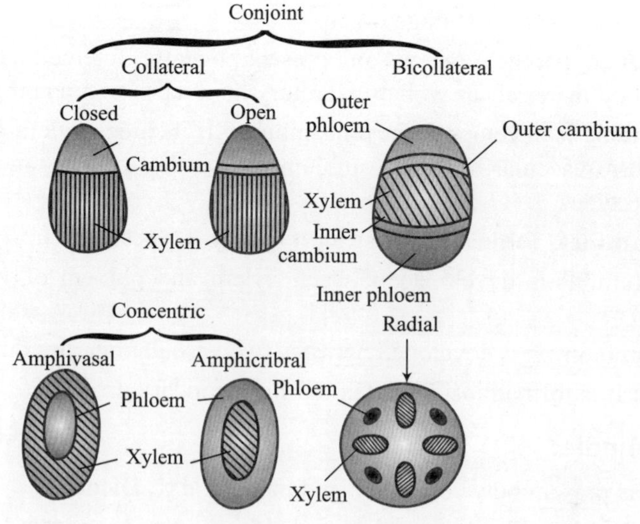

Different types of vascular bundles

Secretory Tissue system: It secretes different substances and performs special functions. It is of two types:

1. **Glandular tissue:** It contains glands which secrete materials. It includes
   (a) **External gland:** It is present on epidermal surface of stem and leaf.
      (i) Glandular hairs: These are present on leaves. e.g., Drosera, Urtica
      (ii) Nectaries: These are present on leaves and thalamus (below ovary).
      (iii) Digestive glands: These secrete proteolytic enzymes. e.g., Insectivorous plants.

(iv) Oil glands: These are present on flower which secretes essential oil.

(v) Hydathodes: These are found along the margins or apices (tip) of leaves which cause guttation. e.g., Garden Nasturtium (*Trepaeolum)*.

(b) Internal gland: It is embedded in tissues. It includes

(i) Mucilaginous glands: These secrete mucous. e.g., leaf of Betel

(ii) Gum glands: These secrete gum. e.g., stem of *Acacia*

(iii) Resin glands: These arc schizogenous in nature on stem. e.g., Resin duct of *Pinus*.

(iv) Oil glands: These are lysigenous in nature which secrete essential oils. e.g., leaves and fruits of citrus plants and *Eucalyptus*.

2. **Laticiferous tissue**: It is thin walled, elongated, branched, multinucleate, tube like structure which contains colourless or milky latex. It is of two types:

(a) Latex cell: It is simple, non-articulated, individual cell which is branched but does not anastomose (form network). e.g., *Calotropis, Euphorbia, Ficus, Nerium, Vinca.*

(b) Latex vessel: It is compound, articulated and multicellular (joined together by dissolution of septa), branched and anstomose (forms network). e.g., *Argemone, Heavea*, Poppy.

Anatomy of Root

Dicot root: The internal structure on transverse section of root shows the following features:

(a) Epidermis: It is single layered, compactly arranged and thin walled. It is not cutinised (without cuticle). It does not contain stomata. It bears unicellular hairs which absorb water and minerals hence epidermis is also called epiblema or rhizodermis or piliferous layer.

(b) Cortex: It consists of several layered, larger thin walled parenchymatous cells with intercellular spaces. In some dicots, one to several layers become suberised below epidermis to form exodermis which is a protective tissue. It does not contain chloroplast. It stores food materials.

(c) Endodermis: It is single layered, closely packed and situated below cortex. It contains thickened band called **Casparian strips** in radial and transverse walls. These strips are made up of suberin. These strips prevent plasmolysis in the endodermis; permit water to pass through symplast of endodermis; prevent leakage of nutrients from vascular tissues; and regulate the flow of water with minerals. A few cells are thin walled and without Casparian strips called passage cells situated opposite to protoxylem. These passage cells help in radial diffusion of water from cortex into xylem.

(d) Pericycle: It is single layered, thick-walled and parenchymatous. It gives rise to lateral root and vascular cambium during secondary growth.

(e) Vascular bundles: These consist of separate alternating strands of xylem and phloem (radial). These strands are usually tetrarch. These are closed due to absence of cambium. These show exarch condition (protoxylem is situated towards periphery and metaxylem towards centre). Conjuctive parenchymatous cells are present between xylem and phloem.

(f) Pith: It is less developed.

Monocot root

(a) Epidermis: It is single layered, compactly arranged, thin-walled. It is not cutinised.

(b) Cortex: In many monocots, exodermis occurs.

(c) **Endodermis:** It is single layered, closely packed and situated below cortex. It contains thickened band called **Casparian strips** in radial and transverse walls. These strips are made up of suberin.

(d) **Pericycle:** It is single layered, thick-walled and parenchymatous. It gives rise to lateral root.

(e) **Vascular bundles:** These are polyarch (containing more than 6 strands of xylem and phloem).

(f) **Pith:** It is well developed and occupied at centre. It is parenchymatous.

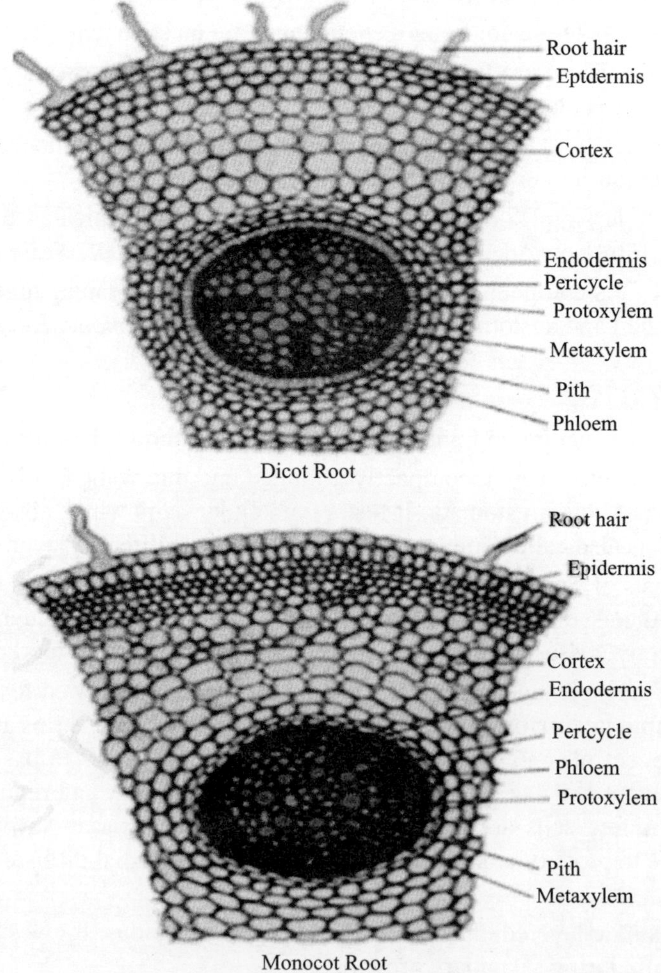

Dicot Root

Monocot Root

Dicot root	Monocot root
Cortex is narrow.	Cortex is wide.
Passage cells are absent in endodermis.	Passage cells are present in endodermis.
Pericycle gives rise to lateral roots and vascular cambium.	Pericycle gives rise to lateral roots only.
Vascular bundles are usually tetrarch.	Vascular bundles are polyarch (≥6).
Pith is not well developed.	Pith is well developed.
Cambium is present.	Cambium is absent
Secondary growth occurs.	Secondary growth does not occur.

Anatomy of Stem

Dicot stem: The internal structure on transverse section of stem shows the following common features:

(a) Epidermis: It is outermost, compactly arranged and single layered. It is heavily cutinised (thick cuticle) to prevent loss of water, and protects inner tissues from pathogens. It bears multicellular hairs having a few stomata.

(b) Hypodermis: It is present below epidermis. It is multilayered and collenchymatous. It provides mechanical support to young stem and also performs photosynthesis, if chloroplasts are present.

(c) Cortex: It is made up of several layered, rounded and parenchymatous cells having intercellular spaces. It bears aerenchyma in aquatic plants. It bears resin duct, latex tubes and reservoir of waste products.

(d) Endodermis: It is single layered, compactly arranged situated just below cortex. It separates the vascular tissues from cortex.

(e) Pericycle: It is single layered and outermost part of stele (central vascular cylinder).

(f) Vascular bundles: These are several in number. Each vascular bundle consists of xylem and phloem. The vascular bundles are ringed and conjoint collateral and open (cambium is present). The endarch condition is seen. In some plants such as cucurbits, bicollateral vascular bundles are seen.

(g) Pith: It is central and innermost part which is made up of parenchymatous cells having large inter cellular spaces. It shows the presence of medullary rays (narrow and vertically elongated bands of parenchymatous cells) extending from pith to periphery. These rays help in radial conduction of food. The pith stores food materials.

Monocot stem:

(a) Epidermis: it is outermost, compactly arranged and single layered. It is heavily cutinised (thick cuticle). It contains stomata.

(b) Hypodermis: It is present below epidermis. It is multilayered, sclerenchymatous and lignified. It provides mechanical strength to stem.

(c) Ground tissue: It constitutes the entire tissue below hypodermis. It is made up of parenchymatous cells rich in food reserves. These cells are larger and more loosely arranged on innerside towards centre compared to peripheral region. It is not diffrentiated into cortex, endodermis, pericycle and pith.

(d) Vascular bundles: These are scattered, conjoint collateral and closed (no cambium) in ground tissues. Each vascular bundle is surrounded by 2-3 layered sclerenchymatous bundle sheath.

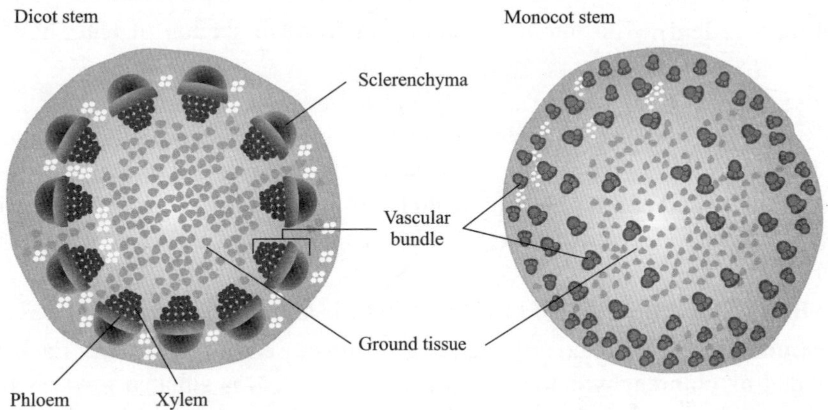

Dicot stem	Monocot Stem
Multicellular hairs are present.	Multicellular hairs are absent.
Stomata is bean or kidney shaped.	Stomata is dumb bell shaped.
Hypodermis is collenchymatous.	Hypodermis is sclerenchymatous.
Ground tissue is differentiated into hypodermis, cortex, endodermis, pericycle, and pith hence these are present.	Ground tissue is not differentiated into hypodermis, cortex, endodermis, pericycle, and pith hence these are absent.
Bundle sheath is absent.	Bundle sheath is present.
Vascular bundles are wedge shaped, of same size, ringed and open.	Vascular bundles are oval, of different size, scattered and closed.
Phloem parenchyma is present.	Phloem parenchyma is absent.
Medullary rays are present between vascular bundles.	Medullary rays are absent.
Schizo-lysigenous cavity is absent.	Schizo-lysigenous cavity is present in protoxylem.
Cambium is present.	Cambium is absent.
Secondary growth occurs.	Secondary growth does not occur.

Anatomy of leaf

Dicot (Dorsiventral) leaf: The internal structure on vertical section of leaf shows the following features:

(a) Epidermis: It consists of upper (adaxial) and lower (abaxial) epidermis. It is compactly arranged and heavily cutinised (thick cuticle) to check the loss of water. it contains variable number of stomata on both surface of leaf. It protects the internal tissues from the damage and infection.

(b) Mesophyll: It is present between two layers of epidermis. It is differentiated into two distinct layers-(i) the palisade cells which are upper, vertically elongated, parallel and closely packed with numerous chloroplast, and (ii) spongy cells which are inner (below palisade cells), rounded and loosely arranged with fewer chloroplast extending into lower epidermis. The palisade cells perform photosynthesis whereas spongy cells allow gaseous exchange between atmosphere and leaf.

(c) Vascular bundles: These are present in midrib and veins. If vein is thicker then size of vascular bundle will be larger. These are conjoint, collateral and closed (no cambium). Xylem is situated towards adaxial (upper) side and phloem towards abaxial (lower) side. Each vascular bundle is surrounded by thick-walled bundle sheath cells which are distinct from mesophyll cells.

Monocot (Isobilateral leaf): The internal structure on vertical section of leaf shows the following features:

(a) Epidermis: It consists of upper (adaxial) and lower (abaxial) epidermis. It is single layered, compactly arranged and heavily cutinised thick cuticle). It contains equal number of stomata on both epidermis. In some grasses, the epidermal cells are arranged in long furrow, larger in size, colourless, thin-walled and highly vacuolated and highly turgid called motor or **bulliform** cells which help in rolling of leaves upwards to minimise the exposed surface in dry weather (scarcity of water).

(b) Mesophyll: It is not differentiated into palisade and spongy cells. It is compactly arranged.

(c) Vascular bundles: These are conjoint, collateral and closed (no cambium). Each vascular bundle is surrounded by chlorenchymatous bundle sheath. Xylem is situated towards upper epidermis and phloem towards lower epidermis. Sclerenchyma is present on both sides of vascular bundle. The vascular bundles are uniform in size in all veins except midrib.

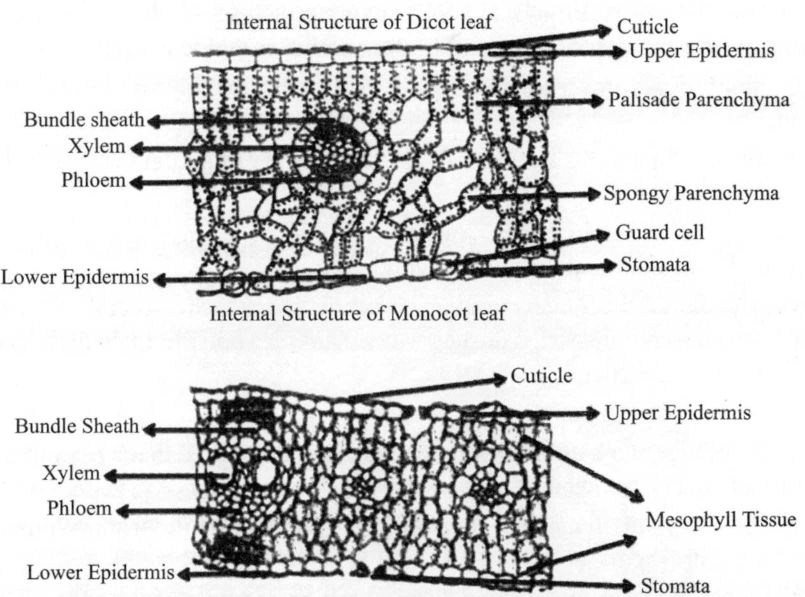

Dicot (Dorsiventral) leaf	Monocot (Isobilateral) leaf
Both surface of leaf is distinct. (upper/ventral/adaxial) and Lower/dorsal/ abasial)	Both surface of leaf is uniform.
Cuticle is thicker at upper surface.	Cuticle is uniform in structure on both surface
Stomata is bean or kidney shaped and more on lower side.	Stomata is dumb bell shaped and equal in number on both sides.
Bulliform cells or Motor cells are absent.	Bulliform cells are present on upper epidermis.
Hypodermis is collenchymatous.	Hypodermis is sclerenchymatous.
Mesophyll cells are differentiated into spongy cells and palisade cells.	Mesophyll cells are not differentiated into spongy cells and palisade cells. All are spongy cells.
Bundle sheath is collenchymatous.	Bundle sheath is sclerenchymatous.
Vascular bundles are dissimilar in size and arranged irregularly due to reticulate venation.	Vascular bundles are similar in size and are arranged in a row due to parallel venation.

Secondary growth

It is increase in width or girth or circumference in perennial gymnosperms and shrubs and trees of angiosperms. It also provides support to increasing weight of the aerial growth. It forms corky bark around tree trunk which protects from cold, heat, infection and abrasion. It adds new conducting tissues to replace old and non-functional ones.

Secondary growth in Dicot root: It starts occurring slightly behind root apex. In this vascular bundle is radial and exarch. It involves the following events:

1. **Origin and activity of vascular cambium:** It involves the following events:
   (a) Separate cambial strips are formed from conjuctive parenchymatous cells which become meristematic just below primary phloem and towards pith.
   (b) The number of cambial strips is equal to the number of phloem strands.
   (c) These cambial strips form secondary xylem inside centripetally and secondary phloem outside centrifugally.
   (d) Several more cambial strips are formed just opposite to protoxylem where the cells of pericycle become meristematic.
   (e) These newly formed cambial strips join the firstly formed cambial strips to form complete but wavy ring of vascular cambium. It is secondary in origin.
   (f) Now wavy cambium becomes circular in outline because the secondary xylem produced by the firstly formed cambial strips pushes wavy cambium outside since these cambial strips had started functioning earlier.
   (g) The circular cambial ring cuts off secondary xylem inside and secondary phloem outside.
   (h) The activity of vascular cambium is much faster on outside so that xylem forms major bulk of root after secondary growth.
   (i) Primary xylem strands remain in their original position while primary phloem, endodermis and cortex get crushed.
   (j) Primary medullary rays are formed from cambium against protoxylem. These are distinct widening radial bands of parenchyma. These are well developed in root compared to stem.

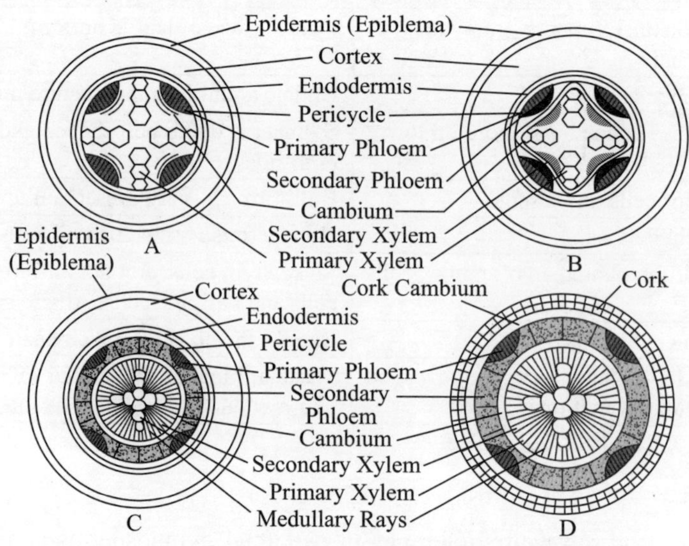

Secondary growth in dicot root. A—D, diagrams showing stages in the secondary growth of a typical dicotyledonous root.

2. **Origin and activity of Cork cambium:** It involves the following events:
   (a) Cork cambium (thin walled and rectangular cells) is formed when the cells of pericycle become meristematic.

(b) Cork cambium (also called phellogen) produces cork (phellem) outside and secondary cortex (phelloderm) inside. Secondary cortex is peripheral (outer) and primary cortex is inner.

(c) Pressure is exerted on peripheral tissues (endodermis, primary cortex and epidermis) so that these tissues get crushed and sloughed off.

(d) Cork cambium gives rise to periderm which is protective layer to replace ruptured epidermis and primary cortex.

(e) Cork is dead due to deposition of suberin. It does not change into bark. Bark is not extensive and forms only a thin covering.

(f) Annual rings are not formed in roots because there is no impact of seasonal variations on roots.

Secondary growth in Dicot stem: Vascular bundle is conjoint, collateral, ringed and open. It involves the following events:

1. **Origin and activity of Vascular cambium:** It involves the following events:

   (a) Fascicular cambium (intrafascicular cambium) is formed between xylem and phloem inside each vascular bundle. It is primary in origin.

   (b) Interfascicular cambium is formed later between different vascular bundles where conjuctive parenchymatous cells become meristematic. It is secondary in origin.

   (c) Both fascicular and interfascicular cambium join together to form complete circular ring. It is partly primary and partly secondary.

   (d) Cambial cells are of two types: fusiform initials and ray initials.

   (e) Fusiform initials are narrow and elongated cells which are formed in vascular cambium. These divide to form secondary xylem inside and secondary phloem outside.

   (f) The amount of secondary xylem produced usually exceeds the amount of secondary phloem because cambial ring is more active towards inner side.

   (g) Ray initials are small and spherical cells which are also formed in vascular cambium. These divide to form elongated parenchymatous cells. These cells accumulate to form medullary rays between secondary xylem and secondary phloem. The medullary rays help in radial conduction.

   (h) Pressure is now exerted by secondary xylem on peripheral tissues (secondary phloem, primary phloem and cortex) which become stretched and later get crushed.

   (i) Epidermis starts dividing to cover the increasing surface area but it gets crushed because of inability to cope with pace of secondary xylem.

   (j) Peripheral tissues need to be replaced to provide protection to stem.

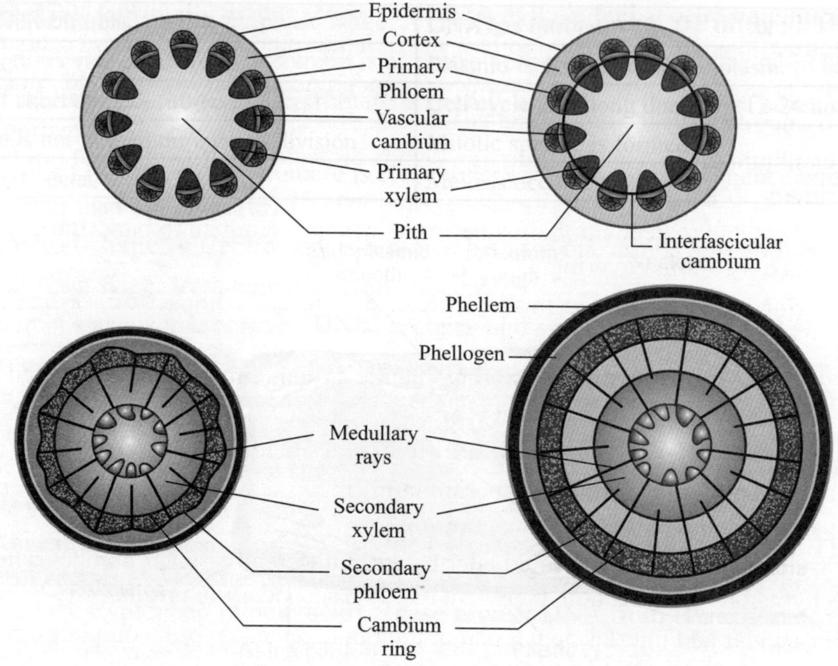

Secondary growth in a dicot stem (diagrammatic)

2. **Origin and activity of Cork cambium:** It involves the following events:
   (a) Cork cambium is formed in outer cortex when permanent parenchymatous cells in outer cortex become meristematic.
   (b) Cork cambium is thick, rectangular and several layered. These cut off cork (phellum) outside and secondary cortex (phelloderm) inside. Thus primary cortex is inner and secondary cortex is peripheral (outer).
   (c) Primary cortex gets crushed and then epidermis gets ruptured.
   (d) Cork is now impervious to water due to deposition of suberin and it cuts off the outer tissues from the supply of water and food materials. Consequently these tissues die and form a part of bark.
   (e) Periderm is later formed which provides to protection to tissues. It gives rise to bark. Periderm consists of the following parts:
      1. Outer cork (phellum): It is dead, prismatic in shape, compact, suberised (made up of suberin), and protective layer.
      2. Middle cork cambium (phellogen): It is living, rectangular, meristematic. Its activity is more on outside.
      3. Inner secondary cortex (phelloderm): It is living, radially flattened, and parenchymatous.

Secondary growth in Dicot Root	Secondary growth in dicot Stem
Vascular cambium is initially wavy and later becomes circular.	It remains circular since beginning.
Vascular cambial ring is secondary in origin as it is formed of conjuctive parenchyma and pericycle.	Vascular cambial ring is completed by both intrafascicular cambium (primary) and interfascicular cambium (secondary).
Vascular rays are wider in the beginning.	Vascular rays are narrower since beginning.
Pericycle gives rise to cork cambium.	Outer layer of cortex gives rise to cork cambium.
Annual rings are not formed.	Annual rings occur commonly.

Differentiation of Wood: As a result of continued secondary growth for several years, the older part of stem is differentiated into two zones:

Sap wood (alburnum)	Heart wood (duramen)
It is peripheral (outer) part.	It is central (inner) part.
It is living having living cells.	It is dead.
It is functional part of secondary xylem.	It is non-functional part of secondary xylem.
Tracheid and vessel do not have any deposition any deposition in their lumen.	Tracheid and vessel have deposition of oils, gums, resins and tannins.
Tracheid and vessel are not plugged by tyloses.	Tracheid and vessel are plugged by tyloses.
It conducts water.	It does not conduct water.
It does not provide mechanical support.	It provides mechanical support.
It is light coloured.	It is dark brown coloured.
It has lower density.	It has higher density.
It is less durable because it is more susceptible to attack by pathogens.	It is more durable because it is less susceptible to attack by pathogens.

Types of wood: On the basis of arrangement of vessels

1. Ring porous: Vessels are arranged in ring. It is advanced. e.g., *Betula*, *Dalbergia, Morus*, Oak (*Quercus*).
2. Diffuse porous: Vessels are uniformly distributed throughout season. It is primitive feature. e.g., Neem, Poplar, *Salix*.

Types of Wood: On the basis of composition

Soft wood	Hard wood
It lacks vessels hence non-porous.	It contains vessels hence porous.
It is made up of 95% tracheids and 5% fibres.	It is made up of 5% tracheids and 95% vessels and fibres.
It is found in gymnosperms	It is found in Angiosperms.

Lightest wood	Balsa (*Ochrorma pyramidale*)
Heaviest wood	*Guaicum officinale*

Bark: It is outermost layer of stem and root of woody plants. It contains all dead tissues outside the vascular cambium. It protects the inner tissues such as phloem from external injuries and prevents loss of water by evaporation. If bark is formed early in season, it is called soft bark. If it is formed late in season then it is called hard bark. If bark is removed then phloem of stem gets damaged first and root dies first due to starvation (food is unable to move downwards hence it will be wasted).

Types of Bark: On the basis of position

1. Inner bark: It is dead tissue between vascular cambium and cork cambium. It contains secondary phloem, primary phloem, endodermis and pericycle.
2. Outer bark (Rhytidome): It is dead tissue outside cork cambium. It contains ruptured epidermis, hypodermis and cork. Outer layer of bark is constantly peel off due to addition of new secondary vascular tissue in interior.

Types of bark: On the basis of arrangement

1. Ring bark: Bark appears in the form of complete ring. e.g., *Betula* (Bhojpatra), *Vitis*.
2. Scaly bark: Bark appears in the form of strip. e.g., *Acacia*, Apple, *Eucalyptus*, Guava, Pine.

Lenticel: It is aerating pore in the bark of trees. It is lens shaped and commonly found in stem. It is filled with loosely arranged, thin walled, rounded, suberised or unsuberised cells called complimentary cells. These cells are formed from loosely arranged phellogen and division of substomatal parenchymatous cells. Pressure is created on epidermis due to increasing number of complimentary cells resulting in rupture.

Functions: (a) It helps in transpiration and exchange of gases.

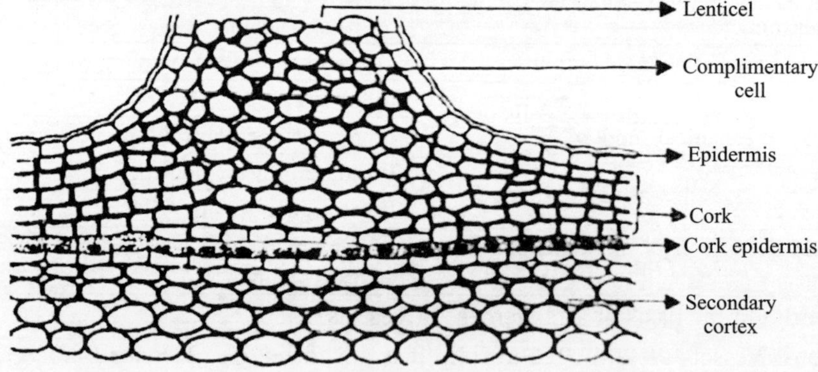

Fig. *Lenticel*

Formation of Secondary structure:

Annual ring or Growth ring: It is the annual or yearly growth appears in the form of distinct ring due to seasonal activity vascular cambium. Hence it is the seasonal increment of wood. One annual ring is formed per year thus numerous rings are formed in successive years in main stem. One annual ring includes one circle of early wood and late wood. The width of ring varies due to fluctuation in atmosphere. If ring is wide then it has got favourable growth condition.

It is developed in temperate areas where climatic conditions are not uniform throughout the year hence there is climatic variations (different seasons have different climate, temperature and rain). In such areas, the activity of cambium is not uniform throughout the year. In spring, there is greater production and activity of foliage leaves. Hence plant needs more water. Thus cambium is more active in spring and forms greater number of vessels with wider lumen. The wood thus formed is early wood.

Dendrochronolgy is study of calculating the age of trees by counting the number of these rings in T.S of stem.

It is not developed in tropical areas, sea coast, desert and equatorial region because uniform climate is not suitable for growth of annual rings all of the year.

Early Wood	Late Wood
It occurs in spring season hence it is known as spring wood.	It occurs in late autumn or winter season hence it is known as autumn wood or winter wood.
It is formed during favourable period of year.	It is formed during unfavourable period.
It constitutes major part.	It constitutes minor part.
Cambium is very active.	Cambium is less active.
Cambium produces large amount of xylary elements which are thin forming a distict concentric ring.	It produces small amount of xylary elements which are thick forming a distinct ring.
Vessel has large lumen.	Vessel has narrow lumen.
Xylem fibres are less in number.	Xylem fibres are abundant in number.
It is lighter in colour.	It is darker in colour.
It has lower density.	It has higher density.

SHORT ANSWER TYPE QUESTIONS

1. What are the salient features of cells of meristematic tissue?
2. Distinguish between primary and secondary meristems.
3. Distinguish between root apex and shoot apex.
4. Explain the quiescent centre concept.
5. Distinguish between collenchyma and sclerenchyma.
6. Distinguish between tracheid and vessel of xylem.
7. Draw a well labelled diagram of phloem.
8. Draw a well labelled diagram of stomata of monocot and dicot leaves.
9. Explain the open and closed vascular bundle.
10. Distinguish between anatomy of dicot root and monocot root.
11. Distinguish between anatomy of dorsiventral leaf and isobilateral leaf
12. Distinguish between sap wood and heart wood.
13. Write notes on annual ring.
14. Write the different components of periderm.
15. Distinguish between secondary growth in dicot root and dicot stem.

LONG ANSWER TYPE QUESTIONS

1. Describe the different types of meristem on the basis of development.
2. Describe the different types of vascular bundles in plants.
3. Distinguish between dicot stem and monocot stem.
4. Describe the origin and activity of cambium during secondary growth in dicot root.
5. Describe the origin and activity of cambium during secondary growth in dicot stem.

MULTIPLE CHOICE QUESTIONS

1. The intercalary meristem in monocots is present at
(a) Base of node (b) Base of internode (c) Base of petiole (d) Everywhere
2. Bast fibres in woody stem belong to
(a) Cork (b) Cortex (c) Xylem (d) Phloem
3. Fascicular cambium found in dicotyledonous stem is a
(a) Primary meristem (b) Intercalary
(c) Apical meristem (d) Secondary meristem
4. Amphivasal vascular bundles are found in
(a) Maize and wheat (b) *Dracaena* and *Yucca*
(c) *Cycas* and *Dryopteris* (d) *Cucurbita* and *Helianthus*
5. Well developed pith is found in
(a) Monocot stem and dicot root (b) Monocot root and dicot stem
(c) Dicot root and dicot stem (d) Monocot root and monocot stem
6. Endodermis of dicot stem is also called
(a) Starch sheath (b) Mesophyll (c) Pith (d) Bundle sheath
7. Phloem parenchyma is absent in
(a) Dicot leaf (b) Monocot stem (c) Dicot stem (d) Dicot root
8. Hypodermis in monocotyledonous stem is
(a) Parenchymatous (b) Collenchymatous
(c) Sclerenchymatous (d) Chlorenchymatous
9. Protoxylem in dorsiventral leaf is directed towards
(a) Adaxial side (b) Abaxial side
(c) Surrounded by phloem (d) Surrounds phloem
10. The correct situation of mesophyll in isobilateral grass leaf is shown by
(a) Palisade towards adaxial surface (b) Palisade towards abaxial surface
(c) Undifferentiated mesophyll (d) Palisade along both the surface
11. Which one of the following tissues is formed in stem from cells cut off by the cambium on its inner side?
(a) Wood fibres (b) Sieve tubes (c) Companion cells (d) Bast fibres
12. In case of dicot root, the cork cambium is derived from
(a) Epidermis (b) Pericycle (c) Cortex (d) Hypodermis
13. The oldest layer of secondary phloem in a dicot stem is located just
(a) outside the vascular cambium (b) inside the primary phloem
(c) outside the secondary xylem (d) inside the vascular cambium
14. Periderm is formed from
(a) Vascular cambium (b) Phellogen
(c) Fascicular cambium (d) Interfascicular cambium

15. As secondary growth proceeds in a tree, the thickness of
    (a) Sap wood increases (b) Heart wood increases
    (c) Both increases (d) Both remain the same
16. The wall of cork cells is mostly impregnated with
    (a) Cutin (b) Suberin (c) Lignin (d) Hemicellulose
17. An example of monocot showing secondary growth in stems is
    (a) *Crocus* (b) *Asparagus* (c) *Yucca* (d) *Lilium*
18. Wilting of plants takes place if
    (a) Few roots are damaged (b) Xylem is blocked
    (c) Phloem is blocked (d) Light is cut to 50%
19. Tyloses are balloon like outgrowth developed by
    (a) Tracheid (b) Vessel
    (c) Xylem fibres (d) Xylem parenchyma
20. A parenchymatous cell which stores ergastic substances is called
    (a) Idioblast (b) Conidioplast (c) Phragmoplast (d) Chromoplast
21. Which of the following would most likely shows totipotency in plants?
    (a) Xylem vessels (b) Meristem (c) Cork (d) Sieve tube
22. Passage cells occur in
    (a) Epidermis (b) Cortex (c) Pericycle (d) Endodermis
23. Which of the following is correctly matched?
    (a) Angular collenchyma- Stem of *Solanum* (b) Lacunate collenchyma- Stem of cucurbits
    (c) Lamellate collenchyma- Stem of sunflower (d) All of the above
24. Sieve tubes differ from sieve cells in
    (a) lacking nuclei (b) having sieve pores at end walls
    (c) being shorter (d) being dead
25. Axillary bud and terminal bud are derived from the activity of
    (a) Lateral meristem (b) Intercalary meristem (c) Apical meristem (d) Parenchyma
26. Grittiness of fruit and seed is due to
    (a) Collenchyma (b) Parenchyma (c) Sclereids (d) Fibres
27. In a dorsivental leaf, protoxylem and metaxylem are located, respectively on
    (a) Abaxial and adaxial sides (b) Adaxial and abaxial sides
    (c) Adaxial and adaxial sides (d) Abaxial and abaxial sides
28. The inner wall of stomata is
    (a) Thinner and elastic (b) thicker and elastic
    (c) Thinner and inelastic (d) Thicker and inelastic
29. The perforated wall is seen in
    (a) Tracheid (b) Vessel (c) Fibre (d) Sclereid

30. Biocollateral vascular bundles have tissue arrangement
    (a) Outer phloem – outer cambium – middle xylem – inner cambium – inner phloem
    (b) Outer xylem – outer cambium – middle phloem – inner cambium – inner xylem
    (c) Outer phloem – outer xylem – middle cambium
    (d) Outer cambium – outer phloem – middle xylem – inner phloem – inner cambium
31. Quiescent centre occurs in
    (a) Shoot apex (b) Root apex
    (c) Both shoot apex and root apex (d) Meristematic tissue
32. Which one is not formed from procambium?
    (a) Xylem (b) Phloem
    (c) Intrafascicular cambium (d) Interfascicular cambium
33. Casparian strips contain deposition of
    (a) Cutin (b) Pectin (c) Suberin (d) Lignin
34. Cells of quiescent centre have lower concentration of
    (a) DNA (b) Proteins (c) RNA (d) All of the above
35. Radial vascular bundles occur in
    (a) Dicot root (b) Dicot stem (c) Monocot stem (d) Monocot leaf
36. Trichomes take part in
    (a) Transpiration and exchange of gases (b) Exudation of water drops
    (c) Protection and reduction (d) Desiccation
37. The cells of collenchyma are
    (a) Dead and hollow (b) Dead with reserve food
    (c) Living with no reserve food (d) Living with protoplasm
38. In barley stem, vascular bundles are
    (a) Closed and radial (b) Open and scattered
    (c) Closed and scattered (d) Open and in a ring
39. The cork cambium, cork and secondary cortex are collectively called
    (a) Phelloderm (b) Phellogen (c) Periderm (d) Phellem
40. Which of the following is nonnucleated at maturity?
    (a) Palisade cell (b) Cortical cell (c) Periderm (d) Companion cell
41. The apical meristem of the root is present
    (a) Only in radicle (b) Only in tap roots
    (c) Only in adventitious roots (d) All of the above
42. Vascular tissues in flowering plants develop from
    (a) Periblem (b) Dermatogens (c) Phellogen (d) Plerome
43. The annular and spirally thickened conducting elements generally develop in the protoxylem when the root of stem is:
    (a) elongating (b) widening (c) differentiating (d) maturing

44. In which of the following phloem parenchyma is absent
    (a) Maize (b) Sunflower (c) Guava (d) Banyan
45. Passage cells are found in
    (a) Dicot stem (b) Monocot root (c) Aerial root (d) Monocot stem
46. Bordered pits occur in
    (a) Sievc tube (b) Sieve cell (c) Vessel (d) Companion cell
47. The maximum harm to a tree occurs due to
    (a) Loss of half of its leaves (b) Loss of half of its branches
    (c) Loss of all of its leaves (d) Loss of all of its bark
48. Palisade parenchyma is absent in leaves of
    (a) Soyabean (b) Grain (c) Sorghum (d) Mustard
49. The dicot root is distinguished from dicot stem by
    (a) Presence of passage cells (b) Position of protoxylem
    (c) Absence of secondary xylem (d) Presence of medullary rays
50. The following events in secondary growth of dicot stem are as follows
    I. formation of circular cambium
    II. formation of cork cambium
    III. formation of fascicular cambium
    IV. formation of interfascicular cambium
    V. formation of wavy cambium
    The proper sequence of above events is:
    (a) I – III – II – IV –V (b) III – IV – V – I – II
    (c) II - IV - V – III - I (d) V – I – II – IV - III
51. The tissues present in annual ring are
    (a) Primary xylem and primary phloem (b) Secondary xylem and secondary phloem
    (c) primary phloem and secondary xylem (d) Secondary xylem
52. Girdling experiment cannot be performed in monocot plants because
    (a) Vascular bundles are scattered (b) Vascular bundles are closed
    (c) Condition of xylem is exarch (d) Protoxylem is situated outside
53. The histogen layer present at the apex of root tip is called
    (a) Dermatogens (b) Calyptrogens (c) Plerome (d) Procambium
54. A leaf primordium grows into the adult leaf lamina by
    (a) Initially apical meristem and later by marginal meristem
    (b) Initially marginal meristem and later by apical meristem
    (c) Apical meristem only
    (d) Lateral meristem
55. The stalk of leaf (petiole) is derived from
    (a) Parenchyma (b) Collenchyma (c) Fibres (d) Sclereids

56. Stomata are present only on upper side of leaf in
(a) *Potamogeton* (b) Apple (c) Water lily (d) *Nerium*
57. If assign board was nailed on the side of tree 5 feet above in year 2000, how high would be sign be in year 2015 if the tree grew 1 inch taller each year
(a) 5 feet long (b) 20 feet long (c) 19 feet long (d) 21 feet long
58. The face of cell wall where lignifications occurs in sclerenchyma is
(a) Inner (b) Outer (c) Radial (d) All of the above
59. The term 'sterome' refers to
(a) All vascular tissues (b) All mechanical tissues
(c) All packing tissues and mechanical tissues (d) All vascular tissues and mechanical tissues
60. Needle like crystals appearing in parenchyma are called
(a) Raphides (b) Cystolith (c) Druses (d) Lithocyst
61. The unthickened space of a pit is called
(a) Pit chamber (b) Pit cavity (c) Pit aperture (d) Pit membrane
62. The sieve area of a sieve plate is comparable to
(a) Perforation plate of vessel (b) Primary pit field
(c) End wall of tracheid (d) All of the above
63. Vascularisation of plants occurs in the manner of
(a) Procambium differentiates first, xylem and then phloem
(b) xylem differentiates first, phloem and then procambium
(c) Procambium differentiates first, phloem and then xylem
(d) Xylem differentiates first, procambium and then phloem
64. Which one of the following is incorrect regarding meristem?
(a) These are thin walled and compactly arranged
(b) These contain plastid
(c) These have no reserve food
(d) These lack intercellular spaces
65. Fascicular cambium is found in
(a) Promeristem (b) Primary meristem (c) Secondary meristem (d) All of the above
66. Histogen theory is applicable for
(a) Root apex (b) Shoot apex
(c) Both root apex and shoot apex (d) None of the above
67. Root cap is formed from
(a) Histogen (b) Plerome (c) Procambium (d) Calyptrogen
68. Interfascicular cambium and cork cambium are formed as a result of
(a) Differentiation (b) Dedifferentiation (c) Undifferentiation (d) Redifferentiation
69. Which of the following is produced as a result of redifferentiation?
(a) Root apex (b) Parenchyma (c) Cork cambium (d) Secondary xylem

70. Xylem and phloem are not associated with
(a) Parenchyma (b) Collenchyma (c) Fibres (d) All of the above
71. Collenchyma is made up of
(a) Cellulose (b) Pectocellulose (c) Lignocellulose (d) Chitin
72. Which of the following thickenings is considered as advanced?
(a) Annular thickening (b) Scalariform thickening
(c) Spiral thickening (d) Pitted thickening
73. Sieve cells are present in
(a) Bryophytes and pteridophytes (b) Pteridophytes and gymnosperms
(c) Gymnosperms and angiosperms (d) All of the above
74. Which of the following cells is present in gymnosperms instead of companion cells?
(a) Complimentary cell (b) Sister cell
(c) Albuminous cell (d) Nonalbuminous cell
75. Multiseriate epidermis is present in
(a) *Ficus* (b) *Nerium* (c) Maize (d) All of the above
76. During night, the stomatal apparatus contains
(a) Guard cell only (b) Guard cell and subsidiary cell
(c) Guard cell, stoma and subsidiary cell (d) Stoma and subsidiary cell
77. Sunken stomata is found in
(a) Hydrophytes (b) Mesophytes (c) Halophytes (d) Xerophytes
78. Multilayered pericycle is present in
(a) Root of *Ficus* (b) Root of onion (c) Root of *Smilax* (d) Root of orchid
79. Mesarch condition of protoxylem is found in
(a) Root of *Cycas* (b) Rhizome of ferns (c) Root of orchid (d) Stem of cucurbits
80. Pith is absent in
(a) Monocot root (b) Monocot stem (c) Dicot root (d) Dicot stem
81. The vascular bundles are wedge shaped and ringed in
(a) Monocot root (b) Monocot stem (c) Dicot root (d) Dicot stem
82. Which of the following is incorrect regarding palisade cells of dicot leaf?
(a) These are elongated (b) These are compactly arranged
(c) These are more photosynthetic (d) These contain less chloroplast
83. Cork cambium in dicot root is formed by
(a) Cortex (b) Endodermis (c) Pericycle (d) Vascular tissue
84. The youngest layer of secondary xylem is situated
(a) Inside vascular cambium
(b) Outside vascular cambium
(c) Outside cork cambium
(d) Outside youngest layer of secondary phloem

85. Tracheary elements of xylem consists of
   (a) Tracheids only (b) Tracheids and vessels
   (c) Vessels and xylem fibres (d) Tracheids, vessels and xylem fibres
86. Complementary cells are associated with
   (a) Lenticels (b) Stomata (c) Cuticle (d) Bark
87. Annual rings are not developed in the trees of
   (a) Tropical areas (b) Sea coast (c) Equatorial region (d) All of the above
88. Silica is abundant in
   (a) Grasses and cereals (b) Grasses and horse tails
   (c) Horse tails and ferns (d) Mosses and ferns
89. Elongation of grass internodes is due to activity of
   (a) Apical meristem (b) Intercalary meristem (c) Lateral meristem (d) All of the above
90. Which of the following tissues becomes dead due to activity of phellogen in dicot root?
   (a) Tissues outside pericycle (b) Tissues outside vascular cambium
   (c) Tissues inside secondary cortex (d) Tissues inside vascular cambium
91. Vascular rays are formed
   (a) Centripetally (b) Centrifugally (c) Both of the above (d) Laterally
92. Annual rings are most conspicuous in
   (a) Tropical forest (b) Subtropical forest
   (c) Temperate deciduous forest (d) All of the above
93. Cambial cells divide in
   (a) Tangential manner (b) Anticlinal manner (c) Radial manner (d) Periclinal manner
94. Annual rings are the bands of
   (a) Secondary xylem and xylem rays (b) Secondary xylem and secondary phloem
   (c) Secondary phloem and medullary rays (d) All of the above
95. Diffuse porous woods grow in
   (a) Tropical areas (b) Alpine areas (c) Temperate areas (d) All of the above
96. Vascular cambium in stem is
   (a) Completely primary (b) Completely secondary
   (c) Partly primary and partly secondary (d) None of the above
97. The bundle sheath cells in Krantz anatomy are characterised by
   (a) Thick wall, no intercellular space and large number of chloroplast
   (b) Thin wall, no intercellular space and large number of chloroplast
   (c) Thick wall, having intercellular space and small number of chloroplast
   (d) Thick wall, having intercellular space and large number of chloroplast
98. Bicollateral vascular bundle is seen in
   (a) Stem of tomato (b) Root of orchid (c) Root of *Ficus* (d) Stem of cucurbits

99. Cambium is absent in
    (a) Radial vascular bundle
    (b) Conjoint vascular bundle
    (c) Concentric vascular bundle
    (d) All of the above
100. Stomata in epidermis consists of
    (a) One guard cell and two subsidiary cells
    (b) Two guard cells and two subsidiary cells
    (c) Two guard cells and several subsidiary cells
    (d) Several guard cells and several subsidiary cells

ANSWERS

1.b	2.d	3.a	4.b	5.b	6.a	7.b	8.c	9.a	10.c
11.a	12.b	13.b	14.b	15.a	16.b	17.c	18.b	19.d	20.a
21.b	22.d	23.d	24.b	25.c	26.c	27.b	28.d	29.b	30.a
31.b	32.d	33.c	34.d	35.a	36.c	37.d	38.c	39.c	40.c
41.d	42.d	43.c	44.a	45.b	46.c	47.d	48.c	49.b	50.b
51.d	52.a	53.b	54.a	55.b	56.c	57.a	58.a	59.b	60.a
61.b	62.b	63.a	64.b	65.b	66.c	67.d	68.b	69.d	70.b
71.b	72.d	73.b	74.c	75.d	76.b	77.d	78.c	79.b	80.b
81.d	82.d	83.c	84.a	85.b	86.a	87.d	88.b	89.b	90.a
91.c	92.c	93.a	94.a	95.a	96.c	97.a	98.d	99.c	100.c

6

Cell: The Unit of Life

Cell biology (formerly called cytology) is the study of structure and function of cell and cell organelles.

Father of Cytology	Robert Hooke
Father of Modern Cytology	Swanson

Discovery:

Robert Hooke discovered the cell in 1665 when he examined the thin slices of bottle cork (outer dead layer of bark) of Oak plant (*Quercus suber*) under his crude microscope. He noticed the presence of several small hexagonal boxes which together appeared as honey comb. Externally he observed only the cell wall that surrounds the protoplast. He coined the term 'Cell' (meaning: a small room) in his book '*Micrographia*'. He considered cell as hollow space (empty cell wall of plant tissue). Cell is the basic structural, functional and biological unit of all known living organisms. A cell is the smallest unit of life that can replicate independently hence called building blocks of life. Now it is not regarded as a hollow space but it is separable unit having living contents called protoplast enclosed by cell wall.

Dutrochet (1824) defined cell is the basis of life.

Corti studied the living content of plant cell.

Rudolf Virchow (1855) proposed 'Cell lineage theory'. It states that cells arise from pre-existing cells ('*Omnis cellula e cellula*'). His proposal established that cell division occurs in the continuity of life.

Matthias Schleiden (German Botanist) and Theodor Schwann (German Zoologist) proposed 'Cell theory' in 1839 and later modified by Rudolf Virchow.

The general unifying principles of **cell theory** include:

1. All living organisms are composed of one or more cells and cell products.
2. All cells arise from pre-existing cells by divisions (not derived from spontaneous generation).
3. All cells are fundamentally similar in chemical composition and metabolic activities.
4. The function of organisms as a whole is the result of activities and interaction of cells.
5. Energy flow occurs within cells.

Exceptions to cell theory:

1. Viruses, viroid and prion: These do not have cellular organisation.
2. Prokaryotes: These do not have nuclei and membrane bound cell organelles.
3. Mammalian RBCs and sieve tube: These lack nuclei.
4. *Rhizopus* and *Vaucheria*: These are multinucleated and without cell differentiation.

This theory is considered as significant because

(a) It shows that cells of different organisms are similar in structure.
(b) It explains that life exists in the form of cells.
(c) It explains that all cells are similar in metabolic activities.
(d) It confirms that life passes from one generation to other in the form of cells.

Organisation of Cell:

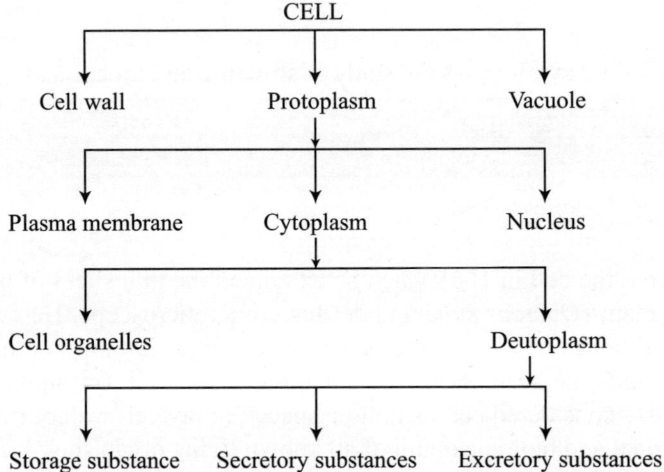

(carbohydrate, proteins, fats) (enzymes, pigments, nectar) (gums, resins, alkaloids,tannins)

Terms:

Protoplast: It is cell without cell wall.

Cytosol (Hyaloplasm or Cytoplasmic matrix): It is liquid part of cytoplasm without cell organelles. It contains scattered ribosome, cytoskeletons, tRNA, enzymes of anaerobic respiration.

Features of Cell:

Number: There is one (unicellular organisms) to indefinite cells (multicellular organisms).

Shape: There is great variation. It is usually rounded but may be may be elongated, polygonal, columnar, discoidal etc. it depends on the function.

Size: It varies from 0.2μm to 20μm. It depends on Nuclear-cytoplasmic ratio (N/C or N: C ratio) and metabolic activity. The N-C ratio is the ratio of size (volume) of the nucleus of a cell to the size (volume) of the cytoplasm of that cell. It indicates the maturity of a cell. As the cell matures (increases its size), the size of nucleus will decrease and surface: volume ratio of cell decreases. Small size is extremely important that allows for rapid uptake and intracellular distribution of nutrients and excretion of wastes due to large surface:volume ratio.

Smallest cell	*Mycoplasma gallisepticum* (0.2μm)
Largest cell	Egg of Ostrich (6 inches in diameter with shell)
Longest cell	Nerve cell (90 cm)
Longest Plant cell	Fibre of Ramie (50 cm)
Longest unicellular plant	*Acetabularia,* a green alga (10 cm)

Cell size (maturity)=1/**efficiency**=1/**metabolic activity**=1/**N:C ratio**=1/Surface:Volume ratio

- Smaller cell (immature cell) = More efficient = More metabolically active = High N:C ratio=High surface: volume ratio. The nucleus of small cell efficiently controls the activity of cytoplasm which is small.
- Larger cell (mature cell) = Less efficient = Less metabolically active = Low N:C ratio = Low surface: volume ratio. The nucleus of large cell does not control the activity of cytoplasm which is very large.

Types of cell: Dougherty (1957) divided cells into three types on the basis of nucleus.

1. Prokaryotic cell: Nucleus is unorganised and undeveloped (primitive) and histone proteins are absent. e.g., Archaebacteria, Bacteria and Cyanobacteria.
2. Mesokaryotic cell: It was coined by Dodge (1964). Nucleus is well developed but histone proteins are absent. e.g., *Peridinium balticum* (a dinoflagellate).
3. Eukaryotic Cell: Nucleus is well developed and histone proteins are present. (Advanced). e.g., Protista, Fungi, Plants and Animals.

Differences between prokaryotic cell and eukaryotic cell

Prokaryotic cell	Eukaryotic cell
Its size is small (between 0.2-5μm)	Its size is large (between 5-100μm).
Flagella do not show 9+2 arrangement and are made up of flagellin protein. They are not enclosed in plasma membrane and show rotatory movement.	Flagella show 9+0 arrangement and are made up of tubulin protein. They are enclosed in plasma membrane and show undulatory movement.
Cell has single envelope organisation (only plasma membrane is present).	Cell has double envelope organisation (plasma membrane and membrane of cell organelles).
Plasma membrane contains respiratory enzymes.	Plasma membrane does not contain respiratory enzymes.
Membrane bound cell organelles are absent.	Membrane bound organelles are present.
Cytoplasm does not show cyclosis or streaming movement.	Cytoplasm shows cyclosis or streaming movement.
Vacuoles are absent.	Vacuoles are present.
Plasma membrane infolds to form mesosome.	Plasma membrane does not infold to form mesosome.
Single thylakoid lies free in cytoplasm.	Several thylakoids are grouped in granum of chloroplast.
Ribosomes are of 70S. These lie free in cytoplasm or join to form polyribosome.	Ribosomes are of 80S. These are free or attached with nuclear membrane and endoplasmic reticulum.
Cytoskeletons are absent.	Cytoskeletons are present.
Endocytosis and exocytosis do not occur.	Endocytosis and exocytosis occur.
Cell is haploid.	Cell is generally diploid but haploid cells are also found.
Nucleus is unorganised and undeveloped.	Nucleus is well developed.
Nuclear membrane and nucleolus are absent.	Nuclear membrane and nucleolus are present.
Nucleoid is equivalent to single chromosome.	Nucleus contains more than one chromosomes.
Histone proteins are absent.	Histone proteins are present.
DNA is small, circular, naked and occurs in cytoplasm	DNA is large, linear (organellar DNA is circular), closed and occurs in nucleus.

DNA does not have introns.	DNA has introns.
Plasmid is found in cytoplasm.	Plasmid is not found in cytoplasm.
Cell cycle is of short duration(30-60 minutes)	Cell cycle is of long duration (12-24 hours).
Mitotic spindle is not formed during cell division.	Mitotic spindle is formed.
Meiosis does not occur.	Meiosis occurs.

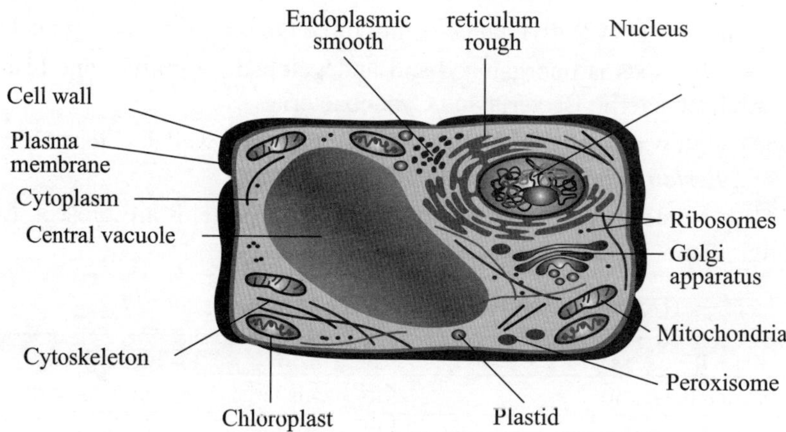

Fig. ***Plant cell***

Differences between Plant cell and Animal cell

Plant Cell	Animal Cell
It is larger in size (>100µm)	It is smaller in size (<100µm).
Cell wall is present.	Cell wall is absent.
Vacuole is single, large, central and permanent.	Vacuoles are more, small, peripheral and temporary.
Contractile vacuoles and food vacuoles are absent.	Contractile vacuoles and food vacuoles are present.
Nucleus is peripheral in position.	Nucleus is central in position.
Mitochondria are lesser in number.	Mitochondria are larger in number.
Cristae of mitochondria are tubular.	Cristae of mitochondria are plate like.
Plastid is present.	Plastid is absent.
Golgi complex contains more (10-20) cisternae.	Golgi complex contains less (3-7) cisternae.
Lysosome is rarely found.	Lysosome is commonly found.
Glyoxysome is present.	Glyoxysome is absent.
The number of chromosome is relatively small.	The number of chromosome is relatively large
The size of chromosome is relatively large.	The size of chromosome is relatively small.
Reserve food is starch and fat.	Reserve food is glycogen and fat.
Mitosis occurs in both haploid and diploid cell.	Mitosis occurs in diploid cell only.
Spindle is without aster (anastral).	Spindle has aster at each pole (amphiastral).
Cytokinesis occurs by cell plate.	Cytokinesis occurs by cell furrow (constriction).

Adjacent cells are connected through plasmodesmata.	Adjacent cells are connected through cell junction.
Tight junction and desmosomes are absent.	Tight junctions and desmosomes are present.
Tissue fluid is absent.	Tissue fluid is present.
It can synthesise all amino acids, coenzymes and vitamins.	It cannot synthesise amino acids, coenzymes and vitamins.
It becomes turgid if placed in hypotonic solution.	It bursts when placed in hypotonic solution.

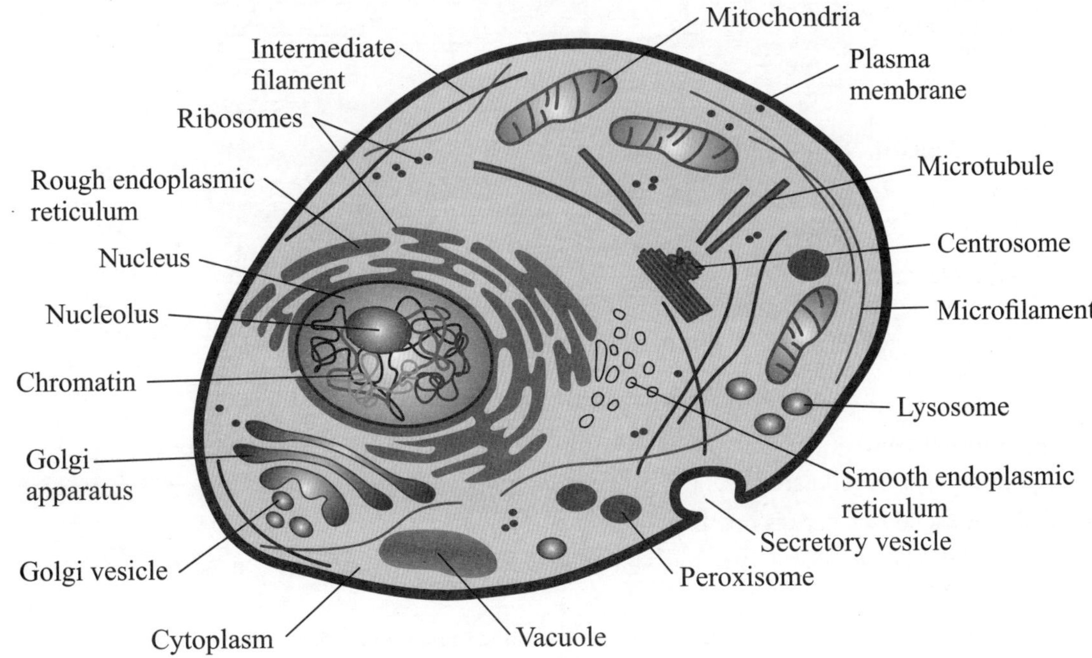

Fig. ***Animal cell***

Bacterial Cell

Bacteria (sing: bacterium) are the simplest, most primitive and microscopic organisms. Antonie Van Leeuwenhoek (1875) discovered bacteria from rain water, sewage water and saliva using single lens microscope of his own design. Bacteria being unicellular consist of the following parts:

1. **Surface appendage:**
   (a) **Flagella:** It is long, filamentous and much thinner than eukaryotic flagella. It protrudes from cell wall. It develops from basal granules present in plasma membrane. It consists of single fibril (single stranded) surrounded by 3-8 fibrils without differentiation of axoneme and sheath. It lacks 9+2 arrangement. It is made up of flagellin protein. It contains basal body, hook and filament (shaft). Basal body has central rod and rings. There are two rings (S ring in peptidoglycan layer and M ring in plasma membrane) in Gram positive bacteria. There are four rings (L ring associated with lipopolysaccharide; P ring associated with peptidoglycan; S ring directly attached to plasma membrane and M ring embedded in plasma membrane) in Gram negative bacteria.

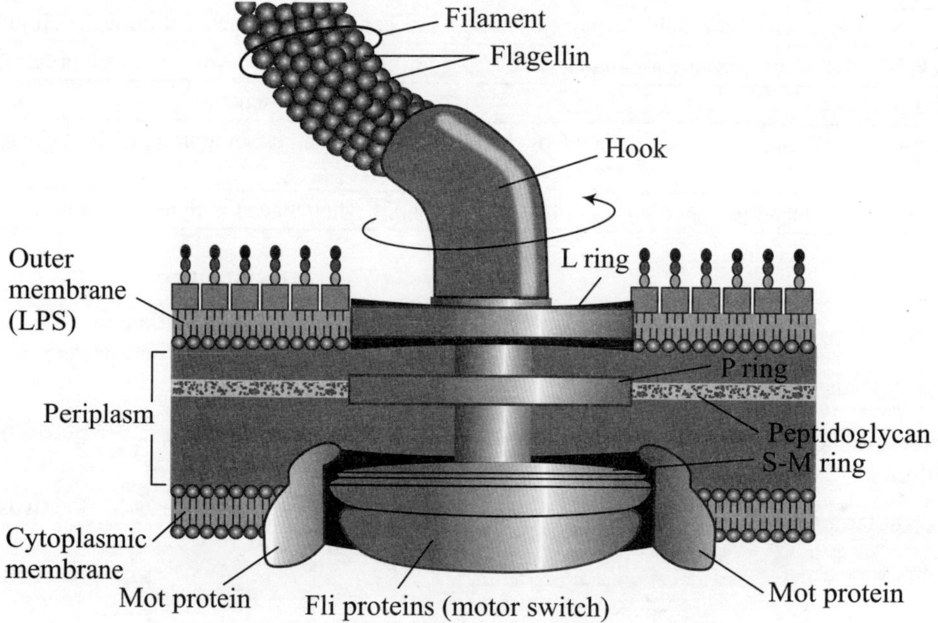

Fig. ***Bacterial flagella (Internal structure)***

Types of flagella:

1. Atrichous: Flagella are absent. e.g., *Lactobacillus, Streptococcus.*
2. Monotrichous: There is one flagellum. e.g., *Vibrio*
3. Amphitrichous: There are two flagella; each at two opposite ends. e.g., *Alkaligenes, Nitrosomonas*
4. Lophotrichous: There is single bundle of flagella at one end only. e.g., *Spirillum, Pseudomonas*
5. Cephalotrichous: There are two bundles of flagella each at both ends. e.g., *Pseudomonas fluorescence.*
6. Peritrichous: There are many flagella all over the surface. e.g., *E. coli, Bacillus, Clostridium, Salmonella.*

Function: It helps in locomotion (60 cell length in per sec so it takes 245 days to reach 1 km.

(a) Pili: These are long, a few (1-4) in number, thick and non-helical appendage. These are found in Gram negative bacteria. These are made up of pilin protein (glucophosphoprotein). Their formation is controlled by plasmid (F-factor). These help in process of conjugation by serving as passage of DNA from donor to recipient cell.

(b) Fimbriae: These are shorter, more (300) in number, and thinner than pili. These are found in Gram negative bacteria. These are made up of pilin protein. Their formation is controlled by nuclear gene. These help in attachment.

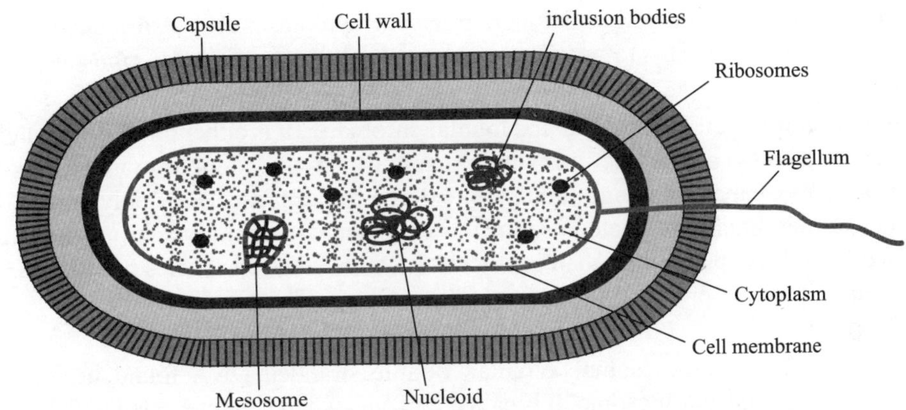

Fig. ***Bacterial cell***

2. **Glycocalyx**: It is outermost mucilage layer. It differs in composition and thickness in different bacteria which are as follows:
   - **(a) Capsule:** It is thick and tough layer made up of polysaccharides.
   - **(b) Slime layer:** It is thin and loose layer made up of polysaccharides.
   - **(c) S-layer:** It is thin and loose layer made up of protein or glycoprotein.

   Functions:
   1. It protects bacteria from toxic materials.
   2. It protects bacteria from bacteriophage (virus).
   3. It protects bacteria from engulfing by eukaryotic cell (macrophage)
   4. It protects bacteria from desiccation or drying by binding water.
   5. It prevents aggregation of cells.
3. **Cell wall:** It is rigid and solid covering made up of peptidoglycan or mucopeptide or murein. Its sugar part forms backbone and made up of alternating units of amino sugars such as N-acetyl glucosamine (NAG) and N-acetyl muramic acid (NAM) joined together by β-1,4 linkage (glycosidic bond). Its protein part contains four amino acids which are attached to NAM. These are D-Alanine, L-alanine, D-glutamic acid and L- lysine (D-aminopimelic acid in case of Gram negative bacteria). It forms structural framework of cell and maintains shape of cell.
4. **Plasma membrane:** It is innermost and trilaminar membrane made up of lipid and proteins. It does not contain sterol but contains pentacyclic sterol (hopanoids) in some bacteria. It has respiratory enzymes at inner face. It is selectively permeable. It takes part in respiration by performing oxidative phosphorylation.
5. **Mesosome:** It is convoluted structure develops from infolding or invagination of plasma membrane. It increases the surface area of plasma membrane for secretory functions. It is confined to peripheral region. It is found in Gram positive bacteria. It is considered analogous to mitochondria due to presence of respiratory enzymes such as dehydrogenase and cytochromes associated with respiration. It takes part in binary fission by forming septum and separation of replicated DNA.
6. **Chromatophore:** It contains single thylakoid which is found freely or diffused in cytoplasm. It contains bacteriochlorophyll and bacterioviridin. It is analogous to chloroplast.

7. **Ribosomes:** These are of 70S and found freely in cytoplasm and also attached with plasma membrane. Their size is 20 x15 nm. These form polyribosome when 4-6 ribosomes are attached to m-RNA.
8. **Nucleoid (Genophore):** It is fibrillar and lobular chromatin. It is either directly attached to plasma membrane or through mesosome. It is single (haploid), circular, double stranded, naked (without histone proteins), supercoiled and tightly folded due to the presence of polyamines. It does not have nuclear membrane and nucleolus thus it is considered as primitive chromosome. It occupies 10-20% of cell. It is 1000 longer than cell itself. It varies between species of bacteria. It can code 2000 to 3000 different proteins. It does not follow Chargaff's rule. *E.coli* has 4300 genes having 4.6 million base pairs.
9. **Plasmid:** It is extra-chromosomal, circular, double stranded DNA found in cytoplasm. It is analogous to bacterial chromosome. It is called episome when it is integrated with bacterial DNA. It undergoes autonomous replication. It is of several types
   (a) F-factor (fertility factor): It helps in conjugation.
   (b) R-factor: It confers resistance to various antibiotics and heavy metals.
   (c) Col-factor: It produces a chemical, colicin which kills closely related species of bacteria or sensitive *E. coli.*
   (d) Ti-plasmid: It induces formation of crown gall (tumour).
10. **Cell inclusions:** These are nonliving, membrane-less and found freely in cytoplasm. These are
   (a) Poly beta hydroxybutyrate (PBH) granule: It stores carbon and energy source.
   (b) Gas vacuole: It stores gases which provides buoyancy and protects from radiations.
   (c) Volutin granule: It is metachromatic granule which stores phosphate.
   (d) Sulphur granule: It picks up H_2S as reducing power in photosynthesis.

Gram Staining Method: It is differential staining technique developed by Christian Gram (1884). This difference is due to differentiation in lipid content and thickness of cell wall.

Procedure: Smear was prepared from sample and then heat dried. There are four stains which are given one by one for about one minute, then washed gently by tap water and observed microscopically.

Stain	Action	Time	Gram positive	Gram negative
Crystal violet	Primary stain	1 min	Violet	Violet
Iodine (0.5%)	Mordant	1 min	Violet	Violet
Absolute alcohol	Decolourising agent	1 min	Violet	Colourless
Safranin	Counter stain	1 min	Violet	Red/pink

Mechanism for Gram positive bacteria	Mechanism for Gram negative bacteria
Crystal violet-iodine (CV-I) complex is formed.	Crystal violet-iodine (CV-I) complex is formed
Low amount of lipid is dissolved by alcohol.	High amount of lipid is dissolved by alcohol.
Small pore is formed in cell wall.	Large pore is formed in cell wall
Pore is closed due to dehydration of proteins.	Pore is not closed due to dehydration of proteins.
CV-I complex is trapped.	CV-I complex is leaked.
Decolourisation does not occur.	Decolourisation occurs.
Thus it remains violet/purple	Thus it appears red/pink

Difference between Gram Positive and Gram negative bacteria

Gram positive	Gram negative
Cell wall is thick, single layered, simple, homogeneous, smooth and rigid.	Cell wall is thin, two layered, complex, heterogeneous, wavy and plastic.
Amount of peptidoglycan is more (90%).	Amount of peptidoglycan is less (10%).
Amount of lipid is low (3-4%).	Amount of lipid (phospholipids, lipoprotein and lipopolysaccharide) is high (25%).
L-lysine is present in peptide.	D-aminopimelic acid is present in peptide.
Sialic acid is absent.	Sialic acid is present.
Techoic acid (acidic polymer of glucose, phosphate and alcohol) is present. It protects from high pH and temperature fluctuation.	Techoic acid is absent.
These are mostly capsulated.	These are mostly non-capsulated.
Cell wall is digested by lysozyme hence protoplast is formed.	Cell wall is resistant to lysozyme.
Porins are absent.	Porins are present
Pili are absent	Pili are present.
Mesosome is present.	Mesosome is absent.
These are more sensitive to penicillin.	These are less sensitive to penicillin.
These are less sensitive to streptomycin.	These are more sensitive to Streptomycin.
A few are pathogenic and produce exotoxins.	Majority are pathogenic and produce endotoxins.
Examples: *Bacillus, Coccus, Clostridium, Mycobacterium* (lesser in number).	*Azotobacter, E.coli, Rhizobium, Salmonella, Vibrio, Xanthomonas* (more in number).

Cell Wall

It is outermost, nonliving, tough, flexible (sometimes rigid) and structural layer outside plasma membrane of cell. It was discovered (observed) and coined by Robert Hooke.

Distribution: It is found in prokaryotes, some protists (diatoms, dinoflagellates and spore of slime moulds), fungi and plants but it is absent in *Mycoplasma*, *Euglena*, somatic phase of slime moulds. protozoans and animals.

Composition: Water (60%), cellulose (15%), hemicellulose (15%), pectin (10%). The composition differs according to cell type and functions. Cellcuose microfibrils form skeletal of cell wall whereas hemicellulose and pectin form matrix of cell wall.

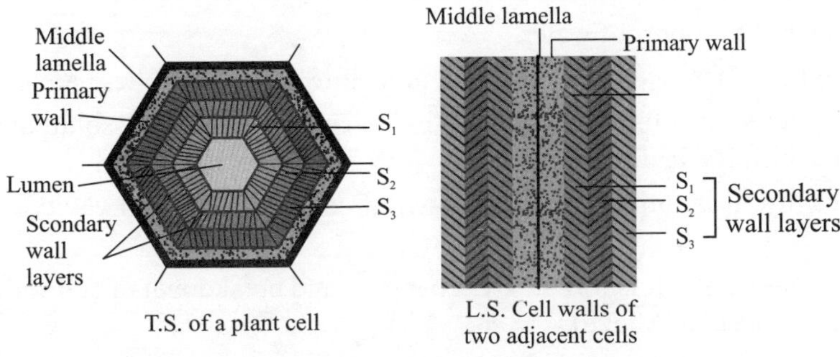

T.S. of a plant cell

L.S. Cell walls of two adjacent cells

Structure: It consists of following layers

1. **Middle lamella:**
   (a) It is outermost, thin and sticky cementing layer between two adjacent cells.
   (b) It is made up of calcium pectate (more) and magnesium pectate.
   (c) It is first layer formed as a result of cytokinesis.
   (d) It ruptures to create intercellular spaces in many tissues.
   (e) It is dissolved or gelatinised by pectinase hence softening of ripened fruits occurs.
2. **Primary wall:**
   (a) It is thin, elastic and single layered present below middle lamella.
   (b) It is made up of cellulose (25%), hemicellullose (25%) in the form of organised microfibrils and pectin (35%).
   (c) It develops by intussusceptions (deposition of wall materials in forms of grains).
   (d) It is stretchable because cell wall microfibrils are transversely oriented (anisotropic).
   (e) It is present in parenchymatous cells and expands in mersitematic cells.
3. **Secondary wall:**
   (a) It is thick, non-elastic (rigid) and three layered.
   (b) It is made up of more cellulose and less hemicellulose. It may contain small amount of lignin, cutin or suberin.
   (c) It develops by accretion (deposition of wall materials in the form of layers).
   (d) It is present in collenchyma, sclerenchyma, xylem tracheid and xylem vessel.
   (e) It is found in hard and woody parts of plants when cell has stopped growing.
   (f) It forms pits because secondary wall is not of uniform thickness.
4. **Tertiary wall:**
   (a) It is present occasionally.
   (b) It is made up of xylan (a type of hemicellulose).
   (c) It is found in tracheid of gymnosperms.

Functions:

1. It protects the protoplasm from mechanical injury by middle lamella.
2. It protects the protoplasm from the attack of pathogens.
3. It provides rigidity or mechanical support to the cell.
4. It maintains the shape of cell due to middle lamella.
5. It prevents cell from bursting (undue expansion) as a result of endosmosis.
6. It helps in apoplast transport of water.

Intercellular spaces: These are spaces formed in mature cells only. These are of three types:

1. Schizogenous cavity: It is formed by separation of cell wall due to dissolution of middle lamella. e.g., Mucilage duct, Resin duct of *Pinus*.
2. Lysigenous cavity: It is formed by breakdown of cell wall. e.g., Oil cavity in citrus plants and *Eucalyptus*.
3. Schizo-lysigenous cavity: It is formed by separation and breakdown of cell wall. e.g., Protoxylem lacunae in monocot stem (Maize).

Plasmodesmata: These are microscopic channels or pores which directly connect the cytoplasm of neighbouring plant cells to each other. These are lined by plasma membrane. These establish living bridges between cells. A typical plant cell may have between 10^3 and 10^5 plasmodesmata.

Protoplasm

It is living substance of cell and performs all vital functions of a cell. It is fundamental material of which living organisms are composed.

Discovery: Felix Dujardin (1835) discovered protoplasm and called it 'Sarcode'. The term 'protoplasm' was coined by Purkinje (1837).

Von Mohl (1844) stated the importance of protoplasm in cell division.

Huxley (1868) defined it as 'Physical basis of life'.

Max Shultze (1861) proposed the 'Protoplasmic theory' in which living matter of cell is protoplasm which is limited by outer membrane and contains nucleus.

Fisher and Hardy (1894) proposed 'colloidal theory'. It states that protoplasm is a heterogeneous polyphasic colloidal system which appears differently-reticular, fibrillar or granular in different phases.

Composition: Water (80%), Protein (15%), lipid (3%), carbohydrate (1%), minerals (1%). Among elements, the amount of Oxygen is very high (62%) followed by Carbon (20%), Hydrogen (10%) and Nitrogen (3%). It is difficult to determine the exact composition because it undergoes continuous change. Thus it is complex crystallo-colloidal mixture where some solute form solution and most of the solutes remain colloidal.

Properties:

(a) It is complex, granular, elastic, colourless and viscous.

(b) It is selectively permeable.

(c) Its pH is slightly alkaline but in plants it is acidic.

(d) Its specific gravity is slightly more than one.

(e) It coagulates on heating up to 60°C or treating with conc. Acid or base.

(f) It shows Brownian movement (zig-zag movement due to random collision of molecules) and Tyndall effect (scattering of light by particles).

(g) It shows intracellular movement like cyclosis or streaming. It occurs in all eukaryotic cells. It depends on viscosity and temperature. It helps in movement of cell organelles and cell inclusions. It helps in formation of pseudopodia in *Amoeba*. Cyclosis is of two types:

  (a) Rotation: It moves in one direction (either clock wise or anticlock- wise) around vacuole inside cell. e.g., Leaf cell of *Hydrilla* and *Vallisneria*.

  (b) Circulation: It moves in different directions around different vacuoles inside cell. e.g., Staminal hairs of *Tradescantia* (=*Rhoeo discolor*).

Vacuole

It is non-cytoplasmic, nonliving and fluid filled reservoir. It was discovered by Spallanzani in plant cell. It is occupied by 30% of the volume of cell. It is found in cells of plant, fungi, some protists and some animals. It is very small or absent in young cells but fuse to form very large and central vacuole in mature cells such as parenchymatous and collenchymatous cells when cells increase in size. It is bounded by single membrane called tonoplast which is lipo-proteinaceous and selectively permeable. Its structure varies according to the requirement of the cell. It varies in different tissues.

It is formed in plant cells because plants cannot produce enough protein to fill their enlarging cells with protoplasm.

It contains watery fluid called cell sap which is generally neutral but slightly acidic at maturity. Cell sap is hypertonic (higher concentration than cytoplasm) which generates turgor pressure in plant cell.

Vacuole is known as storehouse of cell products because cell sap contains inorganic substances such as K^+ (more), Na^+ and organic substances such as acetic acid (more), fumaric acid, amino acids, enzyme, sugar and secondary metabolites such as pigments, alkaloids, latex, tannins, crystals of calcium oxalates). It also contains excretory products not useful for the cell.

Origin: It develops from invagination of plasma membrane. It also develops from endoplasmic reticulum.

Types:

1. Contractile vacuole: It picks up water from surrounding cytoplasm, expand and collapse to throw water outside thus performs excretion and osmoregulation in Protozoa.
2. Food vacuole: It helps in digestion because it contains digestive enzymes.
3. Gas vacuole: It stores metabolic gases and helps in floating in prokaryotes.

Functions:

1. Vacuole performs osmoregulation by maintaining osmotic relation of cell. Thus it helps in absorption of water.
2. It provides colouration of flowers of rose and sunflower; roots of beet due to presence of anthocyanin (red, blue, purple) in cell sap.
3. It stores food reserves such as sucrose, salts, inulin and waste products.
4. It maintains turgor pressure inside the cell.
5. It maintains acidic internal pH.
6. It contains hydrolytic enzymes hence it acts as lysosome in plant cell.
7. It exports unwanted substances.

Plasma membrane

It is thin and film like membrane enclosing the cell. It was discovered and coined by Nageli and Krammer. It is also known as cell membrane or plasmalemma. The term 'Plasmalemma' was coined by Plower. It is considered living because it is an essential part of protoplasm that makes cell alive. It constitutes 5% of all membrane.

Composition:

(a) Lipids (32%)-Phospholipids are common (phosphatidyl serine, phosphotidyl choline (lecithin), p-ethanol amine (cephalin). Others are glycolipid and sterol (cholesterol in animal cell and *Mycoplasma*; phytosterol in plant cell and ergosterol in bacterial cell).

(b) Proteins (42%):

(c) Carbohydrate (6%): It is oligosaccharide associated with glycoprotein (60%) and glycolipid (10%).

(d) Water: 20% by weight.

Types of plasma membrane	Lipid	Protein	Carbohydrate
Human RBC	40%	52%	8%
Nerve cell	80%	20%	-
Outer membrane of mitochondria	47%	51%	2%
Inner membrane of mitochondria	23%	76%	1%

Models:

1. **Overton** (1902) proposed that plasma membrane is composed of thin layers of lipid.
2. **Danielli and Davson** (1935) proposed trilamellar (trilayer) model. Plasma membrane is composed of three layers in which a layer of phospholipid is sandwiched between two layers of proteins.
3. **Robertson** (1959) proposed unit membrane model. Plasma membrane is composed of one light middle layer of lipid which is sandwiched between two dark layers of proteins.

   Thickness: Protein + Lipid + Protein

   $20\ A^0 + 35\ A^0 + 20\ A^o = 75A^0$
4. **Jonathan Singer and Garth Nicolson** (1972) proposed fluid mosaic model. It is most accepted model which explains the selective permeable nature of plasma membrane.

Plasma membrane is composed of a continuous bilayer of phospholipid in which the globular proteins are embedded. It looks like iceberg (proteins) floating in a sea of phospholipids. Plasma membrane is two dimensional quasi-fluid structure where proteins are arranged in mosaic fashion. Proteins are not arranged in continuous layer but they are present separately and arranged asymmetrically. Proteins can diffuse freely in a lateral direction within lipid matrix thus there is fluid movement of proteins and lipids. Thus the lateral movement of proteins make its nature fluid like. Proteins cannot move from one side of membrane to the other side (flip-flop) diffusion. Lipids rarely show flip-flop movement (movement of lipid from inside to outside). The components of membrane (lipids, proteins and oligosaccharides are held in their positions by non-covalent interactions. Lipid molecules are in constant lateral movement and not rigidly locked hence these provide permeability, fluidity and elasticity to membrane. The degree of fluidity depends on the percentage (length) of unsaturated fatty acid present in lipid bilayer and the temperature. Lipids have kinks (double bond in tail) which prevent from close packing of molecules and make the membrane more fluid. Fluidity is due to hydrophobic interaction of lipid and protein.

Lipids are stable and amphipathic molecules. It consists of

(a) **Phosphate heads** are hydrophilic and polar. These face outward and lie at free surface.

(b) **Hydrocarbon tails** are hydrophobic and non-polar containing two long chains of fatty acids. These face inside. Tails remain in contact with each other by Wander Wall's forces hence tail is protected from aqueous environment by its hydrophobic nature. Lipid bilayers are generally impermeable to ions and polar molecules. The arrangement of hydrophilic heads and hydrophobic tails of the lipid bilayer prevents polar solutes (e.g., amino acids, nucleic acids, carbohydrates, proteins and ions) from diffusing across membrane but generally allows for the passive diffusion of hydrophilic molecules.

Proteins are of three types:

(a) **Extrinsic proteins** (peripheral proteins): These constitute 30% of total proteins. These are loosely bound (covalently attached) at the polar surface of lipid head. These are soluble. These can be extracted or removed from membranc. These are involved in cell interaction and cell recognition (cell signalling). e.g., ATPase in mitochondria, spectrin in RBC etc.

(b) **Intrinsic proteins** (Integral protein): These constitute 70% of total proteins. These are deeply penetrated or embedded into lipid bilayer. These are insoluble. These cannot be extracted or removed from membrane due to strong hydrophobic interaction with non-polar hydrophobic tail of lipid. These are involved in transport of polar solute molecules. e.g., porin, glycophorin, rhodopsin in rod cell of retina.

(c) **Tunnel protein**: It is the intrinsic protein which projects on both surface hence known as transmembrane protein. It may be monotopic (exposed only at one surface), bitopic (exposed at both surface) and polytopic (exposed across surface). It provides mechanical framework.

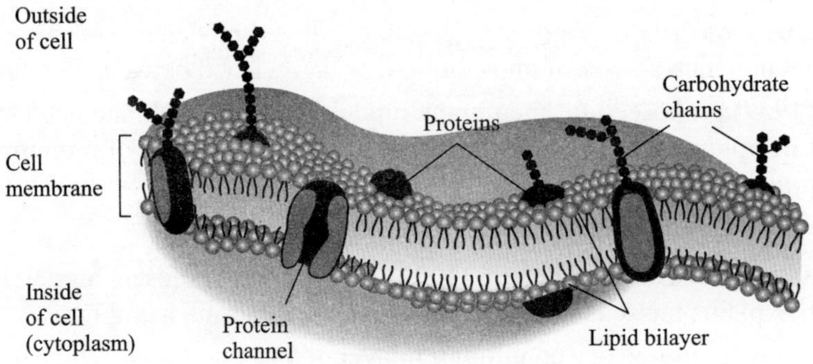

Fig. ***Plasma membrane***

Carbohydrates are in the form of oligosaccharides which are covalently attached to external surface of lipid (called glycolipid) and protein (called glycoprotein). Sialic acid (N-acetylglucosamine or NAG) is terminal sugar of oligosaccharide chain present in glycoprotein of plasma membrane.

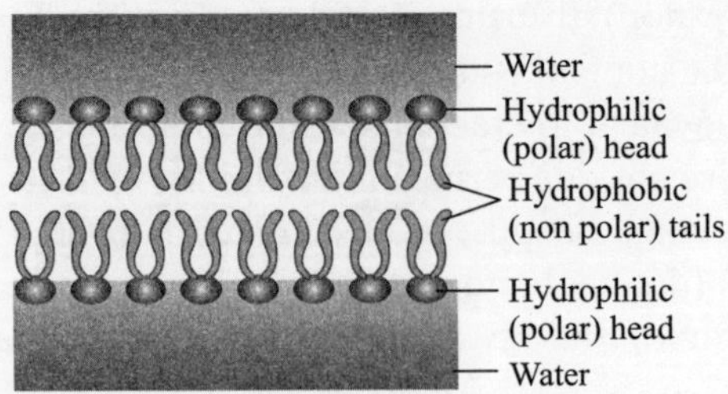

Fig. ***Arrangement of lipids and proteins in Plasma membrane***

Origin: Plasma membrane is developed due to accumulation of surface active substances provided by cytoplasm.

In case of animal cells, plasma membranes of neighbouring cells in tissues are linked to each other at specialised sites called cell junctions. These may be

- Adhering junction (Desmosome): It holds cells together (adhesion).
- Impermeable junction: It seals the cells together.
- Communicating junction: It mediates the passage of small molecules from one interacting cell to the other.

Functions:

1. Plasma membrane transports the materials into and outside the cell. It is selectively or differentially permeable. It takes place by

(a) Simple diffusion: O_2, CO_2, H_2O etc.

(b) Facilitated diffusion: Charged ions, glucose in RBC etc.

(c) Active transport: Salts, glucose and amino acids (up to size of 1-15A^0).

(d) Bulk transport: It occurs in eukaryotic cell. It takes place by **exocytosis** in which cholesterol is removed by carrier vesicle. It also takes place by **endocytosis** which occurs by following two methods;

1. **Pinocytosis (cell drinking)**: It is intake of ions, sugars and amino acids in liquid form by forming small vesicle (100 nm in diameter). It was reported by Lewis.
2. **Phagocytosis (cell eating)**: It is intake of materials in solid form by forming large vesicle (>250 nm in diameter). It was reported by Metchnikoff.

2. It protects cell from injury and provides mechanical support.
3. It maintains the shape of cell and cell organelles.
4. It helps in cell to cell recognition allowing immune system to differentiate between body cells and foreign cells and cell to cell adhesion due to presence of sugar of glycoprotein and glycolipid at surface.
5. It delineates and separates intracellular components from external environment (compartmentalisation).
6. It helps in chemoreception because it contains receptors which recognise specific hormones and convey the flow of information to the interior of cell.
7. It forms the sheath of cilia and flagella.
8. It is involved in plasmolysis.
9. It helps in pseudopodial movement of cells (e.g., WBC, fibroblast, *Amoeba*, Macrophage).
10. It determines the antigenic specificity due to presence of glycophorin on the surface of RBC.
11. It is involved in transmission of nerve impulse.
12. It is involved in oxidative phosphorylation in prokaryotes.

Cytoplasm

It was discovered by Kolliker (1862). It is jelly like semi-fluid general mass of protoplasm. It includes cytosol or cytoplasmic matrix or hyaloplasm which is clear fluid part (pH 7.2) of cytoplasm and does not contain cell organelles.

It exists in two zones-**Ectoplasm** (plasmagel) is dense, gelatinous, outer layer concerned with cell movement and **Endoplasm** (plasmasol) is fluid like and inner part where most of the organelles are found. It contains 90% of water. It is actually crystallo-colloidal complex where some chemicals (glucose and minerals) are present in the form of true solution and others (proteins (maximum), enzymes, fatty acids and some sugars) are present in the form of colloidal solution.

Functions:

1. It is the seat of synthesis of enzymes, proteins, fats and nucleotides.
2. It contains raw materials for the functioning of cell organelles.
3. It is the seat of catabolic activities such as glycolysis, anaerobic respiration, pentose phosphate pathway (PPP).
4. It performs cytoplasmic streaming or cyclosis which distributes various materials inside the cell.

5. It receives products of cell organelles.
6. It is involved in β-oxidation of fats in prokaryotes.

Mitochondria

Mitochondrion (pl: Mitochondria) is hollow, and sac like organelle.

Discovery: It was discovered by Kollicker (1850) in striated muscle of insect and called sarcosome (granule). Flemming (1882) called it 'fila' (thread like). Altman (1890) called it 'bioblast' (a granule). Carl Benda (1898) coined the term 'Mitochondria'. Friedrich Meves (1904) first observed it in cells of a plant, white water lily (*Nymphaea alba*).

Distribution: It is found in eukaryotic cells except in mammalian RBCs. It is absent in prokaryotes.

Size: Its average size is 0.5 μm in diameter. It is the largest cell organelle in animal cell and second largest in plant cell. It approaches bacterial cell. Its smallest size is found in yeast cell and the largest size occurs in oocyte of *Rana pipens.*

Shape: It is usually oval or sausage shaped. It changes due to chemical or osmotic change.

Number: It shows great variation in organism, tissue and cell types. It is one in *Chlorella* (a green alga), *Microasterias* (a green alga), *Trypanosoma* (a protozoan) to 500,000 in flight muscle of birds. Its average number is 1000 per cell. Its number is lesser in plant cell due to the presence of chloroplast which meets the energy requirement. Its number depends on the metabolic requirements of the cell.

Activity: It remains stable in cells performing same function. It undergoes continuous movement within cytoplasm during interphase.

Stain: It can be stained with a vital dye, Janus green. It appears yellow due to pigment, riboflavin.

Composition: Phospholipids (30%), Proteins (70%), RNA (3%), and DNA (1%). It contains 70% of total enzymes of the cell.

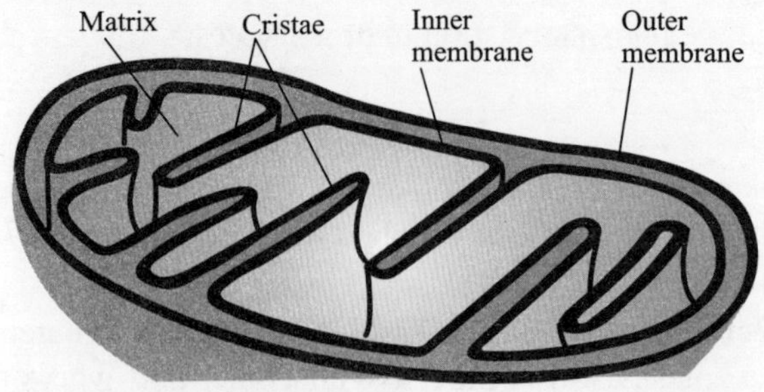

Fig. ***Mitochondria***

Structure: It is double membrane-bound cell organelle. The two membranes have different properties. It consists of following parts:

(a) Outer membrane: Its thickness is $60A^0$. It is smooth and permeable to a number of metabolites because of pores having porins (intrinsic protein). It contains more lipids, and sialic acid (glycoprotein and glycolipid) but less proteins. It contains enzymes such as monoamine oxidase and fatty acid Co-A ligase.

(b) Perichondrial space: It is space between two membranes. It contains enzymes such as adenylate cyclise, nucleoside diphosphokinase, cytochrome c etc. It shows lowest pH.

(c) Inner membrane: Its thickness is $60A^0$. It contains more proteins but less lipids and sialic acid (glycoprotein and glycolipid). It contains succinate dehydrogenase, cytochrome oxidase, ATPase etc.

(d) Cristae: These are several finger-like structures develop from inner membrane. These are tubular in plant cells and plate like in fungal cells and animal cells. The number of cristae indicates the intensity of respiration. Cristae increase the surface area of inner membrane for reactions to occur. Its outer surface is covered with 10^4-10^5 elementary particles or oxysomes. These are also known as subunits of Parson or Racker's particles or Fo-F_1 particlcs. These were discovered by Fernandez Moran (1961). Oxysome is functional unit of mitochondria. These are evenly spread over interval of $100A^0$. Each oxysome consists of head (F_1), stalk and base (F_o). Each oxysome with one of respiratory chain complex constitutes a phosphorylating unit. The whole electron transport system occurs in base and each base is a complex containing a portion of enzymes of respiratory chain. Oxysome is lipo-proteinaceous. The head of oxysome contains a coupling enzyme, ATPase which synthesises ATP in matrix.

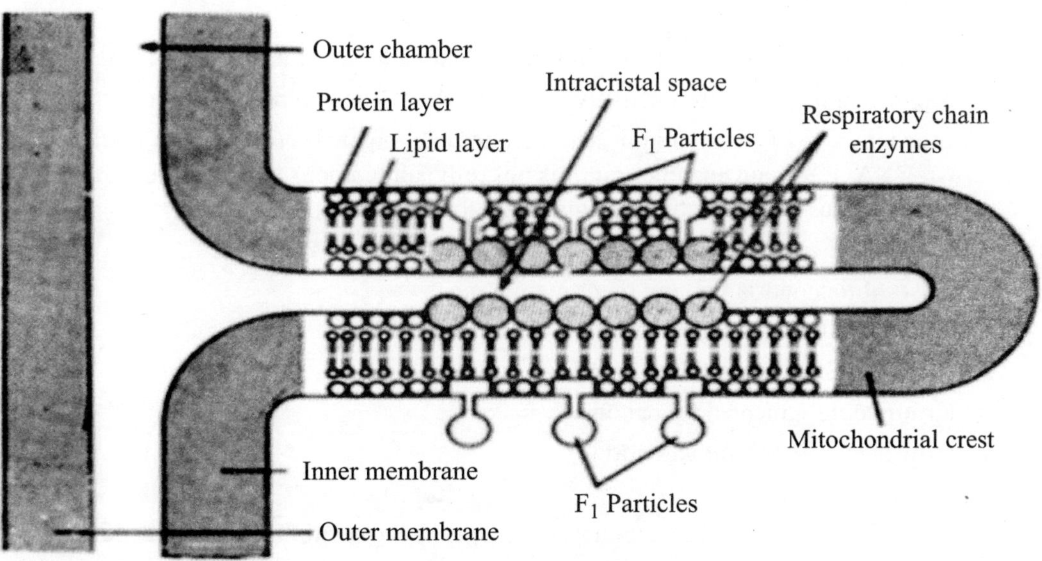

Fig. ***Internal structure of Mitochondria***

(e) Matrix: It is the space enclosed by the inner membrane. It contains dense granules (inorganic salts), ribosomes (ranging from 55S to 70S), DNA, RNA and enzymes of Krebs cycle, fatty acid oxidation, protein synthesis, and nucleic acid synthesis.

Ribosome is smaller than bacterial ribosome and cytoplasmic ribosome. Its subunits are 25 S (containing 12 S r-RNA) and 35S (containing16S r-RNA and 5S r-RNA).

DNA is single, circular, double stranded, naked (without histone proteins). It was discovered by Chevremont and Nass. It is 1% of total DNA of cell. It is rich in G-C ratio. It has 16,569 base pairs encoding 37 genes. It does not have repair mechanism and has higher mutation rate than nuclear DNA.

Origin: Small free living prokaryotes were engulfed (but not consumed) by larger prokaryotes. The two organisms developed a symbiotic relationship over time, the larger organism providing nutrients to smaller organism and in return they got ATP from smaller organism. Eventually larger organism developed into eukaryotic cell and smaller organism into the mitochondria. Thus mitochondria were originally aerobic bacteria like organism which entered accidentally into a host cell of primitive nucleated eukaryote and established with it for the successful endosymbiotic association. The conditions inside host cell were favourable for prokaryote. The bacteria provided increased capacity to produce ATP and provided the host the ability to respire aerobically. Thus mitochondria are also called intracellular prokaryotic parasite or bacterial endosymbiont. Bacteria and mitochondria show resemblances in several respects:

(a) The size of both mitochondria and bacteria is almost same.
(b) Both mitochondria and bacteria undergo binary fission.
(c) Porin protein is present in plasma membrane of both mitochondria and bacteria.
(d) DNA of mitochondria and bacteria is single, circular and double stranded.
(e) Ribosome of mitochondria and bacteria is of 55S close to 70S.
(f) Respiratory enzymes in both are present on inner surface of inner membrane.
(g) Protein synthesis in both is sensitive to streptomycin and chloramphenicol.

Mitochondrion is considered as semiautonomous organelle (cell within cell) because it has own DNA, RNA and enzymes but it is not fully autonomous because it is partially controlled by the nucleus and availability of materials from cytoplasm.

Terms:

1. Mitoplast: Mitochondria without outer membrane.
2. Chondriome: All mitochondria in a cell.
3. Chondriocont: Small rod of mitochondria.
4. Chondriomite: Fragmented mitochondria
5. Chondriosphere: Fused mitochondria

Functions:

1. Mitochondria produces large amount of energy by oxidation of organic substrate such as carbohydrates, proteins and fats. The energy is used in biosynthesis, membrane transport, cell division, movement etc. Hence it is known as power house of cell (as coined by Siekevitz).
2. It forms several intermediates for synthesis of chlorophyll, cytochromes, alkaloids, steroids, pyrimidines etc.
3. It is involved in (β-oxidation) of very long chain of fatty in animal cell.
4. It is involved in synthesis of glutamic acid (an amino acid) from α-ketoglutaric acid, and aspartic acid (amino acid) from oxaloacetic acid.
5. It shows cytoplasmic inheritance (maternal influence).
6. It is involved in cell signalling, cellular differentiation and cell death.

Plastid

The plastid is flat and circular cell organelle found in plant cell and some protista. The term 'plastid' was coined by Haeckel (1861).

Types of Plastids: Schimper divided plastid into three types on the basis of types of pigments.

1. **Leucoplast**: It is the largest and colourless plastid. It does not contain chlorophylls. It stores food in roots, underground stem and seed where light is not available. It is of three types
   (a) Amyloplast: It stores starch (carbohydrate).
   (b) Aleuroplast: It stores proteins.
   (c) Elaioplast: It stores oils and fats.
2. **Chromoplast:** It is second largest and coloured other than green. It contains carotenoids. It is found in petals of flower; and fruits in which colouration is provided by a pigment, anthocyanin which is dissolved in cell sap. e.g., Phaeoplast (brown algae) and rhodoplast (red algae).
3. **Chloroplast:** It is green plastid containing high concentration of chlorophyll.

   Discovery: It was discovered by Julius von Sachs. The term 'chloroplast' was coined by Strasburger.
4. **Distribution:** It may change with the amount of light. When chloroplasts are exposed to sun light, they stack along the anticlinal cell walls to minimise exposure but in the dark they spread out in sheets along the periclinal walls to maximise light absorption.

   Size: Its size is 3-5μm in diameter.. The chloroplast of shade plants (sciophytes) is larger having more grana with more thylakoid area than that of sun plants (heliophytes). It is larger in olyploidy cells than diploid cells.

   Shape: It is biconvex lens or disc or oval shaped in higher plants. Its position and orientation change in relation to the intensity of incident light. Its shape is variable in algae.

Ribbon or spiral shaped	*Spirogyra*
Cup shaped	*Chlamydomonas*
Collar (girdle) shaped	*Ulothrix*
Rounded	*Chlorella*
Reticulate or branch like	*Oedogonium*
Star shaped or stellate	*Zygnema*

   Number: It is present in green parts of a plant. It is usually 20-40 chloroplasts per cell of higher plants. It is one in *Chlamydomonas* and 100 in palisade cell of leaf.

   Composition: Phospholipids (30%), Proteins (50%), Chlorophyll (10%), carotenoids (5%), RNA (2%) and DNA (1%).

   Structure: It is double membrane cell organelle. It consists of following parts:

   (a) Outer membrane: It is highly permeable because of the pores containing porins (intrinsic proteins). It is lipo-proteinaceous. It is rich in lipid and porin protein.

   (b) Periplastidial space: It is space between two membrane. Its thickness is 20 nm.

   (c) Inner membrane: it contains a central space called stroma which is granular and selectively permeable. It is rich in proteins.

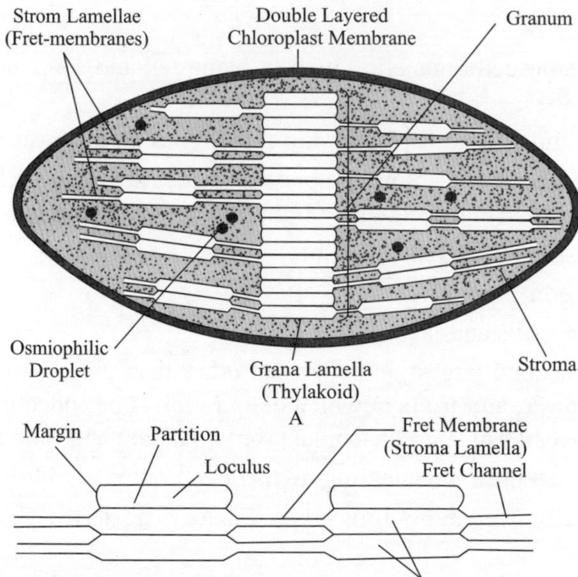

Fig. ***A. chloroplast, B. Thylakoid***

(d) Stroma: It is semi-fluid, alkaline (pH=8) and colloidal complex which contains 50% of soluble proteins. It contains starch granules (pyrenoid in algae), ribosomes, DNA, RNA. The cpDNA is single, circular, double stranded, naked (without histone proteins), heavily stained. It contains more genes than mitochondrial genes. It has high G+C content containing about 150,000 base pairs.

Chloroplast is considered as semiautonomous organelle (cell within cell) because it has own DNA, RNAs, ribosomes but depends on nuclear DNA for synthesis of chlorophyll, carotenoids and enzymes for photosynthesis hence it is not fully autonomous.

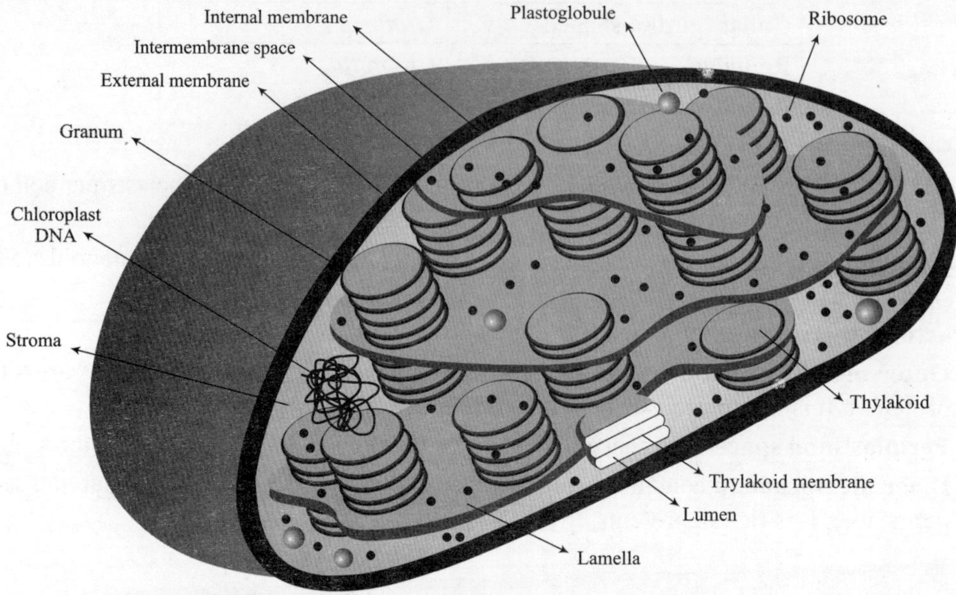

5. **Grana:** It consists of membranous, flattened or circular sac and highly organised structures called thylakoids which are arranged in stacked manner. It develops from lamella. There are 40-60 grana per chloroplast. Grana are absent in algae and bundle sheath cell of C_4 plants. Each granum is made up of 2-100 thylakoids (usually 10-20). Thylakoid is structural unit of chloroplast. The light reaction takes place on the thylakoid membranes. The interior of thylakoid is acidic. Grana are interconnected by a network of tubules called stroma lamellae or intergranal lamellae or Fret. These lamellae are continuous with numerous thylakoids within granum and also with thylakoids in different granum. These lamellae act as channels for transporting various photosynthetic substrates. The inner side of thylakoid membrane contains group of fine, ultrastructural particles called quantasomes (185x155x110A°) arranged in specific manner. These were discovered by Park and Biggin but the term 'quantasome' was coined by Park and Pons. Quantasome is the smallest functional unit of chloroplast and photosynthesis. Each quantasome contains 230 chlorophyll molecules (160 chlorophyll-a and 70 chlorophyll-b) and 50 carotenoids. These photosynthetic pigments are organised into distinct units or particles of protein complex called photosystem or Pigment systems. Quantasome has been resolved into four major complexes such as Pigment system-II, cytochrome b_6-f complex, pigment system I and ATP synthase complex (cf_o-cf_1 or coupling factor).

There are three spaces in chloroplast such as periplastidial space, stromal space and thylakoidal space.

Origin: Chloroplast divides by fission. Like mitochondria it also originates from cyanobacteria through endosymbiotic relationship with eukaryotic cell. Secondary endosymbiosis occurs in which chloroplast was developed after mitochondria. Thus chloroplast is also known as intracellular prokaryotic parasite or bacterial endosymbiont.

Evolution: Chromatophore (Photosynthetic bacteria) → Single thylakoid → Granal thylakoid or Chloroplast (Plants)

Terms:

1. **Chromatophore:** It is the scattered thylakoids in cytoplasm.
2. **Proplastid:** It is colourless, immature (undeveloped) and double membranous structure. It is precursor to plastid. It has the ability to divide and differentiate into various types of plastids.
3. **Etioplast:** It is plastid of etiolated plant (growing in dark) which lacks chlorophylls but contains chlorophylloid (protochlorophyll) and carotenoids.
4. **Plastidome:** All chloroplasts in a cell.

Functions:

1. It performs photosynthesis (light reaction occurs in thylakoid of grana and dark reaction takes place in stroma).
2. It traps solar energy and converts it into chemical energy (ATP and $NADPH_2$).
3. It stores starch permanently in pyrenoid in Algae and temporarily in higher plants.
4. It stores fats in the form of plastoglobuli.
5. It provides colours to flowers and fruits by converting into chromoplast.

Photochemical conversion: All three types of plastids are photochemically interconvertible.

(a) In case of Carrot, leucoplast is converted into chromoplast.
(b) In case of Tomato, leucoplast is firstly changed to chloroplast then to chromoplast.
(c) In case of Chilly and Mango, chloroplast is converted to chromoplast.

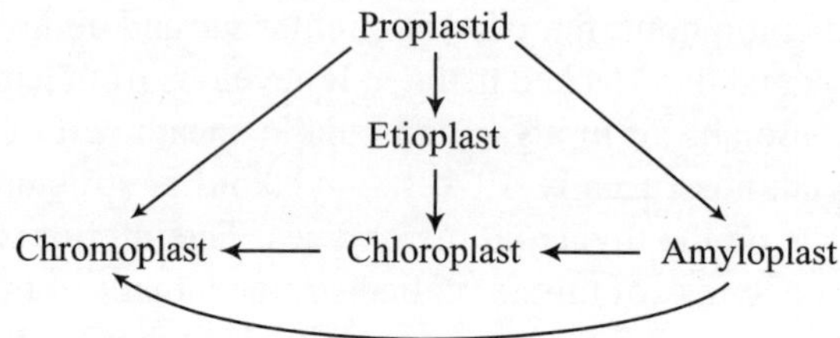

Endoplasmic Reticulum

It was discovered by Porter, Claude and Fullam (1945). The term 'Endoplasmic reticulum' was coined by Porter (1953).

Distribution: It is absent in prokaryotes. It is present in eukaryotic cells except mammalian RBCs and sperm.

It was the first cell organelle to be revealed by electron microscope. It constitutes 40% of endomembrane system (having endoplasmic reticulum, Golgi body, lysosome and vacuole because these structures remain coordinated but these have no coordination with other membrane bound organelles). It is poorly developed during cell division. It is well developed in metabolically active or enzyme secreting cells. It is not a stable structure bccause it undergoes break down and reconstruction. It undergoes partial fragmentation during cell division. It is basophilic in nature due to high content of RNA hence it is called ergastoplasm.

Endoplasmic reticulum of muscle cell	Sarcoplasmic reticulum
Endoplasmic reticulum of retinal cell	Myeloid body
Endoplasmic reticulum of nerve cell	Nissl's granule

Structure: Endoplasmic reticulum divides the intracellular space into two distinct compartments: luminal (inside ER) and extraluminal (cytoplasmic face). It is made up of single membrane containing several channels which are interconnected system. These channels run through cytoplasm. It may be continuous with nucleus and plasma membrane. It consists of following parts:

(a) Cisternae: These are long, flattened, unbranched and sac-like structures which are parallel to each other in bundles. Their size is 400-500A^0.

(b) Tubules: These are tubular and branched or unbranched. These are connected with cisternae or vesicles to form a reticular system. Their diameter is 50-100A^0.

(c) Vesicles: These are small, oval and remain isolated in cytoplasm. Their diameter is 25-500A^0.

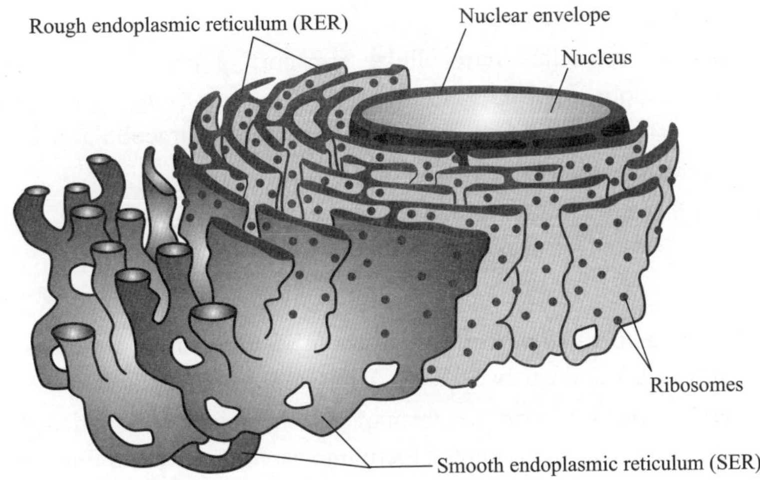

Fig. ***Endoplasmic reticulum***

Types: ER is of two types on the basis of nature of its membrane

Rough Endoplasmic reticulum	Smooth endoplasmic reticulum
It has rough or granular membrane due to presence of ribosomes.	It has smooth or agranular due to absence of ribosomes.
It is firstly formed.	It is later formed.
It develops from nuclear membrane.	It develops from rough endoplasmic reticulum.
It is attached to nucleus.	It is attached to plasma membrane.
It occupies two-third of total ER.	It occupies one-third of total ER
It is mainly made up of cisternae.	It is mainly made up of tubules and vesicles.
It is more stable and internal.	It is less stable and peripheral.
It contains ribophorins which provide attachment to ribosomes.	It does not contain ribophorins.
It contains more proteins and less lipids.	It contains less proteins and more lipids.
It develops lysosomes.	It develops sphaerosome.
Function: Protein synthesis	**Functions:** Fat synthesis except fatty acids Glycogen synthesis Synthesis of steroidal hormones (testosterone, estrogen) in animal cells. Synthesis of visual pigments from vitamin-A in retinal cell. Detoxification of drugs, chemicals, pollutants in liver cell by cytochrome P_{450}

General Functions:

1. It acts as cytoskeleton (endoskeleton) along with microfilaments, microtubules and intermediate filaments by providing mechanical support due to formation of network like structure inside the cell.

2. It keeps all cell organelles in position.
3. It provides a passage for quick intracellular transport of proteins by diffusion via nuclear membrane → nuclear pore → RER → transport vesicle → cis-Golgi → trans Golgi → Secretory vesicle → plasma membrane and thus transports outside for extracellular use.
4. It forms plasmodesmata which joins two adjacent cells.
5. It increases the surface area inside the cell for various enzymatic actions.
5. It gives rise to vacuoles.

Terms:

1. **Transitional Endoplasmic reticulum:** It is ribosome free region of endoplasmic reticulum. It is considered as partly RER and partly SER.
2. **Sarcoplasmic reticulum:** It is modified form of smooth endoplasmic reticulum found in striated muscle. It stores Calcium ions and pumps them into sarcoplasm when muscle fibre is stimulated to permit relaxation of myofibrils following each round of muscle contraction.
4. **Microsomes:** These were discovered by Claude. These are vesicle like artefact reformed from pieces of ER when eukaryotic cells are subjected to homogenisation. These can be concentrated and separated from other cellular debris by differential centrifugation. These are not found in healthy and living cells. These are of two types:
   (a) Smooth microsome: It consists of SER, parts of mitochondria, Golgi apparatus and plasma membrane.
   (b) Rough microsome: It consists of RER only.

Golgi apparatus

It was discovered by Camillo Golgi (1898) in cerebrum of neuron of barn owl by metallic impregnation method (staining with $AgNO_3$) and termed as 'Internal reticular apparatus'. It is also known as Lipochondria or Baker's body or Idiosome or Dalton comlex or Middle man of cell.

Distribution: It is absent in prokaryotes. It is found in eukaryotic cells except mammalian RBCs, sieve tube of phloem, antherozoids of bryophytes and pteridophytes. It is large and more numerous in synthetic and secretory cells.

Its shape, size and number vary because it is constantly being formed, changed, reformed or broken down. It depends on metabolic activities of cell. Its number increases during cell division. It is present anywhere in the cell but mostly adjacent to ER and has similar organisation in one type of cell.

It comes under endomembrane system (endoplasmic reticulum, Golgi apparatus, lysosome, vacuole) which functions in close coordination with one another.

It shows zone of exclusion because it is found in the area of cytoplasm where other cell organelles are absent.

It develops from plasma membrane, smotth endoplasmic reticulum (SER) and nuclear membrane.

Composition: Lipids (phospholipids), proteins and enzymes (glycosyl transferase, thiamine pyrophosphatase).

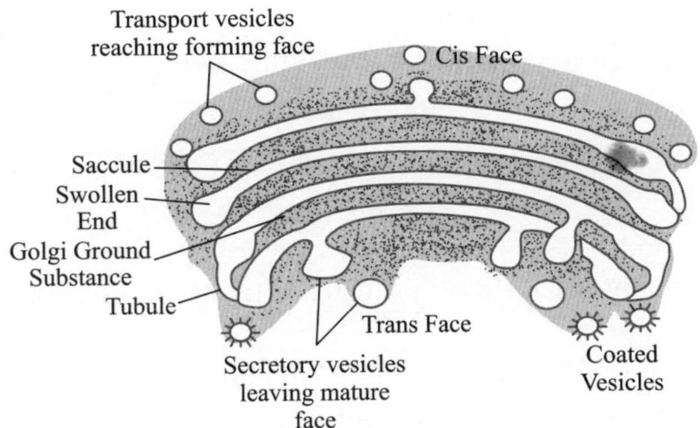

Fig. ***Golgi apparatus***

Structure: It is single membrane bound cell organelle. It consists of the following parts:

(a) Cisternae: These are disc shaped, flattened and membrane bound intercommunicating channels and arranged in parallel stacked manner. One stack contains 3-20 cisternae (usually 6). The cisternae vary in number, shape and organisation in different cells. Cisterna is constant element and functional unit of Golgi body. Cisternae are scattered and net-like in plant cells and each stack of cisternae is called dictyosome. In fungi, dictyosome is unicisternal. Each stack of cisterna is polarised structure because its margin is curved. It has two faces: cis and trans. Both faces differ in thickness, enzyme content and degree of vesicle formation.

Cis-face: It is also called convex face or proximal face or forming face. It is collection of fused, flattened and membrane enclosed discs. It is associated with nuclear membrane or endoplasmic reticulum. It receives materials and vesicles from cytosol and endoplasmic reticulum. It is also called entry face.

Trans-face: It is also called concave face or distal face or maturing face. It is towards plasma membrane. It gives out vesicles and Golgian vacuoles. It is also called exit face.

(b) Tubules: These are flat and tubular structures arising from peripheral area of cisternae.

(c) Vesicles: These are small and sac like structures attached to tubules. These may be smooth vesicles (secretory vesicles) which are numerous and coated vesicles which are less in number.

(d) Golgian vacuoles: These are large, rounded and expanded part of cisternae. These develop on trans face.

Functions:

1. Golgi body is involved in processing, sorting, packaging, transport and release of secretory proteins coming from ER outside the cell (for extracellular use) by exocytosis or reverse pinocytosis along the route (RER→transport vesicle→cis-Golgi→median-Golgi→trans-Golgi→secretory vesicle→Plasma membrane. The proteins are also used for intracellular use. It also performs post translational modification of proteins.
2. It forms glycoprotein by glycosylation of protein.
3. It forms glycolipid by glycosidation of lipid (addition of oligosaccharide to phospholipid of membrane).
4. It helps in formation of cell wall in plants during cytokinesis where cell plate is formed by secreting pectin (polysaccharide).

5. It lubricates the root tip by secreting mucilage.
6. It helps in formation of primary lysosome.
7. It forms acrosome of sperm during spermiogenesis.
8. It forms insulin, melanin and yolk.

Ribosome

Discovery: It was discovered by Palade in animal cell and termed it 'microsome'. The term 'Ribosome' was coined by Claude.

Size: It is the smallest cell organelle. Its size is 20x15 nm in diameter.

Distribution: It is found in both prokaryotes and eukaryotic cells. It is most abundant cell organelle. It is found floating freely in cytoplasm of prokaryotes but in case of eukaryotes, it is found floating freely in cytoplasm as well as attached with nuclear membrane and RER by hydrophobic interaction. It is also found in matrix of mitochondria and stroma of chloroplast of eukaryotic cell.

Number: It is several thousands (20,000) in prokaryotes and millions in cell of eukaryotes.

Origin: It originates from nucleolus.

Synthesis: In prokaryotes, the ribosomes are synthesised in cytoplasm but in eukaryotes, these are synthesised in nucleolus.

Composition: r-RNA (65%) and proteins (35%). Lipid is not present.

Structure: Its structure is not stable as it is constantly being bound and released from the membrane. It is not bounded by unit membrane. Hence it is an organelle within an organelle. The basic structure of ribosome is due to r-RNA. Proteins play secondary role. Ribosome is strongly negative because the positive charges are not enough to neutralise the negative charges of RNA. It consists of two distinct unequal subunits- the smaller one is oval and larger one is dome shaped. The smaller subunit has head, base, platform and cleft. The larger subunit has ridge, valley, protuberance and stalk.

These two subunits are dependent on Mg^{+2} concentrations (0.001M). These two subunits separate if Mg^{+2} concentrations will be less in cytoplasm. These two subunits unite if Mg^{+2} concentrations will be increased ten times. Both subunits are found separately in cytoplasm but rejoin at the time of protein synthesis.

Isolation: Ribosomes are isolated from other cell components by fractionation and ultracentrifugation. These are so small that these are last organelle to be sedimented in a centrifuge. It is measured in Svedberg unit named after Svedberg. It is rate or velocity of sedimenting molecule per unit of gravitational field in dense solution of sucrose. It is a measure of how rapidly a particle (organelle or macromolecule) sediments. The greater is S number, the greater will be rate of sedimentation. The shape, size and density (molecular weight) of ribosome depend on sedimentation coefficient in ultracentrifuge.

Types: Ribosome is of two types on the basis of sedimentation coefficient

70 S	80 S
Its size is smaller.	Its size is larger.
It has lower molecular weight.	It has higher molecular weight.
It is found in prokaryotes.	It is found in eukaryotes.
It is synthesised and assembled in cytoplasm.	It is synthesised and assembled in nucleolus (5S r-RNA in nucleoplasm) and final maturation occurs in cytoplasm.

It has two subunits: 30S and 50S.	It has 40S and 60 S
Three molecules of r-RNA are found →30S has 16 S rRNA (1500 bases). →50S has 23 S (3000 bases) and 5S (120 bases).	Four molecules of rRNA are found. →40S has 18S rRNA (2000bases) →60S has 28S rRNA (5000bases), 5S rRNA (160 bases) and 5.8 S rRNA (120 bases).
rRNA component is 60% in 30 S and 50% in 50 S	rRNA component is 50% in 40 S and 65% in 60 S
Protein content is 21 in 30S and is 31 in 50S	Protein content is 33 in 40 S and is 49 in 60S
The ratio of rRNA and protein is 2:1.	The ratio of rRNA and protein is 1:1.

Functions:

1. Attached ribosomes (ER bound) synthesise secretory proteins, integral membrane proteins, lysosomal proteins for extracellular use (transport). These ribosomes form most of the proteins.
2. Free ribosomes synthesise non-secretory and structural proteins such as soluble cytoplasmic proteins, peripheral proteins, mitochondrial proteins, chlorophyll proteins, nuclear proteins and peroxisomal proteins for intracellular use.

Microbodies

These are very small (<1 μm), rounded cell organelles bounded by single membrane. These are absent in prokaryotes. These include:

1. **Lysosome:** It was discovered by Christian de Duve (1955) in liver cell of rat. The term 'lysosome' was coined by Navikoff. It is also known as pericanalicular body. It is small, rounded and dense body bounded by single membrane. It is absent in prokaryotes but it is present in animal cell of eukaryotes except mammalian RBCs. It is usually absent in plant cells where function of lysosome is performed by vacuoles and sphaerosome. It has been reportedly found in root tip of maize, yeast and *Neurospora.*

   It is named as lysosome because it contains hydrolytic enzymes. It is rich in 60 types of hydrolytic enzymes which function at pH 4.5-5. Some of the enzymes are phosphatase (common and marker enzyme), protease, lipase, sulfatase, glycosidase etc. It contains a special transport protein which pumps H^+ into the cavity of enzymes. Its single membrane is impermeable to substrate of enzymes and resistant to hydrolase. It is also called suicidal bag of cell because it causes destruction of cell once enzymes are released in cell.

   It is polymorphic hence it passes through different forms in the same cell on the basis of function. These forms are as follows:

   1. **Primary lysosme (Storage granule):** It is very small in size and newly formed lysosome. It is formed when enzyme of lysosome is formed on RER and is transported to Golgi apparatuus which buds off in cytoplasm. It contains only one enzyme which fuses with larger acidic vesicle.
   2. **Secondary lysosome (Heterophagosome)**: It is formed when primary lysosome is fused with vesicles such as phagosome or pinosome formed by plasma membrane for endocytosis. These vesicles pass digested food materials into cytoplasm for absorption and secondary lysosome is left with undigested food materials. It contains primary lysosome and pinosome or phagosome.
   3. **Tertiary lysosome (Residual body)**: It is formed when secondary lysosome containing undigested food materials is fused with vesicle formed by plasma membrane for exocytosis to remove debris outside.

4. **Autophagosome (Cytolysosome)**: It is formed by fusion of primary lysosome with dead or worn out cell organelles which are engulfed and later digested by autodigestion. It helps in disposal of cell debris.

Origin: It is formed by joint activity of Golgi apparatus and Endoplasmic reticulum.

GERL= G + ER → Lysosome.

Lysosome stabilisers: These chemicals strengthen lysosome because these prevent enzymes from leaking into cytoplasm and causing destruction of cell organelles. e.g., cortisol, cortisone, cholesterol, heparin, antihistamine.

Lysosome labilisers: These chemicals weaken lysosome because these help release of enzymes outside lysosome by rupturing membrane causing destruction of cell organelles. e.g., Progesterone, testosterone.

Functions:

1. It helps in intracellular digestion of stored substances such as proteins and fats present inside the cell during starvation or fasting and thus it provides nourishment.
2. It helps in extracellular digestion of substances by releasing acid hydrolases outside the cell.
3. It provides immunity to body by phagocytic activity against microbes (viruses and bacteria).
4. It helps in scavenging by removing old, dead and useless organelles.
5. It is engaged in autolysis of old, diseased and useless cell by bursting.
6. It secretes thyroid hormones in epithelial cell of thyroid due to presence of hydrolase enzymes.

2. **Peroxisome:** It was discovered by Tolbert (1960) in liver cell but in plant cell (leaf cell of spinach) it was discovered by de duve It is also called uricosome. It is small, rounded and bounded by single membrane. It is the smallest membrane-bound cell organelle. It is absent in prokaryotes but present in plant and animal cells of eukaryotes. It arises from cisternae of ER. It is around 100 in number and found in close association of RER, mitochondria and chloroplast.

   It contains peroxide producing enzymes such as peroxidase, D-amino acid oxidase and urate oxidase. These oxidise substrate and produce H_2O_2 which is highly reactive, toxic and oxidising agent. It also contains catalase which degrades H_2O_2 to H_2O and O_2. Catalase is the fastest acting enzyme.

   Functions:

   1. It is involved in the process of photorespiration (glycolate cycle or C_2 metabolism) along with mitochondria and chloroplast in cell of C_3 plants.
   2. It carries out lipid metabolism (β-oxidation of long chain fatty acids) in animal cells.

3. **Glyoxisome:** It was discovered by Beevers (1951) in endosperm cell of castor. It is small rounded and bounded by single membrane. It contains enzyme rich matrix and crystalloid core. It is found in plant cell but absent in animal cell. It arises from RER. It occurs in seedlings (germinating seeds) of oil seed plants. It contains isocitrate lyase, malate synthetase, glycolate oxidase and catalase hence it is also called specialised peroxisome.

   Functions:

   1. It is involved in gluconeogenesis or glyoxalate cycle (conversion of fats into glucose).
   2. It breaks down fatty acid into acetyl-CoA (β-oxidation) in cell of oil seeds.

4. **Sphaerosome (Oleosome):** It was discovered by Dangeard and Hanstein (1880). It is small, rounded and bounded by single membrane. It is found in plant cell. It is also known as plant lysosome. It arises from SER. It is abundantly found in endosperm and cotyledon of oil seeds. It is

stained with osmium tetroxide and sudan dye. It contains hydrolytic enzymes such as hydrolase, protease, esterase, phosphatase, ribonuclease etc. and high percentage of lipids (95%).

Function: It synthesises and stores fats (lipid metabolism).

Centriole

It is small, hollow and cylindrical structure (diameter=150 nm).

It was first observed by Von Beneden and coined by Theodor Boveri.

It is not bounded by any membrane. It is absent in prokaryotes, fungi, gymnosperms and angiosperms. It is found in motile cell of lower plants (from algae to pteridophytes) and animal cell except mammalian RBCs.

It is situated in clear area of cytoplasm near interphase nucleus called centrospheres or kinoplasm. The position of centriole determines the position of nucleus. It is present in two sets (pairs) which are at right angles to one another called diplosome. Thus centrioles and centrospheres constitute centrosome. Thus centrosome is an organelle containing two cylindrical structures called centrioles.

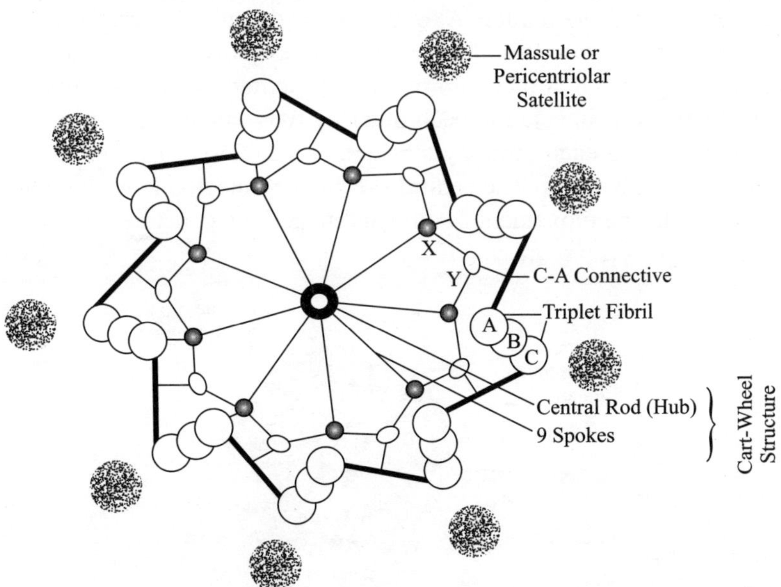

Fig. ***Centriole***

Structure: It is cylindrical structure. Centriole is surrounded by dense, amorphous and protoplasmic sphere in one or more series called massules or perinuclear satellites. These form new centrioles.

Each centriole is made up of 9 sets of peripheral triplets which are arranged parallel to each other. These triplets are evenly spaced showing 9+0 arrangement. Each triplet fibril has 3 subfibres (A-B-C). Each triplet is situated at angle of 40° with radius of cylinder. The adjacent triplets are connected by A-C proteinaceous linkers. The first tubule is connected with third tubule of next. Each subfibre has 13 protofilaments like microtubules. At the centre there is a proteinaceous mass called hub which is connected with 9 triplets by 9 radial spokes. Each radial spoke has a thickening called X before joining with A subfibre. Another thickening called Y is present nearby and attached both to X and A-C linkers by connectives. Thus whole structure looks like cart wheel.

Origin: Centriole is synthesised during S phase but not replicated.

Functions:

1. Centriole gives rise to basal body (9+0) which forms cilia and flagella (9+2).
2. Centriole forms spindle fibres and asters (rays radiating in all direction) during cell division.

Cilia and Flagella

These are fine hair like and contractile appendage of cell. These are outgrowths of plasma membrane and extend outside the cell through pores of cell wall. These are absent in cyanobacteria, rhodophyceae, *Pinus* of gymnosperm, and angiosperm. These may be isokont (of same length) and heterokont (of different length). Flagella show different places of insertion.

(a) Akrokont-Flagella are inserted apically
(b) Opisthokont-Flagella are inserted posteriorly
(c) Pleurokont-Flagella are inserted laterally.

Structure: It is cylindrical and same for both cilia and flagella. It contains core (axoneme or shaft) which is surrounded by plasma membrane. Axoneme is made up of 9 peripheral or radially arranged doublets of microtubules and 2 central singlet of microtubules showing 9+2 arrangement. Each doublet gives rise to a pair of arms which is made up of dyenin (acting as ATPase to provide energy for the movement). The doublets are connected to each other by A-B linkers that are made up of nexin. The central singlet are connected to each other by a bridge and enclosed in a central sheath. Each singlet fibril is made up of 13 protofilaments. The central sheath is connected to each doublet by radial spoke which is made up of nexin. The movement of cilia and flagella is generated by microtubule.

Origin: These arise from basal body.

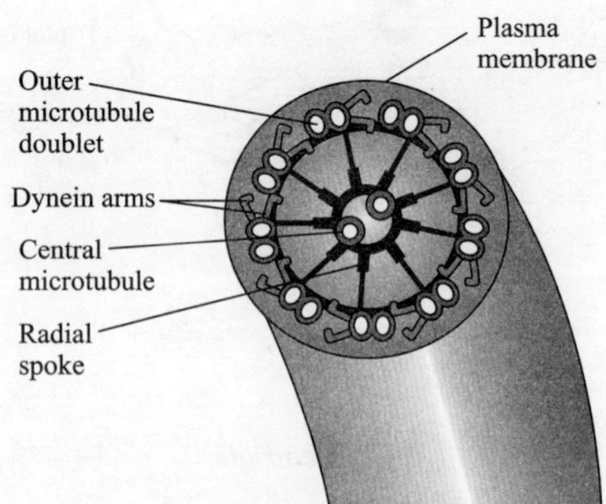

Fig. ***Flagella***

Differences between cilia and flagella

Cilia	Flagella
These are hair like appendages.	These are thread like appendages.
These are present in eukaryotic cells.	These are found in prokaryotic and eukaryotic cells.
These are larger in number (300-14000).	These are lesser in number (1-4).
These are shorter in length.	These are longer.
These are present all over the surface.	These are present at the end of cell.

These beat isochronously or synchronously (simultaneosly) and metachronously (successively).	These beat independently.
These move very fast.	These move very slow.
These show rotational movement like motor.	These show wavelike or undulating movement.
These help in locomotion, feeding, circulation.	These help in locomotion.

Differences between prokaryotic flagella and eukaryotic flagella

Prokaryotic flagella	Eukaryotic flagella
These run in rotary movement.	These run in bending movement.
These use rotary motor.	These use sliding filament system.
These are proton driven due to concentration of gradient set up by cell's metabolism.	These are ATP driven.

Cytoskeletons

These are minute and fibrous structures found in eukaryotic cells. These include:

Microfilaments: These are long, narrow, solid (rod like) unbranched fibrils. These were discovered by Paleviz (1974). These are found just below plasma membrane. These are made up of G-actin proteins. Their thickness is 7 nm. These look like bead due to close helical arrangement of actin. These are contractile. These form extensive networks in cytoplasm.

Functions:

1. These help in cyclosis (cytoplasmic streaming) or cell movement by which transport of nutrients occurs.
2. These help in movement of plasma membrane during endocytosis.
3. These form cell plate and cell furrow.
4. These form pseudopodia.
5. These help in sol-gel transformation.

Intermediate filaments: These are tubular, solid and branched filaments. These are scattered in cytoplasm. These are made up of structural proteins such as desmin, keratin, kinesin etc. These are cross-linked with one another containing 5 protofilaments. These are non-contractile. Their thickness is 10 nm.

Functions:

1. These form nuclear matrix (nuclear lamina) present on the surface of inner membrane of nucleus.
2. These form scaffold (supporting array) for chromatin and cytoplasm.
3. These support all membrane bound cell organelles including nucleus

Microtubules: These are very long, tubular hollow and unbranched fibrils. These were discovered by de Robertis and Franchi (1953) in axon of medullated nerve fibre. These are found in centriole, cilia and flagella. Their wall is made up of 13 parallel protofilaments which are composed of alternate spirals of α and β- tubulin proteins. These are rich in ATPase. These are non-contractile. Their thickness is 25 nm. These require GTP and Ca^{+2} for their assembly and dissembly.

Functions:

1. These help in formation of spindle and cell plate during cell division.

2. These maintain shape, form and rigidity of cell.
3. These help in anaphasic movement of cell.
4. These help in movement of cilia and flagella.
5. These help in intracellular transport of materials.

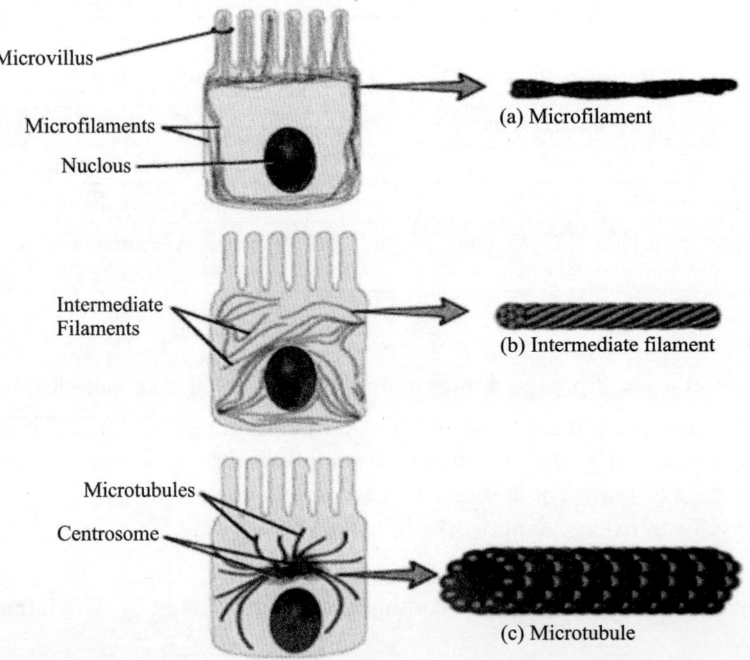

Nucleus

Discovery: It was discovered by Robert Brown (1833) in cell of root of Orchid. It is rounded, dense and highly specialised body. It is studied best at interphase stage. It is considered as component because it is extracytoplasmic.

Size: Its size is 5-25µm. It is large in active cell. It depends on protein and DNA content; metabolic activity and volume of cell.

Shape: It is lens shaped and peripheral in plant cell and rounded and central in animal cell.

Number: It is usually one per cell but there are several exceptions which are as follows.

(a) Binucleate condition (micronucleus and meganucleus) is seen in *Paramecium* but dikaryotic condition is seen in *Puccinia* (rust fungus) containing nuclei from opposite mating types.

(b) Multinucleate condition (syncytium) is seen in epidermis of *Ascaris* where fusion of cells occurs.

(c) Multinucleate condition (coenocyte) is seen in *Rhizopus, Mucor* (fungi), plasmodium of slime moulds and liquid endosperm of coconut where nuclear division occurs without cytokinesis.

Distribution: It is absent in prokaryote but present in eukaryotic cells except mammalian RBCs and sieve tube of phloem. It occupies 10% of volume of cell.

In a cell, there is definite nucleo-cytoplasmic ratio.

Nucleo-cytoplasmic index = Volume of Nucleus/ (Volume of Cell - Volume of nucleus)

Composition: Its dry weight contains protein (90%), DNA (9%) and RNA (1%).

Structure: It is double membrane cell component. It consists of

(a) Nuclear membrane: It is double membrane and porous. It is simple in plants and annulated in animals. It is lipo-proteinaceous in nature. The outer membrane communicates with ER at several points. Outer membrane shows presence of ribosomes at outer surface. Perinuclear space (20 nm) is space between outer and inner membrane. It is filled with fluid. It has 1000-10,000 minute circular pores (size=20-80 nm) which serve as a means of nucleo-cytoplasmic exchange of materials (mRNA, proteins, micronutrients, water, ions etc.) in both direction. It is lined with a fibrous protein network called nuclear lamina which provides shape to nucleus, strength to nuclear membrane and holds nuclear pores in shape. The area of membrane where ribosomes are absent, forms vesicles (blebbing). The inner membrane is smooth.

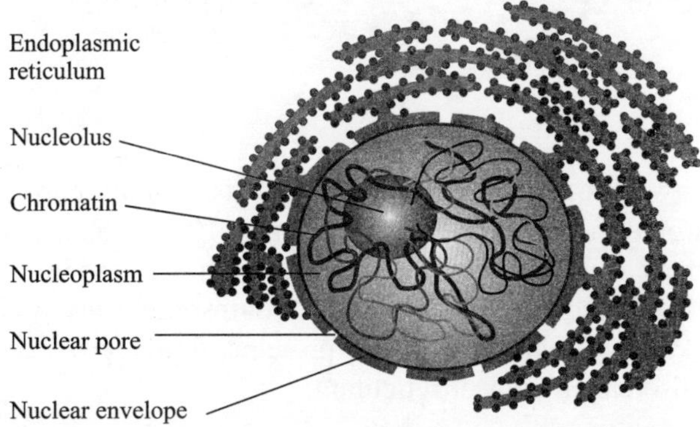

Fig. ***Nucleus***

(b) Nucleoplasm: It is clear, gel like, and granular ground matrix (pH 7.4). It is more dense than cytoplasm. It contains ions, enzymes (polymerases), proteins, water and nucleotides. Its aqueous nature allows ions and organelles to be transported within cell. It is site of enzyme activities and synthesis of DNA, RNA and ribosomal subunits.

(c) Nucleolus: It was discovered by Fontana (1781). The term 'nucleolus' was coined by Bowman. It is rounded and densest structure inside nucleus. It is not bounded by any membrane. It is absent in prokaryotes. It is present in eukaryotic cells except mammalian RBCs, mammalian sperms, young embryo cell, and yeast. It is one in haploid but one to four in diploid cells. It is present anywhere (either at centre or periphery). It is large and more in number in actively (protein) synthesising cells. It occupies one-third area of nucleus. It is found at secondary constriction or nucleolar organising region (NOR) of chromosome. This region contains repetitive DNA which forms r-RNA.

Composition: Nonhistone proteins (65-70%), rRNA (30%), Ca^{+2}

Structure: It has following parts:

(a) Nucleolar chromatin: There are perinucleolar chromatin which surrounds nucleoli and intranucleolar chromatin embedded in matrix.

(b) Amorphous matrix: It is homogeneous containing granules and fibrils. It is rich in RNA polymerase enzyme.

(c) Granules: These are made up of ribonucleoproteins (rRNA and protein=1:2). These are precursor to ribosomes.

(d) Fibrils: These are made up of ribonucleoproteins. This part contains many copies of repetitive DNA which directs synthesis of rRNA.

Function: It is involved in synthesis of rRNA and ribosome. It also stores nucleoproteins.

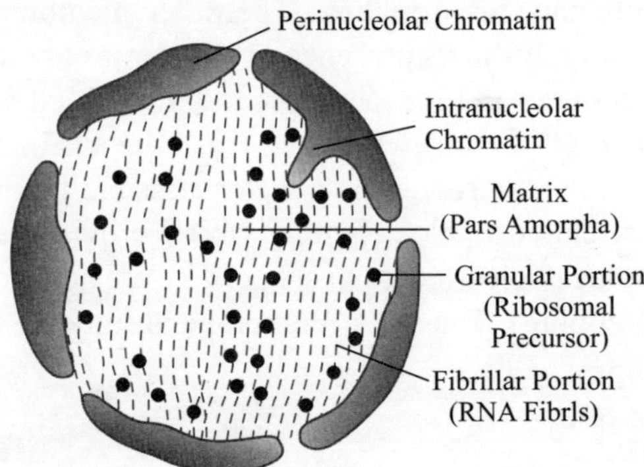

Fig. ***Nucleolus***

(d) Chromatin: It is network of nucleoprotein fibrils. It is actually chromosome in fully extended form. The term 'Chromatin' was coined by Flemming (1979) because it has ability to get stained with basic dyes such as acetocarmine or Fuelgen stain. It is found in interphase of cell cyclc. It s condensed to more tightly coiled threads called chromosomes during M phase of cell cycle (cell division) due to packaging of DNA by histone proteins. Emil Heitz developed staining technique for chromatin. It is also called nuclear reticulum.

Chromatin is differentiated into two regions during interphase.

Euchromatin	Heterochromatin
It is loosely coiled or diffused region of chromosome.	It is tightly coiled or highly condensed region of chromosome.
It gets light stain in interphase.	It gets dark stain in interphase.
It is fibrous and less labile.	It is granular and more labile.
It is present towards centre of nucleus.	It is present towards nuclear membrane.
It is present in large area.	It is present in certain areas.
It contains large amount af DNA	It contains small amount of DNA.
It contains loosely packed DNA hence it is available for transcription and recombination (genetically active).	It contains tightly packed DNA hence it is unavailable for transcription and recombination (genetically inactive).
It is condensed during cell division only.	It remains condensed throughout life cycle.
It replicates during early S-phase.	It replicates later at S-phase.
It does not show heteropycnosis.	It shows heteropycnosis.

Types of Heterochromatin:

1. Constitutive Heterochromatin: The heterochromatin remains permanently in all cell types at all stages of life cycle. It is located near telomeres and centromere. It shows heteropycnosis (difference in thickening between different regions). It contains repetitive sequences.

2. Facultative Heterochromatin: The heterochromatin remains permanently in some special cell types or at some particular stages in life cycle. It can change to euchromatin and will become loose. It is sometimes expressed. e.g., Barr body.

Functions of nucleus:

1. It contains genetic information for metabolism, development and reproduction. **Hammerling's grafting experiment:** John Hammerling performed grafting experiment on *Acetabularia* to prove that nucleus is store house of all hereditary information. *Acetabularia* has base, stalk and cap. Nucleus is present in base. He selected two species: *Actabularia crenulata* and *Acaetabularia mediterranea*. The stalk of one species is grafted on base of another species. The cap formed is similar to species to which base belongs. It proves that shape of cap is determined by nucleus which is present in base.
2. It passes the genetic information to daughter cells.
3. It contains direction for regulating cellular activities through information stored in mRNA.

Chromosome

It is thread like structure found in nucleus. It is organised structure of DNA and protein.

Discovery: It was discovered by Hofmeister in nucleus of pollen mother cell of *Tradescantia*. The term 'Chromosome' was coined by Waldeyer due to its ability to take up colours in basic dyes.

Number: Its number varies. Chromosomes are present in sets (ploid). Haploid has single set of chromosome and diploid has double set of chromosomes inherited from maternal and paternal parent. Animal cell contains larger number of chromosomes than plant cell.

Among animals, the lowest number (2n=2) is found in *Ascaris microcephala* and the highest number (2n=1600) is found in *Aulacantha* (a protozoan).

Among plants, the lowest number (2n=4) is found in *Haplopappus gracilis* and *Brachycome lineariloba* and the highest number (2n=1262) is found in *Ophioglossum reticulatum* (a pteridophyte).

Plants	Number of chromosome	Animals	Number of chromosome
Garden pea, Barley	2n=14	*Drosophila*	2n=8
Onion	2n=16	Housefly, *Coenorhabditis*	2n=12
Neurospora (a fungus)	n=7	Cockroach	2n=24(male), 2n=23 (female)
Broad bean	2n=12	Mosquito	2n=6
Rice, Tomato	2n=24	Silk worm	2n=56
Wheat	2n=42	Frog	2n=26
Maize	2n=20	Mouse	2n=20
Cotton	2n=52	Dog, Chicken	2n=78
Tobacco, Potato	2n=48	Guinea pig, Horse	2n=64
Cabbage, Mustard	2n=18	Rhesus monkey, Rat	2n=42
Sugarcane	2n=80	Man	2n=46
Yeast	2n=32	Gorilla, Chimpanzee	2n=48

Size: Its size varies from 0.5 to 30μm. Its average size is 3μm. It can be measured at metaphase stage. Its smallest size is found in fungi and birds and the largest size is seen in *Trillium*. The plant cell has larger size than animal cell. Among plants, monocots have larger size than dicots. Each human cell contains roughly 2 meters of DNA.

Shape: It can be observed at anaphase stage. A chromosomes shows the different shapes.

(a) Metacentric (V-shaped): Centromere is at centre hence both arms are equal in length.

(b) Submetacentric (L-shaped): Centromere is slightly away from the centre hence one arm (p arm) is shorter than other (q arm).

(c) Acrocentric (J-shaped): Centromere is close to its end (sub-terminal) hence one arm (p arm) is much shorter than other (q arm).

(d) Telocentric (I-shaped): Centromere is terminal because one arm is absent. It is rare.

(Robert Sovian stage: from metacentric to telocentric)

Composition: Chromosome contains DNA, proteins and small amount of RNA.

(a) DNA: It constitutes 40%.The amount of DNA is same in all cells of an organism. The amount is constant in different tissues. The amount is constant within one species but varies from one species to another. The amount is related to the number of chromosome. The amount is proportionate in haploid, diploid and tetraploid.

(b) Proteins (60%): These may be

(a) Histones: These constitute 35%. These are basic and structural proteins. These are highly heterogeneous and synthesised during S-phase. The amount of DNA and histones is constant. These lack tryptophan (amino acid). These protect DNA from the action of nuclease. These are modified by acetylation, methylation and phosphorylation. These are of different types based on difference in sequence of amino acid.

H_1: It is most divergent protein. It is lysine rich. It is the largest in size. It is more easily extracted from chromatin. It is almost species specific in its sequence.

H_2a and H_2b: These are less conserved than H_3 and H_4. These are slightly lysine rich (Lysine is moderately more than arginine).

H_3 and H_4: These are highly conserved. These are arginine rich.

Functions:

1. Histones stabilise the structure of chromosome.
2. These protect DNA from the action of nuclease.
3. These neutralise the anionic charges of phosphate in DNA.
4. These repress the activity of gene.

(2) Non-histones: These constitute 25%. These are acidic proteins. These are synthesised during entire life cycle. These are 15-20 in number. These differ greatly in tissues and species. These are complex and not as observed as histones.

Functions:

1. These stimulate gene activity.
2. These induce the formation of DNA in eukaryotes.

Structure: It is cylindrical double structure which consists of two longitudinal halves called sister chromatids. A chromatid is called half chromosome or daughter chromosome. Each chromatid has two arms-short (p) and long (q). These chromatids are held together at a single point called centromere (also known as primary constriction). The chromosome is narrower in this region. Centromere is functional during cell division while rest part of chromosome is generally inactive. Usually there is single centromere (monocentric) in a chromosome but several centromeres (polycentric) chromosome) are found in diffused condition (holocentric) along the length of a chromosome in *Ascaris*, insects,

Spirogyra. The position of centromere is fixed for a particular chromosome but its position varies in different chromosomes. The position of centromere determines the relative length of arm.

$$\text{Arm ratio} = \frac{\text{Length of longer arm}}{\text{length of short arm}}$$

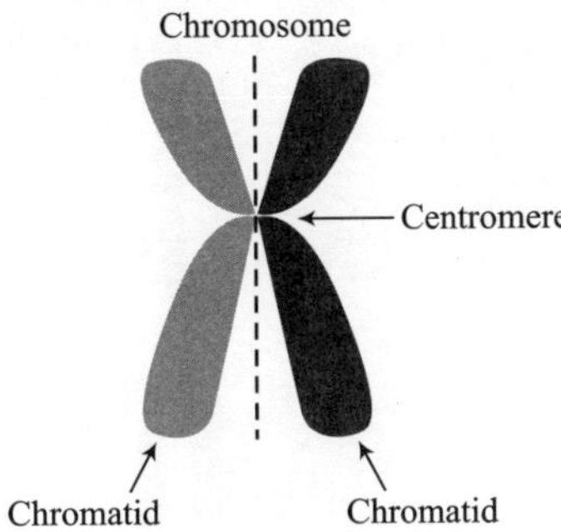

Fig. ***Chromosome***

The arm ratio is high in acrocentric chromosome. Centromere contains highly repetitive DNA called satellite DNA.

The protein disc attached to centromere is called kinetochore. It is C-shaped and two in number. It is actual site of attachment of spindle fibres to the centromere.

Secondary constriction is also observed in some chromosomes at constant position in addition to primary constriction. Secondary constriction lacks centromere and is narrower and metabolically more active than primary constriction. It occurs in distal end and fixed in position hence used as marker**.** Secondary constriction cuts off a small fragment of chromosome at distal end called trabant or satellite (terminal part of a chromosome beyond secondary constriction). It remains attached to the rest of chromosome by a thread of chromatin. Chromosome containing satellite is called satellite chromosome. Chromosomal satellite is different from satellite DNA. The number of SAT (Sine acid Thyminucleinico) chromosome varies in genome. Human somatic cell has 10 SAT chromosome (13th pair, 14th pair, 15th pair, 21st pair and 22nd pair). Secondary constriction is site of formation of nucleolus hence called nucleolar organising region (NOR). The NOR of each SAT chromosome contains several hundred copies of the genes coding for rRNA. The SAT-DNA constitutes about 10% of total DNA present in genome of a species. The NOR is essential for the formation of nucleolus.

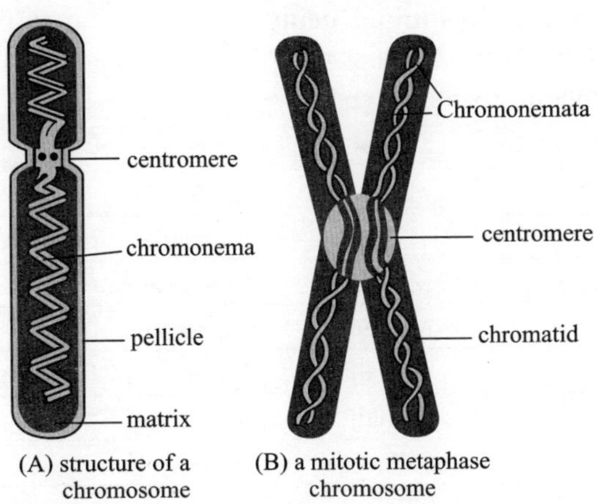

(A) structure of a chromosome (B) a mitotic metaphase chromosome

Each chromatid is composed of an elongated, coiled filament called chromonemata which bears several beaded or gene bearing structures called chromomeres. It lies embedded in matrix which is further enclosed in a proteinaceous sheath called pellicle. It is present throughout the length of chromosome. Each chromonemata consists of a single thread of DNA associated with histones. The number of chromonemata is two in prophase and metaphase but one in anaphase.

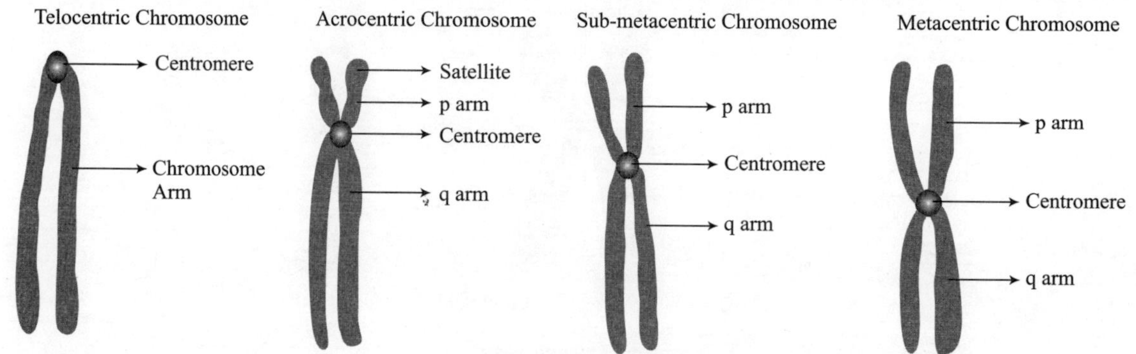

Fig. ***Different shapes of chromosome***

Telomere is the end of a chromosome. It is highly stable. It does not unite with telomere of other chromosome hence it is not sticky and it ensures survival and polarity. It contains short tandem repeat DNA sequence. It also contains terminal single strand which is folded to form a hair pin loop giving it stability. Telomere facilitates the attachment of chromosome with nuclear membrane.

Types of giant chromosomes:

1. **Polytene chromosome:** These were discovered by Balbiani (1881). These are found in the cells of salivary gland of larva of chironomus insect (dipteran). These are also found in gut, trachea, fat body cells and malpighian tubules. These are wide and 1000 times longer than somatic chromosome. These contain a number of light bands and dark bands. The bands are enlarged to form puff or Balbiani rings involved in synthesis of RNA. It contains several chromatids attached with one centromere.
2. **Lampbrush chromosomes:** These were discovered by Ruckert (1892). These are found in diplotene stage of prophase 1 of meiosis in nuclei of amphibian oocyte. These contain loops which are involved in synthesis of RNA and proteins.

Classification of chromosomes in human beings: Tijo and Laven proposed the diploid number of chromosomes in human being (2n=23 pairs or 46). Denver classified all 23 pairs of homologous chromosomes of human beings in seven groups:

Group	Type	Size	Shape
A	1,2,3	Large	Metacentric
B	4,5	Large	Submetacentric
C	6,7,8,8,10,11,12, X	Medium	Submetacentric
D	13,14,15	Medium	Acrocentric
E	16,17,18	Small	Metacentric(16),Submetacentric (17,18)
F	19,20	Small	Metacentric
G	21,22,Y	Smallest (Y)	Acrocentric

Karyotype: It is the complete diploid set of chromosomes in the cells of an organism. It is the arrangement of chromosomes in descending order of size keeping their centromere in a straight line thus the largest chromosome is placed on extreme left and smallest one on extreme right.

It is done in WBC, skin cells and cells of oral mucosa. It is most often studied when the cell is at metaphase of mitosis and all the chromosomes are present as dyads.

Technique: It was prepared by Casperson (1968). It involves the following methods:

1. Banding patterns:
   (a) Q-banding: It uses quinacrine dye. It is A-T rich region of chromosome. It is used for animals.
   (b) C-banding: It uses Giemsa stain. It is part of constitutive heterochromatin. It is used for plants only.
   (c) R-banding: It uses reverse Giemsa stain. It is sulphur deficient region of chromosome. It is used for animals.
   (d) G-banding: It uses Giemsa stain. It is sulphur rich but low G-C content of chromosome. It is used for animals.
2. Fluorescence *In situ* Hybridisation (FISH): It is used for plants and animals using DNA probe.
3. Flow cytometry: It is the latest technique.

Types of karyotype:

Symmetric karyotype	Asymmetric karyotype
It is small difference between the smallest and largest chromosome.	It is large difference between the smallest and largest chromosome.
It contains more metacentric chromosomes.	It contains a few metacentric chromosomes.
It is primitive feature.	It is advanced feature.

Idiogram: It is the arrangement of homologous pairs of identified chromosomes in a series of decreasing size (length). It is diagrammatic representation of karyotype of a species showing morphological features of chromosomes. It is not possible in symmetric karyotype.

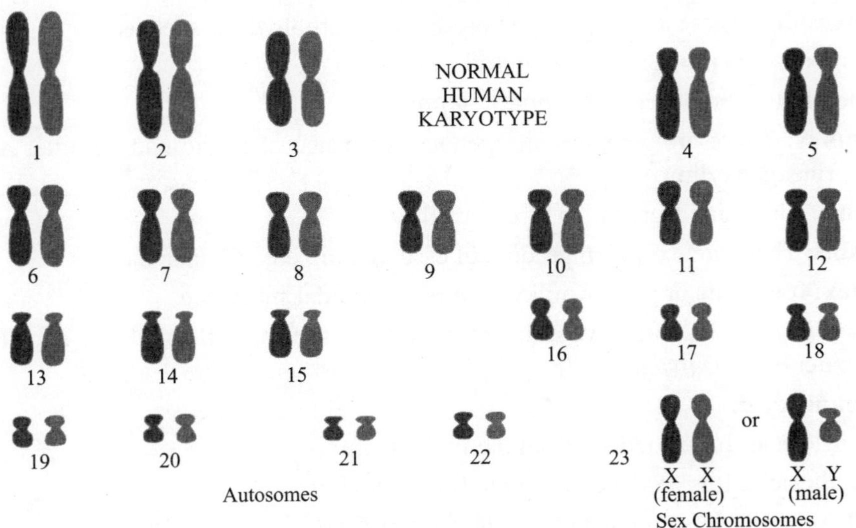

Fig. ***Human karyotype***

Significance:

(a) It provides information of various features of chromosome regarding number, size, shape of chromosome, position of centromere, NOR and secondary constriction and size of satellite bodies.

(b) It detects structural chromosomal aberration.

(c) It indicates the evolutionary relationship among species.

Ergastic substances:

These are non-protoplasmic, non-living and metabolic byproducts present in the cytoplasm, vacuole or cell wall of eukaryotic cell. These are metabolically inactive substances. There are several types of ergastic substances:

1. **Storage substances:** These are carbohydrate, protein and lipids.
   (a) Fructose: It is present in fruit of grapes.
   (b) Sucrose: It is present in root of beet and stem of sugarcane.
   (c) Inulin: It is present in cell sap of *Dahlia*.
   (d) Starch: It is eccentric (deposited only on one side) in potato and concentric (deposited centrally from all side) in maize
2. **Secretory substances**: These are metabolic by-products of plants and not concerned with nutrition.
   (a) Enzymes: Cellulase, Amylase, Protease.
   (b) Pigments: Chlorophyll, Carotenoids, Anthocyanin
   (c) Nectar: This is sweet substance secreted by flower. It is mixture of glucose, fructose and sucrose.
3. **Excretory substance:** These metabolic byproducts are of no use in plants. The excretory organs are absent in plants hence excretory substances are found in cell.
   (a) Alkaloids: These are waste products of metabolism. These are basic, insoluble in water, sour in taste and mostly poisonous. e.g., Atropine, Morphine, Nicotine, Quinine, Reserpine, Thein etc.
   (b) Glucosides: These are degradation products of carbohydrate such as
   Digitoxin: It is produced from *Digitalis purpurea.*
   Saponin: It is produced from fruit of *Sapindus.*
   (c) Tannins: These are complex nitrogenous substance of acidic nature, bitter and astringent. e.g., Haematoxylin
   (d) Gums: These are degradation of cell wall.
   (e) Resins: These are oxidation product of essential oil. e.g., Canada balsam
   (f) Latex: It is white or yellow milky coloured colloidal substance
   (g) Essential oil: These are volatile oil present in special cell or gland. e.g., Rosemary oil, lavender oil, *Eucalyptus* oil
   (h) Organic acid:
   - Citric acid: It is produced from the fruit of lemon
   - Malic acid: It is produced from the fruit of apple
   - Oxalic acid: It is produced from the fruit of *Oxalis*.
   - Tartaric acid: It is produced from the fruit of Tamarind

(i) Minerals: These are crystals in the form of

- Cystolith: It is made up of $CaCO_3$. e.g., *Ficus*
- Raphides (needle like): It is made up of calcium oxalate. e.g., *Eichhornia, Pistia.*
- Sphaeraphides (Druses): It is made up of calcium oxalate. e.g., leaf of *Nerium*
- Silica crystal: It is found in petiole of *Atropa.*
- Silica: It is found in cell wall of grasses and *Eucalyptus*.

SHORT ANSWER TYPE QUESTIONS

1. Write the general unifying principles of cell theory
2. Explain the differentiated cell, dedifferentiated cell and redifferentiated cell.
3. Distinguish between prokaryotic cell and eukaryotic cell.
4. Distinguish between plant cell and animal cell.
5. Write notes on vacuole.
6. Draw a well labelled diagram of fluid mosaic model of plasma membrane.
7. Why mitochondria are considered as semi-autonomous organelle?
8. Discuss the functions of Golgi body.
9. Distinguish between 70S and 80 S ribosomes
10. Discuss the functions of lysosome.
11. Draw a well labelled diagram of centriole.
12. Distinguish between cilia and flagella.
13. Distinguish between heterochromatin and euchromatin.
14. Write notes on nucleolus.
15. Explain the different shapes of chromosome.

LONG ANSWER TYPE QUESTIONS

1. Describe the structure of a typical bacterial cell with well labelled diagram.
2. Describe the structure and functions of chloroplast.
3. Describe the structure and functions of Golgi body.
4. Describe the structure and functions of nucleus.
5. Draw a well labelled diagram of chromosome and describe its different parts.

MULTIPLE CHOICE QUESTIONS

1. Which of the following parts of cell is differentially permeable?
   (a) Cell wall (b) Middle lamella (c) Cell membrane (d) Cuticle
2. The first layer formed in cell wall as a result of cytokinesis is
   (a) Middle lamella (b) Primary wall (c) Secondary wall (d) Tertiary wall
3. The permeability of plasma membrane is provided by
   (a) Extrinsic proteins (b) Intrinsic proteins (c) Lipids (d) Sugar
4. The absorption of dead cells is brought about by
   (a) ER (b) Lysosomes (c) Vacuole (d) Golgi apparatus
5. Thickness of plasma membrane is
   (a) 20 nm (b) 7.5 nm (c) 15 nm (d) 10 nm
6. The examples of semiautonomous organelles are
   (a) Mitochondria and nucleus (b) Chloroplst and Endoplasmic reticulum
   (c) Mitochondria and chloroplast (d) Endoplasmic reticulum and Golgi apparatus
7. Hammerling performed experiments to prove that nucleus was physical basis of heredity while working on
   (a) Neurospora (b) Drosophila (c) Acetabularia (d) Aspergillus
8. A cluster of ribosomes held together on mRNA is called
   (a) Mesosome (b) Dictyosome (c) Polysome (d) Microsome
9. Hydrophobic ends of phospholipid molecules in plasma membrane are
   (a) Polar (b) Neutral (c) Non-polar (d) Bipolar
10. Lysosome along with food content is known as
    (a) Primary lysosome (b) Secondary lysosome (c) Residual body (d) Microbody
11. 'Quantasomes' were discovered by:
    (a) Schimper (b) Strasburger (c) Sachs (d) Park and Biggins
12. Lysosomes occur mostly in:
    (a) Algae (b) Bacteria (c) Plants (d) Animals
13. The resin duct of *Pinus* is a:
    (a) Schizogenous cavity (b) Lysigenous cavity
    (c) Schizo-lysigenous cavity (d) None of the above
14. The correct sequence of cell membrane as per model of singer and Nicolson is:
    (a) L – P – P – L (b) P – P – L – P (c) P – L – L – P (d) L – P – L – P
15. Which of the following chemical constituents is absent in ribosomes?
    (a) Protein (b) Lipid (c) rRNA (d) None of these
16. Rough ER is well developed in cells involved in
    (a) Lipid synthesis (b) Protein synthesis
    (c) Secretory functions (d) Nucleotide synthesis
17. Succinic dehydrogenase enzyme of Krebs cycle is present at
    (a) Outer membrane of mitochondria (b) Inner membrane of mitochondria
    (c) Perichondrial space of mitochondria (d) Matrix of mitochondria

18. Continuity of cytoplasm form cell to cell is maintained through
    (a) Middle lamella
    (b) Plasmodesmata
    (c) Endoplasmic reticulum
    (d) Cell membrane system
19. All are membrane bounded cell organelles except
    (a) Sphaerosomes (b) Mitochondria (c) Ribosomes (d) Lysosomes
20. Golgi apparatus is absent in
    (a) Liver cells (b) Mammalian RBCs (c) Green algae (d) Higher plants
21. Which of the following cell organelle does not belong to microbodies?
    (a) Lysosome (b) Ribosome (c) Peroxisome (d) Glyoxysome
22. The largest membrane bound cell organelle is
    (a) Lysosome (b) Ribosome (c) Peroxisome (d) Glyoxysome
23. Middle lamella of cell wall is composed of
    (a) Lignin
    (b) Suberin
    (c) Calcium pectate
    (d) Calcium phosphate
24. Cellulose and hemicelluloses which are the constituents of cell wall are synthesized by
    (a) Lysosome (b) Micro bodies (c) Smooth ER (d) Golgi apparatus
25. Which of the following cell organelles is common to both Protista and Monera?
    (a) Lysosome (b) Ribosome (c) Chloroplast (d) Vacuole
26. Leucoplast which stores oil is called
    (a) Amyloplast (b) Aleuroneplast (c) Elaioplast (d) Proplast
27. Electron microscope has revealed the presence of
    (a) Leucoplast (b) Chloroplast (c) Chromosome (d) Ribosome
28. Endoplasmic reticulum is attached (continuous) with
    (a) Ribosome
    (b) Golgi body
    (c) Mitochondria
    (d) Nuclear membrane
29. Centrosome is responsible for
    (a) Cell wall formation
    (b) Cell plate formation
    (c) Inhibition of cell division
    (d) Initiation of cell division
30. Cell to cell recognition in cell membrane is by
    (a) Carbohydrate part (b) Lipid part (c) Protein part (d) Nucleotides
31. Isochromosomes (chromosome with identical arms) are formed by
    (a) Pairing of identical chromosomes
    (b) Longitudinal splitting of centromere
    (c) Transverse splitting of centromere
    (d) Disjunction of bivalents
32. Maximum polymorphism is exhibited by
    (a) Lysosomes
    (b) Ribosomes
    (c) Golgi bodies
    (d) Endoplasmic reticulum
33. Peroxisome does not have
    (a) Urate oxidase
    (b) Glycolate oxidase
    (c) Catalase
    (d) D-amino acid oxidase

34. Pinocytosis and phagocytosis are concerned with
 (a) Cell wall (b) Plasma membrane (c) Nucleus (d) Mitochondria
35. The shape of chromosome having a terminal centromere is called
 (a) Metacentric (b) Acrocentric (c) Telocentric (d) Acentric
36. The organelle common in both eukaryotic and prokaryotic cells is
 (a) Microtubules (b) Microbodies (c) Microfilaments (d) Microsome
37. Nucleoli are rich in
 (a) RNA (b) DNA (c) Protein and RNA (d) Carbohydrates
38. The photo respiratory carbon-oxidation cycle involves the interaction of
 (a) Chloroplast (b) Peroxisome (c) Mitochondria (d) All of these
39. Axoneme having 9 + 0 doublet microtubule arrangement is found in
 (a) Cilia (b) Flagella (c) Cilia and flagella (d) Centriole
40. The number of protofilament found in microtubule is
 (a) 5 (b) 9 (c) 10 (d) 13
41. Which of the following structures is an organelle within an organelle?
 (a) Ribosome (b) Peroxisome
 (c) Endoplasmic Reticulum (d) Mesosome
42. Which of the following features is not correct with respect to bacteria?
 (a) Flagella lacks 9+2 arrangement (b) Pilus helps in conjugation
 (c) Fimbriae are involved in attachment (d) Flagella are made up of tubulin protein
43. Which of the following features is appropriate for immature cells?
 (a) These are less metabolically active (b) These are more efficient
 (c) These have low nuclear-cytoplasmic ratio (d) All of the above
44. The number of kinetochore in centromere of chromosome is
 (a) One (b) Two (c) Three (d) Four
45. Which of the following is the densest and membrane less?
 (a) Ribosome (b) Peroxisome (c) Nucleolus (d) Chloroplast
46. The number of mitochondria increases in the cell of
 (a) Dry seeds (b) Dormant seeds (c) Germinating seeds (d) Dead seeds
47. Intracellular compartments are not found in
 (a) Lower plants (b) Eukaryotes (c) Prokaryotes (d) Higher plants
48. The trigger of cell division is
 (a) Peroxisome (b) Ribosome (c) Lysosome (d) Vacuole
49. The pure fractions of cellular components can be isolated by
 (a) Chromatography (b) X-ray diffraction
 (c) Differential centrifugation (d) All of the above
50. Which of the following layers is present nearest to plasma membrane in plant cell?
 (a) Middle lamella (b) Primary wall (c) Secondary wall (d) Tertiary wall

51. Lomasome found in fungi is present in
    (a) Cytoplasm (b) Mitochondria
    (c) Vacuole (d) Between cell wall and plasma membrane
52. Which of the following enzymes catalyses ATP in flagellar membrane?
    (a) Tubulin (b) Nexin (c) Flagellin (d) Dynein
53. In germinating seeds, fatty acids are degraded exclusively in
    (a) Peroxisome (b) Glyoxysome (c) Mitochondria (d) Chloroplast
54. Ribosomal RNA is actively synthesised in
    (a) Ribosome (b) Lysosome (c) Peroxisome (d) Nucleolus
55. A multicellular organism possesses
    (a) Differentiated cells (b) Undifferentiated cells
    (c) Dedifferentiated cells (d) All of the above
56. Which of the following is incorrect regarding bacterial cell?
    (a) Cell wall is made up of NAM and NAG (b) Cutoskeletons are absent
    (c) Ribosomes are of 70 S (d) Nucleoid contains histone proteins.
57. The sugar component of plasma memebrane belongs to
    (a) Monosaccharide (b) Oligosaccharide (c) Polysaccharide (d) All of the above
58. The tail of lipid of plasma membrane is
    (a) Polar and hydrophilic (b) Polar and hydrophobic
    (c) Nonpolar and hydrophilic (d) Nonpolar and hydrophobic
59. Which of the following cell organelles does not belong to endomembrane system?
    (a) Endoplasmic reticulum (b) Golgi complex
    (c) Ribosome (d) Lysosome
60. Triplet fibrils of a centriole are tilted at an angle of
    (a) 40° (b) 45° (c) 60° (d) 65^0
61. Which of the following acts as nucleating centre for the growth of microtubule during formation of new centriole?
    (a) Spindle (b) Massule (c) Aster (d) Chromocentre
62. Plane of cell wall formation in dividing cell is determined by
    (a) Microtubules (b) Microfilaments (c) Microsome (d) All of the above
63. If cells are broken up and sedimented by centrifugation, the new structure formed in one of the cell was
    (a) Centrosome (b) Microsome (c) Ribosome (d) Lomasome
64. Cell theory is not obeyed by
    (a) *E. coli* (b) Bacteriophage (c) *Salmonella* (d) *Nostoc*
65. Lipo-polysaccharide (LPS) is found in
    (a) All Gram positive bacteria (b) All Gram negative bacteria
    (c) Both Gram positive and negative bacteria (d) Viruses
66. Plasma membrane helps in bulk transport in case of
    (a) Active transport (b) Facilitated diffusion (c) Pinocytosis (d) All of the above

67. Mitochondria are stained with
    (a) Acetocarmine (b) Orchill (c) Safranin (d) Janus Green
68. Smooth endoplasmic reticulum cannot perform
    (a) Glycogen synthesis (b) Fatty acid synthesis
    (c) Detoxification of drugs (d) Synthesis of testosterone
69. The processes of glycosylation and glycosidation is performed by
    (a) Mitochondria (b) Endoplasmic reticulum
    (c) Golgi complex (d) Ribosome
70. Multinucleate condition is seen in
    (a) Epidermis of *Ascaris* (b) Mycelia of *Rhizopus*
    (a) Mycelia of *Mucor* (b) All of the above
71. Flagella of prokaryotic and eukaryotic cells differ in
    (a) Type of movement and placement in cell
    (b) Location in cell and mode of functioning
    (c) Microtubular organisation and type of movement
    (d) Microtubular organisation and function
72. Which of the following does not occur during maturation of a proplastid into a chloroplast?
    (a) Increase in size (b) Increase in the number of grana
    (c) Bleaching of chlorophyll (d) Synthesis of pigments
73. Telomeres of eukaryotic chromosomes consist of short sequence of
    (a) Thymine rich repeats (b) Cytosine rich repeats
    (c) Adenine rich repeats (d) Guanine rich repeats
74. Which part of translational modification of proteins does not occur in lumen of ER?
    (a) Glycosylation (b) Formation of disulphide bonds
    (c) Conformational folding (d) Ubiquitination
75. Receptors for steroid hormones are found in
    (a) Cell membrane (b) Cytoplasm (c) Mitochondria (d) Ribosome
76. The maximum size of a cell is limited by
    (a) The need for enough surface area for exchange with environment
    (b) The number of organelles that can be packed inside
    (c) The materials needed to build it
    (d) The amount of flexibility it needs to be able to move
77. Which of the following functional processes results from the presence of protein within plasma membrane?
    (a) Enzymatic activity (b) Signal transduction (c) Intercellular joining (d) All of the above
78. A karyotype would be least likely to show
    (a) An extra chromosome (b) Part of a chromosome duplicated
    (c) A missing chromosome (d) Part of a chromosome turned around

79. The mitochondrial membrane contains a transporter for
    (a) NADH (b) Acetyl CoA (c) GTP (d) ATP
80. The portion of Golgi which contains irregular cisternae and tubules is known as
    (a) Cis-Golgi (b) Trans-Golgi
    (c) Median Golgi (d) Intercisternal Golgi
81. Which of the following does not have in common between chloroplast and mitochondria?
    (a) ATP synthase (b) ETC (c) ATP (d) NADH
82. Euchromatin contains
    (a) Single copy and genetically active DNA
    (b) Repetitive and genetically inactive sequences
    (c) Single copy and genetically inactive DNA
    (d) Repetitive and genetically active sequences
83. Which part of ribosome gets attached to ER?
    (a) Peptidyl site (b) Aminoacyl site (c) Large subunit (d) Small subunit
84. Lysosome contains hydrolytic enzymes which can digest any foreign materials except
    (a) Cellulose (b) Starch (c) Glycogen (d) Lipids
85. Polyribosomes are found in
    (a) Bacteria (b) Fungi (c) Angiosperms (d) Mammals
86. Lysosomes are abundant in
    (a) WBC and osteoblast (b) RBC and spleen (c) Liver and spleen (d) WBC and spleen
87. The middle lamella of plant cells is made up of calcium and magnesium pectate. Thus pectic acid is polymer of
    (a) α-1,4-D glucose (b) β-1,6-D glucose
    (c) α-1,4-D galacturonic acid (d) β-1,4-D galacturonic acid
88. In a cell if acidity of lysosome is lost, then there is loss of
    (a) Phagocytosis of invading bacteria (b) Elevated phosphatise level
    (c) Glycogen degradation (d) No major effect
89. Lipid nature of plasma membrane can be destroyed by
    (a) Hexane (b) Benzene (c) Chloroform (d) NaOH
90. In plasma membrane, the functional and structural roles are played by
    (a) Proteins (b) Lipids (c) Cholesterol (d) Oligosaccharides
91. The most abundant intracellular cation is
    (a) K^+ (b) H^+ (c) Na^+ (d) Ca^+
92. Which of the following is called system of membrane?
    (a) ER (b) Golgi (c) Lysosome (d) Chloroplast
93. Which of the following is not considered as true cell organelle?
    (a) Lysosome (b) Peroxisome
    (c) Endoplasmic reticulum (d) Ribosome

94. Which of the following organelles reduces the number of other cell organelles?
    (a) Ribosome (b) Lysosome
    (c) Endoplasmic reticulum (d) Mesosome
95. A host cell has intracellular bacterial symbionts and the growth rate of bacterial symbiont is always 10% higher than that of the host cell. After 10 generations of the host cell, the density of bacteria in the host cell will increase by
    (a) 10% (b) Two fold (c) Ten fold (d) Hundred fold
96. Mitochondria performs all of the following functions except
    (a) Nucleic acid synthesis (b) Steroid synthesis
    (c) ATP synthesis (d) Polysaccharide degradation
97. In a chloroplast, the highest number of protons occurs in
    (a) Stroma (b) Lumen of thylakoid
    (c) Intermembrane space (d) Antenna complex
98. If a meristematic cell is placed in a medium containing active thymidine, then the radioactivity will be first observed in
    (a) Euchromatin (b) Heterochromatin
    (c) Both simultaneously (d) None of the above
99. In plants, the maximum number of chromosome is found in
    (a) *Haplopappus* (b) *Acetabularia* (c) *Lilium* (d) *Ophioglossum*
100. Centromere is placed at centre in
    (a) Acrocentric chromosome (b) Metacentric chromosome
    (c) Submetacentric chromosome (d) Telocentric chromosome

ANSWERS

1.c	2.a	3.c	4.b	5.b	6.c	7.c	8.c	9.c	10.b
11.d	12.d	13.a	14.c	15.b	16.b	17.b	18.b	19.c	20.b
21.b	22.c	23.c	24.d	25.b	26.c	27.d	28.d	29.d	30.a
31.c	32.a	33.b	34.b	35.c	36.d	37.c	38.b	39.d	40.d
41.a	42.d	43.b	44.b	45.c	46.c	47.b	48.c	49.c	50.c
51.d	52.d	53.b	54.d	55.d	56.d	57.b	58.d	59.c	60.a
61.b	62.a	63.b	64.b	65.b	66.c	67.d	68.b	69.c	70.d
71.a	72.c	73.d	74.d	75.b	76.a	77.d	78.d	79.d	80.c
81.d	82.a	83.c	84.a	85.a	86.d	87.a	88.c	89.b	90.a
91.a	92.a	93.d	94.b	95.b	96.d	97.b	98.a	99.d	100.b

7

Cell Cycle and Cell division

Cell cycle is the series of events that take place between the formation of cell and its division into daughter cells. It is also called cell division cycle. It differs in different organisms and different cell types. It is characteristic of plants and animals. It is 20 minutes in bacteria, 2-3 hours in yeast, 20 hours in onion root tip and 24 hours in human beings.

G_0 phase (Quiescent phase) is a resting phase where the cell has left the cell cycle and has stopped dividing. The cell cycle starts with this phase in response to some stimulus. Non-proliferating or non-dividing cells in multicellular eukaryotes generally enter the quiescent (G_0) state from G_1 and may remain quiescent for long periods of time (as in neuron). This is very common for cells that are fully differentiated but cells are metabolically active. Many cells do not enter G_0 and continue to divide regularly and occasionally throughout an organism's life.

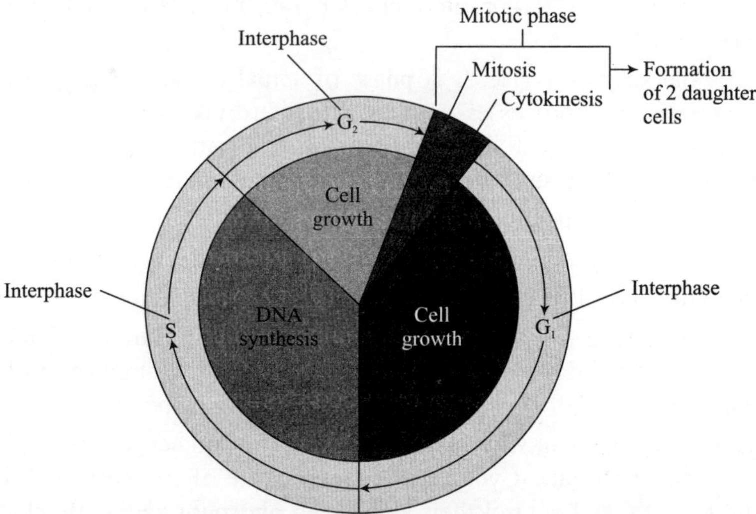

Howard and Pelc (1958) discovered cell cycle in broad bean (*Vicia faba*) and divided into two stages.

1. **Interphase**: It is long and non-dividing phase but it is most active phase. It is period between cytokinesis and beginning of next prophase or it is period between two successive mitotic divisions. Chromosomes lie in the form of chromatin (long and uncoiled). In this phase repair of damaged DNA takes place. It involves following three phases:

   (a) G_I phase (Gap I): It is the longest phase (42%).

   The size of cell and nucleus increases. Cell shows high metabolic rate. Synthesis structural and functional proteins (enzymes) and RNA occurs. Multiplication of cell organelles occurs. It is shorter in unicellular organisms (fast dividing cells) and longer in multicellular organisms (slow dividing cells). The difference in time of cell cycle or rate of cell division is due to variation in length of G_1 phase.

G_1 specific cyclins are synthesised and these activate cyclin dependent kinase (CDK) specific for G_I phase.

First restriction (**R_I**) or major check point: It occurs at late G_1 phase (G_1 to S phase transition). It requires G_1 cyclin and cyclin dependent kinase-2 (CDK-2).

(b) Synthesis (S-phase): It is the second longest phase (33%).

Replication of DNA (doubling the amount) occurs. Synthesis of histone proteins occurs. Growth of nucleus occurs. Duplication of centrioles occurs at late S-phase but completes in G_2 phase.

S-specific cyclins are synthesised and G_1-specific cyclins are destroyed at the end of G_1 phase.

(c) G_2 phase (Gap 2): It is second growth phase. It is the shortest phase in interphase.

The size of cell and nucleus increases. Synthesis of RNA and proteins resumes. Repair of damaged DNA occurs. Synthesis of tubulin proteins occurs that are essential for formation of spindle fibres. Condensation of chromatin into chromosomes occurs. ATP is stored in the cell. Removal of inhibitors occurs. It prepares cell to divide.

G_2 specific cyclins are synthesised and S-specific cyclins are destroyed at the end of S phase. It requires CDK hence called maturation promoting factor which acts as trigger.

Second restriction point (**R_2**) or minor check point: It occurs at late G_2 phase (G_2 to M phase transition).

2. **M phase:** It is the shortest phase. It is phase of actual division. The process of cell division consists of two sequential processes- (i) karyokinesis (division of nucleus) and (ii) cytokinesis (division of cytoplasm). The cell division occurs due to disturbance in nucleo-cytoplasmic ratio or karyoplasmic index. It occurs when nucleo-cytoplasmic index is decreased.

   M-specific cyclins are synthesised.

   After M phase, a cell may either enter fresh cycle or pass into G_0 (quiescent) stage.

Regulation of cell division: It is regulated in the following manner.

Cell division cycle (cdc) gene →Cyclin protein →Cyclin dependent kinase (CDK) →Phosphorylation of histone (H_1) proteins → break down of nuclear membrane→condensation and packaging of chromatin into Chromosome in nucleus→Cell division →Inactivation and disappearance of CDK at late prophase.

Cyclin is a regulatory component whereas CDK is catalytic and acts as protein kinase. Kinase adds phosphate group to target substrate. Cyclin undergoes a cycle of synthesis and degradation in each cycle. Cyclins bind to CDK and control their abilities to phosphorylate particular target protein that triggers for cell division.

Cell cycle depends on nutrients (growth factors), inhibitors and overcrowding.

Types of Cell division

Amitosis:

It was discovered by Robert Remak (1855) in RBC of chick embryo. It is direct nuclear division in which there is no formation of spindle fibres and appearance of chromosome. It rarely occurs. In this, nuclear membrane does not degenerate. Nucleus elongates and constricts in the middle to form two daughter nuclei. Later centripetal constriction of cytoplasm occurs by which two daughter cells are formed. It occurs in prokaryotes, protozoans, yeast, cartilage and foetal membranes of mammals, degenerated, damaged or old tissues.

Mitosis

It is the division of cell in which two genetically identical daughter cells are formed but the number of chromosome and amount of genetic material remain same as that of parent cell. It is direct nuclear division.

It was discovered by Strasburger (1875) in plants but in animals it was discovered by Flemming. The term 'mitosis' was coined by Flemming (1882).

It is common method of cell division. It occurs in all types of cells (haploid, diploid and polyploid) in plants. It occurs in somatic cells. In plants, cell division is confined to root, stem, young leaves, floral buds and cambium but in animals, it is confined to blood cells, outer skin cells and gut epithelium.

Mitosis is best observed in root tip of onion in case of plants because continuous growth of plant occurs throughout their life. The cells of bone marrow, base of nails and skin are also used to study mitosis in case of animals. The total time taken for completion of mitosis is usually from 30 minutes to 3 hours. The duration depends on temperature. As the temperature increases, the duration of mitosis will decrease.

Mitotic apparatus = Aster + Spindle + Centriole = Achromatic figure

In lower plants and animals, mitosis is astral (amphiastral) due to presence of centrioles but in higher plants, it is anastral because asters (rays) are not formed due to absence of centrioles.

Types of mitosis:

1. **Endomitosis**: It is the duplication of chromosomes and DNA without division of nucleus. It leads to the doubling of chromosomes (polyploidy). It is also called endoduplication or endopolyploidy. It leads to polyteny when separation of chromatids does not occur hence there are several chromatids in a chromosome.
2. **Dinomitosis:** It is intranuclear (closed) mitosis in which nuclear membrane persists (does not break down) but spindle fibres are not formed. It occurs in dinoflagellates.
3. **Karyochorisis**: It is intranuclear (closed) mitosis in which nuclear membrane persists and spindle fibres are formed inside nucleus. It occurs in fungi.
4. **Eumitosis:** It is extranuclear mitosis in which nuclear membrane is lost and spindle fibres are formed outside nuclear membrane. It commonly occurs in plants and animals.

Mitosis occurs in four stages which are as follows:

1. **Prophase:**
   (a) It is first, longest and most active phase.
   (b) Cell becomes viscous and refractive.
   (c) Nuclear membrane and nucleolus are present but later start disappearing.
   (d) Chromatin starts condensing into chromosome.
   (e) Chromosome becomes coiled, shortened, thickened and distinct (spireme stage) due to plectonemic coiling (spiralisation or coiling in same direction).
   (f) Chromosome splits longitudinally into two sister chromatids attached with centromere (appearing as double stranded).

2. **Metaphase:**
   (a) Nuclear membrane and nucleolus are absent.
   (b) Chromosome undergoes maximum contraction or thickening or condensation (dehydration).
   (c) Chromosome is clearly visible because plectonemic coiling between sister chromatids is absent.
   (d) Chromosome is arranged at central equatorial plate in one plane (congression) while chromatids extend freely in any direction.
   (e) The spindle fibres get attached to chromosome by kinetochore.
   (f) The structure, size and number of chromosome are observed.
3. **Anaphase:**
   (a) It is the shortest phase.
   (b) Centromere of chromosome divides into two sister chromatids with the division of kinetochore. Each chromatid has its own centromere and is called daughter chromosome.
   (c) Chromatids separate, start repelling each other and reach opposite poles of spindle due to contraction of chromosomal fibres and relaxation of interzonal fibres but connected to each other by interzonal fibres.
   (d) Spindle fibres (made up of tubulin (95%), RNA (3.5%), traces of actin and myosin) are of four types:
      (i) Continuous fibres: from pole to pole
      (ii) Discontinuous fibres: fibres do not reach other pole.
      (iii) Chromosomal fibre: from pole to centromere
      (iii) Interzonal fibre: from one centromere to another.
   (e) Shape of chromosome is observed.
   (f) Spindles of microtubules are polar with their polymerising '+' ends (fast growing) facing centre and non-polymerising '-' ends (slow growing) facing poles.
4. **Telophase:**
   (a) Nuclear membrane and nucleolus reappear.
   (b) Daughter chromosomes uncoil, lengthen and decondense (hydrate) hence finally indistinguishable.
   (c) Spindle fibres disappear.
   (d) Each pole has a nucleus.

Cytokinesis: The division of cytoplasm occurs by **cell plate** in plants. The central part of spindle and intrazonal microtubule grow in size and trap Golgian vacuole of Golgi body to form phragmoplast (barrel shaped). It is vesicle budded out from Golgi complex rich in pectin and line up at centre to form middle lamella which grows centrifugally towards periphery (lateral wall) and fuses with plasma membrane. The content of Golgian vacuole rich in pectin forms middle lamella of cell wall and membrane of cell plate develops plasma membrane of new daughter cell.

In case of animal cells, **cell or cleavage furrow** develops due to contraction of inner surface of plasma membrane and microfilament. This deepens and grows centripetally and finally divides the cytoplasm.

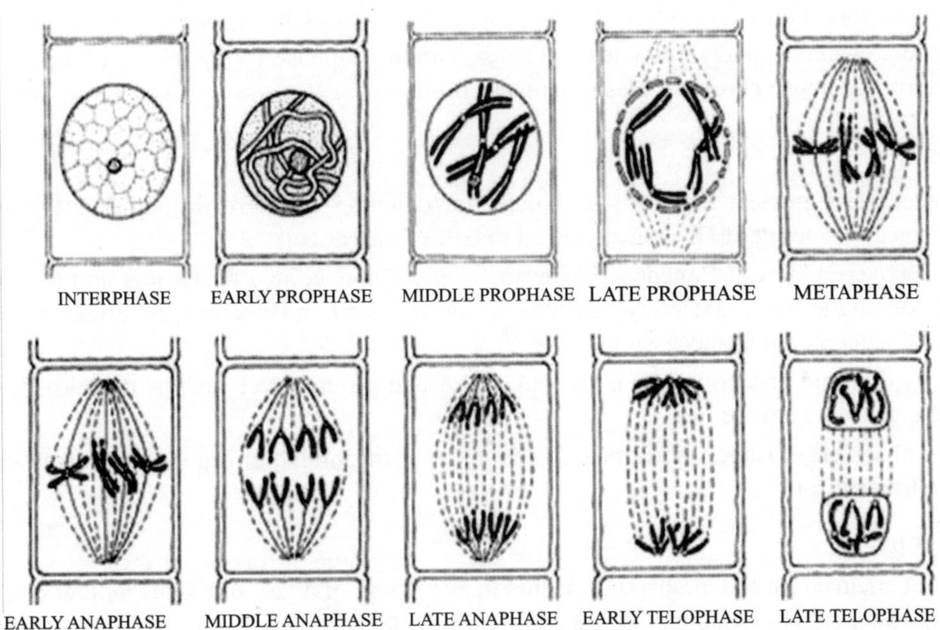

Fig. ***Different stages of mitosis in a plant cell***

Plant cytokinesis	Animal cytokinesis
It takes place in meristematic region.	It takes place at many places throughout body.
Centrioles are absent.	Centrioles are present.
Cytokinesis occurs by cell plate which grows centrifugally	Cytokinesis occurs by cell furrow which grows centripetally.
New cell membrane is formed from vesicles of Golgi body.	New cell membrane is formed from ER.

Significance:

(a) It maintains same genetic constitution (chromosome number) qualitatively and quantitatively.

(b) It maintains the size of cell by establishing nuclear-cytoplasmic ratio.

(c) It is only method of reproduction (multiplication) in unicellular organisms.

(d) It is essential for growth and development in multicellular organisms by forming tissue.

(e) It produces condition of multicellularity.

(f) It helps in repair, regeneration and healing of wounds by forming new cells in some animals.

(h) It leads to cancer due to uncontrolled division.

Mitogens: These are chemical substances which initiate cell division. In plant cells, temperature, auxin and cytokinin etc. are mitogens but in animal cells, the epidermal growth factor, platelet derived growth factor, insulin and lymphokines etc are mitogens.

Mitotic poisons: These are chemical substances which check mitotic division. These include

(a) Chalone: It inhibits cell division.

(b) Mustard gas: It causes agglutination of chromosomes.

(c) Colchicine: It dissolves microtubules hence spindle fibres are not formed at metaphase.

(d) Azide, cyanides, and ribonucleases: These inhibit prophase by inhibiting phosphorylation at interphase hence chromosomes remain as chromatin.

Meiosis

It is two successive nuclear divisions in which four daughter cells are formed thus the number of chromosomes and amount of DNA are reduced to half of parent cell.

It was discovered by Von Beneden. The term 'meiosis' was coined by Farmer and Moore (1905). Flemming coined the terms 'Heterotypic division' for meiosis I and 'Homotypic division' for meiosis II which take place in quick succession.

It occurs in diploid and polyploid cells (but not in odd ploidy level such as triploid). It occurs in reproductive cells or germ cells.

It is best observed in anther of unopened young flower of onion and *Tradescantia*. It is antithetical or opposite to fertilisation.

Types of meiosis

1. Zygotic meiosis (Initial meiosis): It occurs in zygospore of algae and ascus mother cell of fungi.
2. Sporic meiosis (Intermediate meiosis): It occurs in capsule of bryopyhytes; sporangium of pteridophytes; spore mother cell of gymnosperms and angiosperms.
3. Gametic meiosis (Terminal meiosis): It occurs in gametocytes (primary spermatocytes and primary oocytes) of animals. It also occurs in some algae.

Meiosis occurs in two successive stages:

Meiosis-I: It is called reductional or heterotypic division. It includes:

1. Prophase-1: It is the longest phase. It is divided into following five stages:

(a) Leptotene: It is shortest phase.

- Nuclear membrane and nucleolus are present.
- Chromatin fibres condense and coil to form chromosome.
- Chromosome becomes thread like and visible.
- Chromatids are not clear. Each chromosome gets attached at both its end to nuclear membrane called attachment plaque or bouquet stage or synizesis.

(b) Zygotene:

- Chromosomes shorten and thicken.
- Homologous chromosomes (one chromosome of pair comes from maternal side and other from paternal side having identical in shape and size) join and pair laterally (side by side) to form synapsis. It was reported by Montgomery. Synapsis is usually procentric (starting from centromere). It may be proterminal or random.
- Bivalent condition is seen. Both chromosomes of pair have same length, position of centromere and number of genes. Bivalent has four chromatids and two centromeres. Bivalent is preferred term than tetrad.
- Synaptonemal complex (nucleo-proteinaceous complex) is formed which stabilises bivalent. It was discovered by Moses. It consists of two lateral elements (formed by cohesion protein) and one central element (formed by zip protein). It helps in pairing of homologous chromosome.

- A little amount of DNA (0.3%) of each nucleus replicates.

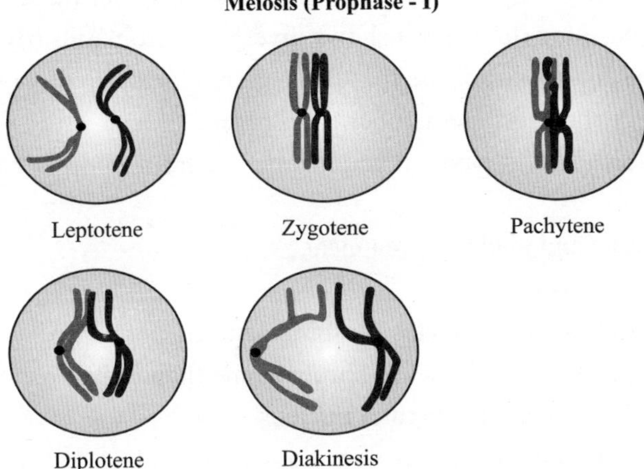

Fig. ***Stages in prophase I of meiosis***

(c) Pachytene:

- Bivalent (tetrad) becomes short and thick.
- Recombination nodule appears between chromosome pairs.
- Crossing over occurs at chiasma (chiasmata is not the cause but the result of crossing over).
- Morphological details of chromosomes are best studied because chromosomes are much longer.

(d) Diplotene:

- It is the longest phase.
- Synaptonemal complex dissolves (desynapsis).
- Bivalent (Tetrad) is clearly visible.
- Chiasma (result of crossing over) is clearly visible (the minimum number of chiasma is one). It was first observed by Janssen in amphibians.
- Chromosomes move apart. (Repulsion of non-sister chromatid)
- Chromosomes become enlarged in size and last for months and years called Lampbrush chromosome as in amphibian oocyte.

(e) Diakinesis:

- Nuclear membrane and nucleolus disappear.
- Terminalisation (shifting at the end) of chiasma occurs.
- Tetrad undergoes maximum thickening or shortening or condensation.
- Repelling of bivalents starts occurring.

2. Metaphase I:

- Two or more equatorial plates are formed.
- Bivalent is arranged at equatorial plate.
- Each equatorial plate will have the total number of chromosome.
- Each chromosome gets attached to spindle pole of its side.

3. Anaphase I:

- Homologous chromosomes separate due to splitting off chiasmata (disjunction).
- Separated chromosome (univalent) shows divergent chromatids (dyad).
- Univalents move towards opposite spindle poles and form two groups of haploid chromosome.
- Real reduction in chromosome number takes place.
- Independent assortment of paternal and maternal chromosome occurs due to disjunction.

4. Telophase I:

- Nuclear membrane and nucleolus reappear.
- Spindle fibres disappear.
- Chromosome uncoils and elongates (despiralisation).
- Daughter nucleus has only one set of chromosome (haploid).
- Amount of DNA is halved in daughter nucleus.

Interkinesis: It occurs in animal cells. It does not occur in plant cell, if it occurs then it is very short. It shows synthesis of RNA and proteins. Replication of DNA does not occur because there is no S phase. Chromosomes are elongated but do not form chromatin.

Meiosis-II: It is known as equational or homotypic division. It occurs in following phases;

1. Prophase-II:

- Nuclear membrane and nucleolus start disappearing.
- Univalent chromosome condenses, shortens or thickens.
- Chromosomes are not plectonemically thickened (compactly twisted).

2. Metaphase-II

- Single equatorial plate is formed at right angle to metaphase-I
- Chromosomes are arranged on equatorial plate.
- Sister chromatids are widely separated.
- Centromere of each chromosome gets attached to spindle pole of its side.

3. Anaphase-II: It is the shortest phase.

- Centromere of each chromosome (univalent) divides.
- Chromatids of each chromosome (univalent) separate.
- Chromatid now becomes daughter chromosome.
- Each daughter chromosome is attached to spindle pole of its side.

4. Telophase-II:

- Nuclear membrane and nucleolus reappear.
- Spindle fibres disappear.
- Daughter chromosome uncoils or elongates or decondenses.
- Chromosome forms a network of chromatin.
- Amount of DNA is halved again.

Cytokinesis: Cell plate is formed and thus four daughter (haploid) cells are formed. These cells resemble each other in the number of chromosomes but differ due to separation of maternal and paternal chromosome as well as formation of new types as a result of crossing over. It is of two types in plant cells.

1. **Simultaneous type:** It takes place after completion of meiosis II resulting in the formation of tetrahedral tetrad in dicot plants. It is considered a primitive feature.
2. **Successive type**: It takes place first after meiosis I and then after completion of meiosis II resulting in the formation of isobilateral tetrad in monocot plants. It is considered an advanced feature.

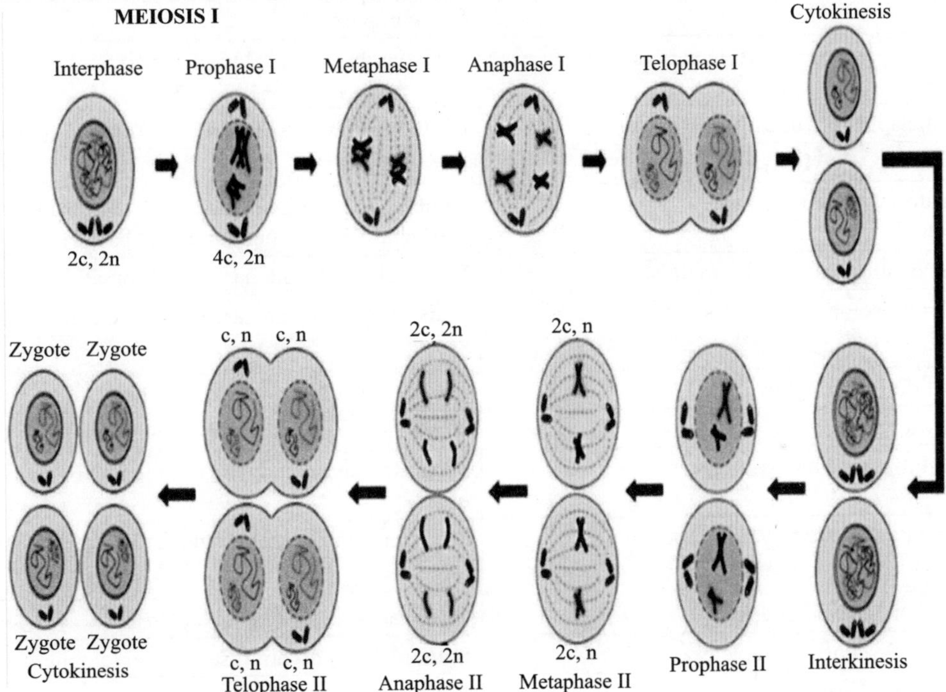

Significance:

(a) Meiosis produces haploid gametes for sexual reproduction.
(b) It brings about variation due to formation of new combination of chromosomes by crossing over.
(c) It causes segregation of two alleles of a gene due to separation of homologous chromosome.
(d) It provides cytological basis of Mendel's law of inheritance.

Differences between Mitosis and Meiosis

Mitosis	Meiosis
It occurs in somatic and reproductive cells.	It occurs in reproductive cells.
It is completed in one stage.	It is completed in two stages.
It is shorter in duration.	It is longer in duration.
It shows cell division only once.	It shows cell division twice.
One parent cell forms two daughter cells.	One parent cell forms four daughter cells.
Each daughter cell contains both homologous chromosomes.	Each daughter cell contains only one of each pair of homologous chromosome.

Replication of DNA occurs in S-phase.	Replication of DNA occurs only in S phase of interphase-I
Replication of DNA occurs once for one cell division.	Replication of DNA occurs once for two cell divisions.
Prophase	
It is simple.	It is complex.
It does not show bouquet stage, synapsis, crossing over and chiasma.	It shows bouquet stage, synapsis, crossing over and chiasma.
Chromosome splits off in early prophase.	Chromosome splits off in late prophase (pachytene).
Metaphase	
Single equatorial plate is formed.	Double and parallel equatorial plates are formed in metaphase-II but single in metaphase-II.
Each chromosome has two chromatids.	Paired chromosome has four chromatids.
Ends of chromosome are towards pole.	Ends of chromosome are towards equatorial plate.
Anaphase	
Centromere divides lengthwise.	Centromere divides only in anaphase-II.
Chromatids separate.	Chromatids separate only in anaphase-II.
Separated chromatids are identical.	Separated chromatids are not identical due to crossing over.
Sister chromatids move towards opposite poles.	Dissimilar chromosomes move towards opposite poles.
Telophase	
It always occurs and is longer.	Telophase-I is shorter.
Spindle fibres disappear completely.	Spindle fibres do not disappear in telophase-I.
It produces interphase nucleus.	It does not enter into interphase.
Nucleolus reappears at telophase.	Nucleolus does not appear in telophase-I.
Results	
Chromosome number remains constant in daughter cells.	Chromosome number is reduced to half in daughter cells.
Each daughter cell is genetically identical to parent cell qualitatively and quantitatively.	Each daughter cell is different from each other qualitatively and different from parent cell qualitatively and quantitatively.

SHORT ANSWER TYPE QUESTIONS

1. What do you mean by quiescent phase
2. Write the functions of S phase.
3. How cell division is regulated?
4. Mention the different types of mitogens and mitotic poisons.
5. Distinguish between metaphase of mitosis and metaphase I of meiosis.
6. Distinguish between anaphase of mitosis and anaphase I of meiosis.
7. Distinguish between anaphase I and anaphase II of meiosis.
8. What do you mean by bivalent?

9. Discuss the importance of mitosis.
10. Distinguish between plant cytokinesis and animal cytokinesis.
11. Discuss the different types of meiosis
12. Discuss the types of cytokinesis in plants.
13. Distinguish between anastral mitosis and amphiastral mitosis.
14. Write notes on amitosis.
15. Discuss the importance of meiosis.

LONG ANSWER TYPE QUESTIONS

1. Describe the different stages of cell cycle.
2. Describe the different types of mitosis.
3. Describe the different stages of mitosis with well labelled diagram.
4. Describe the prophase 1 of meiosis with well labelled diagram
5. Distinguish between mitosis and meiosis

MULTIPLE CHOICE QUESTIONS

1. Anastral mitosis is a feature of
   (a) Higher animals (b) All living organisms (c) Lower animals (d) Higher plants
2. The replication of centrioles occurs during
   (a) Early prophase (b) Late prophase (c) Interphase (d) Late telophase
3. A cell stops growing due to shortage of nutrients at
   (a) G_I phase (b) G_2 phase (c) S phase (d) M phase
4. The most variable length of cell cycle is noticed in
   (a) G_I phase (b) G_2 phase (c) S phase (d) M phase
5. Nuclear membrane is broken down due to activity of
   (a) P protein (b) U protein (c) Kinase (d) Kinetin
6. A cell contains 4n DNA at
   (a) G_I phase (b) S phase (c) G_2 phase (d) M phase
7. Kinetin increases the rate of mitosis by reducing the duration of
   (a) Interphase (b) Metaphase (c) Anaphase (d) Telophase
8. The spindle fibres are composed of
   (a) 3% RNA and 97% tubulin (b) 10% RNA and 90% tubulin
   (c) 40 % RNA and 60% tubulin (d) 60% RNA and 40% tubulin
9. Which stage of prophase-I of meiosis is prolonged?
   (a) Leptotene (b) Zygotene (c) Pachytene (d) Diplotene
10. Replication of chromosomes without division of the nucleus is called
    (a) Dinomitosis (b) Endomitosis (c) Karyochorisis (d) C-Mitosis
11. The spindle in meiosis is oriented at
    (a) acute angle to that in meiosis-I (b) right angle to that in meiosis-I
    (c) obtuse angle to that in meiosis-I (d) Same position to that in meiosis-I
12. The number of DNA in chromosome at G_2 stage is
    (a) 1 (b) 2 (c) 4 (d) 8
13. Interzonal fibres occur in
    (a) Prophase (b) Early metaphase (c) Late metaphase (d) Anaphase
14. The angular divergence of spindle microtubules is decreased at
    (a) Prophase (b) Metaphase (c) Anaphase I (d) Anaphase II
15. The phragmoplast is organised before cell plate at
    (a) Early anaphase (b) Late phase (c) Early telophase (d) Late telophase
16. How many mitotic divisions occur in a cell of root tip to form 128 cells?
    (a) 7 (b) 127 (b) 32 (d) 16
17. How many mitotic divisions are needed for a single cell to make 128 cells?
    (a) 7 (b) 8 (b) 16 (d) 127
18. How many meiotic divisions would be required to produce 101 female gametophytes in an organism?
    (a) 25 (b) 26 (c) 100 (d) 101
19. Chromosomes appear as bead like at
    (a) Leptotene (b) Zygotene (c) Pachytene (d) Diplotene

20. In meiosis, dyad is defined as
    (a) A pair of sister chromatids (b) A pair of non-sister chromatids
    (c) A pair of homologous chromosomes (d) A pair of non-homologous chromosomes
21. During synapsis, the number of threads or chromonemata in each chromosome is
    (a) 1 (b) 2 (c) 4 (d) 8
22. Chiasmata were first discovered by
    (a) Morgan in *Drosophila* (b) Beadle and Tatum in *Neurospora*
    (c) Janssen in amphibians (d) Hammerling in *Acetabularia*
23. The number of chromosome groups at the equatorial plate in metaphase-I of meiosis in a plant with 2n=50 shall be
    (a) 25 (b) 50 (c) 100 (d) 200
24. During meiosis, reduction of chromosome number into half takes place at
    (a) Metaphase I (b) Metaphase II (d) Anaphase I (d) Anaphase II
25. What state of cell cycle is attained when the cell has undergone differentiation?
    (a) G_1 phases (b) G_2 phase (c) S phase (d) G**o** phase
26. The cells of a permanent tissue do not divide because these are arrested at
    (a) G_o stage (b) G_1 stage (c) S stage (d) G_2 stage
27. Colchicine prevents the mitosis of the cells at
    (a) Prophase (b) Metaphase (c) Anaphase (d) Telophase
28. Chromosomes are linear in
    (a) Prophase I (b) Prophase II (c) Interphase (d) Metaphase I
29. If there are 4 chromosomes present during prophase then how many chromosomes are there in each cell at the end of anaphase II?
    (a) 1 (b) 2 (c) 4 (d) 8
30. In an organism if the normal diploid number of chromosomes is 8 then how many chromatids are present in each daughter cell at the end of meiosis I?
    (a) 2 (b) 4 (c) 8 (d) 16
31. Repulsion of homologous chromosomes takes place in
    (a) Zygotene (b) Pachytene (c) Diplotene (d) Diakinesis
32. In which phase of cell cycle, cell division cannot be stopped?
    (a) G_1 phase (b) S phase (c) G_2 phase (d) G_0 phase
33. How many ATPs are required during anaphase to move chromosomes from equator to the poles?
    (a) 15 ATP (b) 30 ATP (c) 36 ATP (d) 38 ATP
34. The nature of coiling of chromatids is
    (a) Plectonemic in both mitosis and meiosis
    (b) Paranemic in both mitosis and meiosis
    (c) Plectonemic in mitosis and paranemic in meiosis
    (d) Paranemic in mitosis and plectonemic in meiosis
35. A little amount of DNA is synthesised at
    (a) Leptotene (b) Zygotene (c) Pachytene (d) Diplotene

36. If the amount of DNA is 20 pg in G_1 stage then how much amount of DNA is formed in each cell at the end of meiosis II?
(a) 5 pg (b) 10 pg (c) 20 pg (d) 40 pg
37. If the amount of DNA in gamete is 10 pg then how much amount of DNA is found at G_2 stage?
(a) 5 pg (b) 10 pg (c) 20 pg (d) 40 pg
38. First restriction point occurs at
(a) Early G_1 stage (b) Late G_1 stage (c) Early G_2 stage (d) Late G_2 stage
39. The stage of polyteny occurs as a result of
(a) Dinomitosis (b) Amitosis (c) Karyochorisis (d) Endomitosis
40. Cytokinesis in plant occurs by cell plate which grows
(a) Laterally (b) Centripetally (c) Centrifugally (d) Randomly
41. Holiday junction is observed during
(a) Mitosis (b) Interphase (c) DNA repair (d) Recombination
42. A cell is post reproductive will remain in
(a) G_1 phase (b) S phase (c) G_2 phase (d) M phase
43. The number of chromatids in a cell in prophase I of meiosis of a person with Turner's syndrome
(a) 23 (b) 25 (c) 47 (d) 90
44. When dividing cells are examined under light microscope, the chromosomes first become visible during
(a) Interphase (b) Prophase (c) Metaphase (d) Anaphase
45. How many chromosomes will be present at mitotic metaphase of diploid cell if its egg has 10 chromosomes?
(a) 10 (b) 20 (c) 40 (d) 80
46. Cells will usually divide if they receive proper signal at
(a) G_1 phase (b) S phase (c) G_2 phase (d) M phase
47. Centrioles separate during
(a) Cytokinesis (b) Prophase I and prophase II
(b) Metaphase I and metaphase II (b) Anaphase I and anaphase II
48. Maize has 10 pairs of chromosome then the number of chromosomes and chromatids at metaphase II are, respectively
(a) 10 and 10 (b) 10 and 20 (c) 20 and 40 (d) 40 and 20
49. Which of the following precedes reformation of the nuclear envelope during M phase of the cycle?
(a) Decondensation from chromosome and reassembly of nuclear lamina
(b) Transcription from chromosome and reassembly of nuclear lamina
(c) Formation of contractile ring and formation of phragmoplast
(d) Formation of contractile ring and transcription from chromosome
50. Which of the following can be seen during interphase with a light microscope?
(a) Nucleosome (b) Introns (c) Euchromatin (d) Heterochromatin
51. When cell has stopped DNA replication fork, then which of the following checkpoints should be predominantly activated?
(a) M phase (b) G_1/ S phase
(c) G_2/ M phase (d) Both G_2/M and M phases

52. A somatic cell that has just completed the S phase of its cycle as compared to gamete of the same species, has
    (a) Twice the number of chromosomes and 4 times the amount of DNA
    (b) Twice the number of chromosomes and 2 times the amount of DNA
    (c) Twice the number of chromosomes and 8 times the amount of DNA
    (d) Same number of chromosomes and 4 times the amount of DNA
53. Cells of certain species of animals have 6 pairs of chromosomes. How many molecules of DNA will remain in a nucleus of these animals?
    (a) 12 (b) 24 (c) 36 (d) 48
54. Nuclear membrane is formed around the groups of daughter chromosomes during the telophase by
    (a) ER (b) Golgi (c) Lysosome (d) Ribosome
55. Segregation of alleles occurs at
    (a) Metaphase I (b) Anaphase I (c) Metaphase II (d) Anaphase II
56. Sister chromatids are
    (a) Created when DNA is replicated (b) Separated during mitosis
    (c) Attached at the centromere prior to division (d) All of the above
57. The nutritional medium was supplied to growing bacteria. If whole nutrient gets depleted in 20 cycles, then at which stage the amount of nutrient was half used?
    (a) At the end of 10 cycles (b) At the end of 19 cycles
    (c) At the beginning of 11 cycles (d) At the beginning of 19 cycles
58. Shape of chromosomes is observed at
    (a) Prophase (b) Metaphase (c) Anaphase (d) Telophase
59. Cell plate occurs in
    (a) Plant cells (b) Animal cells
    (c) Bacterial cells (d) Both bacterial cells and plant cells
60. Dinomitosis occurs in the species of
    (a) Monerans (b) Protists (c) Plants (d) Animals

ANSWERS

1.d	2.c	3.a	4.a	5.c	6.c	7.a	8.a	9.d	10.b
11.b	12.b	13.d	14.d	15.d	16.b	17.a	18.d	19.a	20.a
21.a	22.c	23.a	24.c	25.d	26.b	27.b	28.a	29.b	30.c
31.c	32.b	33.b	34.c	35.b	36.b	37.d	38.b	39.d	40.c
41.d	42.a	43.d	44.b	45.c	46.a	47.b	48.b	49.c	50.d
51. b	52.a	53.b	54.a	55.b	56.d	57.a	58.c	59.a	60.b

8

Transport in Plants

Transport is the translocation of inorganic and organic substances to different parts of plant and also from one part of a cell to another. In nonvascular plants, substances are transported by diffusion but in vascular plants, xylem and phloem transport water and food, respectively.

Means of transport: There are three means of transport.

1. Cellular level transport: It is uptake and release of water and solutes by individual cell.
2. Intercellular or Lateral transport: It is short distance transport of substances from cell to cell by diffusion. It occurs through active or passive process.
3. Longitudinal or Vertical transport: It is long distance transport of sap through xylem (unidirectional) and phloem (bidirectional).

Types of transport:

1. **Simple diffusion**: It is movement of particles (atoms, ions, molecules) of gases, liquid and solids from the region of their high concentration to the region of low concentration until the particles are evenly distributed throughout the available space.

   Features:

   (a) It is spontaneous and passive process but not reversible.
   (b) It is physical and slow process.
   (c) It is due to random kinetic collision of atoms, ions or molecules.
   (d) It depends on free energy or chemical potential (number of molecules per unit volume)
   (e) It operates in any medium.
   (f) It is independent of living system.
   (g) It does not require energy or assistance of proteins for passage.
   (h) Its rate and direction depend on concentration of that substance at different spot.
   (i) It is independent of the presence of other diffusing substance (as O_2 diffuses into leaf and CO_2 diffuses out of leaf simultaneously by stomata in respiration.
   (j) It is only means of gaseous movement within plant body.
   (k) Molecules diffuse only in pure state because molecules have the highest free energy.
   (l) It is very fast in gases because intermolecular attraction is nil but it is very slow in solids because molecules are closely packed.

   Examples:

   (i) Movement of scent molecules of perfume by their kinetic energy without involvement of external force.
   (ii) Colouring of water in beaker with the addition of colouring matter.
   (iii) Spreading of smoke if a firewood is burnt.

Factors:

(a) Temperature: Rate of diffusion increases with rise in temperature

(b) Size and density of diffusing substance: Rate of diffusion decreases with increase in size and density of substance

(c) Diffusion pressure: Rate of diffusion increases with increase in diffusion pressure.

(d) Density of medium: Rate of diffusion is slow if medium is more concentrated and fast if medium is less concentrated.

(e) Concentration gradient: Rate of diffusion increases if difference in concentration of diffusing substance between the two regions will be increased.

Importance:

(a) It is involved in exchange of gases (O_2 and CO_2) between plants and atmosphere through stomata during photosynthesis and respiration.

(b) It releases water in the form of water vapour from leaf to atmosphere through stomata.

(c) It helps in passive absorption of ions of minerals in plants.

(d) It helps in movement of substances over short distance within a cell or between adjacent cells.

(e) It helps in pollination when scent molecules released from flower diffuse in air to attract pollinators such as insects and birds.

2. **Facilitated diffusion**: It is the diffusion of polar and hydrophilic molecules from their high concentration to low concentration with the help of fixed membrane transport protein.

Features:

(a) It is a passive process that moves along concentration gradient.

(b) It does not require energy.

(c) It is very specific as it allows cell to select substances for uptake.

(d) It involves cell membrane as path unlike simple diffusion.

(e) It shows saturation effect as the polar molecules (substances) cannot pass through the non-polar lipid bilayer hence carrier proteins are fully used and then shunted to the other side of protein and to be released there.

(f) It is sensitive to inhibitors because these alter the structure of carrier protein.

Types of transport protein:

(a) Carrier proteins: These bind the selective or specific solute esp. organic molecule (glucose) on one side of membrane, undergo conformational change and release the solute on the opposite side of membrane. These proteins do not form channel but act as enzyme to facilitate the movement. Each type of carrier protein has one or more specific binding sites for its solute (substrate). These involve co-transport when proteins allow transport only if two types of solute move together. These perform in following ways:

(i) Symport: It is co-transport in which both substances cross the membrane in the same direction at the same time.

(ii) Antiport: It is co-transport in which both substances cross the membrane in the opposite direction at the same time.

(iii) Uniport: It is not co-transport in which a substance crosses the membrane independent of other substance.

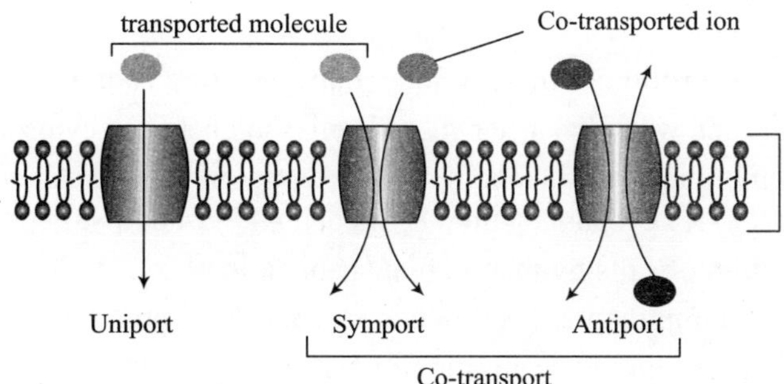

(b) **Channel proteins**: These facilitate diffusion of any inorganic solutes (Na^+, K^+, Cl^-) through pores. These proteins (porins) bind with substrate molecule (solute) then configuration of protein is changed resulting formation of pores (channel) in the membrane. These pores are formed in the outer membrane of mitochondria, chloroplast and some bacteria allowing solutes up to size of small protein to pass through. The channel proteins transport numerous ions much faster than carrier protein. These channels or pores are not permanently open rather they selectively open and close in response to extracellular signals allowing cell to regulate the transport of ions across membrane.

The aquaporin protein of eight different types serves as a water-filled channel which is involved in bulk flow.

3. **Active transport**: It is the transport of substances (ions or salts, glucose, amino acids) from low concentration to high concentration across membrane through proteins of plasma membrane.

Features:

(a) It moves against concentration gradient.
(b) It is very rapid process.
(c) It requires energy and assistance of protein.
(d) It is very specific because specific carrier protein has specific solute.
(e) It shows saturation effect because all proteins are used up when its rate is maximum. The protein works as pump.
(f) It is sensitive to inhibitors (cyanides) which react with protein side chains.

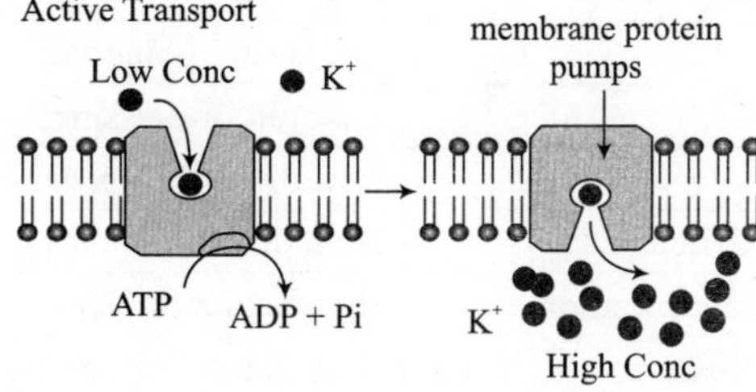

Types of Active transport

1. Primary active transport: The proteins involved are pumps that use ATP.
2. Secondary active transport: It uses potential energy derived from electrochemical gradient.

Importance:

(a) It helps in rapid absorption of mineral ions (nutrients) into root hair cells of plants.

(b) It maintains ionic and water balance between cells; and between cytosol and extracellular fluid.

(c) It maintains membrane potential and resting potential in muscle cell and nerve cell of animals by undergoing Na^+-K^+ exchange pump (For each ATP used, $3Na^+$ and $2K^+$ are pumped).

(d) It helps in excretion of salts by marine plants and animals.

(e) It helps in H^+-K^+ pump in guard cell of stomata and Ca^{+2} pump in RBC.

(f) It prevents the equilibrium concentration of Na^+, K^+, Ca^{+2}, H^+ etc.

Plant water relation:

It is concerned with collection of water from the soil, transport within plant and its loss by evaporation from the leaves. Water is most important constituent of living cells and constitutes more than 90% of protoplasm by volume and weight. It makes 80-90% of fresh weight of herbaceous plants and makes 50% of fresh weight of shrubs and trees (woody plants).

Osmosis: It is the movement of water from high water potential to low water potential through semi-permeable membrane until equilibrium is reached. The term 'Osmosis' was coined by Nollet but it was discovered by Pfeffer.

Features:

(a) It operates only in liquid medium.

(b) It is applicable only to solvent part of solution.

(c) It does not equalise the concentration of solvent on two sides of system.

(d) It develops a comparatively lower pressure up to 100 atm called osmotic pressure.

(e) It is opposed by hydrostatic pressure.

(f) It depends on solute potential.

(g) It also depends on degree of reduction of free energy of one solvent over that of another.

Osmosis is simply described as movement in the following ways:

From	to
Dilute solution	Concentrated solution
High water concentration	Low water concentration
High water potential	Low water potential
High turgor pressure	Low turgor pressure
High diffusion pressure	Low diffusion pressure
Low diffusion pressure deficit (DPD)	High diffusion pressure deficit (DPD)
Low solute or osmotic concentration	High solute or osmotic concentration
Low solute or osmotic pressure	High solute or osmotic pressure
High solute or osmotic potential	Low solute or osmotic potential

Types of Osmosis:

1. **Endosmosis**: It is osmotic entry of water into a cell, organ or system from high water potential to low water potential. As the volume of protoplast increases, it begins to exert pressure against cell wall called turgor pressure. The cell wall being rigid exerts an equal (in magnitude) and opposite

(direction) pressure on protoplast called wall pressure. As the turgor pressure of cell increases, the cell becomes turgid. If a cell is submerged in fresh water, water moves into the cell.

e.g., Swelling of raisins when immersed in water.

2. **Exosmosis**: It is osmotic withdrawal of water from cell, organ or system.

   e.g., shrinkage of grapes when immersed in highly concentrated or sugar solution. If a cell is submerged in salt water, water moves out of the cell.

3. **Reverse osmosis**: It is used in removing salts from saline water as well as extra-purification of water.

Osmotic relation of a plant cell:

1. **Osmotic pressure (OP):** It is maximum actual pressure which can develop in an osmotically active solution when it is separated from its pure water (solvent) by semi-permeable membrane. It is also defined as the pressure required to completely stop the entry of water into an osmotically active solution across a semi-permeable membrane.

   Features:

   (a) It is measured by Osmoscope and Osmometer.

   (b) It is measured in terms of atmosphere, Bar, Pascal (1 atm=1 bar= $1\text{x}10^5$ Pascal=760 mmHg)

   (c) It varies in different plants, organs and tissues.

   (d) It is low in (1-3atm) in hydrophytes, moderate in mesophytes, high in xerophytes and the highest in halophytes (maximum osmotic pressure was recorded in *Atriplex confertifolia*, a halophyte).

   (e) It is lower in roots and higher in upper leaves.

   (f) It is lower in outer cortical cell and higher in inner cortical cell.

   (g) It is 22.4 atm at NTP for ideal molar solution (volume molar solution =1 gram mole of solute dissolved in 1000 cc of solution).

   (h) It is zero in pure water.

   (i) It is opposed by hydrostatic pressure (turgor pressure).

   (j) It is not affected by the addition of sucrose in solution.

   (k) It is numerically equal to osmotic (solute) potential but osmotic potential has negative value.

   Types of solution:

   (a) Hypotonic solution: It is a solution in which concentration or osmotic pressure is less than that of cell sap or another solution.

   (b) Hypertonic solution: It is a solution in which concentration or osmotic pressure is less than that of cell sap or another solution.

   (c) Isotonic solution: It is a solution in which concentration or osmotic pressure is equal to that of cell sap or another solution.

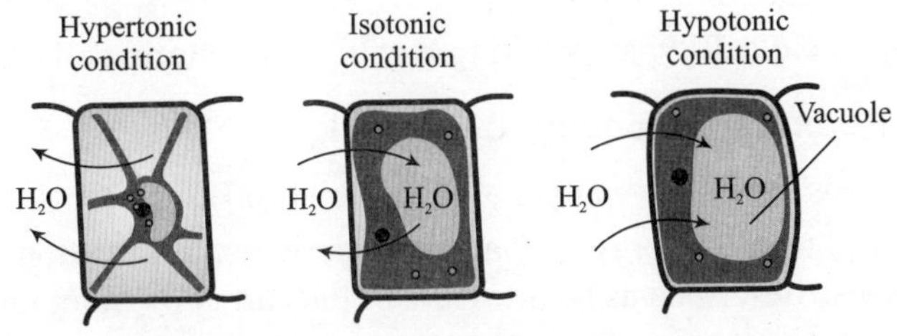

Factors affecting osmotic pressure:

(a) Temperature: Osmotic pressure is directly proportional to temperature.

(b) Concentration of solute particles: OP increases with increase in concentration of solutes because it is colligative property.

(c) Ionisation of solute particles: OP increases with increase in the number of particles.

(d) Hydration of solute particles: OP increases because solutes bind with water hence unable to diffuse.

2. **Turgor pressure (TP):** It is outward hydrostatic pressure exerted by cytoplasm on cell wall due to osmotic diffusion of water into a cell. It causes stretching of cell wall by expanding protoplast. It is zero in flaccid cell and high in fully turgid cell.

Functions:

(a) It maintains the shape of cell.

(b) It maintains the structure of cell organelles.

(c) It helps in opening and closing of stomata.

3. **Wall pressure**: It is inward pressure exerted by cylopiasm upon cytoplasm to counteract the effect of turgor pressure. It is equal but opposite to turgor pressure.

4. **Diffusion pressure deficit (DPD):** It was coined by Meyer (1938).

It is the difference between the diffusion pressure of a solution and pure solvent when both are subjected to same atmospheric pressure or it is amount by which diffusion pressure of a solution is less than that of its solvent so there is loss of chemical free energy of water because of the binding of solvent to solute.

It is also called suction pressure (SP).

It is index of absorbing power of a solution.

It is governed by the relative concentration of solute in a given volume of solution.

It is always positive.

It increases with the addition of solutes in a solution.

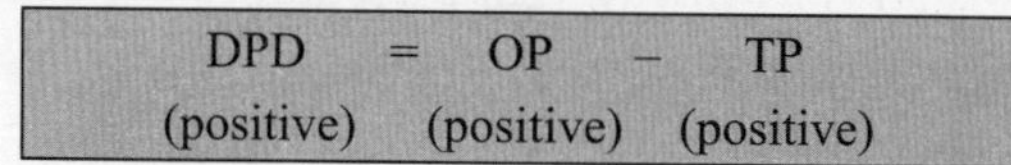

Case:

(a) If a cell is fully flaccid then TP is zero hence DPD = OP. Thus cell absorbs or sucks water with a force equal to osmotic pressure.

(b) If a cell is fully turgid then OP of cell starts decreasing while TP starts increasing then a stage comes when TP = OP hence DPD is zero. The cell has no further capacity to absorb water.

(c) If a cell is plasmolysed then TP is negative hence DPD = OP- (-TP) = OP + TP. Thus DPD is maximum.

The demand of water (from high to low):

Plasmolysed cell > flaccid cell > partially turgid cell > fully turgid cell

Osmosis in terms of Potential (capacity to do work)

1. **Water potential**: It was coined by Slatyer and Taylor (1960).

It is the measure of potential energy in water and derives the direction of movement of water through plants. It is also described as the measure of the ability of water molecules to move freely in solution. It is represented by the Greek letter Psi (ψ).

It is always negative in a solution when pressure potential is less negative than solute potential.

It is zero (highest value) in pure water when pressure potential is equal to solute potential. It increases if more pressure than atmospheric pressure is applied to pure water.

It is negative and equal to solute potential in a solution at atmospheric pressure.

It will be same at equilibrium throughout system.

It is 2.3 Bar in 0.1M solution of glucose.

It is lesser (more negative) in warmer solution than in colder solution.

It will be less (more negative) with the addition of solutes in pure water because solute particles are attracted to water molecules reducing the number of water molecules that can diffuse freely.

Its value plays a great role in plant-water relation in following ways:

$$\psi_w \text{ soil} > \psi_w \text{ root} > \psi_w \text{ stem} > \psi_w \text{ leaf} > \psi_w \text{ atmosphere}$$

Water potential (ψ_w)	= Solute potential (ψ_s)	+ Pressure potential (ψ_p)
(negative)	(negative)	(positive)

2. **Solute (osmotic) potential**: It is the amount of solute by which water potential is reduced. It is the concentration of dissolved solutes in water.

   It is always negative.

   It will be less (more negative) with the addition of solutes in solution.

   It will be more (less negative) with the addition of solvent in solution (due to dilution).

   It will be zero in pure water (solvent).

3. **Pressure potential**: It is turgor pressure (hydrostatic pressure) which is exerted by cytoplasm on cell wall due to osmotic diffusion of water inside cell. It is the effect of turgor pressure on the potential energy of solution. If a pressure greater than atmospheric pressure is applied to pure water or a solution, its water potential increases. If there is positive pressure (as water enters a cell) then water potential increases. If there is negative pressure (as water leaves a cell) then water potential of cell decreases. It increases when solute potential decreases.

   It is usually positive.

   It is zero in pure water kept in an open beaker.

   It is sometimes negative in plasmolysed cell and xylem vessel when water is pulled.

   Cases:

   (a) If a cell is flaccid then water potential is the lowest (ψ_w is more negative) because ψ_p= 0 hence ψ_w= -ψ_s)

   (b) If a cell is partially turgid then water potential is higher (ψ_w is less negative because ψ_s >ψ_p

   (c) If a cell is fully turgid then water potential is the highest (ψ_w=0 because ψ_s= ψ_p). Thus, the net movement of water into the cell is stopped. The cell is in equilibrium with water outside.

Types of membranes on the basis of permeability:

1. Impermeable: It does not allow either solvent or solutes to pass through. e.g., Cutinised cell wall and suberised cell wall.
2. Permeable: It allows both solvent and solute to pass through. e.g., Cellulosic cell wall, lignified cell wall and filter paper.

3. Semipermeable: It allows solvent to pass through but checks solutes. e.g., Animal bladder, egg membrane, cellophane, copper ferricyanide membrane and parchment membrane etc.
4. Selectively or differentially permeable: It allows solvent and some selected solutes to pass through and checks non-selected solutes e.g., Plasma membrane (permeable for gases and water but impermeable for polysaccharides, proteins, phospholipids), biological membrane (membrane of cell organelles), membrane (tonoplast) of vacuole etc.

Importance of osmosis:

(a) It helps in absorption of water by root hairs.
(b) It helps in ascent (upward conduction) of sap (dissolved minerals)
(c) It helps in opening of stomata by providing the turgidity of guard cell.
(d) It helps in distribution of water across the cell and thus turgidity of cell is maintained (70% of total water is found in vacuole of plant).
(e) It provides stability to mitochondria and chloroplast by maintaining their osmotic balance.
(f) It helps in seed germination by maintaining the growth of radicle and plumule.
(g) It helps in autochory (self dispersal of fruits and seeds).
(h) It facilitates shock movement (siesmonasty) like folding and drooping of leaves in *Mimosa pudica* (touch-me-not) and sleep movement (nyctinasty) like folding or rolling of leaves in Bean due to change in turgor pressure of Pulvinus.

Plasmolysis

It is shrinkage of cytoplasm along with plasma membrane from the cell wall due to excessive loss of water from the cell due to osmotic function. It is a reversible process. It was demonstrated best in *Spirogyra*, and leaves of *Tradescantia* and *Rhoeo discolour.* It is a vital process and occurs when a plant cell is placed in hypertonic solution. Water is initially lost from cytoplasm and then from vacuole hence exosmosis occurs. Protoplast is shrunk due to reduction in size of cytoplasm and vacuole. Thus cell becomes plasmolysed or flaccid when water potential (ψ_w) is equal to solute potential (ψ_s) because pressure potential (ψ_p) is negative or zero. Plasmolysis occurs in three stages:

(a) Limiting plasmolysis: The shape of cell slightly changes but cytoplasm does not withdraw from cell wall. The pressure potential (ψ_p) is zero.
(b) Incipient plasmolysis: Protoplast starts shrinking from the corners of cell. The pressure potential (ψ_p) is less negative
(c) Evident Plasmolysis: Protoplast is withdrawn maximum from cell except at one or more places. Hence pressure potential (ψ_p) is more negative.

Case:

(a) If a plasmolysed plant cell is placed in hypotonic solution or pure water then deplasmolysis or endosmosis occurs. Water enters into the cell causing cytoplasm to build up turgor pressure against cell wall. Hence pressure potential (ψ_p) is zero. Now cell comes into original shape.
(b) If a plant cell is placed hypotonic solution or pure water then endosmosis occurs. Cell is enlarged but does not rupture due to rigidity of cell wall. Cell becomes turgid.
(c) If a plant cell is placed in isotonic solution then plasmolysis does not occur because equilibrium is established hence there is no net flow of water.

Plasmolysed cell	Turgid cell
It has low water potential (more negative).	It has high water potential (less negative).
It has low pressure potential (more negative) because protoplast exerts a high positive pressure in cell wall.	It has high pressure potential(less negative) because protoplast exerts a low positive pressure on cell wall
It has low solute potential (more negative).	It has high solute potential(less negative).

Importance:

(a) It proves that cell is living because plasmolysis does not occur in nonliving cell.

(b) It confirms that plasma membrane is semi-permeable and cell wall is permeable and elastic.

(c) It measures osmotic pressure.

(d) It kills the bacteria, spores and weeds by salting the pickles, meat and fish.

(e) It maintains the plant-water relation.

Imbibition:

It is the absorption or uptake of water by hydrophilic solid substance without forming solution. It is a physical process which involves:

(a) Adsorption (attachment of liquid on surface) occurs when a dry material (imbibant) having zero diffusion pressure comes in contact with water (imbibate) having maximum diffusion pressure.

(b) Water potential gradient is established. Water diffuses very rapidly into substance from high water potential to low water potential.

(c) Imbibant gets saturated with water and gradually become equal to that of water thus equilibrium is established.

(d) Some amount of heat energy is released due to compact arrangement of water molecules (water becomes warmer).

(e) Volume of material (imbibant) is increased.

Features:

(a) It does not require semi-permeable membrane.

(b) It develops high imbibition pressure (1000 atm).

(c) It occurs both in living and dead cells.

(d) It is a reversible process.

(e) It requires colloidal and hydrophilic substances but their efficiency to absorb water differ.

(f) It is maximum in agar agar (a phycocolloidal substance) followed by protein (of leguminous seed), starch (of cereals) and minimum in cellulose. That is why proteinaceous (pea) seed swells more than starchy (wheat) seed.

(g) It is not shown by lignin and suberin.

Factors:

(a) Temperature: Rate of imbibition increases with rise in temperature.

(b) Texture of material: Rate of imbibition is more if material is loose.

(c) Pressure: Rate of imbibition decreases if pressure is against imbibant.

(d) Electrolytes: Rate of imbibition decreases in the presence of electrolytes.

Importance:

(a) It helps in absorption of water by young cell of root hair. Hence imbibition is first step in this process.

(b) It helps in germination of seed by absorbing water by seed coat because seed has almost no water.
(c) It helps in growth of seedling.

Absorption of Water

It is a process by which rooted land plants absorb water from soil by root hairs. In plant groups, root hairs are more developed in angiosperms than gymnosperms. Root hairs are also more developed in herbs than shrubs and trees of Angiosperms. Root hair is unicellular and tubular extension of epidermal cell. The cell wall of root hair cell is very thin and two layered; outer layer is made up of pectin and inner layer is made up of cellulose. Root hair increases the available surface area for absorption of water and nutrients (minerals). In case of aquatic plants absorption of water takes place from entire body surface.

Types of water present in soil:

(a) **Capillary water:** It is only available water present in narrow interparticles space or micropores of soil held by capillary forces which work against gravitational forces.
(b) **Gravitational water:** It is unavailable water present in deep strata of soil beyond the reach of root hairs. It is drained downwardly after good rain or irrigation due to the influence of gravitational forces. It is also called run away or runoff water which reaches water body
(c) **Hygroscopic water:** It is unavailable water present in thin film around soil particles held by very strong attractive forces.
(d) **Combined water**: It is unavailable water present attached with chemicals such as hydrated oxides of Silicon, Iron, Aluminium in soil.
(e) **Water vapour**: It is unavailable water present in the form of vapour (gas).

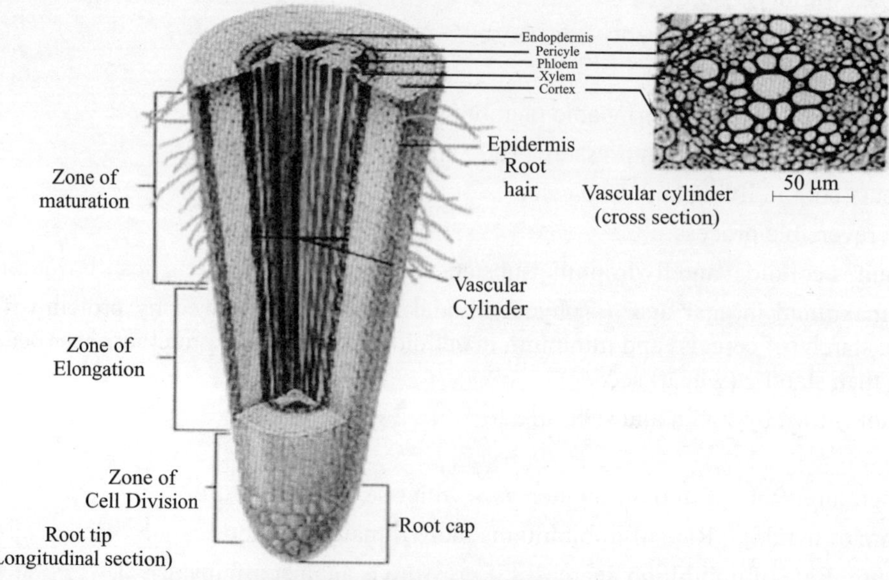

Mechanism of absorption of water: Kramer proposed two mechanisms-active and passive which are as follows:

1. **Process of active absorption and transport**: It was proposed by Priestley (1921).The root cells play active role consuming metabolic energy released through respiration. Root cells behave as

ideal osmotic system through water moves from the soil to the xylem of root along an increasing gradient of DPD (suction pressure, which is real force for water absorption).

It involves **symplast pathway**. The symplast (inner space) is living continuous system of plant which consists of protoplast connected by plasmodesmata. It is a slow process and absorbs small amount of water due to greater resistance of protoplast.

Water is absorbed from soil solution which is dilute having high water potential (-0.3 atm) into root hair cell having cell sap of lower water potential (-3 atm). Water does not enter into cell vacuole. Imbibition occurs in cell of root hair. Endosmosis occurs and water enters into epidermal cell from high water potential to low water potential. Now water enters into cell of cortex, cell of endodermis and pericycle and finally to xylem from high water potential to low water potential. Symplastic movement is aided by cytoplasmic streaming (cyclosis) of individual cell.

2. **Process of passive absorption and transport**: It is mainly due to transpiration. The root cells do not play active role and remain passive.

   It involves both **apoplast pathway** (firstly) as well as **symplast pathway** (later). The apoplast (outer space) is non-living continuous system of plant which consists of cell wall, intercellular spaces and xylem channels. This is called free space. It is a fast process and absorbs large amount of water due to little resistance.

   Water is absorbed from high water potential of soil water (-0.3atm) to low water potential of root hair cell (-3atm). Endosmosis occurs. Water moves from root hair cell to epidermis to cortex through cell wall of intervening cells without entering cytoplasm. Movement of water is through mass flow/bulk flow and it is interrupted in endodermis where it becomes impermeable due to presence of suberised Casparian strips. Water now moves backward and enters into inner cortical cell and then reaches endodermis, pericycle and xylem through symplast involving plasmodesmata. It is the major pathway for the transport of water from the epidermis to xylem of root.

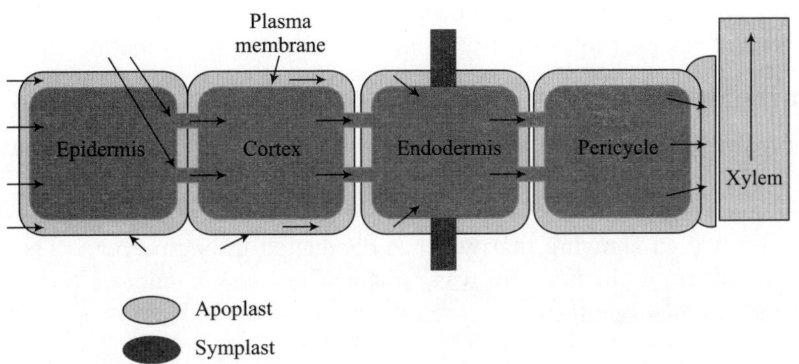

Active absorption	Passive absorption
It absorbs small amount (5%) of water.	It absorbs large amount (95%) of water.
It is slow, vital and unidirectional process.	It is fast, physical and bidirectional process.
It involves symplast pathway.	It involves both apoplast and symplast pathways
It is highly selective and found in some plants only.	It is non-selective and found in all plants.
It occurs in rainy season only.	It occurs throughout the year.

It occurs in slowly transpiring and well-aerated soil.	It occurs in rapidly transpiring plants.
It involves root cells.	It may occur in the absence of roots.
It is influenced by root pressure (positive pressure) developed in roots.	It is influenced by transpiration pull (negative pressure) developed in shoots.
It utilises metabolic energy.	It utilises solar energy.

Factors:

(a) Temperature: The optimum temperature is 20-30°C. Rate of absorption decreases if temperature will be either very low or very high.

(b) Concentration of soil solution: Rate of absorption increases if concentration of soil solution (due to presence of more salts) will be low (high water potential) and decreases if concentration of soil solution will be high (low water potential). Its rate is poor in alkaline soil and marshes.

(c) Aeration of soil (O_2 content): Rate of absorption increases with increase in aeration.

(d) Availability of soil water (Capillary water): Rate of absorption is directly dependent on water availability. Its rate is slowed down when amount of water will be increased in the soil.

Wilting coefficient: It is percentage of water left in soil. Wilting occurs when rate of transpiration is more than rate of absorption of water from soil. It is of two types: (i) **Temporary wilting**: It occurs during day. In night, the rate of transpiration becomes slow hence leaves regain their natural turgid condition. Thus it is the result of drought and (ii) **Permanent wilting**: It occurs under condition of soil water stress. The wilted leaves do not regain their natural state at night and remain as such. Thus it is the result of soil dryness.

Ascent of Sap

It is upward transport of sap (water with dissolved minerals) from root to aerial parts of plant through xylem against gravitational force. Usually sap has less than 1% solute concentration and its pH is 5.0. The ascent of sap is quite rapid process and occurs through xylem with a rate of 75-100 cm/min. The lumen of tracheids and vessels (not wall) of xylem is responsible for ascent of sap.

Evidence in favour of xylem:

Ringing experiment: It was performed by Hartig. He selected two plants and cut under water. In one plant, outer ring of tissue containing phloem (bark) is removed. In another plant, bark is firstly removed, then xylem is removed and then bark is replaced. It was observed that the leaves of plant in which xylem was removed, wilted showing that water is conducted through xylem. There was no wilting of leaves in another plant in which xylem was remained intact. It cannot be performed in monocots because of scattered vascular bundle.

Theories:

1. **Vital theories**: Living cells are involved in ascent of sap.

   (a) Relay pump theory: It was proposed by Godlewsky (1884). Xylem parenchyma and medullary rays are responsible for ascent of sap by pumping water to xylem cells at higher level due to changes in their osmotic pressure.

   (b) Pulsation theory: It was proposed by J.C. Bose (1928). The innermost cortical cells are responsible for ascent of sap. He noticed great vibration in galvanometer when electric current passed in innermost cortical cell.

The above theories were rejected because ascent of sap continued after killing the living cells by poison or chemicals.

2. **Physical theories**: The physical forces of dead cells are responsible for ascent of sap.
   - **(a) Capillary theory**: It was proposed by Boehm (1809). It is due to capillary action (force of surface tension) of xylem vessels.
   - **(b) Imbibition theory**: It was proposed by Unger and Sachs (18878). It is due to imbibition of lumen of vessels and tracheids of xylem.

   The above theories were rejected because it cannot lift water more than 2 metres and further it cannot operate in Gymnosperms where vessels are absent.

3. **Root pressure theory**: It was proposed by Priestley. The root pressure is positive hydrostatic pressure developed in xylem of root due to movement of water from soil into root hairs and from there to xylem. It helps in pushing water to rise in plants. It is measured by Manometer. It is applicable to small sized and soft stemmed plants. This theory was also rejected on the following observations:
   - (a) It is not universal because it does not occur in Gymnosperms.
   - (b) It is not developed in all seasons
   - (c) It is absent in rapidly transpiring plants.
   - (d) It usually occurs at night but maximum ascent of sap occurs in day time.
   - (e) It is not applicable for shrubs and tall trees (because maximum root pressure recorded in plants is 2 atm which can lift water up to a height of 1-2 m).
   - (f) Water can be raised even in absence of roots.

4. **Cohesion Tension or Transpiration pull theory**: It was proposed by Dixon and Jolly (1894). It is widely accepted theory. It is based on
   - (a) Cohesive and adhesive properties of water: Water has very high cohesive force up to 350 atm which develops due to formation of H-bonds between water molecules and adhesive force is developed between water and hydrophilic wall of tracheids and vessels of xylem.
   - (b) Establishment of water column: These cohesive and adhesive forces form water column in xylem elements or it cannot be broken or pulled away from wall of xylem by gravitational forces. Water column has great tensile strength (the ability of getting stretched without breaking) and high capillarity so it is continuous from the root hair cell to transpiring leaf cells.
   - (c) Transpiration pull or tension or suction pressure exerted on water column: It is generated at the top of water column and is transmitted downwards from petiole of leaves, stem and finally reach roots. The transpiration pull or tension of 1 atm is enough to pull water to a height of about 10 m.

Process: The ascent of sap occurs in a well coordinated manner. It involves the following sequential events:

- (a) Water is lost from epidermis of leaf as a result of transpiration through stomata.
- (b) The epidermal cell develops low water potential (water deficit) up to -30 atm.
- (c) The epidermal cell receives water from adjacent mesophyll cell.
- (d) Mesophyll cell now develops low water potential.
- (e) Mesophyll cell draws water from xylem of leaf present on midrib.

(f) Xylem vessels of leaf in turn draw water from xylem of main stem.

(g) A negative pulling pressure is then exerted on stem by all leaves. This combined pressure is called transpiration pull.

(h) Transpiration pull of 20-30 atm is enough to raise water to the tallest trees. It overcomes gravitational pull of 10 atm, resistance of narrow xylem channels, and resistance of living cells of roots that lie in the path of water from soil to xylem.

(i) Transpiration pull or tension is relieved when water is absorbed through roots,

(j) Now water column moves and does not break due to cohesive and adhesive forces.

(k) This is purely physical process and does not require living cells

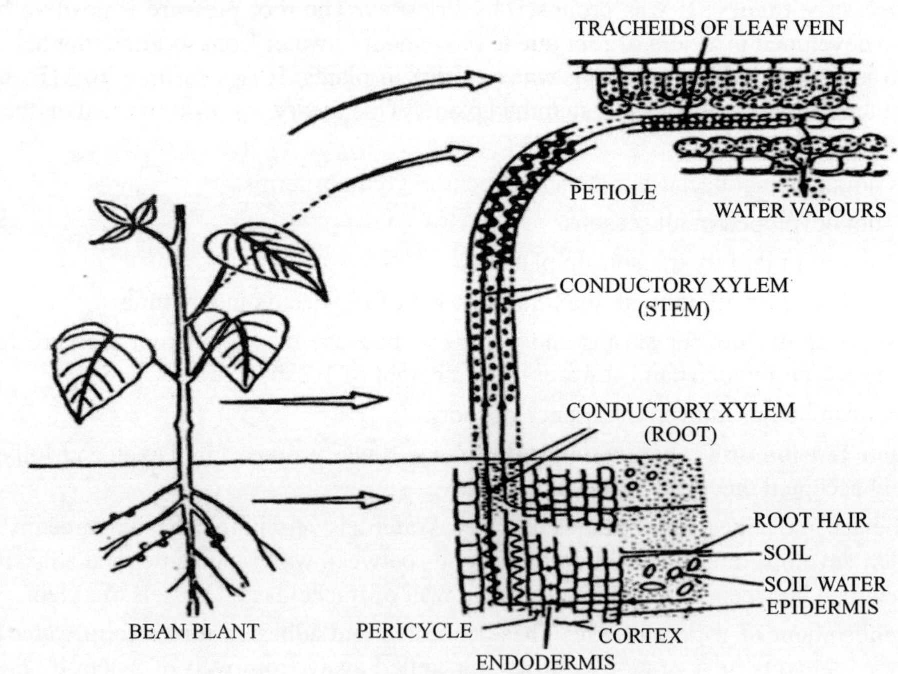

Fig. ***Ascent of sap***

Objections:

(a) There is enough chance of air bubbles in xylem vessels which may break the continuity of the column.

Factors:

(a) Temperature: Rate of ascent of sap increases with rise in temperature

(b) Atmospheric pressure: Rate of ascent of sap decreases with rise in atmospheric pressure.

(c) Atmospheric humidity: Rate of ascent of sap decreases with rise in humidity.

(d) Wind velocity: Rate of ascent of sap increases with increase in wind velocity.

Transpiration

It is the loss of water in the form of vapour (gas) from the aerial organ of plant. It is slow physiological process at the exposed surface of plants. It is universal process occurring in all plants. It is passive process because metabolic energy in the form of ATP is not required for movement of water. It is regulated by guard cells. Out of total absorbed water by land plants, 98% of water evaporates and is lost in atmosphere and rest 2% of water is used for maintenance of their life.

The driving force for transpiration is the difference in water vapour pressure between the leaf air spaces and the external air. It is directly proportional to vapour pressure difference. Transpiration is controlled by several factors such as osmotic pressure and turgor pressure.

Types of transpiration

1. **Stomatal transpiration:** About 90% of transpiration occurs through stomata of young stem and leaves. It occurs during day when stomata open.
2. **Cuticular transpiration**: About 10% of transpiration occurs through cuticle of young stem and leaves by diffusion. It occurs throughout day and night. Actually cuticle checks water loss but permits water vapour to pass by diffusion when cuticle is thin.
3. **Lenticular transpiration:** About a small fraction (0.1%) of transpiration occurs through lenticels in bark and twig of woody (old) dicot stem. It occurs throughout day and night.

Mechanism of stomatal opening:

Potassium pump theory: It was proposed by Levitt. It is widely accepted theory. The opening of stomata is an active process which occurs during day (light). The stomatal opening involves the following events:

(a) Photosynthesis occurs and CO_2 concentration is decreased.
(b) pH of guard cell increases.
(c) Starch of guard cell is converted into glucose phosphate and then to phosphoenol pyruvic acid (PEP) via glycolysis. Then PEP binds with CO_2 by phosphoenol pyruvate carboxylase (PEPcase) to form oxaloacetic acid which is then reduced to Malic acid.
(d) Malic acid dissociates into malate anion and H^+ in guard cell.
(e) H^+ is transported to epidermal cell via subsidiary cell (Efflux of H^+).
(f) K^+ enters into guard cell from epidermal cell via subsidiary cell through H^+-K^+ ion exchange pump in plasma membrane. Influx of K^+ is active process and it is stimulated by cytokinin.
(g) K^+ is balanced by malate anion in guard cell.
(h) Some Cl^- also enters into guard cell to neutralise a small amount of K^+.
(i) Solute concentration (malate anion, K^+, Cl^-) is increased in guard cell.
(j) Water potential is decreased in guard cell.
(k) Water enters into guard cell due to endosmosis.
(l) Turgor pressure is increased in both guard cell. Then the thin outer wall bulges out and forces inner wall into a crescent shape.
(m) The guard cell increases more in length than width due to radial orientation of cellulose microfibrils.
(n) Thus stomata opens.

Periods of stomatal opening:

1. Alfalfa type: Stomata opens throughout day and close at night. e.g., Apple, Grapes, Mustard, Pea, Turnip.
2. Barley type: Stomata opens only for a few hours in day and remain closed in rest time. e.g., Cereals

3. Potato type: Stomata opens whole day and night except for a few hours in evening. e.g., Banana, Onion, Potato
4. Equisetum type: Stomata is always open.

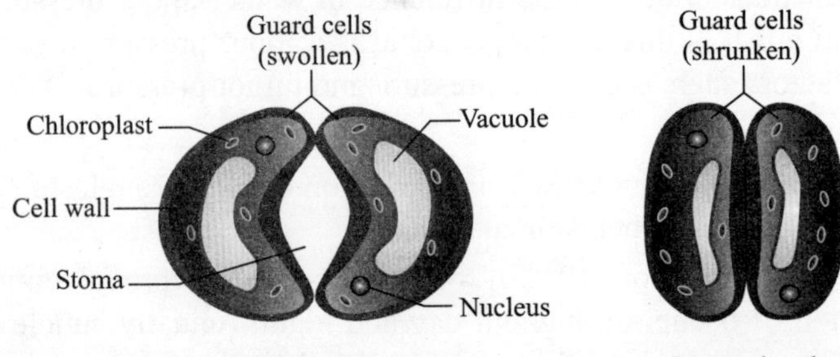

Mechanism of stomatal closing: It occurs during night (darkness). The closure of stomata involves the following steps:

(a) Photosynthesis stops and respiration is continued.
(b) CO_2 concentration is increased.
(c) pH of guard cell is decreased,
(d) Abscisic acid (ABA) becomes functional at low pH.
(e) ABA stops uptake of K^+ by changing diffusion and permeability of guard cell.
(f) H^+ enters into guard cell from epidermal cell.
(g) Malate anion combines with H^+ to form malic acid in guard cell. It is converted into starch.
(h) K^+ is transported from guard cell to subsidiary cell.
(i) Solute concentration (only Cl^-) is decreased in guard cell.
(j) Water potential is increased in guard cell.
(k) Water leaves the guard cell due to exosmosis.
(l) Turgor pressure is decreased. Then the inner elastic wall of guard cell regains its original shape from crescent shape of daytime. Now guard cell becomes flaccid.
(m) Thus stomata closes.

Evidence:

1. Starch is changed to malic acid at the time of stomatal opening.
2. Guard cell of opened stomata has abundant K^+.

Importance:

(a) It removes excess water.
(b) It helps in passive absorption of water and minerals.
(c) It helps in ascent of sap in tall trees.
(d) It produces cooling effect on leaves and prevents heating of leaves.
(e) It helps in development of mechanical tissues.

Disadvantages:

(a) It causes wilting of leaves due to loss of water.
(b) It affects growth due to reduction in availability of water.
(c) It wastes energy while absorbing actively

External factors:

(a) Temperature: Rate of transpiration increases with rise in temperature. Low temperature (0°C) reduces transpiration in light. High temperature (>40°C) reduces transpiration because rate of respiration is increased hence CO_2 is formed.

(b) Light: It is essential for stomatal opening hence rate of transpiration increases. Blue light is more effective than red light because it activates the H^+-ATPase present on plasma membrane of guard cell and further helps in breakdown of starch and malate synthesis.

(c) Atmospheric pressure: Rate of transpiration is inversely proportional to atmospheric pressure.

(d) Relative Humidity (RH): Rate of transpiration is inversely proportional to relative humidity. It increases in dry atmosphere due to low RH. It decreases in humid atmosphere due to high RH (air is already saturated with water hence low water potential gradient).

(e) Wind velocity: Rate of transpiration decreases in high wind due to closure of stomata but increases in slow wind.

(f) Water availability: If water is unavailable, there is no absorption hence there is no transpiration.

Internal factors:

(a) Structure of leaf: A few structures of leaf such as thick cuticle, sunken stomata, waxy coating on leaves, presence of hairs, formation of prickles, spines and scaly leaves check the rate of transpiration.

(b) Number of stomata: Rate of transpiration increases with increase in number of stomata.

(c) Surface area of leaf: Rate of transpiration increases with increase in surface area of leaf.

(d) Root-shoot ratio: Rate of transpiration is directly proportional to R/S ratio. Transpiration is more if absorption of water will be more.

Terms:

Transpiration ratio (TR): It is amount of water transpired per unit of dry matter manufactured by the plant and the number of CO_2 assimilated by photosynthesis.

$$TR = \frac{\text{Amount of water lost}}{\text{Gram dry weight produced}}$$

It is 900/1 in C_3 plants, 300/1 in C_4 plants and 50/1 in CAM plants.

Transpiration flux: It is amount of water transpired by unit area of leaf surface in a unit time.

Stomatal index (SI): It is the correlation between number of stomata and number of epidermal cells per unit area. It is expressed in percentage.

$$SI = \frac{S}{E+S}$$

where S is number of stomata and E is number of epidermal cells.

Antitranspirants:

(a) Film formers: These check diffusion of water vapour. e.g., Silicon oil, Wax

(b) Metabolic inhibitors: These inhibit opening of stomata. e.g., Aspirin, Malic acid, Salicylic acid, Phenyl Mercuric Acetate (PMA) etc.

Measurement:

(a) Porometer: It measures the relative size of stomata.
(b) Potometer: It measures and compares the rate of transpiration.
(c) Cobalt chloride paper method: It is used to compare the rate of transpiration. Water vapour coming out stomata turns blue cobalt chloride paper to pink.

Guttation: It is the loss of water in the form of liquid droplets from the leaves and other parts of injured plants. It was first studied by Bergerstein (1897). It occurs through hydathode. It is a pore in epidermis followed by a large intercellular space, loosely arranged parenchyma called epithem and xylem elements. It is due to root pressure when rate of water absorption is more than the rate of transpiration where sap flows from vein ending, epithem and then hydathode. It is commonly seen in herbaceous plants such as Garden Nasturtium (*Tropaeolum*), Cucurbits, *Colocasia*, Oat, Potato, Tomato. It is not seen in all plants. It regulates water content and turgidity of the leaves as well as the whole plant.

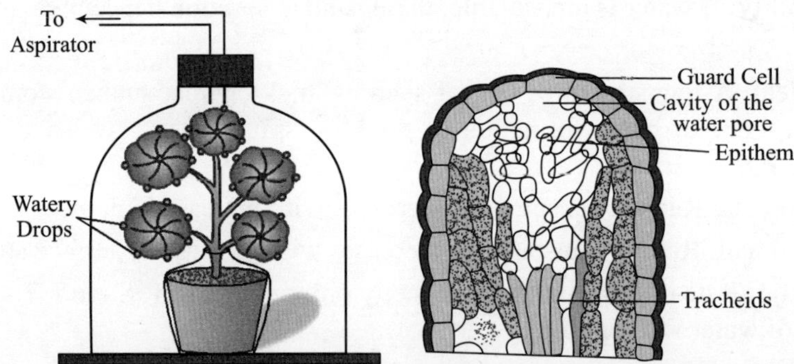

Fig. *Guttation*

Transpiration	Guttation
It occurs in the form of vapour.	It occurs in the form of liquid droplets.
It occurs through stomata, cuticle or lenticels.	It occurs through hydathode
It occurs from general surface of leaf.	It occurs from margin and tip of leaf.
It occurs in day time.	It occurs in night and early morning.
It occurs during dry period.	It occurs during humid period.
It is controlled process.	It s uncontrolled process.
It removes large amount of water.	It removes less amount of water.
It releases pure water.	It releases impure water having organic and inorganic materials.
It causes wilting if excess transpiration occurs.	It does not cause wilting.

Bleeding: It is loss of liquid from injured or cut part of plants anytime. It occurs due to root pressure. It is found in Toddy palm, Rubber plant, *Betula, Vitis*.

Translocation of solutes:

It is transport of organic solute (food) in soluble form from leaf to roots, stem, fruits and seeds through phloem. It takes place by sieve tube of phloem in the form of sucrose. The leaf is the production centre or source whereas root, stem and fruits are consumption centres or sink. It is bidirectional movement along concentration gradient (from high concentration to low concentration) from leaf to storage region and from storage region to growing parts of plant. Its usual speed is 100 cm/hr.

Directions:

1. Downward direction: The organic solutes are transported in downward direction from leaf to root and stem for growth, metabolism, storage and consumption. It is most common.
2. Upward direction: The organic solutes are transported upwards from leaf to stem apex, flower and fruit. In case of deciduous plants where leaves are fallen off the reserved food present in the root and stem helps in growth and development.
3. Radial: The organic solutes are transported from phloem to pith, cortex and epidermis through medullary rays.

Evidences in favour of sugar:

1. Analysis of phloem sap: The chemical analysis of phloem sap (pH=7.5-8.5) confirms that 90% of substance have non-reducing sugar and 10% have other substances such as amino acids, hormones, mannitol, vitamins and K^+. Sugar is present in the form of sucrose which is relatively inactive, but more stable and highly soluble playing with little direct role in metabolism making an ideal transportable sugar. Sucrose is converted back to more active glucose and fructose on reaching consumption centre. Some other sugars such as raffinose (triose), stachyose (tetrose) and verbascose (pentose) etc. are also present in little amount.
2. Tracer technique: It was confirmed through radio-labelling technique where labelled sugar was found to be present in phloem when heavier isotope of carbon was labelled to a leaf during photosynthesis.

Evidences in favour of downward translocation:

1. **Girdling experiment** (Ringing experiment): It was demonstrated by Hartig (1837). In this bark (all tissues outside the vascular cambium including phloem) of a plant was removed from stem below leaf. After sometimes, it was observed that part above ringed portion swelled and part below ringed portion dried and died losing all the stored food. Sugar content was found to be much higher than that of shrivelled or dried part because translocation of sucrose down the plant stopped but upward conduction of water continued. Secondary phloem was found to be involved in translocation of food in downward direction towards root.

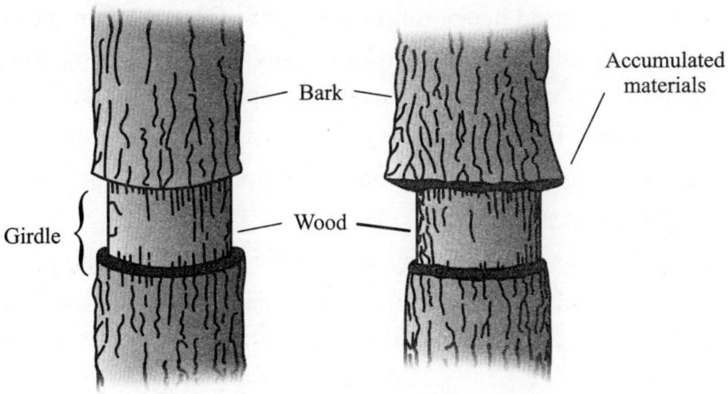

Fig. *Girdling experiment showing downward translocation*

Girdling experiment is not successful in monocots because of scattered vascular bundle. There are several cases which are as follows:

1. If trunk of tree is ringed or girdled then roots die first because roots do not get food.
2. If ringing is done below fruit (including leaves), fruits would be larger in size.
3. If ringing is done between fruits and leaves, fruits would be smaller in size.

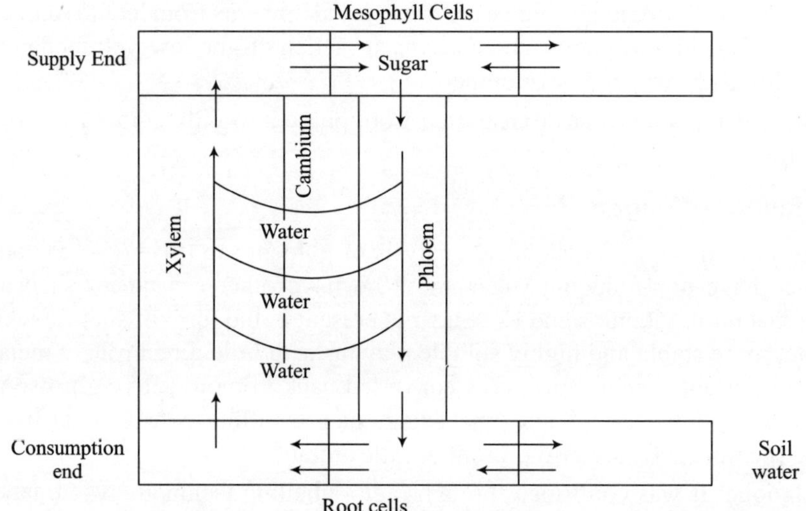

Mechanism: **Mass flow or pressure flow hypothesis**: It was proposed by Munch (1936). The organic solutes are translocated '*en masse*' from source to sink due to turgor pressure gradient between source and sink. It is bidirectional-downwards as well as upwards.

Process:

(a) Glucose is synthesised in mesophyll cell of leaf as a result of photosynthesis.
(b) Glucose is converted into sucrose (oligosaccharide).
(c) Sucrose gets dissolved in cell sap of mesophyll cell.
(d) The solute concentration is increased in mesophyll cell hence water potential is decreased.
(e) Water enters into mesophyll cell from xylem by endosmosis.
(f) Water potential is now increased hence turgor pressure is increased in mesophyll cell.
(g) Turgor pressure forces dissolved sucrose from mesophyll cell to companion cell and then sieve tube of phloem by passive transport. This is called vein loading which is active process.
(h) Solute concentration is now increased and water potential is decreased in phloem.
(i) Water enters into phloem from adjacent xylem by endosmosis (water potential gradient).
(j) Hydrostatic pressure increases in phloem.
(k) Phloem sap (dissolved sucrose) now moves to the area of low osmotic pressure (sink).
(l) Now pressure flow starts when a continuous pressure gradient is established across symplast between mesophyll cell of leaf (source) and root (sink). This is continued as long as vein loading occurs.
(m) Sucrose moves '*en masse*' one sieve tube to next sieve tube in phloem by active transport. This is called phloem unloading or vein unloading.

(n) The organic solutes are either used up or stored in insoluble form at consumption centre (as in root).

(o) Water returns to leaf through apoplast of root, stem and leaf.

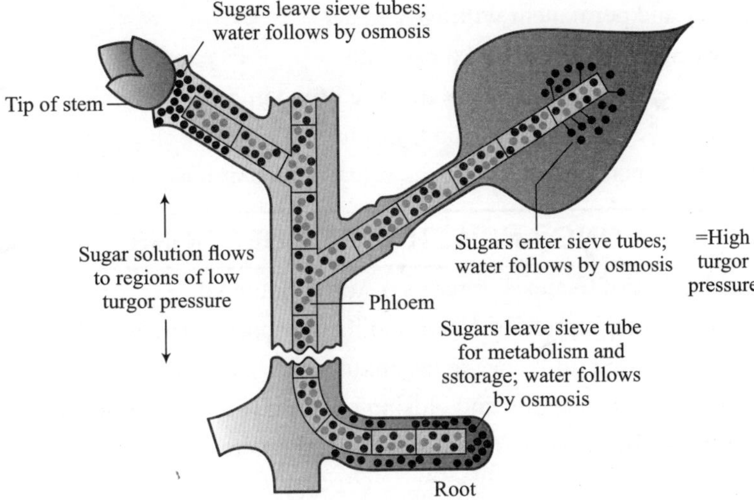

Fig. ***Translocation of solute***

Evidences in favour of Munch's hypothesis:

(a) The sap contains high content of sugar from cut end when a woody or herbaceous plant is girdled.

(b) Positive concentration gradient disappears when plants are defoliated.

(c) The growth hormones move very fast in illuminated leaves than shaded leaves.

Factors:

(a) Temperature: The optimum temperature is 25-30°C. If temperature rises then rate of translocation will increase.

(b) Light: It is essential for translocation.

(c) Oxygen : The rate of translocation is aerobic process

(d) Minerals: Boron stimulates the translocation of sugar.

(e) Inhibitors: Several chemicals such as dinitrophenol, arsenite, fluorides, azide will inhibit the rate of translocation.

SHORT ANSWER TYPE QUESTIONS

1. Distinguish between simple diffusion and facilitated diffusion.
2. Explain the terms symport, antiport and uniport.
3. Write notes on active transport.
4. Define hypotonic, hypertonic and isotonic solutions.
5. Explain Diffusion pressure deficit (DPD).
6. Mention different types of membranes on the basis of permeability.
7. Explain the stages of plasmolysis.

8. Write notes on imbibition.
9. Write notes on root pressure theory of ascent of sap.
10. Distinguish between transpiration and guttation.
11. Explain temporary and permanent wilting.
12. Distinguish between plasmolysed and turgid cell.
13. Explain the ringing experiment to prove the role of xylem in ascent of sap.
14. Discuss the impact of external factors on transpiration.
15. Explain the ringing experiment to prove the role of phloem in ascent of sap.

LONG ANSWER TYPE QUESTIONS

1. Explain the mechanism of osmosis in terms of water potential.
2. Discuss the method of absorption of water and its transport up to xylem of root.
3. Describe the transpiration pull theory in ascent of sap.
4. Discuss the mechanism of opening and closing of stomata in transpiration.
5. Discuss the mass flow hypothesis for translocation of solutes.

MULTIPLE CHOICE QUESTIONS

1. The maximum diffusion pressure occurs in
   (a) A molal solution (b) A molar solution (c) A normal solution (d) Pure solvent
2. Osmosis cannot be demonstrated by a potato osmoscope using a solution of NaCl instead of sugar solution because potato tissue is
   (a) Permeable to Na ion (b) Permaeble to Cl ion
   (c) Permeable to salt solution (d) Impermeable to salt solution
3. When a cell is kept in 0.5 M solution of sucrose, its volume does not alter. If the same cell is placed in 0.5 M solution of NaCl then
   (a) Cell is plasmolysed (b) Cell size will increase
   (c) Cell size would increase (d) There will be no change
4. If cell A with OP =5 atm and TP =4 atm is surrounded by cells with OP =3 atm and TP=1 atm then what will happen
   (a) Water will move from cell A to other cells (b) Water will move from other cells to cell A
   (c) Water will not move (d) None of the above
5. Vein loading occurs due to
   (a) Turgor pressure from mesophyll cell to phloem
   (b) Turgor pressure from phloem to mesophyll cell
   (c) Movement of sucrose from source to sink
   (e) Movement of sucrose from sink to source
6. If a formalin preserved *Spirogyra* plant is placed in a hypertonic solution then what will happen
   (a) It will gain turgidity (b) It will lose turgidity
   (c) It becomes plasmolysed (d) Nothing will happen.
7. If a thin slice of sugar beet is placed is placed in a concentrated solution of NaCl then what will happen
   (a) It will lose water from the cells (b) It will become turgid
   (c) It will absorb water from the salt solution (d) It will neither absorb nor lose water.
8. In seed germination the first phenomenon takes place is
   (a) Plasmolysis (b) Diffusion (c) Osmosis (d) Imbibitions
9. Water enters into root hairs from the soil due to
   (a) Turgor pressure (b) Osmotic pressure (c) Root pressure (d) Suction pressure
10. If a stem of plant is girdled then
    (a) Root dies first (b) Shoot dies first
    (c) Both root and shoot die simultaneously (d) Neither root nor shoot die
11. In hollow hearted trees what happens
    (a) Root dies first (b) Shoot dies first
    (c) Both root and shoot die simultaneously (d) Neither root nor shoot die
12. The maximum harm to a tree occurs when there is
    (a) Loss of half of its leaves (b) Loss of all of its leaves
    (a) Loss of its bark (d) Loss of half of its branches

13. Which of the following plants do not transpire?
    (a) Algae (b) Fungi
    (c) Submerged hydrophytes (d) All of the above
14. Guard cell differs from epidermal cells in having
    (a) Nucleus (b) Mitochondria (c) Ribosomes (d) Chloroplast
15. In guard cells when sugar is converted into starch, then
    (a) Stomatal pore opens partially (b) Stomatal pore opens fully
    (c) Stomatal pore closes completely (d) Stomatal pore remains unchanged
16. Comparative frequency of stomata on a leaf can be determined by
    (a) Porometer (b) Potometer
    (c) Cobalt chloride paper method (d) All of the above
17. Wilting of plant occurs when
    (a) Phloem is blocked (b) Xylem is blocked
    (c) Both xylem and phloem are blocked (d) A few roots are removed.
18. Root pressure is maximum when
    (a) Transpiration is high and absorption is low (b) Transpiration is low and absorption is high
    (c) Both transpiration and absorption are high (d) Both transpiration and absorption are low
19. Guttation occurs if a plant is kept in
    (a) Moist atmosphere (b) Humid soil (c) Dry environment (d) Water
20. Stomata in angiosperms usually do not open during
    (a) Noon (b) Twilight (c) Mid night (d) 10 o'clock
21. If a plant is kept in total darkness, its daily rhythm for opening and closing of stomata
    (a) Stops immediately (b) Continues for some time
    (c) Continues for long time (d) Continues indefinitely
22. Most plant tissues cease to grow when water potential drops to about
    (a) –8 bars (b) –10 bars (c) –15 bars (d) –20 bars
23. Blue light keeps stomata open during the day by
    (a) Stimulating photosynthesis in guard cells
    (b) Promoting movement of sugar into guard cells
    (c) Promoting movement of K^+ into guard cells
    (d) Stimulating formation of K^+ in guard cells.
24. Glycolate induces opening of stomata in
    (a) Presence of O_2 (b) Low CO_2 (c) High CO_2 (d) Absence of CO_2
25. The soil water easily available to the root system is
    (a) Gravitational water (b) Hygroscopic water (c) Capillary water (d) Combined water
26. The water potential and osmotic potential of pure water is, respectively
    (a) 100 and zero (b) Zero and zero (c) 100 and 100 (d) Zero and 100
27. Water potential in the leaf cells is positive during
    (a) Low transpiration (b) Guttation
    (c) Excessive transpiration (d) Excessive absorption

28. If turgidity of a cell surrounded by water increases, the wall pressure will
    (a) Increase (b) Decrease (b) Remain unchanged (d) Fluctuate
29. A plasmolysed cell can be deplasmolysed by placing it in
    (a) Isotonic solution (b) Hypotonic solution
    (c) Hypertonic solution (d) Saturated solution
30. If a cell is kept in hypertonic solution after treating with alcohol, then
    (a) It will shrink (b) It will swell
    (c) It will burst (d) It will remain same
31. If a flower is cut and dipped in dilute NaCl solution then
    (a) Exosmosis will occur (b) Endosmosis will occur
    (c) Shrinkage will occur (d) Nothing will happen
32. Plasmolysis is not observed in boiled plant tissue because
    (a) The cell wall becomes impermeable to water
    (b) The cell wall disintegrates
    (c) The cell membrane disintegrates
    (d) The cell membrane becomes impermeable to water
33. A cell will experience the highest level of endosmosis when it is kept in
    (a) Salt solution (b) Sugar solution (c) Distilled water (d) Protein solution
34. Which of the following is not cotransport?
    (a) Symport (b) Antiport (c) Uniport (d) All of the above
35. Facilitated diffusion is similar to active transport in
    (a) It is very specific (b) It shows saturation effect
    (c) It is sensitive to inhibitors (d) All of the above
36. Which of the following is incorrect regarding water potential?
    (a) It is always maximum in pure water
    (b) It is 2.3 Bar in 0.1 M sugar solution
    (c) It will increase with addition of solutes in pure water
    (d) It is equal to solute potential at atmospheric pressure
37. Which of the following is not selectively permeable membrane?
    (a) Plasma membrane (b) Tonoplast (c) Nuclear membrane (d) Egg membrane
38. Imbibition is highest in
    (a) Agar agar (b) Protein (c) Starch (d) Cellulose
39. Which of the following plant hormones is involved in stomatal opening?
    (a) Abscisic acid (b) Auxin (c) Gibberellin (d) Cytokinin
40. The pH of phloem sap is
    (a) Less than 5 (b) Between 5 and 6 (c) 7 (d) Between 7 and 8
41. As a tree begins transpiring in the morning, tension occurs first in
    (a) Root (b) Stem (c) Leaf (d) All parts of tree
42. The lowest water potential in the xylem is found in
    (a) Root hair (b) Vascular cylinder of root
    (c) Tracheids of stem (d) Leaves

43. If a plant cell is placed in deionised water then the water potential of that cell will become
    (a) More positive because pressure potential becomes more positive
    (b) More positive because pressure potential becomes more negative
    (c) More negative because pressure potential becomes more negative
    (d) Less negative because pressure potential becomes more positive
44. According to pressure flow hypothesis of Munch
    (a) Sugar concentration is highest near sink area
    (b) Water enters the sieve tube by osmosis
    (c) Sugar is transported through phloem as glucose
    (d) Sugar is transported out of sieve tube
45. Sugar moves from sieve tubes into the tissue by
    (a) Diffusion (b) Facilitated diffusion
    (c) Active transport (d) All of the above
46. If one cell is kept in 1M solution of sucrose and another cell is kept in 1M solution of NaCl, then the water potential of cell kept in sucrose solution will be
    (a) Equal to that of cell kept in NaCl solution
    (b) Greater than that of cell kept in NaCl solution
    (c) Lower than that of cell kept in NaCl solution
    (d) No effect
47. Phloem sap consists of mainly
    (a) Water and minerals (b) Water and glucose
    (c) Water and fructose (d) Water and sucrose
48. The smallest amount of pressure needed to stop fluid from moving by osmosis is called
    (a) Water potential (b) Turgor pressure (c) Pressure potential (d) Solute potential
49. The osmotic expansion of a cell kept in water is chiefly regulated by
    (a) Vacuole (b) Mitochondria (c) Chloroplast (d) Ribosome
50. Osmotic potential in plant cell is maintained by
    (a) Proline and β-glycine (b) Histidine
    (c) Lysine (d) Glycine
51. If the concentration of Calcium is 0.3% in a cell and 0.1% in surrounding fluid, then the cell obtains more calcium by
    (a) Diffusion (b) Facilitated diffusion (c) Active transport (d) Osmosis
52. Which of the following solution s has the least osmotic pressure?
    (a) 0.5 M glucose (b) 0.5 M NaCl (c) 0.5 M Na_2SO_4 (d) 0.49 M starch
53. Na^+-K^+ pump is an example of
    (a) Symport (b) Antiport (c) ABC transport (d) Diffusion
54. Guttation is most commonly observed in
    (a) High atmospheric humidity and plenty soil water
    (b) Low atmospheric humidity and plenty soil water
    (c) High atmospheric humidity and little soil water
    (d) Low atmospheric humidity and little soil water

55. Water moving through apoplast from the soil to stele must cross a plasma membrane of
(a) Root hair (b) Cortex (c) Endodermis (d) Vessel
56. If a plasmolysed cell has OP=10 atm, TP= -2 then its DPD will be
(a) 8 (b) 12 (c) 10 (d) –2
57. When sugar is actively transported into a cell then what happens inside the cell?
(a) TP does not change because sugar concentration has no effect on TP
(b) TP increases because sugar concentration directly affects TP
(c) TP decreases because water exits and affects TP
(d) TP increases because water enters and affects TP
58. Water tends to move inside the cell that has
(a) High turgor pressure (b) More positive water potential
(c) More negative water potential (d) Low turgor pressure
59. The guard cells in monocots is
(a) Bean shaped (b) Ribbon shaped
(c) Cup shaped (d) Dumb bell shaped
60. Which of the following modes of transpiration occurs whole day and night?
(a) Stomatal transpiration (b) Cuticular transpiration
(c) Stomatal and Cuticular transpiration (d) Cuticular and Lenticular transpiration
61. The number of aquaporin proteins in water filled channel is
(a) 1 (b) 2 (c) 4 (d) 8
62. What happens to water potential if solutes are added to solution?
(a) It increases (b) It decreases
(c) It remains constant (d) It first decreases then increases
63. Which of the following processes confirms that plasma membrane is semi-permeable?
(a) Osmosis (b) Reverse osmosis (c) Plasmolysis (d) Imbibition
64. Which of the following cells is responsible for ascent of sap according to pulsation theory of J.C.Bose?
(a) Pericycle (b) Endodermis (c) Outer cortical cell (d) Inner cortical cell
65. Which of the following phenomena is involved in ascent of sap according to Dixon and Jolly?
(a) Cohesive and adhesive force (b) Formation of water column
(c) Transpiration pull (d) All of the above
66. Which of the following hormones takes part in stomatal opening?
(a) Auxin (b) Gibberellin (c) Cytokinin (d) Abscicic acid
67. Abscisic acid starts functioning in guard cell of stomata at
(a) Low pH (b) Neutral pH (c) High pH (d) All of the above
68. Which of the light is more effective in transpiration?
(a) Red light (b) Blue light (c) Far red light (d) Violet light
69. Which of the following statements is incorrect regarding guttation?
(a) It occurs through hydathode
(b) It It is loss of water in the form of liquid drop
(c) It is uncontrolled process
(d) It releases pure water

70. Girdling experiment is not successful in monocots because of
    (a) Scattered vascular bundle (b) Absence of cambium
    (c) Parallel venation (d) All of the above
71. The carrier proteins arfe involved in
    (a) Symport (b) Uniport (c) Antiport (d) All of the above
72. There are three cells viz. A, B and C. The water potentials in cell A, B and C are zero, -3 and -6, respectively. What will be the direction of movement of water?
    (a) A→B. B→C, A→C (b) B→A. B→C, A→C
    (c) A→B. C→A, B→C (d) B→C. B→A, A→C
73. Maximum osmotic pressure was observed in
    (a) Mesophytes (b) Hydrophytes (c) Xerophytes (d) Halophytes
74. If a cell is partially turgid, then water potential
    (a) Zero (b) Positive (c) Less negative (d) More negative
75. The demand of water is the highest in
    (a) Fully turgid cell (b) Partially turgid cell (c) Flaccid cell (d) Plasmolysed cell
76. Tonoplast of vacuole is
    (a) Permeable (b) Impermeable
    (c) Semipermeable (d) Selectively permeable
77. Which of the following sequcnces involve in plasmolysis
    (a) Incipient plasmolysis-Limiting plasmolysis-Evident plasmolysis
    (b) Limiting plasmolysis-Incipient plasmolysis-Evident plasmolysis
    (c) Incipient plasmolysis-Evident plasmolysis-Limiting plasmolysis
    (d) Limiting plasmolysis-Evident plasmolysis-Incipient plasmolysis
78. The movement of water is interrupted in endodermis due to the presence of casparian strip which is made up of
    (a) Cutin (b) Lignin (c) Pectin (d) Suberin
79. Root pressure theory for ascent of sap was rejected due to following observations except
    (a) It does not occur in gymnosperm (b) It is absent in rapidly transpiring plants
    (c) It is not applicable for shrubs and tall trees (d) It does not occur throughout day and night
80. Stomatal opening requires
    (a) Efflux of H^+ from guard cell (b) Efflux of K^+ from guard cell
    (c) Efflux of Cl^- from guard cell (d) All of the above

ANSWERS

1.d	2.c	3.a	4.a	5.a	6.d	7.a	8.d	9.d	10.a
11.d	12.c	13.d	14.d	15.c	16.c	17.b	18.b	19.a	20.c
21.b	22.c	23.c	24.b	25.c	26.b	27.a	28.a	29.b	30.d
31.b	32.c	33.c	34.c	35.d	36.c	37.d	38.a	39.d	40.d
41.c	42.d	43.d	44.b	45.c	46.b	47.d	48.d	49.a	50.a
51.c	52.d	53.b	54.a	55.c	56.b	57.d	58.c	59.d	60.d
61.d	62.b	63.c	64.d	65.d	66.c	67.a	68.b	69.d	70.a
71.d	72.a	73.d	74.c	75.d	76.d	77.b	78.d	79.d	80.a

9

Mineral Nutrition

Mineral nutrition is uptake of inorganic minerals present in soil and their use by plants.

Von Helmont (1648) studied on mineral nutrition. He observed that when a dry plant or plant part was heated at high temperature at 600° C, the organic matter of plant dissociated forming ash containing oxides and carbonates of different elements. About 60 elements have been reported in which 30 elements are common to all plants and rest 30 elements vary. There are 17 essential elements playing specific structural and physiological role and 13 nonessential elements out of 30 common elements. Plant ash contains 50% of potassium alone.

The deficiency symptoms may be **acute** when a particular nutrient becomes unavailable to growing plant and **chronic** when a particular nutrient is available but inadequate to the demands of plant.

Essentiality of an element: It was proposed by Arnon and Stout (1939). There are following criteria to know the essentiality of an element.

(a) An element must be essential for growth and reproduction

(b) An element must be specific and not replaceable by another element.

(c) The deficiency of an element causes disorder or disease if it is below critical concentration.

(d) The disorders and disease can by corrected by the availability of that element.

Evidences:

1. **Hydroponics:** It is soilless cultivation or growth of plants by placing the roots in a nutrient solution. It was demonstrated by Julius Von Sachs (1860) and later by Knop (1885). It is generally practised in US and gulf countries.

   Process:

   (a) Firstly nutrient medium (Knop medium or Arnon & Hoagland medium) is prepared. It contains mineral salts and pure water.

   (b) Solution was poured in a jar and covered with black paper to avoid sunlight directly.

   (c) Roots are dipped in nutrient medium.

   (c) Aeration is provided to roots with adequate O_2.

   (d) Then the growth was observed.

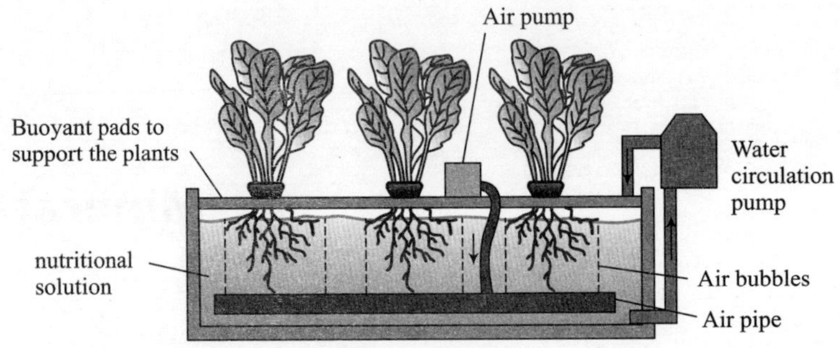

Fig. ***Hydroponics***

Advantages:

(a) It is done to identify the requirement of essential element.

(b) It is done for commercial production of vegetables and flowers out of season.

(c) It is useful in infertile soil or dry soil.

(d) It controls soil borne pathogens.

(e) It avoids problem of weeding.

(f) It conserves water and gives better yield consistently.

Disadvantage: The whole process is costlier.

2. **Aeroponics**: It is technique of growing plants with their roots suspended in plastic vessel bathed in nutrient mist. It was developed by Soilfer Hillel and David Durger. It is done in citrus plants and olive.

Classification of essential elements:

1. Framework elements: These are constituents (cellulose) of cell wall and storage product (starch) of plant. e.g., C, H, O
2. Protoplasmic elements: These are constituent of protoplasm such as chlorophyll, protein and nucleic acid. e.g., C, H, O, N, P, S.
3. Mobile elements: These readily move from old leaves to young leaves and growing tips. These molecules are broken down in old leaves and make them available for mobilisation in young leaves. Their deficiency occurs in old or senescent leaves. e.g., N, P, K, Mg
4. Immobile elements: These cannot move freely from old leaves to young leaves hence their deficiency occurs in young leaves. e.g., Ca, S, Fe
5. Balancing elements: These counteract the toxic effects of other minerals such as As, Hg, Cu by ion balance. e.g., Ca, Mg, K.
6. Storage elements: These are involved in storage of food materials. e.g., C, N, S, P
7. Critical elements: These are usually deficient in soil when present below the critical concentration. e.g., N, P, K.

Types of Essential elements: Most of the elements are present in the form of salt in the soil so the plants consume these elements as ions.

Micronutrients (Trace elements): These are required in small amount (less than 1% of the dry weight or less than10 millimole per kg) in most of the plants. These are generally required for functioning of enzymes. Their slight excess will cause toxicity. The concentration of mineral ion is toxic when there is

10% decrease in dry weight of tissues. This is called critical concentration which varies among different micronutrients. These are 8 in number which are as follows:

1. **Boron** (BO_3^{-3} or $B_4O_7^{-2}$): It is found in leaf and seed.

   Roles:

   (a) It is required for uptake and utilisation of Ca^{+2}.

   (b) It is required for translocation of sugar.

   (c) It helps in cell differentiation, pollen germination and root nodulation.

   Deficiency symptoms:

   (a) Browning of cauliflower

   (b) Brown heart of Turnip

   (c) Heart rot (degeneration of internal tissues) of beet

   (d) Top sickness (necrosis or death of shoot tip) in Tobacco.

   (e) Corky core of apple

2. **Chlorine (Cl^-):** It is found everywhere in plants.

   Roles:

   (a) It maintains solute concentration and cation-anion balance with Na^+ and K^+.

   (b) It helps in photolysis of water during light reaction of photosynthesis by which evolution of O_2 occurs.

   (c) It maintains potential of cell membrane.

   Deficiency symptoms:

   (a) Chlorosis (yellowing of leaves)

   (b) Necrosis (death of tissue)

   (c) Wilting

3. **Copper** (Cu^{+2}). It is found everywhere.

   Roles:

   (a) It helps in chlorophyll synthesis.

   (b) It is constituent of plastocyanin which transfers electrons from cytochrome-f to PS-1 during photosynthesis.

   (c) It is constituent of cytochrome oxidase which transfers electrons during terminal oxidation in respiration.

   (d) It activates tyrosinase, monoamine oxidase, laccase etc.

   Deficiency symptoms:

   (a) Dieback (killing of shoot tip) of Citrus plants.

   (b) Reclamation disease (leaf tip necrosis) of cereals.

   (c) Exanthema (Splitting of bark at several places to exude gum or resin)

4. **Iron** (mostly Fe^{+3} but Fe^{+2} in acidic soil): It is found everywhere. It is required in largest amount among micronutrients.

   Roles:

   (a) It helps in synthesis of chlorophyll but it is not its structural part.

   (b) It is constituent of leghaemoglobin, ferredoxin, flavoprotein (Fe-S) and cytochrome.

(c) It is involved in N_2 fixation, photosynthesis and respiration.

(d) It activates catalase, aconitase, peroxidase, nitrite reductase enzyme etc.

Deficiency symptoms: (a) Chlorosis

5. **Manganese (Mn^{+2}):** It is found in leaf and seed.

Roles:

(a) It helps in photolysis of water by which evolution of O_2 occurs.

(b) It activates nitrite reductase, oxalosuccinic decarboxylase, malic dehydrogenase, hydroxylamine reductase, RNA polymerase etc.

Deficiency symptoms:

(a) Chlorotic and necrotic spot in interveinal areas of leaf

(b) Gray speck(spot) of oat

(c) Marsh spot (cotyledon with brown spot) of pea.

6. **Molybdenum (MoO_2^{+2}):** It is found everywhere. It is required in least quantity among micronutrients.

Roles: It activates nitrate reductase and nitrogenase in N_2 fixation.

Deficiency symptoms:

(a) Whiptail (degeneration of leaf lamina) of cauliflower

(b) Chlorotic mottle in interveinal area of older leaf

(c) Scald (burn) of bean.

(d) Water core in turnip

7. **Nickel (Ni^{+2}):** It is found everywhere.

Roles:

(a) It is component of hydrogenase and urease.

(b) It helps in metabolism of urea.

Deficiency symptoms: (a) Necrosis of leaf tip

8. **Zinc (Zn^{+2}):** It is found everywhere.

Roles:

(a) It helps in biosynthesis of tryptophan and auxin (IAA).

(b) It activates carbonic anhydrase, alcoholic dehydrogenase, carboxylase.

(c) It helps in evolution and utilisation of CO_2.

Deficiency symptoms:

(a) Chlorosis

(b) Mottle (little) leaf in citrus plants

(c) Khaira disease (yellow patches on leaf) of rice

(d) White bud in maize

Meganutrients / Macronutrients: These are required in large amount (greater than 1% of dry weight or greater than10 millimole per kg) in most of the plants. These build up plant body and different cytoplasmic constituents. Their slight excess does not cause toxicity. These are 9 in number which are as follows:

1. **Carbon**: It enters the plant from air in the form of atmospheric CO_2 through diffusion.
   It is most abundant element in organisms on the basis of dry weight
   Roles: It is constituent of biomolecules.
   Deficiency symptoms:
   (a) Stunted growth
   (b) Chlorosis
   (c) Death
2. **Hydrogen**: It is obtained in the form of H_2O
   Roles: It is constituent of biomolecules
   Deficiency symptoms:
   (a) Stunted growth
   (b) Chlorosis
   (c) Death
3. **Oxygen**: It is obtained from air or water in the form of inorganic ions.
   It is most abundant element in organisms on the basis of fresh weight.
   Roles: It is constituent of biomolecules.
   Deficiency symptoms:
   (a) Stunted growth
   (b) Chlorosis
   (c) Death
   Note: Carbon, Hydrogen and Oxygen are non-mineral elements as they are not absorbed from soil instead they are taken up from atmosphere.
4. **Nitrogen (NO_3^-**(mostly), NO_2^- and NH_4^+ in acidic condition). It is found everywhere.
   It is required in largest amount. It is not a mineral but atmospheric gas. It comprises 50% of the dry weight of protoplasm. Plants are unable to make use of nitrogen hence they take up NO_3^- from soil.
   Roles:
   (a) It is constituent of nucleic acid, proteins, hormones, vitamins, cytochromes, chlorophyll, alkaloids, ATP, coenzymes (NAD, FAD, FMN).
   (b) It plays important role in photosynthesis, respiration, protein synthesis, growth, reproduction, heredity etc.
   Deficiency symptoms:
   (a) Chlorosis
   (b) Stunted growth
   (c) Reduction in leaf size
   (d) Purplish leaf due to formation of anthocyanin (a purple pigment)
5. **Phosphorus (**present in soil as **PO_4^{-3}** or **HPO_4^{-2}** and taken up by plants as **H_2PO_4-):** It is found in younger tissue.
   Roles:
   (a) It is constituent of nucleic acid, phospholipid, ATP, NADP, Nucleoprotein, phytic acid etc.

(b) It is involved in cell division.
(c) It is present in cell membrane.
(d) It is involved in energy transfer and phosphorylation reaction.
(f) It is often the limiting factor for growth and biomass production.

Deficiency symptoms:

(a) Poor growth
(b) Dark green colouration of leaves containing necrotic spots.

6. **Sulphur (SO_4^{-2}):** It is found in root tips, stem and young leaves.

   Roles:

   (a) It is constituent of amino acids (cysteine, cystine, methionine), vitamins (Thiamine, Biotin), coenzymes (Co-A, lipoic acid).
   (b) It stabilises the structure of proteins by disulphide linkage.
   (c) It is essential for chloroplast growth and function.

   Deficiency symptoms:

   (a) Chlorosis (starting from younger leaf)
   (b) Tea yellow

7. **Potassium (K^+):** It is found in meristematic tissue (root tip, buds, leaf). It is most common free ion in cell.

   Roles:

   (a) It helps in cation-anion balance in cells.
   (b) It helps in turgidity (water retention) in the cells.
   (c) It helps in opening and closing of stomata.
   (d) It helps in permeability of membrane.
   (e) It activates DNA polymerase.

   Deficiency symptoms:

   (a) Chlorosis
   (b) Necrotic spot at leaf tip

8. **Calcium (Ca^{+2}):** It is found in meristematic and differentiating tissue.

   Roles:

   (a) It is constituent of middle lamella of cell wall.
   (b) It is involved in selective permeability of cell and normal functioning of plasma membrane.
   (c) It helps in development of root and shoot apex (meristem).

   Deficiency symptoms:

   (a) Stunted growth
   (b) Degeneration of root apex and shoot apex (Tip burn).

9. **Magnesium (Mg^{+2}):** It is found in leaf

   Roles:

   (a) It is constituent of chlorophyll.
   (b) It activates hexokinase, phosphorylase, RuBisCO, PEPcase.

(c) It binds subunits of ribosomes in protein synthesis.

Deficiency symptoms:

(a) Interveinal chlorosis

(a) Necrosis

Absorption of minerals: Root absorbs the minerals selectively. Most of the minerals are absorbed by zone of elongation (epiblema) of root and a few minerals are absorbed by zone of maturation (root hairs). The uptake of minerals is also not uniform along the length of root. It is also facilitated by the mycorrrhiza which helps in uptake of phosphate, nitrate and ammonium ion. Absorption of minerals is independent of absorption of water. Minerals are absorbed in two forms: cations and anions. The principal cations associated with clay particles are H^+ in acidic soils and Ca^{+2} in alkaline soils. Minerals are absorbed by two methods:

1. **Passive absorption**: It usually occurs through ion channels which are transmembrane proteins that function as selective pores. It does not require energy. It takes place by the following ways:
   (a) Diffusion: It is most common in passive methods. Mineral ions are diffused along concentration gradient by simple and facilitated diffusion.
   (b) Mass flow/Bulk flow: It was proposed by Krammer (1911). Mineral ions are absorbed along with water under the influence of transpiration.
   (c) Donnan equilibrium: It was proposed by Donnan. Mineral ions are absorbed into the cell across plasma membrane to maintain electrical equilibrium (Donnan equilibrium). The ions get accumulated and firmly attached to inner surface of plasma membrane of cell and to neutralise these, ions of opposite and equal charges enter into the cell passively.

$$Cl^- (K^+ \longrightarrow Cl^- (K^+$$

$$\text{At Donnan equilibrium} = \frac{\text{Concentration of cations inside}}{\text{Concentration of cations outside}} = \frac{\text{Concentration of anions outside}}{\text{Concentration of anions inside}}$$

   (d) Ion exchange: It was proposed by Jenny and Overstreet (1978). Mineral ions are absorbed at root surface and exchanged with similar charged ions of soil solution. Cations are exchanged with H^+ and anions with OH^- or HCO_3^-.

$$\begin{array}{ll} K^+ \mid Br^+ & K^+ \mid Br^+ \\ Cl^- \mid K^+ \longrightarrow & Br^+ \mid K^+ \\ K^+ \mid Br^+ & K^+ \mid Br^+ \end{array}$$

2. **Active absorption**: It occurs against concentration gradient and requires energy by absorbing cells. Most of minerals are absorbed actively unlike water absorption because minerals are present as charged ions in soil solution hence these cannot move across plasma membrane passively and further the concentration of minerals in the soil is lower than the concentration of minerals in the cell so all minerals cannot be absorbed passively by roots.

   It is explained by carrier concept where minerals enter from outer surface of plasma membrane into inner surface of plasma membrane with the help of organic substances (carriers) by forming carrier-ion complex.

   (a) Cytochrome pump theory: It was proposed by Lundegardh and Burstrom (1933). In this, cytochromes act as carriers which are iron containing protein in plasma membrane. The anions are picked up by oxidised cytochrome oxidase and transferred to the other members of chain as they transfer the electrons to next compound. Cytochrome transports electrons from inner surface to outer surface of membrane through electron transport chain and also transports anions from outer surface to inner surface. Cytochromes are involved in active absorption of anions while cations are absorbed passively.

(b) **Protein-Lecithin theory**: It was proposed by Bennet Clark (1956). In this, carrier is protein associated with lecithin (phospholipid). Since carrier is amphoteric hence both cations and anions combine with it. In the transport of ions, lecithin is synthesised and hydrolysed in cyclic manner. Lecithin picks up ions on outer surface during synthesis and releases them on hydrolysis into inner space. ATP is believed to be concerned with the synthesis of at least one of the components of this phosphatide cycle. These are involved in active absorption of both cations and anions.

Transport of Minerals: The minerals are transported from root surface to xylem of root. These show multidirectional transport. Xylem is the pathway for the transport of minerals from the root to higher parts of plant because the analysis of xylem sap shows the presence of mineral salts where these are carried along with water to other parts of plant along transpiration stream. Mineral ions like organic solutes move in upward, downward as well as lateral directions. It follows two pathways:

1. **Apoplast pathway**: It is passive transport of mineral ions across root through diffusion or mass flow. Ions accumulated in root hairs move across cell wall of epidermis, cortex, endodermis, pericycle and reach xylem vessels. Ions enter into pericycle through endodermis which ruptures in the area of lateral roots.
2. **Symplast pathway**: It is active transport of mineral ions. Ions accumulated in root hairs move across cytoplasm of epidermis, cortex, endodermis, pericycle and finally reach xylem vessels through plasmodesmata.

Factors:

1. Temperature: Rate of absorption of minerals increases with rise in temperature but very high temperature decreases the rate due to denaturation of protein carriers.
2. O_2 concentration: Rate of absorption of minerals requires aerobic condition.
3. pH: Low pH increases the absorption of anions and high pH increases the absorption of cations.
4. Age of roots: Rate of absorption of minerals decreases in older roots.

Nonessential elements: These are not essential in most of the plants but some plants require more non-essential elements in addition to essential elements. These include:

1. Sodium: It is found in halophytes and C_4 plants.
2. Gold: It is found in *Equisetum* and mustard.
3. Aluminium: It is found in *Lycopodium* and ferns.
4. Silica: It is found in diatoms and grasses.
5. Iodine: It is found in brown algae.
6. Selenium: It is found in *Astraglus*
7. Gallium: It is found in *Lemna*
8. Cobalt: It is found in Leguminous plants.

Leibig's law of Minimum: If a plant is deficient in anyone nutrient, then it will not grow in spite of adequate amount of other minerals. Or the yield of crop plants is determined by the amount of that essential element which is present in insufficient amount.

Phytotron: It is technique of growing plants in controlled environmental condition (temperature, light, pH).

Nitrogen fixation: It is conversion of free atmospheric nitrogen into inorganic and biologically acceptable nitrogenous compound to make it available for absorption by plants. Nitrogen is abundantly present in atmosphere. It cannot be used directly but has to combined with C, H, O and N to form compounds before being used. It occurs by two methods:

Physico-chemical N_2 fixation: It occurs due to physical processes like lightning or electric discharge which favours combination of gaseous N_2 and O_2 in atmosphere.

$$2NO + O_2 \rightarrow 2NO$$

$$2NO + O_2 \rightarrow 2NO_2$$

$$2NO_2 + H_2O \rightarrow HNO_2 + HNO_3$$

Acids fall on the ground along with rain water and react with alkaline radicals to form water soluble nitrates and nitrites.

$$CaCO_3 + 2HNO_3 \rightarrow Ca(NO_3)_2 + CO_2 + H_2O$$

Nitrates are soluble in water and are directly absorbed by roots of plants. It is less significant because low amount (10%) of nitrogen is fixed due to requirement of rain and lightning.

Biological Nitrogen fixation (Diazotrophy): It was discovered by Winogradsky (1891). It occurs through living organisms (prokaryotes). It occurs at normal atmospheric pressure (1atm) and temperature (around 20°C). It occurs by following methods:

1. **Asymbiotic method**: It occurs by several types of free living organisms which do not form any association with any plants. These organisms are primitive. These fix nitrogen more actively under poor aeration. The whole process is reduction process independent of respiration.
   (a) Aerobic and chemoheterotrophic bacteria: *Azotobacter, Beijerinckia*
   (b) Anaerobic chemoheterotrophic bacteria: *Bacillus, Clostridium, Klebsiella, Aerobacter*
   (c) Anaerobic photoautotrophic bacteria: *Chlorobium, Chromatium*
   (d) Anaerobic photoheterotrophic bacteria: *Rhodospirillum, Rhodopseudomonas*
   (e) Aerobic actinomycetes: *Nocardia*
   (f) Aerobic photoautotrophic Cyanobacteria: *Anabaena, Nostoc, Oscillatoria, Aulosira* (esp. in rice field), *Cylindrospermum* (esp. in maize and sugarcane fields).
2. **Loose symbiotic or Associative method:** It occurs by certain bacteria living in close contact with the roots of higher plants. This is called loose symbiosis. The bacteria live in transition area between soil and roots. The bacteria fix nitrogen and exude a part of fixed N_2 for use by plants. It involves several microbes which are as follows:
   (a) *Azospirillum brasilense* forms loose association with roots of cereals.
   (b) *Azospirillum lipoferum* forms loose association with roots of maize.
   (c) *Azospirillum paspalum* forms loose association with roots of paspalum grass.
   (d) *Beijerinckia* forms loose association with roots of sugarcane.

   These free living microbes do not immediately enrich but after death of plants and animals the fixed nitrogen enters the cycling pool. These involve the following processes:

   Proteolysis: Microbes such as *Clostridium histolyticum* and *Clostridium spirogenes* decompose proteins of dead plants and animals into amino acids.

$$\text{Protein} \xrightarrow{\text{Protease}} \text{Peptides} \xrightarrow{\text{Peptidase}} \text{Amino acid (alanine)}$$

Ammonification: Microbes such as *Bacillus vulagaris, Clostridium, Micrococcus, Actinomycetes, Proteus* etc. cause deamination of amino acids forming ammonia which may either be released to the atmosphere or utilised by the plants in the form of NH_4^+ after changing to ionic form because NH_3 is toxic in gaseous state or oxidised to form nitrates under favourable conditions.

$$\text{Alanine} + \tfrac{1}{2}O_2 \xrightarrow{\text{Alanine deaminase}} \text{Pyruvic acid} + \text{Ammonia}$$

Nitrification: It is oxidation of ammonia to nitrate. Plants depend on soil bacteria to convert NH_3 to nitrates via nitrite. It occurs in aerobic condition. It occurs in two steps:

(a) $2NH_3 + 3O_2 \xrightarrow[\text{Nitrosopira, nitrosocystis}]{\text{Nitrosococus, Nitrosomonas}} 2NO_2^- + 2H^+ + 2H_2O + \text{Energy (84 kcal)}$

(b) $2NO_2^- + O_2 \xrightarrow[\text{Nitrospira, Nitrocystis}]{\text{Nitrococcus, Nitrobacter}} 2NO_3^- + \text{Energy (17 kcal)}$

In this process, energy is released which is used by nitrifying microbes (chemoautotroph) for synthesis of food. Nitrates, thus formed are absorbed by the plants.

Denitrification: Some of nitrates present in soil are reduced to gaseous N_2 which is escaped from soil. It occurs only in anaerobic condition by *Pseudomonas denitrificans, Micrococcus denitrificans, Thiobacillus denitrificans* etc. In this process nitrates serve as e- acceptor. It is undesirable process resulting the loss of fixed nitrogen (nitrates) from the soil without being utilised for any assimilatory process.

$$2NO_32NO_2^- \longrightarrow 2NO \longrightarrow N_2O \longrightarrow N_2$$

Nitrate assimilation: Nitrate is most abundant form of Nitrogen in the soil and is most available to plants. Nitrates are absorbed by the plant roots by active process are directly transported to leaves through transpiration. Nitrates formed in other parts of plant are transported to leaves. Nitrates are not directly used up by the plants hence most of them are reduced to NH_3 and water after absorption in leaves before being converted into organic form (amino acids). It occurs in two steps:

(a) Nitrate is reduced to nitrite in the presence of NADPH (reducing agent as e- donor) and nitrate reductase enzyme with the help of flavoprotein which contains Molybdenum. It occurs mostly in cytosol of leaves.

$$NO_{3-} + 7NADPH + H^+ \xrightarrow{\text{Nitrate reductase, FAD/FMN, Mo}} NO_{2-} + H_2O + NADP^+$$

(b) Nitrite is reduced to ammonia in the presence of NADPH and nitrite reductase with the help of ferredoxin which contains Cu and Fe. It occurs mostly in chloroplast of leaves.

$$2NO_{2-} + 7NADPH + 7H^+ \xrightarrow{\text{Nitrite reductase, Fd, Cu, Fe}} 2NH_3 + 4H_2O + 7NADP^+$$

Overall reaction:

$$NO_3- + 8\ NADPH + 8\ H^+ \longrightarrow NH_3 + 3H_2O + 8NADP^+$$

3. **Symbiotic method**: It occurs when two organisms live together on mutual benefit where host plant supplies carbohydrate to bacteria and in return bacteria supply fixed N_2 to host plant. It involves several microbes such as
   (a) *Rhizobium leguminosarum (=Bacillus radicicola*) forms root nodules on leguminous plants (pea, lentil, sweet pea).
   (b) *Rhizobium phaseoli* forms root nodules on beans.
   (c) *Rhizobium japonicum* forms root nodules on soya bean.
   (d) *Frankia* (an actinomycete) forms nodules in roots of non-leguminous plants such as *Alnus, Casuarina, Dioscorea, Myrica,* rice, maize, cotton etc.

(e) *Nostoc puncteforme* does not form nodule in thallus of *Anthoceros.*

(f) *Anabaena azollae* does not form nodule in leaf of *Azolla.*

(g) *Nostoc puncteforme* and *Anabaena cycadae* do not form nodule in coralloid root of *Cycas.*

(h) *Klebsiella* forms leaf nodule in *Psychotria.*

Mechanism: *Rhizobium* is aerobic, Gram negative, rod-shaped, motile, non-sporulating, chemoheterotoph (saprotroph) and soil borne bacterium. It shows microsymbiosis because one partner (*Rhizobium*) is very small and other (host plant) is large. It involves the following steps:

(a) A number of microbes are present in rhizosphere (close surrounding). Host recognises *Rhizobium* due to susceptibility and specific environmental conditions.

(b) Host plant secretes organic substances or exudates such as Biotin, Thiamine and amino acids by which compatible strains of *Rhizobium* are stimulated over other microbes (chemotactic).

(c) Host plant now secretes a specific plant protein called lectin which attaches Rhizobia on surface of root of host plant.

(d) Rhizobia invade the root by digesting the cell wall of root hair.

(e) *Rhizobium* secretes curling (nod) factors such as cytokinin and polymxin-B by which root hair curls around bacteria.

(f) Tip of curled root hair bends.

(g) *Rhizobium* now penetrates the root hair and grows in the form of infection thread (infolded tubular extension of plasma membrane).

(h) The infection thread grows towards root apex and ramifies in whole cortex where it fuses with plasma membrane of cortical cell releasing Rhizobia in cortical cell.

(i) *Rhizobium* sends mitogenic chemicals that stimulate cell division in cortex.

(k) The infected cortical cells dedifferentiate and start dividing.

(j) Nodule is formed which is stimulated by auxin produced by cortical cell and cytokinin released by bacteria. Nodule is spherical, elongated or flat and protrudes from root surface. It is large and pinkish due to presence of red pigment, Leghaemoglobin (LHb) which is formed during maturation of nodule. Nodule is formed at optimum temperature of 25-30°C and bright light.

(l) *Rhizobium* in nodule is enclosed in a membrane derived from hypertrophied host cortical cell.

(m) *Rhizobium* becomes enlarged and pleomorphic and forms polyhedral bacteroid in the cytoplasm of cortical cell. Bacteroid is swollen, irregular, club shaped, or branched in cytoplasm of cortical cell. Bacteroid is the site of Nitrogen fixation where N_2 is reduced to NH_3.

(n) Bacteroid remains single or in group surrounded by host membrane (peribacteroid membrane). There is a peribacterial space between bacteroid and membrane.

(o) LHb is found outside this space (not inside peribacteriod membrane). It is synthesised by host plant and is located in cytosol of bacteroid infected host cortical cell. The globin protein is encoded by the host plant while haeme group is primarily synthesised by the bacterium and transported in host cytoplasm.

(p) Bacteroid inside nodule produces nitrogenase enzyme that reduces N_2 to NH_3.

(q) The synthesis of NH_3 requires very high amount of energy which is obtained from respiration.

(r) LHb serves as reservoir of O_2 and provides O_2 at low partial pressure (0.5 atm) to bacteroid of nodule for the production of ATP through aerobic respiration for nitrogenase activity. Higher O_2 (>0.5atm) inhibits N_2-fixation due to inactivation of nitrogenase enzyme.

(s) LHb binds the diffusing O_2 to form oxyhaemoglobin to keep away O_2 from nitrogenase and provides protection to nitrogenase and thus LHb creates an O_2 free environment around active site of nitrogenase which does not function in the presence of O_2.

(t) N_2-fixation is reduction process as it functions under anaerobic process. It requires an electron donor which is reduced ferredoxin in most N_2-fixing microbes and 16 molecules of ATP for reducing 1 molecule of N_2. However, the actual reduction of nitrogen molecule without formation of Hydrogen molecule requires only 6 e–.

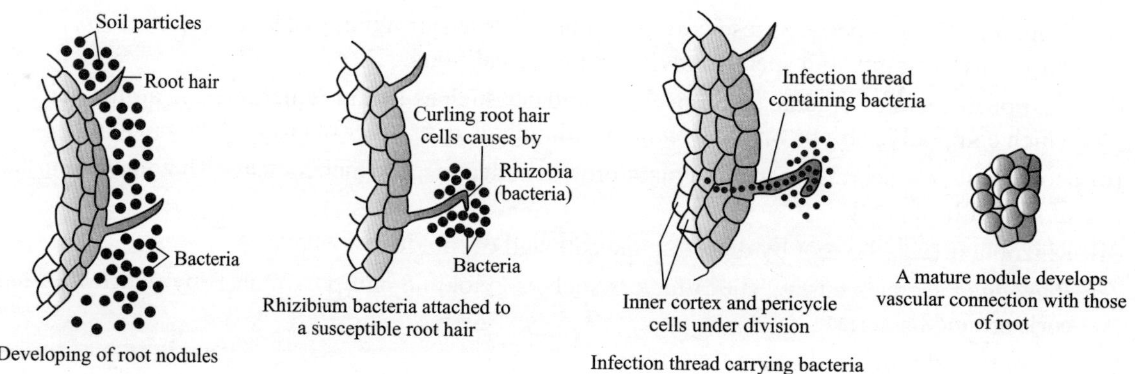

Development of root nodule

$$N_2 + 8e^- + 8H^+ + 16\text{ ATP} \xrightarrow{\text{Nitrogenase}} 2NH_3 + 16\text{ADP} + 16\text{iP} + H_2$$

$$N \equiv N \xrightarrow{2e^-, 2H^+} NH \equiv NH \xrightarrow{2e^-, 2H^+} H_2NNH_2 \xrightarrow{2e^-, 2H^+} 2NH_3$$

Rhizobium living in such nodules gets carbohydrate from host cell and converts N_2 into NH_3 which are again converted into other nitrogenous substances. *Rhizobium* fixes 10 times more than free living N_2 fixing bacteria. When pulse is harvested, the roots are left in the soil and thus root nodules and finally *Rhizobium* are released inside soil.

Nitrogenase Complex

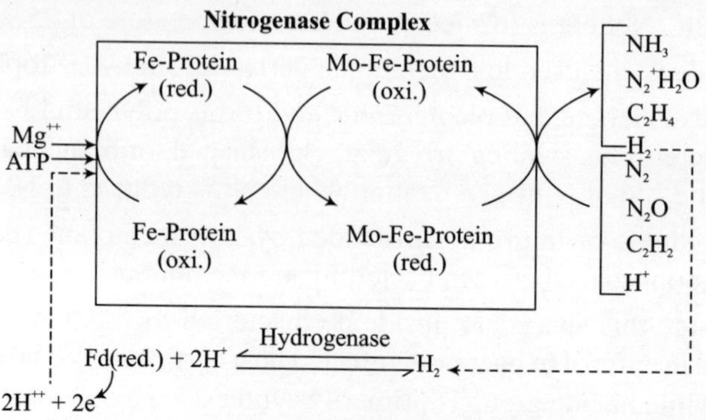

Fate of ammonia or synthesis of amino acids: Plant cells rapidly convert ammonia formed from nitrate assimilation into amino acids in order to avoid ammonium toxicity. The amino acids now form various types of nitrogenous compounds.

(a) Reductive amination: It is conversion of a carbonyl group or keto group of acid to an amine. It occurs in chloroplast of leaves.

NH_3 + α-ketoglutaric acid+$NADPH_2$ $\xrightarrow{\text{Glutamate dehydrogenase}}$ Glutamic acid+ H_2O +NADP

NH_3 + Oxaloacetic acid +$NADPH_2$ $\xrightarrow{\text{Aspartate dehydrogenase}}$ Aspartic acid + H_2O+NADP

(b) Transamination; It is transfer of amino group of one amino acid to keto group of keto acid.

Glutamic acid + Oxaloacetic acid $\xrightarrow{\text{Transaminase}}$ α-ketoglutaric acid + Aspartic acid

Glutamic caid + Pyruvic acid $\xrightarrow{\text{Transaminase}}$ α-ketoglutaric acid + Alanine

(c) Catalytic amidation: It occurs often when the amount of NH_3 available to plant is in excess and becomes poisonous hence it is converted to amides (double keto acids such as glutamine and asparagine). Thus amide formation is protective mechanism. These amino acids such as Glutamine and Asparagine are actually glutamic acid and aspartic acid respectively, with an extra-NH_2 group attached to -COOH group. It requires ATP. Since amides contain more nitrogen than amino acids hence these are transported to other parts of plant esp. shoot through xylem.

NH_3 + Glutamic acid +ATP $\xrightarrow{\text{Glutamine synthetase}}$ Glutamine +ADP + Pi

NH_3 + Aspartic acid +ATP $\xrightarrow{\text{Asparagine synthetase}}$ Asparagine + ADP +Pi

Nitrogen cycle: It is a gaseous cycle by which nitrogen is converted into several nitrogenous chemical compounds carried out mostly by microbes. It involves atmosphere and lithosphere. It circulates in terrestrial and aquatic ecosystems. Nitrogen is present in abundance (78%) in earth's atmosphere. Nitrogen is converted into nitrates (usable form) by physico-chemical method or biological method. The biological method is most significant which is carried out by free-living or symbiotic bacteria. It involves ammonification, nitrification and denitrification.

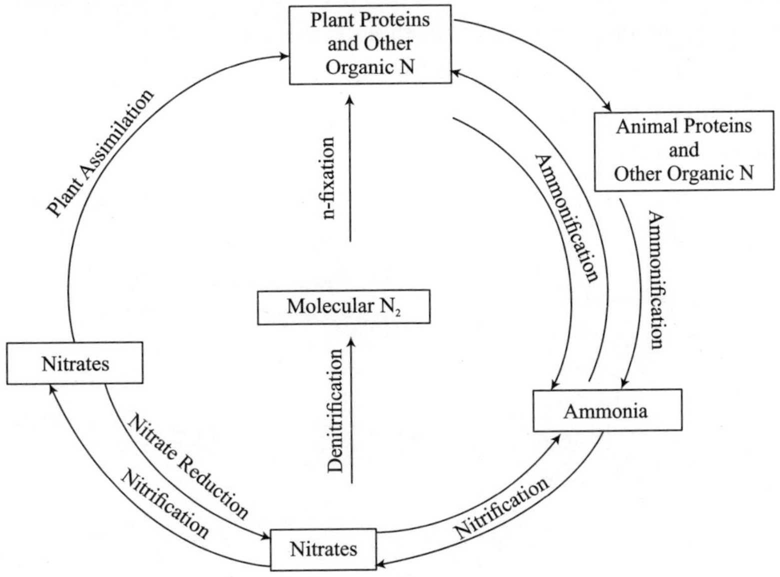

Fig. ***Nitrogen cycle***

SHORT ANSWER TYPE QUESTIONS

1. What are the criteria for the essentiality of an element?
2. What do you mean by hydroponics? What are its advantages?
3. Distinguish between mobile elements and immobile elements?
4. Discuss the roles of Chlorine and manganese in plants?
5. How potassium ion is essential for a plant?
6. Discuss the roles of phosphorus and sulphur in plants.
7. What is Donnan equilibrium?
8. What is difference between asymbiotic and loose symbiotic method of nitrogen fixation?
9. Give the examples of ammonifying bacteria and nitrifying bacteria.
10. What are the features of *Rhizobium* bacteria?
11. Give the examples of nodule forming bacteria.
12. What is Leghaemoglobin? Explain its functions in nitrogen fixation.
13. Discuss the roles of nitrogenise enzyme in nitrogen fixation and also write the final equation.
14. Explain the transamination reaction with examples.
15. Illustrate the nitrogen cycle with diagram.

LONG ANSWER TYPE QUESTIONS

1. What do you mean by micronutrients? Discuss the roles of any five elements.
2. What do you mean by meganutrients? Discuss the roles of any five elements.
3. Discuss the mechanism of active absorption and transport of minerals.
4. Explain the nitrification process and how nitrate is assimilated in plants?
5. Describe the mechanism of symbiotic nitrogen fixation.

MULTIPLE CHOICE QUESTIONS

1. Which of the following elements stimulates the activity of many enzymes but is not a constituent of enzyme?
   (a) Fe (b) K (c) Mg (d) Cu
2. Premature leaf fall is caused due to deficiency of
   (a) Nitrogen (b) Phosphorus (c) Sulphur (d) Magnesium
3. Which of the following elements is obtained when chlorophyll is burnt?
   (a) Fe (b) Mg (c) Fe and Mg (d) Mg and Mn
4. Apple fruit develops internal cork due to deficiency of
   (a) Boron (b) Molybdenum (c) Sulphur (d) Nickel
5. All mineral salts are absorbed in the cell as
   (a) Ions (b) Molecules (c) Atoms (d) All of the above
6. Cell elongation is adversely affected by
   (a) Nitrogen (b) Phosphorus (c) Calcium (d) Magnesium
7. The first stable product of nitrogen fixation is
   (a) Ammonia (b) Nitrate (c) Nitrite (d) Amine
8. *Nif* genes occur in
   (a) E. coli (b) Pseudomonas (c) Agrobacterium (d) Rhizobium
9. The limiting factor in nitrification of soil is
   (a) Light (b) Temperature (c) pH (d) All of the above
10. Which of the following elements is essential for Nitrogen fixation and nitrogen metabolism?
    (a) Zn (b) Mn (c) Mo (d) B
11. Which of the following elements is not absorbed by plants from soil?
    (a) Carbon (b) Hydrogen (c) Nitrogen (d) Boron
12. Plants absorb nitrogen mostly in the form of
    (a) Nitrite (b) Nitrate (c) Ammonia (d) N_2 gas
13. Existence of nitrogen fixing bacteria in soil was discovered by
    (a) Lister (b) Pasteur (c) Koch (d) Winogradsky
14. Mineral salts are absorbed by the roots from soil are in the form of
    (a) Dilute solution (b) Concentrated solution
    (c) Very dilute solution (d) Very concentrated solution
15. Which of the following elements is required to convert nitrate to nitrite?
    (a) Zn (b) Mo (c) Fe (d) B
16. Which of the following microelements is absorbed by foliage?
    (a) Boron (b) Phosphorus (c) Zinc (d) All of the above
17. Which of the following is undesirable while testing the essentiality of an element?
    (a) Tap water (b) Pure salt
    (c) Glazed porcelain container (d) Highly washed sand

18. Deficiency symptoms of elements in plants may vary from
    (a) Individual to individual (b) Species to species
    (c) Genus to genus (d) Family to family
19. Nitrogen fixation by organisms requires
    (a) Alkaline condition (b) Anaerobic condition
    (c) Water free condition (d) Bright sunlight
20. Nitrite ions are reduced to ammonia by ferredoxin in
    (a) Root (b) Stem (c) Leaf (d) All parts of plant
21. Which of the following micronutrient ions required in least amount?
    (a) Fe (b) Zn (c) B (d) Mo
22. Passive absorption of minerals depends on
    (a) Temperature (b) Humidity (c) Metabolic inhibitors (d) All of the above
23. In most higher plants, ammonia is assimilated primarily into
    (a) Asparagine (b) Glutamic acid
    (c) Aspartic acid (d) α-ketoglutaric acid
24. Oxygen absorbed by leghaemoglobin at low O_2 tension in the nodule is transmitted at
    (a) Lower concentration to bacteroid respiratory chain
    (b) Higher concentration to bacteroid respiratory chain
    (c) Lower concentration to root nodule
    (d) Higher concentration to root nodule
25. Nitrogen is translocated through phloem to other parts of plant in the form of
    (a) Glutamine and alpha ketoglutaric acid (b) Asparagine and aspartic acid
    (c) Glycine and glutamine (d) Glutamine and asparagines
26. The main product of biological nitrogen fixation in soil is
    (a) NH_3 (b) NH_4^+ (c) NO_3^- (d) NO_2^-
27. Nitrate reductase is classified as
    (a) Zinc protein (b) Iron protein
    (c) Molybdenum protein (d) Copper protein
28. The common nitrogen fixer in paddy field is
    (a) *Azotobacter* (b) *Azospirillum* (c) *Clostridium* (d) *Klebsiella*
29. The number of ATPs required to reduce 1 molecule of nitrogen into ammonia was
    (a) 8 (b) 16 (c) 32 (d) 64
30. Legume root nodules contain leghaemoglobin which regulates
    (a) Expression of nif genes (b) Dinitrogenase activity
    (c) Oxygen supply (d) Nodule growth
31. Which of the following mineral elements does not appear to be structurally bound in plants?
    (a) Nitrogen (b) Phosphorus (c) Potassium (d) Calcium
32. The enzyme nitrogenase consists of two subunits which have composition respectively being
    (a) Fe protein and Fe-Mo protein (b) Mo protein and Fe-S protein
    (c) Cu protein and Fe protein (d) Fe protein and S protein

33. During symbiotic nitrogen fixation, the recognition of Rhizobium-host interaction involved two classes of molecules. These are
(a) Lectins and lipids (b) Lectins and polysaccharides
(c) Flavonoids and lipids (d) Lectins and alkaloids
34. Which of the following amino acids are formed as a result of catalytic amidation?
(a) Arginine and Asparagine (b) Glutamine and Asparagine
(c) Glycine and Glutamine (d) Glycine and Arginine
35. Some amounts of nitrates are reduced to gaseous nitrogen with the help of
(a) *Rhizobium* (b) *Pseudomonas* (c) *Clostridium* (d) *Nostoc*
36. Which of the following bacteria forms root nodules in non-leguminous plants?
(a) *Rhizobium* (b) *Nocardia* (c) *Azotobacter* (d) *Frankia*
37. Which of the following is incorrectly matched?
(a) Boron-Pollen germination (b) Copper-constituent of plastocyanin
(c) Manganese- Photolysis of water (d) Zinc- Biosynthesis of gibberellin
38. Which of the following is incorrectly matched?
(a) Potassium- Opening and closing of stomata
(b) Sulphur- Constituent of Cysteine and methionine
(c) Calcium-Constituent of middle lamella of cell wall
(d) Magnesium- Development of meristem
39. Which of the following bacteria are aerobic and chemoheterotroph?
(a) *Baciillus* and *Beijerinckia* (b) *Azotobacter* and *Beijerinckia*
(c) *Chlorobium* and *Chromatium* (d) *Clostridium* and *Rhodospirillum*
40. Nitrites are converted into nitrates by
(a) Nitrosococcus (b) Nitrosomònas (c) Nitrobacter (d) Nitrosocystis
41. For the plants, soil is not the source of
(a) Carbon (b) Hydrogen (c) Oxygen (d) All of the above
42. The partial mineral element is
(a) Nitrogen (b) Phosphorus (c) Potassium (d) Iron
43. Which of the following mineral's deficiency causes deficiency of nitrogen?
(a) Sulphur (b) Potassium (c) Manganese (d) Molybdenum
44. Cell division is inhibited due to either lack of or low level of
(a) N but not S (b) S but not K (c) N,S and K (d) K but not S and N
45. Excess of Mn causes deficiency of
(a) Fe (b) Mg (c) Ca (d) All of the above
46. The nodule in a plant root is formed in
(a) Epidermis (b) Cortex (c) Endodermis (d) Pericycle
47. The transported and storage form of nitrogen in plants is
(a) Amino acids (b) Polypeptides (c) α-ketoglutaric acid (d) Amides

48. Which of the following is a prokaryotic autotrophic nitrogen fixing symbiont?
    (a) *Pisum* (b) *Alnus* (c) *Dioscorea* (d) *Cycas*
49. Nitrogen fixation occurs in
    (a) Some bacteria, and legumes (b) Some bacteria, cyanobacteria and legumes
    (c) Some herbaceous plants and legumes (d) All green plants
50. The plants are termed as halophytes if growing in soil having salt concentration above
    (a) 0.1% (b) 0.2% (c) 0.4% (d) 0.5%
51. The essentiality of an element was proposed by
    (a) Sachs (b) Arnon and Stout (c) Bergerstein (d) Levitt
52. Which of the following elements does not belong to immobile element?
    (a) Potassium (b) Calcium (c) Sulphur (d) Iron
53. Which of the following elements is involved in translocation of sugar?
    (a) Cu (b) S (c) K (d) B
54. Plastocyanin is an electron carrier in photosynthesis. It contains
    (a) Cu (b) S (c) Ni (d) B
55. Chlorosis starting from younger leaves is caused by
    (a) Cu (b) S (c) K (d) B
56. Whiptail of cauliflower is caused due to deficiency of
    (a) Manganese (b) Sulphur (c) Potassium (d) Molybdenum
57. Interveinal chlorosis is caused due to deficiency of
    (a) Phosphorus (b) Sulphur (c) Magnesium (d) B
58. The RuBisCO enzyme contains
    (a) Cu (b) S (c) K (d) Mg
59. Which of the following is anaerobic photoautotrophic bacterium?
    (a) *Chlorobium* (b) *Azotobacter* (c) *Azospirillum* (d) *Nocardia*
60. Which of the following bacteria which forms root nodules in non-leguminous plants?
    (a) *Nocardia* (b) *Rhizobium* (c) *Frankia* (d) *Azotobacter*

ANSWERS

1.b	2.a	3.b	4.a	5.d	6.a	7.a	8.d	9.c	10.c
11.a	12.b	13.d	14.c	15.b	16.c	17.a	18.b	19.b	20.c
21.d	22.a	23.b	24.a	25.d	26.b	27.c	28.b	29.b	30.c
31.a	32.b	33.b	34.b	35.c	36.d	37.d	38.d	39.b	40.c
41.a	42.a	43.d	44.c	45.d	46.b	47.d	48.d	49.b	50.d
51.b	52.a	53.d	54.a	55.b	56.d	57.c	58.d	59.a	60.c

10

Photosynthesis in Higher Plants

Photosynthesis is synthesis of food (glucose) by green plants in the presence of light by utilising CO_2 and water and thus releasing O_2 into atmosphere. Photosynthesis is largely responsible for producing and maintaining the oxygen content of the Earth's atmosphere, and supplies all the organic compounds and most of the energy necessary for life on Earth.

It is physico-chemical, endothermic and anabolic process regulated by enzymes. It is ten times faster than respiration. It is a redox process because water is oxidised to form O_2 and CO_2 is reduced to form glucose. It is endergonic process because products have more energy as compared to reactants.

Photosynthesis was first appeared in Cyanobacteria (Blue green algae). Most of photosynthesis (90%) is carried out by algae in ocean and rest (10%) by land plants. Hence these organisms are called photoautotrophs.

Generally photosynthesis is the opposite of cellular respiration. It is the most natural process which contains life on earth. All animals and heterotrophs depend on the green plants for their organic food.

Photosynthesis is mostly oxygenic but in photosynthetic bacteria such as *Chlorobium, Chromatium*, it is anoxygenic (O_2 is not liberated).

It utilises less than 1% (0.2%) of total absorbed water by the plants.

A ray of light incident upon a leaf behaves in three different ways. About 83% of PAR is absorbed, 5% is transmitted and 12% is reflected by the leaf. Out of total absorption only 4%-5% is absorbed by chlorophyll for photosynthesis.

Photosynthetic organisms have three objectives:

1. To capture photon with high efficiency and transfer to reaction centre
2. To convert solar energy to chemical energy
3. To reduce CO_2 to glucose

Landmarks in Photosynthesis

1. **Stephan Hales (1717):** He is father of plant physiology. He proposed that green plants required sunlight to obtain their nutrition from air.
2. **Joseph Priestley (1772):** He concluded that vegetation (plant) purified the air spoilt by burning candle. He observed that a burning candle of a closed jar changed the air inside which would not allow another candle to burn or mouse to live in. But when a twig of Mint plant was placed inside jar then air allowed candle to burn or mouse to live in.

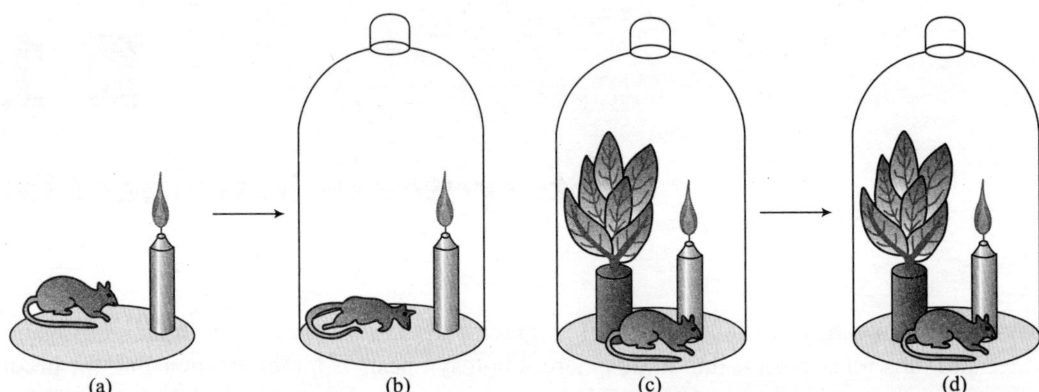

3. **Moll**: He performed half leaf experiment. In this half part of destarched leaf was inserted in air tight bottle containing small quantity of caustic potash (KOH) and other half remained outside bottle and illuminated. Part of leaf outside bottle turned black (positive) when starch test (iodine) was done and the part of inserted leaf did not turn black (negative) because it did not show photosynthesis (formation of starch) due to lack of CO_2. All CO_2 inside bottle was absorbed by KOH. He confirmed that CO_2 was essential for photosynthesis.

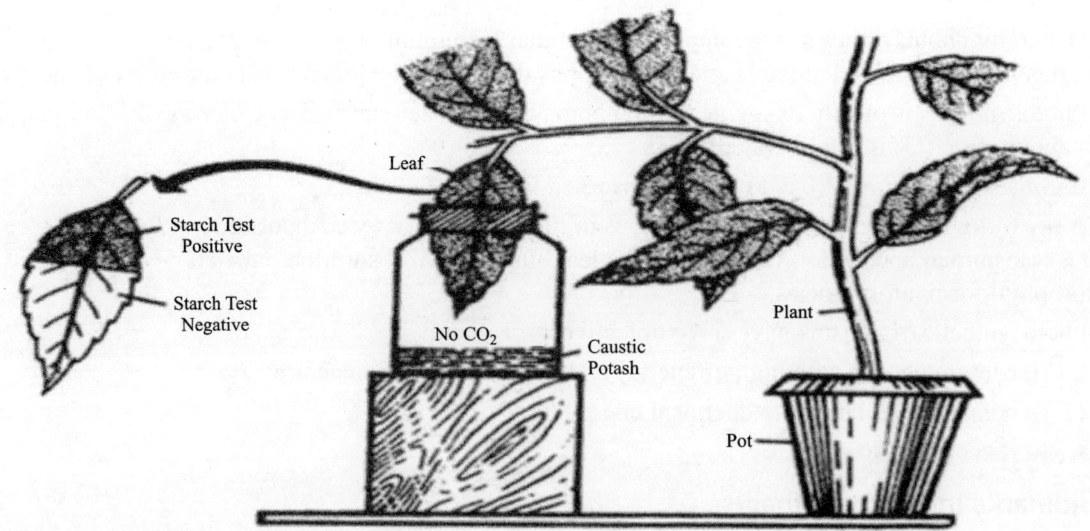

4. **Ingenhousz** (1779): He is real discoverer of photosynthesis. He confirmed that O_2 was released from green plants when twig of *Hydrilla* plant was inserted in funnel and covered by test tube in a beaker and placed in bright sunlight. Small bubbles were formed which are O_2.

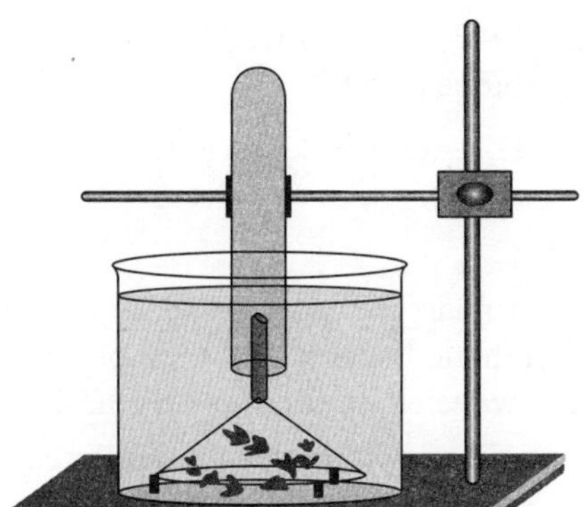

5. **Jean Senebier** (1782): He proposed that increase inCO_2 will increase the evolution of O_2 in green plants.
6. **Theodore de Saussure** (1804): He pointed that green plants require CO_2water and light for photosynthesis. He further proposed that amount of CO_2 absorbed is equal to amount of O_2 evolved.
7. **Pelletier and Caventon** (1818): He discovered the chlorophyll which could be separated from leaf by boiling in alcohol.
8. **Robert Mayer (1845):** He proposed that green plants utilise light energy and convert it into chemical energy of organic matter.
9. **Sachs (1850):** He noticed that green part of plant produces glucose which is stored as starch (first stable product of photosynthesis). He discovered chloroplast in the cell where chlorophyll is located.
10. **Engelmann (1888):** He split light into its components by prism and then illuminated *Cladophora* plant placed in the suspension of aerobic bacteria. He found that bacteria accumulated in the region of red and blue light of split spectrum. He thus discovered the effect of different wavelength of light on photosynthesis. He concluded that chlorophyll was involved in photosynthesis.
11. **Blackman** (1905): He proposed that photosynthesis occurs in two steps (dark reaction along with light reaction). He also proposed "Law of limiting factors".
12. **Warburg** (1920): He introduced *Chlorella* (unicellular, non-motile green alga) as experimental material for photosynthesis.
13. **Van Niel** (1931): He demonstrated that photosynthetic bacteria fixed CO_2 in the presence of H_2S (in case of sulphur photosynthetic bacteria) and H_2 or lactate (in case of non-sulphur bacteria).
$$6CO_2 + 12H_2S \longrightarrow C_6H_{12}O_6 + 6H_2O + 12S$$
He also postulated that green plants evolved O_2 by splitting H_2O (not CO_2).
$$6CO_2 + 12H_2O \longrightarrow C_6H_{12}O_6 + 6H_2O + 6O_2$$
14. **Emerson and Lewis** (1943): They performed red drop and enhancement effect. Photosynthesis is dropped suddenly if red light of higher wavelength (>680 nm or 700 nm) is supplied in *Chlorella*. Photosynthesis is enhanced if red light of shorter wavelength (<680 nm) is supplied in *Chlorella*.
15. **Robert Hill** (1937): He confirmed that isolated chloroplast of green leaves of *Stellaria media* did not bring about photolysis of water and liberate O_2 when illuminated but when reducing agents or electron acceptors (Hill oxidants such as chromate, endophenol, ferricyanide, benzoquinone and ferric oxalate) were added to the illuminated suspension then chloroplast evolved oxygen.

16. **Ruben, Hassid and Kamen** (1947): They proved that O_2 evolved in photosynthesis comes from water labelled using heavier isotope (H_2O^{18}).

$$6CO_2 + 12H_2O \xrightarrow[\text{Chlorophyll}]{\text{Light}} C_6H_{12}O_6 + 6H_2O + 6O_2$$

Photosynthetic pigments:

These are found in thylakoid of chloroplast in plants but these are diffused throughout the cell in case of photosynthetic bacteria (photosynthetic bacteria and Cyanobacteria) because chloroplast is absent. They absorb light energy at specific wavelength for photosynthesis.

1. **Chlorophylls**: These are specialised lipid molecules and green pigments located in lipid part of thylakoid membrane. These were studied by Wistatter, Stoll and Fischer (1912). These are soluble in organic solvent (fat) and insoluble in water. These absorb both shorter waves (blue and violet) and longer waves (orange and red) of visible spectrum. These transmit or reflect green light hence leaves appear to be green.

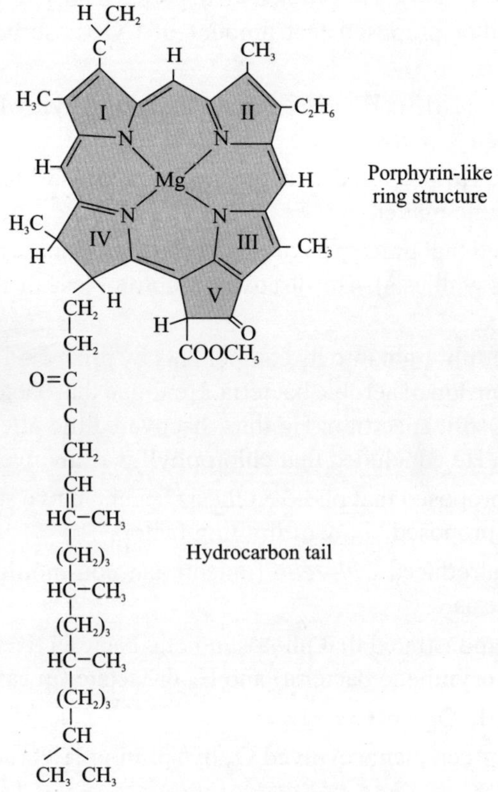

Porphyrin-like ring structure

Hydrocarbon tail

Structure: It is tadpole shaped (square planar) and asymmetrical. It consists of

(a) Porphyrin Head (15x15A^0): It is flat and square structure of alternating single and double bonds containing 4 pyrrole rings which are linked by methine (-CH_2) bridge. Each pyrrole ring is made up of 5 atoms (4C and 1N).There is a non-ionic Mg atom at centre by N atoms of pyrrole rings through two covalent and coordinate bonds. Pyrrole ring contains several side groups which change the properties of pigment. In chlorophyll-a, there is one methyl

($-CH_3$) group at third carbon while in chlorophyll-b, it is replaced by aldehyde (-CHO) group. Chlorophyll without its Magnesium core is colourless and called phaeophytin.

(b) Phytol tail ($20A^0$): It is long chain of carbon and hydrogen atoms (hydrocarbon). It is attached to 7th carbon from head. It anchors the chlorophyll into lipid part (hydrophobic) of thylakoid membrane. In chlorophyll-a, there is only one double bond present in phytol tail.

Synthesis: It is synthesised from succinyl Co-A and glycine. Several intermediates such as aminolevulonic acid, phosphobilinogen, uroporphyrinogen, protoprphyrin, Mg-Protoporphyrin, protochlorophyllide-a and chlorophyllide-a are formed. Finally Chlorophyll-a is formed. Protochlorophyll is converted to chlorophyll by addition of 2H atoms at carbon atoms 7 and 8 of fourth pyrrole ring. Hence protochlorophyll lacks 2 H atoms in one of its pyrrole ring. The etiolated plants contain protochlorophyll.

Types of Chlorophyll: Aronoff and Allen (1906) recognised 9 types of chlorophyll.

(i) Chlorophyll-a (Blue-green): Its formula is $C_{55}H_{72}O_5N_4Mg$. It is most abundant (25%). It is three times more than chlorophyll-b. It absorbs more red light than blue and violet light. It is found in all photosynthetic organisms except bacteria.

(ii) Chlorophyll-b (Yellow-green): Its chemical formula is $C_{55}H_{70}O_6N_4Mg$. It absorbs more blue and violet light than red light. It is found in Euglenoids, Chlorophyceae (green algae) and higher plants.

(iii) Chlorophyll-c (Bluish-green): Its chemical formula is $C_{35}H_{30}O_5N_4Mg$. It does not have phytol chain. It is found in diatoms, dinoflagellates and phaeophyceae (brown algae).

(iv) Chlorophyll-d (Red): Its chemical formula is $C_{54}H_{70}O_6N_4Mg$. It is found in rhodophyceae (red algae)

(v) Chlorophyll-e: It is found in Xanthophyceae (yellow green algae).

(vi) Bacteriochlorophyll-a (Purple): $C_{55}H_{74}O_6N_4Mg$. It is found in purple sulphur bacteria (*Chromatium*).

(vii) Bacteriochlorophyll-b: It is found in purple non-sulphur bacteria (*Rhodospirillum*).

(viii) Bacterioviridin or Chlorobium chlorophyll: It is found in green sulphur bacteria (*Chlorobium*).

(ix) Chlorobium chlorophyll-666:

2. **Carotenoids**: These are formed from basic isoprene units. These are found in algae and higher plants. These are soluble in organic solvents such as chloroform, ether, fats etc. but insoluble in water. These absorb light of shorter wave (blue and violet) and transfer to it to chlorophyll for use in photosynthesis. These are called shield pigments because they provide protection to chlorophyll from photooxidation or photobleaching. These convert nascent oxygen to molecular oxygen.

(a) Carotenes: Its chemical formula is $C_{40}H_{56}$.. These are orange coloured and hydrocarbon (tetraterpenes). These are of several types viz. α, β,γ, flavacene, lycopene (found in tomato) but β-carotene is most common in all plants. These provide red colouration to chilly and potato.

(b) Xanthophylls: Its chemical formula is $C_{40}H_{56}O_2$. These are yellow coloured and oxygen containing carotene. These are of several types such as Lutein (most common), Fucoxanthin (dominant pigment in brown algae or Phaeophyceae) and Violaxanthin.

3. **Phycobillins**: These are noncyclic, linear and open tetrapyrroles attached to proteins in living cells. These are water soluble. These lack Mg atom and phytol tail. These absorb blue and red

light. These are useful in chromatic adaptations. These are of several types such as

(a) c- Phycocyanin (blue): It is found in Cyanophyceae.

(b) Allophycocyanin (blue): It is found in Cyanophyceae and Rhodophyceae.

(c) r-Phycoerythrin (red): It is found in Rhodophyceae.

Ratio of pigments in chloroplast

Chlorophyll-a : Chlorophyll-b = 3:1

Chlorophyll : Carotenoids = 3:1

Carotenes : Xanthophylls = 1 :1

Photoluminescence: It is irradiation of absorbed energy. It is of two types

1. Fluorescence: It occurs instantaneously and it is the immediate emission of long wave radiation. Chlorophylls show an outburst of fluorescence (Katusky effect) during first few months of illumination. These show red fluorescence.
2. Phosphorescence: It occurs after gap and it is the delayed emission of long wave radiation. It continues for sometime after removal of irradiation source.

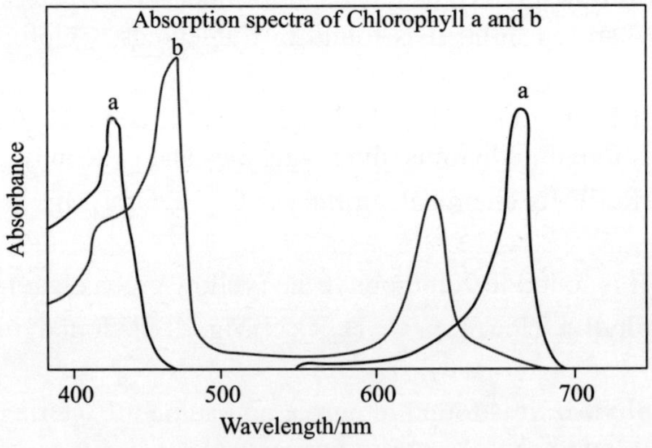

Ability to absorb light of different wavelength

1. Absorption spectrum: It is studied directly. It requires a single exposure to full light. It shows the amount of energy of different wavelength of light absorbed by chlorophyll. Chlorophyll absorbs mostly blue light.
2. Action spectrum: It was studied by Engelmann (1882) in *Cladophora*. It is studied in relation to either CO_2 absorbed or O_2 evolved. It shows the utilisation of light energy (rate of photosynthesis) at different wavelength of light absorbed by a pigment. The rate of photosynthesis is maximum in red light.

Anthocyanin: It is purplish pigment and is present in leaves, flowers and fruits. It gives colouration to leaves. It is soluble in water. It is actually dissolved in cell sap and is not present in cell sap. It does not take part in photosynthesis.

Pigment system (Photosystem): It is network of chlorophyll-a, accessory pigments (carotenoids and phycobillins) and associated proteins on the surface of thylakoid membrane. It is a small group (photosynthetic unit) of 250 pigment molecules and proteins that work together for the reaction centre. It consists of

1. Light Harvesting complex (Antenna complex): It contains 200 chlorophyll molecules which absorb different wavelength of light and transfers to reaction centre by inductive resonance.
2. Core complex: It contains 40-60 chlorophylls, a reaction centre (P-680 / P-700), e^- donors and e^- acceptor. The reaction centre is less than 1%of chlorophyll. The core complex finally performs light reactions.

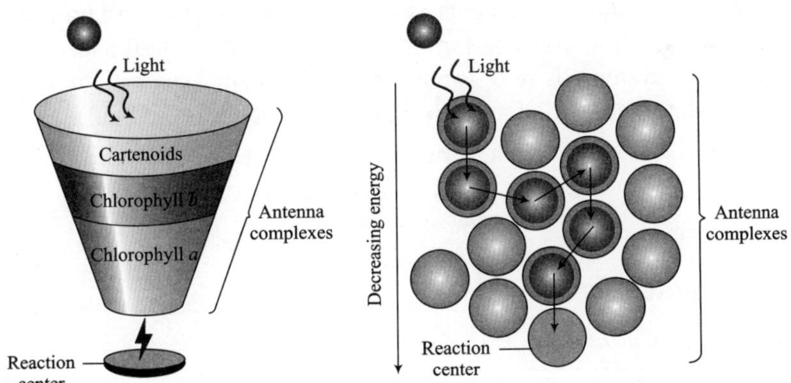

- If Mg atom is removed from Light harvesting complex of Chlorophyll then photosynthetic unit continues to harvest light for reaction centre.
- If Mg atom is removed from chlorophyll of core complex, photosynthesis stops.

Types of Pigment system: There are two types of pigment systems which are structurally distinct but act synergistically.

Pigment system II	Pigment system I
It lies on inner surface of granal thylakoid.	It lies on outer surface of stromal thylakoid and granal thylakoid.
Its reaction centre is P-680.	Its reaction centre is P-700.
It is more fluorescent.	It is less fluorescent.
It absorbs wavelength maximum at 680 nm.	It absorbs wavelength maximum at 700nm.
It is active in red light only.	It is active in both red light and far red light.
It contains less chlorophyll-a than chlorophyll-b.	It contains more chlorophyll-b than chlorophyll-a.
It contains less carotenes than xanthophylls.	It contains more carotenes than xanthophylls.
It performs non-cyclic photophosphorylation.	It performs both cyclic and non-cyclic photophosphorylation.
It is associated with splitting of water hence O_2 is evolved.	It is not associated with splitting of water hence O_2 is not evolved
It generates ATP and $NADPH_2$.	It generates ATP.
It has O_2 as strong oxidising agent because it has tendency to oxidise water molecule	It has ATP as weak oxidising agent.
It has ATP as weak reducing agent.	It has $NADPH_2$ as strong reducing agent because it has tendency to reduce $NADP^+$ molecule

Light reaction (Hill reaction)

It takes place in thylakoid membrane of grana of chloroplast in the presence of light. It is also called photochemical reaction. It is faster process where solar energy is trapped by chlorophyll and is stored in the form of chemical energy (ATP) and reducing power ($NADPH_2$). Both ATP and $NADPH_2$ are collectively called Assimilatory power. It occurs in following steps:

1. **Absorption of solar energy**: The photosynthetic pigments such as chlorophylls, carotenoids and phycobillins absorb solar energy in different regions of visible light.
2. **Transfer of solar energy from antenna pigments to Chlorophyll-a**: Solar energy absorbed by antenna or accessory pigments (pigments other than chlorophyll-a) are transferred by resonance to chlorophyll-a which alone takes part in light reaction. Chlorophyll-a also absorbs light energy directly. In pigment system II, it is P-680 and in pigment system I, it is P-700. Light with wavelength (>680nm) is less efficient than light of wavelength (<680nm).

   Transfer of light energy: Phycoerythrin→Phycocyanin→Carotenoids→Chlorophyll-b →Chlorophyll-a→Reaction centre

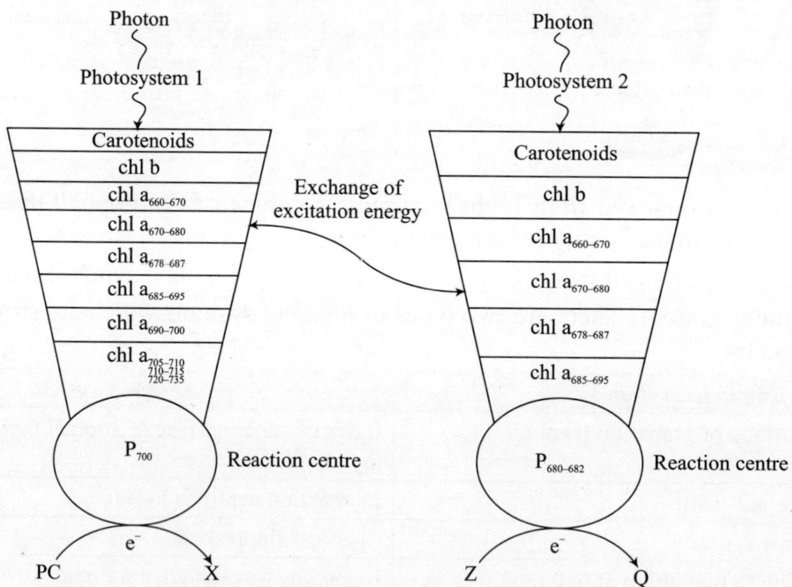

3. **Excitation of Chlorophyll-a**: Both pigment systems (PS-II and PS-I) after receiving a photon (quantum) of light become excited having more energy than ground state energy. After passing through the unstable second singlet state (excited state) and first singlet state, chlorophyll molecules come to metastable triplet state. This form of chlorophyll-a takes part in light reaction. It expels its energy along with an electron and a positive charge comes on chlorophyll-a molecule. The expelled electron contains an extra amount of energy received from light. This is used in the formation of ATP during photosynthesis.

   $$\underset{\text{(Ground state)}}{\text{Chlorophyll-a}} \xrightarrow{\text{light}} \underset{\text{(excited triplet state)}}{\text{Chlorophyll-a}} \longrightarrow \underset{\text{(oxidised)}}{(\text{Chlorophyll-a})^+} + e^-$$

4. **Photolysis (Oxidation) of water:** It is first photochemical reaction. It is mainly concerned with Ps-II and catalysed by Mn^{+2} and Cl^- ions. When Ps-II is active, it is photoactivated then water molecule gets dissociated into H^+ and OH-. After that OH- releases an electron from itself and becomes a free but highly reactive form. The H^+ released in above reaction gets associated with NADP to form $NADPH_2$. The electrons fill the vacancy in chlorophyll molecule created by

photoactivation of chlorophyll. The OH molecules now release O_2 and form H_2O. The photolysis of water maintains the supply of electron. Hence O_2 is released during photosynthesis.

$$4H_2O \rightarrow 4H^+ + 4OH^-$$
$$4OH^- \rightarrow 4OH + 4e^-$$
$$4OH \rightarrow 2H_2O + 4e^- + O_2$$
$$2H_2O \rightarrow 4H^+ + 4e^- + O_2$$

or

$$12H_2O \longrightarrow 24H^+ + 24e^- + 6O_2$$

5. **Photophosphorylation**: It is synthesis of ATP from ADP and iP with the help of light. Arnon (1954) discovered photophosphorylation and observed that isolated chloroplast would synthesise ATP in the presence of light. It is redox process (oxidation and reduction process occur simultaneously) when chlorophyll-a molecule receives a photon of light. It becomes excited and expels the extra energy along with an electron in both pigment systems. Each e- lost is accepted by another molecule, e- acceptor. Thus chlorophyll is oxidised and electron acceptor is reduced. Thus instead of losing its energy, the e- is captured by electron acceptor. This is the conversion of light energy into chemical energy. It occurs in two stages:

   (a) Non-cyclic photophosphorylation: Hill and Bendal proposed Z-scheme for electron transport. In this process, the electrons are not recycled back to chlorophyll molecule and were consumed in reducing $NADPH_2$ and thus a constant supply of water molecule is required for e- and H^+. Thus it is unidirectional movement of electrons. It is carried out by both Ps-II and PS-I. This scheme initially shows uphill transport (from Ps-II to pheophytin), downhill (from pheophytin to PS-I), again uphill (from PS-I to another acceptor, FRS) and finally downhill transport (from FRS to $NADP^+$).

   When sunlight strikes Ps-II, the light harvesting complex receives the radiation first and then passes to other pigment molecules, finally the reaction centre of Ps-II (P-680) by inductive resonance. A photon of light is absorbed by P-680 of chlorophyll-a in PS-II. It gets excited and e- is emitted from it and it causes an electron deficiency which is fulfilled by e- coming from photolysis of water. Light with a wavelength smaller than 680nm is more efficient than wavelength greater than 680 nm.

The e- is trapped by a primary acceptor, pheophytin (a chlorophyll in which Mg atom is replaced by 2H atoms) then it is passed to Plastoquinone (PQ).

The e- is now passed to cytochrome b_6 and cytochrome f (cytochrome complex) where e- loses sufficient energy for the synthesis of ATP. Hence I molecule of ATP is formed by ADP and Pi. The e- is now passed to copper containing protein, plastocyanin (PC) that acts as shuttle between cytochrome f and PS-I. Thus e- reaches PS-I and is accepted eventually by P-700.

At the same time, PS-I also absorbs photon of light and becomes excited to release e- so that e- deficiency occurs in P-700 molecule. The e- deficiency is fulfilled by e- coming from PS-II. The e- is now passed to Ferredoxin Reducing Substance (FRS) which is primary acceptor. The e- is now passed to an iron containing protein, ferredoxin which is natural H acceptor. The e- is transferred to NADP which combines with H^+ released from photolysis of water with the help of NADP reductase to form $NADPH_2$.

$$NADP^+ + 2H^+ + 2e^- \rightarrow NADPH_2$$
$$2NADP^+ + 4H^+ + 4e^- \rightarrow 2NADPH_2$$

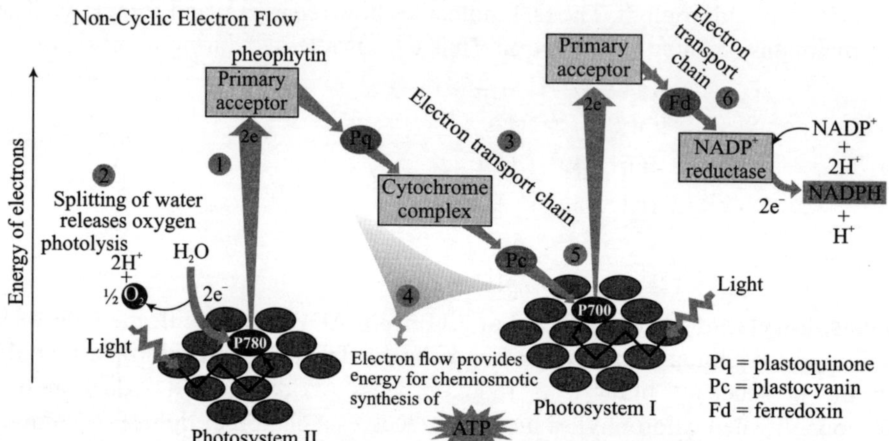

Note:

1. Two e- and two H^+ are required to reduce 1 mole of NADP to $NADPH_2$.
2. Two photons are required to transfer one e-
3. Four e- are required to be transferred to reduce 1 mole of CO_2.
4. Thus 4 mole of P-680 are excited by 4 photons. Similarly, 4 photons will release 4 e- from 4 mole of P-680. Hence photolysis of 2 water molecules results in the production of 4e-, $4H^+$and O_2. Now 8 photons and 2 mole of H_2O are required to reduce 2 mole of NADP+.
5. Quantum yield: It is number of oxygen molecules evolved per photon of light absorbed. Thus it is 1/8=12.5%
6. Quantum requirement: It is the inverse of quantum yield. It is number of photon (8) of light required to for each oxygen molecule evolved.

Cyclic photophosphorylation

In this type, electrons are recycled back to chlorophyll molecule at low energy. It is primitive type and most effective anaerobic phosphorylation mechanism. The electron passes from one component of electron transport system to next, it progressively loses energy so there is lowering of energy gradient. It only occurs when light of wavelength beyond 680 nm is available for excitation; and when level of $NADP^+$ is low and NADPH is high.

A photon of light is absorbed by P-700 of chlorophyll-a in Ps-I. It gets excited and an electron emitted from it is accepted by a primary acceptor, Ferredoxin reducing Substance (FRS).The electron is passed from FRS to plastoquinone (PQ) cytochrome-b_6, cytochrome-f, plastocyanin (PC) and finally back to Chlorophyll-a.

While passing over the FRS, plastoquinone and cytochrome complex, the electron loses sufficient energy to form ATP from ADP and Pi. During this process, protons (H^+) are dragged into thylakoid building a proton gradient which forms ATP by chemiosmosis. It uses light energy to build ATP but it cannot drive dark reaction because it does not produce sufficient ATP just like non-cyclic photophosphorylation. Hence both processes are required for this. It is only process which occurs in certain photosynthetic bacteria.

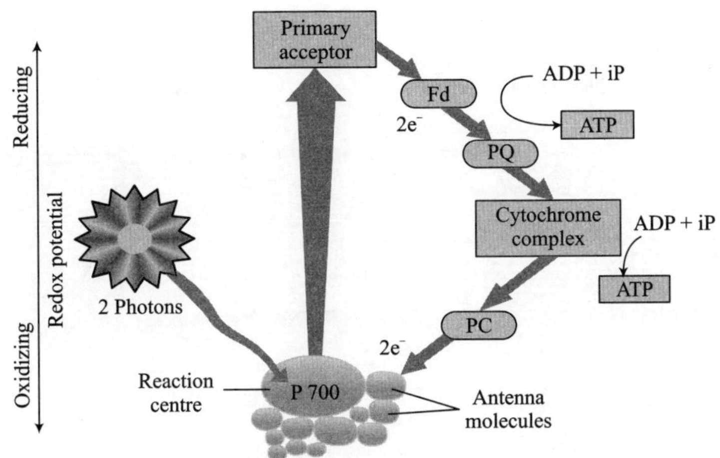

Cyclic photophosyphorylation

Non-cyclic photophosphorylation	Cyclic photophosphorylation
The electron emitted from chlorophyll is not recycled back to same chlorophyll.	The electron emitted from chlorophyll is recycled back to same chlorophyll.
It is unidirectional e- transfer.	It is cyclic e- transfer.
It is carried out by both Ps-I and Ps-II.	It is carried out only by Ps-I.
It occurs in granal thylakoid.	It is carried out by stromal thylakoid (intergranal thylakoid).
It takes place in all plants.	It takes place in photosynthetic bacteria. It occurs in plants as backup to provide additional ATP.
It occurs under optimum light, CO_2 and aerobic condition.	It occurs under low light, low CO_2 and anaerobic condition.
It synthesises 1 ATP and 2 $NADPH_2$.	It synthesises only 2ATP.
It undergoes phosphorylation at one step.	It undergoes phosphorylation at two steps.
It involves photolysis of water and evolution of O_2.	It does not involve photolysis of water and evolution of O_2.
Dichlorophenyldimethyl urea (DCMU) inhibits transport of electrons at plastoquinone stage.	It cannot inhibit transport of electrons.

Chemiosmosis

It was proposed by Peter Mitchell (1961). It is the synthesis of ATP in light reaction by movement of H^+ across selectively permeable membrane along concentration gradient. There are 4 protein complexes in thylakoid membrane.

(a) Ps-II: It is located in stacked region of thylakoid membrane.

(b) Ps-I: It is located in stromal lamellae and in unstacked region of thylakoid.

(c) Cytochrome b_6-f complex: It pumps protons.

(d) ATP synthase (cf_0-cf_1): It is located in stroma lamellae and in unstacked region.

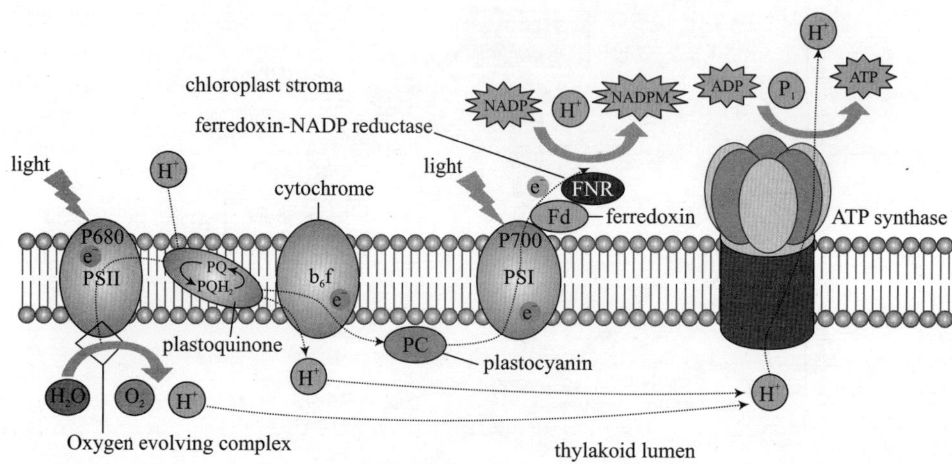

Fig. ***Mechanism of chemiosmosis***

Chemiosmosis requires a membrane, a proton pump, a proton gradient and ATPase. It occurs in following steps:

(a) During electron transport these electrons travel down a electron transport chain through pigment systems, some of energy pump protons (H^+) actively across membrane from stroma to lumen of thylakoid by cytochrome b_6-cytochrome f complex. This occurs when primary acceptor of e- which is located towards the outer side of membrane, transfers its e- not to electron carrier but to H^+ carrier. Thus this molecule removes one H^+ from stroma while transferring an e-. When this molecule passes its electrons to e- carrier present on the inner side of thylakoid membrane, then H^+ is released into lumen of thylakoid.

(b) Water splits on inner side of thylakoid membrane resulting in the accumulation of H^+ in the lumen that contributes to gradient.

(c) $NADP^+$ is reduced to NADPH by NADP reductase enzyme (located on stroma side of thylakoid membrane) by utilising electrons that come from the acceptor of e^- of Ps-I and H^+ from stroma. This also decreases the concentration of H^+ in stroma.

(d) Now the number of H^+ is decreased in stroma (alkaline condition) and increased in lumen of thylakoid (acidic condition).This develops proton gradient (proton motive force) across thylakoid membrane. The gradient is broken down due to movement of H^+ across the membrane to the stroma through trans- membrane channel of the F_0 of ATPase.

(e) The ATPase (ATP synthetase) consists of two metallo-peptide complexes, oneisF_0 (hydrophobic) which is embedded in membrane and forms trans-membrane channel that carries out facilitated diffusion of H^+ across the membrane (from lumen to stroma). The other part is called F_1 (hydrophilic) and projects on the outer surface of thylakoid membrane (in stroma) on the side that forms stroma.

(f) The breakdown of the gradient (or diffusion of protons from lumen to stroma) provides enough energy to cause a conformational change in F_1 particle of ATPase. This energy is used to pump H^+ across the membrane to create gradient or a high concentration of H^+ within lumen of thylakoid.

(g) ATPase has channel that allows diffusion of H^+ back across the membrane. This releases enough energy to activate ATPase enzyme that catalyses the formation of ATP.

(h) Thylakoid membrane is now impermeable to H^+. Hence H^+ can come back from lumen to stroma through channel provided by ATP synthase (cf_0-cf_1). The cf_0 forms a channel across membrane and through which H^+ move from lumen to stroma with the help of proton motive force. This causes synthesis of ATP from ADP (phosphorylation) by cf_1 component of ATP synthase and thus ATP is released into stroma.

$$2H^+ + ADP + Pi \longrightarrow 2H^+ + ATP + H_2O$$

(lumen) (stroma)

The overall equation for **light reaction** is:

$$2H_2O + 2NADP + 3ADP + 3Pi + light \rightarrow 2NADPH + 2H^+ + 3ATP + O_2$$

Anoxygenic photosynthesis	Oxygenic photosynthesis
Photosynthetic pigment is bacteriochlorophyll.	Photosynthetic pigment is chlorophyll-a.
Pigment system II is absent.	Pigment system II is present.
Electron donor is H_2S or H_2 or lactate.	Electron donor is H_2O.
Oxygen is not produced.	Oxygen is produced.
The primary product of energy conversion is ATP only.	The primary products of energy conversion are ATP and $NADPH_2$.

Dark reaction (C_3 cycle)

It is also called biosynthetic pathway in which the assimilatory power (ATP and $NADPH_2$) produced by light reaction is utilised to reduce CO_2 to form glucose by process of carbon fixation.

It was first established by Blackman but later on it was studied in detail by Calvin, Benson and Bassham (1954) who determined the sequence of reactions in *Chlorella* and *Scenedesmus* using autoradiography (C^{14}) and chromatography. It occurs in stroma of chloroplast where all enzymes of dark reaction are found. It does not require light thus it is independent of light hence it can occur in both light and dark.

Process: The dark reaction completes in three steps:

1. **Carboxylation:** It is first process in which atmospheric CO_2 is accepted by Ribulose bisphosphate (5C) in the presence of Ribulose bisphosphate carboxylase oxygenase (RuBisCO) or carboxydismutase to form first stable 3 carbon compound, Phosphoglyceric acid (3-PGA). This process does not utilise ATP AND $NADPH_2$.

   (RuBisCO is most abundant protein which constitutes 50% of total proteins of chloroplast. It is the slowest enzyme located in stroma on outer surface of thylakoid membrane. It requires Mg^{+2} as metal activator. It was crystallised from tobacco leaves).

   $$6CO_2 + 5RuBP \xrightarrow{\text{RuB is CO}} 12(3\text{-PGA})$$

2. **Glycolytic reversal (Reduction):** In this process, 3-PGA undergoes phosphorylation reaction in the presence of phosphoglyceric kinase forming 1,3-diPGA. This reaction utilises 2 ATP per molecule of CO_2 for supplying energy. Then 1,3-diPGA undergoes reduction in the presence of dehydrogenase enzyme forming 3-Phosphoglyceraldehyde (PGAL). This process utilises $2NADPH_2$ for reducing per molecule of CO_2 fixed. The 3-PGAL is key product which is used in the synthesis of both carbohydrates and fats. It does not accumulate hence it is immediately converted to other products such as fructose 1,6-diphosphate, fructose 6-Phosphate and then finally glucose. It seems to be reversal of glycolysis, the only difference is that reducing power is obtained from NADPH (not NADH). Half (50%) of glucose produced is used by plants in cellular

respiration to maintain itself and rest glucose is converted to sucrose for translocation, cellulose for building cell wall and starch for storage.

12(3-PGA) + 12ATP + 12NADPH$_2$ ⟶ 12(PGAL) + 12ADP + 12Pi + 12NADP

5(PGAL) ⟶ 5DHAP

3PGAL + 3DHAP ⟶ 3(Fructose-1,6-diP) ⟶ 3(Fructose 6-P) + 3Pi

1 Fructose-6-P ⟶ Glucose

3. **Regeneration of RuBP**: It is complicated process which requires several enzymes. Simply the large amount of PGAL is converted to RuMP in the presence of kinase and later RuMP is converted to regenerate RuBP in the presence of kinase and Mg^{+2}. This process utilises ATP.

   Since glucose is a 6 carbon compound thus 6 rounds of Calvin cycle are required to synthesise 1 mole of glucose because one carbon atom is reduced in each cycle.

   2Fructose-6-P ⟶ 2(Xylulose-5-P) + 2 (Erythrose-4-P)

   2(Erythrose-4-P) + 2(DHAP ⟶ 2(Sedoheptulose,1,7-diP ⟶ 2(Sedoheptulose-7-P) + 2Pi

   2(Sedoheptulose-7-P) + 2 (3-PGAL) ⟶ 2(Ribose-5-P) + 2 (Xylulose-5-P)

   2(Ribose-5-P) ⟶ 2(Ribulose-5-P)

   4(Xylulose-5-P) ⟶ 4(Ribulose-5-P)

   6(Ribulose-5-P) + 6ATP ⟶ 6RuBP + 6ADP + 6Pi

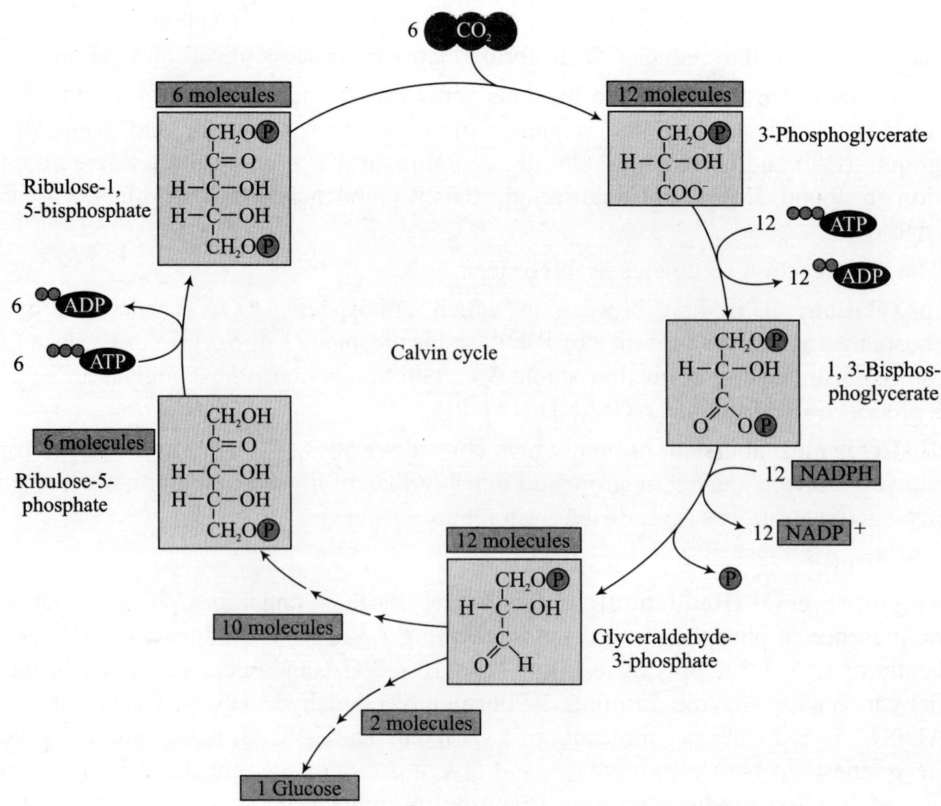

Thermodynamics:

(i) To reduce 1 molecule of CO_2, 8 quanta of light are required.

(ii) To synthesise 1 mole of glucose (from 6 CO_2), 48 quanta of light are required. Therefore, 42 K cal energy is present in one quantum of red light of 680 nm.

(iii) The minimum light energy required for reducing $6CO_2$ to synthesise glucose is 42x48=2016

(iv) Since 1 mole of glucose is oxidised aerobically to yield 686 kcal hence 2016/686=33% of total light energy is utilised in fixing 6molecules of CO_2.

$1CO_2 + 2H_2O + 3ATP + 2NADPH_2 + 1RuBP \longrightarrow (CH_2O)n + H_2O + 3ADP + 3Pi + 2NADP^+ + 2H^+ + 1RuBP$

Overall reaction: $6CO_2 + 12H_2O + 18ATP + 12NADPH + 6RuBP \longrightarrow C_6H_{12}O_6 + 6H_2O + 18ADP + 18Pi + 12NADP^+ + 12\ H^+ + 6RuBP + 6O_2\uparrow$

Conversion of energy:

1. **In C_3 plants,** 18 ATP and 12 $NADPH_2$ are required to synthesise I mole of glucose. 18 ATP = 18 × 7.3 Kcal = 132 Kcal

   12 $NADPH_2$ = 12 × 52 kcal = 624 Kcal

   Total energy required = 132 + 624= 756 kcal

   Since, I mole of glucose produces 686 kcal

   Thus 90% of conversion of energy could be noticed in C_3 plants.

C_4 cycle

It was reported by Kortchak (1964) in leaves of sugarcane but it was studied in detail by Hatch and Slack (1970). It is alternative CO_2 fixation which occurs in tropical and subtropical grasses (monocots) and some dicots of angiosperms. It has been reported in 10 families of angiosperms such as Poaceae, Cyperaceae, Amaranthaceae, Chenopodiaceae, Asteraceae, Euphorbiaceae, Nyctaginaceae, Portulacaeae, Aizoceae, and Zygophyllaceae.

These plants show a very special type of anatomy called Krantz anatomy because of the presence of radially arranged cells of bundle sheath cell around vascular bundle which looks like a ring. The cell of bundle sheath (Krantz cell) is larger in size, thick walled and contain large amount of chloroplast without grana. These cells do not have intercellular space but contain starch. It involves the following steps:

1. **Reactions in mesophyll cell:** The atmospheric CO_2 is accepted by Phosphoenol pyruvic acid (PEP) and undergoes carboxylation in the presence of PEP carboxylase to form first stable 4C compound, oxaloacetic acid (OAA). Later OAA is reduced to malic acid in the presence of malate dehydrogenase enzyme in mesophyll cell. The mesophyll cell remains in contact with stomata.
2. **Reactions in bundle sheath cell:** Malic acid enters into bundle sheath cell through plasmodesmata where it undergoes decarboxylation to form pyruvic acid and thus CO_2 is released. As a result, RuBisCO enzyme functions as carboxylase minimising the activity of oxygenase and thus favours carboxylation. The final fixation of CO_2 (Calvin cycle) occurs in chloroplast of bundle sheath cell where concentration of CO_2 is high but the temperature and O_2 are low. Hence, productivity (yield) is better in C_4 plants.
3. **Reactions in mesophyll cell:** Finally pyruvic acid enters into chloroplast of mesophyll cell through plasmodesmata and reacts to form phosphoenol pyruvic acid in the presence of pyruvate phosphate dikinase enzyme.

These plants are more efficient (greater yield and productivity) because they grow fast at high temperature, high light intensity and under low CO_2 hence these show greater affinity with phosphoenol pyruvate (PEP) at low CO_2 concentration because PEP carboxylase has much higher affinity than RuBP carboxylase and further photorespiration does not occur since it does not accept O_2.

The C_4 plants have had to adjust low level of CO_2 by artificially raising the CO_2 concentration in bundle sheath cell at enzyme site to prevent photorespiration which occurs under low CO_2 and high O_2. The C_4 cycle thus effectively shuttles CO_2 from atmosphere into bundle sheath cell. As a result, the concentration of CO_2 in bundle sheath cell becomes higher than that of atmosphere. This occurs when malic acid of mesophyll cell undergoes decarboxylation to form pyruvic acid in bundle sheath cell to release CO_2. These plants can survive in poor CO_2 condition because there is internal supply of CO_2. High concentration of CO_2 in bundle sheath cell suppresses photorespiration which helps the C_4 plants to photosynthesise more efficiently even at high temperature.

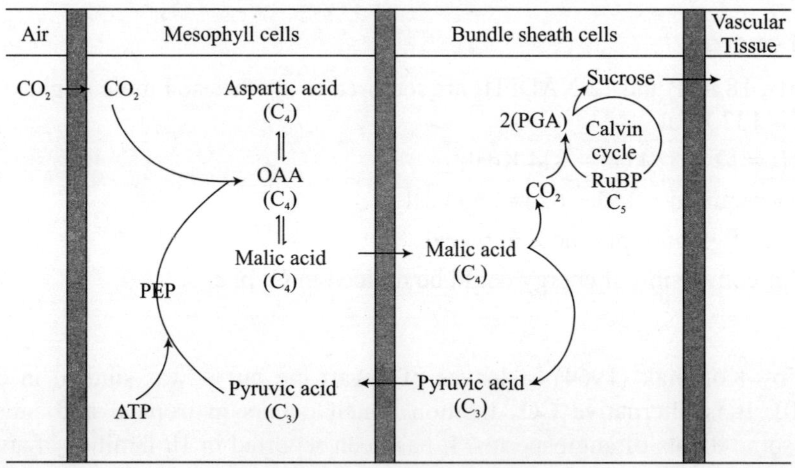

Fig. ***Hatch and Slack pathway***

These plants use more energy (2 more ATP per mole of CO_2 fixed) because this extra energy is required for collecting and transporting CO_2 and hydrogen from mesophyll cell to chloroplast in bundle sheath cell at the site of RuBP carboxylase activity. This is only disadvantage of C_4 plants.

Overall reaction:

$$6CO_2 + 12H_2O + 30ATP + 12NADPH + 6RuBP \longrightarrow$$
$$C_6H_{12}O_6 + 6H_2O + 6O_2 + 30ADP + 30\ Pi + 12NADP^+ + 12H^+ + 6RuBP$$

C_3 plants	C_4 plants
These are more common in dicots than monocots. e.g., rice, wheat, *Atriplex hastata, Euphorbia corallata.* These are found in 90% of plants.	These are more common in monocots than dicots. e.g., maize, sugarcane, *Sorghum, Atriplex rosea, Euphorbia maculata.*
Leaves do not show Krantz anatomy.	Leaves show Krantz antomy
Chloroplast is monomorphic (chloroplast in mesophyll cell is granal).	Chloroplast is dimorphic (chloroplast in mesophyll cell is granal and that of bundle sheath cell is agranal).
It operates through Calvin cycle.	It operates through Calvin cycle as well as Hatch and Slack pathway.

First CO_2 acceptor is RuBP (weak CO_2 acceptor).	First CO_2 acceptor is PEP (strong CO_2 acceptor).
First product is Phosphoglyceric acid (PGA) which is 3 carbon compound	First product is oxaloacetic acid (OAA) which is 4 carbon compound
CO_2 fixation occurs in mesophyll cell.	Initial CO_2 fixation occurs in mesophyll cell and final in bundle sheath cell.
There is one carboxylation process.	There are two carboxylation processes.
CO_2 fixation is slower.	CO_2 fixation is faster.
3ATP and 2NADPH$_2$ are required to fix 1 mol. of CO_2.	5ATP and 2NADPH$_2$ are required to fix 1 mol. of CO_2
Photorespiration occurs.	Photorespiration does not occur.
Rate of net photosynthesis is low.	Rate of net photosynthesis is high (2 to 3 times).
CO_2 compensation point (threshold value) is high (25-100 ppm) due to photorespiration	CO_2 compensation point is low (0-10 ppm).
More CO_2 is released in light than in dark.	Less CO_2 is released in light than in dark.
Photosynthesis is not performed at very low CO_2.	Photosynthesis is performed at very low CO_2.
Optimum temperature is 20°C-25°C for enzymes.	Optimum temperature is 30-45°C for enzymes.
O_2 has inhibitory effect (sensitive) on photosynthesis (Warburg effect).	O_2 has no inhibitory effect (insensitive) on photosynthesis.
These are less efficient (photosynthetic) at high temperature.	These are more efficient (photosynthetic) at high temperature.
These are more efficient (photosynthetic) at low temperature.	These are less efficient (photosynthetic) at low temperature.
It occurs at lower light intensity (1000 foot candles) which is saturation point.	It occurs at higher light intensity and even full sunlight which is not saturation point.
It requires 33% of solar energy to form assimilatory power	It requires 40% of solar energy to form assimilatory power.
These cannot tolerate saline condition.	These tolerate saline condition due to presence of organic acid.
These cannot tolerate drought condition because transpiration ratio is high.	These can tolerate drought condition because transpiration ratio is low.
These are adapted to all climates.	These are only adapted to tropical climate.

Photorespiration

It was discovered by Dicker and Tijo (1959) in tobacco. It is light dependent uptake of O_2 and release of CO_2. In hot and dry conditions, plants close their stomata to prevent water loss. Under these conditions, the CO_2 will decrease and oxygen gas produced by the light reactions of photosynthesis will increase causing an increase of photorespiration. It was observed that photosynthesis was inhibited when there is higher concentration of O_2 and this inhibition was caused due to occurrence of photorespiration. Photorespiration has no relation with normal respiration. Both processes show similarities in uptake of O_2 and release of CO_2. It operates when rate of photosynthesis is ten times higher than respiration due to more availability of O_2. It is a special type of respiration occurring in many green plants (C_3 plants) when these are exposed to strong sunlight.

If temperature is high then oxygenation will be promoted because solubility of CO_2 in water is decreased more rapidly than that of O_2 and further specificity of RuBisCO will be decreased. It occurs when the amount of O_2 is greater than CO_2. The competition occurs between CO_2 and O_2 for RuBisCO. The oxygenase activity increases because O_2 is competitive inhibitor of carboxylase activity of RuBisCO hence affinity of RuBPcase with O_2 increases and with CO_2 decreases. Now it functions as RuBP oxygenase rather carboxylase.

It does not produce ATP, $NADH_2$ and glucose. But overall process is energy consuming process and further there is wasteful loss (or release) of around 25% of photosynthetically fixed CO_2 by utilising ATP when glycine is converted (oxidised) into serine. The loss of carbon is due to dual activity of RuBP carboxylase and oxygenase which can act not only as carboxylase (useful role in C_3 cycle) but also act as oxygenase (enzyme catalysing the direct insertion of molecular O_2 into organic substrate).

It is a cyclic process which occurs in three cell organelles such as chloroplast, peroxisome and mitochondria. It is also called glycolate metabolism or C_2 cycle because the chief metabolite and substrate is glycolate which is two carbon compound.

Reactions in chloroplast:

The first reaction of photorespiration is oxygenation of RuBP in the presence of sufficient O_2 and RuBP oxygenase. It forms 1 mol. of phosphoglycolate (2C) and 1 mol. of Phosphoglyceric acid (3C). if sufficient CO_2 is available then only PGA would have formed. While the PGA is used up in Calvin cycle, the phosphoglycolate is dephosphorylated to form glycolate in chloroplast.

Reactions in peroxisome:

From the chloroplast, the glycolate is diffused to peroxisome where it is oxidised to glyoxalate and H_2O_2. Catalase (Iron containing enzyme) converts H_2O_2 into H_2O. In the peroxisome, the glyoxalate is used to form an amino acid, glycine (2 C).

Reactions in Mitochondria:

Glycine enters mitochondria where 2 mol. of glycine (4C) give rise to 1 mol. of another amino acid, serine (3C) and 1 mol. of CO_2.

Serine is taken up by the peroxisome, and through a series of reactions, it is converted to glycerate. The glycerate leaves the peroxisome and enters the chloroplast, where it is phosphorylated to form phosphoglyceric acid (PGA). The PGA molecule re-enters the Calvin cycle to make carbohydrates but 1 molecule of CO_2 (25%) released in mitochondria during photorespiration has to be refixed.

In other words, 75% of carbon lost in the form of PGA by oxygenation of RuBP is recovered and 25% is lost as release of 1 molecule of CO_2. Thus photorespiration is also called Photosynthetic Carbon Oxidation (PCO) cycle.

It protects the plants from photo-oxidative damage by dissipation of energy of excitation. It also recovers some of carbon from excess glycolate.

Factors: It depends on (i) high temperature, (ii) bright light, (iii) high O_2 and (iv) low CO_2.

Reactions occurring in Mesophyll cell

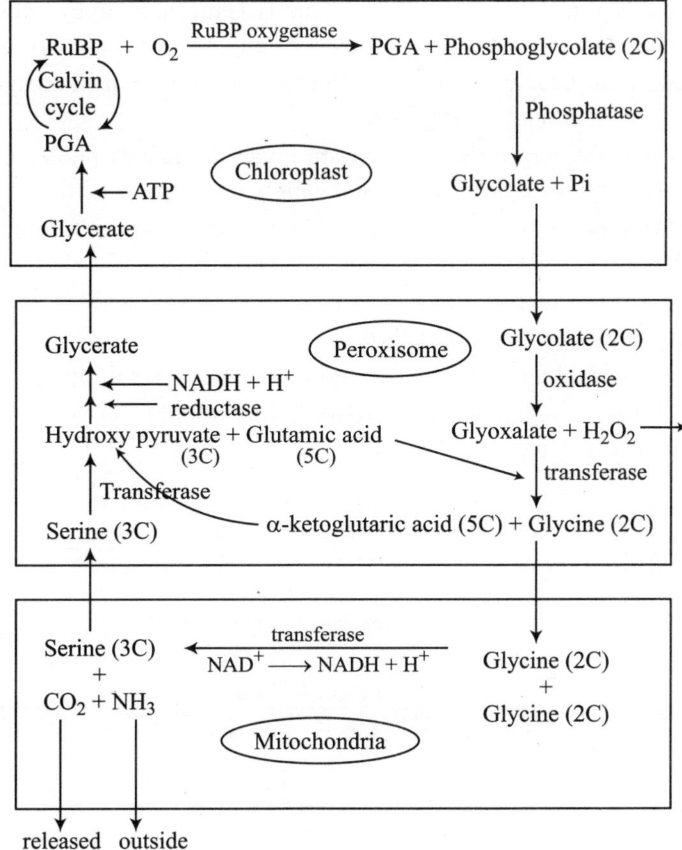

Fig. ***Reactions of Photorespiration***

Terms:

Compensation Point: It is the point or threshold value of CO_2 concentration at which illuminated plant parts stop absorbing CO_2 from environment. At this point, CO_2 fixed in photosynthesis is equal to CO_2 produced in respiration and photorespiration so that there is no net exchange of gases. In other words, it is point in which all O_2 evolved by photosynthesis is used up in respiration and all CO_2 produced in respiration is used for photosynthesis.

Plants cannot survive for long at compensation point because there is net loss of organic matter due to respiration.

It is less in C_4 plants and more in C_3 plants.

It is less in shady plants (sciophytes) and more in sun plants (heliophytes).

Law of limiting factors: It was proposed by Blackman (1905). It states that when a chemical process is affected by several factors then its rate will be determined by the slowest factor. Hence rate of photosynthesis is limited at a given time by one and only one factor which is deficient. The light intensity supplied to a leaf is just sufficient to allow it to utilise 5mg of CO_2 per hour in photosynthesis.

- A denotes no photosynthesis in the absence of CO_2. If CO_2 increases from 0 to 1 mg per hour, then the rate of photosynthesis will increase.

- The rate of photosynthesis, however cannot increase further because of limiting role of CO_2. The rate will increase along the line AB when CO_2 supply is raised to 5mg per hour. The rate does not increase beyond the point B where it becomes constant along the line B-C.
- Any further increase in CO_2 will have no effect upon the rate of photosynthesis because light intensity has now become the limiting factor.
- If light intensity is increased further to medium level, the rate of photosynthesis also increases along the line B-D provided there is increased supply of CO_2.
- Light again becomes the limiting factor at point D when rate of photosynthesis becomes constant along the line D-E.
- Further increase in light to high level, the rate of photosynthesis increases along D-F when sufficient concentration of CO_2 is increased.
- The rate becomes constant along F-G when light alone becomes limiting.
- Hence B, D and F are the points where increase in rate of photosynthesis stops abruptly because of the limiting nature of light.

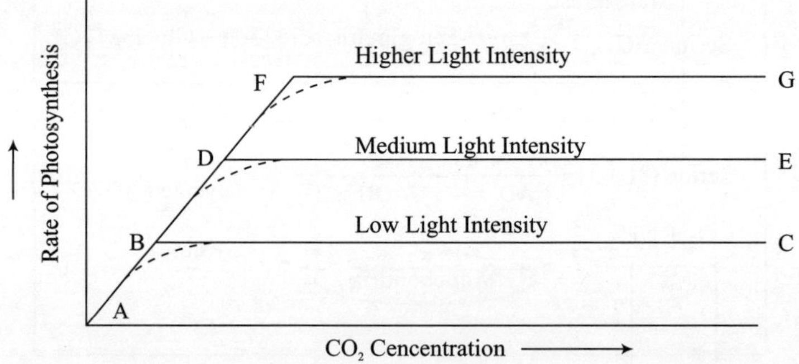

Results:

(i) When all the factors are present but CO_2 concentration is low then rate of photosynthesis is controlled by CO_2 which is limiting.

(ii) Similarly if light intensity is low on cloudy day, it controls the rate of photosynthesis.

(iii) The continued increase of the factor without corresponding increase in the limiting factor would decrease the rate of photosynthesis.

Factors affecting Photosynthesis:

1. Light:

(a) Intensity of light: There are two points:

Light compensation point: It is a point in light intensity at evening where rate of photosynthesis is minimum and equal to the rate of respiration. There is no net CO_2 fixation at compensation point. The low compensation point is indicative of photosynthetic efficiency.

Light saturation point: It is a point in light intensity where rate of photosynthesis is maximum. Above this point, solarisation occurs where rate of photosynthesis starts declining due to photooxidation of chlorophyll.

The rate of photosynthesis starts at low light intensity, starts increasing with increase in light intensity and later decreases with high light intensity due to solarisation.

It is more in intense light than in diffused light. Very low light intensity causes closure of stomata hence entry of CO_2 is reduced thus rate of photosynthesis is reduced.

(b) Quality of light: It occurs in visible light range. It is maximum in white light or sunlight (polychromatic light). It is minimum in green light, average in blue light and maximum in red light. On the other hand, red algae show maximum photosynthesis in green light and brown algae in blue light.

(c) Duration of light: The rate of photosynthesis increases in long continuous light but it is limited by slow rate of dark reaction. Its rate is faster in intermittent light (alternate light and dark period) than in continuous light because essential phase of photosynthesis will occur not only in light but also in dark since assimilatory power is fully consumed in dark reaction. Sugarcane is most efficient converter because it utilises 10% of light.

2. **CO_2:** The rate of photosynthesis is increased from 0.036% (360 ppm) content of atmosphere to 0.9% but starts decreasing above 0.9%. It acts as limiting factor under normal condition of light and temperature (provided other factors are not limiting). CO_2compensation point is that value of atmospheric CO_2 concentration when the rate of photosynthesis is just equivalent to the rate of respiration in the leaf so that there is no net exchange of gases. A plant cannot survive for long at compensation point because there is net loss of organic matter due to respiration of non-green organs and dark respiration. Low CO_2 compensation point in C_4 plants is due to CO_2 concentration mechanism at carboxylation site.
3. **Temperature**: The rate of photosynthesis increases with rise in temperature (optimum 20-25°C for C_3 plants and 30-45°C for C_4 plants) under normal condition of light and CO_2. But the rate decreases due to denaturation of enzymes. Temperature coefficient (Q_{10}) is the ratio of the rate of reaction at a given temperature to the rate at the temperature 10°C lower. Q_{10} is 1 in light reaction where there is no increase in the rate with rise in temperature. Q_{10} is 2 in dark reaction where the rate is doubled with every 10°C rise in temperature.
4. **Water**: It acts as indirect limiting factor. It is the source of H^+ to reduce NADP and release of O_2. It maintains the turgidity of photosynthetic cells; helps in opening of stomata and activates enzymes. If water content of leaves is reduced then stomata closes partially hence entry of CO_2 is reduced by which rate of photosynthesis is decreased.
5. **O_2**: If amount of oxygen is less, the ratio of photosynthesis will be increased. If amount of oxygen is more thus rate will decrease (**Warburg effect**) because high amount of oxygen causes oxidation of chlorophyll and acts as competitive inhibitor of carboxylase activity of RuBisCO. Warburg effect is shown in C_3 plants and insensitive to C_4 plants.
6. **Chlorophyll content:** Rate of photosynthesis increases with increase in content of chlorophyll.
7. **Structure of leaf**: Rate of photosynthesis is affected because entry of CO_2 is dependent on size, position and number of stomata; thickness of cuticle and epidermis; presence of epidermal hairs and amount of mesophyll cells.
8. **Age of plants**: The rate of photosynthesis is slowed down in mature plants.
9. **Accumulation of photosynthetic products**: Rate of photosynthesis is reduced in their presence.

SHORT ANSWER TYPE QUESTIONS

1. Explain Moll's experiment in photosynthesis.
2. Explain Ingenhousz's experiment in photosynthesis.

3. Write the structural formulae of chlorophyll-a and chlorophyll-b and mention their difference.
4. Write notes on carotenoids.
5. Distinguish between PS II and PS I.
6. Distinguish between light harvesting complex and core complex.
7. Distinguish between non-cyclic and cyclic photophosphorylation.
8. Distinguish between anoxygenic photosynthesis and oxygenic photosynthesis.
9. What do you mean by compensation point?
10. Write notes on law of limiting factors.
11. Discuss the role of light in photosynthesis.
12. How photolysis of water is possible in photosynthesis?
13. Write the final equation of Calvin cycle and Hatch-Slack pathway.
14. What do you mean by Krantz anatomy?
15. Why C_4 plants are photosynthetically more efficient?

LONG ANSWER TYPE QUESTIONS

1. Discuss the mechanism of photophposphorylation in light reaction of photosynthesis.
2. Discuss the mechanism of chemiosmotic generation of ATP.
3. Describe the CO_2 fixation in C_3 plants.
4. Describe the CO_2 fixation in Hatch-Slack pathway and distinguish between C_3 plants and C_4 plants.
5. Describe the process of photorespiration in green plants and explain why it is considered a wasteful process?

MULTIPLE CHOICE QUESTIONS

1. Moll while working on photosynthesis confirmed that
   (a) Light is essential (b) CO_2 is essential
   (c) Chlorophyll is essential (b) Water is essential
2. Which of the plant was used by Engelmann to study action spectra?
   (a) *Chlorella* (b) *Mentha* (c) *Chlamydomonas* (c) *Cladophora*
3. Which of the following set of pigments is called antenna pigments?
   (a) All types of chlorophyll (b) Chlorophyll-a and carotenoids
   (c) Chlorophyll-a and b (d) All pigments other than Chlorophyll-a
4. Which of the set of elements is used in photolysis of water?
   (a) Chlorine and Manganese (b) Chlorine and Magnesium
   (c) Manganese and Magnesium (d) Iron and Chlorine
5. How much quanta of light are required to evolve I molecule of O_2?
   (a) 2 (b) 4 (c) 8 (d) 10
6. Which of the following processes occurs in photosystem II?
   (a) CO_2 fixation (b) CO_2 reduction (c) Photolysis of H_2O (d) All of these
7. In which of the following orders, the electron carriers between PS I and PS II are correctly arranged?
   (a) Phe – PQ – Cyt b_6 - Cyt f – PC (b) FRS –PQ – Cyt b_6 - Cyt f – PC
   (c) Phe – PC – Cyt b_6 - Cyt f – PQ (d) Phe – PC – Cyt f - Cyt b – FRS
8. PS II is concerned with
   (a) Photolysis of water (b) Flowering (c) Reduction of CO_2 (d) Release of energy
9. PS I is located on
   (a) Inner surface of thylakoid membrane (b) Outer surface of thylakoid membrane
   (c) Both surfaces of thylakoid membrane (d) Stroma of chloroplast
10. Ferredoxin is a component of
    (a) P680 (b) P700 (c) PS I (d) PS II
11. Cyclic photophosphorylation produces
    (a) $NADPH_2$ (b) ATP and $NADPH_2$
    (c) ATP, $NADPH_2$ and O_2 (d) ATP only
12. ETS associated with photophosphorylation occurs in
    (a) Stroma (b) Mitochondria
    (c) Thylakoid (d) Outer membrane of chloroplast
13. ADPis phosphorylated when three protons are shifted from
    (a) Thylakoid lumen to stroma (b) Stroma to cytosol
    (c) Cytosol to stroma (d) Thylakoid lumen to cytosol
14. Most of experimental work in photosynthesis has been done on
    (a) *Chlamydomonas* (b) *Chlorella* (c) *Hydrilla* (d) *Cladophora*
15. If radioactive CO_2 is given to a C_3 plant of photosynthesis, the radioactivity first appeared in
    (a) 3-Phosphoglyceric acid (b) Oxaloacetic acid
    (c) Aspartic acid (d) Glucose

16. Which of the following is not a step in Calvin cycle?
(a) Carboxylation
(b) Reduction
(c) Regeneration of RuBP
(d) Photophosphorylation
17. During fixation of one molecule of CO_2 in C_3 plants
(a) 3 ATP and 2 $NADPH_2$ are required
(b) 5 ATP and 2 $NADPH_2$ are required
(c) 12 ATP and 12 $NADPH_2$ are required
(d) 18 ATP and 12 $NADPH_2$ are required
18. Dimorphism of chloroplast is found in
(a) C_4 plants
(b) C_3 plants
(c) CAM plants
(d) All of the above
19. Krantz anatomy is found in
(a) Most of monocots
(b) Most of dicots
(c) Algae
(d) All dicots
20. Agranal chloroplasts are found in
(a) Mesophyll of pea leaves
(b) Bundle sheath of mango leaves
(c) Mesophyll of maize leaves
(d) Bundle sheath of sugarcane leaves
21. The inhibitory effect of oxygen in C_3 plants on photosynthesis is called
(a) Warburg's effect
(b) Photo oxidation
(c) Solarisation
(d) Emerson's effect
22. Chloroplasts without grana are known to occur in
(a) Bundle sheath cells of C_4 plants
(b) Bundle sheath cells of CAM plants
(c) Bundle sheath cells of C_3 plants
(d) Mesophyll cells of C_4 plants
23. Which of the following enzymes is not found in C_3 plants?
(a) RUBP carboxylase
(b) PEP carboxylase
(c) NADP reductase
(d) ATP synthase
24. Which of the following cell organelles is involved in photorespiration?
(a) Peroxisome
(b) Golgi body
(c) Endoplasmic reticulum
(d) Lysosome
25. The percentage of light energy fixed in photosynthesis is around
(a) 10
(b) 50
(c) 3
(d) 1
26. Quantasome in chloroplast contains
(a) 200 chlorophyll molecules
(b) 250 chlorophyll molecules
(c) 100 chlorophyll molecules
(d) 300 chlorophyll molecules
27. Which of the following is incorrect regarding anthocyanin?
(a) It is soluble in water
(b) It is purplish pigment dissolved in water
(c) It gives colouration to leaves
(d) It takes part in photosynthesis
28. During daytime, the ratio of CO_2 produced to that of CO_2 consumed is
(a) 1 : 1
(b) 1 : 5
(c) 1 : 10
(d) 1 : 50
29. Photosynthesis is most active in
(a) Red light
(b) Sunlight
(c) Blue light
(d) Green light
30. If half of the leaves of a plant are removed, then the rate of photosynthesis will be
(a) More and the photosynthetic yield will also be more
(c) Less and the photosynthetic yield will also be less
(d) More and the photosynthetic yield will also be less
(d) Less and the photosynthetic yield will also be more

31. Photorespiration is favoured by
    (a) Low O_2 and high CO_2
    (b) Low light and high O_2
    (c) High O_2 and low CO_2
    (d) Low temperature and high CO_2
32. The chloroplast of bundle sheath cell of C_4 plants is
    (a) Large and granal (b) Large and agranal (c) Small and granal (d) Small and agranal
33. Which of the following plants is most efficient converter of solar energy?
    (a) Wheat (b) Maize (c) Soya bean (d) Sugarcane
34. C_4 plants are adapted to
    (a) Hot and dry climate
    (b) Temperate climate
    (c) Cold and dry climate
    (d) Hot and humid climate
35. Which of the pigment systems is inactivated in Red drop?
    (a) Ps I (b) Ps II (c) Both Ps I and Ps II (d) None of the above
36. Glycolate accumulates in chloroplast when there is
    (a) Low CO_2 (b) High CO_2 (c) Bright light (d) Low temperature
37. Photorespiration shows the formation of
    (a) ATP only
    (b) Sugar
    (c) Both ATP and sugar
    (d) Neither ATP nor sugar
38. The ratio between 2 Carbon and 3 Carbon intermediates having $-NH_2$ group formed in photorespiration is
    (a) 1 : 1 (b) 1 : 2 (c) 2 : 1 (d) 3 : 2
39. In photorespiration, the role of peroxisome is to
    (a) Reduce glycolate
    (b) Oxidise glycolate
    (c) Convert glycine into serine
    (d) Oxidise RuBP
40. The porphyrin ring of chlorophyll molecule is a complex ring structure of
    (a) Double bonds
    (b) Triple bonds
    (b) Alternating single and double bonds
    (d) Alternating double and triple bonds
41. Plants do not store carbohydrate in the form of glucose because
    (a) Glucose dissolves in water thereby altering the osmotic balance
    (b) Glucose is an unstable molecule
    (c) Glucose would replace ribose
    (d) Glucose attracts insect herbivores
42. Which of the following portions of chloroplast shows the lowest pH when sunlight falls on chloroplast?
    (a) Stroma
    (b) Cytosol
    (c) Periplastidial space
    (d) Space enclosed by thylakoid membrane
43. C_4 plants are found among
    (a) Poaceae family
    (b) Monocots only
    (c) Dicots only
    (d) Monocots and dicots
44. Protochlorophyll differs from chlorophyll in lacking
    (a) 2 H atom in one of its pyrrole ring
    (b) 4 H atom in one of its pyrrole ring
    (c) 2 H atom in two of its pyrrole ring
    (d) 4 H atom in two of its pyrrole ring

45. Phosphoglyceric acid (PGA) as the first CO_2 fixation product was discovered in photosynthesis of
(a) Algae (b) Vascular cryptogams
(c) Higher plants (d) Bacteria

46. How chlorophyll would appear when seen in red light?
(a) Red (b) Far red (c) Black (d) Green

47. Bacteriochlorophyll differs from chlorophyll-a in having
(a) one pyrrole nucleus with one hydrogen (b) one pyrrole nucleus with two hydrogen
(c) one pyrrole nucleus with three hydrogen (d) one pyrrole nucleus with four hydrogen

48. Emerson effect confirms the existence of
(a) Photorespiration (b) Photophosphorylation
(c) Light and dark reactions (d) Two distinct photochemical reactions

49. The dry weight of a green leaf will be maximum at
(a) Morning (b) Evening
(c) Afternoon (d) Same throughout whole day

50. On a normal sunny day, the rate of photosynthesis is maximum during
(a) Early morning (b) Late morning (c) Mid day (d) Evening

51. RuBP carboxylase was crystallised from
(a) *Chlorella* (b) Leaf of maize (c) Leaf of sugarcane (d) Leaf of Tobacco

52. Seedlings grown in dark are
(a) Shorter than those grow in light (b) Taller than those grow in light
(c) Similar to those grow in light (d) They do not grow at all.

53. The process of photorespiration involves
(a) Chloroplast - Mitochondria - Peroxisome (b) Mitochondria - Peroxisome - Chloroplast
(c) Peroxisome - Mitochondria - Chloroplast (d) Chloroplast - Peroxisome - Mitochondria

54. Which of the following is the most limiting factor under normal condition?
(a) CO_2 (b) Light (c) Temperature (d) Water

55. The number of turns of Calvin cycle required to synthesise one molecule of glucose was
(a) Two (b) Three (c) Four (d) Six

56. Which of the following processes of Calvin cycle does not require any energy?
(a) Carboxylation (b) Reduction
(c) Glycolytic reversal (d) Regeneration of RuBP

57. The first CO_2 acceptor in Hatch-Slack pathway is
(a) RuBP (b) PEP (c) OAA (d) PGA

58. Which of the following pigment systems is involved in bacterial photosynthesis?
(a) Ps I (b) Ps II
(c) Both PS I and PS II (d) None of these

59. Which of the following electron carriers contains copper?
(a) Plastoquinone (b) Plastocyanin (c) Ferredoxin (d) Pheophytin

60. Which of the following pigments is known as shield pigment?
(a) Phycobilin (b) Chlorophyll (c) Carotenoid (d) Anthocyanin

61. The reaction centre consists of
    (a) Single molecule of chlorophyll a
    (b) Single molecule of chlorophyll b
    (c) Light harvesting complex
    (d) Both a and b
62. The electron flow from pigment system II to NADP via ETS and PsI follows
    (a) Uphill→Downhill→Uphill→Downhill
    (b) Downhill→Uphill→Downhill→Uphill
    (c) Both a and b
    (d) None of the above
63. Which of the following materials is not recycled between dark and light reaction?
    (a) NADPH $+H^+$
    (b) ADP
    (c) ATP
    (c) O_2 and CO_2
64. The pH of thylakoid space (lumen) in bright light becomes
    (a) More acidic
    (b) Less acidic
    (c) More alkaline
    (d) Neutral
65. What happens during light reaction of photosynthesis?
    (a) ADP is phosphorylated and NADP reduced
    (b) ADP is hydrolysed and NADP reduced
    (c) ATP is hydrolysed and NADP is reduced
    (d) ATP is hydrolysed and NADP is reduced
66. The cooperation of two pigment systems of the chloroplast is required for
    (a) ATP synthesis
    (b) Reduction of $NADP^+$
    (c) Oxidation of reaction centre of photosystem
    (d) Generation of proton motive force
67. C_4 plants are mostly found in
    (a) Temperate region with more humid condition
    (b) Tropical region with more dry condition
    (c) Temperate region with less humid condition
    (d) Tropical region with more humid condition
68. In Calvin cycle, ATP is required in
    (a) Carboxylation
    (b) Reduction
    (c) Regeneration
    (d) Both b and c
69. If mature leaves were enclosed for a fixed amount of time in a transparent bag that had radioactive CO_2, then maximum radioactivity was found to occur in
    (a) Guard cell of leaves
    (b) Mature leaves
    (c) Actively growing leaves
    (d) Senescent leaves
70. The productivity of crop declines when leaves begin to wilt because
    (a) Stomatal closure prevents entry of CO_2
    (b) Photolysis of water does not occur
    (c) Mesophyll cell becomes flaccid
    (d) Chloroplast decomposes
71. If one plant is grown in the shade of green house and other is grown under shade of trees, then rate of photosynthesis will be
    (a) Greater in the shade of green house
    (b) Lesser in the shade of green house
    (c) Equal in both condition
    (d) Equally low in both conditions.
72. In an experiment on evolution of O_2 by photosynthesising *Hydrilla* plant, a pinch of $NaHCO_3$ is added to water then the rate of photosynthesis will
    (a) Decrease
    (b) Increase
    (c) Remains constant
    (d) First decrease then increase
73. How much oxygen is formed from 264 g. of CO_2 and 216 g. of H_2O?
    (a) 96 g
    (b) 216 g
    (c) 264 g
    (d) 192 g
74. Photosynthesis cannot be operated in
    (a) Red light
    (b) Blue light
    (c) Green light
    (d) Yellow light

75. C_4 plants by decreasing photorespiration and efficiency carrying out photosynthesis can
    (a) Decrease stomatal opening and thereby reduce water loss
    (b) Decrease stomatal opening and thereby reduce leaf temperature
    (c) Increase stomatal opening and thereby increase water loss
    (d) Increase stomatal opening and thereby increase temperature rate
76. The major difference in Ps I and Ps II found in chloroplast is
    (a) Position on lamellae (b) Chlorophyll a
    (c) Position of electron carriers (d) Energy harvesting
77. Light harvesting complex II is located in the
    (a) Thylakoid lumen (b) Stroma
    (c) Thylakoid membrane (d) Outer membrane of chloroplast
78. If Mg atom is removed from core complex of photosystem then
    (a) Photosynthesis will stop
    (b) Photosynthesis will initially decrease then increase
    (c) Photosynthesis will continue with normal rate
    (d) Photosynthesis will initially increase then decrease
79. Which of the following plants is both C_3 and C_4 plant?
    (a) *Atriplex* (b) *Euphorbia* (c) Maize (d) Sugarcane
80. Photophosphorylation was discovered by
    (a) Calvin (b) Blackman (c) Robert Hill (d) Arnon
81. Which of the following plants was used by Priestley to confirm that plant purified the air spoilt by burning candle?
    (a) Mint (b) *Hydrilla* (c) *Cladophora* (d) *Chlorella*
82. Which of the following plants was used by Ingenhousz to confirm the release of oxygen?
    (a) Mint (b) *Hydrilla* (c) *Cladophora* (d) *Chlorella*
83. Each pyrrole ring of chlorophyll head contains
    (a) 4C and 1N atoms (b) 5C and 1N atoms (c) 3C and 2N atoms (d) 4C and 2N atoms
84. Which of the following groups is found in chlorophyll-b?
    (a) $-CH_3$ (b) –CHO (c) –COOH (d) =CO
85. Chlorophyll is synthesised from
    (a) Succinic acid (b) Oxaloacetic acid
    (c) Succinyl Coenzyme A (d) Malic acid
86. Which of the following pigments does not contain phytol chain?
    (a) Chlorophyll-a (b) Chlorophyll-b (c) Chlorophyll-c (d) Chlorophyll-d
87. Which of the following pigments does not involve in photosynthesis?
    (a) Carotene (b) Xanthophylls (c) Anthocyanin (d) Phycoerythrin
88. Which of the following pigments does not belong to antenna pigments?
    (a) Chlorophyll-a (b) Chlorophyll-b (c) Chlorophyll-c (d) Chlorophyll-d
89. Which of the following phosphorylation reactions occurs in bacteria?
    (a) Non-cyclic photophosphorylation (b) Cyclic photophosphorylation
    (c) Substrate level phosphorylation (d) Oxidative phosphorylation

90. C_4 plants perform photosynthesis under
    (a) Low CO_2 (b) High temperature (c) Bright light (d) All of the above
91. The first stable product in C_4 cycle is
    (a) Phosphoglyceric acid (b) Oxaloacetic acid
    (c) Lactic acid (d) Phosphoenol pyruvic acid
92. The first CO_2 acceptor in C_3 plants is RuBP which is
    (a) 3 Carbon compound (b) 4 Carbon compound
    (c) 5 Carbon compound (d) 6 Carbon compound
93. The first reaction in photorespiration is
    (a) Carboxylation (b) Oxygenation (c) Dehydration (d) Reduction
94. The temperature coefficient in light reaction is
    (a) 1 (b) 2 (c) 3 (d) 4
95. Photorespiration does not produce
    (a) ATP (b) $NADPH_2$ (c) Glucose (d) All of the above
96. How many ATPs are required to fix 6 molecules of CO_2 in C_4 plants?
    (a) 18 (b) 24 (c) 30 (d) 36
97. How much amount of glucose formed as a result of photosynthesis will undergo respiration?
    (a) 10% (b) 50% (c) 75% (d) 100%
98. The final CO_2 fixation in Hatch and Slack pathway occurs in
    (a) Mesophyll cell (b) Bundle sheath cell (c) Companion cell (d) Xylem vessel
99. Photorespiration is also known as
    (a) C_2 cycle (b) C_3 cycle (c) C_4 cycle (d) C_5 cycle
100. Chlorophyll absorbs maximum
    (a) Red light (b) Blue light (c) Green light (d) Orange light

ANSWERS

1.b	2.d	3.d	4.a	5.c	6.c	7.a	8.a	9.b	10.c
11.d	12.c	13.a	14.b	15.a	16.d	17.a	18.a	19.a	20.d
21.a	22.a	23.b	24.a	25.d	26.b	27.d	28.c	29.b	30.c
31.c	32.b	33.d	34.d	35.b	36.a	37.d	38.c	39.b	40.c
41.a	42.b	43.d	44.a	45.a	46.c	47.b	48.d	49.c	50.b
51.d	52.b	53.d	54.a	55.d	56.a	57.b	58.a	59.b	60.c
61.a	62.a	63.d	64.a	65.a	66.b	67.b	68.d	69.c	70.a
71.a	72.b	73.d	74.c	75.c	76.d	77.c	78.a	79.a	80.d
81.a	82.b	83.a	84.b	85.a	86.c	87.c	88.a	89.b	90.d
91.b	92.c	93.b	94.a	95.d	96.c	97.b	98.b	99.a	100.b

11

Respiration in Plants

Respiration is an oxidative process which is mainly concerned with the production of energy which is required for all living cells to perform various vital activities. The released energy is stored in the form of ATP in the cells. Respiration is of three types:

1. **External respiration:** It is exchange of oxygen and carbon dioxide between environment and organisms (breathing). It occurs through stomata in land plants, entire plant surface in submerged hydrophytes, general body surface in algae, fungi and bacteria.
2. **Internal respiration:** It is exchange of O_2 and CO_2 between internal body fluids such as blood and individual cell.
3. **Cellular respiration:** It is biochemical oxidation of organic substance esp. glucose resulting in its stepwise break down into simpler ones with gradual and stepwise release of energy (ATP). The energy is utilised for metabolic activities in the cell. It is exothermic process.

$$C_6H_{12}O_6 + 6O_2 \longrightarrow 6CO_2 + 6H_2O + \text{energy.}$$

Respiration occurs whole day and night in all the living cells but the rate is maximum in growing regions like root tip, stem tip, flower bud and germinating seeds. Plants take in free O_2 from the air which enters through the stomata and lenticels and diffuses through the continuous intercellular space to reach the living cell.

During day time when the rate of photosynthesis is higher than the rate of respiration and when both the processes are going on simultaneously, all the oxygen liberated in photosynthesis is not consumed in respiration, whereas all the CO_2 liberated in respiration is completely consumed in photosynthesis and some more CO_2 must enter the plant for the higher rate of photosynthesis. Thus the effects of respiration are completely masked during day time. During night, however there is no photosynthesis and only exchange of gases (O_2 consumed and CO_2 liberated) between the plant and atmosphere is the effect of respiration. Respiration occurs in the liquid medium in a cell. Respiration consumes 50% of sugars formed in photosynthesis. The cell undergoing respiration will prefer carbohydrate (firstly monosaccharide then oligosaccharide and lastly polysaccharide), then fats and lastly proteins when both carbohydrate and fats are unavailable to them.

Types of respiration: Blackman divided respiration into two types on the basis of organic substrate used.

Floating respiration	Protoplasmic respiration
Respiratory substrate is carbohydrate.	Respiratory substrate is protein.
It is common and runs continuously throughout.	It is rare and occurs only for some periods.
It does not produce toxic products.	It produces toxic product.

Types of respiration: On the basis of availability of O_2.

Anaerobic respiration	Aerobic respiration
It does not require O_2 to break down the substrate.	It utilises O_2 to break down the substrate.
It occurs in cytoplasm.	It occurs in both cytoplasm and mitochondria.
It usually occurs in microbes, yeast and parasitic worms.	It occurs in plants and animals.
It involves glycolysis and incomplete breakdown of pyruvic acid.	It involves glycolysis, Krebs cycle and terminal oxidation.
It is less efficient.	It is more (15 times) efficient.
The substrate is incompletely broken and one of the end products is at least organic.	The substrate is completely oxidised and end products are inorganic.
Water is not formed and lesser number (2) of CO_2 is released.	Water is formed and larger number (6) of CO_2 is released.
One molecule of glucose produces 2 ATP on complete oxidation.	One molecule of glucose produces 36/38 ATP.
One molecule of glucose produces less amount of energy (56 K.cal).	One molecule of glucose produces large amount of energy (686 Kcal).

Anaerobic respiration can only continue for a short time in higher plants. Finally the plants die if they are deprived of oxygen for longer time because energy produced in this process is not sufficient for continuing the life processes for long time. Further accumulation of alcohol might reach the level of toxicity. Anaerobic respiration stops when the concentration of alcohol produced reaches about 13-15%.

Both anaerobic and aerobic respiration involves

1. Glycolysis

It is anaerobic breakdown of glucose into two molecules of pyruvic acid. It was discovered by Embden, Mayerhoff and Parnas (1930) hence it is known as EMP pathway. It occurs in cytosol of the cell because all enzymes are found. It occurs in both anaerobic and aerobic respiration. It is an oxygen independent metabolic pathway used by both autotrophs and heterotrophs. It does not require molecular oxygen for any of its reactions but it can take place in the presence of O_2 also hence it involves oxidation without direct use of oxygen.

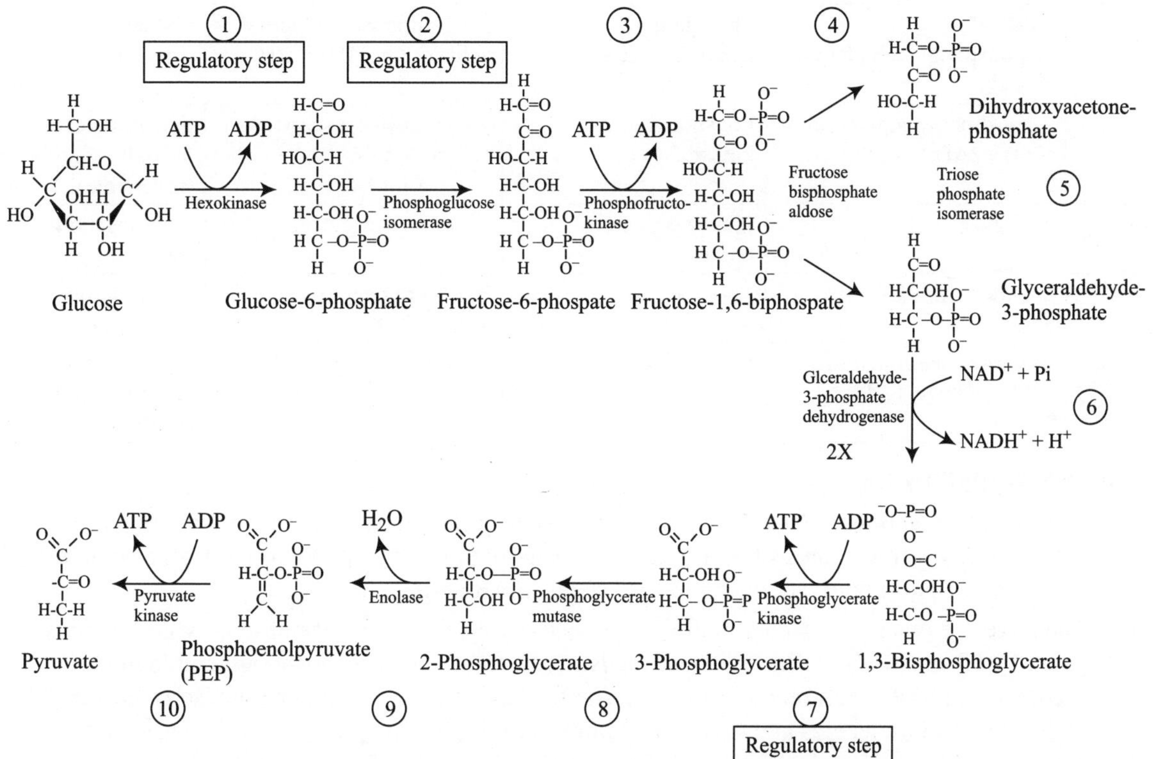

Reactions of glycolysis

Final reaction:

$$C_6H_{12}O_6 + 2ADP + 2Pi + 2NAD^+ \longrightarrow$$
$$2CH_3COCOOH + 2ATP + 2NADH + 2H^+ + 2H_2O$$

Steps involved in Glycolysis:

Steps	Substrate	Enzyme	Products	Reaction
1.	Glucose + **ATP**	Hexokinase, Mg^{+2}	Glucose-6-phosphate + ADP + Pi	Phosphorylation (irreversible)
2.	Glucose -6-phosphate	Glucose phosphate isomerase	Fructose-6-phosphate	Isomerisation (reversible)
3.	Fructose-6-phosphate + **ATP**	Phosphofructokinase, Mg^{+2}	Fructose-1,6-diphosphate + ADP + Pi	Phosphorylation (irreversible)
4.	Fructose-1,6-diphosphate	Aldolase	Dihydroxyacetone phosphate (ketose) + 3-Phosphoglyceraldehyde (aldolase)	Aldol splitting (reversible)
5.	Dihydroxyacetone phosphate	Phosphotriose isomerase	3-Phosphoglyceraldehyde.	Isomerisation (reversible)

6.	2 (3-phosphoglyceraldehyde) + NAD^+	Phosphoglyceric dehydrogenase	2(1,3-diphosphoglyceric acid) + **2 NADH + 2H$^+$**	Oxidation (reversible)
7.	2(1,3-diphosphoglyceric acid) +2ADP + 2 Pi	Phosphoglyceric kinase, Mg^{+2}	2 (3-phosphoglyceric acid) + **2ATP**	Substrate level phosphorylation (reversible)
8.	2(3-phosphoglyceric acid)	Phosphoglycero-mutase	2(2-phosphoglyceric acid)	Isomerisation (reversible)
9.	2(2-diphosphoglyceric acid)	Enolase	2 (Phosphoenol pyruvic acid) + 2 H_2O	Dehydration (reversible)
10.	2 (phosphoenol pyruvic acid) + 2ADP + 2Pi	Pyruvate kinase, Mg^{+2}	2(Pyruvic acid) + **2 ATP**	Substrate level phosphorylation (irreversible)

Features of glycolysis:

(a) The first five reactions (first half of glycolysis) come under preparatory or investment phase since they consume 2 molecules of ATP (one in 1st and one in 3rd step) to convert glucose into two molecules of phosphoglyceraldehyde (3 Carbon compound).

(b) The next five reactions (second half of glycolysis) are called pay-off phase in which there is net gain of ATP and NADH. Since glucose leads to two triose sugars (phosphoglyceraldehyde) in the preparatory phase, each reaction in the pay-off phase occurs twice per glucose molecule. This yields 2NADH molecules and 4ATP molecules (two in 7th step and two in 10th step), leading to a net gain of 2 NADH molecules and 2ATP molecules from the glycolytic pathway per glucose.

(c) The process of glycolysis is turned off when ATP is present in sufficient amount because ATP is allosteric inhibitor of phosphofructokinase enzyme.

(d) Two molecules of Pyruvic acid are formed as products from one molecule og glucose.

(e) CO_2 is not released, hence there is no carboxylation reaction.

Balance sheet of ATP in glycolysis

Consumption of ATP by phosphorylation of glucose at first step: – 1 ATP

Consumption of ATP by phosphorylation of glucose at third step: – 1 ATP

– 2 ATP

Production of ATP:

Substrate level phosphorylation (direct gain) at 7th step: 2 ATP

Substrate level phosphorylation (direct gain) at 10th step: 2 ATP

4 ATP

(Net gain of ATP = 4ATP - 2ATP = 2ATP)

Oxidative phosphorylation (through 2 $NADH_2$) at 6th step: **6 ATP**

Total production of ATP in glycolysis = 6ATP + 2 ATP = 8ATP

Types of phosphorylation process:

1. Photophosphorylation: It occurs in photosynthesis in which light is used.
2. Substrate level phosphorylation: It occurs in the process of respiration where there is direct transfer of a phosphate group from one of substrate to ADP. It does not require O_2 and ETC.

3. Oxidative phosphorylation: It occurs in the process of respiration.

Pasteur effect: It is sudden change from anaerobic respiration to aerobic respiration due to availability of O_2 hence there is reduction in consumption of respiratory substrate. Pasteur observed that when yeast were subjected to aerobic condition then their production of alcohol and consumption of glucose declined due to process of fermentation. He concluded that yeast consumed far more sugar when growing anaerobically rather aerobically.

Fate of Pyruvic acid: Pyruvic acid is a key compound which stands at the junction of anaerobic respiration and aerobic respiration. It shows following three fates which are as follows:

(a) If O_2 is available then pyruvic acid will enter into Krebs cycle via oxidative decarboxylation process.

$$2 \text{ Pyruvic acid} + 6\, O_2 \longrightarrow 6CO_2 + 6H_2O + \text{energy}$$

(b) If O_2 is limited as in microbes, yeast and plant cells then alcoholic fermentation occurs.

(c) If O_2 is absent as in skeletal muscle, liver and microbes then lactic acid fermentation occurs. If lactic acid accumulates in muscle it causes fatigue and pain in muscle. However during rest it is again converted into pyruvic acid which undergoes aerobic respiration.

Fermentation: It is a process in which chemical changes are brought about by the action of enzymes produced by living organisms (microbes and some fungi) on organic substrate (carbohydrate, protein, fats). It can also be defined as 'a process of converting carbohydrate to alcohol or organic acids using microorganisms (bacteria and yeast) under anaerobic condition". The science of fermentation is called zymology. It is less efficient because it produces only 2 ATPs from glucose.

It was discovered by Gay Lussac (1815). He described the production of alcohol from sugar.

$$C_6H_{12}O_6 \longrightarrow 2C_2H_5OH + 2CO_2$$

Pasteur (1875) first of all carried out alcoholic fermentation of yeast to investigate why wine sometime turned distasteful. He showed that fermentation occurs by the action of living microorganisms and concluded that fermentation was catalysed by a vital force called 'ferment' within yeast cell. Thus yeast's glucose consumption decreased under aerobic conditions of fermentation, in comparison to anaerobic effect (Pasteur effect). Later Edward Buchner (1897) demonstrated that the conversion of glucose to ethyl alcohol was possible using a nonliving extract of yeast due to the action of enzymes in the extract. He coined the term 'zymase'. It requires temperature (32-36°C) and pH=4.5-5.0

Types: Fermentation is of several types:

1. **Alcoholic fermentation**: It occurs in limited amount of oxygen. It involves two steps: In first step, pyruvic acid itself becomes acceptor of protons (H^+) and electrons (e-) in the absence of oxygen and undergoes decarboxylation forming acetaldehyde and in second step, acetaldehyde undergoes dehydrogenation forming ethyl alcohol. It is used to produce beer, wine and distilled spirits.

   (a) $2 \text{ Pyruvic acid} \xrightarrow[\text{Thiamine pyrosphosphate (TPP)}]{\text{Decarboxylase, } Mg^{2+}} 2 \text{ Acetaldehyde} + 2CO_2$

   (b) $2 \text{ Acetaldehyde} + 2NADH + 2H^+ \xrightarrow{\text{Alcoholic dehydrogenase}} 2\text{Ethyl alcohol} + 2NAD^+$

2. **Lactic acid fermentation**: It occurs in complete absence of oxygen and pyruvic acid itself becomes acceptor of protons (H^+) and electrons (e-) in the absence of oxygen. In strenuous exercise, when energy demands exceed energy supply, the respiratory chain cannot process all of the H atoms joined by NADH. It produces lactic acid in skeletal muscle, liver and microbes. If lactic acid accumulates in muscle, it will cause fatigue. However during rest the same lactic acid is again converted into pyruvic acid and undergoes aerobic respiration. It is of two types:

(a) **Homolactic**: It is carried out by *Lactobacillus, Streptococcus, Pediococcus* etc. It is used for souring milk and also for preparation of cheese, yoghurt. It involves single step in which pyruvic acid is converted into lactic acid in the presence of lactic dehydrogenase enzyme (it also acts as isoenzyme in which different molecular forms catalysing the same reaction). It occurs in one step.

$$2 \text{ Pyruvic acid} + 2 \text{ NADH} + 2H^+ \xrightarrow[\text{Flavin Mononucleotide (FMN), } Zn^{+2}]{\text{Lactic dehydrogenase}} 2 \text{ Lactic acid} + 2 \text{ NAD}^+$$

(b) **Heterolactic**: It is carried out by *Leuconostoc*. It is used for preparing sauerkraut. It uses Pentose phosphate (PPP) pathway. It produces lactic acid, ethyl alcohol and CO_2 in several steps.

3. **Acetic acid fermentation**: It is carried out by *Acetobacter aceti*. It is used for preparation of vinegar.

$C_6H_{12}O_6 \longrightarrow 2\ C_2H_5OH + 2CO_2$

$C_2H_5OH \xrightarrow{\textit{Acetobacter}} CH_3COOH + H_2O$

4. **Propionic acid fermentation**: It is carried out by *Propionebacterium.*

$C_6H_{12}O_6 \longrightarrow 2\ CH_3CO.COOH \longrightarrow CH_3CHOH.COOH \longrightarrow CH_3CH_2COOH{+}CO_2$

5. **Butyric acid fermentation**: It is carried out by *Clostridium butyricum*

$C_6H_{12}O_6 \longrightarrow CH_3CO.COOH \longrightarrow CH_3CH_2CH_2.COOH{+}2CO_2$

6. **Butanediol fermentation**: It is carried out by *Enterobacter, Erwinia, Klebsiella.*

$C_6H_{12}O_6 \longrightarrow$ 2 Pyruvic acid $\longrightarrow$ Acetolactic acid $\longrightarrow$ Acetoin $\longrightarrow$ 2,3-butanediol

Fermentation and anaerobic respiration are identical because

(a) Glucose is substrate in both processes.
(b) The principal end products are CO_2 and ethyl alcohol.
(c) Zymase occurs commonly.
(d) Pyruvic acid and acetaldehyde are formed as intermediates.

Fermentation	Anaerobic respiration
It is an extracellular process.	It is intracellular process.
It usually occurs in microbes, yeast, root and seed of higher plants, muscles of animals.	It usually occurs in plants and animals.
It occurs in a small (limited) quantity of O_2.	It occurs in complete absence of O_2.

2. Pyruvate oxidation

Pyruvic acid, the end product of glyclolysis will undergo oxidative decarboxylation in mitochondrial matrix. It is also called link reaction or acetylation of pyruvic acid. This reaction is complicated and involves several steps which are as follows:

(a) Pyruvic acid + TPP $\xrightarrow{\text{Pyruvate decarboxylase, } Mg^{2+}}$ TPP-acetate + CO_2

(b) TPP-acetate + lipoic acid $\xrightarrow[\text{Dehydrogenase}]{\text{Acetyl transferase}}$ Acetyl hydrolipoate + TPP

(c) Acetyl hydrolipoate + Co.A $\longrightarrow$ Acetyl Co.A + Reduced lipoic acid

(d) Reduced lipoic acid +NAD$^+$ $\longrightarrow$ Oxidised lipoic acid +NADH$^+$ + H$^+$

Substrates/Reactants	Enzyme	Products	Reaction
Pyruvic acid + Thiamine pyrophosphate + Lipoicacid + CoA + NAD+	Pyruvate dehydrogenase	Acetyl CoA + CO_2 + NADH + H^+	Oxidative decarboxylation

$$2CH_3.CO.COOH + 2CoA + 2NAD^+ \xrightarrow{\text{Pyruvate dehydrogenase}} 2CH_3CO.CoA + 2CO_2 + 2NADH + 2H^+$$

Acetyl CoA (derived from oxidation of pyruvic acid or from β-oxidation of fatty acid) is the only fuel to enter Krebs cycle. It is connecting link between glycolysis and Krebs cycle. With each turn of cycle, one molecule of acetyl CoA is consumed for every molecule of oxaloacetic acid present in mitochondrial matrix and is never generated. Actually it is the oxidation of the acetate part of acetyl CoA that produces CO_2 and water. Two molecules of CO_2 are released since two molecules of pyruvic acid are involved in this step.

Balance sheet of ATP in pyruvate oxidation:

Oxidative phosphorylation through 2 $NADH_2$= 2 × 3 = **6 ATP**

3. Krebs cycle

It is a series of chemical reactions used by all aerobic organisms to release stored energy through oxidation of acetyl CoA. It was investigated by Hans Krebs (1940) when he observed the increase in oxygen consumption in breast muscle of pigeon due to addition of certain chemicals. It occurs in matrix of mitochondria where all enzymes are present except succinic dehydrogenase which is found in inner membrane. It occurs in the presence of oxygen but oxygen has no role in Krebs Cycle.

It is called citric acid cycle because citric acid is first product and is constantly generated in every turn of cycle. It is also called tricarboxylic acid (TCA) cycle because citric acid contains three carboxylic groups.

It is called central metabolic pathway involving oxidation of carbohydrates, proteins and fats into smaller compounds (catabolism) and synthesis of macromolecules (amino acids, chlorophylls, cytochromes etc) from intermediate products (anabolism or anapleurotic reaction). Hence it is called amphibolic process.

It consumes acetyl coenzyme A and water, reduces NAD^+ to NADH and produces 2 molecules of CO_2 as a waste product. At the end of Krebs cycle, the final product is oxaloacetic acid (OAA). This is identical to the OAA that begins the cycle. Now the OAA is ready to accept another acetyl CoA to begin another turn of the cycle.

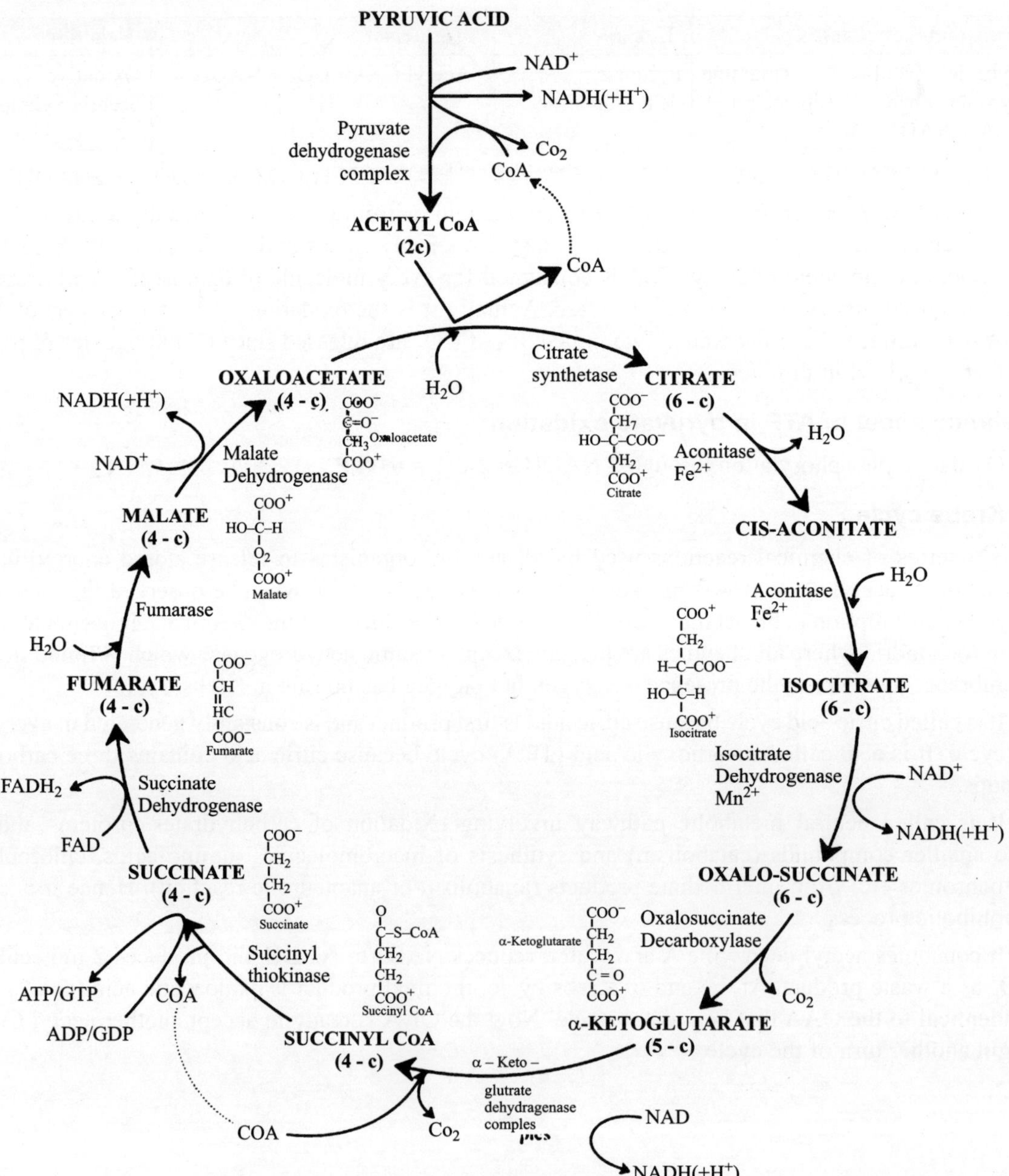

Fig. ***Pyruvate oxidation and Krebs cycle***

Overall reaction of Krebs cycle (single cycle considering one molecule of acetyl Co.A):

$CH_3COCH_2COOH + CH_3CO.CoA + 3NAD^+ + 1\ FAD + 1\ ADP + Pi + 4H_2O \longrightarrow OAA + 2CO_2 + 3NADH + 3H^+ + 1FADH_2 + 1\ ATP + H_2O$

Overall reaction of Krebs cycle (two cycles considering two molecules of acetyl Co.A):

$2\ CH_3COCH_2COOH + 2\ CH_3CO.CoA + 6NAD^+ + 2FAD + 2ADP + 2Pi + 8H_2O \longrightarrow 2OAA + 4CO_2 + 6NADH + 6H^+ + 2FADH_2 + 2\ ATP + 2H_2O$

Steps	Substrate	Enzyme	Products	Types of Reaction
1	Oxaloacetic acid + Acetyl CoA + H_2O	Citrate synthetase	Citric Acid + CoA	Aldol condensation (irreversible)
2	Citric acid	Aconitase	Cis-aconitic acid + H_2O	Dehydration (reversible)
3	Cis-aconitic acid +H_2O	Aconitase	Isocitric acid	Hydration (reversible)
4	Isocitric acid + NAD^+	Isocitrate dehydrogenase	Oxalosuccinic acid +**$NADH^+$ + H^+**	Oxidation
5	Oxalosuccinic acid	Oxalosuccinic decarboxylase	α-ketoglutaric acid + CO_2	Decarboxylation (irreversible)
6	α-ketoglutaric acid + NAD^+ + CoA	α-ketoglutarate dehydrogenase	Succinyl CoA + **$NADH^+$** + **H^+** +CO_2	Oxidative decarboxylation (irreversible)
7	Succinyl Co A + GDP + Pi + H_2O	Thiokinase	Succinic acid + CoA + **GTP** (animal) /**ATP** (plant)	Substrate level phosphorylation
8	Succinic acid + FAD	Succinic dehydrogenase	Fumaric acid + **$FADH_2$**	Oxidation
9	Fumaric acid + H_2O	Fumarase	Malic acid	Hydration
10	Malic acid + NAD^+	Malate dehydrogenase	Oxaloacetic acid + **$NADH^+$** + **H^+**	Oxidation (reversible)

Balance sheet of ATP in Krebs cycle

Substrate level phosphorylation (Direct gain) at 7th step: → 2ATP

Oxidative phosphorylation (Indirect gain)

Through 2 molecules of $NADH_2$ at 4th step: → 6 ATP

Through 2 molecules of $NADH_2$ at 6th step: → 6 ATP

Through 2 molecules of $NADH_2$ at 10th step: → 6 ATP

Thorough 2 molecules of $FADH_2$ at 8th step: → 4ATP

Total production of energy: **24** ATP

Features of Krebs cycle:

(a) Krebs cycle undergoes twice since 2 molecules of acetyl CoA are formed from two molecules of pyruvic acid. Acetyl CoA is consumed in Krebs cycle and is never regenerated.

(b) It produces reducing power which undergoes electron transport chain.

(c) There are three oxidation steps.

(d) In each cycle, 2 molecules of CO_2 (2x2 = 4 CO_2) are released outside mitochondria and lost as waste product. The main purpose of acetyl coenzyme A is to produce two molecules of CO_2.

(e) There are 4 major regulatory enzymes such as citrate synthetase, isocitrate dehydrogenase, α-ketoglutarate dehydrogenase and succinate dehydrogenase.

(f) The first 4 carbon compound involved was oxaloacetic acid.

(g) The first 5 carbon compound formed was α- ketoglutaric acid (first dicarboxylic acid).

(h) α-ketoglutaric acid undergoes both decarboxylation and dehydrogenation reactions.
(i) The first 4 carbon compound formed was succinyl coenzyme A.
(j) Krebs cycle doesnot produce too much direct energy.

Importance:

(a) It produces large amount of energy in controlled manner.
(b) Pyruvic acid produces an amino acid, alanine.
(c) Acetyl Co A is raw material for steroid, alkaloid, carotenoid, terpenoid, gibberellins.
(d) Citric acid is used as precursor for the synthesis of fatty acid.
(e) α-ketoglutaric acid produces an amino acid, glutamic acid and thus helps in transamination reaction.
(f) Succinyl Co-A is involved in the synthesis of chlorophyll, phytochrome, cytochrome, haeme etc.
(h) Malic acid is used as precursor for synthesis of glucose.
(h) Oxaloacetic acid produces an amino acid, aspartic acid which forms alkaloids and pyramidines.

Combined equation of glycolysis, pyruvate oxidation and Krebs cycle

$$\text{Glucose} + 10\ NAD^{+} + 2FAD + 4ADP + 2\ Pi + 6\ O_2 \longrightarrow$$
$$6CO_2 + 10\ NADH^{+} + 10\ H^{+} + 2\ FADH_2 + 4\ ATP + 6\ H_2O$$

Inhibitors:

Fluoroacetate	It acts as aconitase hence synthesis of isocitric acid is inhibited.
Arsenic dehydrogenase	It acts as α-ketoglutaate dehydrogenase hence synthesis of succinyl Co-A is inhibited.
Malonate	It acts as succinic dehydrogenase hence synthesis of fumaric acid is inhibited

Entry of organic materials in different stages of aerobic respiration

1. **Starch**
   ↓amylase
   Glucose ⟶ **Glycolysis**
2. **Glycogen**
   ↓phosphorylase
   Glucose-1-phosphate
   ↓phosphoglucomutase
   Glucose-6-phosphate ⟶ **Glycolysis**
3. **Lactose**
   ↓lactase
   Glucose ⟶ **Glycolysis**
   \+
   Galactose

   UDP Galactose

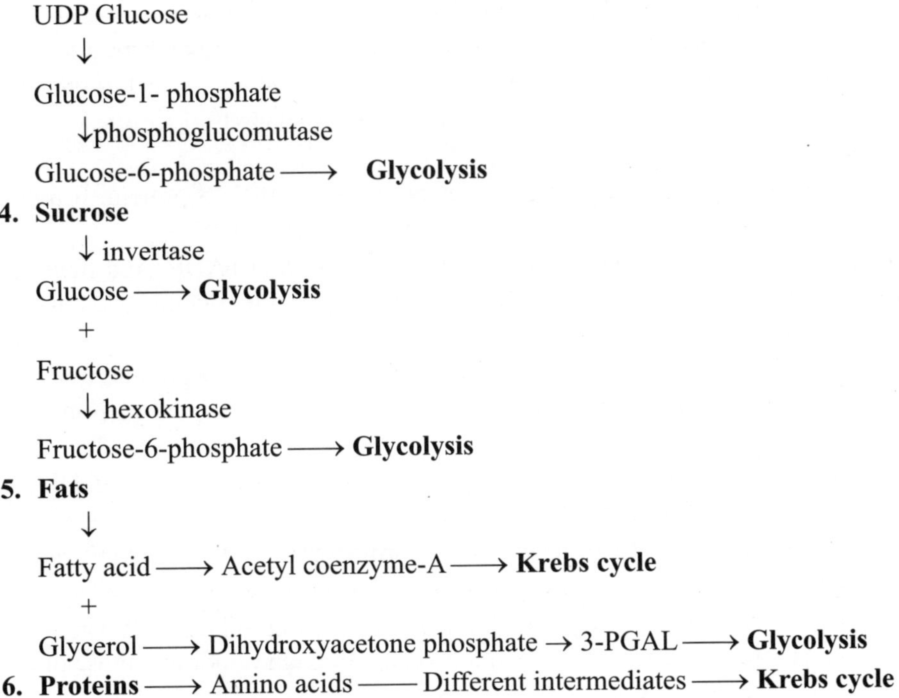

UDP Glucose

↓

Glucose-1- phosphate

↓phosphoglucomutase

Glucose-6-phosphate ⟶ **Glycolysis**

4. **Sucrose**

↓ invertase

Glucose ⟶ **Glycolysis**

\+

Fructose

↓ hexokinase

Fructose-6-phosphate ⟶ **Glycolysis**

5. **Fats**

↓

Fatty acid ⟶ Acetyl coenzyme-A ⟶ **Krebs cycle**

\+

Glycerol ⟶ Dihydroxyacetone phosphate → 3-PGAL ⟶ **Glycolysis**

6. **Proteins** ⟶ Amino acids —— Different intermediates ⟶ **Krebs cycle**

Biosynthesis of amino acids:

(a) Glucose -6-phosphate → Histidine
(b) Phosphoglycerate → Serine → Cysteine, Glycine
(c) Phosphoenolpyruvate → Phenylalanine, Tryptophan, Tyrosine
(d) Pyruvate → Alanine, Valine, Leucine, Isoleucine
(e) Alpha ketoglutaric acid → Glutamic acid → Glutamine, Arginine, Proline
(f) Oxaloacetic acid → Aspartic acid → Asparagine, Methionine, Threonine, Lysine

Electron Transport chain (Respiratory chain)

It is an assembly of electron transporting coenzymes (carriers) to transfer electrons present in inner membrane in redox and sequential manner. It is the last event of aerobic respiration and only part of glucose metabolism that uses atmospheric oxygen and releases maximum energy. It is unidirectional flow of electrons from NADH and $FADH_2$ to molecular oxygen. These electron carriers are arranged according to their increasing positive redox potential. Each electron donor will pass electron to a more electronegative acceptor until e^- are passed to oxygen (the most electronegative and terminal electron acceptor in the chain). In other words, each electron carrier or acceptor has a more positive redox potential (higher electron affinity) than previous one and with each successive transfer, the electrons lose additional free energy which is utilised to generate proton gradient. As the electrons travel through the chain, they move from a higher to lower energy level because each electron carrier holds electron at lower energy level than the previous electron carrier. The transport of e^- releases energy to generate a proton gradient across inner membrane by active pumping of protons into intermembrane space or intracristal space (C side). In other words, the electrons move from low redox potential (to lose electrons) to high redox potential (to accept electrons). Hence it is a process which oxidises NADH and

$FADH_2$ produced in glycolysis, pyruvate oxidation and Krebs cycle and thus traps and releases energy in the form of ATP. The main function of ETC in mitochondria is to provide energy for the synthesis of ATP during oxidative phosphorylation. Most eukaryotic cells have mitochondria which produce ATP from the products of Krebs cycle, fatty acid oxidation and amino acid oxidation.

ETC consists of **5 complexes** of integral membrane protein.

1. Complex-**I** (**NADH dehydrogenase complex**): It contains FMN and Fe-S protein. It transfers e- from NADH to UQ and it pumps $2H^+$ across inner membrane into C-side.
2. Complex-**II** (**Succinate dehydrogenase complex**): It accepts e- from $FADH_2$ and transfers e- from succinate oxidation to UQ (Ubiquinone). It does not pump H^+. It contains FAD and Fe-S protein.
3. Complex-**III** (**Cytochrome b-c_1 complex**): It contains cytochrome b, Fe-S protein, ubiquinone (UQ) and hydroquinone (UQH_2) and cytochrome c_1 (located in C side). It accepts e^- from UQ and transfers to cytochrome-c. It pumps H+ into C side.
4. Complex-**IV**: (**Cytochrome-c oxidase complex**): It accepts e- from cytochrome-c and transfers to reduce molecular O_2 to water. It pumps H^+ into C-side.
5. Complex-**V** (**ATP synthase complex**): It catalyses the formation of ATP from ADP and iP. It contains transmembrane channel through which H^+ diffuse down concentration gradient.

The reduced esenzymes (NADH and $FADH_2$) have hidden potential to produce ATP. The reduced coenzymes of glycolysis enter into inner membrane of mitochondria through C-side and those of Krebs cycle reach inner membrane through m-side.

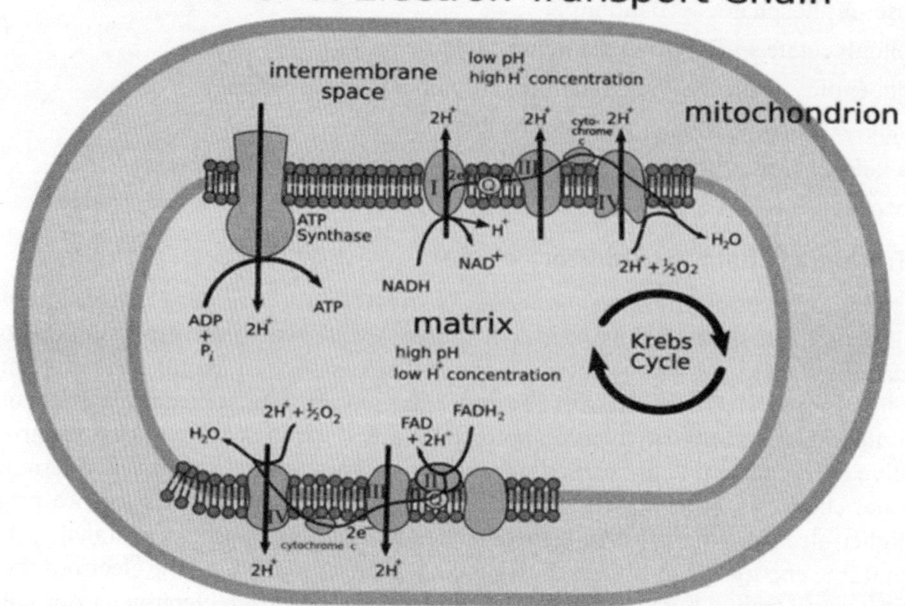

Fig. ***Mechanism of chemiosmosis***

Mechanism: The intermediates such as 3-PGAL, pyruvic acid, isocitric acid, α-ketoglutaric acid, succinic acid and malic acid formed as a result of breakdown of glucose are dehydrogenated (oxidised) hence a pair of H atom is removed from each one of them in the form of reduced coenzyme ($NADH_2$). A pair of H atom dissociates into $2H^+$ and $2e^-$.

($2H \longrightarrow 2H^+ + 2$ e–)

(a) NAD^+ accepts 2 H^+ and 2 e^- and is reduced to NADH and transfers 2 e- and 1 H^+ to FMN. Thus NADH is oxidised back to NAD^+ in Krebs cycle and returns.

(b) FMN accepts 2e- and 1 H^+ from NADH and further accepts 1 H^+ extra from matrix to be reduced to $FMNH_2$. The $FMNH_2$ pumps $2H^+$ to C-side (intracristal side or intermembrane space) and transfers 2e- to Fe-S protein and $FMNH_2$ is oxidised back to FMN.

(c) Like NADH, $FADH_2$ deposits its electrons in the ETC but it does so bypassing the complex I entirely. $FADH_2$ transfers 2 e- to Fe-S protein and then to ubiquinone (UQ).

(d) Fe-S protein transfers 2e- to UQ (ubiquinone), a lipid soluble carrier and mobile link between complex-I and complex-III; and complex II and complex III. It is highly oxidised. It receives 2 H^+ from matrix. These two protons enter into inner membrane because these have greater potential.

(e) UQ transfers 2e- and 2 H^+ to UQH_2 (hydroquinone) which pumps $2H^+$ to C-side from complex III.

(f) UQH_2 (hydroquinone) transfers 2e- to cytochrome b.

(g) Cytochrome- b transfers 2e- to Fe-S protein.

(f) Fe-S protein transfers 2 e- to Cytochrome-$c_{1.}$

(i) Cytochrome-c_1 transfers 2 e- to cytochrome-c (mobile link between complex-III and complex IV of ETC).

(j) Cytochrome-c accepts 2 H^+ from matrix and transfers 2 e- and 2 H^+ to cytochrome-a.

(k) Cytochrome-a pumps 2 H^+ to C-side and transfers 2 e- to cytochrome-a_3 and it is reduced.

(l) Finally reduced cytochrome-a_3 (cytochrome oxidase), a marker enzyme of mitochondria) transfers 2e- to oxygen (final acceptor molecule) to activate it. In this activated state, oxygen will extract the free $2H^+$ from M-side to form water. This is called terminal oxidation.

$1/2O_2 + 2\ e^- + 2\ H^+ \leftrightarrow H_2O$

Thus oxygen is terminal acceptor of e- and most electronegative in ETC. In Krebs cycle, $FADH_2$ is also produced and transfers e- and H^+ to ubiquinone (UQ) directly at later stage in chain bypassing the first stage. All cytochromes have Fe only but cytochrome a_3 has both Fe and Cu. The cytochrome a_3 accepts 2e- through its Fe part and transfers 2e- to oxygen through its Cu component.

Oxidative phosphorylation: It is synthesis of ATP by direct oxidation of substrates or reduced coenzymes such as NADH and $FADH_2$ by e- transport in respiration. It occurs in plasma membrane of prokaryotes but in inner mitochondrial membrane of eukaryotes.

Peter Mitchell proposed chemiosmotic hypothesis where electron transport and ATP synthesis are coupled by a proton gradient. Chemiosmosis is the chemical formation of ATP driven by diffusion force similar to osmosis.

The enzyme complexes such as complex-I, complex-III, and complex IV present in inner membrane act as proton pump which transfers protons (H^+) from M-side to C-side.

The vast majority of energy contained in food material is ultimately used to generate H^+ gradient in inner membrane of mitochondria. This gradient is used to drive the synthesis of ATP.

As the electrons move energetically downhill, the complexes capture the released energy and use it to pump H^+ ions from M side to C side. The H^+ concentration is increased in C side and becomes acidic or positively charged and thus matrix becomes slightly negatively charged. This internal negativity attracts the positively charged H^+ and induces them to re-enter matrix to equalise the pH. Now the inner membrane becomes impermeable to protons. This pumping across inner membrane forms proton motive force which is the electrochemical potential difference between protons in the aqueous phases on different sides of membrane. Proton motive force causes H^+ move back to the matrix passing through ATP synthase which uses the energy of gradient to catalyse the synthesis of ATP from ADP.

The elementary particle or oxysome of mitochondria has three parts: Base (F_0), Stalk and Head (F_1). The base (F_0) is an integral membrane protein complex embedded in inner membrane. It forms proton channel through which H^+ cross the inner membrane. The head (F_1) is peripheral membrane protein complex containing ATP synthase protruding into the matrix. It contains the site for synthesis of ATP from ADP and Pi. The isolated F_1 cannot make ATP from ADP and Pi but it can hydrolyse ATP into ADP and Pi.

For synthesis of each ATP, two protons (H^+) are passed through proton channel formed by F_0 from C-side to M-side along high concentration to low concentration. Since 3 ATPs are produced from complete oxidation of 1 NADH hence 6 protons (3 pairs) are passed from C-side to M-side but in case of oxidation of each $FADH_2$, 2 pairs of protons are passed from C-side to M-side hence 2ATPs are produced because it enters electron transport chain at later stage and transfers e- and H^+ to UQ (ubiquinone) bypassing the complex I.

The energy of proton gradient (as H^+ flows down the gradient from C-side to M-side via inner membrane) is used in attaching an inorganic phosphate (Pi) radical to ADP by high energy bond producing ATP.

$$ADP + Pi \longrightarrow ATP + H_2O$$

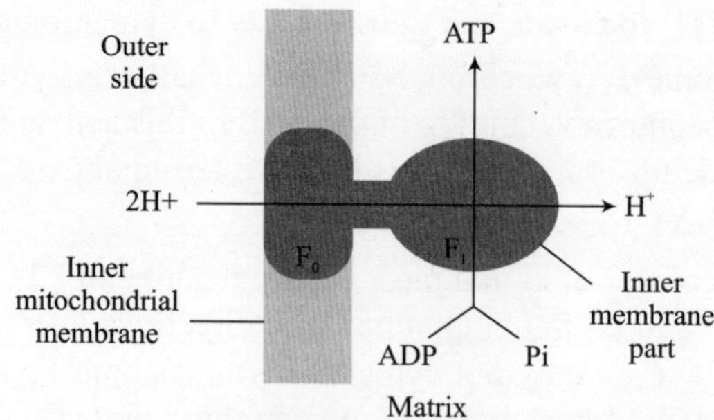

ATP (Adenosine Triphosphate): It was discovered by Lehmann (1920) but its function was discovered by Lipmann. It is immediate source of energy within cell. The hydrolysis of ATP releases energy. Since ATP is found in all living cells hence it is called universal energy carrier or energy currency of cells. It is mobile and transports chemical energy to energy requiring process within the cell. ATP like GTP, CTP and UTP is essential for various activities such as synthesis of carbohydrates, proteins, fats; active transport, translocation of food, growth and streaming of protoplasm etc. ATP is derivative of Adenosine monophosphate (AMP) to which two additional phosphate groups are attached through an anhydride linkage. When first high energy phosphate bond is hydrolysed, 8.9 Kcal energy is released; second phosphate bond is broken, 6.5 Kcal energy is released and if third phosphate group (of AMP) is hydrolysed, then only 3.4 Kcal energy is released. The energy released by the breaking down of phosphate groups in ATP and ADP is much more than the energy released on the hydrolysis of most of the other covalent bonds.

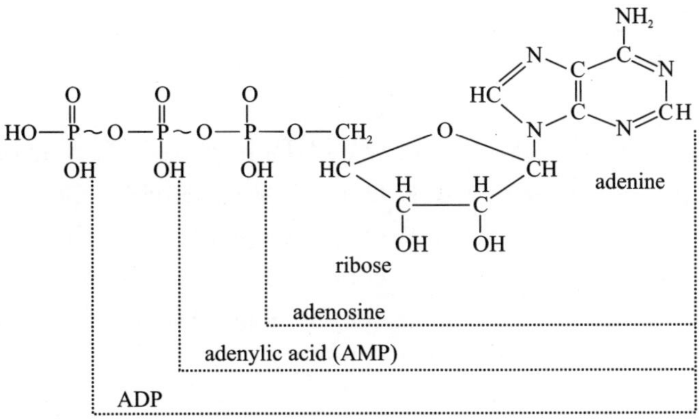

Fig. *ATP molecule*

Its catabolism to ADP yields 8.15 kcal/mol of energy.

$$ATP + H_2O \longrightarrow ADP + Pi + Energy$$

ADP may be rephosphorylated to ATP by respiratory activity or another high energy compound.

$$ADP + Pi \longrightarrow ATP + H_2O$$

The actual number of ATP produced may be lower in certain eukaryotic cells (skeletal muscles) because energy (2ATP) is consumed in transporting the NADH across the mitochondrial membrane hence there is formation of 36 ATP instead of 38 ATP.

The **final equation of aerobic respiration** is as follows:

$$C_6H_{12}O_6 + 6O_2 + 38\ ADP + 38\ Pi \longrightarrow 6CO_2 + 6\ H_2O + 38\ ATP$$

List of reactants consumed and products formed:

Total number of CO_2 produced: **6**

(2 CO_2 in step of Pyruvate oxidation)
(2 CO_2 in 5th step of Krebs cycle)
(2 CO_2 in 6th step of Krebs cycle)

Total number of O_2 consumed: **6**

(1 by ETS in 6th step of Glycolysis)
(1 by ETS in step of Pyruvate oxidation)
(1 by ETS in 4th step of Krebs cycle)
(1 by ETS in 6th step of Krebs cycle)
(1 by ETS in 8th step of Krebs cycle)
(1 by ETS in 10th step of Krebs cycle)

Number of glucose molecule formed: **1**

Total number of H_2O consumed: **10**

(2 H_2O in 6th step of Glycolysis)
(2 H_2O in 1th step of Krebs cycle)
(2 H_2O in 3th step of Krebs cycle)
(2 H_2O in 7^{h} step of Krebs cycle)
(2 H_2O in 9th step of Krebs cycle)

Total number of H_2O produced: **16**

(2 H_2O in 9th step of Glycolysis)
(2 H_2O by ETS in 6th step of Glycolysis)
(2 H_2O by ETS in step of Pyruvate oxidation)
(2 H_2O in 2th step of Krebs cycle)
(2 H_2O by ETS in 4th step of Krebs cycle)
(2 H_2O by ETS in 6th step of Krebs cycle)
(2 H_2O by ETS in 8th step of Krebs cycle
(2 H_2O by ETS in 10th step of Krebs cycle

Net H_2O produced: 16 – 10 = 6 H_2O

Process	Final hydrogen (electron) acceptor	Types of phosphorylation
Fermentation	An organic molecule	Substrate level phosphorylation
Anaerobic respiration	Usually an inorganic substance (nitrate , sulphate or carbonate)	Substrate level and oxidative phosphorylation
Aerobic respiration	Molecular oxygen	Substrate level and oxidative phosphorylation

Inhibitors:

Compound	Use	Effect	Site of action	Action
Cyanide, CO, Azide, H_2S	Poisons	Inhibitors of ETC	Complex IV	These inhibit the ETC by binding more strongly than oxygen to Fe-Cu center in cytochrome-c oxidase thus prevents transfer of electron to oxygen.
Rotenone	Pesticide	Inhibitor of ETC	Complex I	It blocks the transfer of e- from complex I to UQ by blocking the UQ binding site.
Malonates and Oxaloacetate	Poisons	Inhibitors of ETC	Complex II	These are competitive inhibitors of succinic dehydrogenase.
Antimycin A	Piscicide	Inhibitor of ETC	Complex III	It binds to cytochrome-c reductase and thus inhibits oxidation of ubiquinone.
2,4-dinitrophenol	Poison	Uncoupler of oxidative phosphorylation	Inner membrane	It uncouples (dissociates) ETC and phosphorylation reaction. It increases the permeability of mitochondria to H^+. It helps H^+ travel down their gradient and back into M-side without passing through ATP synthase.
Oligomycin	antibiotic	Inhibitor of oxidative phosphorylation	Complex V	It inhibits ATP synthase by blocking the flow of H^+ through F_0 subunit (proton channel).

Factors affecting Respiration

(a) O_2: Rate of respiration is decreased if atmospheric O_2 concentration is below 5% for whole tissue. Rate of respiration is increased with increase in O_2.

(b) CO_2: Rate of respiration is decreased if amount of CO_2 in air is more than usual (0.03%) because stomata will be closed hence O_2 will not enter.

(c) Light: Rate of respiration is increased because light promotes photosynthesis hence organic substance will be formed.

(d) Temperature: Rate of respiration is decreased if temperature is increased from 30^0C. Rate of respiration is increased when temperature is between 0^0C and 30°C (Q_{10}=2).

(e) Water: Rate of respiration increases if there is shortage of water or slight desiccation of tissues.

(f) Inorganic salts: Rate of respiration is increased if plant tissue is transferred from water to salt concentration.

(g) Injury: Rate of respiration is increased if tissue will be injured or wounded.

(h) Age: Rate of respiration is decreased in old stage, steady in growing or meristematic stage and increased (climacteric) in mature fruit.

Efficiency of aerobic respiration:

1ATP produces 8.15 kcal or 34kJ /mol and thus total energy obtained from 38 ATP=38x 8.15=309 Kcal. Glucose stores energy= 686 Kcal

$$\text{Efficiency} = \frac{309}{686} \times 100 = 45\% \text{ and rest 55\% energy is lost as heat.}$$

Respiratory quotient (RQ): It is the ratio of volume of CO_2 evolved to the volume of O_2 consumed in respiration. It is measured by Ganong's Respirometer. It depends on type of substrate.

$$RQ = \frac{\text{Volume of } CO_2 \text{ evolved}}{\text{Volume of } O_2 \text{ consumed}}$$

Case:

1. If substrate is **carbohydrate** (as in germinating cereals; sprouting potatoes, and green leaves kept in dark) then RQ will be one or unity.
   $C_6H_{12}O_6 + 6O_2 \longrightarrow 6\,CO_2 + 6H_2O + \text{energy}$
   RQ = 6/6 = 1 (Unity)
2. If substrate is **protein** as in germinating pulses or **fat** as in germinating oilseeds such as castor, then RQ will be less than one (0.7 in fats and 0.9 in proteins)
   $2\,(C_{51}H_{98}O_6) + 145\,O_2 \longrightarrow 102\,CO_2 + 98\,H_2O + \text{energy}$
   (Tripalmitin)
   RQ = 102/145 = 0.7
   Fats are poorer in oxygen hence it requires more O_2 for complete oxidation.
   RQ is also less than 1 in mixed diet, coloured petals and starved leaves.
3. If substrate is **organic acid (Malic acid or Oxalic Acid)** then RQ will be more than one
   $CH_2\text{-}COOH + 3O_2 \longrightarrow 4\,CO_2 + 3\,H_2O$
   CHOH-COOH
   (Malic acid)
   RQ = 4/3 = 1.3 (more than 1)
   $2\,COOH + O_2 \longrightarrow 4\,CO_2 + 2\,H_2O$
   COOH

(Oxalic acid)

RQ = 4/1= 4 (more than 1)

4. If **anaerobic** respiration occurs as in microbes (bacteria and yeast), deep seated tissue of plants and animals; and fruits then RQ will be infinite

$$C_6H_{12}O_6 \xrightarrow{\text{Zymase}} 2\ C_2H_5OH + 2CO_2$$

RQ = 2/0 = ∞

5. If material is **succulent or CAM** plants (e.g., *Opuntia* and *Bryophyllum)*, RQ will be zero. Glucose is not completely oxidised in CO_2 and H_2O but it is incompletely oxidised or fixed into malic acid. Hence CO_2 is not released during night when stomata is open.

$$2\ C_6H_{12}O_6 + 3O_2 \longrightarrow 3\ \begin{matrix} CH_2\text{-}COOH \\ CHOH.COOH \end{matrix} + 3\ H_2O$$

(Malic acid)

RQ = 0/3 = Zero

Importance:

(a) It determines the type of respiratory substrate.

(b) It determines the type of respiration.

β-oxidation: It is a catabolic process by which very long chain of fatty acids in the for of acyl coenzyme-A are broken down to generate acetyl coenzyme-A which enters krebs cycle.

It occurs in cytoplasm in prokaryotes but takes place in mitochondria in eukaryotes.

C_n acyl Co-A + FAD + NAD^+ + H_2O + Co-B ⟶ C_{n-2} acyl Co-A + $FADH_2$ + NADH + H^+ + Acetyl Co-A

Note:

One turn of **β-oxidation** of fatty acid produces 15 ATPS.

Consumption of ATP = 2ATP and Production of ATP = 17

Net gain = 15 ATPs

β-oxidation of Palmitic acid (16C fatty acid) produces 129 ATP

β-oxidation of stearic acid (18C fatty acid) produces 146 ATP.

SHORT ANSWER TYPE QUESTIONS

1. Distinguish between floating respiration and protoplasmic respiration.
2. Distinguish between anaerobic respiration and aerobic respiration.
3. What are the features og glycolysis?
4. Write notes on Pasteur effect.
5. Distinguish between fermentation and anaerobic respiration.
6. What do you mean by pyruvate oxidation?
7. How different carbohydrates enter into the process of respiration?
8. What are the features of Krebs cycle?
9. Discuss the importance of Krebs cycle.
10. Write the final equations of glycolysis and Krebs cycle.

11. Write notes on ATPase complex.
12. Write notes on ATP.
13. What happens if cyanide will act as inhibitor?
14. What is the efficiency of aerobic respiration?
15. What will happen to RQ in case of anaerobic respiration?

LONG ANSWER TYPE QUESTIONS

1. Describe the different reactions of glycolysis and show the generation of ATP.
2. What do you mean by fermentation? Describe the alcoholic and lactic acid fermentation.
3. Describe the different reactions of Krebs cycle and show the generation of ATP
4. Describe the transport of electrons in inner membrane of mitochondria.
5. Discuss the mechanism of oxidative phosphorylation in mitochondria.

MULTIPLE CHOICE QUESTIONS

1. Formation of phosphoenol pyruvate from 2-phosphoglycerate takes place by
(a) Dehydration (b) Dehydrogenation (c) Oxidation (d) Hydration
2. The number of glucose molecules required to produce 38 ATP molecules under anaerobic condition by a yeast cell is
(a) 2 (b) 4 (c) 19 (d) 38
3. The number of water molecules produced and used in aerobic respiration, respectively, is
(a) 18 and 12 (b) 16 and 10 (c) 12 and 6 (d) 8 and 2
4. The ratio of CO_2 liberated between anaerobic and aerobic respiration is
(a) 3 : 1 (b) 2 : 4 (c) 2 : 2 (d) 1 : 3
5. If glycerol would enter in the Krebs cycle then it will firstly convert to
(a) Acetyl CoA (b) Phosphoglyceraldehyde
(c) Dihydroxyacetone phosphate (d) Pyruvic acid
6. The ratio of NADH formed in glycolysis and that of in Krebs cycle is
(a) 1 : 2 (b) 1 : 3 (c) 1 : 4 (d) 2 : 3
7. The number of CO_2 released during one Krebs cycle is
(a) 2 (b) 3 (c) 4 (d) 6
8. The number of ATP molecules produced as a result of oxidative phosphorylation of glucose in respiration, is
(a) 30 (b) 34 (c) 36 (d) 38
9. The substrate level phosphorylation reaction does not occur?
(a) 1, 3-diphosphoglyceric acid → 3-Phosphoglyceric acid
(b) Pyruvic acid → Acetyl-CoA
(c) Succinyl CoA →Succinic acid
(d) Phosphoenol pyruvic acid → Pyruvic acid
10. The number of ATP molecules produced from the complete oxidation of one molecule of acetyl CoA, is
(a) 12 ATP (b) 15 ATP (c) 36 ATP (d) 38 ATP
11. The coenzyme capable of reducing succinic acid is
(a) FAD (b) FMN (c) NAD (d) NADP
12. The number of carbon atoms present in cis-aconitic acid, malic acid and lactic acid are, respectively
(a) 6, 3 and 3 (b) 6, 4 and 3 (c) 5, 4 and 3 (d) 6, 4 and 2
13. The number of ATP molecules synthesized by substrate level phosphorylation during aerobic respiration of one glucose molecule is:
(a) 8 (b) 6 (c) 4 (d) 2
14. Single turn of citric acid cycle yields
(a) 2 $FADH_2$, 2 NADH and 2 GTP (b) 1 $FADH_2$, 3 NADH and 1 GTP
(c) 2 $FADH_2$, 3 NADH and 1 GTP (d) 2 $FADH_2$, 5 NADH and 1 GTP
15. In Krebs cycle, FAD participates as an electron acceptor during the conversion of
(a) Succinyl CoA to Succinic acid (b) α-ketoglutaric acid to Succinyl CoA
(c) Succinic acid to Fumaric acid (d) Fumaric acid to Malic acid

16. The number of reduced NAD molecules produced in the complete breakdown of one molecule of glucose during aerobic respiration, is
(a) 5 (b) 10 (c) 12 (d) 38
17. In ETS of aerobic respiration, oxygen receives electrons directly from
(a) Ubiquinone (b) Cytochrome-a (c) Cytochrome-a_3 (d) $NADH^+$
18. The sequence of cytochromes in ETS is
(a) Cyt a, b, c, c_1, a (b) Cyt b, c_1, c, a, a_3 (c) Cyt b, a, a_3, c, c_1 (d) Cyt b, c, a_3, a, c_1
19. The respiratory quotient in organic acids is
(a) One (b) More than one (c) Less than one (d) zero
20. The number of ATP molecules formed from one molecule of pyruvic acid in mitochondrial respiration are
(a) 2 (b) 4 (c) 38 (d) 15
21. Which one of the following can respire in anaerobic conditions?
(a) Leaves (b) Stem (c) Flowers (d) Seeds
22. The final electron acceptor in lactic acid fermentation is
(a) Pyruvate (b) NAD (c) Lactic acid (d) Oxygen
23. The terminal electron acceptor in ETS during aerobic respiration
(a) CO_2 (b) O_2 (c) H_2O (c) NADH
24. The metal activator used in lactic acid fermentation is
(a) Mn^{+2} (b) Zn^{+2} (c) Mg^{+2} (d) Cu^{+2}
25. The number of pairs of electrons removed from each glucose molecule is
(a) 6 (b) 10 (c) 12 (d) 18
26. The enzyme, pyruvate dehydrogenase does not contain
(a) NAD (b) Thiamine pyrophosphate
(c) Lipoic acid (d) FAD
27. The number of substrate level phosphorylation reactions in glycolysis, is
(a) One (b) Two (c) Three (d) Four
28. The enzyme released by yeast cells during fermentation is
(a) Oxidase (b) Zymase (c) Invertase (d) Lipase
29. The highest rate of respiration occurs in
(a) Root tip (b) Growing shoot tip (c) Germinating seed (d) Leaf bud
30. In glycolysis, electrons are removed by
(a) ATP (b) NAD
(c) FAD (d) Molecular oxygen
31. In cytochromes, the electrons are picked up and released by
(a) Copper (b) Iron (c) Zinc (d) Molybdenum
32. In mitochondria, the proton gradient required for ATP synthesis develops across the
(a) Outer membrane (b) Inner membrane (c) C-side (d) M-side
33. Chemiosmotic theory of ATP synthesis in mitochondria is based on
(a) Electron gradient (b) Accumulation of Na^+ ions
(c) Accumulation of K^+ ions (d) Proton gradient

34. The number of ATP molecules formed from complete oxidation of fructose-1,6 diphosphate is
(a) 30 (b) 34 (c) 38 (d) 40
35. Excess of ATP inhibits
(a) Pyruvate dehydrogenase (b) Phosphofructokinase
(c) Isomerase (d) Enolase
36. The formation of acetyl Co-A from pyruvic acid occurs as a result of
(a) Substrate level phosphorylation (b) Oxidative decarboxylation
(c) Dehydrogenation (d) Isomerisation
37. Removal of hydrogen and CO_2 from a substrate is called
(a) Dehydrogenation (b) Oxidative decarboxylation
(c) Dehydration (d) Oxidation
38. In Krebs cycle, malic acid hands over hydrogen to
(a) FMN (b) NAD^+ (c) FAD (d) TPP
39. ATP equivalents produced during conversion of succinic acid to fumaric acid, are
(a) 2 (b) 3 (c) 4 (d) 6
40. The universal hydrogen acceptor is
(a) ATP (c) CoA (c) TPP (d) NAD
41. GTP is formed during conversion of
(a) Pyruvic caid to Acetyl CoA (a) Succinyl CoA to Succinic acid
(b) Succinic acid to Fumaric acid (c) Fumaric acid to Malic acd
42. Krebs cycle starts with the formation of six carbon compound by reaction between
(a) Malic acid and Oxaloacetic acid (b) Oxaloacetic acid and Acetyl CoA
(c) Malic acid and Acetyl CoA (d) Puruvic acid and Oxaloacetic acid
43. The electron acceptors in ETS are arranged according to
(a) Decreasing positive potential (b) Increasing positive potential
(c) Increasing negative potential (d) Decreasing pressure potential
44. In electron transport chain , a carrier holds electron at
(a) Lower energy than the previous carrier (b) Higher energy than the previous carrier
(c) Same energy level as the adjacent one (d) None of the above
45. ATP is injected in cyanide poisoning because
(a) It is necessary for Na^+-K^+ pump (b) It breaks down cyanide
(c) It is necessary for cellular function (d) It stops electron movement
46. When a molecule of pyruvic acid is subjected to anaerobic oxidation and produced lactic acid, there is
(a) Gain of 2 ATP (b) Gain of 3 ATP (c) Loss of 2 ATP (d) Loss of 3 ATP
47. Fruits and vegetables are kept in refrigerator
(a) To reduce the rate of respiration (b) To increase the rate of respiration
(c) To reduce the rate of transpiration (d) To promote fermentation
48. CO_2 is not liberated in
(a) Alcoholic fermentation (b) Lactic acid fermentation
(c) Aerobic respiration in plants (d) Aerobic respiration in animals

49. A competitive inhibitor of succinic dehydrogenase is
    (a) Malonate (b) Oxaloacetate (c) Acetate (d) Sulpha drugs
50. Which one of the following cells does not respire?
    (a) RBC (b) Sieve tube cell (c) Cork cell (d) Epidermal cell
51. $FADH_2$ produces lesser amount of ATP and enters into ETS at
    (a) FMN (b) Ubiquinone (c) Cytochrome-b (d) Fe-S
52. Cytochrome-a_3 contains
    (a) Copper and Manganese (b) Iron and Copper
    (c) Iron and Manganese (d) Magnesium and Zinc
53. During alcoholic fermentation of glucose, the number of ATP produced is
    (a) Zero (b) 2 (c) 4 (d) 6
54. The first four carbon compound taking part in Krebs cycle is
    (a) Oxaloacetic acid (b) Fumaric acid (c) Succinic acid (d) Malic acid
55. Glutamic acid, an amino acid is produced from
    (a) Oxaloacetic acid (b) α-ketoglutaric acid (c) Succinic acid (d) Citric acid
56. The chemical nature of cytochrome is
    (a) Metalloflavoprotein (b) Iron containing porphyrin pigment
    (c) Glycoprotein (d) Glycolipid
57. Cyanide resistant respiration occurs in
    (a) Bacteria (b) Viruses (c) Plants (d) Animals
58. An aquatic plant was placed in a test tube containing distilled water and the tube was stoppered. The tube was left outdoors for 24 hours and pH value of water was measured at regular intervals. What is the result?
    (a) pH was lowest just before sunrise (b) pH was highest just before sunrise
    (c) pH was lowest at noon. (d) pH was lowest just before sunset.
59. Cut fruits become brown due to
    (a) Oxidation of tannins and phenolics by enzymes
    (b) Oxidation of sugar by enzymes
    (c) Dehydration of organic compound
    (d) Fermentation
60. The net gain of energy from one molecule of sucrose in aerobic respiration is
    (a) 8 ATP (b) 38 ATP (c) 40 ATP (d) 76 ATP
61. The proton channel of oxysome is
    (a) F_0 particle (b) F_1 particle (c) F_5 particle (d) F_6 particle
62. The ETS in prokaryotes is located at
    (a) Nucleoid (b) Ribosome
    (c) Plasmid (d) Plasma membrane
63. Cyanide is inhibitor to electron transport chain when it prevents transfer of electron to
    (a) FMN (b) Ubiquinone (c) Cytochrome-a (d) Oxygen
64. The number of protons required to produce one molecule of ATP is
    (a) Two (b) Three (c) Four (d) Six

65. Oxaloacetic acid, an intermediate of Krebs cycle, gives rise to
(a) Glutamic acid (b) Glutamine (c) Lysine (d) Aspartic acid
66. The first dicarboxylic acid in Krebs cycle is
(a) Citric acid (b) Alpha ketoglutaric acid
(c) Succinic acid (d) Oxaloacetic acid
67. Lactic acid fermentation requires a coenzyme known as
(a) FMN (b) FAD (c) NAD (d) TPP
68. The electron transport chain occurs in
(a) Cytoplasm (b) Inner membrane of mitochondria
(a) Inter-cristal space of mitochondria (d) Matrix of mitochondria
69. The efficiency of aerobic respiration is
(a) 25% (b) 45% (c) 50% (d) 55%
70. The respiratory quotient in succulent plants is zero because
(a) Glucose is not completely oxidised into CO_2
(b) Glucose is incompletely oxidised into malic acid
(c) CO_2 is not released
(d) All of the above
71. The stoichiometry of anaerobic and aerobic degradation of glucose by yeast are as follows:
$C_6H_{12}O_6 \longrightarrow 2C_2H_5OH + 2CO_2$
$C_6H_{12}O_6 + 6O_2 \longrightarrow 6CO_2 + 6H_2O$
In an experiment,the complete utilisation of 0.5 mol. of glucose, partly under anaerobic and partly under aerobic conditions, yielded 1.8 mole of CO_2. What will be Respiratory quotient for overall combined process?
(a) 0.5 (b) 1 (c) 1.5 (d) 2
72. If temperature is increased above 35°C, then what happens?
(a) Rate of decline of respiration will be earlier than decline of photosynthesis
(b) Rate of decline of photosynthesis will be earlier than decline of respiration
(c) Both decline simultaneously
(d) Both runs usually
73. If a leaf is kept in sugar solution then the rate of respiration will
(a) Increase
(b) Decrease
(c) First decrease then increase
(d) Remains unchanged
74. In glycolysis, during oxidation electrons are removed by
(a) ATP (b) 3-PGAL
(c) NAD+ (d) Molecular oxygen
75. In the absence of oxygen, cells capable of fermentation
(a) Accumulate glucose (b) Oxidise NADH+ to NAD+
(c) Accumulate pyruvate (d) Do not produce ATP

76. Fermentation is performed by
(a) All microbes
(b) All fungi and microbes
(c) Some fungi and some microbes
(d) Yeast only
77. Which of the following steps is associated with substrate level phosphorylation?
(a) 1,3 diPGA → 3PGA
(b) PEP → Pyruvic acid
(c) Succinyl CoA → Succinic acid
(d) All of the above
78. What percentage of ATP is formed by substrate level phosphorylation in glycolysis?
(a) 25%
(b) 100%
(c) 10%
(d) 50%
79. The inner membrane of mitochondria is permeable to
(a) ATP
(b) Glucose
(c) Fructose
(d) Sucrose
80. How many ATPs are produced by the complete oxidation of two molecules of isocitrate by ETS only?
(a) 22
(b) 24
(c) 15
(d) 30
81. Which of the following is first compound which is common for both glucose and fructose in glycolysis?
(a) Glucose-6-phosphate
(b) Fructose-6-phosphate
(c) Fructose-1,6-diphosphate
(d) Fructose-1,6-phosphate
82. The enzyme which occurs in all eukaryotes and shows great homology is
(a) Cytochrome b
(b) Cytochrome c
(c) Cytochrome a
(d) Cytochrome a_3
83. Complex III receives electrons directly from
(a) Ubiquinone
(b) Cytochrome c
(c) $NADH^+$
(d) $FADH_2$
84. ATP synthase complexes are found in
(a) Outer membrane of mitochondria
(b) Inner membrane of mitochondria
(c) Attached to the cristae
(d) Floating in the matrix
85. Which of the following is most conserved in all eukaryotes and acts as respiratory pigment and accepts electron?
(a) Ferridoxin
(b) Cytochrome c
(c) Cytochrome c oxidase
(d) NAD dehydrogenase
86. In aerobic respiration, the release of redox equivalents as H atom for the first time in mitochondrial matrix is catalysed by
(a) PGAL dehydrogenase
(b) Isocitrate dehydrogenase
(c) Pyruvate dehydrogenase
(d) Alpha ketoglutarate dehydrogenase
87. Energy metabolism may have evolved when early prokaryotes
(a) Used up the oxygen in their environment
(b) Began producing poisonous waste products
(c) Used up the ATP in their environment
(d) Began producing significant amount of oxygen
88. If organic substrate is malic acid then RQ will be
(a) Zero
(b) 1
(c) Less than 1
(d) Greater than 1
89. Flavin mononucleotide (FMN) requires
(a) Mg
(b) Zn
(c) Mn
(d) Fe

90. The net gain of ATP in glycolysis is
    (a) 2 (b) 4 (c) 6 (d) 8
91. Succinate dehydrogenase enzyme occurs in
    (a) Outer membrane of mitochondria (b) Inner membrane of mitochondria
    (c) Mitochondrial matrix (d) Perichondrial space
92. The respiratory substrate in protoplasmic respiration is
    (a) Carbohydrate (b) Fats
    (c) Both carbohydrate and fats (d) Protein
93. How many isomerisation reactions occur in glycolysis?
    (a) One (b) Two (c) Three (d) Four
94. Which of the following acids accumulates during muscular fatigue?
    (a) Pyruvic acid (b) Oxaloacetic acid
    (c) Phosphoglyceric acid (d) Lactic acid
95. How many cofactors are required during pyruvate oxidation?
    (a) Two (b) Three (c) Four (d) Five
96. Which of the following cofactors reacts first with pyruvic acid during pyruvate oxidation?
    (a) Coenzyme A (b) Lipoic acid
    (c) NAD (d) Thiamine pyrophosphate
97. The first 4 carbon compound formed in Krebs cycle was
    (a) Citric acid (b) Cis-aconitic acid
    (c) Alpha ketoglutaric acid (d) Succinyl coenzyme A
98. Which of the following electron carriers is mobile link between complex III and complex IV?
    (a) Cytochrome-b (b) Cytochrome-c (c) Ubiquinone (d) Cytochrome-a_3
99. In aerobic respiration, the action of 2,4-dinitrophenol is
    (a) Uncoupler of oxidative phosphorylation (b) Inhibitor of Electron transport chain
    (c) Inhibitor of Oxidative phosphorylation (d) It does not act
100. The respiratory quotient in germinating castor seed is
    (a) One (b) Less than one (c) More than one (d) Zero

ANSWERS

1.a	2.c	3.b	4.d	5.c	6.b	7.a	8.b	9.b	10.a
11.a	12.b	13.b	14.b	15.c	16.b	17.c	18.b	19.b	20.d
21.d	22.c	23.b	24.b	25.c	26.d	27.d	28.b	29.c	30.b
31.a	32.b	33.d	34.d	35.b	36.b	37.b	38.b	39.c	40.d
41.b	42.b	43.b	44.a	45.c	46.d	47.a	48.b	49.a	50.c
51.b	52.b	53.b	54.a	55.b	56.b	57.c	58.a	59.a	60.d
61.a	62.d	63.d	64.a	65.d	66.b	67.a	68.b	69.b	70.d
71.c	72.a	73.a	74.a	75.b	76.c	77.d	78.b	79.a	80.b
81.b	82.d	83.a	84.c	85.b	86.c	87.c	88.d	89.b	90.a
91.b	92.d	93.c	94.d	95.d	96.d	97.d	98.b	99.a	100.b

12

Plant Growth and Development

Development: It is multiphasic sequence of qualitative structural change in structure and functioning of cell, tissue, organ or organism towards a higher or more complex state or it is the conversion of one phase to the next. In plants it starts from seed germination and ends with senescence. It follows as such:

$$\text{Meristematic cell} \xrightarrow{\text{Cell division, Plasmatic growth, Elongation, Differentiation, Maturation}} \text{Mature cell}$$

The development in plants involves the following stages:

(a) Juvenile stage: It starts with seedling phase and ends in vegetative stage.
(b) Reproductive stage: It shows budding phase.
(c) Ripening stage: It shows flowering phase.
(d) Senescent stage: It is terminal phase where death occurs.

Growth

It is permanent and irreversible increase in number, length, area (size), and volume (mass) of an individual cell, organ and its parts. In this, anabolic processes dominate over catabolic process. Growth is a fundamental property of living organisms but it is not defining feature. In plants, limited growth occurs in leaves, flowers and fruits but unlimited growth occurs in root and stem. In vascular plants, growth begins from a single-celled zygote. Now zygote divides to form embryo and then the cells are organised to form embryo in seed. Once the embryo germinates from its seed then it produces vegetative organs (root, stem and leaf). In plants, growth may be primary (increase in length) and secondary (increase in girth or width) due to lateral meristems (vascular cambium and cork cambium).

Parameters of Growth:

(a) Number: Bacteria, yeast and algae increase their number.
(b) Length: Growth of pollen tube
(c) Size: Water melon
(d) Surface area: Leaf
(e) Volume: Storage organs, Fruits.
(f) Weight (fresh or dry): Storage organs, Fruits.

Measurement of Growth:

(a) Direct method: It measures dry weight, volume and number of cells.
(b) Horizontal or Traveller's microscope: It measures growth on length.
(c) Auxanometer: It measures growth in length.
(d) Crescograph: It was developed by J.C. Bose. It measures primary growth very accurately in seconds.

Phases of growth: Growth involves three phases:

1. **Phase of Cell division** (Formative phase): Cells are small in size, isodiametric and metabolically highly active. Mitotic division occurs continuously in apical meristem (root apex and shoot apex) hence the number of daughter cells is increased. Some of the daughter cells retain the meristematic activity, while others enter into next phase-phase of elongation. The resulting cells have dense cytoplasm, small vacuole and a distinct centrally placed nucleus.
2. **Phase of cell elongation**: It is maximum behind root and shoot apex. It may be linear or in all direction (isodiametric cells) hence there is irreversible increase in size of cells due to increase in volume of cytoplasm. Due to high metabolic rate, the solute concentration will be high in their cell sap hence water potential is decreased by which water enters into cells. The elongation occurs by synthesising protoplasm, absorbing water (hydration) and adding new cell wall to make it permanent. The small vacuoles unite to form large centrally placed vacuole. Due to endosmosis, size of vacuole increases and it exerts turgor pressure on cell wall. The cell wall being elastic expands and cell gets elongated.
3. **Phase of Cell differentiation**: Cells become mature and specialised by permanent change in structure (shape and size) and physiological activities such as absorption of water, conduction of water by xylem, translocation of food by phloem, and photosynthesis by mesophyll cell. After differentiation, mature cells do not grow further and remain unchanged until death. The differentiated cells form different kinds of simple and complex tissues which form different functions.

Rates of Growth:

1. **Arithmetic growth**: It occurs at a constant rate from the very beginning and progresses arithmetically (2…3….4…6…). In this one cell divides forming two daughter cells as a result of mitotic divisions and only one daughter cell divides continuously while other cell undergoes differentiation where cell becomes mature, permanent and loses its capacity to divide.

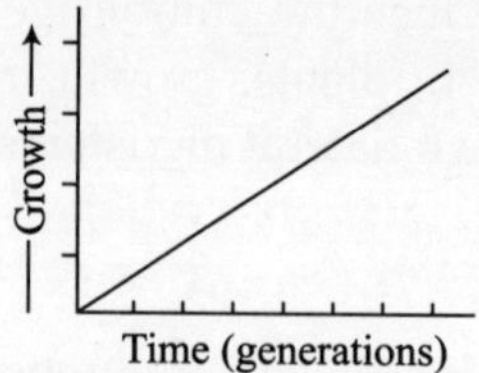

Fig. ***Arithmetic growth curve***

It occurs in root and stem elongating at constant rate. It shows a linear curve. Mathematically, it is expressed as

$Lt = Lo + rt$

Lt = Length (growth) of organ after time 't'

Lo = Length (growth) of organ at the beginning

rt = Rate of growth or elongation per unit time.

2. **Geometric growth**: It occurs when every cell divides and all daughter cells grow and divide again in geometrical manner (2….4…8…16….32….)

It occurs in many higher plants and in unicellular organisms when grown in nutrient rich medium. It involves as such

(a) Lag phase: It is initial phase in which slow or little growth occurs because organism is in the process of mobilising resources.

(b) Log phase: It is phase in which active or rapid growth occurs exponentially and reaches maximum because there is considerable increase in volume of cell sap along with cell wall materials.

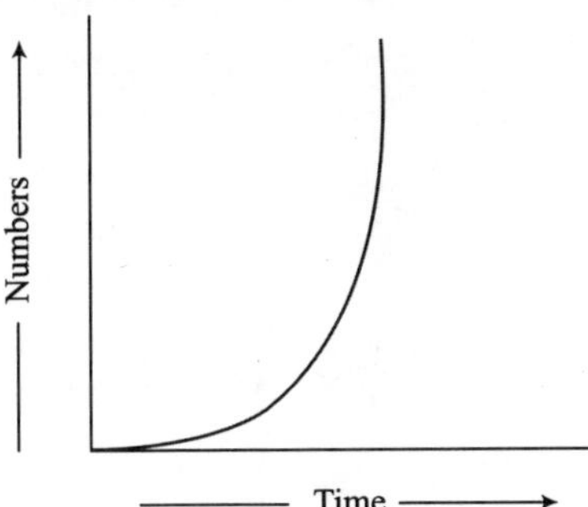

Mathematically, it is expressed as

$W_1 = W_0 = e^{rt}$

W_1 = Final size (weight), Wo = Initial size at beginning of growth

e = Base of natural log; rt = Rate of growth

3. **Sigmoid growth:** It shows when geometric growth cannot be sustained for long because some cells die due to limited nutrient availability by which growth slows down leading to stationary phase. It is characteristic feature of living organisms in their natural environment. It involves:

(a) Lag phase: It corresponds to the phase of cell division. It is initial phase in which slow or little growth occurs because organism is in the process of mobilising resources.

(b) Log phase: It corresponds to the phase of cell elongation. It is phase in which active or rapid growth occurs exponentially and reaches maximum because there is considerable increase in volume of cell sap along with cell wall materials.

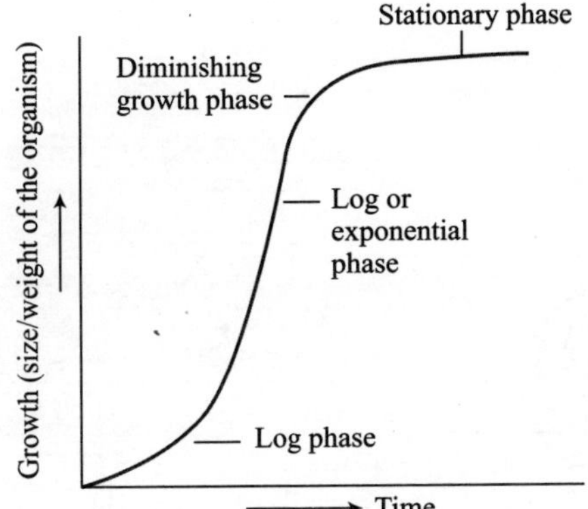

(c) Stationary phase: It corresponds to the phase of differentiation. It starts with inflexion point where growth starts declining. It is a phase in which size of body remains fairly constant because growth has stopped due to shortage of food, space and accumulation of toxins.

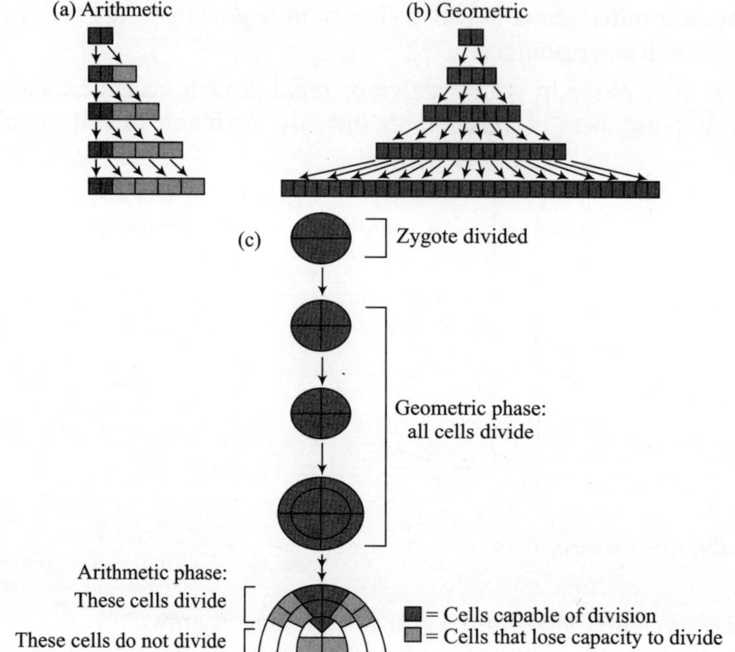

Fig. ***Diagrammatic representation of: (a) Arithmetic (b) Geometric growth and (c) Stages during embryo development showing geometric and arithematic phase***

4. **Absolute growth**: It is actual or total growth per unit time.
5. **Relative growth:** It is growth of given system per unit time. It is expressed as:
   (a) If initial size of leaf A is 5 cm^2 and leaf grows up to 10 cm^2
   (b) If initial size of leaf B is 50cm^2 and leaf grows up to 55 cm^2.
   Thus, in both cases, absolute growth rate is same (5cm^2) but relative growth rate in leaf A is 100% (faster) and in leaf B it is 10% (slower).

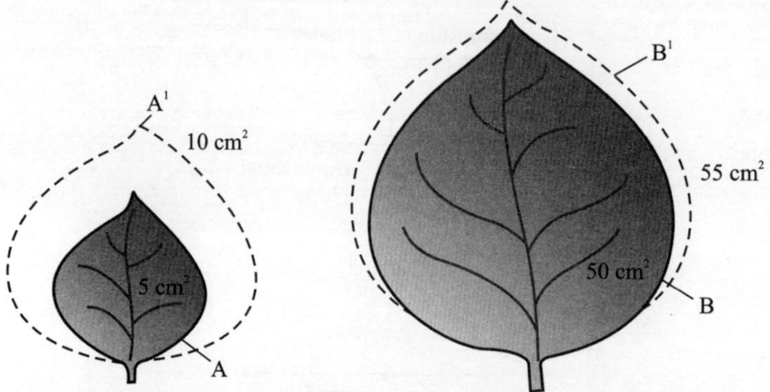

Diagrammatic comparison of absolute and relative growth rates. Both leaves A and B have increased their area by 5 cm^2 in a given time to produce A', B' leaves

Factors affecting growth:

(a) Light: Growth will be maximum in the presence of red light because it is must for photosynthesis. If light is absent then it will cause etiolation.
(b) Temperature: The optimum temperature is 20-30^{0}C. It varies from species to species. Low temperature inhibits growth due to inactivation of enzymes and high temperature also inhibits growth due to denaturation of enzymes and excess transpiration.
(c) Water: Its availability favours growth because it is essential for the activity of enzymes.
(d) Phytohormones: Growth is increased in the presence of auxin and gibberellins,
(e) Carbohydrate / Protein (C/N) ratio: Presence of more carbohydrates compared to nitrogenous compounds favours good growth.
(f) Anabolism/catabolism (A/C) ratio: Growth is more if A/C ratio will be more.
(g) Soil nutrients: Meganutrients are essential raw materials for growth and micronutrients are essential for the activity of different enzymes responsible for growth.
(h) Oxygen: It must be present for the growth of higher plants.

Plasticity: It is the ability of plants to follow different pathways in response to environment or to develop different structure. Plant growth is dependent on environment. e.g., Heterophylly (presence of two types of leaf on the same plant). It is exhibited by the following ways:

(a) The leaves of juvenile plant of cotton, coriander, larkspur (*Delphinium*), mustard, Pea are different in shape than those in mature plants.
(b) The aerial and floating leaves of butter cup (*Ranunculus*) are generally broad and unlobed while submerged leaves are ribbon shaped, linear and much dissected

Phytohormones:

The term 'phytohormone' was coined by Thimann (1948). These are organic compounds and are capable of influencing various physiological activities in plants. These are required in small quantities at very low concentration. When two hormones act together and perform same action then it is said to be synergistic action of hormones.

Auxin

It was the first plant hormone to be discovered.

Discovery: Charles Darwin (1880) reported the existence of auxin while demonstrating of bending of coleoptiles in canary grass (*Phalaris canariensis*) towards light. He observed

(i) If tip was exposed towards light then coleoptile, the tip observed bending towards light
(ii) If coleoptiles tip was covered with opaque tin foil cap, no bending was observed
(iii) If coleoptile was removed then bending was not noticed because stimulus was not perceived.

Finally it was proposed that stimulus was transmitted from tip to subapical or subterminal position where it caused differential growth resulting in bending.

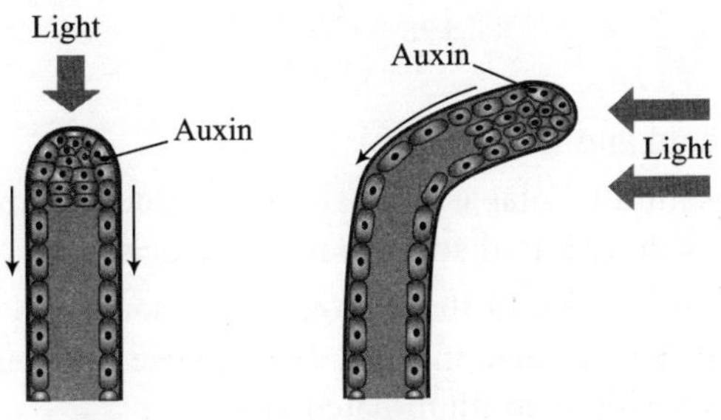

Frits Went (1928) discovered auxin while studying '***Avena* curvature test**'. He placed freshly cut coleoptiles tips of oat on small block of agar for some time. The agar blocks (A and B) were then placed asymmetrically on the coleoptiles stumps for 2 hours in dark. He noticed that coleoptiles containing tips have certain substance which could diffuse down (basipetally) into agar blocks. He termed this substance 'auxin'. He also found that the degree of curvature of coleoptiles is proportional to the amount of active substance present in agar blocks. This could be used to estimate the amount of auxin in any tissue. This test came to be known as *Avena* curvature test.

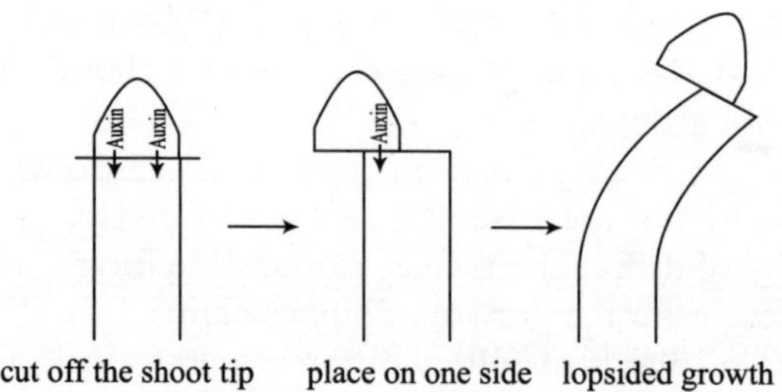

Isolation:

(i) Kogl and Haagen Smit (1931) isolated auxin-a (Auxentriolic acid=$C_{18}H_{32}O_5$) from human urine of pellagra patient.

(ii) They isolated another substance and called it auxin-b (Auxenlinic acid=$C_{18}H_{30}O_4$) from corn germ oil.

(iii) They isolated another auxin and called it Auxin-c or heteroauxin (indole-3-Acetic acid=$C_{10}H_9O_2N$) from human urine.

Biosynthesis: Zn^{+2} is required for synthesis of Indole Acetic Acid (IAA).

Tryptophan (Amino acid)
↓ decarboxylase
Tyramine
↓ aminooxidase
Indole-3-acetaldehyde
↓ dehydrogenase
Indole Acetic Acid (IAA)

O
OH
N
H

Indole-3-Acetic Acid
(an Auxin)

Nature: Auxins are weak acid and water soluble.

Structure: It shows an unsaturated planar aromatic ring structure and a carboxyl side chain. More than 100 types of auxin have been reported so far. These are classified on the basis of function only.

Distribution: Auxins are widely found throughout plant body. These are found in shoot apex in highest amount=10 ppm), root apex, coleoptile tip, bud. These are found throughout day and night. These are found more on darker side than illuminated side.

Transport: Auxins show polar transport where they move from stem apex to root (basipetal) and also from root to shoot (acropetal). The basipetal transport is dominant because higher concentration is found in stem apex. Their speed is slow (6.4-26 mm/hour) and move against concentration gradient.

Bioassay: *Avena* curvature test

Types: Auxins are of two types: Natural and Synthetic

Natural auxins: These are Indole acetic acid, Indole pyruvic acid, Indole butyric acid.

Roles of Auxin or Indole Acetic acid (IAA):

(a) Cell division: It stimulates cell division in vascular cambium resulting in secondary growth.

(b) Cell elongation: It stimulates the growth and elongation of stem by causing cell wall to elongate in more than one direction. It occurs in three ways: (i) with increase in osmotic pressure; (ii) with decrease in wall pressure, and (iii) with increase in permeability of cytoplasm to water (endosmosis).

(c) Cell differentiation: Low auxin induces phloem differentiation and high auxin induces xylem differentiation.

(d) Phototropism: It promotes of bending of shoot towards unilateral light. The differential growth occurs due to lateral distribution of auxin when ight stimulus falls (Cholodny-Went theory).

(e) Geotropism: If growing plant is placed horizontally then it stimulates stem (shoot curves upwards) and inhibits root (to become vertical again). Its low concentration stimulates growth of root.

(f) Apical dominance: It promotes apical dominance (dominance of apical or terminal bud by inhibiting the growth of lateral bud). If apical bud is cut or excised then lateral bud will start growing. The apical dominance was common in plants having excurrent habit (palms, conifers, *Polyalthia*).

(g) Abscission: It prevents organised shedding or fall of leaves, flowers and fruits by preventing the formation of abscission layer at the base of petiole or pedicel.

(h) It prevents or delays senescence.

(i) Parthenocarpy: It promotes development of seed lees fruits by stimulating the swelling of ovary wall before fertilisation.

(j) Respiration: It stimulates respiration indirectly through increased supply of ADP by rapidly utilising ATP in expanding cell.

(k) Secretion of Ethylene: Its higher concentration induces the synthesis of ethylene.

(l) Sex expression: It induces femaleness in flower.

Roles of Indole Pyruvic acid (IPA):

(a) It induces rooting in cut parts (cuttings)

(b) It induces setting of natural fruits

Role of Indole Butyric Acid (IBA): It is natural as well as synthetic auxin.

(a) It induces rooting in cut parts (cuttings)

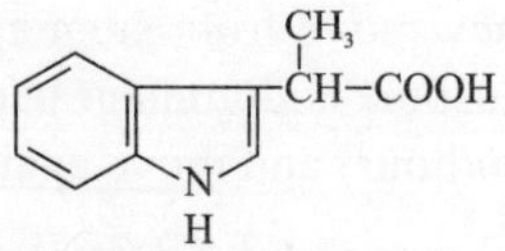

CH_3 | $CH-COOH$ N H

α-Indole propionic acid (IPA)

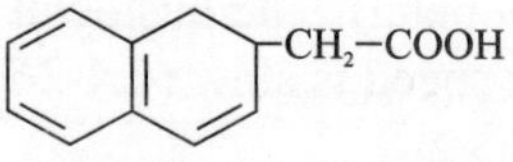

$-CH_2-COOH$

β-NAA

$-CH_2-CH_2-CH_2-COOH$ N H

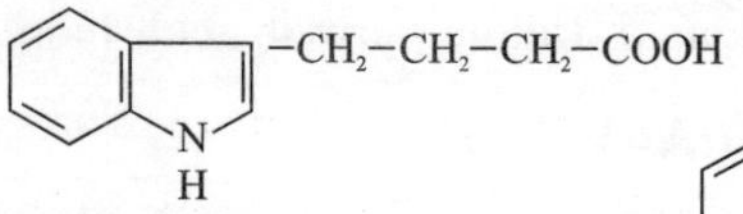

α-Indolebutyric acid (IBA)

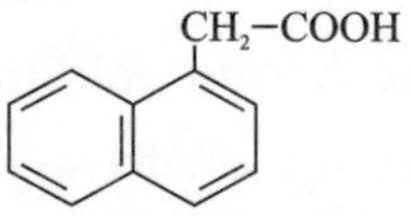

CH_2-COOH

α-NAA

Synthetic auxins: These are more stable and active than natural auxins because these are not destroyed by enzymes which do not exist in plants. These move in all directions.

Roles of Naphthalene acetic acid (NAA):

(a) It induces rooting in cut parts (cuttings)
(b) It induces flowering in Litchi, Pineapple
(c) It induces setting of natural fruits
(d) It prevents lodging (falling)
(e) Its methyl ester prevents sprouting of potatoes.

Roles of 2,4-Dichlorophenoxyacetic acid (2,4-D):

(a) It kills broad leaved weeds (usually dicots) hence it is powerful weedicide.
(b) It induces flowering in Litchi, Pineapple.
(c) It prevents abscission of fruits.

Roles of 2, 4, 5-Trichlorophenoxyacetic acid (2,4,5-T or Dioxin):

(a) The mixture of 2, 4-D and 2, 4, 5-T is called agent orange which causes defoliation of forests (used as chemical warfare in Vietnam war by US).

Other synthetic auxins are 2-methyl 4-chloro Phenoxy acetic acid (MCPA) and Indole-3-Propionic acid.

Synthetic antiauxins: Triiodobenzoic acid (TIBA) , p-chlorophenoxy Isobutyric acid (PCIB) and Naphthylphthalamic acid (NPA).

Gibberellin:

Discovery: The existence of Gibberellin was first noticed by Kurosawa (1926) while working on rice field. He observed that some rice seedlings grew taller and pale in colour than the others. He further observed that such seedlings were infected by a fungus, *Gibberella fujikuroi* (the perfect stage of an imperfect fungus, *Fusarium monoliforme*) which caused a disease called bakanae disease (foolish seedling disease) because seedlings grew foolishly so tall that they resulted into death of plants.

He observed that a sterile filtrate of fungus could stimulate the unusual growth in plants and caused the symptoms of bakanae disease in normal seedlings.

Isolation:

(i) Yabuta and Hayashi (1930): They isolated the substance from culture filtrate of *Gibberella fujikuroi* and named it 'Gibberellin'.

(ii) Yabuta and Sumiki (1938): They isolated gibberellins in crystalline form from the culture of fungus.

Biosynthesis: It occurs in plastid. It is synthesised through isoprenoid pathway of terpene starting from acetyl CoA. It is synthesised mainly in apical tissue.

Acetyl CoA → Acetoacetyl CoA → Mevalonic acid → Mevalonate Pyrophosphate → Isopentyl pyrophosphate → Formaacyl pyrophosphate → Transgeranyl pyrophosphate → Copalyl pyrophosphate → Kaurene → Kaurenoic acid → Hydroxykaurenoic acid → Gibberellic acid.

Nature: It is weak acid and chemically it is gibberellic acid.

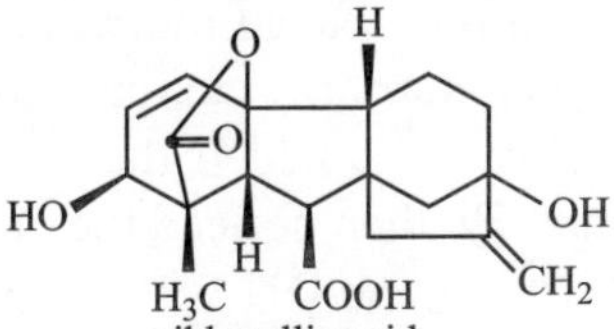

gibberellic acid

Structure: More than hundred types of gibberellins have been reported so far. The most common is GA_3 which is the simplest and first to be discovered. Its chemical formula is $C_{18}H_{22}O_6$. It shows gibbane ring and has tetracyclic terpenoid structure (a complex of group of chemicals related to lipids) having 4 isoprene units. The individual members differ from each other in their side attachments of various groups such as –OH group and $–CH_3$ group. All gibberellins are classified on the basis of structure and function

Distribution: It is found everywhere in plants but more in young leaf and seed. The reproductive parts of plant show greater amount than vegetative parts.

Transport: Its transport is acropetal as well as basipetal but it is not polar. It can be transported from the region of synthesis by simple diffusion, or through xylem and phloem.

Bioassay: (i) Dwarf Pea test (ii) Barley endosperm test

Roles:

(a) Gibberellin stimulates the elongation of stem (internode) in genetically dwarf plants by causing cell elongation and cell division esp. in intercalary meristem of stem. Thus it restores the normal size and growth to genetically dwarf plant.

(b) It causes bolting (elongation of stem into floral axis) in plants which show rosette habit (retarded growth of internode and profuse development of leaves) such as cabbage, carrot, beet etc. If cabbage head is kept under warm night then it will return to rosette habit. It helps in leaf expansion.

(c) It induces *de novo* (new) synthesis of hydrolytic enzymes (α-amylase, protease, lipase) which are mobilised to endosperm during seed germination to breakdown (hydrolyse) the stored food reserve.

(d) It substitutes the requirement of cold treatment (vernalisation) and induce flowering in one year bypassing the winter season in biennial plants.

(e) It induces seed germination by hastening maturity period hence it breaks seed dormancy.

(f) It induces parthenocarpy in apple, pear and also increases the size of fruit (apple) and bunch length in grapes esp. by applying Pomalin (Benzyl adenine+GA_4+GA_7).
(g) It prevents or delays senescence.
(h) It stimulates flowering in long day plants and inhibits in short day plants.
(i) It stimulates fruit setting in apple.
(j) It stimulates sex expression by inducing maleness in flowers.
(k) It induces malting process from barley in brewing industry.

Cytokinin:

Discovery: Skoog (1954) reported a substance in coconut milk which stimulated cell division while working on Tobacco pith culture.

Isolation: Miller (1955) while working on tobacco pith culture wanted it to grow indefinitely. He added several growth substances, vitamins, nutrients etc. in culture medium but these substances did not show any growth of tobacco. By chance he added the contents of old bottle of DNA kept for several years in his lab to culture medium. He then noticed the growth of tobacco pith callus for longer period but he did not get positive result from fresh DNA. He proposed that active substance could be a degraded product of DNA. He then isolated this substance by autoclaving the old DNA and found that this is a compound of purine.

He obtained similar result with DNA of yeast and sperm of Herring and named it kinetin due to its role in cell division. Later, kinetin was identified as 6-furfurylaminopurine. Kinetin is derivative of Adenine which bears furfuryl substituent at 9th position which shifted to 6th position of adenine during autoclaving of DNA.

Letham (1964) isolated first naturally occurring cytokinin from endosperm of young or unripe maize grain (kernel) and named it zeatin later it was identified as benzyl amino purine (BAP), a substituted adenine.

Nature: It is a weak base.

Biosynthesis: It is synthesised from isopentenyl adenosine monophosphate. The root tip is the usual site for synthesis of cytokinin.

Structure: The most common cytokinins are zeatin and isopentenyl adenine.

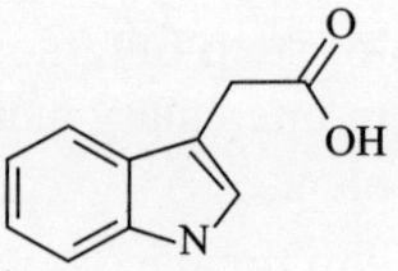

O
OH
N

Cytokinin

Distribution: It is mostly found in meristematic and growing tissues such as embryo and endosperm of seed (more), developing roots and fruits.

Transport: It is translocated from root to shoots by xylem

Bioassay: (i) Tobacco pith culture (ii) Retardation of leaf senescence

Roles:

(a) Cytokinin stimulates cell division in the presence of auxin. The ratio of cytokinin and auxin control cell differentiation or morphogenesis. These never act alone hence no response occurs if auxin and cytokinin acts alone.

- If low amount of cytokinin and high amount of auxin are applied then root develops.
- If high amount of cytokinin and low amount of auxin are applied then shoot (bud, stem and leaf) develops.
- If both cytokinin and auxin are in equal amount then cell divides but fails to differentiate and produce callus (undifferentiated, unorganised and irregular mass of cells).

(b) It counteracts apical dominance by inducing growth of lateral buds if cytokinin is directly applied on axillary bud.

(c) It delays senescence of plants by promoting protein synthesis, mobilising nutrients or resources and preventing degradation of chlorophyll and protein (Richmond-Lang effect). It promotes the conversion of etioplast into chloroplast.

(d) It increases the shelf life of vegetables by several days.

(e) It induces seed germination and thus breaks seed dormancy.

(f) It induces parthenocarpy.

(g) It induces flowering in short day plants.

(h) It stimulates development of chloroplast (synthesis of anthocyanin) in callus and leaves.

(i) It helps in opening of stomata.

(j) It induces sex expression by inducing femaleness in flowers.

(k) It promotes cell expansion in leaves and seeds.

(l) It helps plants to develop resistance against pathogens.

(m) It helps to produce new leaves, lateral shoot and adventitious roots.

Abscisic acid

It is plant growth regulator (inhibitor) and antagonistic to growth promoter esp. Gibberellin.

Discovery: Addicott (1963) discovered it from old cotton fruit and named it dormin which induced dormancy when applied to bud. Wareing (1964) discovered it from young cotton fruit and named it Abscissin-II which caused abscission. Both these substances were found to be Abscissic acid (ABA).

Distribution: It is found everywhere in plants but more in fruit and seed. It is produced in some fungi, algae, bryophytes and in all vascular plants.

Synthesis: It is synthesised in chloroplast of mesophyll cell of leaves by two processes:

Direct synthesis: It is produced from mevalonic acid.

Mevalonic caid → Isopentenyl pyrophosphate → Farnesyl pyrophosphate→ABA

Indirect synthesis: It is produced from xanthophyl (a carotenoid).

Xanthophyll → Violaxanthin → xanthoin→ABA

Transport: It is transported through xylem and phloem.

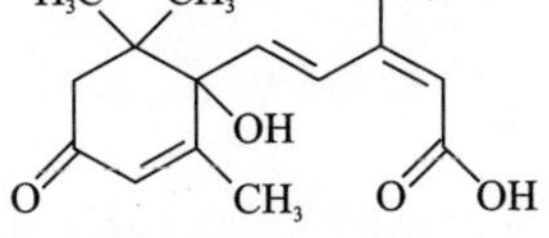

H_3C CH_3 CH_3 OH O CH_3 O OH

Fig. ***Abscisic acid***

Structure: It is dextrorotatory and cis-sesquiterpene (15-carbon).

Roles:

(a) It helps plants to cope up with environmental stress conditions such as drought, water logging, frost (cold) and saline condition. Thus it is called stress hormone.

(b) It causes stomatal closure by inhibiting K^+ uptake in guard cell. Thus it is called anti-transpirant. It prevents water loss under severe drought.

(c) It stimulates abscission and senescence of leaves, flowers and fruits.

(d) It inhibits seed germination and induces bud and seed dormancy (to withstand desiccation and other unfavourable conditions)

(e) It induces parthenocarpy.

(f) It stimulates positive geotropism

(g) It inhibits mitosis in vascular cambium and synthesis of RNA.

Ethylene

It is only gaseous natural plant growth regulator (inhibitor).

Discovery: Crocker reported the release of volatile substance from ripened orange which hastened the ripening of unripe banana. Later it was confirmed by Cousins that the volatile substance was ethylene using the technique of Gas chromatography.

Distribution: It is found in root, stem, leaf, flower, fruit and seed of all higher plants. It is emitted from ripening fruits and wounded organs.

Synthesis: It is produced from an amino acid, methionine in all tissues. Its synthesis is induced by high concentration of auxin but ethylene lowers the level of auxin in tissue (feedback mechanism). The shoot apex is an active site for ethylene biosynthesis.

Transport: It is a gas hence moves by diffusion from its site of synthesis. It is not easily soluble in aqueous medium and volatilises into intercellular space from where it is released in the outer atmosphere.

Structure: It is simple structure. $H_2C = CH_2$

Bioassay: Triple response

Ethephon (2-chloro ethyl phosphonic acid): It releases ethylene within plants when it is sprayed in aqueous medium.

Roles:

(a) It induces the ripening of climacteric fruits such as apple, banana, mango with rise in rate of respiration.

(b) It induces seed germination and sprouting of potatoes hence it breaks seed or bud dormancy.

(c) It stimulates the senescence and abscission of leaves, flowers and fruits.

(d) It stimulates horizontal (transverse) elongation of root and stem but inhibits vertical (longitudinal) elongation. Hence the root and stem look swollen.

(e) It induces flowering and fruit setting in pineapple.

(f) It induces growth of adventitious root and formation of root hair thus increasing the absorption surface.

(g) It induces epinasty (downward bending of leaves due to more growth on upper side than lower side).

(h) It induces triple response (reduced stem elongation, increased stem thickness and tightening of apical hook) on pea seedling.
(i) It induces sex expression by inducing femaleness in flowers of cucumber.
(j) It induces sleep disease (enrolling of petals in opened flowers).

Other growth regulators (inhibitors):

(a) Jasmonic acid: It is phytohormone which stimulates synthesis of phytoalexins.
(b) Brassinosteroid: It is steroidal hormone which inhibits growth of root and stimulates shoot elongation.
(c) Salicylic acid,
(d) Morphactin,
(e) Coumarin and
(f) Paraascorbic acid

Other synthetic regulators (inhibitors):

(a) Malic hydrazide:
(b) Phosphon-D:

Physiology of flowering

The formation of flower is transitional phase in life cycle of plant. During flowering vegetative shoot apex is converted into reproductive shoot apex. It is controlled by two factors:

1. Photoperiodism

It is the response of plants towards flowering to changes in the relative length of day and night.

It was discovered by **Garner and Allard** (1920). They studied the effect of temperature, nutrition, and soil conditions on these plants but none of these was found to regulate the flowering. They found Maryland mammoth (mutant) variety of tobacco plant would flower only if it was exposed to a number of short days. In autumn, this occurred naturally but in summer, this did not flower even after achieving much height until and unless it was provided a number of short days of 7 hours in a green house. When they examined more plants for this behaviour, they found that plants differed in their requirements for day length.

The relative length of dark period (scotoperiod) and light period (photoperiod) in a day varies from place to place and from season to season.

Flowering occurs irradiation of light was applied in hours, days and weeks. In annuals, flowering is primarily affected by photoperiod and secondarily by temperature but in biennials, flowering is affected by temperature. Moonlight and lightning do not play any role in flowering.

Types of plants: There are five types of plants which are as follows:

1. **Short day plants (SDPs):** These plants flower in autumn, winter or early spring in northern hemisphere when dark period is longer than critical dark period or these plants flower when the length of day is shorter than their critical photoperiod. *e.g., Chrysanthemum*, Cocklebur (*Xanthium*), Coffee, Cotton, Cannabis, Dahlia, Rice, Soya bean, Sugarcane, Sweet potato, Strawberry, Tobacco (Maryland mammoth variety). These plants exhibit following responses:

   (a) If dark period is prolonged then flowering will occur.

(b) If darkness is given in photoperiod then flowering will occur.

(c) If red light is given in dark period then flowering will not occur.

(d) If far red light is subsequently given in dark period just after red light then flowering will occur. The type of last treatment determines the type of response.

(e) Red light (torch/flash light) shows maximum inhibition of flowering at the middle of dark period.

(f) Gibberellic acid will not induce flowering.

(g) If short photoperiod and short dark period are given then flowering will not occur.

2. **Long day plants (LDPs):** These plants flower in late spring or early summer in northern hemisphere when photoperiod is longer than critical photoperiod or these plants flower when the length of day is longer than their critical photoperiod. e.g., Barley, Henbane, Lettuce, Oat, Pea, Radish, Snapdragon, Spinach, Sugar beet, Turnip, Wheat, *Salvia* etc.

(a) If photoperiod is prolonged then flowering will occur.

(b) If a flash of red light is given in dark period then flowering will occur.

(c) If far red light is subsequently given in dark period just after red light then flowering will not occur.

(d) If darkness is given in photoperiod then flowering will not occur.

(e) Gibberellic acid will induce flowering.

(f) If short photoperiod and short dark period are given then flowering will not occur.

3. **Day Neutral plants (Intermediate plants):** These plants will flower in any photoperiods. They do not show any response to day length. These plants flower after attaining a certain overall development stage. e.g., Balsam, Bean, Cotton, Cucumber, Maize, Pepper, Potato, Rose, Sunflower, Tobacco, Tomato etc.

4. **Long short day plants**: These plants require long photoperiod for initiation of flower but short photoperiod is required for blooming. These plants flower between summer and autumn. e.g., *Bryophyllum*, Night Jasmine (*Cestrum*).

5. **Short Long day plants**: These plants require short photoperiod for initiation of flower but long photoperiod is required for blooming. These plants flower between spring and summer. e.g., *Campanula*, Rye.

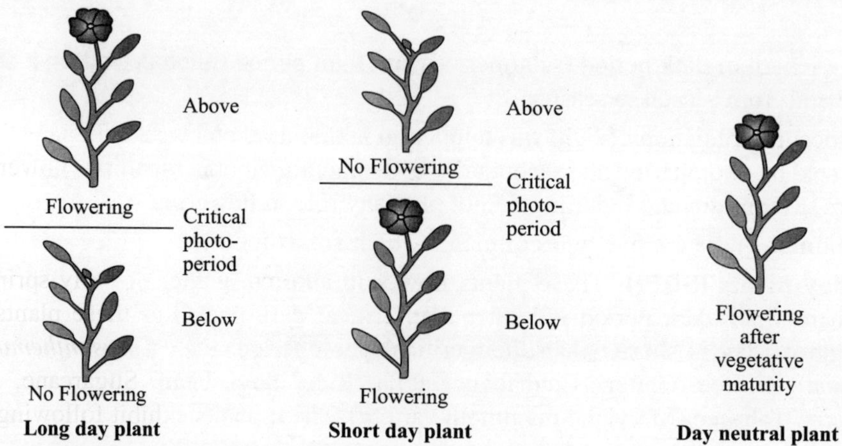

Fig. *Photoperiodism: Long day, short day and day neutral plants*

Photoperiodic induction: It is the persistence of photoperiodism after effect. It generally occurs when plant has achieved certain minimum vegetative growth. An appropriate photoperiod in 24 hours cycle constitutes one inductive cycle. If a plant after receiving sufficient inductive cycles is subsequently placed under unfavourable photoperiod, it will still flower. If the number of inductive cycles is increased, early flowering occurs. The minimum number of photoperiods required for induction varies from one (*Xanthium*) to 25 (*Plantago*). Leaf perceives the light stimulus for flowering in all plants. If all leaves are removed (complete defoliation) then flowering does not occur even after getting proper photoperiod and dark period. Even one leaf or a part (1/8th) of it is sufficient for flowering.

Photoinductive cycle results in the formation of flower forming hormone in leaf known as florigen which is passed from leaf to apex.

Chailakhyan proposed the existence of **florigen** but it has not been isolated so far. He removed the leaves of upper half of *Chrysanthemum* (SDP) but left the leaves on lower half untouched. He then exposed the lower half to short days and simultaneously exposed the upper defoliated parts to long days. Plants started flowering. He again reversed the above process in which lower half was exposed to long day and upper defoliated half to short days. Plants did not flower. He concluded that day length does not have any effect directly on bud but causes the leaves to manufacture a hormone that moves from leaf to bud and induces flowering. Florigen is produced by the joint activity if leaf and growing plants. It induces only flowering. Florigen is same through different groups of plants requiring different photoperiod. The stimulus or florigen responsible for flowering appears to be non-specific and the same for SDP, LDP and DNP because graft of induced twig of one plant into an uninduced twig of another plant causes the latter to flower thus photoperiodic stimulus can be transmitted from one branch of plant to another. The effect of photoperiodic stimulus is highly localised if a branch of SDP was exposed to short day and another to long day then flowering occurs in former and does not occur in latter. The effect of photoperiodic stimulus is systemic if a single leaf of SDP was exposed to short day and rest of plant to long day, flowering occurs in the whole plant.

Phytochrome: It is bluish green pigment which is involved in flowering of SDPs and LDPs. It is primary photoreceptor for photoperiodic stimulus in leaves. It is located in cell membrane. It is found in every plant parts. It has been found in wide range of plants from algae to angiosperms.

It exists in two forms: Red light absorbing form (P_R) and Far red light absorbing form (P_{FR}). Both these forms are photochemically interconvertible. Direct sunlight contains more red light than far red light hence P_R form is converted into P_{FR} form which is present in leaves during day. But in shade or at sunset, there is more far red light than red light hence P_{FR} form is converted into P_R form as night approaches thus P_R form is present in leaves.

P_R form (inactive in cytosol) is converted into P_{FR} form on absorbing red light.

P_{FR} form (active and attached to membrane) is converted into P_R form on absorbing far red light or slowly in darkness.

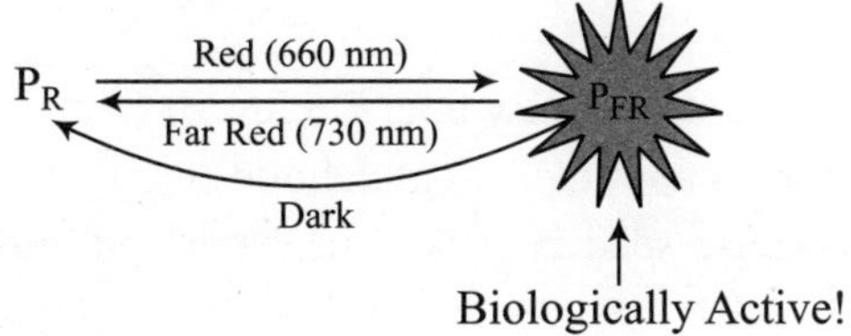

For SDP, Pr/Pfr >1 for flowering

For LDP, Pfr/Pr >1 for flowering

Phytochrome mediated processes:

(a) Flowering (b) seed germination (c) spore germination (d) bud dormancy (e) morphogenesis (f) Leaf abscission (g) Chloroplast movement (plastid orientation) (h) Succulency (i) growth of seedlings (j) sex expression (k) Elongation of root, stem and leaf

Importance of Flowering:

(a) It is useful in keeping plants in reproductive stage to yield more flowers and fruits.

(b) It is useful in keeping some plants in vegetative stage where vegetative parts such as tubers, rhizomes or corms are needed.

(c) It is useful in laying out gardens, orchards and planning crop pattern.

(d) It is useful in hybridisation experiment to cross breed in plants which normally flower in different season.

2. Vernalisation

It is a method of accelerating the flowering ability or shortening the juvenile or vegetative phase of biennials or winter annual by exposing their soaked seeds to low temperature for a week. It was discovered by Lysenko.

It was studied in winter variety of cereals (wheat, oat, barley). There are two varieties of cereals; (i) winter variety (sown in winter and flowering occurs in following summer), and (ii) spring variety (sown in spring and flowering occurs in next summer). If winter variety is sown in spring, it shows vegetative growth but does not flower in summer. It blooms (flowers) only in next spring after passing through winter season. If same variety is exposed to low temperature (0°C-4°C) for a week and sown in spring and then it flowers in following summer.

In nature, plants requiring vernalisation commonly behave as biennials (carrot, cabbage, sugar beet). The biennials complete their life cycle in two growing seasons of two years. They grow vegetatively in first year and produce flowers (monocarpic) in second year of growth. These plants fulfill their cold treatment during winter. If biennials are maintained in green house without cold treatment, these will grow vegetatively for years.

It occurs in temperate and some tropical plants. It is shown in SDPs, LDPs and DNPs.

Perception of stimulus: The stimulus for vernalisation is perceived by active apical meristem (shoot apex and bud). **Melcher** reported the existence of a flowering hormone called **vernalin** in the cells of meristem by grafting experiment. Henbane apex may induce flowering in non-vernalised henbane apex if the two are joined together. Thus flowering stimulus can translocate from vernalised plant to non-vernalised plant through graft union.

Devernalisation: If a vernalised seed/ plant is kept at high temperature, then the effect of low temperature is completely removed. The devernalised plant can be again vernalised by low temperature treatment. It is a reversible process.

Factors:

(a) Temperature (0-4°C): Its duration is few hours to few days.

(b) Oxygen: It is aerobic process and requires metabolic energy.

(c) Water: It is essential because seed is soaked in water and allowed to germinate and then low temperature treatment is done.

(d) Actively dividing cell

(e) Proper nourishment

Photoperiodism	Vernalisation
It is a response of relative lengths of light and dard period	it is the acceleration of the ability to flower by chilling treatment
The stimulus is perceived by mature leaves	The stimulus is perceived by active meristems
It not only prepares the plant for flowing but also intimates flowering	it itself does not induce flowering but prepares the plant for flowering
It is mediated through florigen	It is mediated through vernalin
The effect of photoperiods cannot be reversed	The effect of vernalization can be reversed by exposing the plant to high temperature
The photoperioodic can be transmitted from one plant to the other through grafting	The vernalization stimulus cannot be passed on from on plant to the other through grafting (except in Henbane)
The photoperiodic stimulus is interrupted by dard periods	The vernalization stimulus must be applied continuously

Importance:

(a) It increases the yield.

(b) It increases resistance to cold (frost) and fungal disease.

Senescence

It is deteriorative process which naturally terminates the functional life of organism. It is death of plant or plant parts. It is called phytogerontology. It occurs after reproductive maturity. It shows the following **indications**:

(a) Gradual breakdown of chlorophyll causing yellowing

(b) Reduced photosynthesis due to changes in chloroplast

(c) Reduced rate of respiration.

(d) Stoppage of translocation of solutes.

(e) Decreased production of biosynthetic enzymes

(f) Decreased production of auxin, cytokinin.

(g) Increased production of Abscissic acid, ethylene.

(h) Accumulation of toxic products.

(i) Reduced RNA and protein content.

(j) Deterioration in membranes and cell organelles (chloroplast is first organelle to be deteriorated).

Types:

1. Whole plant senescence: Whole plant dies after seed production. e.g., Rice, Wheat, Gram, Mustard (annual and monocarpic), Henbane (biennial and monocarpic), Bamboos, *Metroxylon* (perennial and monocarpic).
2. Shoot (top) senescence: The above ground part of shoot dies every year after producing flowers and fruits but underground parts survive and give rise to new shoots. e.g., Banana, Ginger, Gladiolus, Daffodil (*Narcissus*).
3. Sequential senescence: The older organ (leaf) and branches and flowers die but apical meristem continues to produce leaves and buds. e.g., Eucalyptus, Mango (many perennials).
4. Simultaneous (Synchronous) senescence: All leaves die and shed at the same time in autumn. e.g., *Dalbergia*, Elm, Maple (deciduous plants).

SHORT ANSWER TYPE QUESTIONS

1. What are the stages of development in plants?
2. Distinguish between arithmetic and geometric growth.
3. Explain the different phases of sigmoid growth.
4. Explain the plasticity in plants with examples.
5. Explain the *Avena* curvature test with respect to auxin.
6. What is full form of 2,4-D and mention its two roles?
7. What is bakanae disease with respect to the discovery of gibberellin?
8. Explain the roles of cytokinin in morphogenesis.
9. Discuss the five roles of abscisic acid (ABA).
10. Discuss the roles of gaseous growth regulator.
11. How photoperiodism was discovered in plants?
12. Distinguish between the short day plants and long day plants with examples.
13. What are the salient features of phytochrome?
14. Give the examples of phytochrome mediated processes.
15. Explain the process of vernalisation.

LONG ANSWER TYPE QUESTIONS

1. What is difference between growth and development? Explain the different phases of growth.
2. Describe the roles of auxin in plants.
3. Describe the important roles of gibberellin in plants.
4. Discuss the roles of phytochrome in flowering and how photoperiodism differs from vernalisation?
5. What are the indications of senescence in plants and mention its various types?

MULTIPLE CHOICE QUESTIONS

1. Which of the following ions is pulled out in apoplast by auxin during growth?
   (a) H^+ (b) K^+ (c) Ca^{+2} (d) Mg^{+2}
2. If a plant is subjected to continuous red light then
   (a) Phytochrome is destroyed (b) Phytochrome is increased
   (c) Phytochrome is decreased (d) Phytochrome remains in equilibrium
3. Which of the following combinations of gases is suitable for fruit ripening?
   (a) 80% CO_2 and 20% C_2H_4 (b) 80% C_2H_4 and 20% CO_2
   (c) 80% CH_4 and 20% CO_2 (d) 80% CO_2 and 20% O_2
4. If a tree flowers thrice in a year (Oct, Jan and July) then it must be
   (a) Both photosensitive and thermosensitive (b) Both photoinsensitive and thermoinsensitive
   (c) Photosensitive but thermoinsensitive (d) Thermosensitive but photoinsensitive
5. Leaves show maximum expansion in
   (a) Red light (b) Green light (c) Blue light (d) Violet light
6. Which of the following instrument can be used to record plant growth by seconds?
   (a) Arc indicator (b) Arc auxanometer (d) Space marker disc (d) Crescograph
7. The S-shaped growth curve and grand period of growth may change with
   (a) Sudden fluctuation in light intensity (b) Change in temperature
   (c) Fluctuation in humidity (d) These remain unaffected
8. Leaf fall occurs as abscission layer is formed when the content of
   (a) Auxin decreases (b) Auxin increases
   (c) ABA decreases (d) Gibberellin decreases
9. Exogenous application of gibberellins induces male flowers on genetically female plants in
   (a) *Coccinia* (b) *Cucurbita* (c) *Cucumis* (d) All of the above
10. Which of the following hormones is sprayed in sugarcane to increase production of sugar?
    (a) Auxin (b) Gibberellin (c) Cytokinin (d) Ethylene
11. When two hormones act together and contribute to same function hence this action is said to be
    (a) Synonymous (b) Antagonistic (c) Symbiotic (d) Synergistic
12. Phytochrome is closely related to
    (a) Carotenoid (b) Phycocyanin-c
    (c) Chlorophyll-e (d) Bacteriochlorophyll
13. If a plant is given only visible light then its growth will
    (a) Decrease (b) Increase
    (c) Firstly decrease then increase (d) Remain unaffected
14. Biological clock in plants is controlled by
    (a) Phytochrome (b) Cryptochrome (c) Cytochrome (d) All of the above
15. RNA formation is induced by
    (a) Florigen (b) Kinetin (c) Auxin (d) Gibberellin

16. Pruning of plants promotes branching because the axillary buds get sensitised to
 (a) IAA (b) GA (c) Kinetin (d) Ethylene
17. Which of the following is the motivative force for growth?
 (a) Root pressure (b) Turgor pressure
 (c) Atmospheric pressure (d) Osmotic pressure
18. Maximum amount of growth in root occurs
 (a) At apex (b) Behind apex
 (c) In the presence of light (d) In the presence of soil
19. An apple tree can be made to produce larger fruits than usual by
 (a) Applying hormones (b) Thinning of blossoms
 (c) Decapitating the stem apex (d) Defoliating
20. Which of the following is incorrect with respect to auxin?
 (a) Zn is essential for its synthesis (b) Tryptophan is precursor to auxin
 (c) It is abundantly produced in shoot apex (d) It cannot be synthetic
21. Phototropism in shoot is attributed to
 (a) Auxin (b) Gibberellins (c) Cytokinin (d) ABA
22. Phototropic curvature is promoted by
 (a) Red light (b) Blue light (c) Green light (d) Violet light
23. If a growing plant is decapitated then
 (a) Leaves become yellow (b) Axillary buds are inactivated
 (c) Axillary buds are activated (d) Leaves become small
24. In decapitated plants, axillary buds are activated because
 (a) These get more light (b) Amount of auxin is increased
 (c) Amount of food material is increased (d) Amount of cytokinin is increased
25. Pruning makes the hedge plant dense because
 (a) Injury induces dense growth
 (b) Apical dominance is removed
 (c) Root sprouts additional branches
 (d) Pruning removes shade and allow germination of new seedling to impart a dense growth
26. Reactivation of cambial cells in spring season is caused by
 (a) Auxin (b) Gibberellins (c) Cytokinin (d) Ethylene
27. Which of the following hormones is concerned chiefly with root initiation?
 (a) IBA (b) GA (c) ABA (d) Kinetin
28. Parthenocarpic fruits can be produced by the application of
 (a) IBA (b) 2,4-D (c) IAA (d) All of these
29. If the seedling is grown in darkness then
 (a) It is of same size as that in grown in light (b) It is much healthier than that grown in light
 (c) It is similar to that grown in light (d) It is taller than that grown in light

30. A genetically dwarf plant can be converted into tall plant by the use of
    (a) Kinetin (b) GA_3 (c) IAA (d) 2,4-D
31. The habit of cabbage plant can be changed drastically by the application of
    (a) IAA (b) GA_3 (c) ABA (d) 2,4-D
32. Which of the following plant materials is widely used in the preparation of culture medium?
    (a) *Pinus longifolia* (b) *Cocos nucifera*
    (c) *Colchicum autumnale* (d) *Cycas revoluta*
33. Richmond-Lang effect is shown by
    (a) Auxin (b) Gibberellin (c) Cytokinin (d) ABA
34. Ethylene is formed from the amino acid
    (a) Glycine (b) Methionine (d) Tryptophan (d) Lysine
35. Sleep disease of flowers is caused by
    (a) IAA (b) Ethylene (c) ABA (d) IBA
36. Pineapple can be made to flower in off season by
    (a) Zeatin (b) Ethylene (c) Temperature (d) Short days
37. Which of the following plant hormone is known as stress hormone?
    (a) Gibberellin (b) Kinetin (c) Auxin (d) ABA
38. Green organs are turned yellow by application of
    (a) IAA (b) IBA (c) 2,4-D (d) ABA
39. Phytochrome is found in
    (a) Algae (b) Fungi
    (c) Gymnosperm (d) All vascular plants
40. Which of the following pigment is present in two forms?
    (a) Cytochrome (b) Florigen (c) Phytochrome (d) Vernalin
41. In the red and far red responses of plants, it is the
    (a) First treatment that counts (b) Last treatment that counts
    (c) Middle treatment that counts (d) Sum total of all the treatments that count
42. When the dark period of short day plant is interrupted by a brief exposure of light then the plant
    (a) Flowers immediately (b) Turns into a long day plant
    (c) Will not flower at all (d) Gives more flowers
43. The flowering in short day plants does not take place when
    (a) The dark period is interrupted by white or red light
    (b) An interrupted critical photoperiod occurs
    (c) The critical photoperiod is interrupted by light
    (d) Dark period is interrupted by red light followed by far red light
44. Indole acetic acid (IAA) is applied to the root tip of a seedling in high concentration .then the growth of root would be
    (a) Inhibited (b) Increased (c) Decreased (d) Unaffected

45. A short day plant is grafted on a long day plant. The composite is kept in short day conditions. Flowering would be induced in the branches of
(a) Short day plant (b) Long day plant (c) Both (d) None
46. Indole acetic acid (IAA) occurs in the human urine especially in person suffering from
(a) Diabetes (b) AIDS (c) Pellagra (d) Typhoid
47. Which of the following hormones is involved in polar transport in plants?
(a) Auxin (b) Gibberellin (c) Cytokinin (d) ABA
48. Gibberellins are more effective when applied to
(a) Intact plants (b) Coleoptiles (c) Internodes (d) Roots
49. Gibberellic acid induced synthesis of α-amylase in barley aleurone is
(a) Stimulated by ABA (b) Reversibly inhibited by ABA
(c) Irreversibly inhibited by ABA (d) Not affected by ABA
50. Which of the following hormone shows maximum concentration in root tip?
(a) Auxin (b) Gibberellins (c) Cytokinin (d) ABA
51. Which of the following hormone shows maximum concentration in lateral buds?
(a) Auxin (b) Gibberellin (c) Cytokinin (d) ABA
52. Maximum concentration of gibberellins is found in
(a) Young leaves (b) Elongating stem (c) Mature leaves (d) Mature stem
53. Differentiation of shoot is controlled by
(a) High gibberellin: auxin ratio (b) High gibberellin: cytokinin ratio
(c) High auxin: cytokinin ratio (d) High cytokinin : auxin ratio
54. Bud dormancy can be induced by
(a) IAA (b) GA (c) ABA (d) Ethylene
55. Which of the following hormone is not found in plants?
(a) 2,4-D (b) GA (c) IAA (d) ABA
56. Which of the following hormone is used in preserving fruits?
(a) IAA (b) 2,4-D (c) Gibberellins (d) Cytokinin
57. Which of the following hormone causes seasonal leaf fall
(a) Auxin (b) Cytokinin (c) Gibberellins (d) Ethylene
58. The cutting of the upper portion of the shoot tip
(a) Delays growth of plant (b) Growth becomes faster
(c) Growth is not affected (d) None of the above
59. The photoperiodic stimulus is perceived by
(a) Leaves (b) Flowers (c) Buds (d) Meristem
60. ABA is synthesised from
(a) Chlorophyll (b) Terpene (c) Xanthophyll (d) Anthocyanin
61. Seed dormancy can be broken by
(a) IAA and ABA (b) IAA and Ethylene
(c) GA_3 and IAA (d) GA_3 and Ethylene

62. The correct sequence of the developmental process in a plant cell is
    (a) Plasmatic growth, Differentiation, Maturation, Senescence
    (b) Differentiation, Plasmatic growth, Maturation, Senescence
    (c) Differentiation, Maturation, Plasmatic growth, Senescence
    (d) Plasmatic growth, Maturation, Differentiation, senescence
63. Cytokinin and ethylene have opposite effects on
    (a) Elongation of stem (b) Leaf senescence
    (d) Dormancy (d) Lateral swelling of stem
64. The removal of shoot tips (decapitation) means loss of apical dominance and it is applied in
    (a) Tea plantation (b) Hedge making (c) Bolting (d) Both a and b
65. A sigmoid growth curve is characteristic of
    (a) Bacteria growing in culture medium (b) Organisms growing in natural habitat
    (c) All cell, tissues and organs (d) All of the above
66. In plants, IAA causes cell elongation due to
    (a) Increase in pH of cytoplasm (b) Decrease in pH of cytoplasm
    (c) Increase in pH of apoplast (d) All of the above
67. Gibberellin activates α-amylase enzyme activity in
    (a) Endosperm (b) Aleurone layer (c) Cotyledon (d) Embryo
68. Plant hormones act by affecting the activities of
    (a) Genes (b) Enzymes
    (c) Genes and enzymes (d) Genes, enzymes and membranes
69. Once a flower is pollinated, changes occur that make it less attractive to insects. Its petals shrivel and fall off. Pollination must
    (a) Increase the output of cytokinin in the flower
    (b) Block the flow of auxin from the root
    (c) Trigger the release of ethylene in the flower
    (d) Increase the formation of phytochrome which sets the biological clock
70. Long day plants flower especially when
    (a) Light period is greater than some critical period
    (b) Light period is lesser than some critical period
    (c) Dark period is lesser than some critical period
    (d) Light period is greater than some critical period
71. Arithmatical growth shows
    (a) Straight line curve (b) J-shaped curve (c) S-shaped curve (d) Parabolic curve
72. *Avena* curvature test is associated with
    (a) Ethylene (b) Gibberellin (c) Cytokinin (d) Auxin
73. Triple response is shown by
    (a) Auxin (b) Gibberellin (c) Cytokinin (d) Ethylene

74. Cytokinin is chemically known as
(a) 2-furfuryl amino purine (b) 2-furfuryl amino pyrimidine
(c) 6-furfuryl amino purine (d) 6-furfuryl amino pyrimidine
75. Abscisic acid is indirectly synthesised from
(a) Mevalonic acid (b) Acetyl coenzyme A (c) Xanthophyll (d) Tryptophan
76. The size of apple is increased by the application of
(a) Indole acetic acid (b) Naphthalene acetic acid
(c) Gibberellic acid (d) Abscisic acid
77. Which one of the following hormones is not associated with sex expression?
(a) Auxin (b) Gibberellin (c) Cytokinin (d) ABA
78. Which of the following types of gibberellin is most common?
(a) GA_1 (b) GA_2 (c) GA_3 (c) GA_4
79. Which of the following base of DNA is present in cytokinin?
(a) Adenine (b) Guanine (c) Thymine (d) Cytosine
80. The perception of stimulus for vernalisation is found in
(a) Leaf (b) Flower (d) Shoot apex (d) All of the above

ANSWERS

1.a	2.c	3.b	4.d	5.d	6.d	7.d	8.a	9.c	10.b
11.d	12.b	13.a	14.a	15.b	16.a	17.b	18.b	19.b	20.d
21.a	22.b	23.c	24.d	25.b	26.a	27.a	28.c	29.d	30.b
31.b	32.b	33.c	34.b	35.b	36.b	37.d	38.d	39.d	40.c
41.b	42.c	43.a	44.a	45.c	46.c	47.a	48.c	49.b	50.c
51.c	52.a	53.d	54.c	55.a	56.d	57.a	58.a	59.a	60.c
61.d	62.a	63.b	64.d	65.b	66.b	67.b	68.d	69.c	70.a
71.a	72.d	73.d	74.c	75.c	76.c	77.d	78.c	79.a	80.c

13

Reproduction in Organisms

Reproduction is a process by which new individuals of a species are formed from the pre-existing ones. It is one of the characteristic features of living organisms but is not defining feature because there are several organisms which do not reproduce. It maintains the continuity of life.

Life span of some organisms

Organisms	Life span (Years)
Dog	20
Cat	30
Lion	35
Horse	60
Elephant	60
Cow	25
Bat	15
Pigeon	35
Ostrich	50
Parrot	140
Crocodile	60
Tortoise	125
Drosophila	30 days
Rose	5-7
Banyan tree	200

Reproduction takes place by the following methods:

Vegetative reproduction

It is method of reproduction in which new plants are produced from vegetative parts either by fragmentation accidentally or through external force. The newly formed individuals are genetically identical (clone) to parent plant. Each member of clone is called ramet. It is common in flowering plants (wild and domesticated plants). It also occurs in algae, bryophytes and pteridophytes. It occurs by natural and artificial methods:

Natural methods: The vegetative parts naturally form new plant once these are detached from parent plants.

1. **Root**: It develops adventitious buds when placed in soil and then leafy shoots (slip) are grown. The slips are separated and planted to form new plants.

(a) Tap root: It develops adventitious buds to form new plants. e.g., *Albizzia, Dalbergia*, Guava, Murraya, Poplar etc.

(a) Adventitious root : It develops adventitious buds to form new plants. e.g., *Asparagus*, Dahlia, *Ipomoea, Tapioca, Tinospora*, Yam.

2. **Stem**: Its modified part develops buds which form new plants.
   (a) Rhizome: e.g.,Ginger, Turmeric, Banana, *Adiantum,* Water fern (*Marsilea*)
   (b) Corm: e.g., *Amorphophallus, Colocasia, Crocus, Gladiolus*
   (c) Tuber: e.g., Potato, Jerusalem artichoke
   (d) Bulb: e.g., Garlic, Onion, *Narcissus*
   (e) Runner: e.g., *Cynodon, Centella, Oxalis*
   (f) Stolon: e.g., Strawberry, *Vallisneria*
   (h) Offset: e.g., *Eichhornia, Pistia*
   (h) Sucker: e.g., *Chrysanthemum*, Mint
   (i) Phylloclade: e.g.,*Opuntia*
   (j) Bulbil: Fleshy axillary bud: e.g., *Dioscorea*, Lily
   Fleshy terminal bud: e.g., *Agave*, Pineapple
3. **Leaf**:It develops buds to form new plants. e.g., Walking fern (*Adiantum*), *Begonia, Bryophyllum, Kalanchoe, Saintpaulia, Streptocarpus*.
4. **Turion**: It forms fleshy buds in aquatic plants e.g., *Potamogeton, Utricularia*

Artificial methods: The vegetative cut parts are dipped in root promoting phytohormones such as Indole butyric acid (IBA) and Naphthalene acetic acid (NAA). It is generally performed in spring and rainy seasons.

1. **Cutting:**
   (a) Root cuttings: e.g.,Blackberry, Raspberry
   (b) Stem cuttings: e.g.,Carnation , Cotton, Coffee, Tea, Croton, Citrus plants (Lemon and Orange), Sugarcane, Rose, Grapes
   (c) Leaf cuttings: It is cut into two or three parts and planted in vertical position. e.g., *Sansevieria* (Snake plant).
2. **Layering**: It is rooting-cutting type where adventitious roots are induced to develop on soft stem while it is still attached to parent plant. It is performed on one year old basal shoot branches during early rainy season.
   (a) Tip layering: In this, tip of a branch is inserted in the soil where root formation is induced and later shoot tip develops. e.g., Raspberry
   (b) Serpentine layering: Inthis, the basal branch is inserted at several places hence several plants develop. e.g., *Clematis*
   (c) Mound (stool) layering: In this stem is cut just one inch above soil in early spring and covered with mound (soil). Later new side branches (shoots) containing roots develop. These shoots are separated and planted to produce new plants. e.g., *Clematis*, Grapes, Jasmine, Strawberry (herbaceous plants).

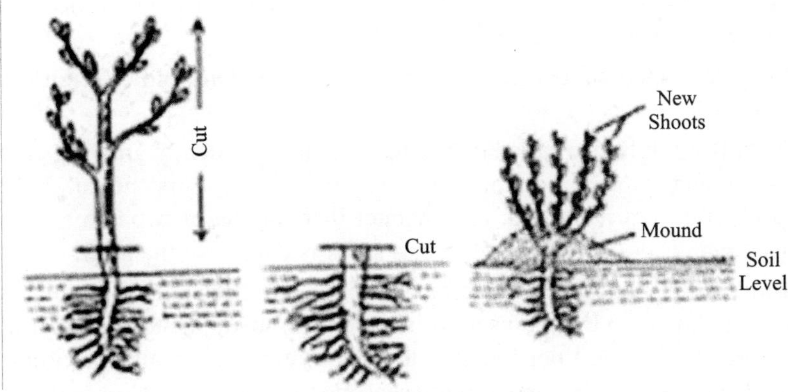

Fig. *Mound layering*

(d) **Air layering (Gootee)**: In this ring of bark is removed or girdled. The cut part is covered with grafting clay (cow dung, finely divided hays and clay=1:1:2). Root promoting hormones and water are added. It is then wrapped in polythene bag from preventing from desiccation and left for 2-3 months. After that, roots are developed below cut ends of shoot and then branch is cut below it and planted. e.g., *Bougainvillea*, China rose, Litchi, Lemon, Orange, Pomegranate. It is commonly used for shrubs and trees.

This is not possible in monocot due to scattered vascular bundle hence water and food supply will be harmed.

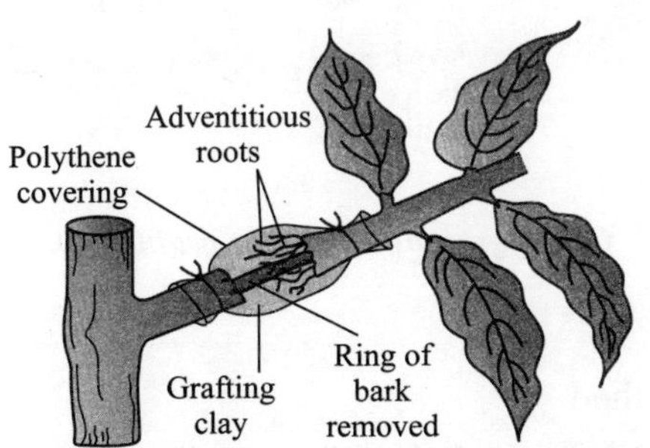

Fig. *Air layering*

3. **Grafting**: It is technique of forming composite plant by joining stock and scion of two different plants of close or related species. It is done in those plants where cambium is present (e.g., gymnosperms and dicots). It employs stock which is efficient root system of plant of disease resistant or wild variety and scion which is shoot system having one or more buds. Stock is always older than scion. Both stock and scion are fixed and united in such a manner that cambia of the two come into contact. Callus is formed by cambium. Thus it is mixing of traits of two varieties without hybridisation. It is generally performed in Apple, Lemon, Orange, Mango, Plum, Peach, Pear, Rose, Rubber etc.

   (a) **Tongue grafting**: Oblique cut occurs on both stock and scion. The diameter of both stock and scion is same.

   (b) **Crown grafting**: Several V-shaped cut occurs on scion and these are inserted on several cuts on the side of stock. The diameter of stock is much greater than scion.

(c) **Wedge grafting**: V-shaped cut occurs on stock and wedge shaped cut occurs on scion. The diameter of stock and scion is same.

(d) **Side grafting**: V-shaped cut occurs on stock on one side. One end of scion is sharpened and inserted into stock.

(e) **Approach grafting**: It is also called inarching. Two independently growing plants are brought near. The stock and scion of both are cut at same level for a distance of 2.5-5 cm. Scion is fixed on stock after removing bark in a manner that cambia of two come in contact. After a few months, Scion is cut below the graft and stock is cut above the graft.

(f) **Bud grafting**: In this, bud is taken as scion with a small portion of bark having intact cambium. The living tissue of scion bud is inserted into incision made in stock so that it reaches up to cambium of stock. The bud portion should remain exposed in air. The joint is thoroughly sealed using bandage. This is done in rainy season to ensure survival of bud. After one month, bud starts growing and sprouts. The stock is cut above the level of graft. e.g., Apple, Rose, Peach

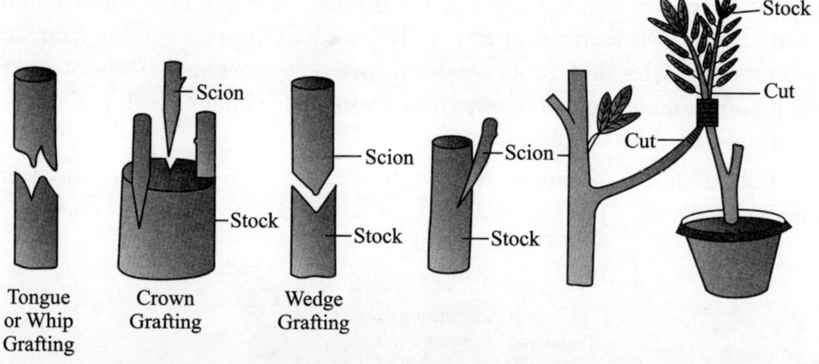

Fig. ***A, Tongue or whip grafting, B. Crown grafting, C. Wedge grafting, D. Side grafting, E. Approach grafting.***

Advantages:

(a) It is cheaper and easier method.
(b) It multiplies superior varieties with weak roots.
(c) It preserves god qualities of plants for long time.
(d) It maintains purity of plants (without variation) by retaining original plant.
(e) It is only method of multiplication in seedless plants such as Banana, Grapes, Fig, Orange, Sugarcane, Pineapple, Rose etc.
(f) It is done in those plants where the number of viable seeds is less.
(g) It is performed in those plants where seeds remain dormant.

Disadvantages:

(a) The vegetatively produced plants are less adapted to changed environment.
(b) The vegetative parts are easily decayed and infected with viruses, bacteria and fungi.
(c) It cannot produce new varieties.

Asexual reproduction

It is a type of reproduction in which new individuals develop from a single parent without involvement of gametes. It is more primitive and simpler than sexual reproduction.

Features:

(a) It involves only a single parent.
(b) It occurs usually in unicellular organisms (e.g., Monerans and Protists) and also in Fungi, plants and some animals.
(c) It can occur with or without gamete formation but there is no fusion of gametes.
(d) It involves only mitotic division.
(e) It shows rapid multiplication because gamete formation, fertilisation and embryo development do not occur.
(f) It does not form zygote.

Types:

1. **Binary fission**: In this, the organism divides into two smaller and equal sized daughter individuals. Parental cell does not die as it continues to exist in the form of daughter cells. The division of cytoplasm is equal. It is of following types on the basis of plane of division.
   (a) Transverse binary fission: In this, DNA replicates while attached to chromosome. Mesosome divides and new mesosomes get attached to daughter DNA. Finally transverse septum develops and grows centripetally. e.g., Bacteria, Diatoms, *Paramecium, Planaria*
   (b) Longitudinal binary fission: In this, nucleus divides and then cytoplasm divides lengthwise (longitudinally). e.g., *Euglena, Vorticella*

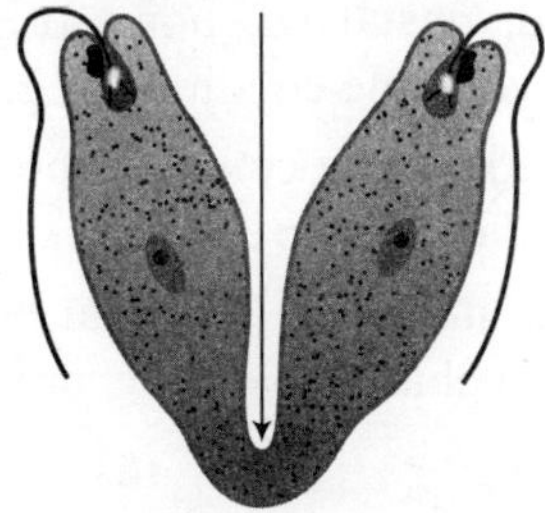

Fig. ***Euglena***

   (c) Irregular fission: In this, cytoplasm divides in any planes but always at right angle. e.g., *Amoeba*
   (d) Oblique fission: In this, cytoplasm divides obliquely. e.g., *Ceratium, Gonyaulax*
2. **Multiple fission**: In this, parental cell divides mitotically several times producing many daughter cells. e.g., *Amoeba, Plasmodium*

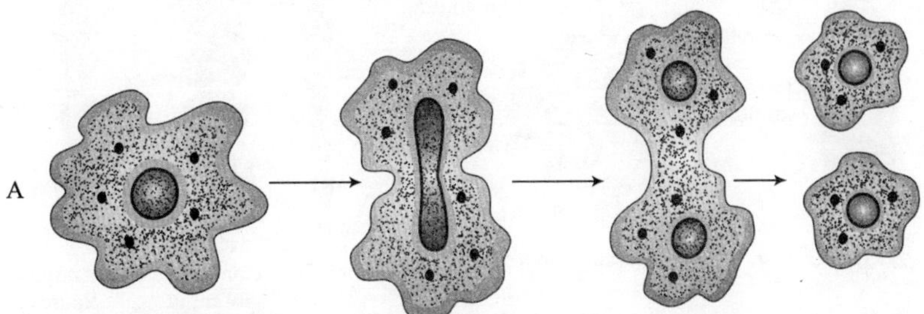

3. **Plasmotomy**: In this, multinucleate parent cell undergoes cytokinesis to form two multinucleate daughter cells without karyokinesis. The daughter cells regain the same number of nuclei by further nuclear division. e.g., *Opalina, Pelomyxa*
4. **Budding:** A small bud is produced by cell wall on one side. This bud later grows in size. Parental cell remains intact and division of cytoplasm is unequal. Parent nucleus divides and one of daughter nuclei shifts into bud. Bud grows further, constricts at base and finally separate to form new cell. e.g., Yeast, *Hydra* etc.

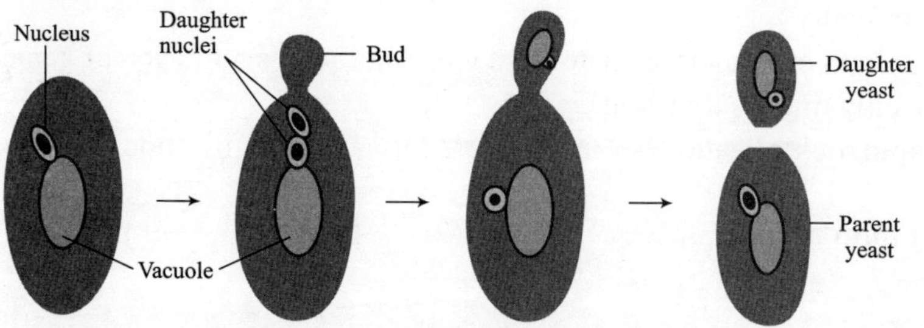

Fig. ***Yeast undergoing budding***

5. **Sporulation:** It takes place by specialised reproductive bodies called spores. These are small, unicellular and uninucleate body released from parent body. These are involved in reproduction, dispersal and perennation. These are of following types.
   (a) Zoospore: It is oval, unstalked, motile (flagellate-2 or 4), endogenous (developed in sporangium) and formed during favourable condition. e.g., *Chlamydomonas,* and *Ulothrix* (among algae); *Pythium* and *Saprolegnia* (among fungi).
   (b) Sporangiospore: It is rounded, unstalked, non-motile (non-flagellate), endogenous, most common and formed during favourable condition. e.g., *Mucor*, *Rhizopus*
   (c) Conidium: It is rounded, stalked having conidiophore, non-motile, exogenous (borne singly or in chain) and formed during favourable condition.
   (d) Chlamydospore: It is rounded, unstalked (terminal or intercalary), non-motile thick walled and developed directly from hyphal cell during unfavourable condition. It is regarded as resting spore, e.g., *Agaricus.*
   (e) Oidia (Arthropspore): It is rounded, unstalked, non-motile, thin-walled and developed in the condition of excess water, sugar and salt during unfavourable condition. e.g., *Agaricus*
   (f) Aplanospore: It is non-motile and thin walled. e.g., *Spirogyra, Ulothrix*

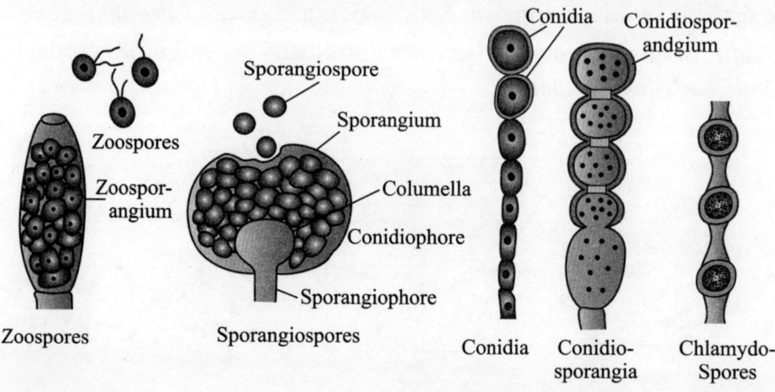

6. **Gemmulation:** It occurs in sponges. Gemmule is formed when aggregates of cells (archaeocytes) become isolated at the surface of sponges and surrounded by a protective covering. It germinates to form new sponge on the onset of favourable condition.
7. **Regeneration:** It is formation of new individual from a fragment when the body of the parent breaks into small pieces. e.g., *Hydra*, Star fish

Advantages: It forms genetically similar daughter individuals (clones) because there is no variation.

Disadvantages:

(a) It makes organisms less adaptable to changed environment.

(b) It does not show variation among the offsprings hence there is no role in evolution.

(c) It results in overcrowding due to rapid multiplication.

(d) It produces weaker organisms.

Sexual reproduction

It is the development of new individuals through the formation and fusion of male and female gametes. These individuals are different from the parents and also amongst themselves. It is called amphimixis. Rudolf Camerarius (1694) discovered sexual reproduction in plants.

Features:

(a) It involves both parents but not always.

(b) It involves both meiosis (during sporogenesis) and mitosis (during embryogenesis).

(c) It requires formation of gametes hence zygote is formed.

(d) It is complicated and slow process.

(e) It forms genetically different daughter individuals from parents.

Phases in life cycle: There are three phases in the life span of all such organisms.

1. **Vegetative or Juvenile phase**: It is period of growth starting from seed germination and up to reproductive maturity. It bears only foliar leaves. Its duration varies in different organisms. It may last from a few weeks to several years (Perennial plants).
2. **Reproductive or maturity phase**: It develops floral structure which is sign of sexual maturity. Its duration varies in different organisms. Sexually there are two types of plants

   (a) Monocarpic: These plants bear flower and fruit only once in their life time. e.g., all annuals, all biennials (Beet, Cabbage, Carrot, Henbane, Lettuce, Radish, Turnip) and some perennials (*Agave*, *Bambusa*). *Strobilus kunthiana* (bearing flowers after 12 years).

   (b) Polycarpic: These plants bear flowers and fruits every year. e.g., Perennials, China rose (bears flowers throughout the year).
3. **Senescent phase**: It is terminal and irreversible state of ageing. It shows yellowing and falling of leaves.

   Thus annuals and biennials have distinct vegetative, reproductive phase and senescent phases but in perennial plants, these phases are not well marked.

Events in sexual reproduction: The sexually reproducing organisms follow regular sequential events right from the formation of gametes. These are as follows:

1. **Gametogenesis:** It is a process by which two types of gametes (male and female) are formed. It is pre-fertilization event. These gametes are always haploid in nature but the parent body may be

haploid or diploid. These are produced as a result of **meiosis** in case of diploid individuals such as pteridophytes, gymnosperms, angiosperms and most of the animals including human beings or **mitosis** in case of haploid individuals such as fungi, algae and bryophytes. In case of some algae, the two gametes are morphologically similar (**isogametes** or homogametes) but in most of sexually reproducing organisms, the gametes are morphologically dissimilar (heterogametes or **anisogametes**). The male gamete is known as antherozoids or sperms and the female gamete is called egg.

Types of Plants: On the basis of sexuality

(i) Monoecious (Homothallic**):** It is a condition in which male and female sex organs are present in same plant. It is always bisexual in case of fungi but in case of flowering plants, it may be bisexual (presence of male and female sex organs in same flower. e.g., most common) and unisexual (presence of male and female sex organs in separate flowers. e.g., Cucurbits, Maize, coconut etc.

(ii) Dioecious (Heterothallic**):** It is a condition in which male and female sex organs are present in separate plants. In flowering plants, it is always unisexual **(**presence of male and female sex organs in separate flowers) where male flower is called staminate and female flower is known as pistillate. e.g., Date palm, *Mulberry*, Papaya.

Types of animals: On the basis of sexuality

(i) Hermaphrodite: It is a condition in which both male and female sex organs are present in same animal. e.g., sponges, earthworm, tapeworm, leech.

(ii) Unisexual (sexually dimorphism): It is a condition in which male and female sex organs are present in separate animals. e.g., most of animals including human beings.

2. **Gametic transfer:** It is also pre-fertilisation event and occurs in following four ways:

(a) In case of some algae and few fungi, both male and female gametes are motile (flagellate) and float on water.

(b) In case of algae, sex organs are unicellular and non-jacketed but in bryophytes, and pteridophytes, the sex organs (antheridium and archegonium) are multicellular and jacketed and transfer of gametes requires water because male gamete is motile (biflagellate in bryophytes and multiflagellate in pteridophytes) and reaches towards female gamete which is non-motile (unflagellate) and stationary. It has been noticed that in spite of availability of water, male gamete fails to reach female gamete that is why the number of male gametes produced is much more than the number of female gametes produced.

(c) In case of seed plants (gymnosperms, angiosperms), the pollination is effective mechanism for pollen grains (carrying non-motile male gametes) of anther to stigma of carpel. Pollination may be self (autogamy) in case of bisexual flowers of monoecious plants as well as cross (geitonogamy) in case of unisexual flowers of monoecious plants and cross (xenogamy) in case of unisexual flowers of dioecious plants.

(d) In case of sexually dimorphic (unisexual animals where male and female gametes are formed in different animals, a special complex mechanism is evolved for transfer of gametes for fertilisation.

(e) In case of mammals, the transfer of gametes occurs by mating ehere male individual releases sperms into reproductive tract of female and further semen of male and secretion of female reproductive tract provide a medium for sperms to move and fertilise the eggs.

3. **Fertilisation:** It is a process in which fusion of gametes (syngamy) takes place resulting in the formation of zygote.

   Types of syngamy: On the basis of place of occurrence

   (a) External fertilisation (exogamy): It occurs outside the body of organism mostly in water. e.g., Most of algae, fish and amphibians.

   (b) Internal fertilisation (endogamy): It occurs inside body of organism. e.g., Most of the fungi, plants other than algae (bryophytes, pteridophytes, gymnosperms and angiosperms) and animals (reptiles, birds and mammals).

   Types of syngamy: On the basis of structure of fusing gametes

   (a) Isogamy: It is fusion of morphologically similar but physiologically dissimilar gametes. It is considered primitive. e.g., algae *(e.g.,Spirogyra, Ulothrix, Oedogonium, Chlamydomonas debaryanum* etc*),* fungi (e.g., *Synchytrium, Rhizopus* etc).

   (b) Anisogamy: It is fusion of morphologically and physiologically dissimilar gametes. e.g., *Chlamydomonas braunii*, Fungi, Frogs, Mammals

   (c) Oogamy: It is fusion of smaller, motile and active male gamete with larger, non-motile and passive female gamete but both are formed in sex organs. It is more advanced. e.g., *Chlamydomonas coccifera*, *Volvox, Chara* (among chlorophyceae), phaeophyceae, rhodophyceae, fungi, and higher plants.

   (d) Hologamy: It is fusion of two organisms which act as gamete. e.g., yeast

   Parthenogenesis: It is the development of embryo from unfertilised egg. The female gamete (egg) forms new individuals without undergoing fertilisation. It occurs in some organisms. It is of three types:

   (a) Arrhenotoky**:** In this, only males are produced. e.g., rotifers, drones of honey bees, wasp, turkey (bird) etc.

   (b) Thelytoky**:** In this, only females are produced. e.g., rock lizard (*Lacerta saxicola*), blind snake (*Typhlina*) etc.

   (c) Amphitoky**:** In this, the individuals of any sex are produced. e.g., aphid

4. **Zygote formation:** It is pόst-fertilisation event. Zygote is diploid in nature in all sexually reproducing organisms. Zygote is link between two successive generations. In some algae and fungi, zygote develops a thick wall that is resistant to desiccation and damage during unfavourable condition. It undergoes resting period before germination. In most of algae (showing haplontic life cycle), the zygote undergoes meiotic division rather mitosis to form haploid spores that grow in haploid individuals (gametophytic plants). In animals, zygote starts developing just after its formation.

5. **Embryo formation (Embryogenesis):** It is the development of embryo from zygote which undergoes mitotic division and then differentiation. Embryo is formed in all plant groups except algae. The number of cells in embryo will be increased by cell division while differentiation form specialised tissues and later organs which finally establish an organism. The post-fertilisation stages include withering and falling off sepals, petals and stamens; development of seed from ovule and development of fruit from ovary. The fruit develops a thick wall called pericarp and seed germinates to form new plant.

Types of animals:

1. **Oviparous:** These animals lay fertilised eggs covered by tough calcareous shells in safer place outside their body. These eggs are incubated for a certain period of time after which young ones hatch by breaking the shell. e.g., most fish, amphibians, reptiles, birds and a few mammals (monotremes).
2. **Viviparous:** These animals give birth to young ones because fertilised egg (zygote) develops into embryo within uterus of mother. The embryo gets nourishment from the blood of mother through placenta. Their eggs are very small which remain inside their bodies. e.g., Majority of mammals including human beings.
3. **Ovoviviparous:** These animals instead of laying eggs produce eggs that develop within the mother's body. The eggs are hatched and remain inside body of mother for a certain period of time and are nourished there but not by placenta. Finally the young ones are born alive. e.g., some species of shark.

Advantages of sexual reproduction

(a) It helps plants adapt to changed environment.
(b) It helps in evolution of species by causing variation.
(c) It results in multiplication and perpetuation of species.
(d) It increases vigour and vitality among organisms.

Disadvantages:

(a) It is slow process.
(b) It is based on chance.

Terms:

(a) Parthenogamy: It is the union of two incompatible gametes like two female gametes.
(b) Parthenoapogamy: It is fusion of vegetative nuclei.

SHORT ANSWER TYPE QUESTIONS

1. Mention the different modified part of stem in asexual reproduction with examples
2. Mention the different cuttings of plant in artificial method of vegetative reproduction.
3. Distinguish between mound layering and air layering.
4. Write notes on bud grafting.
5. Distinguish between conidium and zoospore.
6. Mention the disadvantages of of asexual reproduction.
7. Distinguish between monocarpic condition and plycarpic condition.
8. Distinguish between isogamy and oogamy.
9. Write notes on parthenogenesis.
10. Write notes on ovoviviparous type of animals.

LONG ANSWER TYPE QUESTIONS

1. Describe the different techniques of grafting. What are its advantages and disadvantages?

5. What are the features of asexual reproduction? Describe its different methods.

3. Discuss the events taking place in sexual reproduction.

MULTIPLE CHOICE QUESTIONS

1. Which of the following modified part of stem is used as vegetative propagule in banana?
   (a) Rhizome (b) Stolon (c) Sucker (d) Offset
2. The fleshy axillary bud is vegetative propagule in
   (a) *Dioscorea* and Lily (b) *Agave* and Lily
   (c) *Agave* and pineapple (d) *Dioscorea* and pineapple
3. Which of the following set of phytohormones is used in rooting of cuttings?
   (a) IAA and IBA (b) IBA and IPA (c) IAA and NAA (d) IBA and NAA
4. Leaf develops buds to form new plants in
   (a) *Begonia* and *Discorea* (b) *Begonia* and *Bryophyllum*
   (c) *Dioscorea* and *Bryophyllum* (d) *Agave* and *Dioscorea*
5. Grafting is not possible in monocots due to
   (a) Presence of cambium (b) Presence of scattered vascular bundles
   (c) Absence of cambium (d) All of the above
6. Which of the following set of terms is used in grafting?
   (a) Stock and trunk (b) Scion and trunk
   (c) Stock and scion (d) Scion and propagule
7. Which of the following binary fissions occurs in bacteria?
   (a) Longitudinal binary fission (b) Transverse binary fission
   (c) Irregular fission (d) Multiple fission
8. Longitudinal binary fission occurs in
   (a) *Paramoecium* (b) *Opalina* (c) *Plasmodium* (d) *Euglena*
9. Budding occurs in
   (a) Yeast and *Planaria* (b) *Hydra* and *Planaria*
   (c) Yeast and *Hydra* (d) *Euglena* and *Yeast*
10. Gemmule in Sponges is an aggregation of
    (a) Choanocytes (b) Myocytes (c) Archaeocytes (d) Thesocytes
11. Which of the following spores is stalked, non-motile and exogenous?
    (a) Zoospore (b) Sporangiospore (c) Conidium (d) Chlamydospore
12. Which of the following algae shows isogamy, anisogamy and oogamy?
    (a) *Volvox* (b) *Chara* (c) *Spirogyra* (d) *Chlamydomonas*
13. Which of the following types of syngamy is most advanced?
    (a) Isogamy (b) Anisogamy
    (c) Physiological anisogamy (d) Oogamy
14. Sexual reproduction in plants was discovered by
    (a) Rudolf Virchow (b) Rudolf Camerarius (c) Strasburger (d) Hartig
15. Which of the following spores is thick-walled and develops under unfavourable condition?
    (a) Conidium (b) Zoospore (c) Sporangiospore (d) Chlamydospore
16. The features of aplanospore are
    (a) Thin walled and motile (b) Thin walled and non-motile
    (c) Thick-walled and motile (d) Thick-walled and non-motile

17. External fertilisation occurs in
    (a) Most of fishes (b) Most of fungi
    (c) Most of bryophytes (d) Most of angiosperms
18. The vegetative phase, reproductive phase and senescent phase are not well marked in
    (a) Annuals (b) Biennials
    (c) Perennials (d) Annuals and biennials
19. Monocarpic condition is shown by
    (a) All annuals and biennials (b) All biennials and perennials
    (c) Perennials only (d) All of the above
20. Parthenogenesis occurs in
    (a) Queen of honeybees (b) Worker of honeybees
    (c) Drone of honeybees (d) All of the above
21. Which types of parthenogenesis shows only male individuals?
    (a) Arrhenotoky (b) Thelytoky (c) Amphitoky (d) All of the above
22. Quadriflagellate sperms are seen in
    (a) *Spirogyra* (b) *Chlamydomonas* (c) *Chara* (d) *Ulothrix*
23. Which of the following sequence of events occur in sexual reproduction in fungi?
    (a) Karyogamy-Plasmogamy-Meiosis (b) Plasmogamy-Karyogamy-Meiosis
    (c) Meiosis-Plasmogamy-Karyogamy (d) Plasmogamy-Meiosis-Karyogamy
24. Which of the following perennial plants produces flowers once in its lifetime?
    (a) Mango (b) *Bambusa* (c) Orange (d) Fig
25. Asexual reproduction in monerans and protists takes place by
    (a) Budding (b) Binary fission (c) Conidia (d) Multiple fission
26. Meiocyte is observed in
    (a) Cyanobacteria (b) Human egg (c) Dicot plant (d) Fern
27. The terms homothallic and monoecious are used to denote
    (a) Unisexual condition (b) Bisexual condition (d) Staminate flowers (d) Pistillate flowers
28. Under unfavourable conditions, adventitious buds present on the surface of tuberous roots produce
    (a) Seeds (b) Adventitious roots
    (c) Axillary bud (d) Nodes and internodes
29. Which of the following is pre-fertilisation event in flowering plants?
    (a) Transfer of pollen grains (b) Development of embryo
    (c) Development of endosperm (d) Development of seed
30. Which types of parthenogenesis is shown by drones of honey bees?
    (a) Arrhenotoky (b) Thelytoky (b) Amphitoky (d) All of the above

ANSWERS

1.a	2.a	3.d	4.b	5.c	6.c	7.b	8.d	9.c	10.c
11.c	12.d	13.d	14.b	15.d	16.b	17.a	18.c	19.a	20.c
21.c	22.d	23.b	24.b	25.b	26.a	27.b	28.b	29.a	30.a

14

Sexual Reproduction in Flowering Plants

The sexual reproduction in higher plants involves development of male and female gametes produced from respective sex organs. In angiosperms, male gametes are produced from pollen grain (microsopore while female gametes are produced inside embryo sac.

Flower is specialised condensed or modified shoot because it has nodes and internodes. It is reproductive organ of angiosperms. It is meant for sexual reproduction.

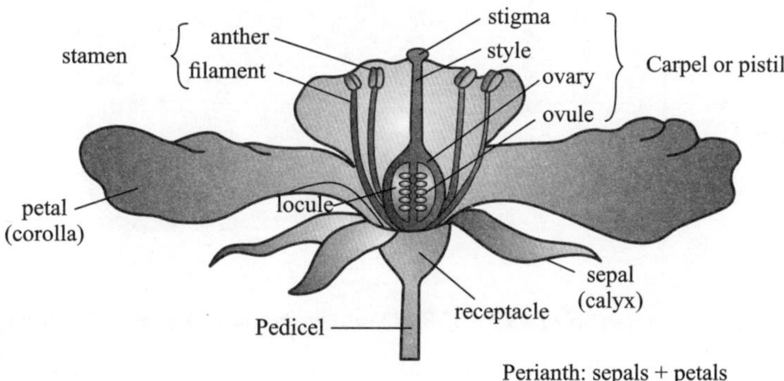

Parts of flower:

Accessory outer whorl: It does not take direct part in sexual reproduction. It contains calyx and corolla.

1. Calyx: Its unit is sepal. It is green, lower and smaller. It protects other inner floral parts.
2. Corolla: Its unit is petal. It is coloured other than green, upper and larger. It attracts pollinators and protects inner floral parts.

Essential inner whorl: It is fertile whorl and takes part in sexual reproduction. It contains androecium and gynoecium.

3. Androecium: Its unit is stamen. It is male reproductive organ concerned with production of microspores or pollens. It consists of broad terminal anther, connective and cylindrical filament. The proximal end of filament is attached to the thalamus or the petal of a flower.
4. Gynoecium: Its unit is carpel. It is female reproductive organ concerned with production of megaspores. It consists of basal swollen ovary, elongated style and terminal stigma.

Development of male gametophyte

It involves the following events in stamen.

Development of anther

It is fertile part of stamen. It is usually dithecous (two lobes having four sporangia). Both lobes of anther are attached to each other by sterile parenchymatous tissue called connective but connective is absent in monothecous stamen. Anther is monothecous (one lobe having two sporangia) in members of Malvaceae family such as cotton, lady's finger. Anther consists of only one microsporangium in *Arceuthobium* (smallest parasitic plant).

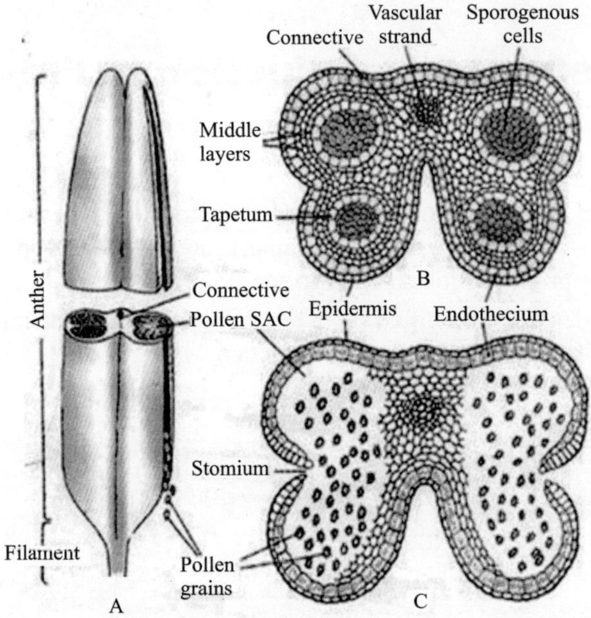

Fig. ***Structure of anther: A (LS of anther), B (TS of young anther) and C (TS of mature anther)***

Structure: Each anther lobe has two long and cylindrical microsporangia or pollen sacs. Now four sporangia of anther are situated at four corners of anther. These run parallel to one another. The development of microsporangia is eusporangiate type in which microsporangia develop hypodermally from a group of archesporial cells. The archesporial cells divide periclinally (parallel to outer wall of anther lobe) forming outer parietal cell (towards epidermis) and mass of inner sporogenous cell. The parietal cell undergoes a series of periclinal and anticlinal divisions forming 3-5 concentric layers of anther wall which consists of epidermis, endothecium, middle layers and tapetum. The outer three wall layers are protective and also help in dehiscence of anther by which pollens are released.

(a) Epidermis: It is firstly formed, outermost and one layered. It becomes stretched, flattened and finally shrivel off at maturity. It is protective layer.

(b) Endothecium: It is sub-epidermal, one layered and radially elongated. It develops fibrous thickening containing α-cellulose in radial and tangential walls. It develops maximum at the time of dehiscence. It is hygroscopic. The cells of endothecium become thin walled and lack fibrous thickening at the place of dehiscence called stomium. It helps in dehiscence of anther at maturity due to presence of fibrous bands, differential expansion of outer and inner tangential wall and hygroscopic nature.

(c) Middle layer: It is short lived (ephemeral) and 1-3 layered. It is absent in *Lemna* and *Najas*. It disintegrates at maturity to provide nourishment to growing microspore mother cell.

(d) Tapetum: It is innermost, one layered and viscous. It is rich in dense cytoplasm and more than one prominent nuclei. It possesses mitochondria, plastids, dictyosomes and a number of spherical lipid bodies called Ubisch bodies. It surrounds sporogenous cell. It attains maximum development at pollen tetrad stage. It breaks down at the time of maturity. It either becomes multinucleate (coenocytic) without cytokinesis or its nucleus becomes polyploid due to endomitosis (duplication of chromosomes due to separation of chromatids without division of nucleus). It is of two types:

(i) Glandular or secretory tapetum: It is most common type. It remains *in situ* (original position) throughout microspore development. It secretes sporopollenin, pollenkit (oily and sticky substance of lipid and carotenoids) and compatibility proteins. It helps in ornamentation (external thickening) of exine of pollen grain due to Ubisch bodies.

(ii) Amoeboid or Invasive tapetum: It is short-lived. Its cells break down at early stage and their entire protoplasts move towards centre to nourish microspores where it joins to form plasmodium or periplasmodium. It is considered primitive and is found in *Alisma*, *Typha, Potamogeton, Tradescantia* etc.

Functions:

(a) It secretes enzymes, hormones and nutrients to provide nutrition to the developing pollen grains.

(b) It secretes callase enzyme which dissolves callose (β-1, 3-glucan) layer of pollen tetrad to separate pollen grains.

Microsporogenesis: It is formation of haploid microspores (pollen grains) from diploid microspore mother cell by meiotic division inside pollen sac (microsporangium) of anther.

Primary sporogenous cells increase their numbers due to divisions and fill the entire cavity of microsporangium. These are transformed into microspore mother cells or pollen mother cells which are diploid. Initially it is covered by callose and later separates. The separated microspore mother cell undergoes meiosis to produce a tetrad of microspore or pollen grains. Callose binds all four microspores into tetrad.

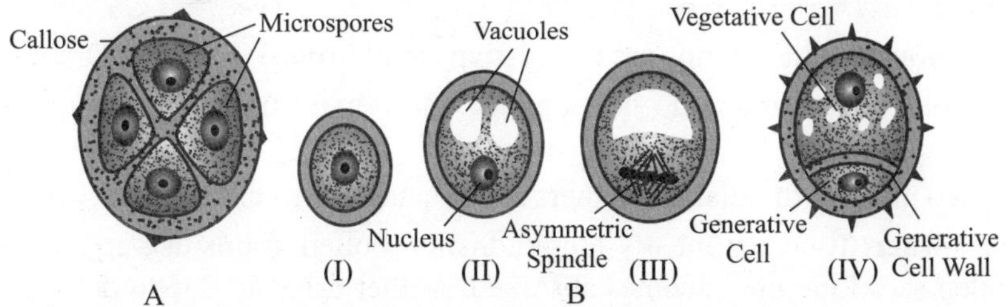

Microsporogenesis. A, a microspore tetrad, B, a microspore maturing into a pollen grain.

$$\text{Primary sporogenous cell} = \text{Microspore mother cell} \xrightarrow{\text{meiosis}} \text{Microspore tetrad}$$

Types of tetrad:

1. Tetrahedral: It shows **simultaneous** type of cytokinesis (first and second cytokinesis occurs after completion of meiosis). It is primitive feature. It is most common and found in dicots.
2. Isobilateral: It shows **successive** type of cytokinesis (first cytokinesis occurs after meiosis I and second cytokinesis after meiosis II). It is advanced feature. It is found in monocots.
3. Decussate: It is found in *Atriplex, Crocus, Magnolia* etc.
4. Linear: It occurs in *Halophila.*
5. T-shaped: It is found in *Aristolachia, Butomopsis*.

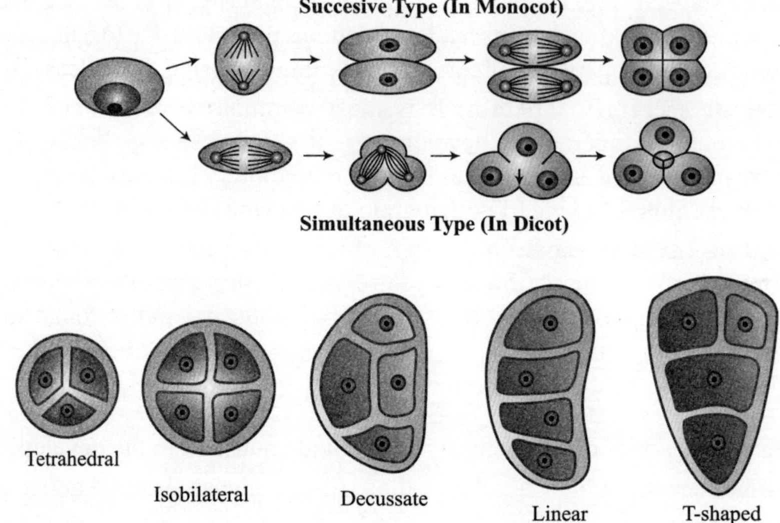

Peculiarities:

1. Out of four microspores, three degenerate and one remains functional in case of *Cyperus*. Hence one meiotic division produces one functional pollen grain.
2. All five types of tetrad are seen in a species of *Aristolachia* and *Butomopsis*.
3. Polyspory condition (occurrence of more than 4 spores esp.11) is seen in *Cuscuta*.
4. Compound pollen grains (all the four microspores do not separate) are seen in *Juncus, Jatropha, Drimys, Dracaena, Drosera, Typha* etc.
5. Masulae (compound pollen grains are attached in small group) is seen in Orchid.
6. Pollinia (all four pollen grains do not separate and remain united in a single sac) are seen in Orchid.
7. Translator (two pollinia of adjacent anthers are attached) is found in *Calotropis*.

Dehiscence of anther: It occurs mostly longitudinally. Pollen grains dry up and become powdery inside mature pollen sac. Tapetum becomes absorbed. Anther dries up due to dehydration. The sterile partition wall between two pollen sacs of anther lobe gets destroyed forming single chamber. Dead cells of endothecium contract from outer thin wall due to further drying so that anther lobe forms longitudinal concavities. Slit occurs lengthwise on anther. Anther lobe ruptures in region of stomium. These microspores after their release from tetrad are called pollen grain. Thus thousands of pollen grains are dispersed. The pollen grain represents the male gametophyte.

Exceptions:

1. Transverse dehiscence: *Althaea, Ocimum*
2. Porous dehiscence: *Solanum*
3. Valvular dehiscence (by lifting surface layers): *Berberis*
4. Irregular dehiscence: *Najas*
5. Gun powder mechanism (by bursting): *Urtica*
6. Fly trap mechanism: *Aristolachia*

Structure of Pollen grain: Pollen grain is generally rounded having size 25-50μm. Its shape, size, colour and design vary from species to species. It is rich in nutrients (proteins=7-26%, carbohydrate=24-46%, and fat=1-15%) and water (7-16%). Study of pollen grains is called Palynology.

Father of Palynology	Erdtman
Father of Indian Palynology	P.K.K.Nair
Smallest pollen	*Myosotis*
Largest pollen	*Mirabilis* (250μm)
Longest pollen	*Zostera*
Largest number of pollen	Maize (25 million)

Pollen grain consists of two layers:

1. **Exine**: It is outermost, thick and cuticularised. It is made up of sporopollenin (a fatty substance). It is further divided into ectexine (sexine) and endexine (nexine). Ectexine is made up of three layers: outer discontinuous tectum, middle discontinuous baculum and inner continuous foot layer. It has clear aperture called germ pore where sporopollenin is absent. Germ pore is developed in those areas where exine is either absent or thin. It has three germ pores (tricolpate) in dicots and one germ pore (monocolpate) in monocots.

   Exine can tolerate high temperature, acids, alkali and hazardous environment. It cannot be degraded by any enzymes. It can be preserved for long period during fossilisation due to presence of sporopollenin which is resistant to physical and biological decomposition.

   Exine can exhibit fascinating arrays of patterns and designs due to tectum. These are of taxonomic importance.

   Exine is spiny as well as covered by a yellowish, viscous, sticky and oily coating called pollenkit in entomophilous (insect pollinated) pollen grains. Pollenkit is chiefly made up of lipids and carotenoids. It is secreted by tapetum. Pollenkit helps in attachment of pollens to the bodies of insects, acts as insect attractant and protects pollens from UV radiations.

2. **Intine:** It is innermost, thin and delicate. It is made up of pecto-cellulose.

Zonocolpate	*Arachis, Argemone*
Zonocolporate	*Cleome, Lathyrus*
Zonoporate	*Datura*
Pantoporate	*Hibiscus, Mirabilis*

Microgametogenesis: It is development of male gamete from pollen grain (microspore). Microspore (partially developed) is first cell of male gametophyte and represents immature male gametophyte. The male gametophyte is developed inside microsporangium (pollen sac) of anther. It is called precocious development. It is more or less uniform in all flowering plants.

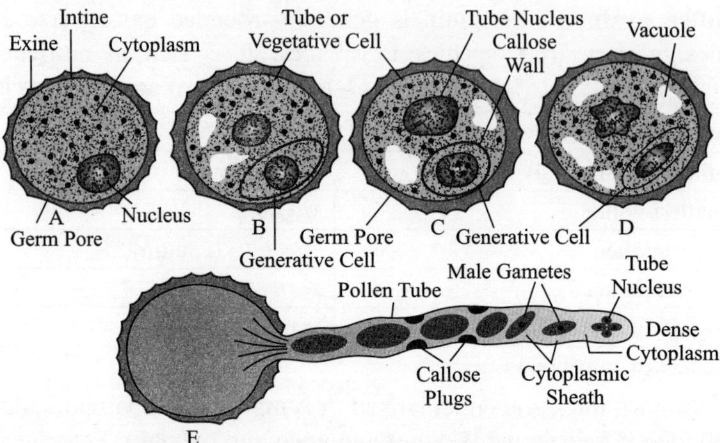

Fig. ***Germination of pollen grain and formation of male gametophyte in an angiosperm***

Microspore is small, unicellular, uninucleate, haploid and large in numbers. Freshly formed microspore has rich cytoplasm and centrally placed prominent nucleus. It grows in size due to flow of nutrients nd development of vacuole.

The nucleus of microspore divides mitotically forming larger tube nucleus and smaller generative nucleus. Both surround cytoplasm and become larger vegetative cell and smaller and spindle-shaped generative cell. The vegetative cell has large nucleus and vacuolated cytoplasm which is rich in food reserve. The generative cell has thin dense cytoplasm surrounding a prominent nucleus. A layer of callose is deposited around generative cell in pollen tube to separate the generative cell from pollen grain wall.

Pollen grain sheds at 2-celled stage in dicots (70% of angiosperms). This takes place after the release of pollen from anther. In this case, generative cell divides after pollen has landed on stigma or upon its germination.

The generative cell again divides mitotically forming two non-motile and amoeboid male gametes. The male gametes surround cytoplasm and form male cells. The pollen tube is developed from germ pore from intine. Pollen grains shed at 3-celled stage in monocots (30% of angiosperms) and *Cycas* and 4-5 celled stage in *Pinus*. The vegetative cell enlarges and descends to the tip of pollen tube where it later degenerates.

Pollen viability: It is the ability of pollen grains to germinate. Its duration is variable from 30 minutes (very small) in cereals or flowers pollinated in bud condition to several months in legumes. It is longest in *Typha* (300 days). It depends on temperature, humidity and genetic potentiality of species. It can be stored for years in liquid nitrogen (-196°C) in pollen banks (cryopreservation) for future use in plant breeding programme.

Pollen products: These are in the form of tablets, creams and syrup used as food supplement by atheletes because the pollens are rich in nutrients.

Allergic Pollens: The pollen grains of *Amaranthus,* Castor, *Chenopodium,* Carrot grass (*Parthenium*), *Prosopis, Sorghum* cause allergy and respiratory trouble.

Pollen bank: Pollen grains are stored in liquid Nitrogen (-196°C).

Pollen or male sterility: It is caused by

(a) Anther suppression, abortion, phyllody.
(b) Failure of anther dehiscence
(c) Aberrant meiosis or gametophyte development
(d) Premature dissolution of callose
(e) Fungal infection

Development of female gametophyte: Embryo sac is known as female gametophyte. It is formed from megaspore. It involves the following events in ovary:

Development of ovule: Ovule is integumented megasporangium present inside ovary. It is attached with placenta. It develops into seed after fertilisation. It is generally oval in shape. There are several types of ovule based on shape and orientation.

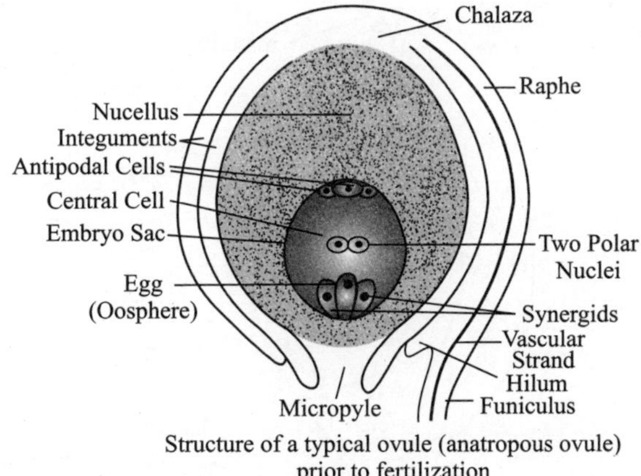

Structure of a typical ovule (anatropous ovule) prior to fertilization

1. Anatropous: It is most common. It is called inverted ovule because nucellus is rotated by 180^0 and micropyle is on lower side and comes close to funicle due to unilateral growth of ovule. It has following parts:

(a) Nucellus: It is main body of ovule. It has abundant reserve food hence it provides nutrition to developing embryo. It may be

- Crassinucellate: Nucellus is massive and well developed. It is many layered because parietal cell undergoes division. It is seen in *Casuarina*, members of Polypetalae subclass of Dicots; Monocots.
- Tenuinucellate: Nucellus is thin and ill-developed. It remains single layered because parietal cell does not divide. It is seen in members of Gamopetalae subclass of Dicots.

(b) Chalaza: It is basal part of ovule and present just opposite to micropyle or at the junction of nucellus and integument.

(c) Micropyle: It is narrow opening at one end of ovule. It is formed by integuments.

(d) Funicle: It is stalk of ovule. It remains attached with placenta of ovary. It contains vascular strand for supply of nutrients to ovule.

(e) Integument: It is protective layer and covers the ovule except at micropyle. It arises from chalaza. It is of following types:

1. Ategmic: Integument is absent. e.g., *Loranthus, Santalum*

2. Unitegmic: There is one integument. e.g., Gymnosperms, Gamopetalae of dicots (Solanaceae, Asteraceae).
3. Bitegmic: There are two integuments. e.g., Polypetalae of dicots (Brassicaceae, Malvaceae) and monocots.
4. Tritegmic: There are three integuments. e.g., *Asphodelus*, Litchi.

(f) **Hilum**: It is point of attachment of funicle with nucellus of ovule.

(g) **Raphe**: It is a ridge when funicle gets fused with the nucellus on one side.

(h) **Embryo sac:** It is female gametophyte and lie at the centre inside nucellus.

2. Orthotropous (Atropous): It is erect ovule in which micropyle and funicle are in one straight vertical line. It is simplest and most primitive. e.g., Gymnosperms, Betel, *Polygonum*.

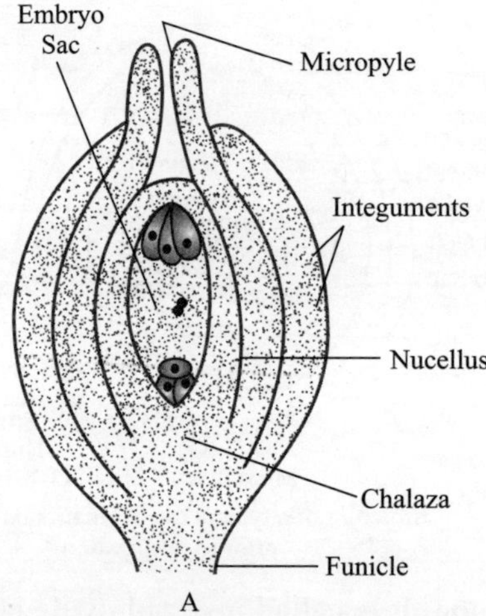

Fig. Orthotropous ovule

3. Hemitropous (Hemianatropous): It is transverse ovule in which micropyle and funicle lie at right angle. It is intermediate between anatropous and orthotropous ovule. e, g., *Primula, Ranunculus*.

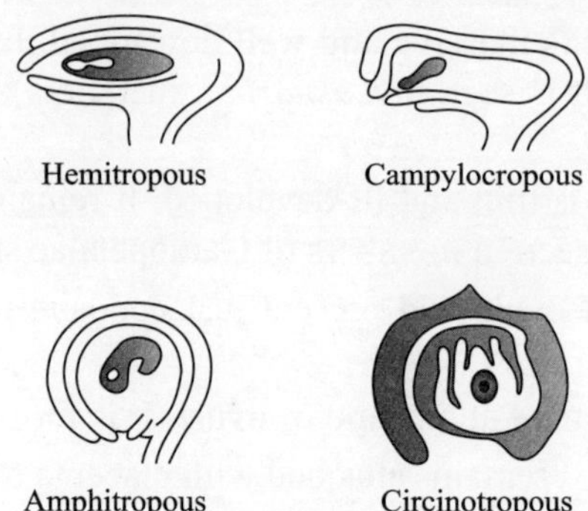

4. Amphitropous: Both embryo sac and nucellus are horse shoe shaped. e.g., *Alisma*, *Lemna,* Poppy.

5. Campylotropous: Nucellus is curved hence micropyle and chalaza lie on either side of funicle. e.g., *Capparis, Chenopodium,* Members of Brassicaceae, Fabaceae.

6. Circinotropous: Micropyle faces upwards but funicle is large and coiled around ovule by more than 360°. It is initially orthotropous but later become anatropous due to unilateral growth of funicle. e.g., *Opuntia*

Barriers:

1. **Hypostase**: It is group of lignified (thick walled) cells present below embryo sac. It arises from cells of nucellus. It acts as barrier or boundary for growing embryo sac and thus it prevents it from pushing into base of ovule. It also maintains water balance. It is seen in *Agave*, Lily, *Aristolachia* etc.
2. **Epistase:** It is group of lignified (thick walled) cells present above embryo sac. It arises from cells of nucellus. It forms a cap like structure.

Megasporogenesis: It is formation of haploid megaspore from diploid megaspore mother cell by meiotic division inside the ovule (megasporangium).

Archesporial cell is differentiated in hypodermis from any cell of nucellus towards micropyle. It grows in size and develops dense cytoplasm with prominent nucleus. Archesporial cell divides mitotically (periclinally) forming outer parietal cell and inner primary sporogenous cell. Parietal cell divides one or more times and forms parietal tissue. Primary sporogenous cell commonly functions as megaspore mother cell. It shows deposition of callose in its wall but absent in *Polygonum* type of embryo sac. Megaspore mother cell (diploid) undergoes meiotic division forming linear tetrad of haploid megaspores. At this stage, two integuments develop from base of ovule.

Megagametogenesis: It is development of female gamete from megaspore. Megaspore is the first cell of female gametophyte (embryo sac). Female gametophyte like male gametophyte is reduced and dependent on nucellus (sporophytic tissue). Its development is endosporous (within megaspore).

Panchanan Maheshwari classified embryo sac into three types on the basis of (a) number of megaspore involved in the development of embryo sac, (b) number of divisions occurring in nucleus of megaspore and (c) organisation of nuclei in mature embryo sac.

1. **Monosporic Embryosac**: It develops from a single functional megaspore and as such all the nuclei present are genetically identical. It is of two types:
   (a) ***Polygonum* type**: It was discovered by Strasburger (1879) in *Polygonum divaricatum.* It is most common and primitive. It develops from chalazal megaspore (lowermost or fourth from micropyle).

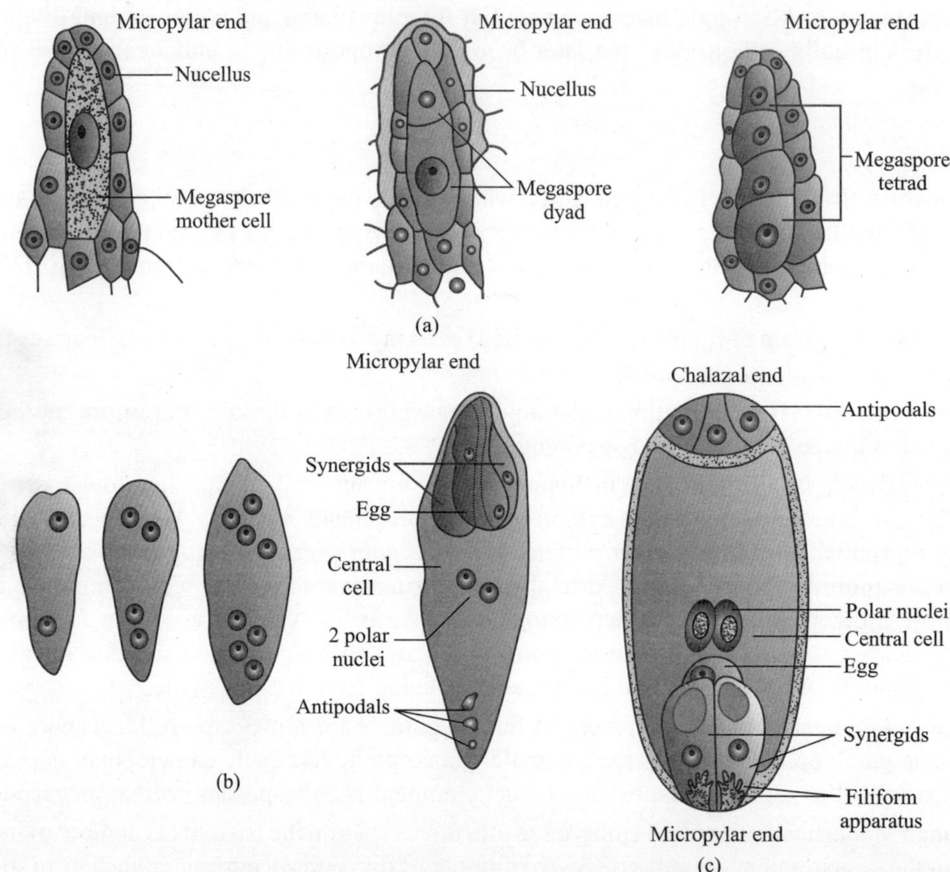

Fig. ***Development of Polygonum type of embryo sac***

Functional megaspore undergoes successive three mitotic divisions without cytokinesis (free nuclear type in which karyokinesis is not followed by cytokinesis) forming 8 nucleated embryo sac in which four nuclei are present in chalazal end and four nuclei at micropylar end. Now reorganisation of nuclei occurs. One nucleus from both ends moves to middle forming two polar nuclei which unite to form secondary nucleus. The remaining three at chalazal end forms antipodal cells and rest three at micropylar end forms egg apparatus in which two are synergid cells and one is egg cell. Thus embryo sac is 7-celled containing 8 nuclei.

Embryo sac is embedded in nucellus towards micropylar half of ovule. It is polar in organisation in which cytoplasmic elements are arranged towards chalaza and vacuolated cytoplasm towards micropyle. Embryo sac contains following type of cells.

(a) Central cell: It is the largest cell and mother cell of endosperm. It contains two polar nuclei (later unite to form secondary nucleus or definitive nucleus) which are present at centre. It has highly vacuolated cytoplasm.

(b) Antipodal cells: These are three in number and present at chalazal end. These are rich in nutrients. These provide nutrition to developing embryo sac and endosperm. These degenerate before or during or after fertilisation.

(c) **Egg**: It is one in number and present at micropylar end. It is partly embraced by synergid cells at lateral side. The nucleus is much larger than synergid cell and lies towards chalazal end and vacuole present towards micropylar end. Egg shows common walls with two synergid cells and central cell.

(d) **Synergid cells**: These are two in number and lie side by side of egg. The nucleus is present at centre. It develops several filiform apparatus from cell wall to distribute nutrients in embryo sac, to secrete chemotropic substance for attracting or directing pollen tube into synergids towards egg for fertilisation and to provide mechanical strength to synergid cells. The synergid cells usually degenerate before fertilisation and hence degenerated synergids (either one or both) form suitable space for pollen tube discharge.

(b) *Oenothera* type: It was discovered by Geerts (1908) in *Oenothera lamarckiana*. It develops from micropylar megaspore. It undergoes two successive mitotic divisions without cytokinesis forming four nucleated embryo sac. There is no antipodal cell. Polar nucleus is one hence endosperm is diploid in this case.

2. **Bisporic embryo sac**: It is formed from two of the four megaspores. It is of two types:
   (a) *Allium* type: It was discovered by Strasburger in *Allium fistulosum.* It develops from chalazal dyad cell and is 8 nucleated.
   (b) *Endymion* type: It was discovered by Swamy in *Schisandra chinensis*. It develops from micropylar dyad cell and is 8 nucleated.
3. **Tetrasporic Embryo sac**: All four megaspore nuclei remain in cytoplasm forming coenomegaspore because megaspore mother cell undergoes meiotic division which is not followed by cytokinesis. It is more heterogeneous because four products of meiosis are genetically different. It is of several types:
   (a) *Peperomia* type: It is 16 nucleated.
   (b) *Paenaea* type: It is 16 nucleated.
   (c) *Drusa* type: It is 16 nucleated.
   (d) *Fritillaria* type: It is 8 nucleated.
   (e) *Plumbagella* type: It is 4 nucleated.
   (f) *Plumbago* type: It is 8 nucleated.
   (g) *Adoxa* type: It is 8 nucleated.

Terms:

1. Bambacioni effect: It is fusion of spindles of three chalazal nuclei in the coenomegaspore to form a triploid nucleus.
2. Nemec phenomenon: Nemec (1898) reported 8 nucleated embryo sac in pollen grain of *Hyacinthus orientalis* (a member of Liliaceae).

Pollination

It is transfer of pollen grains from anther to stigma. It occurs before fertilisation. It is exhibited by gymnosperms and angiosperms. It is essential for fertilisation, fruit and seed formation. It has freed the angiosperms from dependence on water. It also determines the suitability of pollen for carrying out sexual reproduction. Broadly it is of types

1. **Self-pollination (Autogamy):** It is transfer of pollen grains from anther to stigma within same flower. It is possible when anther and stigma are close to each other. It takes place when there is synchrony between pollen release and stigma receptivity. It occurs usually by wind or touch. It does not require external agents for carrying pollen grains. If self pollination is continued then it may lead to inbreeding depression.

   Adaptations to ensure self-pollination or to check cross-polllination (Inbreeding devices)

   (a) Bisexuality: It is a condition in which androecium and gynoecium are present in same flower. e.g., Members of Poaceae

   (b) Homogamy: It is a condition in which both anther and stigma of open (chasmogamous) bisexual flowers mature at same time. e.g., Periwinkle (*Catharanthus*), *Mirabilis*

   (c) Cleistogamy: It is a condition in which both anther and stigma of bisexual flowers remain closed (never opened) in some plants throughout their life cycle. It forms seed setting in spite of lack of pollinators. These plants show both chasmogamous (open) and cleistogamous (closed) flowers. e.g., *Commelina, Oxalis*, Common pansy (*Viola*).

   (d) Bud pollination: Both anther and stigma of bisexual flowers mature or open before anthesis (opening of bud).

   Advantages:

   (a) It maintains parental character throughout (purity of race).

   (b) It maintains pure lines for hybridisation experiment.

   (c) It eliminates bad recessive characters.

   (d) It saves pollen grains because it does not require large number of pollen grains.

   Disadvantages:

   (a) It decreases the chance of producing healthier and new varieties.

   (b) It causes loss of vigour and strength in progeny after every generation.

   (c) It reduces the adaptability in changed environment.

2. **Cross pollination (Allogamy):** It is the transfer of pollen grains from one flower to another. It is of two types:

   (a) Geitonogamy: It is transfer of pollen grains from anther of a flower to the stigma of another flower of the same plant. It is referred as genetically self pollination and ecologically (functionally) cross pollination.

   (b) Xenogamy: It is transfer of pollen grains from anther of one plant to stigma of another plant of same species. It is referred as genetically and ecologically cross pollination.

Adaptations to ensure cross pollination or to check self pollination: (Outbreeding devices)

(a) Unisexuality (Dicliny): It is the condition in which androecium and gynoecium are present on separate flowers. It is seen in some monoecious plants where male and female flowers are present separately on same plant ensuring only geitonogamy. e.g., Maize, Castor, Coconut, Cucurbits and all dioecious plants where male and female flowers are present on two separate plants ensuring only xenogamy. e.g., Date palm, Mulberry, Papaya, *Asparagus, Cannabis, Dioscorea, Fragaria, Populus, Rumex, Smilax, Vallisneria.*

(b) Self sterility (Self Incompatibilty): It is the failure of pollen grains to germinate on stigma after falling. It ensures only xenogamy. It may be between the plants of the same species or different species. It is of two types:

- Sporophytic self incompatibility: Rejection of pollen is controlled by the genotype of the sporophytic tissue of the plant from which pollen is derived.
- Gametophytic self incompatibility: Rejection of pollen is controlled by the genotype of male gametophyte. e.g., Tea, Tobacco, *Petunia*, Potato, Radish, *Malva, Passiflora*

(c) **Dichogamy**: It is the condition in which anther and stigma of a bisexual flower mature at different times. It may be

Protandry: Anther matures and sheds pollens earlier than stigma. e.g., Cotton, Marigold, Rose, *Salvia*, Sunflower

Protogyny: Stigma matures earlier (ten days before) than anther. e.g., *Aristolachia, Gloriosa, Ficus, Magnolia, Michelia, Plantago, Polyalthia.*

(d) **Heterostyly**: It is the condition in which stamen, style and stigma of bisexual flowers are of different length. Flowers may be

- **Dimorphic**: Two types of flowers are found. In one type, short stamen, and long carpel or pistil (Pin eyed) are seen and in another type, long stamen and short carpel or pistil (thrum eyed) are seen. e.g., Primrose, *Linum.*
- **Trimorphic**: Three types of flowers are found. In one type, carpel is short and stamens are long or intermediate; in second type, carpel is intermediate, and stamens are short or long; and in third type, carpel is long, and stamens are short or intermediate. e.g., *Lythrum, Oxalis*

(e) **Herkogamy**: It is a condition in which a natural barrier develops between anther and stigma to check self pollination and to ensure cross pollination. e.g., Gynostegium disc develops in *Calotropis*; Pollinia formation occurs in Orchid; Stigma lies inside flap in Pansy (*Viola tricolor*); Anther lies in corolla basket in *Kalmia*.

Advantages of cross pollination:

(a) It produces new varieties by bringing about genetic recombination.
(b) It produces healthy and disease resistant plants due to hybrid vigour.
(c) It increases yield.
(d) It makes the organisms better fitted or more adaptable in changed environment.
(e) It produces large, more and viable seeds.

Disadvantages of cross pollination:

(a) It requires large number of pollen grains.
(b) It does not eliminate undesirable characters.
(c) It wastes lots of energy and food materials in unnecessary adaptations.

Agents of pollination:

1. **Anemophily:** It occurs by wind. It is non-directional hence it is wasteful. Flowers are small, colourless, odourless and nectarless but packed in inflorescence. Anthers are well exposed. Pollen grains are small, dry, light, non-sticky and produced in large numbers so they are blown away. Stigma is large, well exposed and feathery (hairy) to catch pollen. e.g., Gymnosperms, Cereals, Grasses, Bamboo, Sugarcane, Coconut palm, Date palm, Papaya, *Cannabis*.
2. **Hydrophily**: It occurs by water. It is not very common and occurs only in few aquatic plants esp. monocots. Flowers are small, colourless, odourless and nectarless. Pollen grains are light, dry,

unwettable due to mucilaginous covering and released on water surface. Stigma is long, sticky, unwettable due to mucilaginous covering and released on water surface. It is of two types:

(a) **Hypohydrophily:** It occurs below water surface. It is seen in *Ceratophyllum, Najas, Zostera*. The male flowers of *Ceratophyllum* (monoecious) breaks at base, rises upwards, reaches to water surface and undergoes dehiscence. The liberated pollen grains germinate and start sinking. They come in contact with large and sticky swaying stigma of female flowers while sinking in water. Thus pollination takes place.

(b) **Epihydrophily:** It occurs on water surface. It is seen in *Vallisneria* (Tape grass/Ribbon weed). The male flowers of *Vallisneria* (dioecious) break from spadix and float on water surface but remain attached to female plant by long stalk. Female flowers have long sticky stigma and while floating male flowers come close to female flowers where anther bursts to release pollen. Thus pollination occurs.

3. **Entomophily:** It occurs by insects such as bees (most common), butterfly, moth, wasp, fly. The flowers are large, brightly coloured, odour full (fragrant), nectarfull. The pollen grains are rough and sticky due to pollenkit. Stigma is rough, sticky and unbranched. There are several mechanisms which are as follows:

(a) **Turn pipe /Lever mechanism:** It occurs by bees in *Salvia* plant. *Salvia* is bisexual and protandrous. It has bilipped or bilabiate corolla. A bee visits young flower and sits on lower lip in search of nectar. The pollen grains are deposited upon the back of bee. The pollen dusted bee visits older flower and its back rubs mature stigma thus bringing about pollination.

(b) **Trapdoor mechanism:** It occurs by Gall wasp (*Blastophaga*) in Fig. Wasp enters inside hypanthodium (inflorescence) to lay eggs inside. As wasp comes out of hypanthodium, it pushes pollen of male flower situated near the opening on female flower located at the base. Thus pollination occurs.

(c) **Pseudocopulation:** It occurs by hairy wasp (*Colpa aureus*) in *Ophrys* speculum (Orchid). *Ophrys* bears flowers which resemble the female wasp in colour, odour and appearance. Male wasp matures first and leaves the burrow before the females come out for mating. The inexperienced males try to pseudocopulate the *Ophrys* flowers presuming them as their female partners. Wasp carries pollinia fom one flower to another and brings about pollination. Coevolution (evolutionary modification of organisms in response to other organisms) occurs because *Ophrys* flower coevolves to maintain the similarity of its petal to female wasp.

(d) **Obligatory Symbiotic relationship:** It occurs by Pronuba moth (*Tageticula*) in *Yucca*. Coevolution occurs. Moth cannot complete the life cycle without association of *Yucca* and in turn, yucca has no other pollinator. Female moth visits flowers at night due to fragrance. It collects pollen grains in the form of balls and it deposits the egg in ovary of flower. After that it climbs to the top of style and pushes pollen ball into the hollow of stigma to bring about pollination.

4. **Ornithophily:** It occurs by birds (Crow, Parrot, Sunbird, Humming bird). Flowers are large, leathery, brightly coloured (yellow), odourless, and nectar full. e.g., *Agave, Bignonia*, Red Silk Cotton (*Bombax*), *Butea*, Bottle brush (*Callistemon*), Coral tree (*Erythrina*), *Sterlitzia.*

5. **Cheiropterophily:** It occurs by bats. Flowers are large, dull coloured, odourfull (fermenting or fruiting odour) and nectar full. Pollen grains are large in number. e.g., *Anthocephalus*, Baobab tree (*Adansonia*), *Bauhinia*, Sausage tree (*Kigella*).

6. **Malacophily**: It occurs by snail. e.g., *Colocasia*, *Lemna,* Lily

7. **Myrmecophily**: It occurs by ants. e.g., Guava, Litchi, Mango

8. **Ophiophily**: It occurs by snake. e.g., *Arisaemia, Michelia, Santalum*
9. **Cantherophily**: It occurs by Beetles. e.g., members of Rubiaceae

Pollen-Pistil interaction: It is the interaction between chemicals of pollen grains and those present on stigma. The interaction determines whether a pollen grain will germinate or not. It is dynamic process. It is essential for fertilisation.

The exine of Pollen grain is composed of sporopollennin. The cavities present in exine layer are filled with specific proteins (enzymes), glycoproteins, lipids and certain allergens derived from tapetum of anther during microsporogenesis. These chemicals carry some sporophytic recognition factors since these are derived mainly from tapetum. The intine of pollen grain, on the other hand, contains carbohydrate and proteins of gametophytic origin.

Stigma of angiosperms is wet as well as dry. Wet stigma secretes free floating amino acids, proteins and lipids, and dry stigma contains a hydrated layer of carbohydrate, proteins and lipids of sporophytic origin. Both these types of chemicals interact for the process of fertilisation.

Fertilisation

It is the fusion of male and female gamete to form zygote. Zygote is the first cell of sporophyte. It was discovered by Strasburger (1894) in *Monotrapa*.

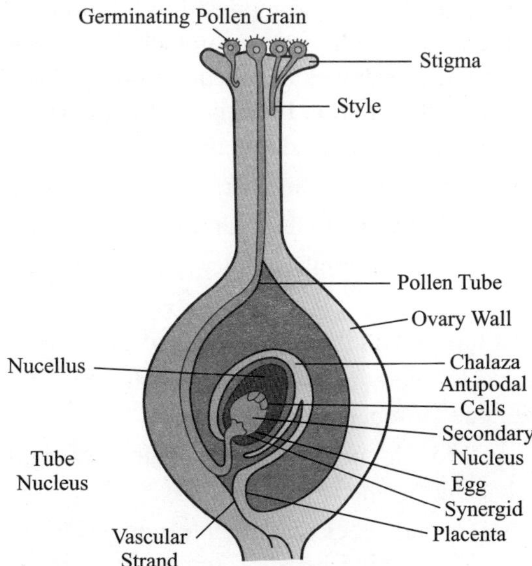

Fig. ***Mechanism of fertilisation***

Process:

(a) Pollen grains (2-celled stage in most of the plants) after maturing are transferred by different agencies fall on stigma. Only right pollen belonging to same species would be able to germinate and wrong or incompatible pollens are rejected by the stigma so they are unable to germinate and thus they perish. Pollen grains absorbs moisture from stigma and swells.

(b) Stigma secretes substances which protect pollen as well as stigma.

(c) Pollen grain germinates by rupturing exine to release pollen tube from intine through germ pore. High Relative humidity is first essential condition for pollen germination. Pollen tube was first observed by Amici in *Portulaca*. Pollen tube wall is three layered; outer pectin layer, middle pecto-cellulose layer and inner callose layer. Such fertilisation through pollen tube is called siphonogamous. Usually one pollen tube (monosiphonous) develops from a pollen grain.

(d) Growth of pollen tube is unidirectional and controlled by Ca^{+2}, a chemotropic agent present on stigma. The generative cell divides and forms two male gametes during growth of pollen tube in the stigma. In some plants in which pollen grains sheds at 3-celled stage, the pollen tube carries two male gametes from the very beginning. Pollen tube passes down to stigma and pollen grain remains on stigma. It grows initially at the expense of energy and food materials stored in pollen tube.

(e) Further growth of pollen tube occurs at the expense of cell of stigma and transmitting tissue of style. Pollen tube secretes pectinase and hydrolytic enzymes to move through style. Pollen tube moves intercellularly and chemotropically and apically along the concentration gradient of Ca^{+2}-borate-inositol-sugar complex. Boron prevents bursting of pollen. Style is either solid (dicots) or hollow.

(f) Pollen tube does not reach the ovary in a straight line. The pollen tube grows near style and curls to the bottom of the ovary and then near receptacle. The pollen tube then breaks into the ovule through the micropyle and then the micropyle bursts into the embryo sac.

(g) Growth of pollen tube in embryo sac is now guided by obturator. It develops from funicule of ovule. It is bridge on which pollen tube grows to reach ovule. It secretes surface exudates for nutrition of pollen tube and also provides chemical and mechanical guidance to pollen tube. It disappears after fertilisation. The duration between formation of pollen tube and its reaching to embryo sac varies between a few minutes to few months.

(h) The entry of pollen tube occurs in three manners:

- **Porogamy:** Pollen tube usually enters through micropyle. It is most common.
- **Mesogamy**: Pollen tube enters through middle part of integument (e.g., cucurbits) or funicle (e.g.,*Pistachia*).
- **Chalazogamy**: Pollen tube enters through chalaza. e.g., *Casuarina*

(i) Pollen tube is now attracted by secretions of synergid cells which develop filiform apparatus. The synergid cells degenerate due to impact of pollen tube.

(j) Pollen tube usually passes through one of the degenerated synergid cells. The tip of pollen tube enlarges and ruptures releasing two male gametes.

(k) Syngamy (true fertilisation) occurs when one of male gametes fertilises with egg and thus zygote is formed. It takes much longer time because egg is in inactive state. Syngamy (generative fertilisation) is of three types.

(a) Pre-mitotic: Fusion of male and female gamete occurs immediately. e.g., Asteraceae, Poaceae

(b) Intermediate: Fusion of male and female gamete is incomplete. It is most common.

(c) Post-mitotic: The nuclei of both male and female gamete remain in contact for a while and fuse after when both nuclei enter into division. e.g., *Lilium, Fritillaria*

Post fertilisation changes:

Before fertilisation	After fertilisation
Calyx, corolla and stamen	These fall off .
Style and stigma	These shrivel and wither away.

Ovary	It forms fruit.
Ovary wall	It develops pericarp.
Ovule	It forms seed.
Outer integument	It is converted into testa of seed.
Inner integument	It is converted into tegmen of seed.
Funicle	It forms stalk of seed.
Nucellus	Remnant of nucellus is present called perisperm. e.g., beet, black pepper, *Nymphaea*
Egg	It forms Zygote.
Two polar nuclei (Secondary nuclus)	It forms Endosperm.

Double fertilisation: It was discovered by Nawaschin (1898) in *Lilium* and *Fritillaria*. It occurs when another male gamete enters into central cell and fuses with two polar nuclei (secondary nucleus) forming primary endosperm nucleus (3n) which later develops into endosperm. This process is also called triple fusion or pseudo-fertilisation or trophomixis. It is characteristic feature of angiosperms. It takes shorter time because central cell is in active state. Triple fusion provides a stimulus to resume growth and form nutritive tissue to provide nutrition to developing embryo. Further it ensures that nutritive tissue is formed only when formation of embryo has taken place by syngamy.

Thus in the process of double fertilisation five nuclei are involved.

Syngamy = Male gamete + Egg → Zygote (2n) → Embryo

Triple fusion = Male gamete + 2 polar nuclei → Primary endosperm nucleus(3n) → Endosperm

Endosperm

It is nutritive tissue which is produced as a result of triple fusion. It develops from the central cell of embryo sac. It is characteristic feature of angiosperms except Orchidaceae, Podostomaceae and Trapaceae. In angiosperms, it is triploid because it is formed after fertilisation. It is haploid in gymnosperms because endosperm is represented by female gametophyte which is formed before fertilisation. It is starchy in cereals; proteinaceous in legumes (pulses) and oily in Castor and Coconut.

Types of Endosperm: On the basis of development:

1. **Nuclear endosperm:** It is most common and found in 56% of angiosperms.

   The primary endosperm nucleus (PEN) undergoes first and subsequent mitotic divisions without cytokinesis (wall formation). Thus, a large number of nuclei remain free in the central cell of embryosac. Now a large vacuole appears at the centre in central cell that pushes all the nuclei to peripheral cytoplasm. Later cytoplasm thickens hence size of vacuole is decreased and finally it disappears. Multinucleate cytoplasm is formed. Large sized nuclei are aggregated at micropylar and chalazal end and small sized nuclei are present at centre.

   Zygote undergoes resting period which is longer in nuclear endosperm.

   e.g., Coconut milk, Cereals, *Capsella*, Cotton, Sunflower

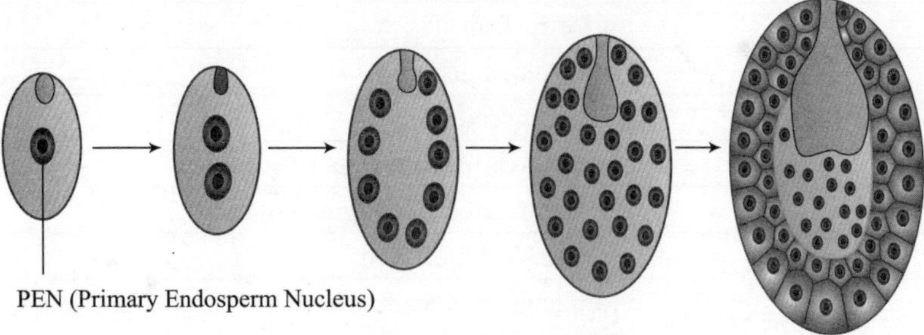

Fig. ***Nuclear endosperm***

2. **Cellular endosperm:** It is less common and found in 25% of angiosperms.

   The primary endosperm nucleus undergoes first and subsequent mitotic divisions accompanied by cytokinesis (with wall formation). The first division is usually transverse but subsequent divisions may be transverse or longitudinal or oblique. Finally all cells are uniformly present in central cell. It also shows the occurrence of haustoria.

   Zygote undergoes resting period which is shorter in cellular endosperm.

   e.g.,Coconut meal (copra or white kernel), *Adoxa*, Balsam, *Datura*, *Magnolia, Peperomia, Petunia, Utricularia* etc.

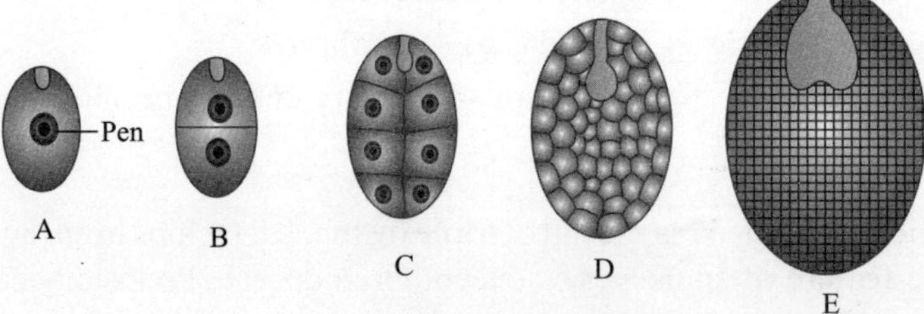

Fig. ***Cellular endosperm***

3. **Helobial endosperm:** It is very less common (19%). It is intermediate between nuclear and cellular endosperm. The primary endosperm nucleus undergoes first division which is transverse followed by cytokinesis (wall formation) to form larger micropylar cell and smaller chalazal cell. Micropylar cell undergoes further division without cytokinesis. Chalazal cell is either remain undivided or may undergo a few divisions to form haustoria.

   e.g., *Asphodelus, Saxifraga* and Helobiales order of Monocots.

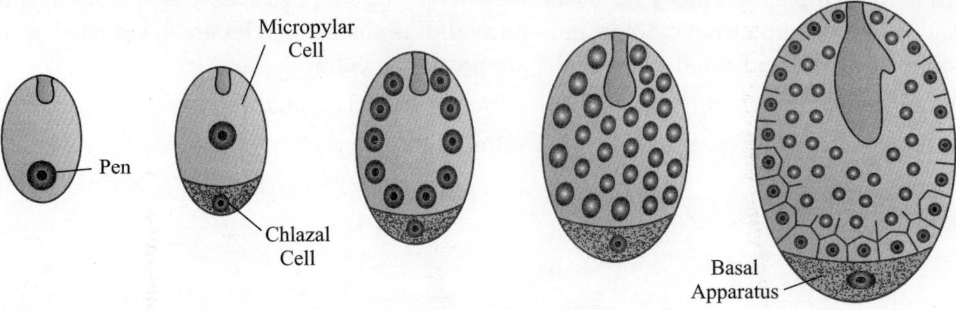

Fig. ***Helobial endosperm***

Types of endosperm:

1. **Ruminate endosperm**: The surface of endosperm is uneven due to ridges or furrows. It is formed due to uneven hard seed coat outer to soft endosperm. It is homogeneous. e.g., Members of Annonaceae, Aristolachiaceae, Arecaceae, Myristicaceae, Rubiaceae.
2. **Mosaic endosperm:** The surface of endosperm contains patches of different colours. It is heterogeneous. e.g., Maize

Ploidy level of endosperm

1. Triploid: n + 2n = 3n. It is most common.
2. Diploid: n + n = 2n. It is found in *Oenothera*
3. Pentaploid: n + 4n = 5n. It is found in *Penea, Plumbago, Plumbaginella, Fritillaria*
4. Nanoploid: n + 8n = 9n. It is found in Peperomia

Other terms

1. **Xenia**: It was coined by Focke. It is direct effect of pollen on endosperm (structure inside embryo sac except embryo). It is seen in maize.
2. **Metaxenia:** It was coined by Swingle. It is effect of pollen on tissue outside embryo sac or endosperm. It is seen in date Palm.

Fate of Endosperm:

1. It is either consumed by developing embryo before seed maturation.
2. It may persist in the seed hence seed is endospermic (albuminous).

Functions:

1. It provides nutrition to developing embryo.
2. It stores carbohydrate, protein and fats which are used in the establishment of seedling during seed germination.

Embryogenesis

It is the development of embryo from zygote. It occurs when endosperm development is complete. It occurs at the micropylar end of embryo sac where zygote is situated. Zygote undergoes first division which is transverse forming terminal cell and basal cell. Panchanan Maheshwari recognised five types of embryos on the basis of plane of division of terminal cell or apical cell (embryonal cell) in 2-celled embryo.

1. The terminal cell of 2-celled proembryo divides longitudinally
   (a) **Crucifer or Onagrad type**: The terminal cell takes part in the development of embryo but basal cell plays a minor role.
   (b) Asterad type: Both terminal cell and basal cell are involved.
2. The terminal cell of 2-celled proembryo divides transversally.
   (a) Caryophyllad type: The terminal cell is involved in the development of embryo but basal cell does not undergo further division.
   (b) Solanad type: The terminal cell is involved in the development of embryo but basal cell forms only suspensor.
   (c) Chenopodiad type: Both terminal cell and basal cell are involved in the development of embryo.

Development of Dicot embryo (Crucifer type): It is typically studied in shepherd's purse (*Capsella bursa-pastoris*). It involves following steps:

(a) Zygote increases its size and secretes its own cellulosic cell wall.

(b) Zygote divides transversally but unequally forming larger basal cell and smaller terminal cell towards micropylar end.

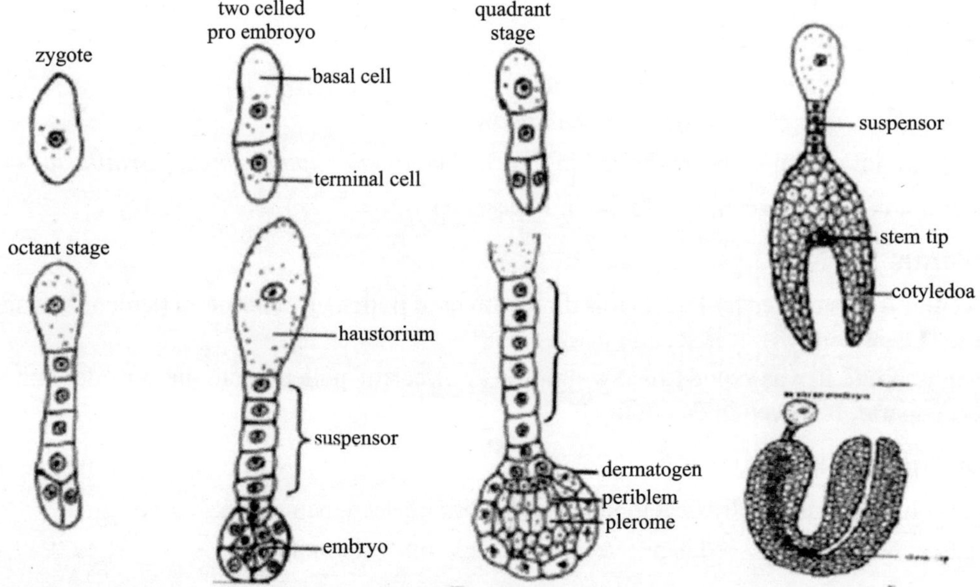

Fig. ***Developmental stages in dicot embryo***

(a) Basal cell divides transversally three times forming 8-celled suspensor. The first cell of suspensor towards chalaza becomes swollen and functions as haustorium, Suspensor is short lived structure and is concerned with nutrition of proembryo. Suspensor pushes the embryo in the endosperm. The last cell of suspensor towards chalaza is called hypophysis. It gives rise to root apex which later forms radicle and root cap.

(b) Terminal cell divides longitudinally twice forming 4-celled quadrant. The quadrant again divides transversally forming 8-celled octant. The octant divides periclinally forming two tiers of octant stage-epibasal (terminal) and hypobasal (near suspensor). The epibasal cells form two cotyledons and plumule. The hypobasal cells form hypocotyl except at its tip.

(c) Embryo is initially circular (globular) and undifferentiated. It changes to heart shaped then torpedo shaped and finally become mature.

(d) Mature embryo consists of embryonal axis and two cotyledons. The part of axis below cotyledon is called hypocotyl which at lower end terminates into radicle (embryonic root). The part of axis above cotyledon forms epicotyl which terminates into plumule (embryonic shoot).

(e) The development of embryo is endoscopic (embryo is pushed inside the interior of embryo sac or female gametophyte.

Development of Monocot embryo: It involves the following steps:

(a) Zygote divides transversally forming terminal cell and basal cell.

(b) Basal cell towards chalazal end does not divide further and gives rise to large and swollen vesicular suspensor which functions as haustoria to draw nutrition from endosperm.

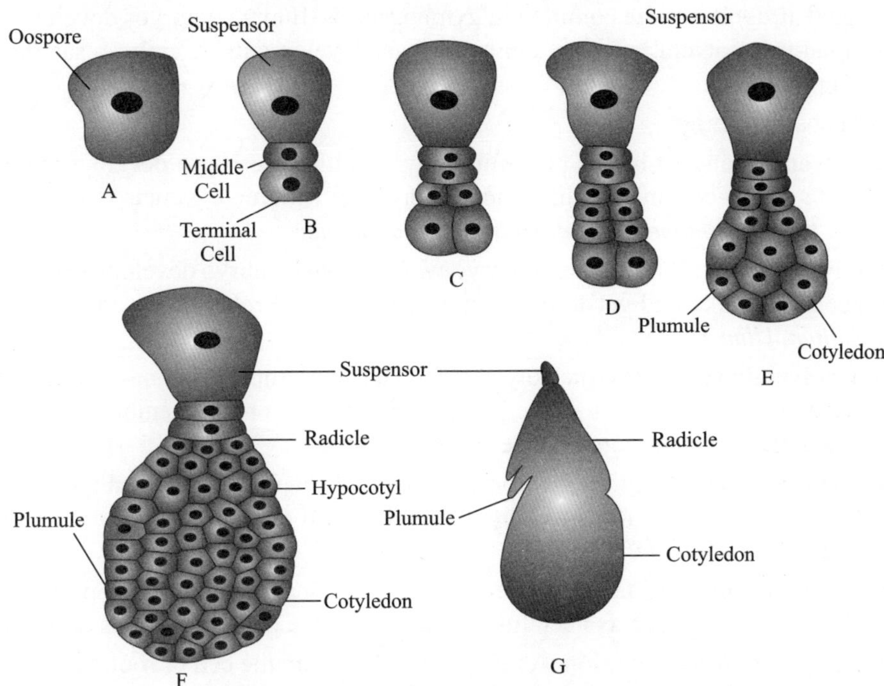

Fig. ***Developmental stages in monocot embryo***

(c) Terminal cell divides transversally forming two cells-top cell and middle cell. The top cell undergoes longitudinal division and subsequently a series of divisions forming plumule and one cotyledon. The middle cell undergoes transverse division and later several divisions forming hypocotyls and radicle. It also adds a few cells to suspensor.

(d) Cotyledon called as scutellum (shield-shaped) grows rapidly and pushes plumule laterally on one side. The part of axis above the level of attachment of scutellum is called epicotyl.

(e) Radicle is covered by a sheath called coleorhiza and plumule is covered by a sheath called coleoptile.

Dicot embryo	Monocot embryo
Both terminal cell and basal cell form embryo.	Only terminal cell forms embryo.
Basal cell divides to form 8-celled suspensor.	Basal cell does not divide further and forms 1-celled suspensor.
First division of terminal cell is longitudinal.	First division of terminal cell is transverse.
It has two lateral cotyledons.	It has one terminal cotyledon.
Plumule is terminal or distal between two cotyledons.	Plumule is lateral due to excessive growth of cotyledon.
Scutellum is absent (its homologous structure is cotyledon).	Scutellum is present.
Radicle is not covered by coleorhiza (absent).	Radicle is covered by coleorhiza.
Plumule is not covered by coleoptile (absent).	Plumule is covered by coleoptile.

Polyembryony

It is the presence of more than one embryo in one ovule/seed. It was discovered by Leeuwenhoek (1719) in Orange (Citrus).It is more common in gymnosperms. If extra embryos develop in the embryo sac then it is called true polyembryony. if extra embryos develop outside embryo sac then it is called false polyembryony.

Types: On the basis of origin

1. **Simple Polyembryony:** It is true polyembryony. Additional embryo develops from one or both synergid cells which become egg like and get fertilised with male gamete released by additional pollen tube. e.g., *Argemone, Aristolachia, Poa, Sagittaria.*
2. **Mixed polyembryony**: It is true polyembryony. Additional embryo develops from antipodal cells which become egg like and get fertilised with male gamete released by additional pollen tube. e.g., *Paspalum, Ulmus.*
3. **Cleavage polyembryony:** It is true polyemryony. It is the simplest and most common. Additional embryo develops by splitting of single proembryo into two or more embryos or by cleavage of zygote. It is more common in gymnosperms. e.g., orchids
4. **Multiple embryo sac**: Additional embryo develops because several embryo sacs are formed by differentiation of same megaspore mother cell or two or more megaspore mother cell. e.g., Brinjal, *Casuarina, Hydrilla.*
5. **Adventive embryony**: It is false polyembryony. Additional embryo develops by differentiation of diploid cells of ovule such as nucellus and integument. e.g., Citrus plants, Mango, *Opuntia.*
6. **Induced polyembryony**: Additional embryo develops from the cell of root, leaf, pollen grains in culture medium.

Causes:

1. Necrohormone theory: It was proposed by Haberlandt. The degenerative cells of nucellus act as stimulus to adjacent cells to divide and form adventives embryo.
2. Recessive gene theory: It was proposed by Kappert. One or more recessive genes cause polyembryony.

Importance:

(a) It produces uniform parental type of seedlings which are free from disease and produce better yield.

(b) It is better than those produced from cuttings (vegetative reproduction).

Apomixis

It is asexual production of seed (embryo) without fertilisation. The term 'Apomixis' was coined by Hans Winkler (1907). It occurs in members of Rutaceae, Asteraceae, Rosaceae, Poaceae. It is more common in polyploids than diploids.

Types:

1. **Nonrecurrent apomixes**: Meiosis occurs hence the nucleus of embryo is haploid. It is of two types:
   (a) Haploid apogamy: It is the development of haploid embryo from any cell of embryo sac such as synergid and antipodal cells. e.g., Tobacco, Orchid.
   (b) Haploid parthenogenesis: It is the development of haploid embryo from unfertilised egg. e.g., Banana, *Datura*, Maize.

2. **Recuurent apomixis**: Meiosis does not occur hence the nucleus of embryo is diploid. It is of two types:
   - **(a) Generative apospory (Diplospory or Agamospermy)**: It is the development of diploid embryo from megaspore mother cell (diploid). e.g., *Parthenium*
   - **(b) Somatic apospory (Adventive embryony or sporophyte budding)**: It is the development of diploid embryo from the diploid cell of ovule such as nucellus (e.g., *Acalypha*, Citrus plants, Mango, *Opuntia*) or integument (e.g., *Allium, Potentilla*).

Importance:

(a) It produces genetically uniform plants.

(b) It produces multiplication of desired characters in hybrids because there is no segregation and recombination due to absence of meiosis.

(c) It produces virus free scion materials.

(d) It bypasses the production of seeds.

(e) It provides expression of characters like hybrid vigour.

(f) It produces viable pollen.

(g) It has great scope in breeding experiment.

Parthenocarpy

It is production of seedless (sterile) fruit without pollination and fertilisation. The term 'Parthenocarpy' was coined by Noll (1902). It is commonly found in those plants which have large number of ovules per ovary such as Citrus plants, Grapes, Pineapple, Melon, Fig etc.

Types: Nitsch (1963) recognised four categories of parthenocarpy.

1. Natural Parthenocarpy: Parthenocarpic fruit occurs naturally. e.g., Banana, Grapes, Orange, Pineapple.
2. Genetic parthenocarpy: It occurs due to mutation in axillary bud or hybridisation (crossing) e.g., Navel orange, Cucurbits, Grapes.
3. Artificial or chemical parthenocarpy: It occurs when auxin was applied after anthesis thus ovary starts maturing or ripening. e.g., Tomato. It also occurs when gibberellin was applied at anthesis hence ovary starts ripening before fertilisation. e.g., Grapes
4. Environmental parthenocarpy: It occurs when sexual reproduction is checked due to low temperature, fog or frost hence sex organs do not function. e.g., Chilly, Pear, Olive.
5. Physiological parthenocarpy: It occurs due to delayed pollination or zygotic sterility.

Advantages:

(a) Seedless fruits have increased proportion of edible parts than normal fruits.

(b) These are widely used in fruit processing industry for preparing jam and jelly.

Disadvantage:

Parthenocarpy is undesirable in those plants where seeds are eaten such as Pomegranate, Cashew, *Pistachio*, Walnut.

SHORT ANSWER TYPE QUESTIONS

1. Write notes on essential whorl of flower.
2. Distinguish between glandular and amoeboid tapetum.
3. Distinguish between tetrahedral spore tetrad and isobilateral spore tetrad.
4. Define homogamy and cleistogamy with respect to self pollination.
5. Write notes on pollen viability.
6. Draw a well labelled diagram of anatropous ovule.
7. What are the advantages of cross pollination?
8. Distinguish between epihydrophily and hypohydrophily.
9. Write notes on pseudocopulation.
10. Write notes on pollen-pistil interaction.
11. Write notes on egg apparatus.
12. Discuss the post fertilisation changes in flower.
13. Write notes on double fertilisation.
14. Distinguish between dicot embryo and monocot embryo.
15. Distinguish between true polyembryony and false polyembryony.

LONG ANSWER TYPE QUESTIONS

1. Describe the development of anther in floerr.
2. Discuss the microsporogenesis and microgametogenesis in flower.
3. Describe the development of ovule in flower.
4. Discuss the megasporogenesis and megagametogenesis in flower.
5. What is cross pollination? Mention the different adaptations to ensure cross pollination.
6. Discuss the process of fertilisation with ell labelled diagram.
7. Desribe the development and types of endosperm.
8. Describe the development of dicot embryo.
9. Mention the different types of polyembryony with examples.
10. Mention the different types of apomixis with examples.

MULTIPLE CHOICE QUESTIONS

1. How many nuclei are involved in double fertilisation if tetraploid male plant is crossed with hexaploid female plant?
   (a) 7 (b) 9 (c) 11 (d) 13
2. In anatropous ovule the micropyle is
   (a) in straight line with funicle (b) at right angles with funicle
   (c) at 45° with funicle (d) side by side with funicle
3. Caruncle is derived from:
   (a) Peduncle (b) Cotyledon (c) Integument (d) Funicle
4. Modified shoot with highly condensed internodes is:
   (a) Leaf (b) Flower (c) Microsporophyll (d) Megasporophyll
5. Anthers are generally:
   (a) Monosporangiate (b) Bisporangiate (c) Tetrasporangiate (d) Trisporangiate
6. Pollen grains are preserved for long periods in fossils due to
   (a) Cellulose (b) Sporopollenin (c) Carotenoids (d) Ubisch bodies
7. Obturator in ovule is generally derived from
   (a) Placenta (b) Funiculus (c) Integument (d) Nucellus
8. The constant feature of embryo sac is
   (a) Synergids (b) Antipodals (c) Egg (d) Polar nuclei
9. Filiform apparatus is the feature of
   (a) Egg (b) Synergid cells (c) Antipodal cell (d) Zygote
10. The 2 polar nuclei of embryo sac are derived from
    (a) Both from micropylar end (b) Both from chalazal end
    (c) One each from micropylar and chalazal end (d) None of the above
11. The flowers which expose their mature anthers and stigma to the pollinating agents are called
    (a) Cleistogamous (b) Autogamous (c) Pseudogamous (d) Chasmogamous
12. The number of nuclei involved in double fertilization is
    (a) 3 (b) 4 (c) 5 (d) 7
13. The pollen-pistil rejection reaction in a plant showing gametophytic self-incompatibility usually occurs
    (a) on the stigma (b) in the style
    (c) in the ovary (d) during gametic union
14. Which type of spore tetrad is formed as aresult of successive type of cytokinesis?
    (a) Isobilateral (b) Tetrahedral (c) Decussate (d) T-shaped
15. The ovule found in gymnosperms is
    (a) Anatropous (b) Orthotropous (c) Amphitropous (d) Hemitropous
16. Perisperm is remaining part of
    (a) Endosperm (b) Ovule (c) Nucellus (d) Integument

17. How many nuclei are present in fully developed male gametophyte of flowering plants?
    (a) 1 (b) 2 (c) 3 (d) 4
18. Development of embryo sac from any cell of nucellus is called:
    (a) Apogamy (b) Apospory (c) Parthenogenesis (d) Diplospory
19. Tapetal cells in anther are:
    (a) Haploid (b) Polyploid (c) Diploid (d) Triploid
20. The stamen of flower represents:
    (a) Microporangium (b) Megasporangium (c) Megasorophyll (d) Microsporophyll
21. Endosperm in coconut is:
    (a) Cellular (b) Nuclear
    (c) Both cellular and nuclear (d) Neither cellular nor nuclear
22. In adventive embryony, embryos develop from
    (a) Endosperm (b) Synergids (c) Nucellus (d) Egg
23. When a diploid female plant is crossed with a tetraploid male plant, the ploidy of endosperm cells in the resulting seed is
    (a) Diploid (b) Triploid (c) Tetraploid (d) Pentaploid
24. At which stage, one can distinguish between nuclear and cellular type of endosperm?
    (a) Stage following division of primary endosperm nucleus
    (b) Mature stage of endosperm
    (c) When embryo has started divisions
    (d) At heart shaped stage of embryo
25. Megasorangium with its protective layers called integuments is called
    (a) Ovary (b) Ovule (c) Funicle (d) Chalaza
26. Morphologically and physiologically specialized outer layers of endosperm of cereals constitute:
    (a) Tapetel tissue (b) Perisperm (c) Aleurone layers (d) Chalazosperm
27. Exine of pollen is formed by activity of
    (a) Tapetum (b) Endothelium (c) Middle layers (d) Endothecium
28. Cells of tapetum are characterized by
    (a) Free cell formation (b) Meiotic divisions
    (c) Amitosis and endomitosis (d) Mitosis and endomitosis
29. Which is a chemotropic agent in gynoecium?
    (a) K^+ ions (b) Ca^{2+} ions (c) Mg^{2+} ions (d) Na^+ ions
30. Tectum, baculum and foot layer are found in
    (a) Pollen wall (b) Integuments of ovule
    (c) Pollen kitt (d) Embryo sac
31. If the number of chromosomes in egg cell is 8, then the number of chromosomes in endosperm is
    (a) 8 (b) 12 (c) 16 (d) 24
32. A close relation between flower and pollinating agent is best exhibited by
    (a) Cocos (b) Salvia (c) Yucca (d) Avena

33. If there are 4 cells in anther, what will be the number of pollen grains?
    (a) 8 (b) 4 (c) 16 (d) 12
34. In flowering plants, the male gametes are formed by the:
    (a) Generative cell (b) Vegetative cell
    (c) Pollen tube (d) Uninucleate microspore
35. Adventive embryony in Citrus is due to
    (a) Nucellus (b) Integument (c) Zygotic embryo (d) Fertilized egg
36. In a mature embryo sac, the central cell is
    (a) Binucleate (b) Single nucleate (c) Tetranucleate (d) Eight nucleate
37. The point of attachment of ovule to the funicle is known as:
    (a) Raphe (b) Chalaza (c) Micropyle (d) Hilum
38. Covering of microspores tetrad is made of:
    (a) Callose (b) Pectocellulose (c) Sporopollenin (d) Pectin
39. The aleurone layer in maize grain is especially rich in:
    (a) Starch (b) Lipid (c) Hormone (d) Protein
40. The type of pollen grains found in insect pollinated flowers is
    (a) Hygroscopic (b) Rough and sticky
    (c) Smooth and dry (d) Heavy and coloured
41. Which of the following pairs of plant parts are both haploid?
    (a) Nucellus and antipodal cells (b) Antipodal cells and megaspore mother cell
    (c) Nucellus and primary endosperm nucleus (d) Antipodal cell and egg nucleus
42. In fully organised *Polygonum* type of embryo sac, the ratio of haploid, diploid and triploid nuclei is
    (a) 3 : 1 : 3 (b) 6 : 0 : 1 (c) 6 : 1 : 0 (d) 3 : 2 : 3
43. The direction of micropyle in anatropous ovule is
    (a) Upward (b) Downward (c) Right (d) Left
44. Cellular totipotency is demonstrated by
    (a) Gymnosperms (b) All plants (c) Eukaryotes (d) Bacteria
45. Apomictic embryos in citrus plant arise from
    (a) Synergids (b) Maternal sporophytic tissue
    (c) Antipodals (d) Egg cell
46. Even in absence of pollinating agents, seed setting is assured in
    (a) *Commelina* (b) *Zostera* (c) *Salvia* (d) Fig
47. An organic substance that can withstand environmental extremes and cannot be degraded by any enzyme is:
    (a) Cuticle (b) Sporopollenin (c) Lignin (d) Cellulose
48. In the embryos of a typical dicot and monocot, true homologous structures are
    (a) Coleorhiza and coleoptile (b) Coleoptile and scutellum
    (c) Cotyledons and scutellum (d) Hypocotyl and radicle

49. Anthesis is a phenomenon which refers to:
    (a) Reception of pollen by stigma (b) Formation of pollen
    (c) Development of anther (d) Opening of flower bud
50. How many meiotic divisions are required for formation of 100 seeds?
    (a) 25 (b) 50 (c) 125 (d) 400
51. How many pollen grains (microspores) are produced by each head inflorescence of Asteraceae family having 10 actinomorphic flowers if each anther produces 20 pollen grains?
    (a) 500 (b) 800 (c) 1000 (d) 1200
52. A bilobed dithecous anther has 100 microspore mother cells per microsporangium, then how many male gametophyte s are produced from this anther
    (a) 400 (b) 800 (c) 1200 (d) 1600
53. The arrangement of megaspore tetrad in angiosperms is
    (a) Linear (b) Tetrahedral (c) Isobilateral (d) T-shaped
54. The number of female nuclei involved in double fertilisation is
    (a) 1 (b) 2 (c) 3 (d) 4
55. Which of the following parts of gynoecium determines the compatible nature of pollen grain?
    (a) Ovary (b) Pistil (c) Egg (d) Ovule
56. In angiosperms, the functional megaspore develops into
    (a) Ovule (b) Embryo sac (c) Endosperm (d) Pollen sac
57. Long ribbon-like pollen grains are seen in
    (a) Wind pollinated plants (b) Bird pollinated plants
    (c) Gymnposperms (d) All of the above
58. Both autogamy and geitonogamy are prevented in
    (a) Papaya (b) Castor (c) Maize (d) Cucurbits
59. A pollen mother cell undergoes microsporogenesis and forms 4 microspores. Each of these forms 2 male gametes. How many meiosis and mitosis are required for the formation of 4 male gametophytes from a pollen mother cell?
    (a) 1 meiosis and 4 mitosis (b) 2 meiosis and 4 mitosis
    (c) 1 meiosis and 8 mitosis (d) 2 meiosis and 4 mitosis
60. One celled suspensor is found in
    (a) *Triticum* (b) *Mangifera* (c) *Pisum* (d) *Solanum*
61. Which of the plants does not show clear cut vegetative, reproductive and senescent phases?
    (a) Annuals (b) Biennials (c) Perennials (d) Both a and b
62. The proximal part of filament of stamen is attached to
    (a) Thalamus or sepal (b) Thalamus or petal (c) Pedicel (d) Ovary
63. In a typical embryo sac, the nuclei are arranged as
    (a) 3 + 2 + 3 (b) 3 + 3 + 2 (c) 2 + 3 + 3 (d) 2 + 4 + 2
64. Long silky hairs coming out of cob of maize are meant for
    (a) Fruit dispersal (b) Attracting insects
    (c) Protecting seeds (d) Long styles and stigma

65. Emasculation is not required when flowers are
 (a) Unisexual (b) Bisexual (c) Intersexual (d) Both b and c
66. Perisperm differs from endosperm in
 (a) Having no reserve food
 (b) Being a diploid tissue
 (c) Being a haploid tissue
 (d) Formation by fusion of secondary nucleus with several sperms
67. The number of male gametes formed by 16 microspore mother cell is
 (a) 16 (b) 32 (c) 64 (d) 128
68. Emasculation is not required for artificial hybridisation in
 (a) Castor (b) Maize (c) Papaya (d) All of the above
69. Aleurone layer is outermost layer of endosperm in cereals. It helps in
 (a) Protection of endosperm (b) Growth of endosperm
 (c) Mobilisation of reserve food in endosperm (d) Accumulation of reserve food in endosperm
70. Plants growing in harsh environment such as desert, sand dunes and arctic tundra often reproduce vegetatively because
 (a) There are few animals for pollination
 (b) Fruits would freeze or dry out in such condition
 (c) Vegetative reproduction is not as risky as making seeds
 (d) These are members of such families which reproduce asexually.
71. Egg apparatus consists of
 (a) One egg cell and one synergid cell (b) One egg cell and two synergid cells
 (c) Two egg cells and one synergid cell (d) One egg cell and several synergid cells
72. Unisexuality of flowers prevents
 (a) Autogamy but not geitonogamy (b) Autogamy and geitonogamy
 (c) Allogamy and geitonogamy (d) Geitonogamy but not xenogamy
73. Which of the following floral parts forms pericarp after fertilisation?
 (a) Outer integument (b) Inner integument (c) Nucellus (d) Ovary wall
74. Male gametophyte with least number of cells is present in
 (a) *Funaria* (b) *Pteris* (c) *Pinus* (d) *Pisum*
75. Pollen grains can be stored for several years in liquid nitrogen having temperature of
 (a) –50°C (b) –100°C (c) –150°C (d) –196°C
76. A dioecious flowering plant prevents both
 (a) Autogamy and xenogamy (b) Autogamy and geitonogamy
 (c) Geitonogamy and xenogamy (d) None of the above
77. Pseudocopulation occurs between
 (a) *Salvia* and Bee (b) *Ficus* and Gall wasp
 (c) *Yucca* and Pronuba moth (d) *Ophrys* and Hairy wasp

78. If a diploid male plant and tetraploid female plant are crossed, the ploidy level of endosperm shall be
    (a) Triploid (b) Tetraploid (c) Pentaploid (d) Hexaploid
79. Which of the following ovule shows straight line of micropyle and nucellus?
    (a) Anatropous (b) Orthotropous (c) Hemitropous (d) Amphitropous
80. Translator condition of pollen grains is seen in
    (a) Orchids (b) Cucurbits (c) *Calotropis* (d) *Mirabilis*

ANSWERS

1.d	2.b	3.c	4.b	5.c	6.b	7.b	8.a	9.b	10.c
11.d	12.c	13.b	14.a	15.b	16.c	17.c	18.b	19.b	20.d
21.c	22.c	23.c	24.a	25.b	26.c	27.a	28.d	29.b	30.a
31.d	32.c	33.c	34.a	35.a	36.a	37.d	38.a	39.d	40.b
41.d	42.a	43.b	44.b	45.b	46.a	47.b	48.c	49.d	50.c
51.c	52.d	53.a	54.c	55.b	56.b	57.b	58.a	59.c	60.a
61.c	62.b	63.a	64.d	65.a	66.b	67.d	68.d	69.c	70.c
71.b	72.a	73.d	74.d	75.d	76.b	77.d	78.c	79.b	80.c

15

Principles of Inheritance and Variation

Genetics is the study of inheritance and variation of characters from one generation to another. The term 'Genetics' was coined by William Bateson.

Classical genetics: It is study of phenotype leading to isolation of genes and study of DNA sequences.

Reverse Genetics: It is study of DNA sequences leading to determination of unknown phenotypic effect and its introduction in plant or animal.

Father of Classical Genetics	Gregor Johann Mendel
Father of Modern Genetics	Bateson
Father of Experimental Genetics	Thomas Hunt Morgan
Father of Cytogenetics	Muller
Father of Human and Biochemical Genetics	Archibald Garrod

Heredity is the transmission of characters from one generation (parents) to another (offspring) and fixed for a particular individual. It is based on 'like begets like'. Thus human beings are like human beings in overall characteristics. Although the offspring differ from their parents but they resemble much and inherit the parental characters.

Early theories of heredity:

1. **Pythagoras** (500 BC) proposed **vapour theory**. Each organ of male produced moist vapour during coitus which formed the body parts of embryo.
2. **Aristotle** (450 BC) proposed **fluid theory**. Men's semen was highly purified blood and woman's menstrual fluid (female semen) was not as pure as male semen. Both fluids combined during intercourse. Female semen provided inert substance for the formation of embryo and male semen provided form and vitality to embryo. Any defect in mixing up the fluids results in missing of characters of one or both parents.
3. **Empedocles** proposed that both parents contribute semen to the formation of embryo. Therefore, children show the features of both parents.
4. **Leeuwenhoek** (1677) observed the human sperm for the first time and termed it 'animalcules'. He proposed that each sperm has the capacity to develop into a new individual when introduced into the womb of woman where it could get nourishment. This was rejected because maternal characters also inherit in offspring.
5. **John Swammerdam** (1679) proposed **preformation theory**. The organism is already present or preformed in the sperm or egg in miniature form called homunculus which enlarged to form embryo.

6. **Maupertius** proposed **particulate theory**. Particles are found in the semen of each animal and the union of particles of both parents results into formation of new individual.
7. **Wolff** proposed **theory of epigenesis**. The organs of body were developed gradually from undifferentiated living materials but sperm and egg contains some definite substances.
8. **Charles Darwin** proposed **theory of pangenesis**. Pangenes or minute gemmules are produced in every cell, tissue and organ. All the pangenes of body collect in the gametes via blood and are passed on to the zygote where they guide the growth of different parts of the embryo to form an offspring. Hence offspring has a blend (mixture) of the characters of both parents. Several objections are as follows:
   (a) **Weismann** (1892) proposed **theory of continuity of germplasm.** He observed that the tail of mice was cut and studied up to 22 generations. He found that complete tail structure was still inherited. (if tail was removed then pangenes for the tail will not be available so the tail should not have developed in next generation). He assumed that germplasm is immortal and it alone is passed from generation to generation. Somatoplasm that forms the body is mortal and perishes when organism dies.
   (b) Characters do not blend or mix and differ from those of parents. Some characters of distant parent (grandparent) are expressed in the offspring. It shows that characters may remain unexpressed for one or more generations but do not blend.
   (c) The character of sex does not blend itself in unisexual organisms. Such organism can either be male or female.
   (d) **Kolreuter** (1760) obtained fertile interspecific hybrids in Tobacco. These do not resemble either of parents. Hybrids were self-pollinated. Some offspring resembled the hybrid while remaining resembled one of the grandparents in different characters. This proves that traits have particulate nature and remain discrete.

Variation: It is the degree of difference in the progeny or between progeny and parent. It is also referred to a single difference in a trait. The difference in characters of offspring mainly depends on unique process of crossing over which occurs during meiosis. Variation occurs commonly in sexually reproducing organisms. It does not occur in identical twins and asexually reproducing organisms. Not all organisms are similar to their parents. Certain characters appear to belong to neither of parents. These dissimilarities are called variations. Variations are found in all characters and in every direction. Variation becomes fewer if there is intensive breeding (mating between closely related organisms).

Types of variation: On the basis of part

1. Meristic variation: It is change in number. e.g., Polydactyly
2. Substantive variation: It is change in appearance (shape, size, colour, weight).

Types of Variation: On the basis of nature

1. Phenotypic variation: It is observable.
2. Genotypic variation: It is constitutive.

Types of variation: On the basis of cells affected

1. **Somatic variations:** These occur in somatic or body cells of organisms. These are also called modification or acquired characters because these are got by individual after birth during its life time. These are non-inheritable because these disappear with the death of individual. These occur due to impact of environmental factors such as **light** (If there is no light, plants become etiolated showing longer internodes, thinner and broader leaves) and in the presence of light, plants develop

more mechanical tissues and thicker smaller leaves), **temperature** (in low temperature, root system is not well developed, metabolic activity is reduced and rate of transpiration is decreased and in high temperature, root system is well developed, metabolic activity is increased and rate of transpiration is increased.) and **water** (in scarcity of water, rate of transpiration is reduced, retentivity of water (succulence) occurs, and there is formation of spines, small leaves, sunken stomata and thick coating on leaves and in the abundance of water there is luxuriant growth in plants). The somatic variation also occurs due to use and disuse of organs, and conscious efforts.

2. **Germinal variations:** These occur in germ cells or reproductive cells. These are heritable. These help organisms adapt in a particular environment. These are of two types:
   (a) **Continuous (quantitative) variations:** These are fluctuating in nature. These are already present in different organisms or races of a species. These are caused by **recombination** which occurs due to segregation of chromosomes during gamete formation in meiosis, random fertilisation (fusion of gametes) and crossing over in meiosis.
   (b) **Discontinuous (qualitative) variation:** These are sudden or unpredictable in nature. These form basis of evolution of new species. These are caused by **mutation** which occurs due to change in base sequence (structure) of gene, structural aberration in chromosomes and numerical aberration in chromosomes.

Significance of variations:

(a) Variations make organisms better fitted in struggle for existence.
(b) These help individuals adapt in changed environment.
(c) These provide raw materials for evolution of species.
(d) These develop new traits.
(e) These improve the races of plants and animals with respect to resistance, yield and growth.

Mendelian inheritance

Gregor Johann Mendel worked for 7 years from 1856 to 1863. He conducted experiments of hybridisation on 29000 pea plants. He read out his results of observations in Natural History Society of Brunn in 1856. He published his research paper in German language entitled "Experiments on Plant Hybridisation" in 1866 in Annual Proceedings of Natural History Society of Brunn. He also worked on Hawk weed (*Hieracium*) and Hyacinth Bean (*Dolichos lablab*) but did not get success. He died in 1884.

Mendel's experiments:

1. **Selection of pure parental plant**: Mendel selected Garden pea (*Pisum sativum*, 2n=14) because
   (a) Pea plant has distinct, well defined and contrasting characters.
   (b) It is bisexual and self-pollinated hence it is true breeding (purity of same trait or character is maintained from generation to generation).
   (c) It is much easier to cross pollinate (hybridise) the flowers by removing anthers (emasculation) before the maturity of pollen grains.
   (d) It is annual hence it has short life span.
   (e) It is easy to grow, maintain and handle.
   (f) It produces large number of seeds in one generation.

He selected seven traits each having two forms or variants of pure or true breeding variety of pea (initially he took 34 traits, then 22 traits and finally 7 traits). All the selected traits had easily distinguishable alternate traits. The genes for 7 pairs of characters (traits) are such that some are present on same chromosome but not linked. Thus these seven traits are situated only on four chromosomes (1st, 4th, 5th and 7th) as confirmed by Blixt.

Trait/character	Dominant form	Recessive form	Position on chromosome
Seed colour	Yellow (Y)	Green (y)	1
Seed shape	Round (R)	Wrinkled (r)	7
Flower colour	Purple (P)	White (p)	1
Flower position	Axial (A)	Terminal (a)	4
Pod colour	Green (G)	Yellow (g)	5
Pod shape	Inflated (I)	Constricted (i)	4
Plant height	Tall (T)	Short (t)	4

2. **Hybridisation of Pure pea plants for obtaining F_1 generation**: He performed reciprocal crosses between plants having contrasting form of a character such as tall and short. The flower on plant (tall) was emasculated (anthers were removed) and pollens from the flower of another plant (short) were dusted over the stigma of tall plant. Now flower was covered with paper bag to avoid entry of unwanted pollen. The seeds formed by crossing the plants were collected and sown. Hybrid plants (first filial or F_1) were produced next year.
3. **Self-pollination of Hybrid plants for obtaining F_2 generation**: Self-pollination was done in F_1 hybrid plants. The flowers were covered in paper bags to keep out foreign pollens. The seeds produced were collected and sown. Later plants of next generation (second filial or F_2) were produced and thus it was observed that seeds of tall plants produced tall plants and yellow seeds produced plants of yellow seeds and so on.

Findings of the experiments:

1. F_1plants of reciprocal crosses were similar.
2. F_1 plants were not intermediate between the two contrasting forms of a trait. Rather they resembled one parent in having a single alternate form.
3. Both the parental forms of a trait are expressed in F_2 generation.
4. One form of a trait which did not appear in F_1generation must be hidden or unexpressed.
5. The organism should possess two factors of each trait (**Principle of paired factors**).
6. Out of two factors (alleles) representing the particular trait, one is dominant and expresses itself in the hybrid or F_1 generation. The other factor or allele is recessive and does not show its effect (**Principle of dominance**).
7. There is no mixing of the two factors in the hybrid.
8. The two factors segregate or separate at the time of gamete formation and pass into different gametes. A gamete contains one factor of a pair. Thus Mendel predicted the occurrence of meiosis long before it was discovered. The gametes fuse randomly during fertilisation so that factors come together in new generation and express themselves.
9. The two forms of a trait appear in F_2 generation in the ratio of three dominant to one recessive, 3:1(monohybrid ratio) and out of dominant individuals, one is pure breeding (homozygous

dominant) and two are hybrids(heterozygous dominant). This is possible only when the two factors of each trait segregated during gamete formation (principle of segregation).

10. In a dihybrid cross, four types of plants are formed in F_2 generation, in which there are two parental and two recombinants. The ratio is 9 (both dominant): 3(one dominant second recessive): 3 (one recessive second dominant) : 1 (both recessive). It is dihybrid ratio. This shows that the factors or alleles of two traits assort independently (**Principle of independent assortment**).

Basic Features of Heredity

(a) All traits have alternative or contrasting forms or variants.

(b) Traits are developed in the individuals by discrete particulate entities which do not blend or change.

(c) One alternate form of a trait may be expressed more often than other.

(d) Alternate form of a trait may remain hidden for one or more generations and then reappear later in unchanged state.

(e) Particular form of two or more traits may occur together in one generation and separate in next generation.

(f) One alternate forms of trait present together in an individual is expressed more often than the other.

Basis of heredity:

1. **Physical basis**: Mendel proposed that heredity was controlled by factors (now called as genes) which are present in all the cells and are transferred to next generation through gametes. These factors or genes are physical basis of heredity.
2. **Chemical basis**: It is confirmed that heredity is controlled by genes which are situated in the chromosomes present in the nucleus of gametes. The genes are made up of DNA and protein. DNA functions as genetic material and forms chemical basis of heredity. Genes code for proteins which develops trait of organism.

Rediscovery of Mendel's work: His works remained unnoticed and unappreciated for 34 years due to following reasons:

(a) He was himself not sure of his findings because he did not get success in his earlier works.

(b) He proposed concept of stable and unblending discrete unit factor.

(c) He failed to prove the physical existence of factors.

(d) He worked ahead of his time because at that time Biologists were impressed with the theories of Darwin (Origin of Species,1859).

(e) He introduced statistical analysis which was totally new to that time.

(f) He failed to project his findings due to limited circulation of journal and lack of communication.

But the works of Mendel were found to be correct due to independent findings of these three scientists who were credited with the rediscovery of Mendelism.

1. **Hugo de Vries** of Holland worked on evening primrose (*Oenothera lamarckiana*).
2. **Carl Correns** of Germany worked on maize, and
3. **Erich Von Tschermak** of Austria worked on different flowering plants.

Finally Correns assigned Mendel's postulates the status of law. Bateson and Punnet found Mendel's works to be of universal application including animals.

Terms: Bateson coined the terms allele, F_1, F_2, homozygous, heterozygous, epistasis etc. Johannsen worked on princess variety of red kidney bean or rajma (*Phaseolus vulgaris*). He coined the terms gene, genotype and phenotype.

1. Unit factor: It is a gene which controls a trait.
2. Trait: It is inherited character or feature.
3. Allele or Allelomorph: It is two contrasting forms of a factor (gene) or it is any two forms of a gene present on a same locus in homologous pair of chromosomes and controlling same trait (height of plant) but producing different effects (tall or short).
4. Phenotype: It is external physical appearance of an individual.
5. Genotype: It is genetic constitution of an individual.
6. Dominant: It is a trait that will appear in the offspring. It will appear if a dominant allele is present.
7. Recessive: It is a trait that will appear either in the absence of dominant allele or in the presence of identical allele.
8. Homozygous: It is diploid condition where two alleles of a gene for a particular character are similar. It possesses dominant alleles or recessive alleles.
9. Heterozygous: It is diploid condition where two alleles of a gene for a particular character are dissimilar. It possesses both dominant and recessive alleles.
10. Hemizygous: It is a diploid condition where one allele of a gene is missing.
11. Cross: It is mating between two individuals leading to fusion of gametes.
12. Monohybrid cross: It is a cross involving one pair of contrasting character.
13. Dihybrid cross: It is a cross involving two pairs of characters.
14. Hybrid: It is an offspring produced from a cross between two genetically different plants.
15. Wild allele: It is original, common or widespread in population and dominant in expression.
16. Mutant allele: It is altered version, less common in population and recessive in expression.
17. Epistasis: It is a type of gene interaction when one gene masks or modifies the expression of another gene at distant locus. The former gene is called epistatic gene and latter gene is known as hypostatic gene. It is intergenic or non-allelic.
18. Punnet square: It is checker board (simple box like) developed by Punnet (1927). In this, female gametes are arranged vertically and male gametes are arranged horizontally. It shows the result of a cross between two organisms.
19. Phenocopy: It is the different appearance (phenotypes) of plants in different seasons of year such as growing season, flowering season, fruiting season or season of leaf fall. It is not hereditary.

Mendel's laws of Inheritance:

1. **Law of paired factors**: Each character is controlled by a gene that has at least two alleles. It shows monogenic inheritance. These two alleles (factors) are present on two homologous chromosomes at the same locus. These may represent the same (homozygous TT in case of pure tall plant, tt in case of pure dwarf plant) or different expression (heterozygous Tt in case of hybrid tall plant) of same trait.

   Exceptions: (a) Multiple alleles (b) Polygenic inheritance

2. **Law of dominance**: If two organisms of contrasting characters are crossed then only one form (dominant) of character is expressed in the offspring and the other form (recessive) remains unexpressed.

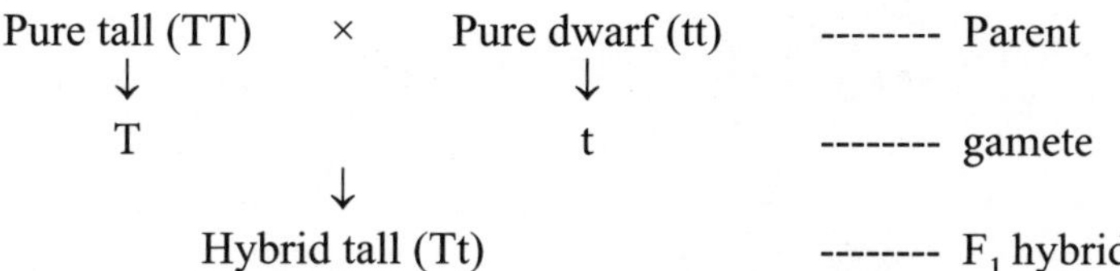

This law explains that individuals of F_1 express character of one parent. This law is not universally acceptable.

Exceptions: (a) Incomplete dominance (b) Codominance

3. **Law of segregation:** The factors (two alleles of a gene) of each character segregate or separate during gamete formation hence each gamete receives only one factor (allele) for each character.

It was explained by monohybrid cross.

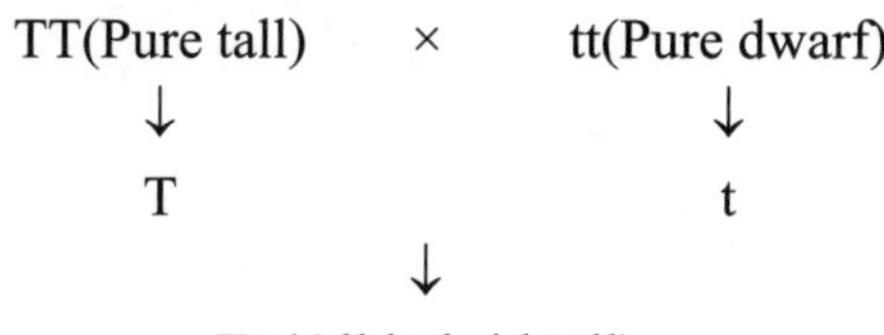

Selfing of F_1 hybrids (Tt x Tt)

Female/Male gametes	T	t
T	TT	Tt
t	Tt	tt

F_2 phenotypic ratio = 3 (tall) : 1 (short)

F_2 genotypic ratio = 1 (TT) : 2 (Tt) : 1 (tt)

(homozygous dominant) (heterozygous dominant) (homozygous recessive)

(pure tall) (hybrid tall) (pure dwarf)

This law is universally acceptable. It is applicable for diploid organisms because they form haploid gametes to reproduce sexually. It also explains that hereditary factors are discrete and do not blend when present together.

Bateson called this law as 'Principle of purity of gametes' because segregation of two Mendelian factors of a trait results in gamete receiving only one factor (allele) out of a pair. As a result, gametes are always pure for a character. It is also known as law of non-mixing of alleles.

In haploid organisms such as bacteria, algae and most fungi, the characters are governed by only one allele hence law of dominance and law of segregation do not apply for them. In these organisms, monohybrid ratio is 1:1 instead of 3:1 and dihybrid ratio is 1:1:1:1 instead of 9:3:3:1.

4. **Law of independent assortment**: When two individuals different from each other in two or more characters are crossed then the factors (alleles) of each character are distributed or inherited into the gametes independently of the factors of another character and anyone of a pair of characteristic may combine with anyone of another pair.

It is explained by dihybrid cross.

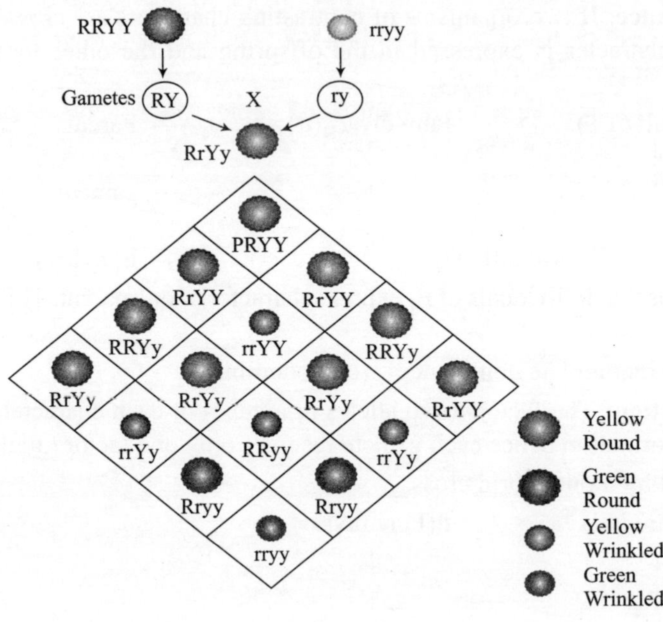

F_2 phenotypic ratio:	9	:	3	:	3	:	1
Number of Phenotypes = 4	Round yellow		Round green		Wrinkled yellow		Wrinkled green
F_2 genotypic ratio:	1 : 2	:	1 : 2	:	4 : 2	:	1 : 2 : 1

Number of genotypes = 9 (RRYY, RRYy, RRyy, RrYY, RrYy, Rryy, rrYY, rrYy rryy)

It produces four types of plants in F_2 generation. There are two or more parental types which show that the factors of each of two characters assort independently of others as if the other pair of factors are not present. The outcome of dihybrid cross makes it very clear that segregation of seed colour is independent of seed shape and both parental and new combinations of characters appear in F_2offspring. This law clearly states that genes for one trait are not inherited together with another trait.

It is applicable for two or more than two characters. It is applied to genes present far apart on same chromosome or unlinked genes present on different (nonhomologous) chromosomes. It does not apply for genes present on the same chromosome.

This is universally unacceptable.

Exception: Linkage

Mendelian crosses:

1. **Reciprocal cross:** It is a cross which involves the same trait but carried by sexes opposite to those in original or first cross. When two parents are crossed in such a way that in one experiment, it is used as male/female and in other experiment it is female/male but the result is same.

   TT(male) × tt (female) TT (female) × tt (male)

   ↓ ↓

   Tt Tt

   Reciprocal cross does not affect those characters which are controlled by nuclear genes. It affects those characters controlled by cytoplasmic genes.

2. **Back cross**: It is a cross between F_1 hybrid and either of the parents (homozygous dominant or homozygous recessive). It is performed to find out the heterozygosity of any genotype. It is also called top cross. It is of two types:
   (a) Out cross: It is a cross between F_1 hybrid and homozygous dominant parent.

Tt × TT

Female/male gametes	T	T
T	TT	TT
t	Tt	Tt

Results: All are tall (50% are pure tall and 50% are hybrid tall).

It is used by breeders to bring useful characters in homozygous condition.

   (b) Test cross: It is a cross between F_1 hybrid and homozygous recessive parent.

Tt × tt

Female/male gametes	t	t
T	Tt	Tt
t	tt	tt

Results: It is monohybrid test cross in which 50% are hybrid tall and 50% are short.

It helps to know the genotype of organisms. It is done to determine whether dominant character is due to homozygous or heterozygous condition. It is performed to find out the heterozygosity of F_1 individual.

It distinguishes (pure) homozygous dominant from (hybrid) heterozygous dominant.

Thus all test crosses are back cross but all back crosses are not test cross.

Monohybrid test cross ratio (Aa × aa)	1 : 1
Dihybrid test cross ratio (AaBb × aabb)	1 : 1 : 1 : 1
Trihybrid test cross ratio (AaBbCc × aabbcc)	1 : 1 : 1 : 1 : 1 : 1 : 1 : 1

Account of Mendel's Work

Experiment	Number of gametes = 2^n = n = number of heterozygote	Number of zygotes = $(\text{Gamete})^2$	Number of phenotype = 2^n where n = number of heterozygotes	Number of genotype = 3^n where n = number of heterozygote	Phenotypic ratio	Genotypic ratio
Monohybrid cross (Aa × Aa)	$2^n = 2^1 = 2$	$2^2 = 4$	$2^n = 2^1 = 2$	$3^n = 3^1 = 3$	3 : 1	1 : 2 : 1
Dihybrid cross (AaBb × AaBb)	$2^2 = 4$	$4^2 = 16$	$2^2 = 4$	$3^2 = 9$	9 : 3 : 3 : 1	1 : 2 : 1 : 2 : 4 : 2 : 1 : 2 : 1
Trihybrid cross (AaBbCc × AaBbCc)	$2^3 = 8$	$8^2 = 64$	$2^3 = 8$	$3^3 = 27$	27 : 9 : 9 : 9 : 3 : 3 : 3 : 1	$(1:2:1)^3$

Mendel got success due to following reasons which are as follows:

1. He selected only pure or true breeding varieties of garden pea. He took 2 years (1857-59) for checking that his experimental materials are true breeding.
2. He identified very clear contrasting characters in Pea plant.

3. He chose those traits which did not show linkage, incomplete dominance or any type of non-allelic gene interaction.
4. He concentrated on the inheritance of one particular trait with contrasting form at one time.
5. He studied the inheritance for three or more generations.
6. He performed reciprocal crosses.
7. He took care to avoid contamination from pollen grains brought about by insects.
8. He did not explain the variation found in his results but left them as such, e.g., linkage of flower colour and seed colour.
9. He maintained the complete and accurate record of every cross, subsequent selfing and number of seeds produced.
10. He used statistical methods and law of probability for analysing his results.

Exceptions to Mendelism or Post-Mendelian discoveries:

1. Incomplete dominance: It is a phenomenon in which both the alleles fail to express in F_1 hybrid or it is the failure of one allele to dominate other in heterozygous condition. It comes under intragenic interaction.

Examples:

(a) **Flower colour of 4'oclock plant (*Mirabilis jalapa*) and Snapdragon or Dog flower (*Antirrhinum majus*).** It was discovered by Correns (1903).

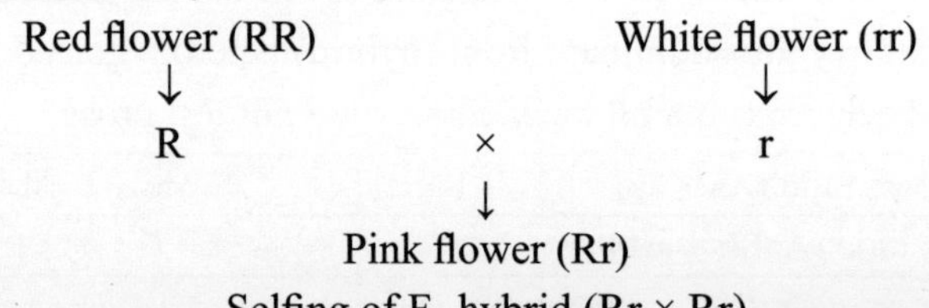

Female/male gametes	R	r
R	RR (red)	Rr (pink)
r	Rr (pink)	rr (white)

The genes for red and white colour do not actually mix in F_1 hybrid (Pink colour) as both parental characters (red and white) reappear due to segregation in F_2 plants. Hence it is not called blending inheritance on the basis of different F_2 progeny.

Findings:

(i) Hybrid has new phenotype (Pink) which is intermediate between those of two alleles. Hence it is called partial or semi-dominance. But the effect of one of the allele is more pronounced.

(ii) There is no specific gene for pink colour.

(iii) Phenotypic ratio is 1 (red) : 2 (Pink) : 1 (white) which differs from Mendelian ratio (3:1).

(iv) Genotypic ratio is 1 (RR) : 2 (Rr) : 1 (rr) which is same as Mendelian ratio.

(b) **Starch synthesis (size) in seed of pea.** It is not correlated with shape of seed. The homozygous dominant (BB) shows round shape and large size of grain, homozygous recessive (bb) shows wrinkled shape and small size of grain but heterozygous dominant (Bb) shows round shape and intermediate size of grain. Hence size of grain shows incomplete dominance.

Round seed (large grain) × Wrinkled seed (Small grain)
BB bb
↓
Round seed (intermediate size of grain)
Bb

(c) Size of leaf in Snapdragon (*Antirrhinum majus*)

Intermediate leaf

(d) Feather colour of Andalusian fowl

Blue plumage

2. Codominance: It is phenomenon where both alleles express themselves fully, equally, independently and simultaneously in heterozygous (F_1 hybrid) condition. It does not show dominant-recessive relationship or Mendelian interaction. It is also called mosaic inheritance.

Examples:

(a) (Coat (skin) colour of cattle.

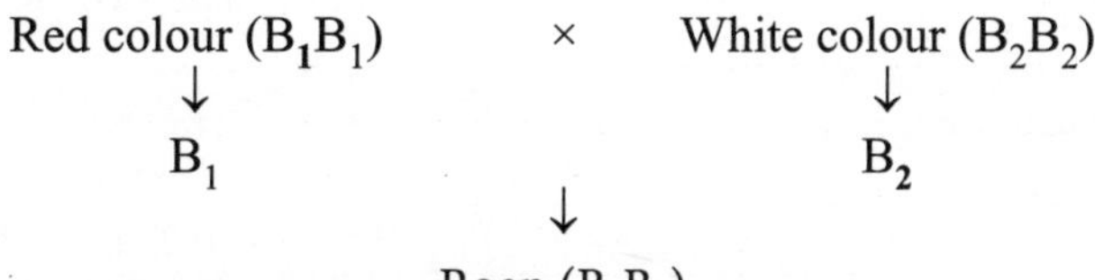

Roan (B_1B_2)

Inbreeding of F_1 hybrid ($B_1B_2 \times B_1B_2$)

Female/male gametes	B_1	B_2
B_1	B_1B_1 (Red)	B_1B_2 (Roan)
B_2	B_1B_2 (Roan)	B_2B_2 (White)

Findings:

(i) Both alleles produce phenotype in F_1 hybrid.

(ii) Phenotypic ratio: 1 (Red) : 2 (Roan) : 1 (White)

(iii) Genotypic ratio: 1 (B_1B_1) : 2 (B_1B_2) : 1 (B_2B_2)

(b) Sickle cell anaemia of human beings: The carrier individuals (heterozygotes) show codominance because HbA and HbS alleles are present. These are more tolerant of malaria than normal homozygotes.

Normal haemoglobin (Hb^AHb^A) × Sickle celled haemoglobin (Hb^SHb^S)
↓ ↓
Hb^A Hb^S
↓
Carrier (Hb^AHb^S)

Crossing of carrier individuals ($Hb^AHb^S \times Hb^AHb^S$

Female/male gametes	Hb^A	Hb^S
Hb^A	Hb^AHb^A (normal)	Hb^AHb^S (carrier)
Hb^S	Hb^AHb^S (carrier)	Hb^SHb^S (death)

(c) **MN** blood group individuals and **AB** blood group individuals.

3. Non-allelic or Intergenic interaction: According to Mendel, genes are functioning independently of each other, i.e. each of seven traits considered was controlled by a single gene. But many traits of an organism are determined by the complex contribution of many different genes. When two or more different genes (non-allelic) influence the outcome of a single trait, this is called gene interaction. In this two or more independent genes present on same or different chromosomes interact to produce a new expression. There are different forms of gene interaction:

Each gene pair affecting a same character:

(a) Complimentary genes (Duplicate recessive epistasis gene): There is complete dominance at both gene pairs, but either recessive homozygous gene is epistatic to the effects of the other gene. It occurs when recessive alleles at either of two loci mask the expression of dominant allele at the two loci.

Example: **Flower colour of Sweet pea** (*Lathyrus odoratus*)

Gene pair A: Purple dominant over white

Gene pair B:Colour dominant over white

Phenotypic classes	Genotypes
9 purple	AABB (1), AABb (2), AaBB (2), AaBb (4)
7 white	AAbb (1), Aabb(2), aaBB (1), aaBb (2), aabb (1)

Epistasis: aa is epistatic to B,b; bb is epistatic to A,a

(b) Supplementary genes (Recessive epistasis gene): There is complete dominance at both gene pairs but one gene, when homozygous recessive, is epistatic to the other.

Example:**Skin (coat) colour of Mouse**

Gene pair A: Colour is dominant over albino

Gene pair B: Agouti colour is dominant over black

Phenotypic classes	Genotypes
9 agouti	AABB (1), AABb (2), AaBB (2), AaBb (4)
3 black	AAbb (1), Aabb (2)
4 albino	aaBB (1), aaBb (2), aabb (1)

Epistasis: aa is epistatic to B,b

(c) Masking gene (Dominant epistasis gene): There is complete dominance at both gene pairs, but one gene, when dominant, is epistatic to the other gene.

Example: **Fruit colour in squash** (*Cucurbita pepo*).

Gene pair A: White is dominant over colour

Gene pair B: Yellow colour is dominant over green

Phenotypic classes	Genotypes
12 white	AABB(1),AABb(2), AaBB(2),AaBb(4), AAbb(1), Aabb(2)
3 yellow	aaBB(1), aaBb(2),
1 green	Aabb(1)

Epistasis: Ais epistatic to B,b

(d) Duplicate dominant epistasis gene: There is complete dominance at both gene pairs, but either gene, when dominant, is epistatic to the other gene. It occurs when a dominant allele at either of two loci masks the expression of recessive alleles at the two loci.

Example: **Shape of seed of shepherd's purse** (*Capsella bursa-pastoris*)

Gene pair A: Triangular shape is dominant over oval

Gene pair B: Triangular is dominant over oval.

Phenotypic classes	Genotypes
15 triangular	AABB (1),AABb (2), AaBB (2), AaBb (4), AAbb (1), Aabb (2), aaBB (1), aaBb (2)
1 oval	Aabb (1)

Epistasis: A is epistatic to B,b ; B is epistatic to A,a

4. Lethality: It is a condition in which the alleles created by mutations in these genes cause the death of organism. There are several types which are as follows:

(a) Absolute lethality: Death occurs in embryonic state.

(b) Sub-lethality: Death occurs at later stage before maturity.

(c) Delayed lethality: Death occurs after sexual maturity.

(d) Conditional lethality: Death occurs under certain environmental condition (temperature) only.

(e) Semilethality: Death occurs in only some individuals of population.

Lethal alleles are of two types:

(a) Dominant lethal alleles: These show lethal effects in homozygous condition. A single recessive allele in heterozygous condition can produce normal phenotype and survive but only structural changes occur. These are rare.

(b) Recessive lethal alleles: These show lethal effects only in homozygous condition.

Examples:

(a) Coat (skin) colour in mice. It was reported by Lucien Cuenot (1905). It is an example of dominant lethal.

Inbreeding of yellow mice (Yy x Yy)

Female/male gametes	Y	y
Y	YY(death)	Yy (yellow)
y	Yy (yellow)	yy (brown)

Findings:

(i) Death occurs in embryonic state (absolute lethality).

(ii) Lethal genes cause modification of 3:1 ratio to 2:1

(iii) Phenotypic ratio= 2 (yellow) : 1 (brown)

(b) Leaf colour of Snapdragon (*Antirrhinum majus*). It was reported by E. Baur (1907).
Selfing of golden leaved plant (Gg x Gg)

Female/male gametes	G	g
G	GG (death)	Gg (golden)
g	Gg (golden)	gg (green)

Findings:

(i) Death occurs before germination because it lacked the ability to make chlorophyll.

(ii) Lethal genes cause modification of 3 : 1 ratio to 2 : 1

(iii) Phenotypic ratio = 2 (golden) : 1 (green)

It is also observed in dexter gene in cattle, achondroplastic dwarfism in human being and xeroderma pigmentosum.

5. Pleiotropy: It is the ability of a gene to influence two or more unrelated phenotypic effects as it influences a number of characters simultaneously. It is against Mendelian interaction where a single gene controls a single trait or phenotype or character. It is not necessary that the characters are equally influenced. Sometimes, the effect of gene is more evident in case of one trait (major effect) and less evident in case of other traits (secondary effect). These unrelated traits are caused by the same protein produced by the same allele.

Examples:

(a) Pea: A gene controls the colour of flower, and the same gene also controls the colour of seed and red spot in leaves.

(b) Cotton: A gene controls the size of long fibre (lint), and the same gene also controls the size of plant, size of boll and number of seed.

(c) Drosophila: A recessive gene controls the size of vestigial wing, and the same gene also controls the length of body, bristles of wing, white colour of eye and structure of reproductive organs.

(d) Human beings: A recessive gene causes **sickle cell anaemia** and the same gene also causes increased resistance to malaria; effect on the development of tissues and organs (bone, heart, liver, and lung).

A recessive gene causes **phenylketonuria,** and the same gene also causes mental retardation, reduced hair and skin pigmentation.

A recessive gene causes **Marfan syndrome**. The same gene also causes abnormally tall height; thin fingers and toes; dislocation of lens of eye; heart problems.

Other examples of pleiotropy are albinism and schizophrenia.

6. Multiple alleles: It is the presence of more than two alleles of a gene. These alleles are located on same gene locus of same or homologous chromosomes. These produce different alternatives of a single trait. These alleles show dominance-recessive behaviour or codominance. These do not show crossing over because these are present on same locus. These are found in population.

Examples:

(a) ABO Blood group of human beings: There are four types of blood groups in human beings-A, B, O and AB. Landsteiner discovered A, B and O blood groups but de Castello and Steini (1902) discovered AB blood group.

Bernstein (1925) reported the blood group inheritance. Blood groups are determined by two types of antigens (A and B) present on the plasma membrane surface of RBC. The antigens occur on oligosaccharide rich head region of glycophorin.

Person of blood group A has antigen A; group B has antigen B; group AB has both antigens A and B while group O does not have antigen.

There are three immunogen alleles: I^A, I^B and I^o (symbolised as I, an abbreviation of Isohaemoagglutination) occupying the same locus determine the presence, absence and type of antigens. I^A forms antigen A; I^B forms antigen B while I^O does not form any antigen. All three alleles develop by mutation from a single ancestor. Now three allele of gene present on 9th chromosome are known.

First allele I^A codes for enzyme that adds a molecule, N-acetylglucosamine to glycoprotein resulting in 'A' blood group.

Second allele, I^B codes for enzyme that adds a molecule, galactose to glycoprotein resulting in 'B' blood group.

Third allele, I^O codes for non-functional enzyme that cannot add any sugar to glycoprotein.

I^A and I^B are dominant alleles and I^O is recessive allele. Both I^A and I^B are dominant over I^O but not over each other ($I^A = I^B > I^O$).

Case: Human being is a diploid organism, each person has any two of three alleles, one from each parent.

1. If both I^A and I^B are present in an individual then both express their own type of antigens in the presence of each other and person is said to be having AB blood group. Hence, both I^A and I^B behave as codominant alleles.
2. If both I^A and I^O are present in an individual, then only I^A will be expressed because I^O does not produce sugar. It shows dominant-recessive pattern of Mendelian inheritance.
3. If both I^B and I^O are present in an individual, then only I^B will be expressed because I^O does not produce sugar. It shows dominant-recessive pattern of Mendelian inheritance.

Blood group/ phenotype	Genotype	Antigen	Antibody	Reaction with anti-A	Reaction with anti-B	Compatible blood groups
A	I^AI^A, I^AI^O	A	b	+	-	A, O
B	I^BI^B, I^BI^O	B	a	-	+	B, O
O	I^OI^O	-	ab	-	-	A,B,O and AB
AB	I^AI^B	AB	-	+	+	O

Allele of first parent	Allele of second parent	Genotype of offspring	Blood group of offspring
I^A	I^A	I^AI^A	A
I^A	I^B	I^AI^B	AB (universal recipient)
I^A	I^O	I^AI^O	A
I^B	I^B	I^BI^B	B
I^B	I^O	I^BI^O	B
I^O	I^O	I^OI^O	O (universal donor)

The possible genotypes are determined by formula.

G = n (n + 1)/2 where n= number of alleles.

G = 3(3 + 1)/2 = 6 genotypes

The possible phenotypes are determined by formula

P = n + 1 where n = number of alleles.

P = 3 + 1 = 4 phenotypes (A,B,O and AB).

The phenotypes are tested by two antisera; ant-A and anti-B.

Determination of blood groups:

Blood group of parents	Genotypes	Possible blood group of offspring	Impossible blood group of offspring
A × A	$I^AI^O \times I^AI^O$	A,O	B,AB
A × B	$I^AI^O \times I^BI^O$	A,B,O,AB	--
A × O	$I^AI^O \times I^OI^O$	A,O	B,AB
A × AB	$I^AI^O \times I^AI^B$	A,B,AB	O
B × B	$I^BI^O \times I^BI^O$	B,O	A,AB
B × O	$I^BI^O \times I^OI^O$	B,O	A,AB
B × AB	$I^BI^O \times I^AI^B$	A,B,AB	O
O × O	$I^OI^O \times I^OI^O$	O	A,B,AB
O × AB	$I^OI^O \times I^AI^B$	A,B	O,AB
AB × AB	$I^AI^B \times I^AI^B$	A,B,AB	O

(b) **Coat (skin) colour of Rabbit**: It is controlled by four alleles. The possible genotypes are 10 and bpossible phenotypes are 5.

(c) **Eye colour of Drosophila**:

7. Polygenic inheritance: It is the inheritance of character controlled by two or more genes being expressed at one time. It was discovered by Nilsson-Ehle who studied the inheritance of grain colour in wheat. It is also known as multiple factors inheritance because it is additive effect of two or more genes on a single phenotype. Such genes which are non-allelic and effect the phenotype of a single quantitative trait, are called polygenes or cumulative genes.

Features:

(a) It produces wide range of phenotypes (variants/shades) hence called quantitative inheritance.

(b) It involves many genes at several different loci.

(c) It shows F_1 individual which is intermediate between two parents.

(d) The intensity of character depends on the number of dominant alleles.

(e) It shows continuous variation in the phenotype.

(f) It can be easily influenced by environment.

(g) The dominant alleles have cumulative effect and full trait is shown when all dominant alleles are present.

(h) The frequency of the trait follows the shape of bell curve.

(i) It is also expressed in metric trait expressed in length, weight and number.

(j) It controls skin colour, eye colour, hair colour, height, weight, body shape, intelligence, mechanical skills, personality and tendency to develop disease.

(k) Polygenes originate due to duplication of a part of chromosome, increase in the number of chromosome (polyploidy) and mutation.

Examples:

1. Skin colour in human beings: It was reported by Davenport (1913) in negro and white population of US.

(a) If skin colour is controlled by **two gene pairs** (each with 2 alleles). These genes are present at two different loci and each dominant gene is responsible for synthesis of fixed amount of pigment, melanin. The amount of melanin produced depends on the number of dominant gene.

Marriage between Negro and white

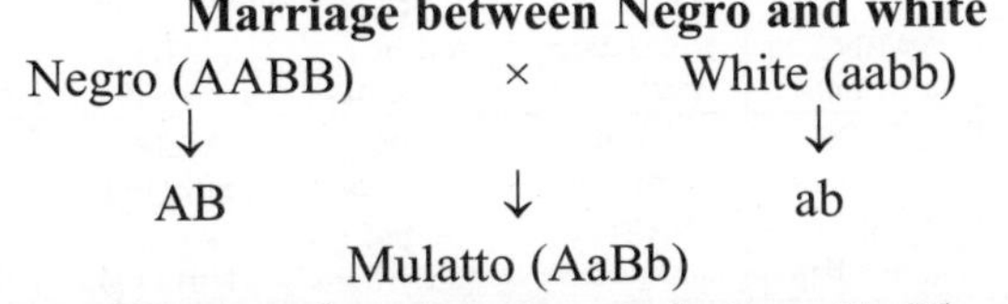

Marriage between two mulattoes (AaBb × AaBb)

Female / male gametes	AB	Ab	aB	ab
AB	AABB (Negro)	AABb (darker than Mulatto)	AaBB (darker than Mulatto)	AaBb (Mulatto)
Ab	AABb (darker than Mulatto)	AAbb (Mulatto)	AaBb (Mulatto	Aabb (lighter than Mulatto)
aB	AaBB (darker than Mulatto)	AaBb (Mulatto)	aaBB (Mulatto)	aaBb (lighter than Mulatto)
ab	AaBb (Mulatto)	Aabb (lighter than Mulatto)	aaBb (lighter than Mulatto)	Aabb (White)

Results:

(a) The number of intermediate type increases with increase in number of polygenes but the number of parental types remains same.

(b) Phenotypic ratio= **1** (negro): **4**(darker than mulatto): **6**(mulatto): **4**(lighter than mulatto) : **1**(white).

(c) The number of dominant alleles is 4(negro), 3 (darker than mulatto), 2 (mulatto) 1(lighter than mulatto) and 0 (white).

(d) Number of phenotypes: $2n + 1$ or $(2 \times 2) + 1 = 5$ where n=number of genes

(b) If skin colour is controlled by **3 pairs of genes**: Now it is considered that 3 pairs of genes are responsible for skin colour in human beings. The likely conditions are as such:

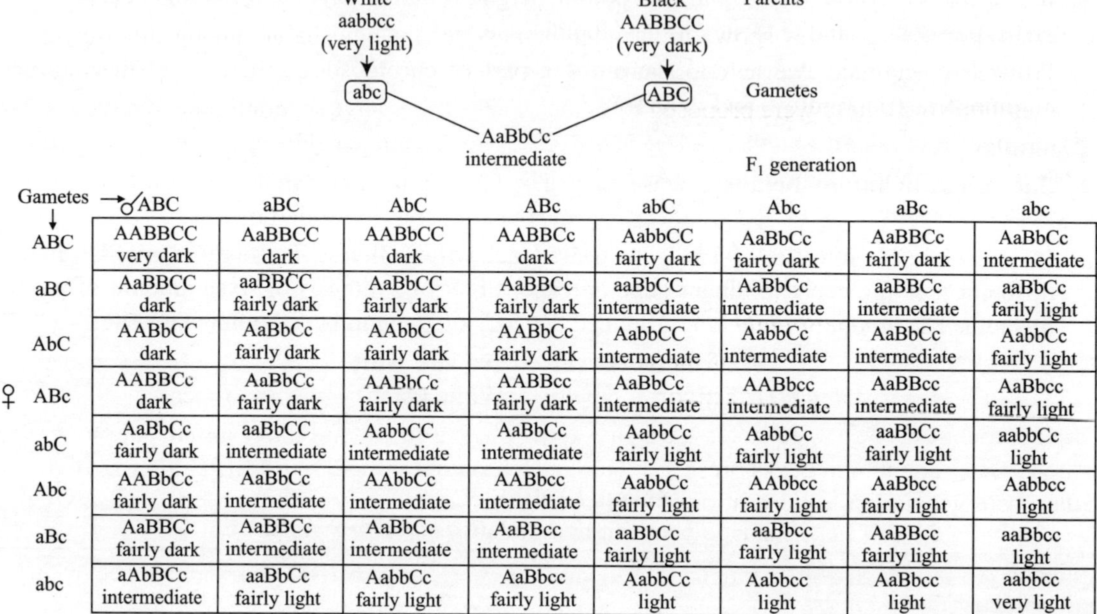

Gametes ♂ → / ♀ ↓	ABC	aBC	AbC	ABc	abC	Abc	aBc	abc
ABC	AABBCC very dark	AaBBCC dark	AABbCC dark	AABBCc dark	AabbCC fairty dark	AaBbCc fairty dark	AaBBCc fairly dark	AaBbCc intermediate
aBC	AaBBCC dark	aaBBCC fairly dark	AaBbCC fairly dark	AaBBCc fairly dark	aaBbCC intermediate	AaBbCc intermediate	aaBBCc intermediate	aaBbCc farily light
AbC	AABbCC dark	AaBbCc fairly dark	AAbbCC fairly dark	AABbCc fairly dark	AabbCC intermediate	AabbCc intermediate	AaBbCc intermediate	AabbCc fairly light
ABc	AABBCc dark	AaBbCc fairly dark	AABbCc fairly dark	AABBcc fairly dark	AaBbCc intermediate	AABbcc intermediate	AaBBcc intermediate	AaBbcc fairly light
abC	AaBbCc fairly dark	aaBbCC intermediate	AabbCC intermediate	AaBbCc intermediate	AabbCc fairly light	AabbCc fairly light	aaBbCc fairly light	aabbCc light
Abc	AABbCc fairly dark	AaBbCc intermediate	AAbbCc intermediate	AABbcc intermediate	AabbCc fairly light	AAbbcc fairly light	AaBbcc fairly light	Aabbcc light
aBc	AaBBCc fairly dark	AaBBCc intermediate	AaBbCc intermediate	AaBBcc intermediate	aaBbCc fairly light	aaBbcc fairly light	AaBBcc fairly light	aaBbcc light
abc	aAbBCc intermediate	aaBbCc fairly light	AabbCc fairly light	AaBbcc fairly light	AabbCc light	Aabbcc light	AaBbcc light	aabbcc very light

Phenotypes: Very Dark (Black)-1, Dark-6, Fairly Dark 15, Intermediate-20, Fairly light-15, Light-6, Very light(White)-1

Fig. ***Quantitative inheritance of skin colour in human beings***

Results:

(a) The number of intermediate type increases with increase in number of polygenes but the number of parental types remains same.

(b) Phenotypic ratio= **1**(very dark/negro) : **6** (darker than Mulatto) : **15** (fairly darker than mulatto) :**20** (intermediate/Mulatto) :**15** (fairly lighter than mulatto) : **6** (lighter than mulatto) : 1(very light/white).

(c) Number of phenotypes= 2n +1 or (2x3) +1=7

2. Kernel (grain) colour in wheat: It was reported by Nilsson-Ehle (1909). It is controlled by 2 gene pairs (each with 2 alleles).

Phenotypic ratio:1 (red) : 4 darker than intermediate) : 6 (intermediate) : 4(lighter than intermediates) : 1 (colourless).

3. Corolla length of *Nicotiana longiflora*: It was reported by Emerson and East (1916).

Qualitative genetics	Quantitative genetics
It deals with the inheritance of traits of kind such as colour, form, structure etc.	It deals with the inheritance of traits of degree such as height, weight, number etc.
It shows discontinuous variation	It shows continuous variation
Each qualitative trait is governed by two or more alleles of a single gene.	Each quantitative trait is governed by many non-allelic genes or polygenes.
The phenotypic expression of a gene is not influenced by environment.	The phenotypic expression of a gene is influenced by environment.
It is concerned with individual.	It is concerned with population.
Its analysis is done by counts and ratios.	Its analysis is done by statistical methods.

Terms:

(i) Isoalleles: These alleles show similar phenotypes but distinguishable among themselves after testing *Drosophila* shows several alleles for red eye colour.

(ii) Pseudoalleles: These were proposed by Fisher. These genes have separate gene loci lying side by side but often inherit together. These show close linkage and can be separated by rare crossing over. These are shown by

(a) Rh blood group of Drosophila. Rh antigen production is controlled by a set of 3 genes (CDE) that act as one gene. When any gene is dominant, the person is Rh- and when all the genes are recessive (cde), the person is Rh-.

(b) Eye colour of *Drosophila*

8. Linkage: It is the tendency of the genes situated in the same chromosome to stay together in hereditary transmission. It is an exception to Mendel's law of independent assortment.

There are so many genes present in a chromosome. The chromosomes move as units during meiosis into the gametes and all the genes present in the same chromosome will segregate as a group and travel together from generation to generation. The genes located in the same chromosome are called linked genes and those present in different chromosomes are called unlinked genes.

Linkage was explained by Morgan (1919) in *Drosophila* where red eyes and grey body colour occur in same individual because the genes of red eyes and grey body colour are located in the same chromosome and hence move together in the gamete and the next generation.

Types of linkage: Morgan divided linkage into two types:

1. Complete linkage: The genes remain linked together in the same chromosome for two or more generations and do not assort independently. It is shown by following cross.

Red eyed & normal winged *Drosophila* × purple eyed & vestigial winged *Drosophila*

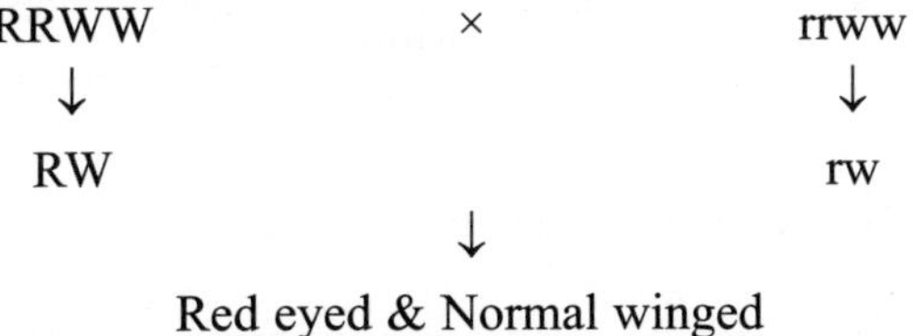

Red eyed & Normal winged

RW/rw

Test crossing (F_1 hybrid male with homozygous double recessive female)

RW/rw (male) × rrww(female)

↓

Female/male gametes	rw	Ratio	Phenotypes
RW	RWrw	1 (50%)	Red eyed & normal winged (parental)
rw	rwrw	1(50%)	Purple eyed & vestigial winged (parental)

Results:

(a) It shows that red eye colour and normal wing are inherited together because these genes are linked together.

(b) Purple eye colour and vestigial wing are inherited together because these genes are linked together.

(c) Only parental combinations are formed in the ratio of 1:1(50% each)

(d) Recombinants or non-parental types are not formed.

(e) There is wide departure from dihybrid test cross ratio(1:1:1:1) to 1:1 and dihybrid ratio (9:3:3:1) to 3:1.

2. Incomplete linkage: The genes do not remain linked on the same chromosome but separate due to crossing over to form new combinations. It is shown by following cross.

Red eyed & normal winged *Drosophila* × purple eyed & vestigial winged *Drosophila.*

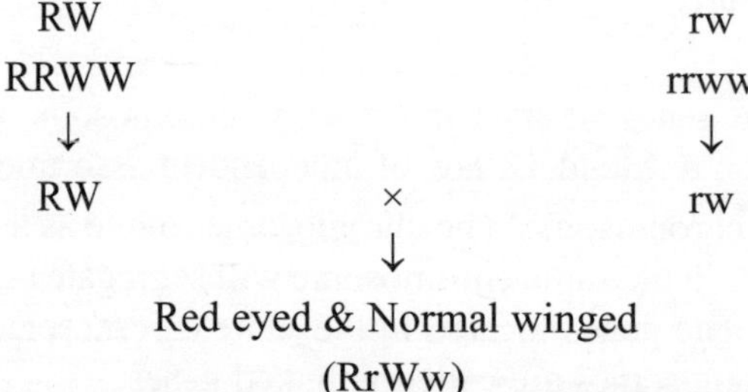

Test crossing (F_I hybrid female with homozygous double recessive male)

RrWw (female) × rrww (male)

↓

Female/male gametes	rw	Ratio	Phenotypes
RW	RrWw	7 (Parental)	Red eyed & normal winged
Rw	Rrww	1(Recombinant)	Red eyed & vestigial winged
rW	rrWw	1(Recombinant)	Purple eyed & normal winged
rw	rrww	7(Parental)	Purpleeyed & vestigial winged

Results:

(a) The genes are incompletely or partially linked hence parental progenies are more (>50%) than recombinants (<50%).

Complete linkage	Incomplete linkage
The genes closely located in the same chromosome remain linked in hereditary transmission and donot separate.	The genes distantly located in the same chromosome remain separate.
Crossing over does not occur between closely linked genes.	Crossing over occurs.
It produces only parental progenies.	It produces more (>50%) parental progenies than less (<50%) recombinants.
It occurs rarely. (e.g., Male *Drosophila* and female silk worm).	It commonly occurs.

3. Unlinked genes: The two genes are not linked because these are located in different chromosomes hence assort independently. It is shown by following cross.

Red eyed & normal winged *Drosophila* × Purple eyed & vestigial winged *Drosophila*

Red eyed & Normal winged

RrWw

Test crossing (F_I hybrid female with homozygous double recessive male)

RrWw (female) × rrww (male)

↓

Female/male gametes	rw	Ratio	Phenotypes
RW	RrWw	1(25%)	Red eyed & normal winged (parental)
Rw	Rrww	1 (25%)	Red eyed & vestigial winged(recombinant)
rW	rrWw	1(25%)	Purple eyed & normal winged(recombinant)
rw	rrww	1(25%)	Purpleeyed & vestigial winged(Parental)

Results:

(a) Four types of individuals are formed in equal proportions (25% each).

(b) Out of four progenies there are two parental and two recombinants.

Principles of linkage: It was discovered by Bateson and Punnet (1906) on sweet pea (*Lathyrus odoratus*) where two pairs of alleles donot show independent assortment. They explained linkage on the basis of

1. Coupling phase: It is observed when two dominant or recessive alleles coming together from same parent tend to enter the same gamete and inherit together. It is explained by following cross.

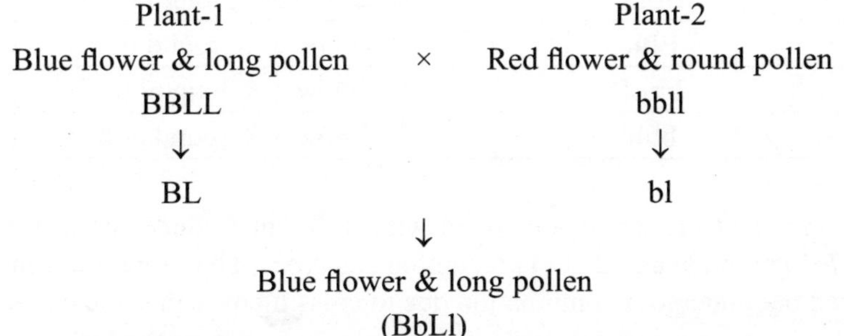

Test crossing (Blue flower & long pollen with red flower & round pollen)

BbLl × bbll

Female/male gametes	bl	Ratio	Phenotypes
BL	BbLl	7	Blue flower & long Pollen
Bl	Bbll	1	Blue flower & round pollen
bL	bbLl	1	Red flower & long pollen
bl	bbll	7	Red flower & round pollen

Results:

(a) Both parental and recombinant progenies are found with different frequencies in the ratio of 7 : 1 : 1 : 7. There are 14(7 + 7) parental and 2 (1 + 1) recombinants types. Therefore, the genes are linked but undergo recombination due to crossing over in some cases.

(b) Traits were coupled and not easily assorted in an independent manner.

(c) If genes of two characters are completely linked, then only parental progenies should have in the ratio of 1 : 1 (50% each).

(d) If genes of two characters are unlinked then four types of progenies should have in the ratio of 1 : 1 : 1 : 1 (25% each).

2. Repulsion phase: It is observed when two dominant and recessive alleles coming from different parent tend to enter the different gametes to remain separate in the offspring. The separation of linked genes occurs due to exchange of segments between a homologous pair of chromosome during meiosis.

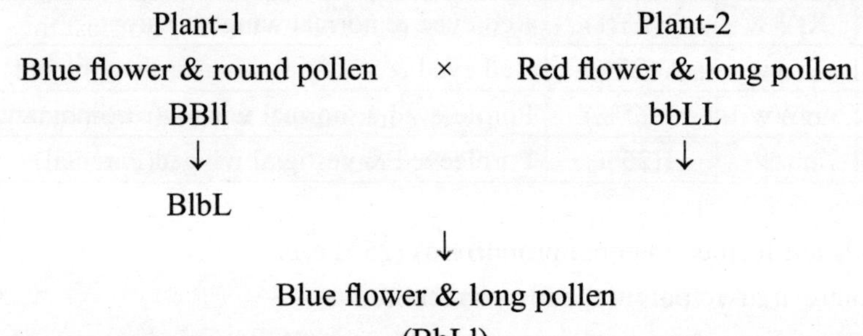

Test crossing (Blue flower & long pollen with red flower & round pollen)

BbLl × bbll

Female/male gametes	bl	Ratio	Phenotypes
BL	BbLl	1	Blue flower & long Pollen
Bl	Bbll	7	Blue flower & round pollen
bL	bbLl	7	Red flower & long pollen
bl	bbll	1	Red flower & round pollen

Results:

(a) Both parental and recombinant progenies are found with different frequencies in the ratio of 1:7:7:1. There are 14(7+7) parental and 2 (1+1) recombinants types. Therefore, the genes of two characters are not linked but undergo recombination due to crossing over in some cases.

Morgan suggested that two genes present on same chromosome are in coupling phase and when present on two different homologous chromosomes are in repulsion phase.

Arrangement of linked genes on chromosome: If there are 2 pairs of alleles (BbLl), then the possible arrangement may be

1. **Cis-arrangement**: It represents coupling phase. It shows BL/bl.

   Two dominant genes are present on same homologous chromosome.

B	**L**

   Two recessive genes are present on the other member of homologous chromosome

b	**l**

2. **Trans-arrangement**: It represents repulsion phase. It shows Bl/bL.

   One dominant and one recessive allele are located on one member of homologous pair of chromosome.

B	l

   One dominant and one recessive allele are located on another member of homologous pair of chromosome.

b	L

Contributions of Morgan:

1. He proposed classical theory of linkage.
   (a) Genes are located in chromosome and arranged in linear order.
   (b) Strength of linkage is determined by the distance between linked genes in chromosome.
   Strength of linkage = 1/distance between genes
   (c) Linked genes staying together in hereditary transmission are located in same chromosome.
   (d) Genes tend to stay in parental combination except for occasional crossing over.
2. He verified chromosomal theory of inheritance.
3. He discovered the linkage and distinguished between linked and unlinked genes.
4. He discovered the basis for variation due to sexual reproduction.
5. He proposed chiasma type hypothesis showing that formation of chiasmata shows the occurrence of crossing over.
6. He along with Sturtevant proposed that frequency of crossing over (recombination) between two linked genes is directly proportional to distance between two genes (1% recombination= 1 Centimorgan or 1 map unit)
6. He proposed criss-cross inheritance.

Importance:

1. Linkage maintains specific parental character because it reduces recombination.
2. It maintains good character of newly developed variety.
3. It provides strong proof for linear arrangement of genes on chromosome.

Disadvantages:

1. It dilutes the use of desirable trait if undesirable character is also present on same chromosome.
2. It disallows the plant and animal breeder to combine all desirable traits in a single variety.

Linkage group: It is a linearly arranged group of linked genes which are normally inherited together. It is equal to the haploid number of chromosomes found in a species and corresponds to the number of chromosomes in its one genome. It was discovered by Sutton (1903). In other words, it represents all the chromosomes if organism is haploid (fungi) or homologous pairs if organism is diploid. All the genes present on a particular chromosome form linkage group. The size of linkage group is directly proportional to the size of chromosomes. If number of chromosomes is less then linkage group will be small. It proves that genes are present on chromosome.

Diploid organisms	Number of chromosome	Number of linkage group
Pea	2n = 14	7
Wheat	2n = 42	21
Apple	2n = 34	17
Drosophila	2n = 8	4
Frog	2n = 26	13
Human	2n = 46	22 + XX = 23 in female 22 + X + Y = 24 in male
Haploid organisms		
Neurospora	n = 7	7
Mucor	n = 2	2
E.coli	n = 1 (nucleoid)	1

Linkage map (chromosome map or gene map): It is linear graphic representation of the sequence and the relative distance of various genes present in a chromosome. It was given by Sturtevant. It does not depict the physical distance among genes because of interference of one crossing over and heterochromatin.

The linkage map is easily detected in haploids (*Neurospora*) because

(a) There is single allele per locus hence products of recombination can be studied without any complication due to dominance.

(b) It is easy to analyse the products of single meiosis because these are linearly arranged in the form of ordered tetrad so it is possible to distinguish between first division segregation and second division segregation.

Significance of Linkage Map:

1. It locates the different genes in specific chromosome.
2. It shows linear arrangement of genes.
3. It shows the strength of linkage as well as occurrence and number of recombination.
4. It predicts the effect of loss of chromosome segment.

Problems:

1. If two heterozygous tall pea plants having genotypes (Tt) is crossed. What is the probability that two out of five plants will be dwarf ?

   Ans: It is solved by binomial expression

   $$\text{Probabilty} = \frac{n!(p^x q^{n-x})}{x!(n-x)!}$$

   n = Total number of plants = 5

   x = Number of events in one category (dwarf plant) = 2

   p = Individual probability of x (dwarf plant) = 1/4: obtained from cross (Tt × Tt)

   q = Individual probability of other category (tall plant) = 3/4: obtained from cross (Tt × Tt)

! = symbol of factorial (products of integar from down to one.

$$\frac{5!(1/4)^2 \times (3/4)^{5-2}}{2!(5-2)}$$

or $\frac{5 \times 4 \times 3 \times 2 \times 1}{2 \times 1(3 \times 2 \times 1)} \times \frac{1}{16} \times \frac{3 \times 3 \times 3}{4 \times 4 \times 4} = 0.26 = 26\%$

2. A heterozygous pea plant that is tall with yellow seeds (TtYy) is allowed to self fertilise then what is the probability that an offspring will be either tall with yellow seeds, tall with green seeds or dwarf with yellow seeds?

   It is solved by sum rule.

   According to dihybrid cross, Plants with tall and yellow seed = 9/16

   Plants with tall and green seeds = 3/16

   Plants with dwarf and yellow seeds = 3/16

   Sum rule = 9/16 + 3/16 + 3/16 = **15/16**

Crossing Over

It is mutual exchange of genetic material between two non-sister chromatids (adjacent paternal and maternal) of homologous chromosome.

It was discovered by Janssen (1909).

The term 'crossing over' was coined by Morgan.

Its physical basis was first demonstrated by Creighton and Mclintock.

Features:

(a) It is the separation of linked genes. If crossing over occurs, the genes separate and pass into different gametes and offspring. Thus linkage and crossing over are antagonistic to each other.

(b) It takes place at pachytene stage of prophase-I of meiosis.

(c) It occurs at chiasma at four stranded stage as proved by the 2:2:2:2 arrangement of ascospores in ascus of *Neurospora*.

(d) It occurs in germinal cells.

(e) It involves breakage and reunion at certain point.

(f) Usually it involves two of the four chromatids of two homologous chromosomes.

(g) Its value varies from 0 to 50%.

(h) Its value can be worked out from test cross progeny.

Proof: C.B. Hutchinson (1922) took colour and shape of grain (seed) of maize. He performed following crosses:

1. **Coupling cross**: He crossed a pure breeding coloured and smooth (rounded) grained variety of maize with pure breeding colourless and shrunken variety of maize.

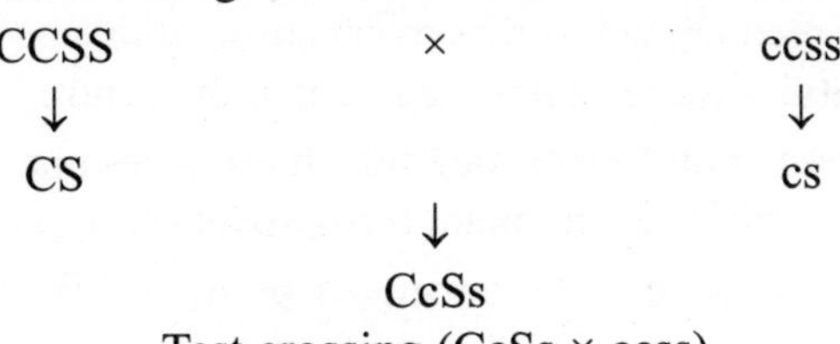

Test crossing (CcSs × ccss)

Gametes	cs	Ratio	Phenotypes
CS	CS/cs	48%	Coloured and smooth (parental)
Cs	Cs/cs	2%	Coloured and shrunken (recombinant)
cS	cS/cs	2%	Colourless and smooth (recombinant)
cs	cs/cs	48%	Colourless and shrunken (parental)

Results:

(a) Four types of progenies are produced in the ratio of 48:2:2:48.

(b) The two genes are linked in the same chromosome but 4% recombinants (2+2) occur by exchanging segments. Strong linkage exists between seed colour and seed shape.

(c) The large deviation of the observed F_2 population from the expected segregation occurred because separation in one pair of alleles is not independent of the separation in the other pair of alleles.

(d) The data shows that the two pairs of genes have not assorted independently.

2. **Repulsion cross**: He crossed a pure breeding coloured and shrunken variety of maize with pure breeding colourless and smooth variety of maize.

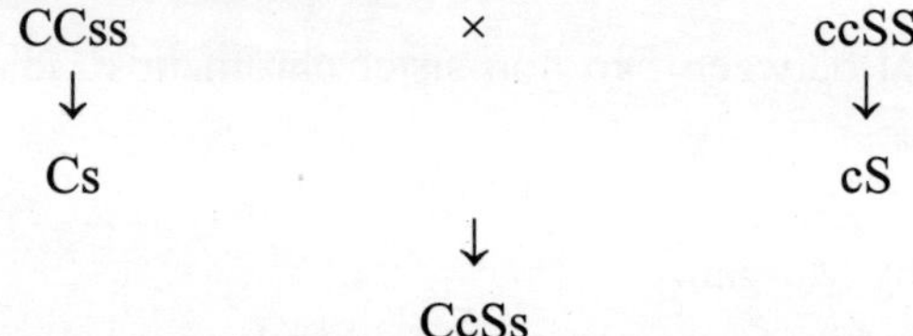

Crossing of F_1 hybrid with homozygous recessive parent

(CcSs × ccss)

Gametes	cs	Ratio	Phenotypes
CS	CS/cs	48%	Coloured and smooth (parental)
Cs	Cs/cs	2%	Coloured and shrunken (recombinant)
cS	cS/cs	2%	Colourless and smooth (recombinant)
cs	cs/cs	48%	Colourless and shrunken (parental)

2% Same results were observed as those of coupling cross.

Types of Crossing over:

1. **Single crossing over**: It occurs at one point (chiasma). It produces two parental type (non-cross over) chromatids and two recombinant (cross over) chromatids. Each recombinant chromatid acquires a new segment at the end.
2. **Double crossing over**: It occurs at two points (chiasma). It is two types:
   (a) Reciprocal: It involves the same non-sister chromatids. It produces two parental type (non-cross over) chromatids and two recombinant (cross over) chromatids just like single crossing over. Each recombinant chromatid acquires a new segment at the centre.

   (b) Complementary: It involves three or all four chromatids. It produces one or no parental (non-cross over) chromatid and three or four recombinant (cross-over) chromatids.
3. **Multiple crossing over**: It occurs at several (3 or more) points (chiasma) randomly. It is uncommon.

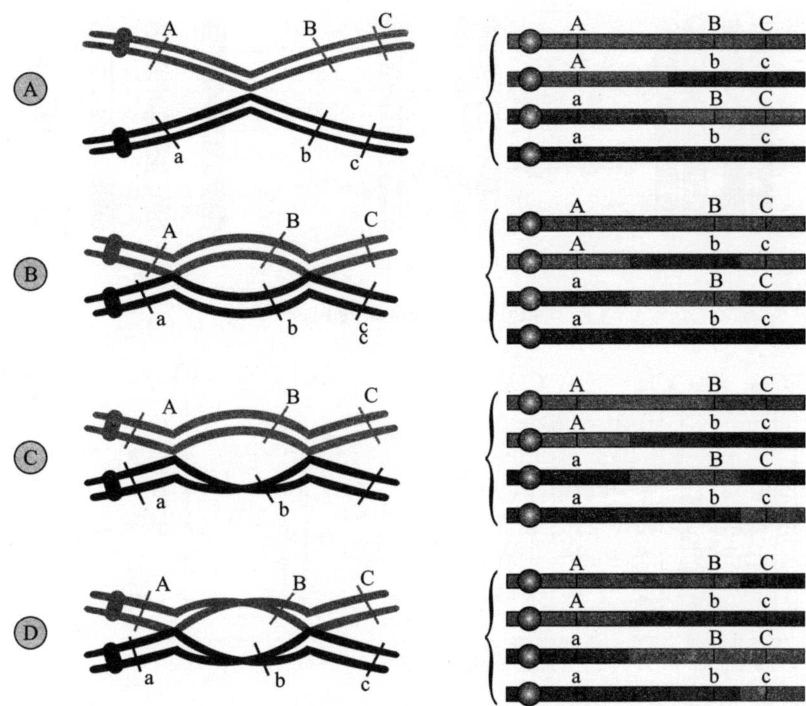

Fig. ***(a) Single crossing over; (b) Double crossing over (reciprocal); (c) and (d) Double crossing over (complementary)***

Mechanism:

(a) The pairing of homologous chromosomes (synapsis) takes place at zygotene stage.

(b) Each member of paired (synapsed) chromosome divides longitudinally into two chromatids forming tetrad.

(c) Chiasma is formed where crossing over occurs. Chiasma may be terminal (at the end of pairing chromatids) or interstitial (at the middle).

(d) Crossing over occurs between non-sister chromatids of homologous chromosome.

(e) Non-sister chromatid breaks at corresponding point which is facilitated by endonuclease enzyme.

(f) The broken part of one chromatid joins the other broken part of other non-sister chromatid by ligase enzyme.

(g) Exchange of chromatids containing genes takes place.

Hybrid DNA model: It was proposed by Holliday. It is simple model. In this only one strand in each of two DNA duplexes belonging to non-sister chromatids (from homologues) breaks. The break would occur in strands having same polarity. The single strands released from these breaks then pair crosswise with unbroken strands by complimentary base pairing. This results in the formation of hybrid DNA segments.

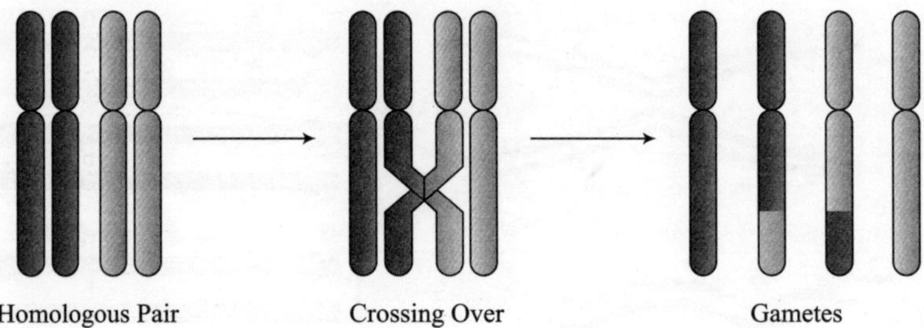

Fig. ***Single crossing over***

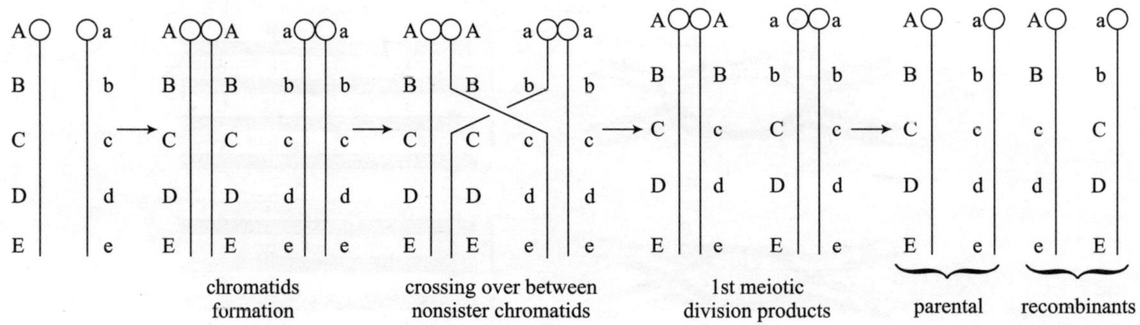

Fig. ***Crossing over at 4-stranded stage***

Factors affecting crossing over:

1. Distance: Frequency of crossing over increases with increase in distance between genes.
2. Temperature: Frequency of crossing over increases with rise in temperature (from 22^0C).
3. X-rays: Frequency of crossing over increases with X-ray irradiation.
4. Chemicals: Frequency of crossing over increases with treatment of chemicals.
5. Heterochromatin: Frequency of crossing over decreases with the presence of heterochromatin.
6. Location: Frequency of crossing over decreases at the end (telomere) and centromere.
7. Age: Frequency of crossing over decreases with increase in age.
8. Sex: Frequency of crossing over is less in male mammals. Crossing over is not seen in male *Drosophila* and female silk worm.
9. Structural changes: Frequency of crossing over reduces if translocation and inversion take place.
10. Interference: One crossing over reduces the occurrence of another in adjacent area.

Interference: It is a phenomenon in which one crossing over reduces the occurrence of another in adjacent area.

The term 'interference' was coined by Muller (1911).

It varies from chromosome to chromosome and within a chromosome from one point to another. It is maximum near the centromere and telomere (at the end of chromosome).It is affected by distance between the genes on chromosome. Lesser the gene distance greater is the interference and vice-versa. It may be

(a) Positive interference: It is most common in prokaryotes and eukaryotes. Its value varies from zero to one when observed value (from data) is lesser than expected value (to be calculated).

(b) Negative interference: The presence of crossing over in one region enhances the chance of crossing over in another adjacent region. It occurs in bacteriophage, *Aspergillus* etc.

Recombination frequency: It is directly related to distance between genes. The unit of crossing over has been termed as centiMorgan (cM) as in eukaryotes or map unit as in prokaryotes. One centiMorgan is equivalent to 1% crossing over. If 1 % Crossing over means 1% centimorgan or 1 map unit, 10% crossing over means decimorgan or 10 map unit and 100 % crossing over will be Morgan or 100 map unit. A distance of 1 cM between two genes means that the genes are separated to different chromosomes on average one per 100 meiosis. When chiasma is organised in between two gene loci, only 50% meiotic products shall be crossovers and 50% non-crossovers. Thus, the chiasma frequency is twice the frequency of cross over products i.e., chiasma % = 2 (cross over %) or crossover % = 1/2 (chiasma). The percentage of crossing over is calculated by test cross.

$$\text{Cross over value (\%)} = \frac{\text{Number of recombinants in test cross}}{\text{Total number of individuals}} \times 100$$

The maximum cross over value (COV) or recombination frequency is 50% between the two genes hence these genes are linked. It is main parameter to measure map distance. Each cross over event results in 50% recombinant gametes and 50% parental gametes. The larger the COV, the less accurate to measure map distance. It is noted that map distance and recombination frequency do not correspond because over long distance, map distance will exceed even 100 but recombination frequency between any two genes will never exceed 50. It is calculated by 2-point or 3-point test cross.

Coefficient of coincidence (COI): It is opposite to interference. It is expressed as the ratio between the actual number of double crossing over and the expected number of such double crossing over. It measures interference.

$$\text{COI} = \frac{\text{\% actual (observed) double crossing over}}{\text{\% expected double corssing over}} \times 100$$

$$\text{COI} = 1 / \text{Interference}$$

If double crossing over is absent, then coincidence is zero and interference is 100%.

If double crossing over is equal to expected value, then coincidence is 100% and interference is zero (impossible).

In practice, double crossing over is always less than expected value.

Significance:

(a) It produces new traits due to formation of new chromosomes in the gametes and develops variation in newly formed individuals.

(b) It produces new and useful varieties of crop plants and animals.

(c) It gives clear proof of linear arrangement of genes on chromosomes.

(d) It locates the point of genes on chromosome.

(e) It helps in preparing chromosome / gene mapping.

Cytoplasmic inheritance:

It is transmission of certain characters by the genes outside the nucleus. It was discovered by Carl Correns (1908). It is found in most eukaryotes and is commonly known to occur in cytoplasmic organelles such as mitochondria and chloroplast due to the presence of plasmagenes. It is also called extranuclear inheritance. It is a form of non-Mendelian inheritance.

The total self-replicating hereditary material present in cytoplasm is called plasmon. It is exhibited by the following examples:

(a) **Plastid inheritance** in 4 o'clock plant (Mirabilis jalapa): Correns reported the three types of leaves such as green, white or variegated (green and white). He performed the following crosses in which offsprings resembled the female branch.

Leaf phenotype of branch (male plant)	Leaf phenotype of branch (female plant)	Leaf phenotype of progeny (F_1)
Green	Green White Variegated	Green White Variegated (green and white)
White	Green White Variegated	Green White Variegated (green and white)
Variegated	Green White Variegated	Green White Variegated (green and white)

(b) Kappa particles in *Paramoecium*: It was reported by sonneborn.

(c) Shell coiling in snail (*Limnaea*): There are two types of coiling: Dextral (towards right side) is dominant and Sinistral (towards left side) is recessive.

Male (dextral) × Female (sinistral) → F_1 progeny (sinistral)

Male (sinistral) × Female (dextral) → F_1 progeny (dextral)

(d) Cytoplasmic male sterility in maize:

(e) Mitochondrial inheritance in petites of yeast: It was studied by Euphrussi.

Cytoplasmic inheritance	Nuclear inheritance
It is controlled by plasma genes present in cytoplasm.	It is controlled by nuclear genes present at specific loci in chromosome.
It shows the transmission of traits by female parent only (uniparental or maternal).	It shows the transmission of traits by both male and female parent (biparental).
It shows different results of reciprocal crosses.	It shows same results of reciprocal crosses.
It does not show segregation in F_2 and subsequent generations.	It shows segregation.

Chromosomal theory of inheritance:

It was proposed by Walter Sutton and Theodor Boveri (1904). It was elaborated by Morgan, Sturtevant and Bridges. They observed a close similarity and parallel behaviour between the transmission of hereditary characters (genes) and chromosomes while passing through one generation to next through gametes.

1. Male and female gametes play an equal role in transmission of hereditary characters.
2. Gametes serve as bridge between two successive generations.
3. Hereditary information is present in nucleus because only nucleus of sperm unites with egg.

4. The specific number of Chromosome in nucleus carries the hereditary characters.
5. Both chromosomes and genes occur in pairs in somatic or diploid cells.
6. A gamete contains only one chromosome of a type and only one of two alleles of a character (A+X or A+Y).
7. A sort of parallelism is observed between Mendelian factors and chromosomes.
8. Numerous genes are present in chromosome hence chromosome is carrier of Mendelian factors.
9. The paired condition of both chromosomes as well as Mendelian factors is restored during fertilisation.
10. The homologous chromosomes synapse (pair) during meiosis and then separate or segregate independently into different cells.
11. Sex is determined by sex chromosome.
12. A wide variety of phenotypes in living organisms shows that there are far more numerous characters than chromosomes. Hence each chromosome carries several genes on it.
13. The structural and functional changes in organisms are caused due to chromosomal aberration.

Sex determination:

It is the establishment of male and female individuals or male and female organ of an individual. It differs from sex differentiation which involves development of male and female sex organs as well as male and female secondary sexual characters. It involves

Primary signal (environment, hormone, gene) → sex determining genes → sex differentiating genes

Tijo and Levan (1956) found that human cells have 23 pairs or 46 chromosomes where 22 homologous pairs or 44 chromosomes are called autosomes and the last or 23rd pair are called sex chromosome. There are two sex chromosomes such as X and Y (XX in females and XY in males).

X chromosome was discovered by Henking in male bug and Y chromosome was discovered by Nettie Stevens (1906) in beetles.

Henking (1891) firstly proposed the connection of chromosome with sex while working on a hemipteran insect, *Pyrrhocoris apterus* and found that half of daughter cells contained one element more than the other during anaphase-I of meiosis in spermatogenesis.

McClung proposed that X chromosome was connected with sex determination.

Sex chromosome (X and Y) has two regions: (i) homologous region which carries incompletely sex linked genes because they undergo crossing over. This region helps in pairing and (ii) non-homologous region or differential region which carries completely sex linked genes as they do not undergo crossing over.

Sex of offspring in the animals with two types of sperms, is determined at the time of fertilisation by chance event (by the kind of sperm that fuses with egg) whereas in animals with two types of eggs, sex is not determined but is already fixed during meiosis before fertilisation.

Types of sex determination:

1. Progamic: Sex is determined before fertilisation. e.g., Drones of honey bees
2. Syngamic: Sex is determined during fertilisation. e.g., Plants and animals.
3. Epigamic: Sex is determined after fertilisation. e.g., Workers and Queen of honey bees.

Different controls of sex determination:

1. **Hormonal control:**
   (a) ***Bonellia viridis*** (marine annelidian worm): Basically all larvae are cytologically and genetically similar hence sexually undifferentiated. If a particular larva settles near proboscis of adult female, it will become male individual and lives in uterus of female. If it develops free in water, it will become female individual. If a partly developed male is detached from proboscis then it will become intersex. It is clear that proboscis of adult female secretes a substance (hormone) which inhibits femaleness and induces larva to differentiate into males.
   (b) The **Crew's hen** (female fowl): It is an example of sex reversal reported by Crew (1923). If a fertile hen produces offspring which change to fully fertile male (cock) due to damaged ovary in hen. Ovary in female secretes a male suppressing hormone hence in the absence of ovary, testis develops. In fact, when functional ovary is destroyed or removed, the steroid cells of adrenal cortex become active and provoke the development of rudimentary gonad into testis. Both adrenal cortex and testis secrete large amount of male hormones which inhibit the activity of female hormone cells of pituitary. Thus female sex is reversed into male sex due to sex reversal phenomenon.
   (c) The **Free martin**: It is twin of cattle in which male is normal and female is sterile. It is a case of different misalignment of primary and secondary sexual differentiation. This abnormal development is only found in females that have been a member of a pair of mixed sex twins. The embryonic blood supplies of twins in these spaces are fused in uterus and hence XY blood cells and hormones can enter the circulation of female twin. This is sufficient to force the gonads to develop into testes and partially to block normal female development even though the animal is genetically XX. The male twin is not affected by the presence of XX cells in its blood system.
2. **Environmental control:**
   (a) *Crepidula fornicata* (slipper snail): It becomes male in the company of female and becomes female if reared alone.
   (b) *Ophryotrocha* (annelidian worm): It is male at younger stage and becomes female at later stage.
   (c) Turtles: Low temperature (25^0C) induces maleness and high temperature (35^0C) for femaleness.
   (d) Crocodiles and Snakes: Low temperature induces femaleness and high temperature for maleness.
3. **Chromosomal theory**: It was proposed by Wilson and Stevens (1906). It is most common. In diploid organisms, there are several autosomes and two sex chromosomes. In one sex, two sex chromosomes are homomorphic (XX) and in other sex, two sex chromosomes are heteromorphic (XY). Sex is determined at the time of fertilisation. This theory proposed that one sex of any species produced two types of gametes, each is responsible for the formation of one type of sex. The presence of sex determining region of Y chromosome causes the development of testis in male. Thus, the sex is determined by sperm.

1. **Male heterogamety**: In this, male individual produces two different types of gametes. It shows two types sex chromosomal conditions:

(a) XY type (XX-XY type): It is called lygaeus type because it was firstly discovered in lygaeus insects. e.g., Human being, *Drosophila* and dioecious plants (*Coccinia, Melandrium*). The key to sex determination in humans is the SRY (sex region on Y) gene located on the short arm of the Y chromosome. If Y chromosome is present then the individualwill be male. If it is absent then female will be formed.

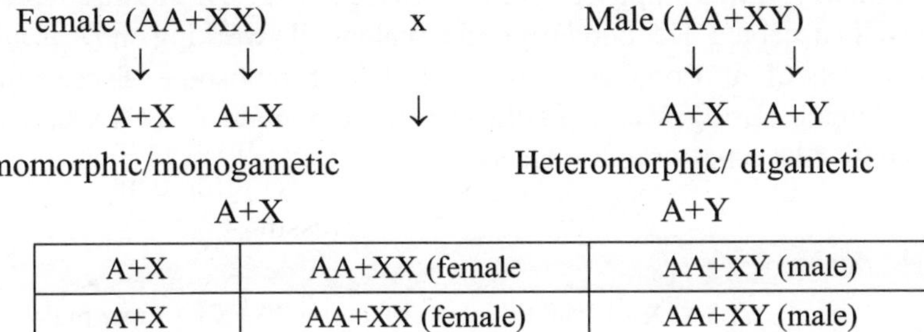

A+X	AA+XX (female	AA+XY (male)
A+X	AA+XX (female)	AA+XY (male)

Sex ratio in offspring is 1:1

(b) XO type (XX-XO type): It is called protenor type because it was firstly discovered in protenor insects. e.g., *Ascaris*, bed bug, cockroach, grass hopper, plants (*Dioscorea, Vallisneria*) etc.

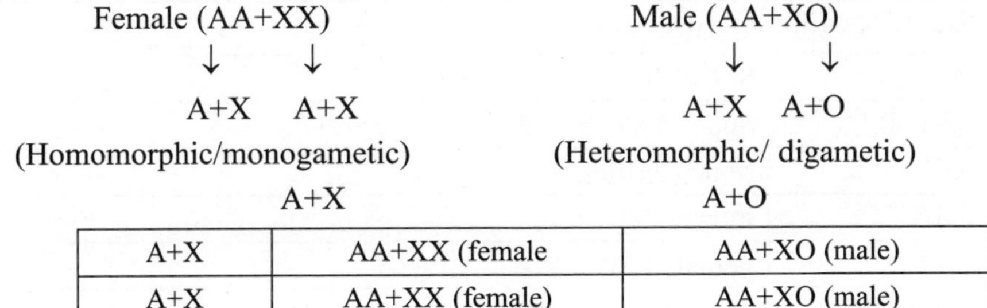

A+X	AA+XX (female	AA+XO (male)
A+X	AA+XX (female)	AA+XO (male)

Sex ratio in offspring is 1:1.

2. **Female heterogamety:** In this, female individual produces two different types of gametes. Thus, the sex is determined by the egg.

(a) ZW type (ZZ-ZW Type): e.g., Fish, reptiles, birds, *Fragaria* (plant).

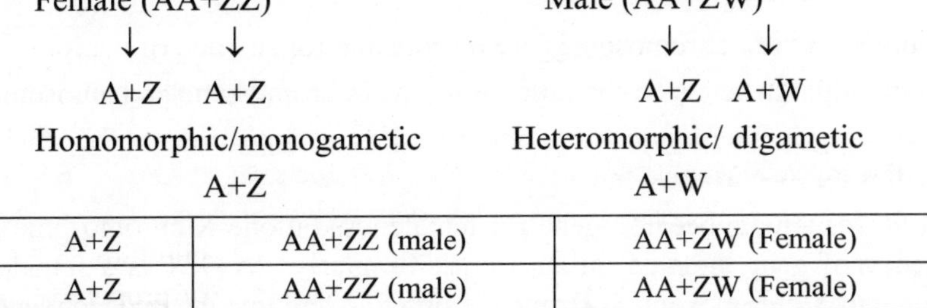

A+Z	AA+ZZ (male)	AA+ZW (Female)
A+Z	AA+ZZ (male)	AA+ZW (Female)

Sex ratio in offspring is 1:1

(b) ZO type (ZZ-ZO Type): e.g., Moth, butterfly.

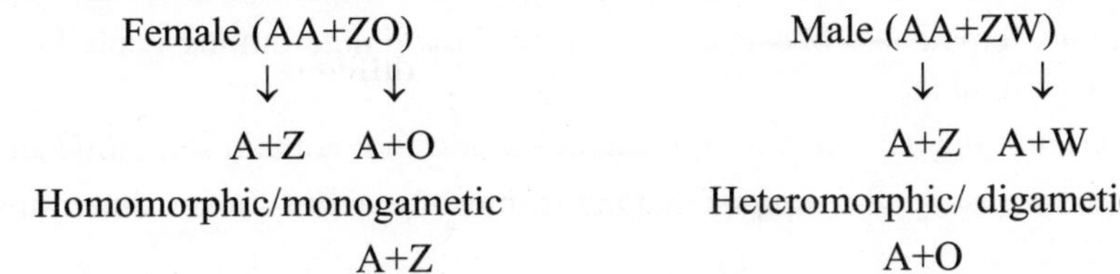

A+Z	AA+ZZ (male)	AA+ZO (Female)
A+Z	AA+ZZ (male)	AA+ZO (Female)

Sex ratio in offspring is 1:1.

4. **Genic balance theory**: It was proposed by Calvin Bridges 1926). It was discovered on *Drosophila*. It is not universally accepted. He found triploid female while working on *Drosophila*. The triploid female has three sets of autosome and three sets of X chromosome. These triploid females are normal fertile female. Bridges crossed these triploid female (3A+XXX) with diploid male (2A+XY) and noticed 8 types of individuals.

3A + XXX x 2A + XY

Femalc/male gametes	A+X	A+Y
2A+X	3A+XX (Triploid intersex)	3A+XY (Super male)
A+XX	2A+XXX (Super female)	2A+XXY (Diploid female)
2A+XX	3A+XXX (Triploid female)	3A+XXY (Triploid intersex)
A+X	2A+XX (Diploid female)	2A+XY (Diploid male)

X/A ratio determines different types of individuals.

X/A ratio	Sex of Individuals
Less than 0.5	Super male or meta male (sterile)
0.5	Male (fertile)
0.5-1	Intersex
1	Female (fertile)
More than 1	Super female or metafemale (sterile)

- **Intersex**: It is neither male nor female but sterile intermediate between male and female. It contains XX chromosome in some cells and has XY chromosomes in some other cells. It has genital that seems to be between usual male and female type. Girl has large clitoris without vaginal opening and boy has small penis with divided scrotum looking like labia.

Findings:

(i) Both autosomes and sex chromosomes are responsible for sex determination.

(ii) Sex of an individual depends on the ratio between X chromosome and autosome.

(iii) The gene showing inclination towards maleness lies on autosome while gene showing towards femaleness lies on X- chromosome.

(iv) 2A+XY produces a male because gene for femaleness on one X chromosome cannot overcome male tendency of gene situated on autosome. Similarly, 2A+XX is a female because female character on two X chromosome is strong enough to overcome the male tendency found in genes on 2A. Thus one X is more powerful than one A (autosome).

(v) Autosomes show maleness. In male, autosomal material is more than in female compared to X.

(vi) X chromosome shows femaleness because of carrying female determining gene, Sxl. So higher number of X produces female.

(vi) Y chromosome does not play significant role because it does not contain sex influencing genes.

(viii) The presence of intersex is a definite proof that autosomes also carry for sex determination.

Sex determination in Honey bee (Haplodiploidy): It occurs in Honey bees, ants, wasps where female is diploid (2n=32) and male is haploid (n=16). In honey bees there are three types of individuals.

1. Queen is diploid, fertile and female developed from fertilised egg.
2. Drone is haploid, fertile and male developed from unfertilised egg
3. Worker is diploid, sterile and female developed from fertilised egg.

In this, sex is determined by the number of sets of chromosomes an individual receives.

Female (2n=32) Male (n=16)
↓---meiosis---↓ ↓---mitosis
Egg (n=16) Egg (n=16) ------×---------- Sperm (n=16)
↓parthenogenesis ↓
Male (n=16) Female (2n=32)
(Drone) **(Queen, Worker)**

Gynandromorph: It is abnormal feature (sex mosaic) where individual has right half of body male and left half female. It commonly occurs in *Drosophila* and Silk worm. It is developed due to accidental loss of X chromosome from somatic cell (2A+XX) at metaphase of first mitotic division of zygote after fertilisation. Hence, one of the two cells of 2-celled proembryo will have 2A+XX with X/A=1 and other cell will have 2A+X with X/A=0.5

Sex determination in plants: Mostly plants are monoecious and some are dioecious. R.P. Roy performed experiment on *Coccinia* and Warmke conducted on *Melandrium*. The findings are as follows:

(a) The ratio of X and Y determines sex as in human beings,
(b) Autosomes have no visible effect on sex determination.
(c) If Y chromosome is present then individual is male.
(d) If Y chromosome is absent then individual is female.
(e) If ratio of X and Y is 4:1 then hermaphrodite plant is produced.

Drosophila as experimental material for genetic research

(a) It is more suitable than pea.
(b) It is easy to be reared and bred on synthetic medium under laboratory condition.
(c) It has short life span (two weeks) so numerous generations are obtained.
(d) It shows a number of clearly visible contrasting traits.
(e) It has smaller number (2n=8) of chromosomes.
(f) It has heteromorphic (XY) sex chromosome in male in and the transmission of heteromorphic chromosome can be easily studied from one generation to another.

Mutation:

It is sudden change in phenotype due to qualitative or quantitative change in the composition of gene or chromosome.

The term 'Mutation' was coined by Hugo de Vries (1900) for sudden, large and discontinuous change in evening primrose (*Oenothera lamarckiana*). He observed the sudden appearance of several new and different plants on breeding (self-pollination).

Darwin called it 'sports'.

Morgan studied mutation scientifically with the discovery of White eye mutant of *Drosophila*.

Bateson called it discontinuous variation. Haploids are more suitable than diploids for study of mutation because they allow expression of recessive mutation in F_1 generation.

Features:

(a) Mutation occurs in any cells or tissues at any time.
(b) It occurs in all living organisms including viruses.
(c) It may affect the structure and function of organisms.
(d) It produces mutant or changed expression.
(e) It occurs throughout the genome all the time but some areas are more prone to mutation called hot spot.
(f) Most of the mutations are harmful.
(g) Most of the mutations are recessive.
(h) It occurs during any developmental stages of organisms.
(i) It occurs randomly in both autosomes and sex chromosomes.
(j) It shows pleiotropic effect (single mutant allele affects more than one character).
(k) It is recurrent (the same mutation may occur again and again).
(l) It produces visible or detectable changes (macromutation) as well as invisible or insignificant changes (micromutation).
(m) It may be forward (wild or normal allele is changed to mutant allele and backward (reverse) mutation (mutation or altered allele is changed to wild allele).
(n) It alters the morphological features of organisms such as colour, shape, size etc. and prevents the production of a biochemical in organisms.
(o) It shows lethal condition (kills all the individuals), sublethal (kills most of the individuals), vital (does not kill the individuals), subvital (kills some of the individuals) and supervital (increases the survival of the individuals).

Types: On the basis of types of cell

1. **Somatic mutation:** It occurs in somatic cell of organism. It is non-inheritable. It disappears with the death of organism. It affects the organisms partially. It is insignificant because it is limited to certain cells. e.g., white spot in red eye of *Drosophila*.
2. **Germinal mutation:** It occurs in reproductive cells. It is heritable. It does not disappear with the death of organism. It affects the whole organisms. It is significant in evolution. e.g., short leg in Ancon sheep.

Types: On the basis of magnitude of effect

1. **Dominant mutation:** It occurs when mutated gene is dominant to wild type. It is expressed in both homozygous and heterozygous condition. It is rare hence its frequency is low. It is easy to identify. e.g., white eye and notch wing of *Drosophila*
2. **Recessive mutation:** It occurs when mutated gene is recessive to wild type. It is expressed only in homozygous condition. It is produced immediately. e.g., bar eye (narrow) of *Drosophila*

Types: On the basis of amino acid replacement

1. **Missense mutation:** It occurs when one of the bases of a codon is replaced by another base. It occurs in first base of codon (original sequence).

5′-AUG CCU **AGA** GGG CAA-3′ (Original sequence)

Met Pro ↓Arg Gly Gln

3′-AUG CCU **GGA** GGG CAA-3′ (Mutated sequence)

Met Pro Gly Gly Gln

It alters the sequence of amino acid in a normal protein. It is tolerated by most of protein. It causes deleterious effect if it takes place in parts of proteins which are important for structure or function. It will produce mutant phenotype (phenotypic effect).

2. **Nonsense mutation**: It occurs when any base of a codon is changed in a manner to form any one of terminating codons (UAA,UAG and UGA).

   5′-AUG CCU **UGU** GGG CAA-3′ (Original sequence)

   Met Pro Cys ↓ Gly Gln

   3′-AUG CCU **UGA** GGG CAA-3′ (Mutated sequence)

   Met Pro **Stop** Gly Gln

   It terminates the elongation of polypeptide chain hence there is formation off shortened protein. It has serious effect on the activity of encoded protein. It will produce mutant phenotype (phenotypic effect).

3. **Same sense (Silent) mutation**: It occurs when one of the bases of a codon is replaced by another one but there is no alteration in amino acid sequence of protein because the changed codon specifies the same amino acid. It occurs in third base of codon.

   5′-AUG CCU UGU **GGG** CAA-3′ (Original sequence)

   Met Pro Cys Gly↓ Gln

   3′-AUG CCU UGU **GGA** CAA-3′ (mutated sequence)

   Met Pro Cys Gly Gln

   It is no longer functional and cannot be detected unless the nucleic acid sequencing technique is applied and genome is screened. It has no effect on encoded protein and does not produce mutant phenotype.

Types: On the basis of origin

1. **Spontaneous mutation**: It occurs automatically without any apparent reason under natural conditions. It is also called background mutation. It is heritable. It occurs rarely. Its frequency or rate varies but it can be measured. It is higher (10^{-4}-10^{-6} per gene per generation) in eukaryotes but lower (10^{-5}-10^{-7} per gene per generation) in prokaryotes. The large gene provides a large target and tends to mutate more frequently. It is caused by the following means:
   (a) UV rays of solar radiation present in the environment.
   (b) Error in DNA replication due to DNA polymerase enzyme.
   (c) Integration of transposons and insertion sequence at specific sites in genome.
   (d) Methylation of cytosine base of DNA at 5th position by methylase enzyme followed by spontaneous deamination.

   e.g., White eyed male *Drosophila*; and sudden occurrence of new plants and animals (sports).

2. **Induced mutation**: It occurs artificially due to effect of either physical agents or chemicals (mutagens). It causes change in normal sequence of bases of DNA hence it is also known as **gene or point mutation**. It is basically intragenic. Its unit is muton and it consists of one or many nucleotide pairs.

Physical mutagens:

Ionising radiation	Particulates-α ray, β-ray, fast or thermal neutron Non-particulates- X ray, gamma ray, cosmic ray
Non-ionising radiation	UV rays

Chemical mutagens: These are more hazardous than physical mutagens. Auerbach firstly reported the mutagenic action of a chemical (mustard gas).

Base analogues	5-bromouracil, 5-chlorouracil, 5-iodouracil, 2-aminopurine, 2,6-diaminopurine, 5- bromodeoxyuridine
Alkylating agents	Ethyl methane sulphonate (EMS), Ethyl ethane sulphonate (EES), Methyl methane sulphonate (MMS), Dimetyl sulphonate (DMS), N-nitrosoguanidine (NTG), Mustard gas (nitrogen, sulphur).
Deaminating agents	Nitrous acid, hydroxyl amine
Intercalating agents	Acridine orange (dye)-Acriflavin (for bacteria and plants), proflavin (for viruses), Ethidium bromide
Inorganic compounds	Ferrous salts, Manganese salts
Organic compounds	Benzopyrene, Formaldehyde, organic peroxides (for fungi), Lysergic acid Dimethylamide (LSD), Aflatoxins

Induced mutation by UV ray: It is non-ionising radiation with mutagenic properties. It is present in solar radiation or produced by mercury vapour lamp or tube. Its wavelength is1000-4000 A^0 and it is poorly penetrating.

It is highly mutagenic in microbes where DNA is not covered by UV absorbing material. It can penetrate only the surface layer of cells in multicellular organisms.

It is specifically absorbed maximum by bases of DNA at wavelength 2540 A° which is mutagenic wavelength.

It causes direct effect on DNA by forming pyrimidine dimmer. It causes molecular vibration or promotion of electrons to high energy level within thymine (target molecule). It becomes reactive and joins covalently with another thymine located nearby in same DNA strand and forms thymine dimer at 2600 A°. Normal condition is restored at 2400A^0.

Thymine dimer causes mis-replication due to bend in DNA duplex. Hence DNA polymerase cannot catalyse DNA replication in the region containing dimer. This causes strand separation leading to distortion. Finally mutation occurs when cell tries to repair damage using the error prone repair system called SOS repair system.

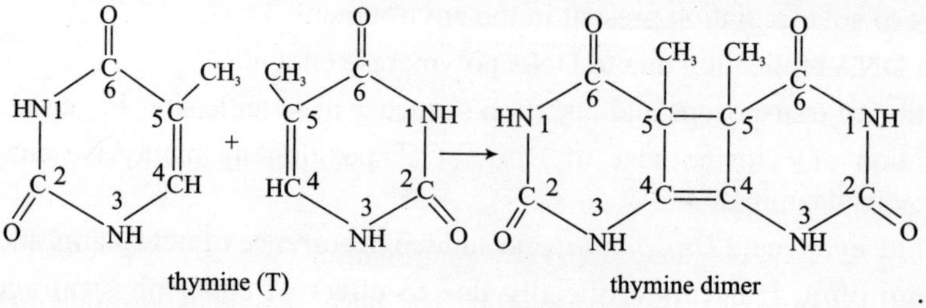

thymine (T) → thymine dimer

It also causes cytosine dimer and Thymine-cytosine dimer.

It also causes another direct effect cytosine hydrate due to addition of an –H atom at 6th position and –OH atom at 5th position. Hence the double bond between 5th and 6th positions is disrupted.

Induced mutation caused by X-ray: H. J. Muller (1927) discovered the mutagenic effect of X-ray on Drosophila. Stadler discovered it on Barley and maize. X-ray is produced by X-ray tube. Its wavelength is 0.1-10 A° and highly penetrating. It can cause mutation in seeds of plants in which embryo is deep inside. Its frequency depends on radiation dose (measured in Roentgen unit). It causes direct major effects on DNA in several ways such as change in a base or loss of a base or disruption of H-bond between complimentary strand of DNA or break of single and double strand in DNA and cross linkage of chromosomal proteins.

The efficiency of radiation causing mutation depends on stage of cell division, temperature, plant or animal species, O_2 concentration and moisture content.

Types of Gene mutation:

1. **Substitution mutation**: It occurs when one base in DNA is replaced by different base. It is most common. It is also called copy error mutation because it occurs during DNA replication due to error in base pairing. It is of two types:
   (a) **Transition mutation**: It occurs when one base of purine is replaced by another base of purine or when one base of pyrimidine is replaced by another base of pyrimidine. It is more common. There are four types of possible transitions:
   Adenine is replaced by Guanine (A→G)
   Guanine is replaced by Adenine (G→A)
   Thymine is replaced by Cytosine (T→C)
   Cytosine is replaced by Thymine (C→T)

Transition mutation caused by oxidative deamination: Nitrous acid is best known chemical mutagen which causes oxidative deamination by converting amino group of a base to keto group.

HNO_2 deaminates Adenine to hypoxanthine; cytosine to uracil; and Guanine to Xanthine. The frequency of deamination is maximum in adenine, then in cytosine and minimum in guanine. HNO_2 cannot deaminate Thymine.

$$\text{Adenine} \xrightarrow[\text{Deamination}]{HNO_2} \text{Hypoxanthine}$$

Adenine is deaminated to hypoxanthine.
Hypoxanthine is guanine like and pairs with cytosine rather than Thymine.
Cytosine pairs with Guanine in next replication.

```
                     A-T (normal)
A-T ──→ H-T---------↕                    H-C
↑                     H-C-------------------------↕
↑                                        G-C (mutant)
↑←-------------------------<--------------------------------←|
```

Thus base sequence (A-T) is changed to G-C by deamination.

Transition mutation caused by Base analogue: The base analogues are unnatural bases whose molecular structure is similar to nitrogenous base in DNA. These rare derivatives of normal base with altered base pairing properties and cause direct mutagenesis. As a result these are readily incorporated in DNA during replication. The higher frequency of pairing error is due to higher electronegativity of bromine in bromouracil as compared to $-CH_3$ group in Thymine. The analog fits perfectly in DNA. This brings about mutation in two different ways:

Mutation by mis-incorporation or error.

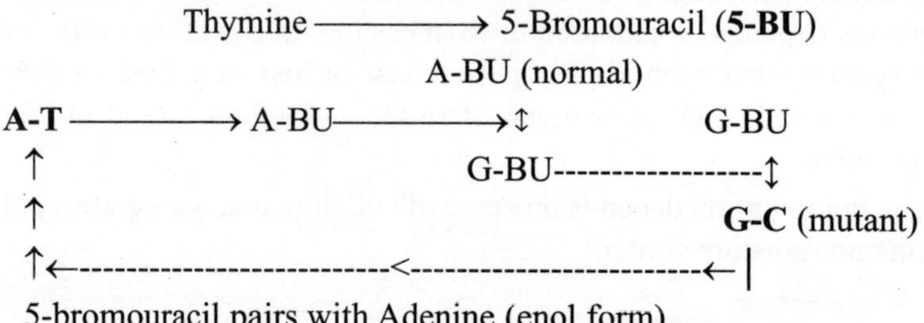

5-bromouracil pairs with Adenine (enol form).

5-BU pairs with guanine in next replication. Guanine will pair with cytosine in next replication. Hence, base sequence (A-T) is changed to G-C by base analogue.

Mutation by mis-replication:

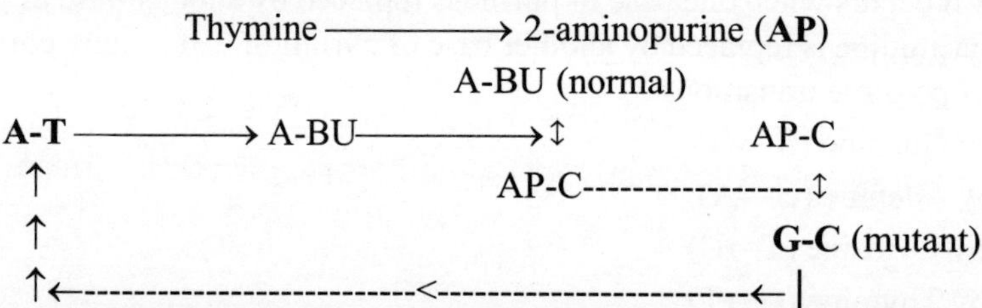

AP pairs with thymine. AP pairs with cytosine in next replication. Cytosine will pair with guanine in next replication. Hence, base sequence (A-T) is changed to G-C by base analogue.

Transition mutation caused by tautomerisation: The tautomerism is the ability of a molecule to exist in more than one chemical forms. The normal bases of DNA are usually in keto form but due to tautomerisation these bases can be transformed into the rare enol forms. Thus each base of DNA can exist in two states: normal and rare (abnormal). The rare state can exist in each base of DNA due to rearrangement or redistribution of electrons and protons in the molecule. This is tautomeric shift. Thus, tautomerism is a phenomenon where the original keto form of base is replaced by its imino or enol form. These tautomeric shifts change the H-bonding characteristic of base. Hence bases are converted into less stable state.

Adenine (amino form) ⟶ Rare adenine (imino form tautomer)

Guanine (keto form) ⟶ Rare guanine (tautomer)

Thymine ⟶ Rare thymine (tautomer)

Cytosine ⟶ Rare cytosine (tautomer)

If the base is in rare state, it will not pair with normal partner.

Rare adenine will pair with cytosine but not with normal thymine

Rare guanine will pair with thymine but not with normal cytosine

Rare thymine will pair with guanine but not with normal adenine.

Rare cytosine will pair with adenine but not with normal guanine

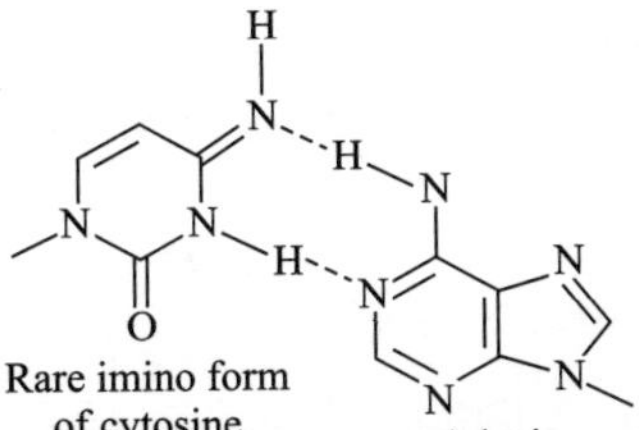

Rare imino form of cytosine — Adenine

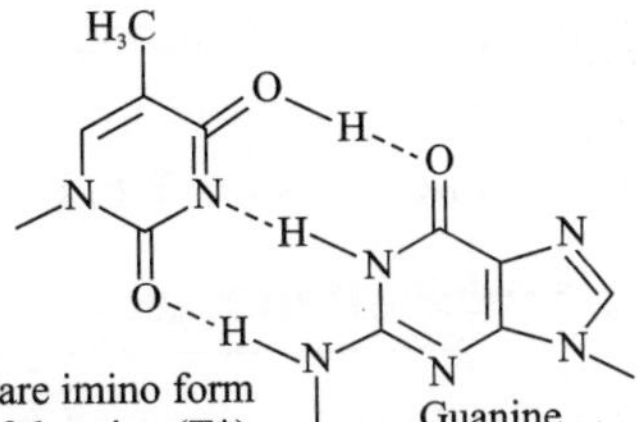

Rare imino form of thymine (T*) — Guanine

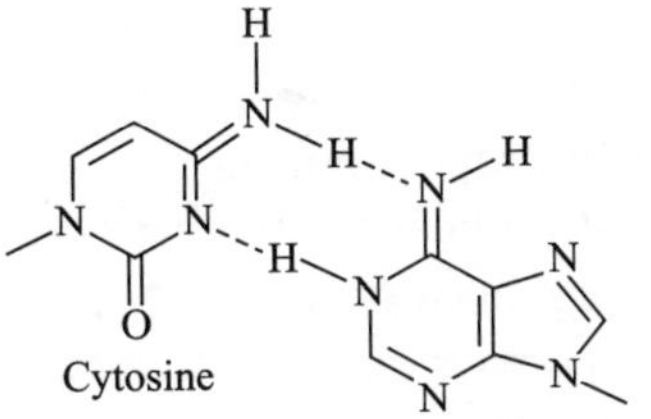

Cytosine — Rare imino form of adenine (A*)

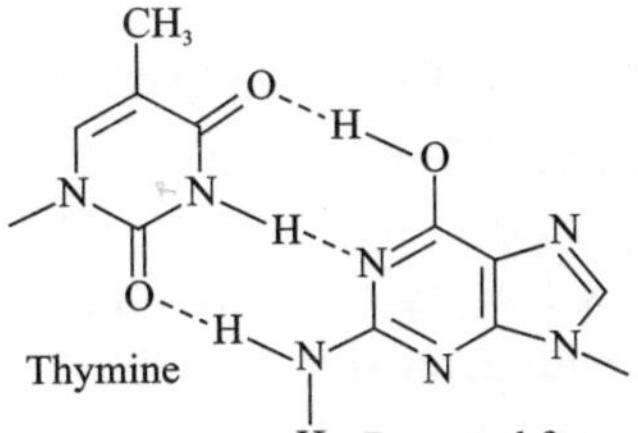

Thymine — Rare enol form of guanine (G*)

Such pairing is called forbidden base pairing. It has been found that such situation is not stable. In next replication, rare state may change to normal state. But once a forbidden base pair is formed, it will include a changed base. This will result in two kinds of DNA; one is normal DNA and other is mutated DNA which has undergone a base pair substitution.

(b) Transversion mutation: It occurs when one base of purine is replaced by another base of pyrimidine or vice-versa. It shows more noticeable effects than transition mutation. There are eight types of possible transversions.

Adenine is replaced by Thymine (A→T)
Thymine is replaced by Adenine (T→A)
Adenine is replaced by Cytosine (A→C)
Cytosine is replaced by Adenine (C→A)
Guanine is replaced by Thymine (G→T)
Thymine is replaced by Guanine (T→G)
Guanine is replaced by Cytosine (G→C)
Cytosine is replaced by Guanine (C→G)

Transversion mutation by alkylating agents: Ethyl methane sulphonate (EMS) as alkylating agent donates alkyl group on the 7th position of Nitrogen atom of Guanine in DNA. As a result, either miscoding through abnormal base pairing occurs or depurination (through gap formation) due to excision of guanine takes place. The gap may then interferes with DNA replication or may incorporate wrong base.

$$\text{Guanine} \xrightarrow{\text{EMS}} \text{7 - Ethyl Guanine}$$

Guanine will become 7-Ethyl guanine (alkylated base).

The 7-Ethyl guanine will mispair with Thymine instead of pairing with Cytosine.

The 7-EG cause pairing mistake during DNA replication.

```
                              7-EG-C
G-C———————> 7-EG——————————↕
↓                             7-EG-T
|-----------------------------------↑
```

2. **Frame shift mutation:** It is caused by addition or deletion of one or more nitrogenous base pairs in DNA or mRNA because this mutation shifts the reading frame from the site of change onwards. It is called gibberish mutation. It is much more harmful than substitution mutation. It occurs by intercalating agents (acridine dyes) which are flat, three ringed compound, similar in shape to base (purine) and can be intercalated between base pairs. The replication of intercalated DNA results in addition or deletion of one to few base pairs which produce frame shift mutation. It takes place by the following two methods:
   - **(a) By deletion:** When acridine dye is inserted during replication in new chain, this does not allow any base to pair with it. Therefore, in next replication, the base pair will be missing from DNA leading to shortening and changed organisation in DNA. The new strand will replicate with deleted base and DNA will be formed with one base less than normal.

     Original sequence: CAT **A**GT GAC GCA
     ↑deletion of **Adenine** base
     Modified sequence: CAT GTG ACG CA
     (unchanged) frame shifts towards right→
   - **(b) By addition (insertion):** When acridine dye is inserted during replication in new chain, then a base is inserted opposite to acridine in new strand. Therefore, the complementary base (X′) will pair with X leading to elongation of DNA and changed organisation in DNA.

     Original sequence: CATAGT GAC GCA
     ↑insertion of **Thymine** base
     Modified sequence: CAT TAG TGA CGC CA
     (Unchanged)→frame shifts towards left

Results:

1. One deletion may be neutralised by one addition or vice versa provided this occurs at the same place or very close to each other.
2. Deletion or addition of one or two bases results in a new sequence of codons which may code for entirely different amino acids. This results on a drastic change in the protein synthesised which is usually non-functional.
3. Deletion or addition of three bases may disturb only a few frames and shift by three nucleotides. The resulting protein is normal except that it lacks one amino acid or may contain extra amino acid.

Examples of disease caused by frameshift mutation:

(a) Huntington's disease
(b) Myotonic dystrophy
(c) Kennedy disease

Detection:

(a) Dominant lethal mutation is detected by Russel's test

(b) Sex-linked recessive mutation is detected by attached X method.

(c) Sex-linked lethal mutation is detected by CIB method.

Roles of mutation:

(a) It will produce better qualities of plants and animals.

(b) It will develop new species with the help of natural selection, isolation, genetic drift, migration.

(c) It will make plants adaptable in changed environment.

(d) It will give rise to new mutant strains of microbes for better yield of antibiotics, better fermenting ability.

(e) It will allow the genes to be identified and studied. e.g., white eye locus was identified only after the white eyed mutant of *Drosophila*.

(f) It permits the fine structure and analysis of gene, the relationship between genes and proteins as well as some features of genetic code.

Structural chromosomal aberration (Chromosomal mutation)

These are produced due to breakage on chromosome and later joining in chromosome segment. These aberrations occur very rarely in nature in the following ways:

1. **Deletion**: It is loss of one or more genes due to loss of a part of chromosome. It was first structural aberration detected by Bridges on X-chromosome of *Drosophila*. It can be detected at the time of synapsis. It may be homozygous deletion if both chromosomes are involved. It may be heterozygous deletion if only one chromosome is involved. Large heterozygous deletion inhibits synapsis hence causes non-disjunction and lower fertility. It is of two types:

   (a) Terminal: In this, last segment or telomere is broken or lost due to single break. It is rare.

   (b) Intercalary: In this, any middle segment between telomere and centromere is lost or broken due to two breaks in G_1 period. It is most common.

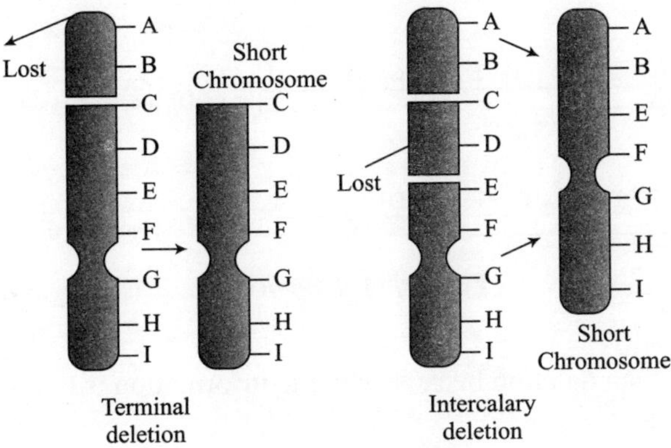

Fig. ***Types of deletion***

Genetic effects:

(a) A deletion of small size is not harmful but of large size is always harmful.

(b) It is less harmful in polyploid species.

(c) It helps plant survive if entire chromosomes are deleted. It is observed in monosomic and nullisomic wheat and *Datura*.

(d) It forms characteristic loop.

(e) It produces sterile pollen.

(f) It produces pseudo-dominance when a recessive allele behaves like dominant allele.

(g) Crossing over does not occur in deleted part of chromosome. It was confirmed by Benzer on genetic analysis of rII locus of T_4 phage of *E.coli*.

(h) Deletion of 3c7 band of X chromosome in *Drosophila* gives rise to Notch wing.

(i) Deletion of half part of short arm (p) of 5th chromosome (sub-metacentric) causes cat's cry syndrome.

(h) Deletion of some part of long arm (q) of 22nd chromosome (acrocentric) or Philadelphia chromosome causes Chronic Myeloid Leukaemia (CML) and deleted part is translocated to 9th chromosome.

2. **Duplication:** It is addition of one or more genes due to repetition of a part of chromosome. It involves breakage and reunion of a chromosome segment. It occurs due to abnormal events in crossing over. It occurs in following ways:

(a) Tandem: It is repetition of adjacent part.

(b) Reverse tandem: It is repetition of reversal of duplicated part.

(c) Displaced homobrachial: It is repetition of displaced position on same arm of chromosome.

(d) Displaced heterobrachial: It is repetition of displaced position on different arm of chromosome.

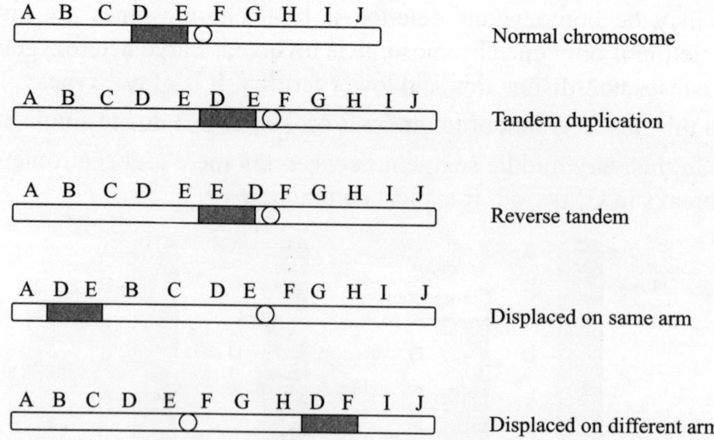

Fig. ***Types of duplication***

Genetic effect:

(a) It is less harmful than deletion because genetic information is neither lost nor altered.

(b) It occurs frequently.

(c) It protects the organisms from lethal mutation.

(d) It is beneficial and hence basis of evolution of new genes because there is increase in DNA content per cell. Hence duplication provides additional genes which mutate in new form while older genes synthesise necessary enzymes for normal activities.

(e) It can be detected if it is large due to presence of loop.

(f) Tandem duplication of 16 A region of X chromosome in *Drosophila* produces bar eye (narrow eye) which is smaller than wild type. This is called position effect (phenotypic changes in expression of genes which are due to differences in the position of genes) detected in salivary gland chromosome.

(g) Reverse tandem duplication of 4th chromosome produces eyeless dominant.

(h) Tandem duplication of 3rd chromosome causes confluence (thickening of veins).

(i) Duplication of centromere does not survive because the two centromeres may orientate to opposite poles at division breaking the chromosome.

4. **Translocation**: It is transfer of a part of chromosome from one chromosome to non-homologous chromosome. It was discovered by Bridges while working on *Drosophila.* It occurs due to irregularities during crossing over. It is of two types:

   (a) Nonreciprocal translocation: It is transfer of a part of one chromosome to another non-homologous chromosome. In this, one chromosome becomes deficient and other gains from other non-homologous chromosome. It rarely occurs. It is of two types:

   (i) Terminal: It involves single break in the chromosome. The broken part is directly added to other non-homologous chromosome. It is rare.

   (ii) Intercalary: It involves three breaks. The broken part is directly added within break produced in non-homologous chromosome.

   (b) Reciprocal translocation: It is mutual transfer of part of chromosome between pairs of non-homologous chromosome. It occurs at single break in each of the two non-homologous chromosomes. It is of two types:

   (i) Reciprocal homozygous: Both the homologous chromosomes exchange parts with the two homologues of another pair.

   (ii) Reciprocal heterozygous: Only one member of different homologous pairs exchanges their parts.

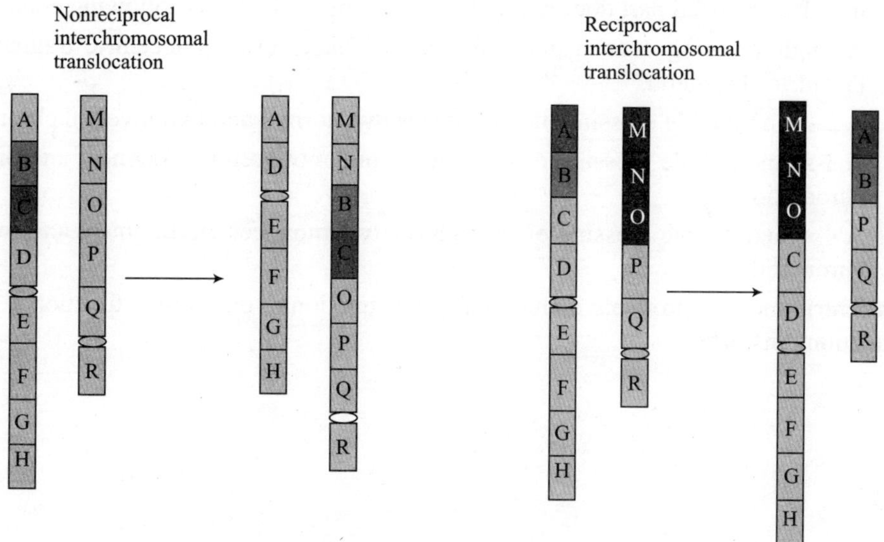

Fig. ***Types of Translocation***

Genetic effects:

(a) Translocation gives rise to new chromosomes and new linkage groups.

(b) It forms isochromosome.

(c) Reciprocal translocation forms multivalent rings in meiosis.

(d) Reciprocal translocation from 22nd to 9th chromosome causes chronic myeloid leukaemia (CML).

(e) Reciprocal translocation causes Down syndrome in human beings.

(f) Translocation heterozygotes in plants may produce fewer seeds than normal and such plants appear sterile.

Robertsonian translocation: It is a condition in which two telocentric or acrocentric chromosomes with no significant genetic material on short arm fuse at centromere to give one metacentric or sub-metacentric chromosome. It results in reduction in chromosome number.

4. **Inversion:** It is the case of reversal of order of genes due to reversal of a part of chromosome by 180°. It is an example of rearrangement of genes. It was first reported by Sturtevant in *Drosophila*. It does not change the amount of DNA unlike deletion and duplication but changes the linear sequence of genetic information. It occurs due to formation of loop on a chromosome then two breaks occur at intersection of loop followed by the joining of intercalary part in reverse order. It is of two types:

(a) Pericentric inversion: It is less common. It is heterobrachial because both breaks occur in both arms. The inverted part of chromosome includes centromere. It may change the location of centromere. It shows following conditions:

- If crossing over takes place within loop, the resulting chromatids include half of chromatid with extra genes and deficiencies rendering it non-functional.
- If two breaks are not equidistant from the centromere, then this will change the shape of chromosome (metacentric chromosome may become sub-metacentric or vice-versa).

(b) Paracentric inversion: It is more common. It is homobrachial because both breaks occur in one arm. The inverted part does not include centromere. It shows following conditions:

- A single crossing over in inversion loop produces one monocentric fragment and one dicentric chromatid.
- A 2-strand double crossing over produces two normal and two inverted chromatids.
- A 3-strand double crossing over produces one monocentric fragment and one dicentric chromatid.
- A 4-strand double crossing over produces two monocentric fragment and two dicentric chromatids.

Thus, paracentric meiotic products include half non-functional, one fourth functional normal and one fourth functional inverted.

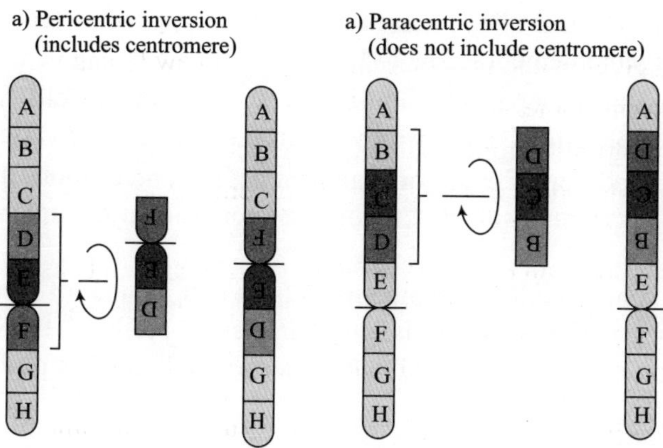

Genetic effects:

(a) It reduces the rate of crossing over and acts as crossing over suppressor because only unrecombined parental chromosomal arrangement of inverted region can be transmitted.

(b) It causes male or pollen sterility which suppresses the transcription of genes.

(c) It does not change the nature of genes.

(d) It shows position effect when genes are transferred from euchromatic segment to heterochromatic segment

(e) It maintains the heterozygosity in the species.

(f) It confirms the evidence of crossing over in which only two chromatids are inverted in crossing over.

Numerical Chromosomal Aberration

These are produced when number of chromosomes change from diploid state. These are of two types:

1. Aneuploidy: It is loss or gain of one or a few chromosomes in somatic chromosome. It was discovered by bridges. It occurs mostly due to non-disjunction of chromosomes during mitosis and meiosis. Aneuploids rarely survive and are less viable than euploids. It is always deleterious due to genetic imbalance.

(a) Hypoploidy: It is loss of chromosome. It is of three types:

(i) Monosomy: It is loss of one chromosome from its somatic number. Its chromosome complement is 2n-1. It causes a major genetic imbalance which is not tolerated by diploid species. It is generally weaker than normal forms. e.g., Turner syndrome (44+X).

(ii) Double monosomy: It is loss of two non-homologous chromosomes (each from different pair). Its chromosome complement is 2n-1-1.

(iii)Nullisomy: It is loss of a pair of homologous chromosome. Its chromosome complement is 2n-2. It is viable and survives in polyploids. It produces deleterious effect. e.g., hexaploid wheat (*Triticum aestivum*).

(b) Hyperpoloidy: It is gain of chromosome. It is of three types:

(i) Trisomy: It is gain of one extra chromosome. Its chromosome complement is 2n+1. It was firstly obtained by Blakeslee in jimson weed (*Datura stramonium*). It occurs spontaneously due to production of n+1 type of gametes as a result of rare non-disjunction of a bivalent. e.g., Down syndrome, Patau syndrome, Klinefelter syndrome.

(ii) Double trisomy: It is gain of two non-homologous chromosome (each from a different pairs). Its chromosomal complement is 2n+1+1.

(iii) Tetrasomy: It is gain of a pair of homologous chromosome. Its chromosomal complement is 2n+2. It is more deleterious than trisomy because it shows more variation.

2. Euploidy: It is the condition of a cell or organism having one or more than one complete set of chromosomes excluding sex chromosomes. Euploidy is more tolerated in plants than in animals. It is of following types:

(a) Haploidy: It is the condition in which organisms have half number of chromosomes as compared to diploid or normal organisms. Haploids can be artificially produced by X-ray treatment, delayed pollination, temperature treatment, colchicine treatment and through pollen or anther culture.

Genetic effects:

(a) Haploids show reduction in vigour.

(b) These show reduction in size of vegetative as well as reproductive organs.

(c) These are used in crop improvement because chromosome doubling of haploids produce diploid plants as new varieties) in short span of time.

(d) These are useful in isolation of mutants.

(b) Monoploidy: It is the condition in which organisms have single basic set of chromosome. In diploid organisms (most plants and animals) the haploid and monoploid numbers are the same. However, in certain polyploidy plants such as wheat, the chromosome numbers of haploid and monoploid are different. It is observed in wheat as such:

2n (diploid) = 6X (hexaploid) = 42 (wheat)

or X (monoploid) = 42/6 = 7

(c) Polyploidy: It is the condition in which organisms have more than two sets of chromosomes. These are more common in plants than animals because mostly plants are monoecious. Among animals, polyploidy is very common in amphibians and reptiles. It may be

Natural polyploidy: It occurs naturally due to non-disjunction (non-separation) of chromosomes at anaphase or due to absence of spindle formation.

Artificial polyploidy: It is produced by

(a) Heat or cold treatment to flowers at the first division of zygote

(b) Irradiation of X-rays to vegetative and floral buds.

(c) Chemicals (colchicine) treatment to vegetative and floral buds. Colchicine is a chemical obtained from seed and corm of autumn crocus (*Colchicum autumnale*). It can be obtained either by soaking by seeds and axillary bud or injecting at seedling stage. It arrests and breaks spindle so that cell division occurs without cytokinesis leading to doubling of chromosome number.

Types of Polyploidy:

1. Autopolyploidy: It is obtained by multiplication of same basic set of chromosome but all chromosomes come from same species. Natural autopolyploids are rare. e.g., doob grass (*Cynodon dactylon*) is autotriploid. It is of two types:

(a) Autotriploids: These are produced from a cross between autodiploid and autotetraploid.

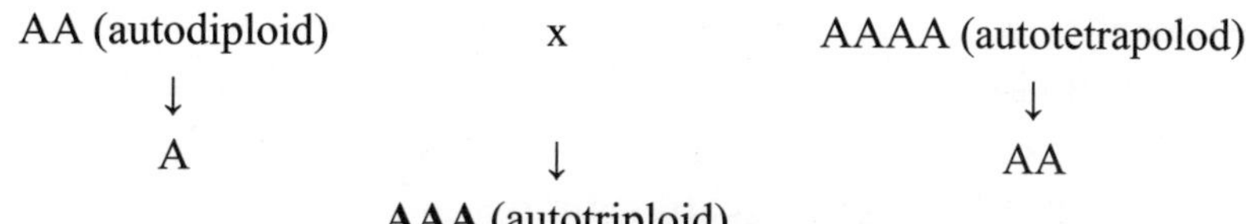

AAA (autotriploid)

These are produced by union of haploid gamete and diploid gamete.

A × AA ⟶ **AAA** (autotriploid)

Features:

(i) These are 3 sets of homologous chromosomes hence trivalents are formed but do not disjoin normally and either disjoin 2:1 chromosomes to two poles or will disjoin 1:1 leading one chromosome as laggard.

(ii) These are sterile because odd chromosomes do not form synapsis because there is no homologue to pair.

(iii) These are more vigorous and leafy.

(iv) These do not form seeds hence reproduce vegetatively. These are economically important in those plants where seed is not commercially important. E.g., Orange, lemon, water melon, grapes

(b) Autotetraploids: These are produced by doubling of autodiploid genome due to colchicine treatment.

AA (autodiploid)
↑←Colchicine treatment
AAAA (autotetraploid)

These are also produced by union of unreduced diploid gametes produced in the absence of meiosis or abnormal meiosis in diploid zygote.

AA × AA⟶**AAAA** (autotetraploid)

Features:

(i) These are 4 sets of homologous chromosomes hence quadrivalents are produced which disjoin in a normal 2:2 manner giving diploid gametes.

(ii) These are fertile because each gamete gets a balanced set of chromosome.

(iii) They are vigorous and highly disease resistant.

(iv) These have showy flowers, larger fruit and seeds and greater vitamin-c content.

(v) These form seeds. These are produced in those plants where larger leaf, flower, fruit, seed are commercially important. e.g., Marigold, lily, snapdragon, primrose.

2. Allopolyploidy: It is obtained by multiplication of hybrid. It is produced either spontaneously by natural hybridisation between two different species followed by spontaneous chromosome doubling or it can be artificially produced crossing two different species followed by colchicines treatment of F_1 hybrid. The alloployploids have better adaptability than diploids. Most of the present day crop plants are alloployploids. It is of two types:

(a) Segmental: It is obtained by hybridisation of two related species hence genesof their chromosome are similar to each other.

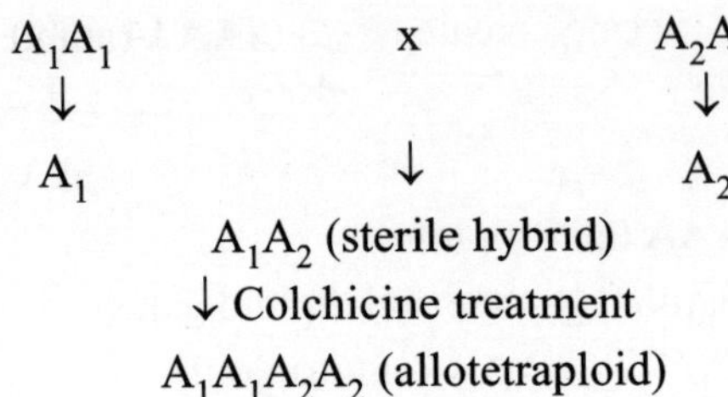

(b) True or genomic: It is obtained by hybridisation of two distantly related species hence chromosomes are different (non-homologous) so the pairing does not occur in F_1 hybrid which becomes sterile.

AA × BB
↓ ↓
A B
↓
AB (sterile hybrid)
↓Colchicine treatment
AABB (allotetraploid)

Examples of new varieties:

1. **Raphanobrassica:** It is intergeneric cross between Radish (*Raphanus sativus*) and cabbage (*Brassica oleracea* var. *capitata*). It was developed by Karpechenko (1928).

Raphanus sativus (2n = 18) x *Brassica oleracea* (2n=18)
↓ ↓
n = 9 n = 9
↓
Sterile hybrid (n+n=2n=18)
↓ Colchicine treatment
(Raphanobrassica) (2n=36=4x) or Allotetraploid or amphidiploids (fertile)

2. **Modern wheat (Common bread wheat):** It was developed by McFadden, Sears and Hitoshi Kihara (1946).

Triticum dicoccum (2n = 28 = 4x) × *Aegilops squarrosa* (Goat grass) (2n = 14 = 2x)
↓ ↓
n = 14 = 2x n = 7 = x
↓
Sterile hybrid (2n=21=3x)
↓Colchicine treatment
Triticum aestivum (2n = 42 = 6x)- Allohexaploid or amphitriploid (fertile)

3. **Triticale:** It is manmade cereal.

Triticum durum (2n = 28 = 4x) × Secale cereale (Rye) (2n = 14 = 2x)

↓ ↓

n = 14 = 2x n = 7 = x

↓

Sterile hybrid (2n=21=3x)

↓Colchicine treatment

Triticale (2n=42=6x) - Allohexaploid or amphitriploid (fertile)

4. **New world cotton:**

Gossypium herbaceum (2n = 26) × *Gossypium raimondii* (2n=26)

(Old world cotton) ↓ ↓ (new world cotton)

n=13 n=13

↓

Sterile hybrid (2n=26(13+13)

↓colchicine treatment

Gossypium hirsutum (2n=52(26+26) -allotetraploid or amphidiploid (fertile)

(Upland or new world cotton)

5. **Autoallopolypolidy:** It is obtained by union of autoplyploid and allopolyploid.

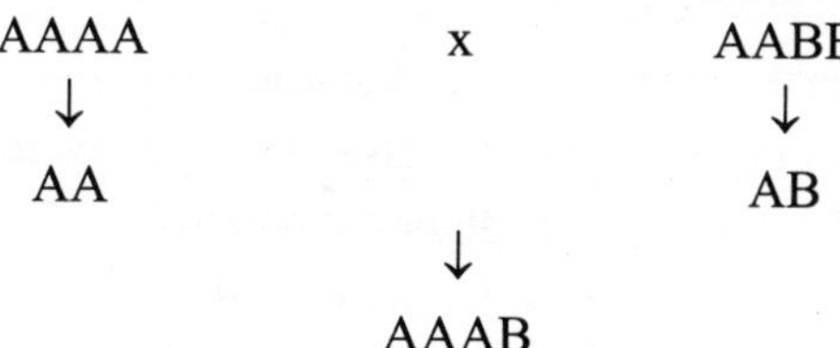

AAAA x AABB

↓ ↓

AA AB

↓

AAAB

↓colchicine treatment

AAAAAABB (autoallooctaploid)

e.g., Sun flower (*Helianthus annus*)

Advantages of polyploidy:

(a) It produces new varieties of plant which are better in quality.
(b) Polyploidy increases the size of different parts of plant giving higher yield.
(c) It increases the size of cell hence larger cell improves the texture of fruits and vegetables.
(d) It increases the vigour and disease resistance.
(e) It increases the vitamin-c content.

Disadvantages:

(a) Polyploids are slow to evolve because they cannot show rapid response to selection.
(b) These show slower growth rate due to slower rate of cell division.
(c) These show low resistance to cold and frost.
(d) These can develop sterility.
(e) These show delayed flowering.

Sex linked Inheritance (Sex linkage)

It is transmission of sex linked characters and their determining genes borne on sex chromosomes from generation to generation.

It was discovered by Thomas Hunt Morgan (1910) on fruit fly (*Drosophila melanogaster*) when he studied the inheritance of red-white eye colour trait. He observed when a white eyed male appeared in a culture of red eyed males.

Traits are of two types: autosomal traits and sex-linked traits

1. **Autosomal traits**: The determining genes are found on autosomes. These show same results in reciprocal crosses. A single recessive allele does not express its effect. These are equally affected in both sexes. There are always two alleles in both sexes. These show direct (non criss-cross) inheritance from parents to offspring.
2. **Sex linked traits:** The determining genes are found on sex chromosomes. All sex linked traits are inherited together. e.g., Haemophilia, colour blindness.

Features:

(a) Most of the sex linked traits are recessive.

(b) These traits show different results in reciprocal crosses.

(c) These traits are more common in males than in females.

(d) These are mostly harmful.

(e) These show mostly criss-cross inheritance. It is of two types:

Diagynic: Father passes X-linked allele of trait to daughter who passes to son.

Father ⟶ daughter ⟶ Son

Diandric: Mother passes X-linked allele of a trait to son who passes to daughter.

Mother ⟶ son ⟶ Daughter

Proof: It was proved by Morgan while working on *Drososphila*.

The chromosome complement in *Drosophila* is XX (female) and XY (male). Its eye colour is sex-linked. Allele of eye colour gene is located on X chromosome and there is no allele in Y chromosome. Female expresses sex linked recessive trait when it has two genes for it while male expresses even it has a single gene. The normal eye colour is red and is dominant over the mutant white eye colour which is recessive. The inheritance of X-linked eye colour in *Drosophila* is shown in the following two crosses:

First cross: Red eyed female × White eyed male

If a homozygous red eyed female is mated with a hemizygous (having a single allele for a trait) white eyed male, then all F_1 individuals are red eyed irrespective of sex. It shows that the gene for white colour is recessive to red colour.

X^RX^R × X^rY (hemizygous/1-allelic)

	X^r	Y
X^R	X^RX^r (red eyed female)	X^RY (red eyed male)
X^R	X^RX^r (red eyed female)	X^RY (red eyed male)

Now red eyed female and red eyed male of F_1 are inter-crossed.

X^RX^r × X^RY

	X^R	Y
X^R	X^RX^R (red eyed female)	X^RY (red eyed male)
X^r	X^RX^r (red eyed female)	X^rY (white eyed male)

Results:

Red eyed individuals : White eyed individuals
3 : 1

Red eyed female : Red eyed male : White eyed male
2 : 1 : 1

In this cross, white eyed females were not observed. Morgan concluded that the gene for eye colour is located on X-chromosome. *Drosophila* will have white eye colour only when both X-chromosomes contain gene for white eye (homozygous recessive).

The above cross shows that a recessive X-linked trait follows criss-cross inheritance (the trait is transmitted from father to grandson through daughters). The latter are carriers because they have a trait but do not express it.

Second cross: White eyed female × Red eyed male

If a homozygous white eyed female is mated with a hemizygous (having single allele for a trait) red eyed male, then all F_1 individuals show that all females are red eyed and all males are white eyed. This cross showed that eye colour and sex were linked.

X^rX^r × X^RY (hemizygous/1-allelic)

	X^R	Y
X^r	X^RX^r (red eyed female)	X^rY (whit eyed male)
X^r	X^RX^r (red eyed female)	X^rY (white eyed male)

Red eyed female : White eyed male
2 : 2

Now red eyed female and white eyed male of F_1 are intercrossed.

X^RX^r × X^rY

	X^r	Y
X^R	X^RX^r (red eyed female)	X^RY (red eyed male)
X^r	X^rX^r (white eyed female)	X^rY (white eyed male)

Result:

Red eyed female : White eyed male : Red eyed male : White eyed male
1 : 1 : 1 : 1

The results of this cross are entirely different from the previous cross. In this cross, both red eye and white eye colour are in equal proportions in both the sexes of F_2 generation.

Morgan finally concluded that eye colour gene is linked to sex and is situated on X-chromosome.

The above cross clearly shows criss-cross inheritance (father transmits his dominant sex-linked trait (red eyed) to his grandsons through daughter).

Thus both crosses show that the pattern of inheritance of eye colour trait exactly parallels the behaviour of X-chromosomes carrying the alleles of eye colour gene. This proves that eye colour gene (mutant white eyed allele or normal red eyed allele) lies on X-chromosome.

Types of sex linked traits:

(a) **X-linked traits:** These traits are determined by homologous region of X chromosome. These do not have alleles on Y-chromosome. These occur more in males than females because there is no homologues of this gene on Y chromosome. Males suffer due to single recessive allele but females will suffer when two recessive alleles are situated.

Man cannot be carrier. He is always diseased. Thus he is hemizygous having one copy of X chromosome for trait because his Y chromosome carries no defective allele.

Woman is homozygous for recessive disorders because her both X chromosomes carry defective alleles hence she is diseased and passes the allele of sex linked trait to both her sons and daughters.

Woman is also heterozygous for recessive disorders because her only one X chromosome carries defective allele hence she is carrier (appears normal).

These traits or disorders show criss-cross inheritance which occurs in two ways:

It is explained in the following five cases:

Case 1: Marriage between a normal man and a carrier woman

XY × X^cX

	X^c	X
X	X^cX (carrier daughter)	XX(normal daughter)
Y	X^cY (diseased son)	XY (normal son)

Results: 50% daughters are carrier and 50% are normal.

50% sons are diseased and 50% are normal.

Case 2: Marriage between a normal man and a diseased woman

XY × X^cX^c

	X^c	X^c
X	X^cX (carrier daughter)	X^cX(carrier daughter)
Y	X^cY (diseased son)	X^cY (diseased son)

Results: 100% daughters are carriers.

100% sons are diseased.

Case 3: Marriage between a diseased man and a normal woman

X^cY × XX

	X	X
X^c	X^cX (carrier daughter)	X^cX(carrier daughter)
Y	XY (normal son)	XY (normal son)

Results: 100% daughters are carriers.

100% sons are normal.

Case 4: Marriage between a diseased man and a carrier woman

X^cY × X^cX

	X^c	X
X^c	X^cX^c(diseased daughter)	X^cX(carrier daughter)
Y	X^cY (diseased son)	XY (normal son)

Results: 50% daughters are diseased and 50% are carrier.

50% sons are diseased and 50% are normal.

Case 5: Marriage between a diseased man and a diseased woman

X^cY × X^cX^c

	X^c	X^c
X^c	X^cX^c(diseased daughter)	X^cXc(diseaseddaughter)
Y	X^cY (diseased son)	XcY (diseased son)

Results: 100% daughters are diseased.

100% sons are diseased.

(b) **Y-linked traits:** These traits are determined by holandric genes present on non-homologous region of Y-chromosome only. These do not have alleles on X-chromosome. These are expressed only in males but not expressed in females. These are rare because Y chromosome is very small and does not contain many genes. These show non criss-cross inheritance.

Holandric: These are inherited straight from father to son.

Father → Son → Grandson. (male to male).

e.g., Hypertrichosis(excessive growth of hairs on ears), Testis determining factor(TDF), Keratoderma dissipatum (thickened skin on extremities), Webbed toes in humans, porcupine skin.

(c) **X-Y linked traits:** These are determined by the genes which are localised in homologous region (the part which synapses) of X and Y chromosome. e.g., Xeroderma pigmentosum (patches of pigments on skin and skin cancer).

Sex limited traits: These are autosomal inherited traits caused by a single gene and expressed only in particular sex but their genes are also found in other sex. e.g., Development of Ovary and breast, secretion of milk; feminine musculature and voice in human female; and beard, stout musculature and deep voice in human male; Brilliant plumage in male peacock and Antler in male deer.

Sex influenced traits: These are autosomal inherited traits observed more frequently in one sex than in other hence these are influenced differently in male and female due to sex hormones. e.g., beard, moustache, deep voice (low pitch), cleft palate (fissure in roof of mouth), forelock (area of skin without pigment), hair lip (a fissure in upper lip), stammering or stuttering (speak haltingly), pattern baldness (male show baldness if he has at least one dominant allele (Bb) and female shows baldness if she has both dominant alleles (BB). The baldness is summarised as such:

Genotypes for	**Male**	**Female**
Baldness	BB, Bb	BB
Normal	bb	Bb, bb

Genetic disorders: These occur in both sex chromosome and autosomes.

Sex-linked disorders: These occur due to gene mutation in sex chromosomes. These are of two types:

1. **Sex-linked (X-linked) recessive disorders**: The mutated gene is recessive. These are caused by defective gene on X-chromosome. These are found more in male than female. These include:
   (a) **Haemophilia**: It was discovered by John Cotto. The affected individual is unable to clot blood properly from a minor cut hence such individual may die of blood loss hence called **Bleeder's disease**. It is quite common in royal families of Europe. The disease spread to them through children of Queen Victoria (1819-1901). The ancestors of queen did not possess the disease. It appears that gene for haemophilia developed either in germ cell of her father or herself through mutation. It is of two types:

   Haemophilia A: It occurs due to mutation of Anti-Haemophilic Globulin (AHG) or Factor VIII gene hence factor VIII is not formed. It is more common (one in every ten thousand in men and one in every ten millions in women).

   Haemophilia B: It occurs due to mutation of Plasma Thromboplastin Factor or Factor IX gene hence factor IX is not formed. It is called Christmas disease. It is less common.

   (b) **Colour blindness**: It was reported by Dalton and described by Horner (1876). The affected individual is unable to distinguish between red and green colours due to absence of some colour sensitive cells of retina. Red colour blindness is called protanopia and green colour blindness is called deuteronopia.

   (c) **Duchenne Muscular Dystrophy**: The mutated gene on X-chromosome does not produce a protein, dystrophin. The affected individual suffers from muscular degeneration due to absence of dystrophin protein. The muscular contraction is inhibited because muscle cell fails to release Calcium in the absence of dystrophin. The muscular degeneration leads to loss of skeletal muscle control.

   (d) **Glucose-6-Phosphate dehydrogenase deficiency(G6PD) Syndrome**: The mutated gene on X-chromosome does not produce Glucose-6-phosphate dehydrogenase which is essential for carrying out Hexose Monophosphate Shunt (HMS) in RBC. The haemoglobin crystallises and RBC ruptures (haemolysis) in the absence of G-6-PD which is used to form glutathione which protects Hb and RBC. The affected individual suffers from haemolytic anaemia, yellowing of skin and white of eye (jaundice). It is also known as favism. Such individuals cannot suffer from malaria because they cannot have plasmodium in their RBC.

2. **Sex-linked (X-linked) dominant disorders**: The mutated gene is dominant which is present on X-chromosome so only one copy of allele is enough to cause disorder when inherited from parent. These occur more in females than in males. Males are diseased when defective allele is present only on X-chromosome. Males cannot be diseased if mother is not diseased. Females are diseased when defective allele is present on one X-chromosome. These include:
   (a) Chondrodysplasia punctata
   (b) Hypophosphatemic ricket
   (c) Retinitis pigmentosa

Autosomal disorders: These occur due to gene mutation in autosome. These are of two types:

1. **Autosomal recessive disorders:** The mutated gene is recessive. These occur in both male and female. Homozygous dominant individuals are normal (AA) and heterozygous dominant individuals are carriers (Aa) but homozygous recessive individuals are diseased (aa). These include

(a) Alkaptonuria (AKU): It was first recessive disorder to be discovered by Garrod. It is defect in amino acid metabolism. It occurs due to mutation of gene present on 11th chromosome by which homogentisic acid oxidase is not produced. The affected individuals cannot convert homogentisic acid into acetoacetic acid due to lack of homogentisic acid oxidase enzyme. Thus homogentisic acid (alkapton) is accumulated in blood and is excreted in large amount in urine. The urine turns black after coming in contact with air due to oxidation of homogentisic acid.

(b) Albinism: It is defect in amino acid metabolism. It occurs due to mutation of gene present on 11th chromosome by which tyrosinase enzyme is not produced. The affected individuals cannot convert tyrosine into melanin due to lack of tyrosinase enzyme. Hence melanin (black pigment) is not formed in skin, hairs and iris of eye. The albinos have poor vision but lead normal life.

(c) Phenylketonuria (PKU): It is defect in amino acid metabolism. It occurs due to substitution mutation of gene present on 12th chromosome by which phenylalanine hydroxylase enzyme is not produced. The affected individuals cannot convert phenylalanine into tyrosine due to lack of phenylalanine hydroxylase enzyme. Thus phenylalanine is accumulated in blood and some of it is passed from urine. The affected child is normal at birth but within a few weeks, the level of phenylalanine starts rising in plasma and by age of six months, pale skin, walking disability and severe mental retardation occur. It occurs one in five thousand births.

(d) Sickle cell anaemia: It is hereditary blood disorder. It occurs due to transversion type of substitution mutation of gene present on 11th chromosome. The mutated gene (Hb^SHB^S) causes single base change in amino acid sequence of β-chain hence glutamic acid (hydrophilic) is changed to valine (hydrophobic) at 6th position of β-chain. Valine being hydrophobic forms hydrophobic bond with complimentary sites of other globin molecules due to polymerisation. It causes Hb molecules to aggregate, forms fibres after collapsing and distorts the shape of RBC (biconcave) into sickle shaped.

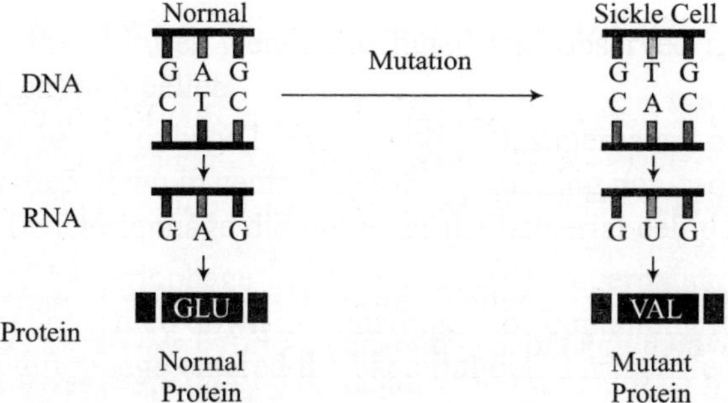

It occurs under low oxygen tension (at high altitude and during strenuous exercise), temperature changes and stress. It is found in population of equatorial region (tropical Africa) and also in American blacks whose ancestors belonged to Africa.

Sickle shaped RBC causes secondary effects such as severe anaemia (because it has tendency to clot and rupture), reduction in blood circulation and damage to spleen, liver and brain.

It is controlled by two alleles. There are following three genotypes:

- Hb^AHb^A: It is homozygous dominant and normal.
- Hb^AHb^S: It is heterozygous dominant and normal but carrier. It causes periodic discomfort.

- Hb^SHb^S: It is homozygous recessive and diseased. The affected individuals die before maturity. Natural selection keeps the lethal sickle allele (HbS) at a slow frequency in the population because affected individuals die before attaining the age of 20 without producing offspring.

(e) Thalassemia: It was reported by Cooley (1925). It is inherited blood disorder. It occurs due to either mutation or deletion of gene controlling the formation of globin chain of haemoglobin (made up of 4 chains-2 α and 2β) by which rate synthesis of one of globin chain is reduced causing anaemia.

Types: On the basis of types of globin chain affected

(a) Alpha thalassemia: In this production of alpha chain is reduced and that of beta chain is increased. It is controlled by two genes (HBA-1 and HBA-2) having four alleles on 16th chromosome. There are four cases:

(a) If mutation occurs and there is one defective allele then it causes thalassemia silent. The genotype is ●ααα.

(b) If mutation occurs and there are two defective alleles then it causes thalassemia minor (large number of microcytic RBC and lesser number of Hb). The genotype is ●●αα.

(c) If mutation occurs and there are three defective alleles then it causes Haemoglobin bart (no Hb) in infants and haemoglobin H (unstable Hb) in adults. The genotype is ●●●α.

(d) If mutation occurs and there are all four defective alleles then it causes thalassemia major (severe haemolytic anaemia, pale skin, fussiness). It is also called Cooley's anaemia. The genotype is ●●●●.

(b) Beta thalassemia: In this, the production of beta chain is decreased. It is controlled by single gene (HBB) with two alleles present on 11th chromosome.

(a) If mutation occurs and there is defective allele then it causes thalassemia minor. The genotype is ●β.

(b) If mutation occurs and there are two defective alleles then it causes thalassemia major. The genotype is ●●.

(c) Cystic fibrosis: It is hereditary disease. It occurs due to mutation in a gene present on 7th chromosome. This causes inability to transport salts hence mucous thickens on lung, pancreas, liver, kidney, intestine. The affected individuals contain high level of Na^+ and Cl- in their sweat.

(d) Galactosemia: It is rare metabolic disorder. It occurs due to mutation of gene present on 9th chromosome by which galactose-1-phosphate uridyltransferase enzyme is not produces. The affected individual fails to convert or metabolise galactose into glucose due to absence of this enzyme. Hence the level of galactose-1-phosphate is increased in various tissue and becomes toxic. It causes hepatomegaly (enlarged liver), cirrhosis, renal failure in child and becomes fatal at the age of 3 years.

(e) Gaucher's disease: It was reported by Phillipe Gaucher (1882). It occurs due to mutation in a gene present on 1st chromosome by which glucocerebrosidase enzyme is not produced. The affected individual shows accumulation of glucocerebroside (Sphingolipid) in WBC of blood, spleen, liver, kidney and brain. It causes anaemia, low platelet count, enlarged spleen and limb bones.

(f) Tay-Sach disease (Infantile amourotic idiocy): It was reported by Warren Tay (1881). It occurs due to mutation of in a gene present on 15th chromosome by which hexosaminidase is not produced. The affected individual shows abnormal accumulation of gangliosides in nerve cell of brain and spinal cord. It causes degeneration of Central nervous system (brain and spinal cord) and becomes fatal at the age of 4 years.

2. **Autosomal dominant disorders:** The mutated gene is dominant. Only one mutated gene is sufficient for disorders. These occur in both male and female. Homozygous dominant individuals are diseased (AA) and heterozygous dominant individuals are diseased (Aa) but homozygous recessive individuals are normal (aa). These include:
   - (a) **Adactyly**: In this, either fingers or toes or both are absent.
   - (b) **Brachydactyly**: In this, fingers or toes are abnormally short.
   - (c) **Polydactyly**: In this, extra finger or toe are found.
   - (d) **Achondroplasia**: In this, bones do not grow much causing dwarfism.
   - (e) **Huntington chorea**: It occurs due to mutation of gene present on 4th chromosome. The affected individual shows gradual deterioration of muscles and brain causing uncontrollable shaking (dance like movement) and loss of cognition (thinking ability).
   - (f) **Myotonic muscular dystrophy**: It occurs due to mutation of gene present on 19th chromosome. The affected individual shows weakness and wasting of muscles. It occurs in adulthood.
   - (g) **Marfan syndrome**: It occurs due to mutation of a gene present on 15th chromosome.
   - (h) **Neurofibromatosis**:It occurs due to mutation of a gene present on 22nd chromosome. It causes non-malignant brain tumour.

Disorders due to gene incompatibility:

1. **Rh-factor incompatibility**: Landsteiner and Weiner(1940) discovered an antigen or protein called Rh factor in blood of Rhesus monkey. Formation of Rh antigen is controlled by dominant gene (R) and its absence by recessive gene (r). The genotypes of Rh+ individuals are RR and Rr and those of Rh- individuals are rr.

   It causes Erythroblastosis foetalis in which death of foetus occurs within womb or just after birth.

   Generally human blood is devoid of Rh antibodies. But it has been noticed that on transfusion of blood of a Rh^+ person to Rh- person. The recipient develops Rh antibodies in its blood plasma. If Rh^+ blood is transfused for the second time it causes agglutination and leads to death of Rh- person.

The father of Rh affected foetus is Rh^+ and mother is Rh-. The child inherits the Rh^+ trait from the father. A few Rh^+ red blood corpuscles of foetus in womb enter in the blood of mother where they develop Rh antibodies. As mother's blood is Rh-, it causes no harm to her. These Rh antibodies along with the mother's blood on reaching the foetal circulation cause clamping of foetal RBCs or agglutination reaction. The first child is somehow born normal because by that time the number of antibodies in mother's blood remain lesser but they increase with successive pregnancies. So a marriage between Rh^+ man and Rh- woman should be avoided.

Man	Woman	Types of biological marriage
Rh^+	Rh^+	Compatible marriage
Rh-	Rh-	Compatible marriage
Rh-	Rh^+	Compatible marriage
Rh^+	Rh-	Incompatible marriage

Chromosomal abnormalities:

These occur in chromosomes. These are of two types:

1. **Autosomal abnormalities:** These occur in autosomes. These occur in both male and female. These can be detected by karyotyping.

   (a) Down syndrome (Mongolism or Mongolian idiocy): It was reported by Langdon Down (1866). It was first syndrome to be reported. It is caused by the presence of extra chromosome in 21st chromosome due to trisomy. It also occurs due to Robertsonian translocation or mosaicism. Trisomy is developed due to non-disjunction hence n+1 male or n+1female gametes are produced. Non-disjunction is meiotic error. It occurs when a pair of homologous chromosome fails to separate or segregate during anaphase-I of meiosis-I. This is called primary non-disjunction which is more common. If it occurs when sister chromatids fail to separate during anaphase-II of meiosis-II. This is called secondary non-disjunction. As a result, one daughter cell receives both the chromosomes of a pair and the other without a chromosome of that pair. It also occurs in mitosis after fertilisation forming mosaicism.

Male gametes	Female gametes	Offspring
22 + X	**23** + X	**45** + XX (female having 47 chromosomes)
23+ Y	22 + X	**45** + XY (male having 47 chromosomes)

   It is more common in females. Its frequency of occurrence is one in 750.

   Symptoms: Rounded face, broad forehead, flat hand or broad palm (palm crease), asymmetrical skull, open mouth, slant eye, epicanthus (folded eyelid), protruding tongue, short neck and fingers, defective heart, undeveloped genitalia and gonads, mental retardation.

   (b) Edward syndrome: It was reported by Edward (1960). It is caused by the presence of extra chromosome number in 18th chromosome due to trisomy.

Male gametes	Female gametes	Offspring
22 + X	**23** + X	**45** + XX (female having 47 chromosomes)
23+ Y	22 + X	**45** + XY (male having 47 chromosomes)

   It is more common in females. It is fatal at the age of 3-4 months. Its frequency of occurrence is one in 5000.

   Symptoms: Small mouth, jaws and nose; short stature, syndactyly (fingers tightly clenched against palm of hand), defective heart, mental retardation etc.

   (c) Patau syndrome (D-syndrome): It was reported by Patau (1960). It is caused by the presence of extra chromosome number in 13th chromosome due to trisomy.

Male gametes	Female gametes	Offspring
22 + X	**23** + X	**45** + XX (female having 47 chromosomes)
23+ Y	22 + X	**45** + XY (male having 47 chromosomes)

   It is more common in females. It is fatal at the age of 3-4 months. Its frequency of occurrence is one in 5000.

   Symptoms: Small head and chin; Deformed ears, defective eyes, cleft palate, syndactyly, mental retardation.

   (d) Cri-Du-chat syndrome: It was reported by Lejeune (1963) in France. It is caused by deletion of half part in short arm of 5th chromosome (submetacentric). It is very rare.

   Symptoms: Small head, receding chin, widely spaced eyes, children cry like cat.

(e) **Chronic myeloid leukaemia** (Myelogenous leukaemia): It was reported in Philadelphia in 1959. It is caused by deletion of some part in distal region of long arm of 22nd chromosome or Philadelphia chromosome (acrocentric).

Symptoms: Excess formation of granular WBC in blood, reduced formation of RBC, severe anaemia.

2. **Sex chromosomal abnormalities**: These occur in sex chromosome.

(a) **Turner syndrome**: It was reported by Turner (1988).It is caused by the absence of one sex chromosome in female due to monosomy. It occurs due to fusion of normal egg (22+X) with abnormal sperm(22+O) without Y chromosome or fusion of abnormal agg (22 + O) without *x* chromosome with normal sperm.

Egg	Sperm	Offspring
22 + X	22 + O (abnormal)	44 + XO (sterile female having 45 chromosome
22 + O	22 + X	44 + XO (sterile femal having 45 chromosome

It is sex chromatin negative. Its frequency of occurrence is one in 10,000.

Symptoms: Webbed skin of neck; heavy neck muscles and narrow hip (masculine feature); short stature; undeveloped breast; small uterus; reduced ovary; no egg and menstruation; hearing impairment.

(b) **Klinefelter syndrome:** It was reported by Harry Klinefelter (1942). It is caused by the presence of one extra sex chromosome in male due to trisomy in either sperm or egg. It occurs due to fusion of normal egg (22+X) with abnormal sperm (22+XY) having extra Y chromosome or fusion of abnormal egg with extra X chromosome with normal sperm.

Egg	Sperm	Offspring
22 + X (normal)	22 + XY (abnormal sperm)	44 + **XXY** (sterile male having 47 chromosomes)
22+**XX** (abnormal)	22+ Y (normal)	44 + **XXY** (sterile male having 47 chromosomes)

It is sex chromatin positive. Its frequency of occurrence is one in 500.

Symptoms: Feminine features such as less muscle, body hairs and gynaecomastia (enlarged breast tissue), long leg, broad hip, small penis, small testis, no sperm and reduced sex drive.

(c) **Super female (Metafemale)**: It is caused.by the presence of one or two extra X chromosome in females due to trisomy. It occurs due to fusion of abnormal egg (22+XX) having extra X chromosome with normal sperm.

Egg	Sperm	Offspring
22 + XX **(abnormal)**	22 + X (normal)	44 + **XXX** (super female having 47 chromosomes

Symptoms: Abnormal height, abnormal sexual development, mental retardation.

(d) **Super male (metamale):** It is caused by the presence of one or two extra Y chromosome in males due to trisomy. It occurs due to fusion of normal egg (22+X) with abnormal sperm (22+YY) having extra Y chromosome. It is also known as Jacob's syndrome or criminal syndrome.

Egg	Sperm	Offspring
22+X (normal)	22 + YY (abnormal)	44 + **XYY** (super male having 47 chromosomes

Symptoms: Abnormal height, aggressive attitude, criminal bent of mind, mental retardation.

Approaches to study human genetics:

In human beings there are several traits which follow the same genetic principles of plants and animals. Around 200 traits are hereditary in human beings in which there are dominant and recessive traits. Most of the traits are transmitted in simple Mendelian manner, others show post-Mendelian principles of genetics such as codominance, polygenes, sex-linked genes etc.

Character(Trait)	Dominant	Recessive
Hair	Dark, curly and abundant	Light, straight and less
Skin colour	Normal	Albinism
Eyes	Brown and large	Blue/green/hazel and small
Colour vision	Normal	Colour blind
Ears	Long,wide and free	Short, narrow and fused
Nose	Broad	Narrow
Nostril	Wide	Narrow
Cheeks	Dimpled	Normal
Chin	Cleft chin	Smooth chin (no cleft)
Lips	Broad and thick	Thin
Handedness	Right handed	Left handed
Length of toes	2nd toe longer	1st toe longer
Hearing and speech	Normal	Deaf
Vision	Astigmatism	Normal
Stature	Short (dwarf)	Tall
Baldness	Baldness in male	No baldness in female
Polydactyly	Six fingers	Five fingers (normal)
Tongue rolling	Ability to roll	Inability to roll
PTC tasting	Ability to taste	Inability to taste
Blood groups	A,B and AB	O
Blood clotting	Normal	Haemophilia
Rh factor	Positive	Negative

The inheritance in human beings is very complicated because

(a) It is controlled by two or more than two pairs of genes (not controlled by a single pair of gene).

(b) A person can observe three or four generations.

(c) It has mixed ancestry due to marriage which combines the entirely different genetic characters of two families in children. Hence few traits are pure.

(d) The generation time is long (often 20 years or more).

(e) The number of progenies is small.

(f) Human traits are easily influenced by environment.

(g) Human has large number of chromosomes (46).

(h) Human cannot be subjected to controlled mating.

There are different approaches to study human genetics:

Population genetics: It involves the study of distribution of traits and estimation of gene frequency in population of organisms. It is explained by

1. **Hardy and Weinberg law**: The alleles and genotype frequencies in a population will remain constant from generation to generation in the absence of other evolutionary influences. They proposed an equation

$$p^2 + 2pq + q^2 = 1$$

or

$$p + q = 1$$

or

$$(p + q)^2 = 1$$

p = dominant allele

p^2 = frequency of homozygous dominant individual

q = recessive allele,

q^2 = frequency of homozygous recessive individual

$2pq$ = frequency of heterozygous individual

The seven assumptions underlying Hardy-Weinberg equilibrium are as follows:

(a) Organisms are diploid

(b) Only sexual reproduction occurs

(c) Generations are non-overlapping

(d) Mating is random

(e) Population is large

(f) Allele frequencies are equal in the sexes.

(g) There is no migration, mutation or selection.

2. **Pedigree analysis:** It is a diagram showing the occurrence and appearance of inheritance of certain genetic traits (autosomal or sex chromosomal or dominant or recessive or X-linked or Y-linked) of organism and its ancestors. It is a type of genetic analysis in which trait is traced through several generations of a family to determine how a trait is inherited. It includes law of probability, chance of difference in realised ratio based on smallness of progeny, elimination of alternatives and inheritance of genes which follow Mendelian features.

   Symbols: There are several symbols showing different conditions in a family.

□ = male
[2] = two males
↗□ = proband male
deceased male
■ = affected male
affected by history
examined male
fraternal twins
identical twins (male)
[A] = adopted male

○ = female
(2) = two females
↗○ = proband female
deceased female
● = affected female
affected by history (female)
○ = examined female
⊙ = carrier
○ ○ = fraternal twins
identical twins (female)
(A) = adopted female

◇ = sex unknown
◇ = lived one day
stillbirth
miscarriage
(P) = pregnancy
□—○ = marriage
□═○ = consanguineous marriage
extramarital mating
divorce or separation
no children

Tips to solving the pedigree problems:

(a) Firstly check cytoplasmic inheritance
(b) Then check Y-linked inheritance
(c) Then check X-linked dominant / autosomal dominant inheritance
(c) Finally check X-linked recessive / autosomal recessive inheritance

Basic facts:

(a) When one or both the parents are affected (diseased): It is often a dominant
(b) Both normal parents producing normal offspring: It is always dominant
(c) Both normal parents producing diseased offspring: It is always a recessive

Diseased = Affected; Normal=Unaffected

Types of traits:

Autosomal recessive traits: (AA = Normal, Aa = Carrier and aa = Diseased or affected)

(Alkaptonuria, albinism, phenylketonuria, Sickle cell anaemia, Thalassaemia, Cystic fibrosis)

(a) These occur in both sexes with almost equal frequency.
(b) Disease skips any one generation.
(c) Diseased mother can produce normal children.
(d) Diseased individual has no affected father.
(e) Normal parents (must have carriers) can produce diseased children
(f) If both parents are heterozygous (carriers), then ¼ of the offspring will be diseased and ¾ of offspring will remain normal.
(g) Both the diseased parents cannot have normal offspring.

Autosomal dominant traits: (AA=Diseased, Aa =Diseased and aa=Normal)
(Polydactyly, Huntington chorea, Myotonic dystrophy)

(a) These occur in both sexes with almost equal frequency.
(b) Both sexes transmit the disease to their offspring.
(c) Disease occurs in every generation (affected males and females).
(d) Diseased father transmits disease to his sons.
(e) Diseased individual has atleast one diseased parent.
(f) Normal parents do not transmit the disease.
(g) Diseased parents can transmit the disease in ¾ offspring but ¼ offspring will remain normal.
(h) When one parent is diseased (heterozygous) and the other parent is normal, ½ of the offspring will be diseased.

Sex-linked (X-linked) recessive traits: (Colour blindness, Haemophilia, Duchenne muscular dystrophy)

(a) Disease may skip any one generation.
(b) These are more common in males than females.
(c) Diseased mother will transmit disease to all her sons.
(d) Diseased father will transmit disease to all his daughters.
(e) Disease is never passed from father to son.
(f) Diseased sons are usually born to normal mothers.
(g) Diseased father has carrier daughters.
(h) If diseased mother has one son normal and another son diseased then it cannot be X-linked recessive.

Sex-linked (X-linked) dominant traits: (Hypophosphatemic ricket, Retinitis pigmentosa)

(a) Disease does not skip any generation.
(b) These are more common in females than in males.
(c) Diseased father transmits disease in all daughters. If any daughter is normal then it cannot be X-linked dominant.
(d) Diseased father never transmits disease to sons.
(e) Diseased mother transmits disease to sons.
(f) Diseased daughters must have either diseased mother or diseased father.

Y-linked traits: (Hypertrichosis)

(a) Disease is passed from father to all his sons only.
(b) All females will remain normal.
(c) Disease does not skip any generation.

Maternal or cytoplasmic influence:

(a) Diseased mother transmits disease in all children (sons and daughters).
(b) Normal mother has normal children.

(c) If normal mother has at least one offspring diseased then it cannot be cytoplasmic inheritance.

(d) If diseased mother has normal offspring then it cannot be cytoplasmic inheritance.

A few examples are as follows:

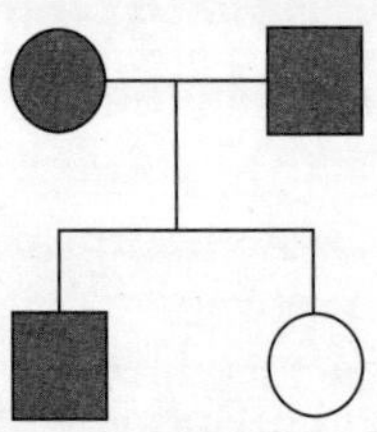

Autosomal Dominant

Cannot be recessive asaffected parents could not have an unaffected offspring

Parents Must be heterozygous

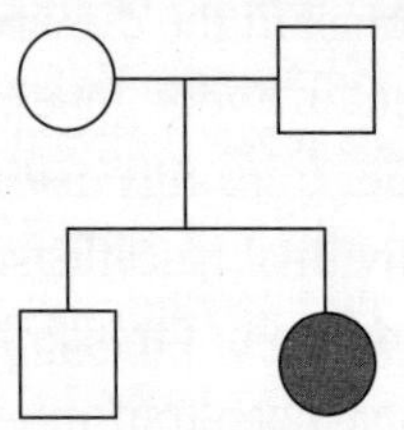

Autosomal Recessive

Cannot be dominant as unaffected parents could not have an affected offspring

Parents Must be heterozygous

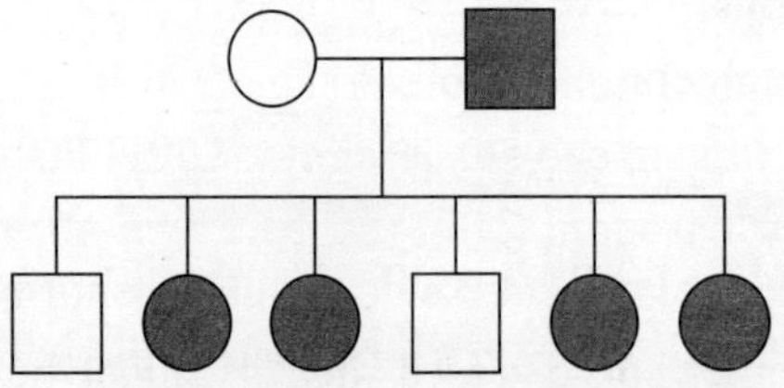

X-Linked Dominant

Mode of inheritance cannot be confirmed

However, 100% incidence of affected daughters from an affected father suggests X-linked dominant inheritance

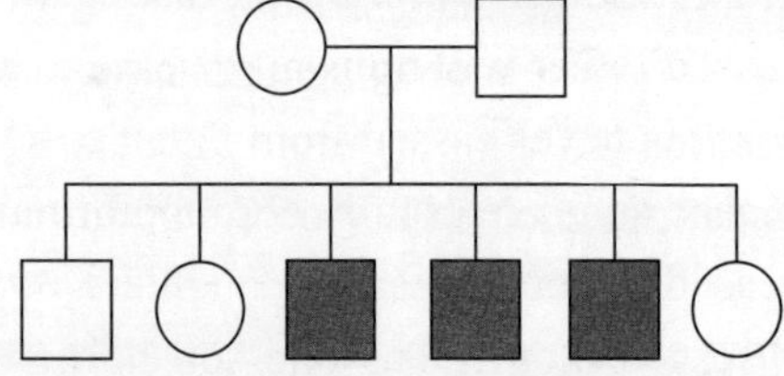

X-Linked Recessive

Sex linkage cannot be confirmed

However, high incidence of affected sons suggests X-linked recessive inheritance

Importance:

(a) It can locate the origin of trait in the ancestor.

(b) It indicates the possibility of children with genetic defects.

(c) It can indicate the harm in the marriage of close relatives.

(d) It is used for genetic counselling.

(e) It is used in medical research.

Examples

1. Solve the following pedigree problem

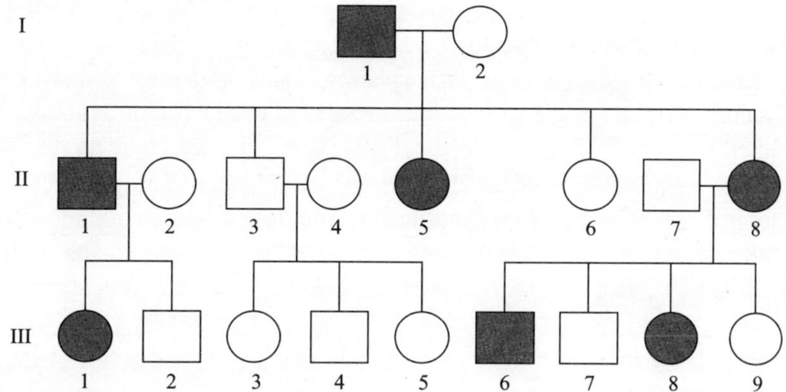

2. Solve the following pedigree problem

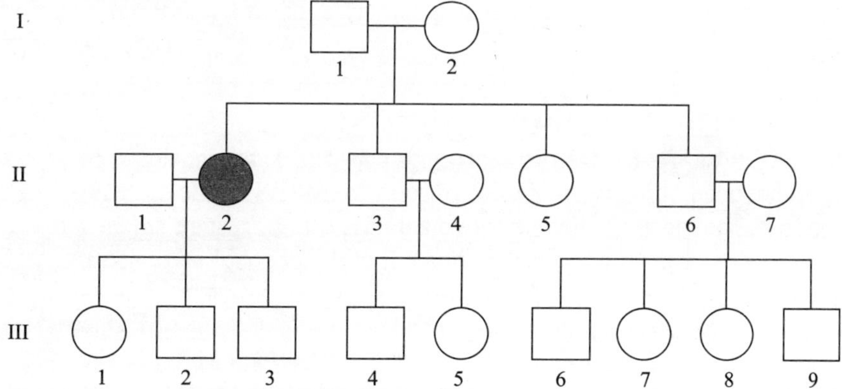

3. Solve the following pedigree problem

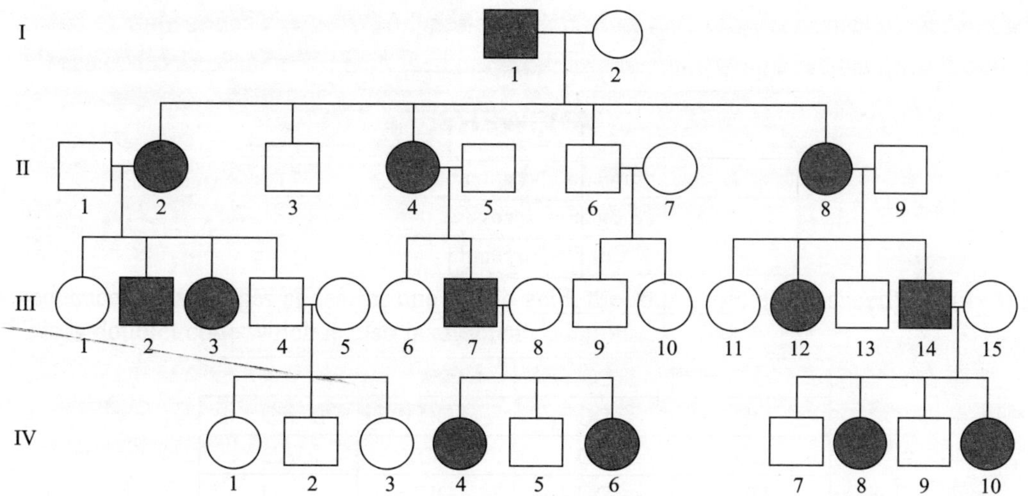

4. Solve the following pedigree problem

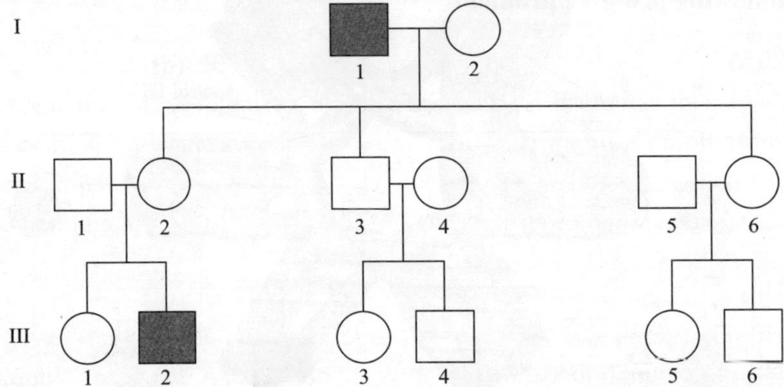

5. Solve the following pedigree problem

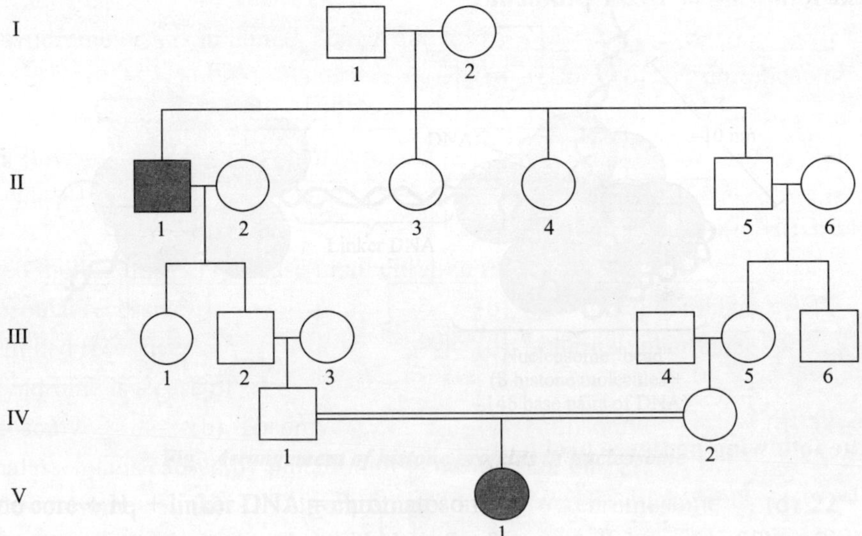

ANSWERS

1.	Autosomal dominant
2.	Autosomal recessive
3.	X-linked dominant
4.	X-linked recessive
5.	Autosomal recessive

SHORT ANSWER TYPE QUESTIONS

1. Distinguish between somatic variation and germinal variation.
2. Discuss the significance of variation.
3. Why Mendel selected garden pea?
4. What are seven traits of Mendel?
5. Who rediscovered Mendelism?
6. Explain reciprocal cross.
7. Distinguish between back cross and test cross.
8. Distinguish between incomplete dominance and codominance.
9. Explain lethal gene with examples.
10. Write notes on pleiotropy.
11. Distinguish between complete linkage and incomplete linkage.
12. Explain interference and coincidence.
13. Explain the environmental control of sex determination with reference to *Bonellia.*
14. What are the contributions of Morgan in genetics?
15. What are the features of mutation?
16. Distinguish between somatic mutation and germinal mutation.
17. Write notes on chemical mutagens.
18. Discuss the role of ultraviolet rays in causing mutation.
19. Distinguish between transition mutation and transversion mutation.
20. Write notes on frame shift mutation.
21. Explain deletion, duplication, translocation and inversion.
22. Distinguish between paracentric inversion and pericentric inversion.
23. Explain monosomy and trisomy.
24. How colchicine causes polyploidy?
25. Distinguish between autotriploid and autotetraploid.
26. How *Raphanobrassica* is developed?
27. Mention the examples of autosomal recessive disorders.
28. Explain sex limited traits and sex influenced traits with examples.
29. Explain the criss cross inheritance.
30. Write notes on haemophilia.
31. Distinguish between autosomal inheritance and sex linked inheritance.
32. How Down syndrome is developed?
33. How Turner syndrome is developed?
34. Distinguish between super female and super male.
35. How autosomal recessive and sex linked disorders are located in a family on the basis of pedigree analysis.

LONG ANSWER TYPE QUESTIONS

1. Discuss the Mendel's laws of inheritance and explain the dihybrid cross.
2. Discuss the multiple alleles with reference to blood group inheritance.
3. Explain the polygenic inheritance with reference to skin colour of human beings.
4. Discuss the chromosomal theory of inheritance.
5. Explain linkage in sweet pea with reference to coupling phase and repulsion phase.
6. What is crossing over? Describe its types, mechanism and significance.
7. Discuss the chromosomal theory of sex determination.
8. Describe the molecular mechanism of gene mutation.
9. Discuss the proof of sex linked inheritance.
10. Discuss the occurrence of sickle cell anaemia and thalassemia.

MULTIPLE CHOICE QUESTIONS

1. Mendel worked on plants for seven years from
   (a) 1850 to 1857 (b) 1854 to 1861 (c) 1856 to 1863 (d) 1863 to 1870
2. A cell is heterozygous at 3 gene loci. How many different types of gametes can it form?
   (a) 2 (b) 3 (c) 6 (d) 8
3. Red and tall dominant character hybrid plant is crossed with recessive white dwarf plant (RrTt × rrtt). What will be the ratio of respective four combination Red tall, red dwarf, white tall and white dwarf in next generation?
   (a) 9 : 3 : 3 : 1 (b) 9 : 3 : 4 : 0 (c) 15 : 1 : 0 : 0 (d) 4 : 4 : 4 : 4
4. A double heterozygous tall plant with yellow colour of seed is selfed. The ratio of dwarf plants with green cotyledons is
   (a) 1/4 (b) 1/8 (c) 1/16 (d) 3/16
5. The recombination frequency between the genes A and B; A and C; B and C are 9%, 17% and 26% respectively. The sequence of genes A, B, C on the chromosome is
   (a) ABC (b) BAC
   (c) ACB (d) information is insufficient
6. A and B genes are linked. What should be the genotype of progeny in a cross between AB/ab and ab/ab ?
   (a) AAbb and aabb (b) AaBb and aabb (c) AABB and aabb (d) aaBB and aabb
7. If haploid number of chromosomes is 10, what would be the number in monosomic and tetrasomic?
   (a) 10 and 40 (b) 19 and 22 (c) 9 and 22 (d) 19 and 40
8. If all the sons of a couple are colour blind then
   (a) Mother is colour blind (b) Mother is carrier and father is colour blind
   (c) Mother is normal and father is colour blind (d) Mother is carrier and father is normal
9. In *Drosophila*, white eye colour is recessive X-linked traits while red eye colour is dominant. A while eyed female is crossed with red eye male. The female offspring with red eye colour would be:
   (a) 0% (b) 25% (c) 50% (d) 100%
10. How many phenotypes are formed in skin colour inheritance of human being if it is controlled by 3 pairs of genes?
    (a) 5 (b) 7 (c) 9 (d) 10
11. A trihybrid cross is made between two yeasts, both having genotypes AaBbCc. The proportion of offspring having genotype aaBbcc will be
    (a) 1/64 (b) 1/32 (c) 1/16 (d) 3/64
12. From a cross AABb × aaBb, the genotypes AaBB : AaBb : Aabb : aabb will be obtained in the ratio of
    (a) 2 : 1 : 1 : 0 (b) 1 : 2 : 1 : 0 (c) 1 : 2 : 1 : 1 (d) 1 : 2 : 0 : 1
13. Two genes 'a' and 'b' present on the same chromosome show 12% recombination. The distance between them is
    (a) 0.12 map units (b) 1.2 map units (c) 12 map units (d) 120 map units

14. Pure homozygous offspring in a dihybrid cross in F_2 generation will be
(a) ¼ (b) 1/8 (c) 3/8 (d) 1/16
15. Hemophilic man marries a normal woman. Their offspring will be:
(a) All normal (b) All haemophilic
(c) All boys haemophilic (d) All girls haemophilic
16. The phenotypic ratio obtained by crossing the F_1 dihybrid with recessive parent is:
(a) 9 : 3: 3 : 1 (b) 3 : 1
(c) 1 : 1 : 1 : 1 (d) 1 : 1 : 1 : 1 : 1 : 1: 1 : 1
17. How many types of gametes are obtained from a plant of genotype AABbccDd?
(a) 2 (b) 4 (c) 4 (d) 8
18. The ratio of parental to recombinants in dihybrid ratio phenotypically is
(a) 6 : 10 (b) 10 : 6 (c) 12 : 4 (d) 9 : 7
19. The number of hybrids dominant of both traits in dihybrid ratio is
(a) 4 (b) 6 (c) 12 (d) 16
20. The possible phenotypes and genotypes for ABO blood groups are
(a) 4 and 6 (b) 6 and 4 (c) 9 and 6 (d) 16 and 10
21. Wheat plant is 6x = 42, then the number of chromosomes in monosomic, haploid and monoploid condition is, respectively
(a) 43, 21 and 7 (b) 41, 21 and 7 (c) 15, 7 and 7 (d) 13, 7 and 7
22. Probability of male child of haemophilic father and normal mother becoming haemophilic is
(a) 0% (b) 25% (c) 50% (d) 100%
23. For MN blood group system, the frequencies of M and N alleles and 0.7 and 0.3, respectively. The expected frequency of MN blood group bearing organism is likely to be:
(a) 40 % (b) 21 % (c) 42 % (d) 49 %
24. From a cross AaBB X aaBB the genotypic ratio in F_1 generation will be obtained in terms of
(a) 3 AaBB : 1aaBB (b) All AaBB : No aaBB
(c) 1AaBB : 1 aaBB (d) 1 AaBB : 3 aaBB
25. What proportion of the offspring of a cross AABBcc x AaBbCc will be completely heterozygous if genes segregate individually?
(a) ¼ (b) 1/16 (c) ½ (d) 1/8
26. A man with blood group B marries a woman with blood group A and the impossible blood group in children will be
(a) A and B only (b) B and AB only (c) A and AB only (d) None
27. In a cross between genotype AB and ++, 650 out of 1000 individuals were of parental type. The distance between A and B is
(a) 35 map units (b) 45 map units (c) 30 map units (d) 15 map units
28. Sickle cell anaemia is an example of
(a) Codominance (b) Pleiotropy (c) Sublethality (d) All of the above

29. The four daughter cells derived from a single meiosis differ from each other due to
    (a) Difference in chromosome number
    (b) Crossing over only
    (c) Independent assortment of chromosomes only
    (d) Crossing over as well as independent assortment of chromosomes
30. If a boy's father has haemophilia and his mother has one gene for haemophilia; what is the chance that the boy will inherit the disease?
    (a) 25% (b) 50% (c) 75% (d) 100%
31. Mr. X is heterozygous for one autosomal gene pair Bb and he carries a recessive X-linked gene d. What proportion of his sperms will be bd?
    (a) 0 (b) 1/2 (c) 1/4 (d) 1/8
32. Which chromosome set is found in male grasshopper?
    (a) XY (b) X (c) YY (d) XX
33. The amino acid substituted in sickle cell anaemia is
    (a) Valine for glutamic acid in the beta chain (b) Glutamic acid for valine in the alpha chain
    (c) Glutamnic acid for valine in the beta chian (d) Valine for glutamic acid in the alpha chain
34. Heterozygous tall (Tt) is crossed with homozygous tall (TT). Percentage of heterozygous tall in the progeny would be
    (a) 25% (b) 50% (c) 75% (d) 100%
35. A human male produces sperms with the genotypes AB, Ab, aB, and ab pertaining to two diallelic characters in equal proportions. What is the corresponding genotype of this person?
    (a) Aabb (b) AABb (c) AABB (d) AaBb
36. The genes controlling the seven pea characters studied by Mendel are now known to be located on how many different chromosomes?
    (a) Seven (b) Six (c) Five (d) Four
37. In order to find out the different types of gametes produced by a pea plant having the genotype AaBb it should be crossed to a plant with the genotype
    (a) AABB (b) AaBb (c) aabb (d) aaBB
38. In pea plants, yellow seeds are dominant to green. If a heterozygous yellow seeded plant is crossed with a green seeded plant, what ratio of yellow and green seeded plants would you expected in F_1 generation?
    (a) 9 : 1 (b) 1 : 3 (c) 3 : 1 (d) 1 : 1
39. In a dihybrid cross between AABB and aabb, the ratio of AABB, AABb, aaBb, aabb in F_2 generation is:
    (a) 9 : 3 : 3 : 1 (b) 1 : 1 : 1 : 1 (c) 1 : 2 : 2 : 1 (d) 1 : 1 : 2 : 2
40. What is the ratio of one pair of contrasting characters in F_2 of a dihybrid cross?
    (a) 1 : 2 : 1 (b) 3 : 1 (c) 9 : 3 : 3 : 1 (d) 1 : 4 : 6 : 4 : 1
41. If two genes 'a' and 'b' are linked and shows 20% recombination, the proportion of gametes produced in F_1 by a dihybrid ++/ab derived from a cross between ++/++ and ab/ab will be
    (a) ++ 40, ab 40, +a 10, +b10 (b) ++ 20, ab 20, +a 40, +b40
    (c) ++ 40, ab 10, +a 10, +b40 (d) ++ 10, ab 10, +a 40, +b40

42. If a – b is 10% and b - c is 20% crossing over, then the frequency of double crossing over in a – c will be
(a) 1% (b) 2% (c) 4% (d) 8%
43. Allele 'A' is dominant over allele 'a' and results in black pigmentation of skin. In a mating of Aa × Aa, 6 offspring are produced then the probability of all having black pigmentation is
(a) 0.25 (b) 0.50 (c) 0.75 (d) 1.0
44. A plant of F_1 generation has genotype AABbCC. On selfing of this plant, the phenotypic ratio in F_2 generation will be
(a) 3 : 1 (b) 1 : 1 : 1 : 1 (c) 1 : 2 : 1 (d) 9 : 3 : 3: 1
45. In population of 1000 individuals, the gene frequency of dwarf plants is 0.4, and then the number of heterozygotes will be
(a) 480 (b) 360 (c) 160 (d) 600
46. Two alternative alleles, the dominant 'A' and recessive 'a' are in equilibrium is population the frequency of A is 0.6. The percentage of individuals showing the dominant trait in the population should be:
(a) 36% (b) 60% (c) 75% (d) 84%
47. If one gamete in 5 carries a recessive allele then the frequency of homozygous recessive genotype in a population at Hardy-Weinberg equilibrium is:
(a) 4% (b) 10% (c) 16% (d) 20%
48. The coupling ratio in incomplete linkage is:
(a) 1 : 7 : 7 : 1 (b) 7 : 1 : 1 : 7 (c) 1 : 1 : 1 : 1 (d) 9 : 3 : 3 : 1
49. In autosomal recessive traits, the genotype of normal but carrier individual is:
(a) AA (b) Aa (c) aa (d) Aa and AA
50. If skin colour is controlled by 2 pairs of gens and the number of individuals in a population is 960 then the number of mulattoes in the population is:
(a) 160 (b) 360 (c) 720 (d) 420
51. The probability of the offspring of genotype aaBbCCdd in a cross between two plants of genotypes AabbCcDd and AaBbCcdd is
(a) 1/8 (b) 1/16 (c) 1/32 (d) 1/64
52. A heterozygous Pea plant having genotype TtYy (tall with yellow seeds) is allowed to self fertilise. What is the probability that an offspring will be either tall with yellow seeds, tall with green seeds or dwarf with yellow seeds?
(a) 64% (b) 74% (c) 84% (d) 94%
53. If two heterozygous plants are crossed and genes are linked then the distribution of phenotypes in F_1 generation for a dihybrid cross will be
(a) 1:1 (b) 3:1 (c) 1:2:1 (d) 1:1:1:1
54. Starch synthesis in Pea seed is controlled by one gene having 2 alleles, B and b. BB produces large sized starch grain while bb produces small sized starch grain. BB seeds are round and bb seeds are wrinkled. Its heterozygote produces
(a) Large sized starch grain and wrinkled seeds (b) Small sized starch grain and wrinkled seeds
(c) Medium sized starch grain and wrinkled seed (d) Medium sized starch grain and round seed

55. If two heterozygous tall pea plants (Tt) are crossed then what is the probability that out of three plants picked up at random, two would be dwarf and one would be tall?
(a) 3/8 (b) 3/16 (c) 9/16 (d) 9/64

56. If two heterozygous tall pea plants (Tt) are crossed then what is the probability that out of five plants will be dwarf?
(a) 16% (b) 26% (c) 36% (d) 46%

57. When a red grain variety of wheat is crossed with another white grain variety of wheat, F_1 hybrid is produced. On selfing F_1 hybrid, F_2 generation is produced. How many offspring of F_2 resemble phenotypically to their parents (grain colour) if it is controlled by3 gene pairs?
(a) 1/16 (b) 2/64 (c) 9/64 (d) 3/64

58. An individual with 'cd' genes is crossed with wild type '++'. On test crossing the progenies were found to be '+c' 105, '+d' 115 , 'cd' 880 and '++' 900. The distance between two 'cd' genes is
(a) 11 map unit (b) 22 map unit (c) 20 map unit (d) 50 map unit.

59. A species of plant A with 20 chromosomes is crossed with another species B with 16 chromosome. When chromosome number in zygote is doubled then it produces a fertile hybrid C. The number of chromosome in each cell of endosperm is
(a) 36 (b) 54 (c) 90 (d) 108

60. Colour blindness is due to sex linked recessive gene. A survey of 500 men from a local population revealed that 20 were colour blind. What percentage of female in this population would be expected to be normal?
(a) 36 (b) 48 (c) 96 (d) 99.84

61. Albinism is known to be due to autosomal recessive mutation. The first child of couple with normal skin pigmentation was an albino. What is the probability of their second child to be albino?
(a) 100% (b) 50% (c) 25% (d) 75%

62. One map unit corresponds to the physical length of of the chromosome in which one cross-over occurs in every
(a) One meiosis (b) Two meiosis (c) Ten meiosis (d) Hundred meiosis

63. Controlled sex mechanism occurs in
(a) *Equisetum* (b) *Luffa* (c) *Sphaerocarpus* (d) *Melandrium*

64. The epistatic effect in which the dihybrid cross ratio 9:3:3:1 between AaBb x AaBb is modified, is the
(a) Interaction between two alleles of the same loci
(b) Interaction between two alleles of different loci
(c) Dominance of one allele on another allele of the same loci
(d) Dominance of one allele on another allele of both loci

65. If Mendel had studied the 7 traits using a plant with 12 chromosomes instead of 14 then what would have been different?
(a) He could have mapped the chromosomes
(b) He would have discovered incomplete dominance
(c) He would have discovered sex linkage
(d) He would not have discovered the law of independent assortment

66. Mental retardation in man increases with
(a) Decrease in X chromosome (b) Increase in X chromosome
(c) Decrease in Y chromosome (d) Increase in Y chromosome

67. *Drosophila melanogaster* has
(a) 2 pairs of autosomes and 1 pairs of allosome (b) 3 pairs of autosomes and 1 pairs of allosome
(c) 4 pairs of autosomes and 1 pairs of allosome (d) 5 pairs of autosomes and 1 pairs of allosome

68. The formation of multivalent at meiosis in a diploid organism is due to
(a) Reciprocal translocation (b) Non-reciprocal translocation
(c) Inversion (d) Deletion

69. One crossing over results in the formation of
(a) 2 parental gametes and 2 recombinants (b) 4 parental gametes
(c) 4 recombinants (d) 2 recombinants

70. Which of the following base is responsible for hot spots for spontaneous point mutation?
(a) Guanine (b) Adenine (c) 5-Bromouracil (d) 5-Methylcytosine

71. A F_1 hybrid having the genotype Ab/aB produces 8% of cross over gametes AB and ab, then the distance between A and B is
(a) 8 cM (b) 16 cM (c) 24 cM (d) 4 cM

72. In a family, father has a blood group A and mother has B. Their children show 50% probability for a blood group AB indicating that
(a) Father is heterozygous (b) Father is homozygous
(c) Mother is heterozygous (d) Either of parent is heterozygous

73. Deletion of half part of 5th chromosome causes
(a) Down syndrome (b) Turner syndrome
(c) Klinefelter syndrome (d) Cat like cry syndrome

74. Rate of crossing over is reduced by
(a) Deletion (b) Duplication (c) Translocation (d) Inversion

75. Raphanobrassica is a classical example of
(a) Autotriploid (b) Autotetraploid (c) Allotriploid (d) Allotetraploid

76. Triticale is manmade cereal developed from
(a) *Triticum durum* and *Aegilops squarrosa* (b) *Triticum aestivum* and *Secale cereale*
(c) *Triticum turgidum* and *Aegilops squarrosa* (d) *Triticum durum* and *Secale cereale*

77. Haemophilia A occurs due to lack of
(a) Factor VI (b) Factor VI (c) Factor VIII (d) Factor 1X

78. Albinism is an autosomal recessive metabolic disorder occurs due lack of
(a) Arginase (b) Tyrosinase
(c) Homogentisic acid oxidase (d) Phenylalanine hydroxylase

79. If red flowered and wrinkled seeded pea plant is crossed with white flowered and round seeded pea plant, then F_1 generation plants are obtained. If selfing of F_1 generation then F_2 generation plants are obtained. How many plants will be with intermediate size of starch grain in F_2 generation?
(a) 9/16 (b) 6/16 (c) 8/16 (d) 1/16

80. In a population of 1000 individuals, 360 belongs to genotype AA, 480 to Aa and the remaining 160 to aa. Based on this data, the frequency of allele 'A' in the population is
    (a) 200 (b) 400 (c) 600 (d) 800
81. A man is heterozygous for autosomal gene pairs (A and b) and is hemizygous for haemophilic gene C. What proportion of his sperms will be bx^c?
    (a) ½ (b) ¼ (c) 1/8 (d) 1/16
82. A couple has 8 children (all boys). What is the chance that their 9th child will be another boy?
    (a) ½ (b) ¼ (c) 1/6 (d) 1/8
83. A couple went to a doctor and reported that both of them are 'carriers' for a particular disorder, their first child is suffering from this disorder and they are expecting their second child. What is the probability that the new child would be affected by the same disorder?
    (a) 25% (b) 50% (c) 75% (d) 100%
84. Multiple alleles are present at
    (a) different loci in the same chromosome
    (b) same loci in the different chromosome
    (c) same locus within the homologous chromosome
    (d) different loci in the different chromosome
85. A man and woman who both appear normal have a child together who has sickle cell anaemia. The woman becomes pregnant again and is told that she is carrying fraternal twins. What is the probability that both of the couple's twins will develop sickle cell anaemia?
    (a) 1/2 (b) 1/4 (c) 1/16 (d) 9/16
86. The genotype of baldness in female is
    (a) BB (b) Bb (c) bb (d) Both BB and Bb
87. Which of the following is incorrect regarding chromosomal abnormalities?
    (a) Epicanthus is seen in Down syndrome
    (b) Klinefelter syndrome occurs due to trisomy of sex chromosome
    (c) Cat like cry syndrome occurs due to trisomy of 5th chromosome
    (d) Turner syndrome is sex chromatin negative
88. A woman heterozygous for colour blindness marries a colour blind man. What is the ratio of carrier daughter, colour blind daughter, normal sons and colour blind sons
    (a) 1:2:1:2 (b) 1:1:1:2 (c) 1:1:1:1 (d) 1:2:2:1
89. Sickle cell anaemia is an example of
    (a) Codominance (b) Sub-lethal gene (c) Pleiotropy (d) All of the above
90. The total number of chromosome in individuals suffering from Turner's syndrome is
    (a) 44 (b) 45 (c) 47 (d) 48
91. The diseased offspring cannot be found in the marriage of
    (a) Normal man and diseased woman (b) Diseased man and normal woman
    (c) Diseased man and carrier woman (d) Diseased man and diseased woman
92. Which of the following laws of Mendel is not applicable for linkage?
    (a) Law of unit factor (b) Law of dominance
    (c) Law of segregation (d) Law of independent assortment

93. A red flowered snapdragon plant is crossed with pink flowered plant then 200 plants are obtained. How many of them are red flowered?
(a) 50 (b) 100 (c) 150 (d) 200

94. A father has AB blood group and mother's blood group is untraced. They have three children. One has 'A' blood group while second has 'B' blood group. What is the possibility of occurrence of 'O' blood group in third offspring?
(a) 0% (b) 25% (c) 50% (d) 75%

95. If parents has free ear lobes and the offspring has attached ear lobes, then the parents must be
(a) Homozygous (b) Heterozygous (c) Hemizygous (d) Codominant

96. Children suffering from PKU are given food low in phenylalanine and supplemented with tyrosine because
(a) They are unable to utilise phenylalanine (b) They do not require phenylalanine
(c) They have increased tyrosine anabolism (d) They have increased tyrosine catabolism

97. Which of the following Mendel's laws is universally acceptable?
(a) Law of unit factor (b) Law of dominance
(c) Law of segregation (d) Law of independent assortment

98. Incomplete dominance was observed in
(a) *Mirabilis* and *Solanum* (b) *Solanum* and *Lilium*
(c) *Lilium* and *Antirrhinum* (d) *Antirrhinum* and *Mirabilis*

99. Complete linkage is observed in
(a) Male *Drosophila* and male silk worm (b) Male *Drosophila* and female silk worm
(c) Female *Drosophila* and male silk worm (d) Female *Drosophila* and female silk worm

100. Hormonal control of sex determination is observed in
(a) Bonellia (b) The Crew's hen (c) Free martin (d) All of the above

101. Haplodiploidy condition is observed in
(a) Honey bees (b) *Drosophila* (c) Human beings (d) Silk worm

102. Which of the following is incorrectly matched?
(a) Non-ionising radiation-UV rays (b) Deaminating agent- Nitrous acid
(c) Alkylating agent- Benzopyrene (d) Intercalating agent-Acridine dye

103. The number of linkage groups in *Pisum sativum* is
(a) 7 (b) 14 (c) 10 (d) 20

104. Thymine dimer is formed in induced mutation is due to
(a) X-ray (b) UV ray (c) Acridine dye (d) Nitrous acid

105. The number of transversion mutation is
(a) 2 (b) 4 (c) 6 (d) 8

106. Frame shift mutation is caused by
(a) Deletion and modification of base pair (b) Deletion and insertion of base pair
(c) Insertion and replacement of base pairs (d) Duplication and inversion of base pairs

107. Gene dosage depends on
   (a) Total number of genes
   (b) Ploidy level of the species
   (c) Frequency of induced mutations
   (d) Number of chromosomes involved in reciprocal translocation

108. Which one of the following alterations in DNA is likely to be lethal?
   (a) Substitution of adenine for cytosine (b) Substitution of methyl cytosine for cytosine
   (c) Deletion of three nucleotides (d) Insertion of one nucleotide

109. Which one of the following phenomena leads to aneuplody?
   (a) Improper division of centromere (b) Desynapsis
   (c) Non-disjunction (d) Disjunction

110. The spontaneous mutations in nature are
   (a) Recessive and lethal (b) Dominant and beneficial
   (c) Dominant and lethal (d) Recessive and beneficial

111. Mendel formulated the law of purity of gametes on the basis of
   (a) Dihybrid cross (b) Monohybrid cross (c) Back cross (d) Test cross

112. Which of the following genotypes is an example of true dihybrid condition?
   (a) Ttrr (b) Ttrr (c) TtRr (d) TTRr

113. If 120 plants are produced on crossing pure red and pure white flowered pea plants, then the ratio of offspring will be
   (a) 90 red : 30 white (b) 30 red : 90 white (c) 60 red : 60 white (d) All red

114. How many types of gametes are produced from the organism with genotype AABBCC?
   (a) 1 (b) 2 (c) 4 (d) 8

115. Two allelic genes are located on the
   (a) Same chromosome (b) Two homologous chromosomes
   (c) Two non-homologous chromosomes (d) Any two chromosome

116. Mrs. X has an autosomal gene pair 'Bb' and she contains X-linked gene 'd' on her both X chromosomes. The percentage of gamete containing 'bd' genes is
   (a) 50% (b) 25% (c) 75% (d) 100%

117. A dihybrid plant on self pollination produced 400 phenotypes with 9 types of genotypes. The number of seeds having genotype TtRr is
   (a) 50 (b) 150 (c) 200 (d) 100

118. In a cross, Aabb x aaBb, the genotypic ratio is
   (a) 1 : 2 : 1 (b) 1 : 1 : 1 : 1 (c) 3 : 1 (d) 9 : 3 : 3 : 1

119. The somatic cell of a human male contains single Barr body then the genetic composition of the person would be
   (a) XY (b) XO (c) XXY (d) XYY

120. In which animals, sex is determined by female individual?
   (a) Human (b) Drosophila (c) Bird (d) Grasshopper

121. If a normal woman marries an albino man and their children are half albino, half normal then the woman is

(a) Homozygous normal
(b) Heterozygous normal
(c) Homozygous recessive
(d) Homozygous dominant

122. The probability of four son to a couple is

(a) 1/4 (b) 1/8 (c) 1/16 (d) 1/32

123. The probability of three daughters to a couple in three children is

(a) 1/4 (b) 1/8 (c) 1/16 (d) 3/8

124. A tobacco plant heterozygous for albinism is self-pollinated and 1200 seeds are subsequently germinated. How many seedlings would have the parental genotypes?

(a) 300 (b) 600 (c) 900 (d) 1200

125. Down syndrome is caused by an extra copy of chromosome number 21. The percentage of affected offspring produced by affected mother and normal father is

(a) 25% (b) 50% (c) 75% (d) 100%

126. Albinism in human being is controlled by a recessive gene 'a'. If both parents were known to carrier (Aa) for albinism then what will be the chance of 1 normal and 3 albino in a family of four?

(a) 12/256 (b) 81/256 (c) 54/256 (d) 100/256

127. Which of the following does not bring variation at individual level?

(a) Gene mutation
(b) Fertilisation
(c) Meiosis and crossing over
(d) Chromosomal aberration and hybridisation

128. If a girl born with sickle cell anaemia, what must be true?

(a) Only father has sickle cell anaemia
(b) Only mother carried sickle cell anaemia
(c) Only mother had sickle cell anaemia
(d) Both parents carried sickle cell anaemia

129. Linkage decreases the frequency of

(a) Hybrid (b) Dominant allele (c) Recessive allele (d) Both b and c

130. Hypophosphatemia is caused by an X-linked dominant gene in human beings. If a hypophosphatemic man marries a normal woman then which of their children will have hypophosphatemia?

(a) All of the daughters
(b) Half of their daughters
(c) All of the sons
(d) Half of their sons

131. Physically 1 map unit on linkage maps can be defined as length of chromosomes in which average crossover formed during 50 cells undergoing meiosis is

(a) 1 (b) 25 (c) 50 (d) 100

132. If two recessive mutations are alleles of different genes, then F_1 progeny are wild type. This is due to

(a) Incomplete dominance
(b) Complementation
(c) Supplementation
(d) Codominance

133. Which of the following chromosomal alterations causes most drastic consequences?

(a) Deletion (b) Duplication (c) Translocation (d) Inversion

134. On a pedigree having the inheritance of PKU, a horizontal line joins a black square and a half black circle. What fraction of this couple's children would you expect to suffer from PKU ?

(a) None (b) 1/2 (c) 1/4 (d) 3/4

135. Mr. X and Mrs. Z are apparently normal but their daughter was born with Alkaptonuria. The probability of their next child being born with alkaptonuria is

(a) 0 (b) 1/2 (c) 1/4 (d) 2/3

136. Duchenne Muscular Dystrophy is caused by a sex-linked recessive allele. Its victims are almost boys who die before the age of 20. This disorder is never seen in girls because

(a) Sex linked traits are never seen in girls
(b) The allele is carried on Y-chromosome
(c) Non-disjunction occurs in males but not in females
(d) A female must have two copies of genes to express it

137. Disorders involving unusual number of sex chromosomes show that maleness is caused by the

(a) Presence of x chromosome (b) Presence of Y chromosome
(c) Absence of X chromosome (d) Absence of Y chromosome

138. In heterozygous cell, the gene of recessive allele

(a) Is always turned off
(b) May be expressed but product is quickly degraded
(c) May be expressed without measurable effect
(d) Is deleted

139. When a plant of chromosomal type 'aa' pollinates a plant of 'AA'. What will be the chromosomal constitution of embryo and endosperm in the resulting seeds?

(a) Diploid zygote of type 'Aa' and triploid endosperm of type 'AAa'
(b) Diploid zygote of type 'aa' and triploid endosperm of type 'Aaa'
(c) Diploid zygote of type 'AA' and triploid endosperm of type 'AAa'
(d) Diploid zygote of type 'Aa' and triploid endosperm of type 'aaa'

140. Reciprocal crosses does not yield equal result if characters are

(a) X-linked (b) Autosomal (c) Extrachromosomal (d) None

141. Dosage compensation in case of human beings is achieved by

(a) Hyperactivation of X chromosome (b) Hyperactivation of Y chromosome
(c) Heterochromatisation of X chromosome (d) Heterochromatisation of Y chromosome

142. Which of the following plants is used as model to study developmental genetics of zygotic embryogenesis?

(a) *Daucus carota* (b) *Arabidopsis thaliana*
(c) *Zea mays* (d) *Nicotiana tabacum*

143. Which of the following genotypes would produce the greatest variety of gametes of the alleles assorted independently?

(a) aaBBCcDd (b) aabbCCDD (c) AaBbCCDd (d) AABBCCDD

144. If bacteria possesses two copies of certain genes then they are referred as

(a) Merodiploid (b) Euploid (c) Polyploid (d) Amphidiploid

145. Mendelian inheritance is obeyed by

(a) Quantitative traits (b) Transposons
(c) Organelles (d) Gene for vertical transfer of disease

146. A woman with albino father marries a albino man. The proportion of her progeny is
(a) All normal
(b) All albino
(c) One normal and one albino
(d) One normal and two albino

147. A man of blood group 'A' marries a woman of blood group 'AB' then which type of progeny would indicate that man is heterozygous 'A'?
(a) AB
(b) A
(c) O
(d) B

148. The weight of fruits in a plant is determined by the number of dominant alleles of a certain number of genes. If seven categories of weight of fruit were found, then how many gene sites would be involved?
(a) 2
(b) 3
(b) 4
(d) 5

149. The ratio of heterozygous, homozygous and colour blind hemizygous offspring of a colour blind man and carrier woman is
(a) 1 : 1 : 2
(b) 1 : 1 : 1
(c) 2 : 1 : 1
(d) 1 : 2 : 1

150. If Mendel had studied 7 pairs of characters in a plant with 12 chromosomes instead of 14 then he could not have discovered
(a) Dominance
(b) Independent assortment
(c) Linkage
(d) Crossing over

151. The number of plants produced from a dihybrid plant with incomplete linkage on test cross, is
(a) 1
(b) 2
(c) 4
(d) 8

152. The number of linkage group in nucleoid of bacteria is
(a) 1
(b) 2
(c) 3
(d) 4

153. Test cross cannot be performed in
(a) Pea
(b) Man
(c) Cat
(d) Neurospora

154. The most suitable medium for culturing *Drosophila* is
(a) Agar agar
(b) Moist bread
(c) Ripe banana
(d) Dung

155. In a random mating population, the frequency of disease causing recessive allele is 80%. What would be the frequency of carrier individuals in population?
(a) 32%
(b) 64%
(c) 16%
(d) 100%

156. A plant is heterozygous (Bb) produces 2 gametes such as B and b. The probability of 'b' gamete fertilising B or b is
(a) Zero
(b) 1
(c) 1/2
(d) 1/4

157. Albinism is determined by a recessive gene in man. The presence of albinism in 50% children born to a couple proves that
(a) Both parents are heterozygous for albinism
(b) Father is homozygous normal and mother is heterozygous
(c) Father is homozygous for albinism but mother is heterozygous
(d) Both parents are homozygous

158. In a population the frequency of a homozygous recessive disease is 16% then the frequency of dominant allele would be
(a) 0.16
(b) 0.6
(c) 0.4
(d) 0.84

159. Both transition and transversion types mutation are induced by

(a) Methyl methane sulphonate (b) Nitrous acid

(c) Base analogue (d) Acridine dye

160. There are two genes A and B which are linked. What shall be the genotype of progeny in a cross between AB/ab and ab/ab?

(a) AAbb and aabb (b) AaBb and aabb (c) AABB and aabb (d) aaBB and aabb

161. A colour blind woman with 'A' blood group marries a man having 'AB' blood group and normal parent. The couple has 2 daughters and 2 sons. The possibility of normal visioned daughter with 'B' blood group will be

(a) 25% (b) 50% (c) 75% (d) 100%

162. In sickle cell anaemia, substitution mutation occurs between

(a) A neutral amino acid and an acidic amino acid

(b) A acidic amino acid and an neutral amino acid

(c) A neutral amino acid and an basic amino acid

(d) A neutral amino acid and an aromatic amino acid

163. Mr. Z is suffering from hypertrichosis and PKU. His father is heterozygous for PKU. The probability of Mr. Z's sperm having one recessive autosomal allele and holandric gene is

(a) 1/2 (b) 1/4 (c) 3/8 (d) 3/4

164. A diseased man marries a normal woman. They have 3 diseased daughters and 5 normal sons. The gene for this disease is

(a) Sex linked recessive (b) Sex linked dominant

(c) Sex limited trait (d) Autosomal dominant

165. How many different genotypes can be expected in F_2 of the cross AABB x aabb when genes are completely linked (coupled) and inherited independently, respectively?

(a) 2 and 4 (b) 3 and 9 (c) 4 and 8 (d) 5 and 9

166. In a graft, the stock has 40 chromosomes and scion has 20 chromosomes. The number of chromosome in root and egg, respectively

(a) 20 and 10 (b) 40 and 10 (c) 10 and 20 (d) 10 and 40

167. Which of the following is a set of discontinuous variation?

(a) Blood group and sex (b) Hair colour and intelligence

(c) Height and excellence in maths (d) Skin colour and mechanical skill

168. In grasshopper, rosy body colour is caused by a recessive mutation. The wild type body colour is green. If the gene for body colour is on X chromosome then what kind of progeny would be obtained from a mating between a rosy female and wild type male?

(a) All female progeny will be green and male will be rosy

(b) 50% female progeny will be green and 50% male will be rosy

(c) All offspring will be green irrespective of sex

(d) All offspring will be rosy irrespective of sex

169. In a population of 6000 individuals, an autosomal recessive disorder is present in 60 individuals. What is the number of homozygous normal (DD) offspring in population?

(a) 4860 (b) 5469 (c) 1000 (d) 2000

170. In a plant, a quantitative character is controlled by 2 polygenes and each dominant allele produces 10 cm height and basic height of plant is 30 cm. What is the height of plant having all dominant alleles?
(a) 40 (b) 50 (c) 60 (d) 70

171. In humans, XX males and XY females are rare but such cases are due to
(a) Deletion of X chromosome (b) Deletion of Y chromosome
(c) XY translocation (d) Duplication of X chromosome

172. A single base pair mutation occurs in strand CATTACCG, its complimentary strand had a sequence GTTATGGC. It means mutation has changed base pair
(a) A→T (b) T→T (c) T→A (d) A→A

173. In human females there is inactivation of one X chromosome for dosage compensation due to
(a) Methylation (b) Acetylation (c) Phosphorylation (d) Formylation

174. Which of the following shows diseased male in pedigree analysis?
(a) Black square (b) Black circle (c) Clear circle (d) Clear square

175. Edward syndrome occurs in human beings due to
(a) Monosomy of 21st chromosome (b) Trisomy of 21st chromosome
(c) Monosomy of 18th chromosome (d) Trisomy of 18th chromosome

176. Two pink flowered four o'clock plants are self fertilised then what will be the probability of first 3 plants appearing to be red?
(a) 1/4 (b) 1/8 (c) 1/16 (d) 1/64

177. If diseased mother transfers disease in all children then it is termed as
(a) Autosomal recessive (b) Autosomal dominant
(c) Sex linked recessive (d) Maternal inheritance

178. Turner syndrome is a case of
(a) Monosomy (b) Trisomy (c) Nullisomy (d) Tetrasomy

179. Alpha thalassemia is caused by mutation of genes present on
(a) 9th chromosome (b) 11th chromosome (c) 16th chromosome (d) 22nd chromosome

180. Folded eye lid or epicanthus is seen in
(a) Edward syndrome (b) Down syndrome
(c) Turner syndrome (d) Klinefelter syndrome

181. Who first proposed that X chromosome was involved in sex determination?
(a) Morgan (b) Henking (c) McClung (d) Sturtevant

182. Chromosomal theory of inheritance was proposed by
(a) Morgan (b) Sutton and Boveri
(c) Janssen (d) Wilson and Stevens

183. Cytoplasmic or plastid inheritance was reported in
(a) *Mirabilis jalapa* (b) *Oenothera lamarckiana*
(c) *Pisum sativum* (d) *Drosophila melanogaster*

184. Which of the following is incorrect about crossing over?
(a) It increases with increase in distance between genes
(b) It increases with irradiation of X-ray
(c) It increase due to interference
(d) It decreases at telomere

185. If the blood groups of your parents are O and AB then the impossible blood group in offspring will be
(a) O and AB (b) O and A (c) O and B (d) A and AB

186. Which of the following blood groups does not have antigen?
(a) A (b) B (c) AB (d) O

187. Codominance was observed in
(a) Flower colour in *Mirabilis jalapa* (b) Skin colour in cattle
(c) Leaf size in *Antirrhinum ajacis* (d) Skin colour in mice

188. Which of the following is sex chromatin negative?
(a) Turner syndrome (b) Klinefelter syndrome
(c) Down syndrome (d) Super male

189. The offspring showing chromosome complement of 44+XXY suffers from
(a) Turner syndrome (b) Klinefelter syndrome
(c) Down syndrome (d) Super male

190. Which of the following is Y-linked traits?
(a) Hypertrichosis (b) Testis determining factor
(c) Porcupine skin (d) All of the above

191. If a diseased man marries to a carrier woman, then
(a) 50% sons will be diseased (b) 25% sons will be diseased
(c) 75% sons will be diseased (d) 100% sons will be diseased

192. The chromosome complement of nullisomy condition is
(a) 2n-1 (b) 2n-2 (c) 2n+1 (d) 2n+2

193. Frame shift mutation is caused by
(a) Nitrous acid (b) Bromouracil (c) Acridine dye (d) All of the above

194. 5-bromouracil is a base analog which causes gene mutation by
(a) Mis-replication (b) Mis-incorporation (c) Tautomerisation (d) Dimer formation

195. The frequency of oxidative deamination for causing transition mutation is maximum in
(a) Adenine (b) Guanine (c) Thymine (d) Cytosine

196. Who discovered mutagenic effect of X-ray on *Drosophila*?
(a) Moragn (b) Muller (c) Sturtevant (d) Hugo de Vries

197. Which of the following is incorrect with respect to mutation?
(a) It occurs in higher plants and animals (b) It is recessive
(c) It shows hot spots (d) It is mostly harmful

198. The large, sudden and discontinuous change was firstly noticed bu Hugo de Vries in
   (a) *Drosophila melanogaster* (b) *Mirabilis jalapa*
   (c) *Oenothera lamarckiana* (d) *Pisum sativum*

199. If X/A ratio is 0.5 then it will be
   (a) Super female (b) Fertile female (c) Intersex (d) Fertile male

200. Which of the following combinations of chromosome is shown by moth and butterfly?
   (a) XX-XY type (b) XX-XO type (c) ZZ-ZW type (d) ZZ-ZO type

ANSWERS

1.c	2.d	3.d	4.c	5.b	6.b	7.b	8.a	9.d	10.b
11.b	12.b	13.c	14.a	15.a	16.c	17.c	18.b	19.a	20.a
21.b	22.a	23.c	24.c	25.d	26.d	27.a	28.d	29.d	30.c
31.c	32.b	33.a	34.b	35.d	36.d	37.c	38.d	39.c	40.b
41.a	42.b	43.c	44.a	45.a	46.d	47.a	48.b	49.b	50.b
51.d	52.d	53.b	54.d	55.d	56.b	57.b	58.a	59.b	60.d
61.c	62.a	63.d	64.d	65.d	66.b	67.b	68.a	69.a	70.d
71.b	72.b	73.d	74.d	75.d	76.d	77.c	78.b	79.c	80.c
81.c	82.a	83.a	84.c	85.c	86.a	87.c	88.c	89.d	90.b
91.b	92.d	93.b	94.a	95.b	96.a	97.c	98.d	99.b	100.d
101.a	102.c	103.a	104.b	105.d	106.b	107.d	108.d	109.c	110.a
111.a	112.c	113.d	114.a	115.b	116.a	117.d	118.b	119.c	120.c
121.b	122.c	123.b	124.b	125.b	126.a	127.d	128.d	129.a	130.a
131.c	132.c	133.a	134.b	135.c	136.d	137.b	138.b	139.a	140.c
141.c	142.b	143.c	144.d	145.a	146.c	147.d	148.b	149.b	150.b
151.c	152.a	153.d	154.c	155.a	156.c	157.c	158.d	159.a	160.b
161.a	162.a	163.a	164.b	165.b	166.b	167.a	168.a	169.a	170.d
171.c	172.a	173.a	174.a	175.d	176.d	177.d	178.a	179.c	180.b
181.c	182.b	183.c	184.c	185.a	186.d	187.b	188.a	189.b	190.d
191.a	192.b	193.c	194.b	195.a	196.b	197.a	198.c	199.b	200.d

16

Molecular Basis of Inheritance

Genetic material is a chemical substance which stores biological information in a coded form, and transmits it to next generation where it is expressed in offspring.

Features of a molecule to fulfil the criteria of genetic material:

1. The genetic material should be present in nucleus of every cell and its amount should be same in all somatic cells of a given species.
2. It should be structurally and chemically stable.
3. It should be able to transmit Mendelian characters to next germination.
4. It should undergo replication (doubling the amount) and
5. It should undergo mutation.

The search for genetic material:

In fact the works for the principle of inheritance by Mendel and discovery of nuclein by Miescher were almost at the same time but to discover and prove DNA as genetic material took very long time. Further the works of Sutton and Morgan confined it (genetic material) to chromosome located in the nucleus of most cells. Chromosome contains DNA as well as protein and was considered as structure for hereditary transmission. It was not clear whether protein or DNA was genetic material. Until 1949, proteins were considered to be playing an informational role. In the mean time, Mirsky and Ris (1949) reported that all cells of an organism have constant amount of DNA and different amount of proteins. This favoured DNA as genetic material.

Evidences in favour of DNA as genetic material:

(a) It is present in nucleus of every cell and its amount is same in all somatic cells of a given species.
(b) It undergoes replication in S-phase of interphase just before cell division.
(c) It passes its copies to next generation where it expresses its genetic information.
(d) It shows mutation occasionally.
(e) It stores biological information in coded form as in triplet code.
(f) It controls various life processes.

Experimental evidences in favour of DNA:

1. Transformation experiment: It was discovered by Frederick Griffith (1928) on *Streptococcus pneumoniae* (earlier known as *Diplococcus pneumoniae*), a bacterium that causes pneumonia in mammals. It is transfer of naked DNA into bacteria. There are two different strains of this bacterium.

(a) S-strain: It is virulent strain which is enclosed by capsule (coat) and forms smooth and shiny colony when grown on agar medium. It is harmful (pathogenic) and causes pneumonia.

(b) **R-strain:** It is non-virulent strain which is not enclosed by capsule (coat) and forms rough colony when grown on agar medium. It is harmless (non-pathogenic) and does not cause pneumonia.

Both S and R strain have several subtypes such as S-I, S-II, S-III and R-I, R-II, R-III, respectively. Griffith tested the virulence of two strains such as S-III and R-II in the following four ways:

(a) Strain S-III (Live, virulent) $\xrightarrow{\text{Injected into mice}}$ mice died of pneumonia

(b) Strain S-III (heat killed at 80°C,virulent) $\xrightarrow{\text{Injected into mice}}$ mice alive

(c) StrainR-II(Live,non-virulent) $\xrightarrow{\text{Injected into mice}}$ mice alive

(d) Strain S-III (heat killed, virulent) + Strain R II (live, non-virulent) $\longrightarrow$ mice died of pneumonia

Griffith found live S-strain of bacterium in the blood of dead mice. He now concluded that a substance (**transforming principle**) had entered the living RII strain from dead III strain which changed (transformed) the RII strain into S-III strain. This was permanent genetic change as S III strain continued to produce similar cells. It was not possible to revert back to RII strain.

He had no idea about what component (carbohydrate or protein or DNA) caused virulence.

Oswald Avery, Colin Macleod and Maclyn McCarty (1944) conducted the experiment of Griffith in vitro (test tube) to identify the component responsible for this transformation. They separated the heat killed filtrate (extract) of S III strain of bacterium into carbohydrate, protein and DNA fraction which were purified and added each fraction into separate culture tube containing R-II strain and observed.

Mixture	Result	Observations
Carbohydrate fraction of S-III strain + RII strain	R-II strain	No transformation
Protein fraction of S-III strain + RII strain	RII strain	No transformation
DNA fraction of S-III strain + RII strain	SIII strain	Transformation occurs
DNA fraction of S-III strain + DNase + RII strain	RII strain	No transformation

It was noticed that only the culture containing DNA fraction of strain SIII transformed RII strain to S III strain. This confirmed that DNA was the transforming agent. When DNA was treated with deoxyribonuclease (an enzyme that digests DNA) then transformation of RII strain did not occur.

2. Blender Experiment: It was performed by Alfred Hershey and Martha Chase (1952) using T_2 bacteriophage (virus) to prove that DNA was the genetic material in the viruses because phage DNA enters into host cell and works as genetic material.

They labelled protein and DNA with different isotopes. They developed the two cultures of *E .coli*. In one culture, a radioisotope S^{35} was labelled into sulphur containing amino acids (cysteine and methionine) of protein and in another culture, a radioisotope P^{32} was labelled into nucleotides of DNA.

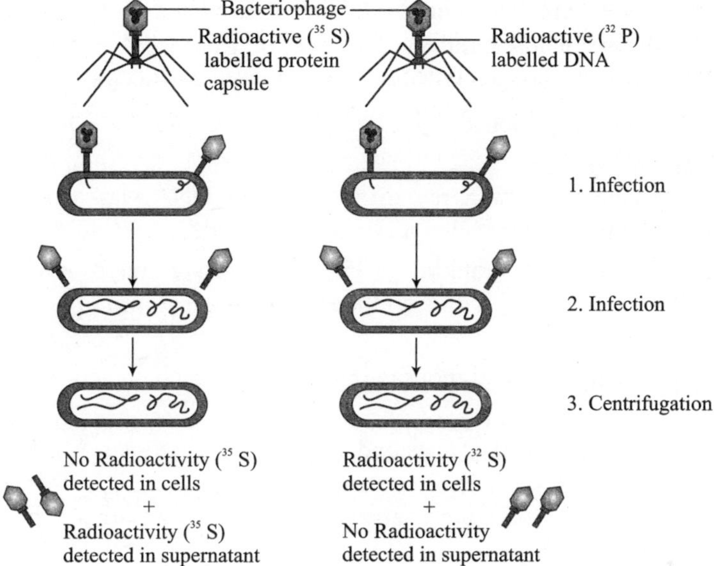

Fig. ***Blender experiment***

Process:

(a) Infection: The phages (viruses) were used separately to infect *E. coli* (bacterium). They introduced T_2 bacteriophages in both cultures and infection was allowed to run for 10 minutes. Multiplication started.

(b) Blending: Both the cultures were gently shaken in a blender at 10,000 rpm to separate ghost or capsid (empty phages) sticking to the surface of *E. coli* or to break contact between viruses and bacteria.

(c) Centrifugation: Both the cultures were centrifuged and observed the level of radioactivity in supernatant (suspension) and pellet.

In one culture (protein of bacteriophage labelled with S^{35}), supernatant contained radioactive empty bacteriophage and pellet settled down in centrifuge tube containing non-radioactive heavier bacteria. It showed that protein coat of bacteriophage did not enter the bacterial cell during infection.

In other culture (DNA of bacteriophage labelled with P^{32}), supernatant contained non-radioactive empty bacteriophage and pellet settled down in centrifuge tube containing radioactive heavier bacteria. It showed that DNA of bacteriophage entered the bacterial cell during infection.

Now it was confirmed that sulphur did not form a part of DNA and phosphorus only formed part of DNA hence P^{32} radioactive was found in *E. coli* infected with P^{32} phage. This proved that protein did not enter bacteria from bacteriophage and only DNA passed from phage to bacteria.

Structure of DNA

Nucleic acids are highly complex, negatively charged, acids, and largest macro molecule. There are two types of nucleic acids found in living organisms. DNA first isolated by Friedrich Miescher (1869) from nucleus of pus cell and sperm of Salmon and termed it 'nuclein'.

Altman (1899) coined the term 'Nucleinic acid' and then called it 'Nucleic acid' because it was found in nucleus and it showed acidic nature.

Fischer (1880) discovered the presence of two types of bases-purine and pyrimidine.

Levene (1910) discovered deoxyribose sugar and phosphoric acid in it. He characterised four types of nucleotides present in DNA.

Astbury proposed through his first X-ray diffraction that DNA was a polynucleotide in which nucleotide was arranged perpendicular to long axis of a molecule.

Maurice Wilkins and Rosalind Franklin (1953) proposed through X-ray photograph (x-ray crystallography) that DNA was a helix with a width of 2 nm and one turn of helix is 3.4 nm having ten layers of stacked bases.

DNA molecule is a polymer of several thousand nucleotides. Each nucleotide consists of sugar, base and phosphate.

Sugar: It is deoxyribose (pentose). At second carbon atom, oxygen is absent hence OH group is replaced by H atom but in RNA it is ribose sugar. The 5th and 3rd carbon atoms participate in phosphodiester bond while 1st carbon atom is always occupied by an organic base. In DNA there are 3 OH groups in deoxyribose sugar.

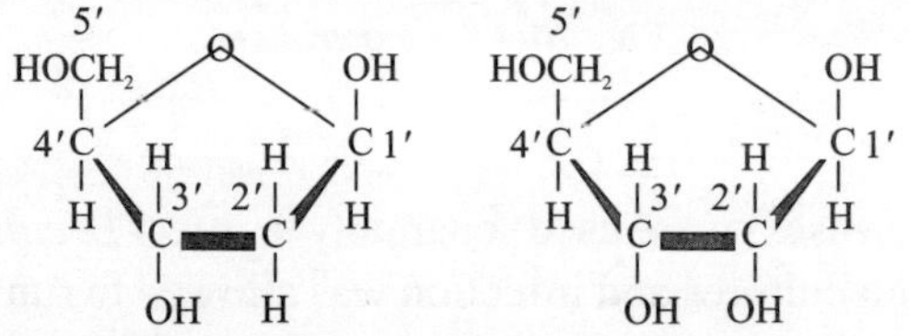

Deoxyribose

Ribose

Nitrogenous bases: These are hydrophobic. There are two types of bases which are heterocyclic aromatic compounds.

Purines: These are double ringed and nine member (5 Carbon and 4 Nitrogen atoms) compound. These have high molecular weight. In these N atom is located at 1st, 3rd, 7th and 9th positions of carbon. These are of two types:

1. Adenine (6- aminopurine)
2. Guanine (2 –amino, 6- oxypurine)

Pyrimidines: These are single ringed and six member (4 carbon and 2 Nitrogen atoms) compound. These have low molecular weight. In these N atom is present at 1st and 3rd positions of carbon. These are of three types:

1. Thymine (5-methyl 2,4-dioxypyrimidine)
2. Uracil (2, 4-dioxypyrimidine): It is present in RNA in place of Thymine.
3. Cytosine (4-amino, 2-oxypyrimidine).

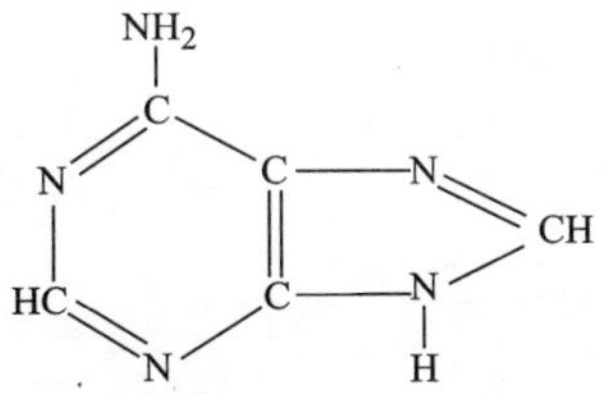

Adenine

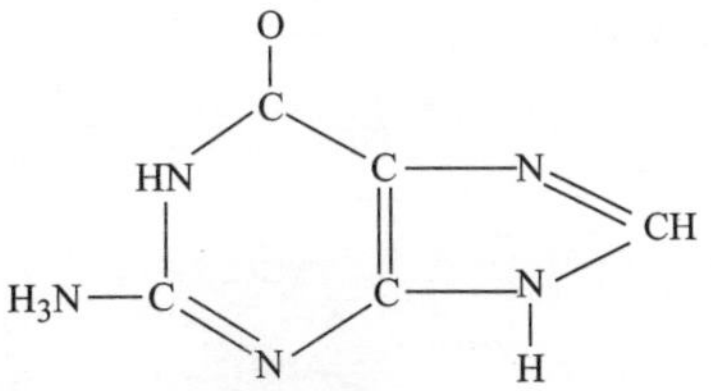

Guanine

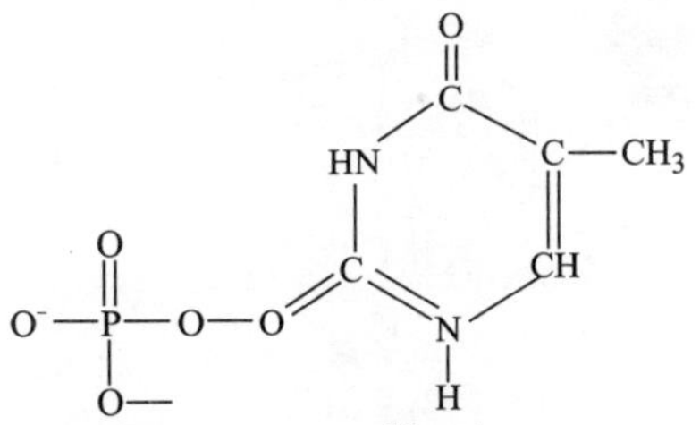

Phosphate

Thymine

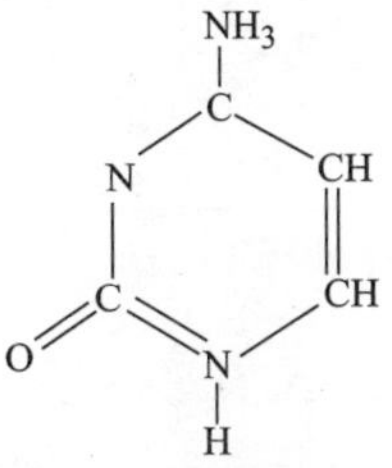

Cytosine

Phosphate: It is present as phosphate group of orthophosphoric acid (H_3PO_4).

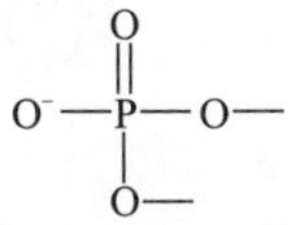

Phosphate

It alternates with sugar. It has three reactive OH groups in which two are involved in sugar phosphate backbone with the help of phosphodiester bond. Each phosphate group is joined to OH group of 3rd carbon of one sugar at one side and joined to OH group of 5th carbon of another sugar of next nucleotide by phosphodiester bond (5′ C-O-P-O-C 3′). It gives acidic nature to DNA. It is used as linker in strand due to multiple bond facility.

Nucleosides: These are formed from sugar and base through β-N-glycosidic bond (C-N covalent bond). It is condensation reaction and water is released.

$$\text{Base + Sugar} \longrightarrow \text{Deoxyriboside} + H_2O$$

Glycosidic bond (or glycosylic bond) develops between 1st carbon of sugar and N atom of 9th position in case of purines; and between 1st carbon of sugar and N atom of 1st position in case of pyrimidines. There are four nucleosides in DNA:

Bases	Sugar	Nucleosides
Adenine +	Deoxyribose →	Deoxyadenosine
Guanine +	Deoxyribose →	Deoxyguanosine
Thymine +	Deoxyribose →	Deoxythymidine
Cytosine +	Deoxyribose →	Deoxycytidine

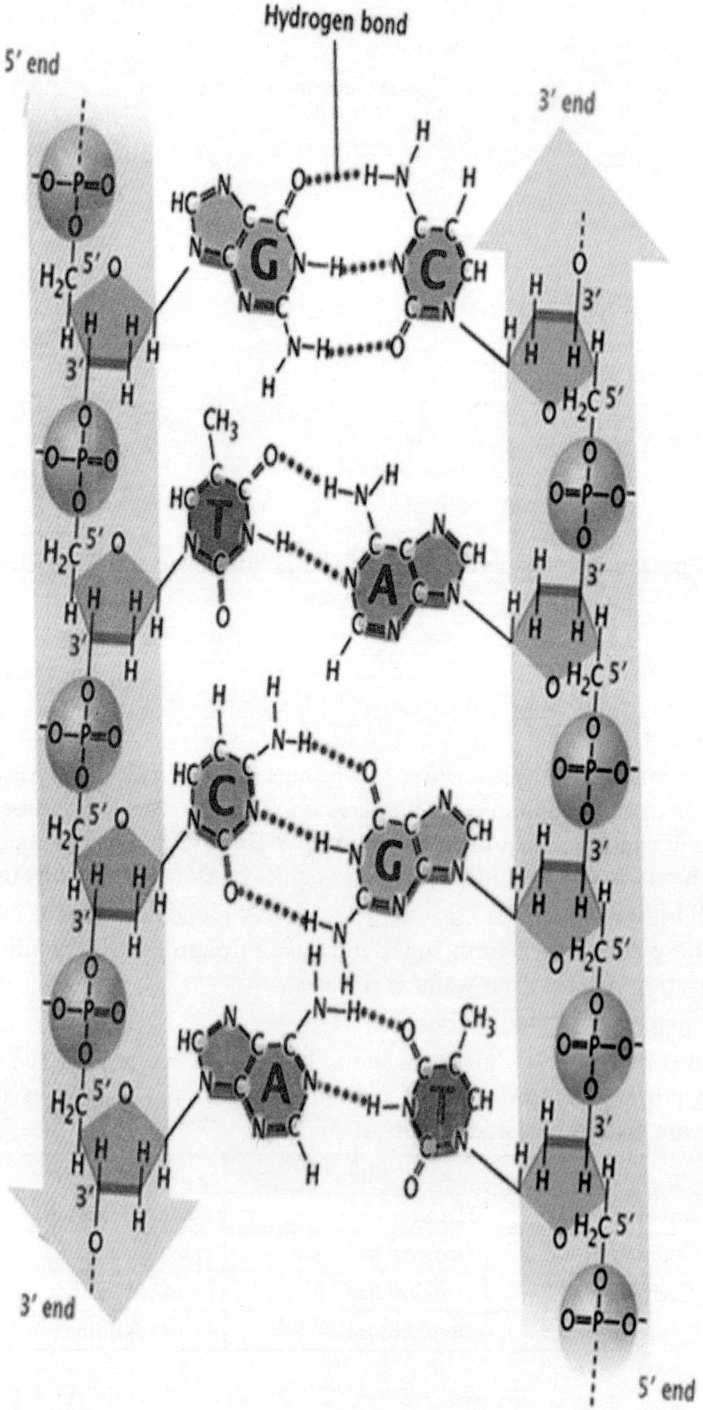

Fig. ***Antiparallel nature of DNA***

Nucleotides: These are formed from sugar, base and phosphate through phosphodiester bond. It is also condensation reaction and water is released.

Phosphodiester bond develops between –OH group of 3rd carbon of sugar of one nucleotide and 5th carbon of another sugar linked with phosphate group. Thus, nucleotides of a strand are linked to each other through phosphodiester bond (3′-C-O-P-O-C-5′) between sugar (3′) of one nucleotide and phosphate (5′) of other nucleotide. At 3′-end of strand, 3rd carbon of sugar is present and usually it has a free –OH group. At 5′-end of strand, 5th carbon of sugar is present and usually it has a phosphate group.

There are four nucleotides in DNA:

Nucleosides	Phosphate group	Nucleotides
Deoxyadenosine +	Phosphate →	Deoxyadenylate (Deoxyadenosine-5- monophosphate)
Deoxyguanosone +	Phosphate →	Deoxyguanylate (Deoxyguanosine -5-monophosphate)
Deoxythymidine +	Phosphate →	Deoxythymidylate (Deoxythymidine-5-monophosphate)
Deoxycytidine +	Phosphate →	Deoxycytidylate (Deoxycytidine-5-monophosphate)

Features:

1. **Polarity:** The polynucleotide chains show polarity (direction). One end of each DNA strand is called 5′ end. The last deoxyribose unit at this end has carbon atom at 5th position. The other end of strand is called 3′ end. The last deoxyribose unit at this end has carbon atom at 3rd position.

   There are two ends.

   (a) **5′:** Phosphorus group of 5th Carbon of sugar is free but its 3rd Carbon is not free because it is attached with P atom forming phosphodiester bond.

   (b) **3′:** Phosphorus group of 5th Carbon of sugar is not free because it forms phosphodiester bond but its 3rd carbon is free because its OH group does not form phosphodiester bond.

2. **Variety:** Only four types of base pairs are involved in the formation of DNA molecule. These base pairs may occur in any sequence and these may be any number of sequences in a molecule. This gives an infinite variety to DNA. Thus each individual has a specific sequence of base pairs in its DNA.

3. **Complementarity of base pairs:** The two polynucleotide chains of DNA are not identical but complimentary to each other with respect to base pairs. This is due to restrictions on base pairings. A base pair most consist of a purine and a pyrimidine. There are two reasons for this limitations:

   (a) The space available between the two sugar phosphate chains of DNA (2nm) can accommodate one purine and one pyrimidine. If there are two purines then the space will be too wide. If there are two pyrimidines then the space will be too narrow. In these cases proper hydrogen bonding will not occur.

   (b) A specific purine base pairs with a specific pyrimidine base because of a perfect match between H-donor and acceptor sites on two bases. Only adenine and thymine; and also guanine and cytosine have the proper spatial arrangements to form correct H-bonding. Adenine and Cytosine and also guanine and thymine do not form proper H-bonding and cannot align correctly thus cannot pair because these combinations are too large and too small to fit inside helix.

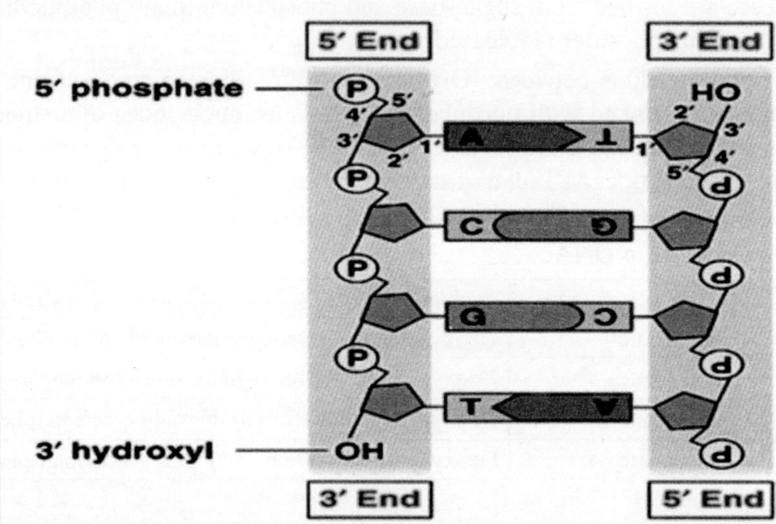

Double helical model: James Watson and Francis Crick (1953) proposed first three dimensional double helical model of DNA from X-ray photograph of Maurice Wilkins and Rosalind Franklin and Edwin Chargaff's analytical data. It could explain the storage of genetic information and DNA replication. There are several features which are as follows:

1. DNA consists of two polynucleotide chains.
2. Two chains spiral plectonemically around central axis to form a pair of right handed helix.
3. A right handed helix turns in a downward clockwise direction.

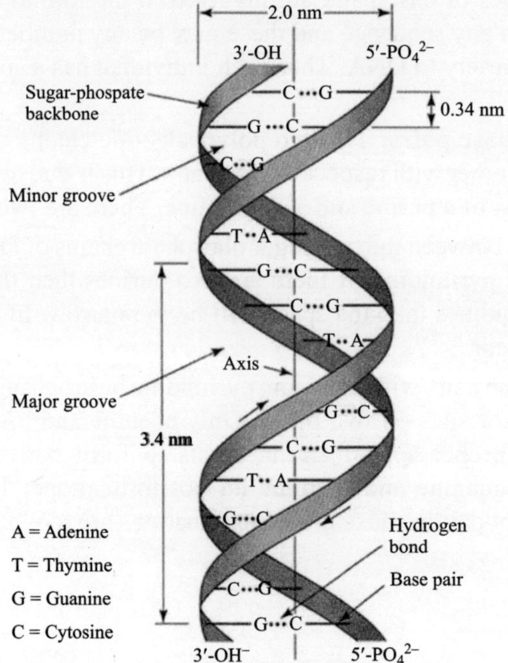

Fig. *DNA double helix*

1. Two chains run in opposite (antiparallel) directions. One chain is aligned in $5' \rightarrow 3'$ direction, its partner must be aligned in $3' \rightarrow 5'$ direction. This antiparallel direction is essential for hydrogen bond formation between base pairs.
2. Sugar phosphate makes the backbone of double helix. Each chain has alternate sugar phosphate back bone. Phosphates are on outside of helix and bases inside.
3. The bases project at right angles to the axis of double helix.
4. Hydrophobic interaction and Vander Wall's forces between the planar bases provide stability to DNA.
5. Two chains are held together by H-bonds (in dotted line) between each base of one chain and an associate base of other chain. The length of H-bond is $3A^0$. Adenine and thymine base pairs are joined by two H bonds whereas guanine and cytosine base pairs are joined by three H-bonds. Thus there are four possible base pairs in DNA: A-T, T-A, G-C and C-G. There is an average of 25 H-bonds within each complete turn of the double helix.
6. The distance from the P atom of backbone to centre of axis is $10A^0$.Thus the width (diameter) of double helix is $20A^0$. The width between two backbones is constant and equal to the width of base pairs (width of a purine + width of a pyrimidine).
7. The association of two purines would extend beyond the specified width and the association of two pyrimidines would not be sufficiently wide.
8. The sequence of one chain determines the sequence in the other. Thus, two chains are said to be complimentary.
9. Chargaff proposed base equivalence rule where the amount of purine (adenine + guanine) is equal to that of pyrimidine (thymine + cytosine).

   $A + G = T + C$ or $A + G / T + C = 1$ for double stranded DNA.

   or $A + G + T + C = 100\%$.

   Hence, $A = T$ or $A/T = 1$; $G = C$ or $G/C = 1$.

   The deoxyribose sugar and phosphate residues occur in equal proportion.
10. The base pairs are spaced at interval of $3.4\ A^0$. Thus one complete turn of double helix containing 10 base pairs has $34\ A^0$.
11. The space between adjacent turns of helix form two grooves of different width alternatively-a wide major groove ($22\ A^0$) and narrow minor groove ($12\ A^0$). The major groove is generated by larger angular distance between sugars and minor groove is formed by smaller angular distance between sugars. In major groove, the edges of the bases are more accessible than in minor groove. In these grooves, specific proteins interact with DNA molecule.

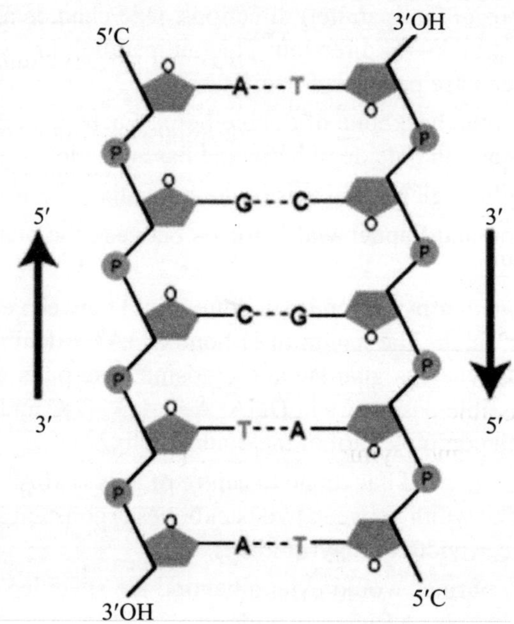

Fig. *Hydrogen bonds in DNA*

Polymorphism in DNA: There are several forms of DNA. Recently Z- DNA was discovered by Alexander Rich. It is found in polytene chromosome of salivary gland of *Drosophila*. Its function is to regulate gene expression. The differences between different forms of DNA are as such:

Features	B form	Z form
Stability	Stable	Unstable
Diameter of helix	20A^0	18 A^0
Number of base pairs per turn of helix	10	12
Distance between two base pairs	3.4 A^0	3.7 A^0
Pitch (one turn) of helix	34 A^0	45 A^0
Vertical rise per base pair	3.4 A^0	5.7 A^0
Base pair tilt to axis of helix	6.3^0	7^0
Rotation per base pair (angle of helical twist)	360/10 = 36^0	720/12 = 60^0
Direction of helix coiling	Right handed	Left handed
Glycoside torsion angle	Anti	Syn (purine) and anti (pyrimidine)
Sugar puckering	3′-exo	3′-endo
Nature of phosphate backbone	Helical	Zig-zag
Distance of phosphate from axis	10 A^0	9 A^0
Repeating unit	One base pair (mononucleotide)	Two base pairs (Dinucleotide)

Base ratio: It is A+T/G+C. It is variable (between 0.4 and 1.9) and constant for a species. It is low (less than 1) in primitive organisms (prokaryotes) and high (more than 1) in advanced organisms (eukaryotes). It can be used to identify the source of DNA and also helps in classification of organisms.

DNA (Deoxyribonucleic acid)	RNA (Ribonucleic acid)
It contains deoxyribose sugar.	It contains ribose sugar.
It contains adenine, guanine, thymine and cytosine.	It contains adenine, guanine, uracil (in place of thymine) and cytosine.
It is usually double stranded.	It is usually single stranded.
It is of one type.	It is of three types.
It is usually genetic material.	It is genetic material in some viruses only.
It occurs mostly in nucleus and small amount in mitochondria and chloroplast.	It occurs mostly in cytoplasm and small amount in nucleus.
It originates later (from RNA).	It originates first (primitive).
It contains millions of nucleotides	It contains fewer (70-12000) nucleotides.
It has same amount of adenine and thymine; and guanine and cytosine.	It has different amount of adenine and uracil; and guanine and cytosine
It follows Chargaff's rule.	It does not follow Chargaff's rule.
It is Fuelgen positive.	It is Fuelgen negative.
It does not contain unusual bases.	It contains usual bases.
It is more stable due to presence of thymine.	It is less stable due to presence of uracil which causes degradation. It also contains 2′-OH group
It is less reactive.	It is more reactive due to presence of 2′-OH group.
It undergoes mutation at slow rate.	It undergoes mutation at fast rate.
Its amount is fixed in each cell of a species.	Its amount is variable in different cells.
It has base pairing throughout the length.	It has base pairing only in hair pin and helical region.
It replicates to form DNA.	It does not replicate to form DNA.
It transcribes to form RNA.	It does not transcribe.
Its renaturation after melting is slow.	Its ranaturation after melting is fast.
It has longer span.	It has shorter span.

Prokaryotic DNA	Eukaryotic DNA
It occurs in cytoplasm.	It occurs in nucleus, mitochondria and chloroplast.
It is circular.	Nuclear DNA is linear but DNA of mitochondria and chloroplast is circular.
Its amount is low.	Its amount is more.
It has smaller part of DNA which is non-functional.	It has greater part of DNA which is non-functional.
It has larger part of DNA which is functional.	It has smaller part of DNA which is functional.
It does not contain introns.	It contains introns.
It can code for few proteins.	It can code for more proteins.

Circular DNA	Linear DNA
It is found in prokaryotes; and organelles of eukaryotes (mitochondria and chloroplast).	It is found in eukaryotes.
It exists as a single chromosome.	It exists in a number of chromosomes.
It does not have histone proteins.	It contains histone proteins

Packaging of DNA double helix

It is the accommodation of long sized DNA in small area through compaction or packaging. It occurs in both prokaryotes and eukaryotes. A single molecule of DNA within a chromosome may be as long as 8.5 cm (3.3 inches) hence it has to be supercoiled into a very complex shape.

DNA packaging in prokaryotes: The length of DNA helix in *E. coli* (haploid organism) is very long as compared to body size. It is calculated as

Length of DNA = Distance between base pairs × Total number of base pairs

0.34 nm × 4.6 million base pairs

$0.34 \times 10^{-9} \times 4.6 \times 10^{6} = 1.56$ mm

Prokaryotic DNA is not scattered throughout the cell in spite of absence of nucleus instead it is present in cytoplasm of cell having size of 1 μm in supercoiled stage. This compaction in the form of loop is called nucleoid or prochromosome or genophore. The long-sized DNA is supercoiled (coiled and recoiled) with the help of RNAs and basic nonhistone protein like polyamines. These proteins are positively charged thus hold firmly the negatively charged DNA.

DNA packaging in eukaryotes: It is very complicated in eukaryotes because the length of DNA is greater than the dimension of nucleus (10^{-6} m). The length of DNA varies not only in different organisms but also in different tissues or even in different regions within same cell of organism. The length of DNA helix in human being (diploid organism) is very long as compared to body size. It is calculated as

Length of DNA = distance between base pairs × Total number of base pairs

0.34 nm × 3.3 billion base pairs (diploid)

$0.34 \times 10^{-9} \times 2\ (3.3 \times 10^{9}) = 2.2$ m

A human diploid cell having 46 chromosomes has 5.6 picogram of DNA hence it contains 174 cm because one pg is equivalent to 31 cm of DNA helix. All DNA in our cells is almost 60 billion km long.

Nucleosome or solenoid model: It was proposed by Kornberg and Thomas (1974). It is widely accepted model.

Chromatin is composed of a repeating unit called nucleosome (nu particle). Nucleosome is ellipsoid or oblate particle about size of 6 x 11nm. It is first order of packaging. It contains an average of 200 base pairs of DNA. The length of DNA associated with nucleosome core is affected by H_1 histone and non-histone proteins. The length of DNA suggests that DNA should be coiled around the surface of nucleosome core particle. It further suggests that histones do not cover DNA but DNA is wrapped around histone core particle. Nucleosome consists of the following components:

(a) Core: It contains a histone octamer (8) composed of two molecules of each of histones H_2a, H_2b, H_3 and H_4 stacked one on top of other. The histone octamer forms a wedge shaped (barrel shaped) disc around which 140 bp long DNA is wrapped outside in 1¾ (1.8) turns in a left hand direction. Presence of histones is necessary for the formation of inner core of nucleosome. The positively charged end of histone proteins are towards outside to attract and hold firmly the negatively charged strand of DNA. This segment of DNA in chromatin is nuclease resistant. It forms beads on a string connected by sequences of linker DNA.

(b) Linker DNA: It contains 8 to 114 base pairs (average 60). Its size is 14nm. It connects with core particles. It is nuclease susceptible. It is additional DNA required to make up the 200 bp nucleosome.

(c) H_1 histone: It contains one molecule in nucleosome. It is mostly located outside linker DNA or between the nucleosomes or at central hole of solenoid. It is large in size which is due to presence of additional C-terminal tail stabilising DNA between the cores of nucleosome. It seals the DNA in nucleosome by binding at the site where DNA enters and leaves the nucleosome. Thus, H_1 helps in grouping of nucleosome. It is most readily removed at moderate salt concentration from chromatin which yields a beads-on- a- string fiber due to expansion of 10 nm nucleosome fiber. If it is removed completely then it does not affect the structure of chromosome. At low salt concentration, H_1 remains bound hence beads are joined together into a more compact structure.

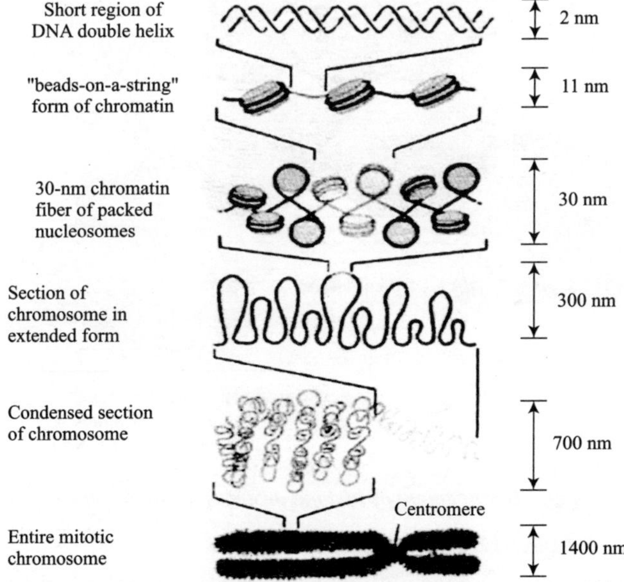

Fig. ***Sequence of packaging***

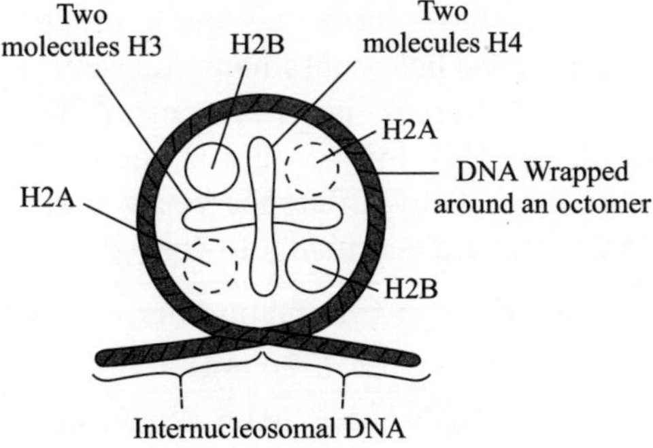

Fig. ***DNA packaging in nucleosome***

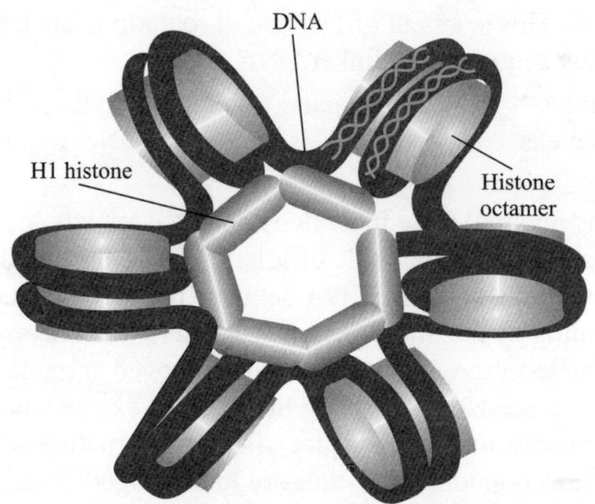

Fig. ***A single solenoid***

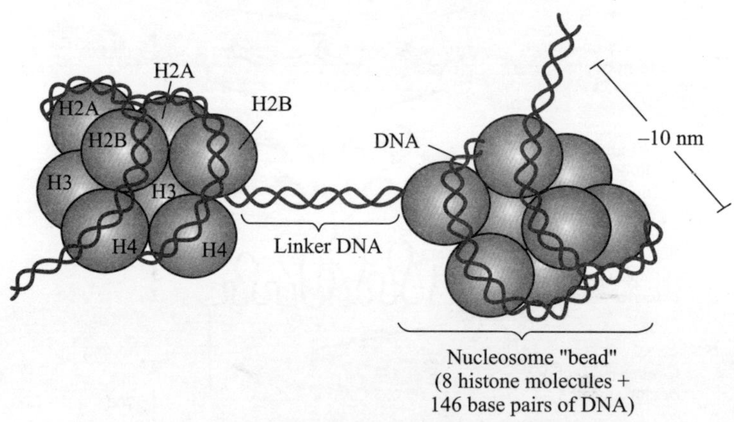

Fig. ***Arrangement of histone proteins in nucleosome***

Nucleosome core + H_1 + linker DNA = chromatosome

Nucleosomes are arranged edge to edge with their flat faces parallel to axis of fiber. Six nucleosomes undergo supercoiling or condensation to form solenoid (30nm diameter) per turn. Solenoid is second order of packaging. H_1 histone is required for its organisation. Solenoid is basic constituent of both interphase and mitotic prophase. It undergoes further coiling to produce chromatin fiber which shows euchromatin (loosely packed structure) and heterochromatin (densely packed structure). If chromatin is digested with endonuclease enzyme, DNA breaks into fragments of 200 bp in which core of nucleosome has 140 bp and linker part has 60 bp. This 140 bp core is common to all types and represents the basic level of organisation of chromatin. Only difference lies in that of linker DNA. Now chromatin undergoes further coiling to form chromatid and finally to chromosome.

DNA ⟶	Nucleosome ⟶	Solenoid ⟶	Chromatin fiber ⟶	Chromatid ⟶	Chromosome
(2 nm)	(10 nm)	(30 nm)	(300 nm)	(700 nm)	(1400 nm)

Supercoiling of DNA: It is coiling of DNA axis upon itself caused by change in linking number from relaxed closed circle. DNA can be supercoiled whether it is circular or linear or having fixed ends. It is negative when DNA is twisted in the opposite direction of unwinding of double helix.

Helix is untwisted. It is positive when DNA is twisted in the same direction as that of DNA helix. In nature most DNA has slight negative supercoiling. The negatively supercoiled DNA is easier to unwind (required for replication and transcription) than relaxed DNA. The supercoiled DNA is more compact than relaxed DNA.

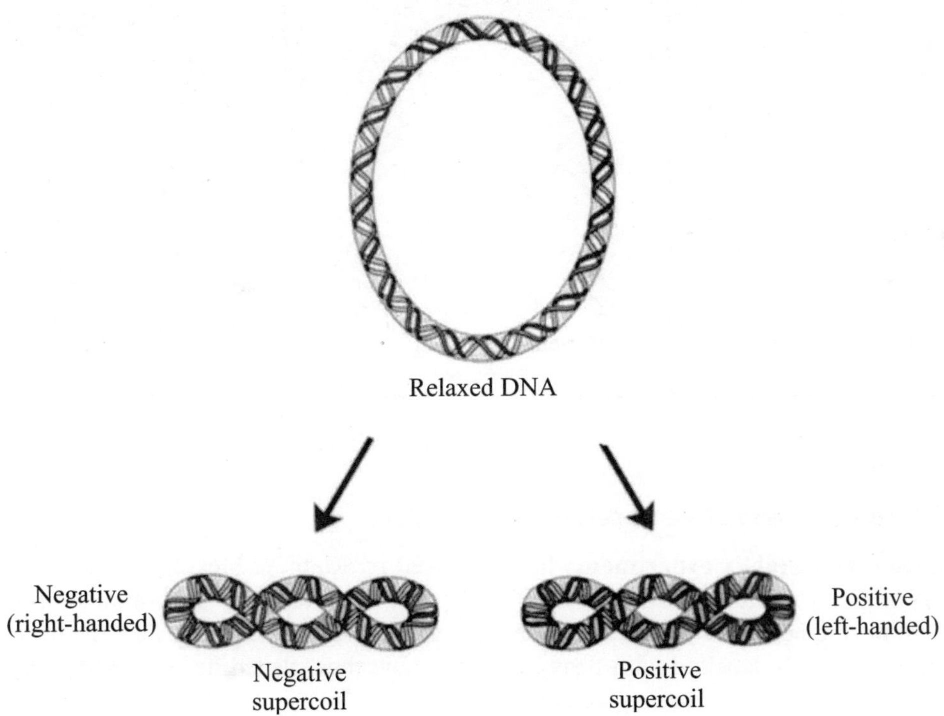

Fig. ***Types of supercoiling***

Replication of DNA

It is a process in which amount of DNA gets doubled or DNA makes its exact copies. It takes place continuously in cytoplasm in prokaryotes and during S-phase of interphase nucleus of eukaryotes. It is bidirectional (usually) as well as unidirectional. There are three modes of replication:

A. Conservative replication: One daughter DNA contains both parental (old) strands and another daughter DNA has both newly synthesised strands.
B. Semi conservative replication: Each daughter DNA contains one parental (old) strand and one newly synthesised strand. It is universally accepted.
C. Dispersive replication: Each daughter DNA contains both parental (old) and newly synthesised strands in dispersed condition when parental DNA is broken into several fragments, each fragment would replicate and the old and new fragments would recombine randomly.

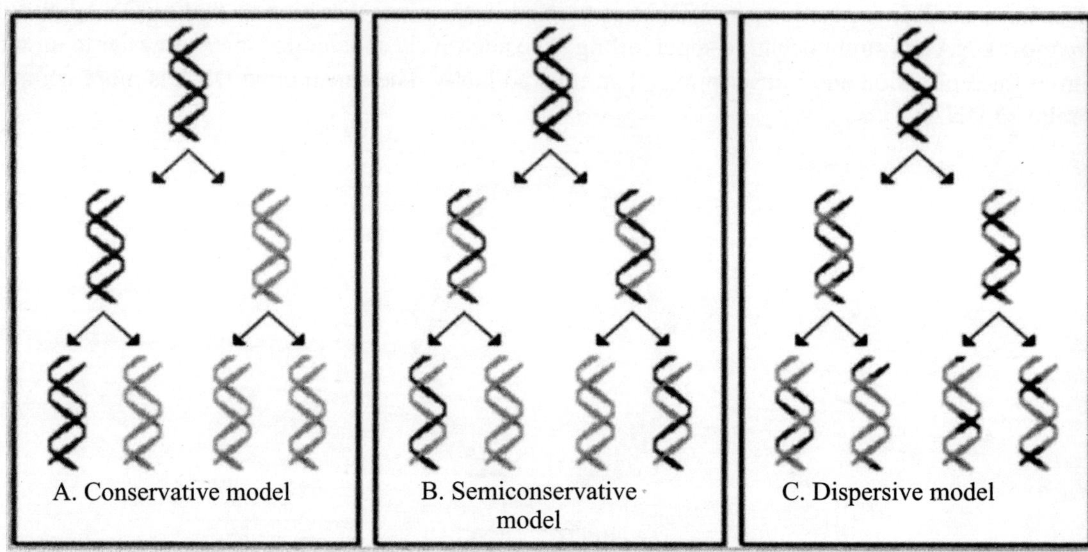

Fig. ***Modes of DNA replication***

Proof of semi-conservative replication of DNA

1. **Messelson and Stahl's experiment:** It was proved by Mathew Messelson and Franklin Stahl (1958) in *E.coli* in the presence of N^{15}. They grew *E.coli* for many generations (up to 14) in a medium having heavy isotope of Nitrogen (N^{15}) in the presence of NH_4Cl till the bacterial DNA became completely labelled with heavy isotope. Now they placed heavy labelled bacteria in a fresh medium containing normal isotope of nitrogen (N^{14}).

   DNA was extracted, purified and centrifuged at high speed (50,000 rpm) for 40-72 hours. They found that all DNA of first generation (replication) was hybrid (N^{15}-N^{14}).

   The daughter cells were allowed to divide again in N^{14} medium. They found two hybrid DNA (N^{15}-N^{14}) and two light DNA (N^{14}-N^{14}).

   Subsequent generations showed increase in number of light DNA and decrease in number of hybrid DNA. There was no heavy DNA.

   Heavy and light DNA can be separated by equilibrium density gradient centrifugation using CsCl. The same CsCl is centrifuged at high speed (50,000 rpm) for 40-72 hours. This leads to formation of a linear gradient of increasing density from top to bottom of centrifuge tube.

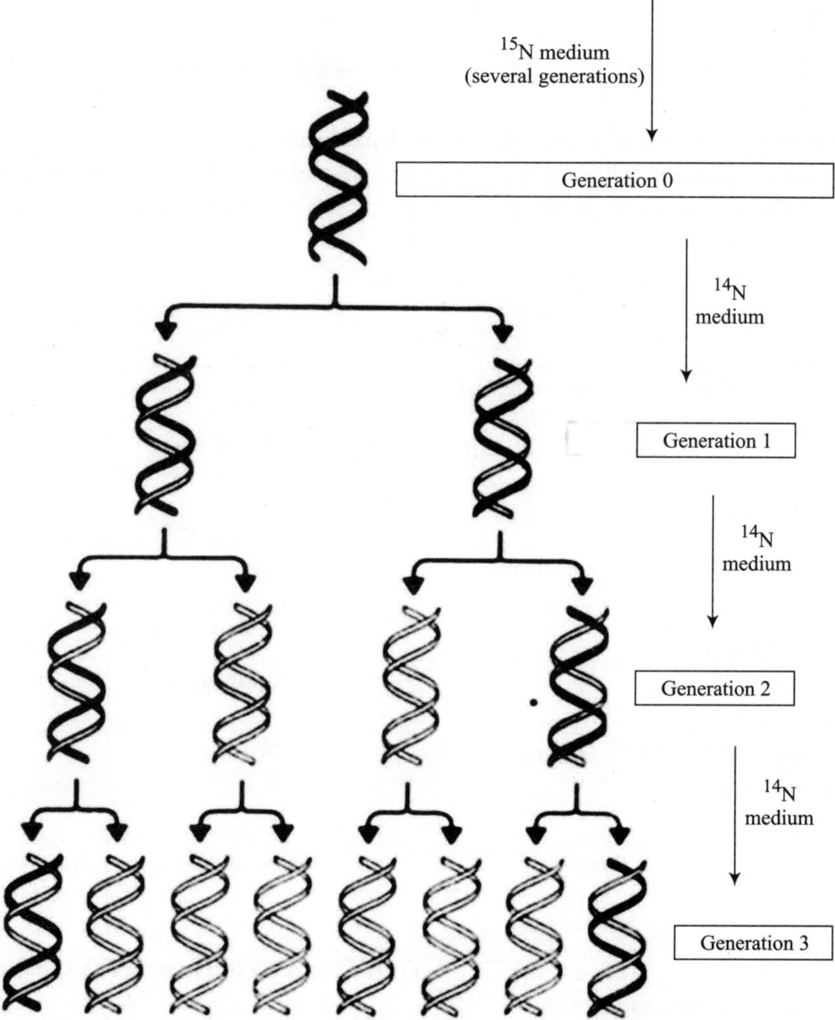

Fig. ***Experiment to prove semiconservative replication***

2. **Cairn's autoradiography experiment:** In this, heavy isotope of hydrogen (H^3) was used to replace thymine with radioactive (tritiated) thymidine in DNA molecule.
3. **Taylor's experiment:** He proved semi-conservative mode of chromosome replication in eukaryotes using radioactive (tritiated) thymidine in the root tip cell of broad bean (*Vicia faba*). He treated root tip with radioactive thymidine to label the DNA. The root tips were grown in the normal medium. In the first generation, both chromatids were labelled and in the second generation, one chromatid of each chromosome was labelled and the other one was normal.

Mechanism of semi-conservative replication: It is complex and multistep process. It is fast (750-1000 bp/sec) in prokaryotes and slow (50-100 bp /sec) in eukaryotes but human genome can be copied only in a few hours because there are multiple initiation sites (many replication forks). *E. coli* takes 38 minutes to replicate at the rate of 2000 bp/sec. The chance of error is very low because DNA is required to preserve the meaning of genetic message from generation to generation.

Requirements:

(a) DNA template: It uses single stranded or double stranded but ssDNA is more efficient because it does not wind.

(b) DNA dependent DNA polymerase enzyme: It was discovered by Kornberg (1956). It is highly efficient enzyme because it has ability to polymerise some 200 bp/sec. It is fast and catalyses the reactions with full accuracy. There are three DNA polymerase enzymes in prokaryotes and five types in eukaryotes.

(c) Primer: It is RNA chain with a free 3′-OH group.

Properties	DNA polymerase I	DNA polymerase II	DNA polymerase III
Molecular weight	High (109,000 dalton)	Higher (120,000 dalton)	Highest (>250,000 dalton)
Nucleotides polymerised at 37°/min	Up to 1000	50	15000
Affinity for nucleotides	Low	Low	High
5′→3′ polymerase activity in Klenow fragment (large)	occurs	Occurs	Occurs
5′→3′ exonuclease activity in smaller fragment	Occurs	Does not occur	Does not occur
3′→5′ exonuclease activity in larger fragment (Klenow fragment)	Occurs	Occurs	Occurs
Function	Repair of DNA and excision of RNA primer	Repair of DNA	Replication of DNA, proof reading and editing.

(d) Deoxyribonucleotides: These are d ATP, d GTP, d TTP, d CTP. These act as substrate and provide energy for polymerisation reaction.

(e) Helicase: It is ATP dependent enzyme.

(f) Single strand binding (SSB) proteins

(g) Superhelix relaxing protein

(h) Ligase: It forms covalent phosphate bond between 3′-OH and 5′-phosphate group.

Process:

(a) DNA replication starts at specific point called origin. In prokaryotes, there is only one origin (*Ori*-c) hence it occurs in both direction. The origin is 245 bp long and has generally A-T rich weak pairs. But in eukaryotes, there are several origins due to giant molecule hence it is unidirectional.

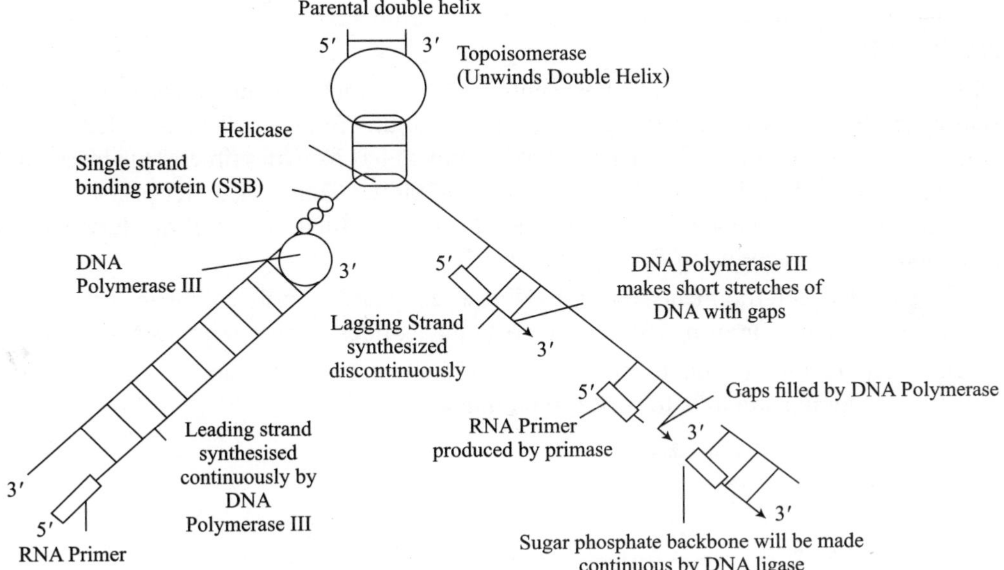

Fig. *Mechanism of DNA replication*

(b) Replication starts with a nick or break or incision on one of the two parental strands by endonuclease enzyme.

(c) Unwinding of DNA double helix occurs by helicase enzyme which breaks H-bonds between base pairs using ATP as source of energy.

(d) Unwinding of helix is promoted by single strand binding (SSB) protein which binds preferentially to single strand of DNA so that normal base of helix is available for replicating enzyme. It also stabilises ssDNA and further prevents from reannealing of strands.

(e) Replication fork (Y-shaped) moves down due to unwinding of helix.

(f) A tension or strain or torque is developed at distal end of DNA. This tension is relieved by superhelix relaxing protein (gyrase) in bacteria and Swivelling protein (topoisomerase) in eukaryotes. This protein has nick and joining activity. Initially it causes nick on one strand of DNA by which rotates over the other to relax the strain or coiling. This strand is joined by ligase enzyme. Now dsDNA becomes normal.

(g) RNA primer is formed which starts the synthesis of DNA. Primer is short fragment (5-10 nucleotides) of RNA which is synthesised by template DNA at 5′-end (fork end of other strand) by primase enzyme (a modified RNA polymerase). Primer is formed because DNA polymerase cannot start the synthesis of a new strand of DNA. It is required to which polynucleotides are added.

(h) Synthesis of DNA occurs from $3'\rightarrow5'$ end on template DNA or from $5'\rightarrow3'$ end on newly synthesised strand by primer. Thus synthesis starts from 3′-end of DNA.

(i) Elongation of new strand occurs by DNA polymerase III which adds deoxyribonucleotides (dATP, dGTP, dTTP, dCTP) to 3′-OH group of last nucleotides of RNA primer. The synthesis occurs in $5'\rightarrow3'$ direction. This is called leading strand which grows continuously as a single piece. Now primer attachéd with 5′-end is removed by $5'\rightarrow3'$ exonuclease activity of DNA polymerase I.

The gap left by removal of primer is filled by complimentary nucleotides by 5′→3′ polymerase activity of DNA polymerase-I.

(j) Okazaki fragments are short DNA fragments are formed on the parental strand in 5′→3′ direction starting from primer. A separate primer is formed for this fragment. This is called lagging strand which grows discontinuously. Each Okazaki fragment has its own primer. This fragment consists of 1000-2000 nucleotides in prokaryotes and 100-200 in eukaryotes. Later primer is removed by DNA polymerase I. The gap left by removal of primer is filled by complimentary nucleotides by 5′→3′ polymerase activity of DNA polymerase I.

(k) Joining of Okazaki fragments occur by ligase enzyme to form complete strand by binding covalently by replacing primers with deoxyribonucleotides. Ligase requires NAD activity.

Thus DNA replication is said to be semi-discontinuous because leading strand is synthesised continuously and lagging strand is formed discontinuously.

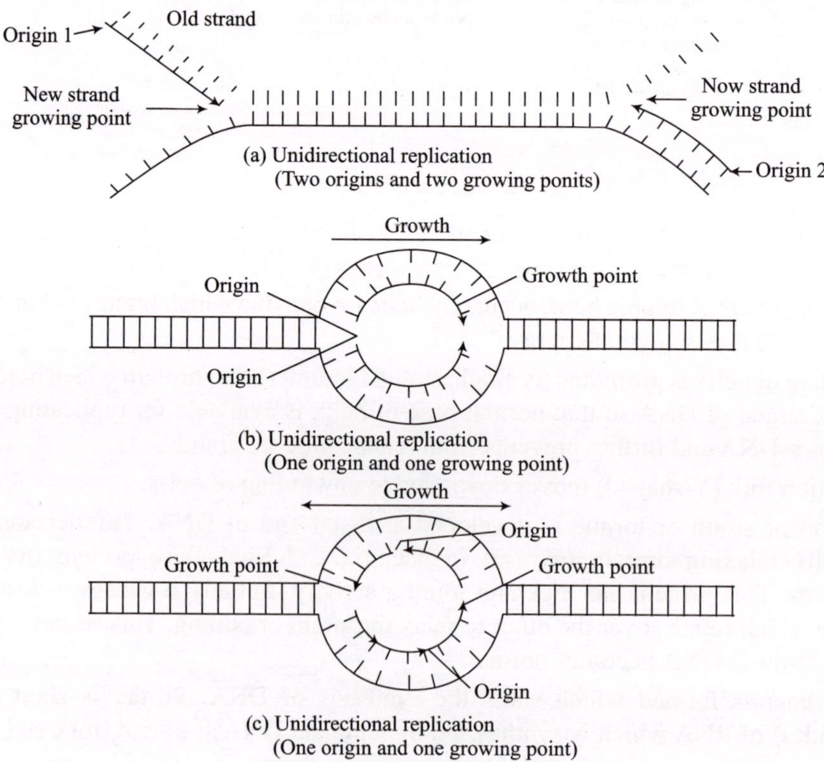

Fig. ***Direction of replication***

The above replication is unidirectional in which one strand is stationary and other strand will move with replication. In this, single replication fork is formed.

The bidirectional replication also occurs when both strands will be moving with replication and none of the strands are stationary. In this two replication forks are formed that move in opposite direction.

Replication of circular DNA in bacteria: The circular DNA has no free ends hence unwinding of helix produces supercoiling of DNA which prevents replication fork from progressing. This problem is overcome by two types of DNA topoisomerase enzyme:

1. DNA topoisomerase I: It produces a transient break in polynucleotide backbone of one of DNA strands a short distance ahead of replication fork. It allows DNA to rotate freely around other intact strand removing supercoiling. The enzyme then rejoins the ends of broken strand.
2. DNA topoisomerase II: It causes transient breaks in both strands of one of DNA molecule. It allows other DNA molecule to pass through then separates the two interlocked circular daughter DNA after completion of replication. This enzyme then rejoins the ends of broken strands.

Terms:

Proof reading: It is removal of mismatched nucleotides by 3′→5′ exonuclease activity of DNA polymerase I and replacement by correct nucleotides.

Nick translation: It is a method in which DNA is nicked in vitro by Klenow fragment (large) of DNA polymerase which is used to label DNA with molecule with radioactive nucleotide.

RNA

Ribonucleic acid is long, unbranched polymer of nucleotides. It is usually single stranded (double stranded in Reo viruses) and laevorotatory. It consists of nitrogenous bases (adenine, guanine, uracil and cytosine), ribose sugar and a phosphate molecule. RNA may form secondary structures such as stem loop and hairpin. RNA has four nucleosides which are as follows:

Base	Sugar	Nucleoside
Adenine +	Ribose →	Adenosine
Guanine +	Ribosc →	Guanosine
Thymine +	Ribose →	Thymidine
Cytosine +	Ribose →	Cytidine

There are four nucleotides in RNA which are as follows:

Nucleoside	Phosphate	Nucleotide
Adenosine +	Phosphate →	Adenylate (Adenosine-5-monophosphate)
Guanosine +	Phosphate →	Guanylate (Guanosine -5-monophosphate)
Uridine +	Phosphate →	Uridylate (Uridine-5-monophosphate)
Cytidine +	Phosphate →	Cytidylate (Cytidine-5-monophosphate)

It is present in cytoplasm and nucleus of all living cells. It is unstable, more reactive and labile than DNA and easily degradable by enzymes or chemicals due to its ribofuranose ring structure. Its unstability is due to the presence of 2′-OH group of ribose sugar, uracil and enzymatic properties. It occurs as single stranded hence it does not show a long regular helical structure like DNA. Inside RNA forms relatively globular conformation in which local regions of helical structure are formed by intramolecular H-bonding and base stacking within single nucleotide chain. It acts as adaptor and plays catalytic role.

RNA world hypothesis: It was proposed by Alexander Rich (1962) but the term was coined by Walter Gilbert (1986). It was supported by Carl Woese (1969). RNA was first form of genetic material to originate. It places RNA at centre-stage when life originated. Later life evolved to use DNA and proteins due to relative instability and poor catalytic properties of RNA. There are a few evidences which are as follows:

(a) RNA is genetic material in some viruses. It stores and transmits genetic information.

(b) It has ability to duplicate itself by ribozyme.

(c) Primer RNA is required for initiation of DNA synthesis (replication).

(d) RNA of *Tetrahymena heterophila* (a ciliate) acts as ribozyme which has catalytic ability and plays a pivotal role in peptide bond formation. It is also involved in splicing (removal of introns).

(e) The different types of RNA (mRNA, rRNA and tRNA) are required for synthesis of protein.

(f) RNA can catalyse the synthesis of a nucleotide, 4-thiouridine from 4-thiouracil and activated ribose.

(g) Ribonucleotide is present in ATP in all living cells.

(h) Ribonucleotides give rise to deoxyribonucleotides to make DNA.

(i) RNA has the ability to form double helices similar to DNA and tertiary structure similar to those of proteins.

Therefore, DNA has evolved from RNA with certain chemical modifications because RNA was very unstable due to its high reactivity. DNA is stable and being double stranded with complimentary strands, it resists changes as it evolved its repair process.

RNA world	--------	Transition world	--------	Present world
(Nucleotides → RNA)		(RNA → DNA)		(DNA → RNA → Protein)

RNA as genetic material: Fraenkel-Conrat and Singer (1957) demonstrated that RNA is genetic material in some viruses. They separated TMV particle into RNA and protein. They developed technique for forming reconstructed viruses containing protein from one mutant strain of TMV and RNA from another or vice-versa. Such hybrid viruses or chimeras were then allowed to infect tobacco leaves and the progeny was examined. In all cases, progeny was phenotypically and genotypically identical like parent RNA from where RNA had been obtained, not the parental protein type.

It is evident from these findings that only RNA (not protein) of TMV has the ability to cause disease and the type of protein present in the virus particle is determined by RNA.

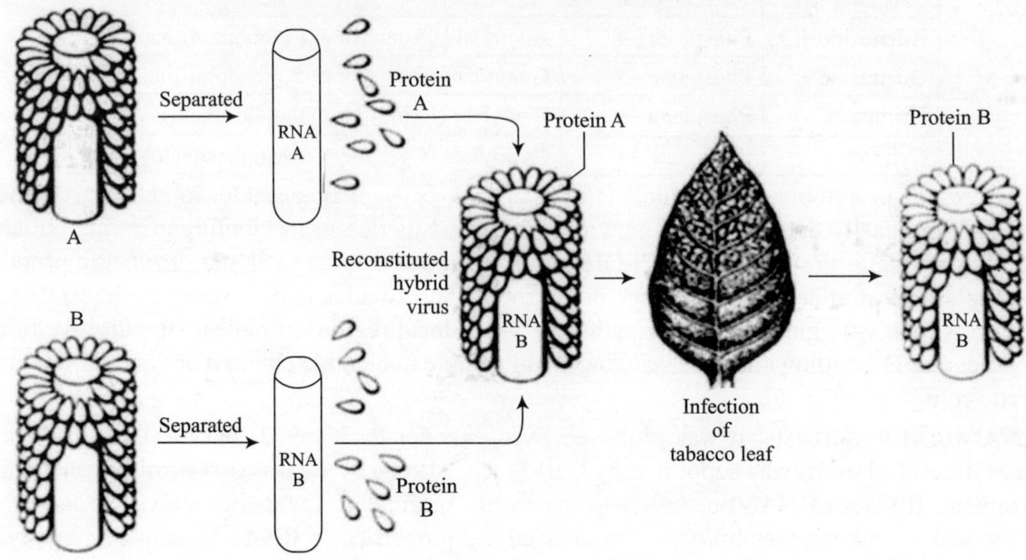

Fig. ***Experiment to prove RNA as genetic material***

In organisms, where genetic information is contained in and transmitted through DNA, RNA does not serve as genetic material in spite of good amount. This non-genetic RNA is synthesised on DNA template.

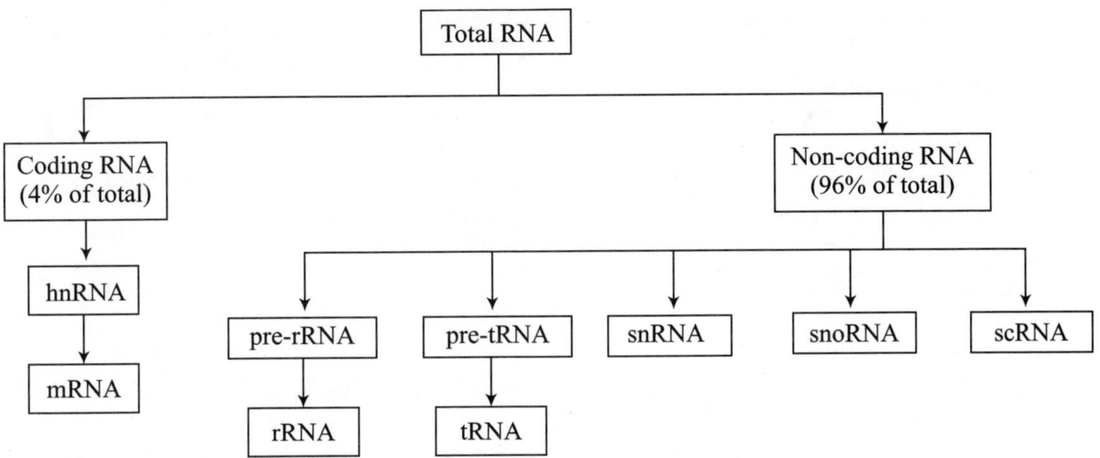

Types of non-genetic RNA: There are three types which are as follows:

1. **Messenger RNA**: The term 'mRNA' was coined by Jacob and Monod (1961).
   (a) It is the longest, linear, single stranded and of high molecular weight (500,0000).
   (b) It is synthesised in nucleus on template DNA.
   (c) It is unstable, short-lived (half life=2 min) in bacteria.
   (d) It is metabolically stable and long lived (half life=6hr) in eukaryotes.
   (e) It constitutes 5% of total cellular RNA because it constantly undergoes break down by ribonuclease enzyme.
   (f) It is most heterogeneous in size and stability.
   (g) Its sedimentation coefficient is 8S.
   (h) It is complimentary to sense strand of DNA.
   (i) It combines with several ribosomes to form polysome or polyribosome.
   (i) It passes from nucleus to cytoplasm for protein synthesis.

Types:

(a) Polycistronic mRNA: It is coded by several adjacent cistron. It occurs in prokaryotes. It is metabolically unstable and is destroyed while metabolic process is on.

(b) Monocistronic mRNA: It is coded by one gene (cistron). It occurs in eukaryotes. It is metabolically stable as it normally waits for the completion process.

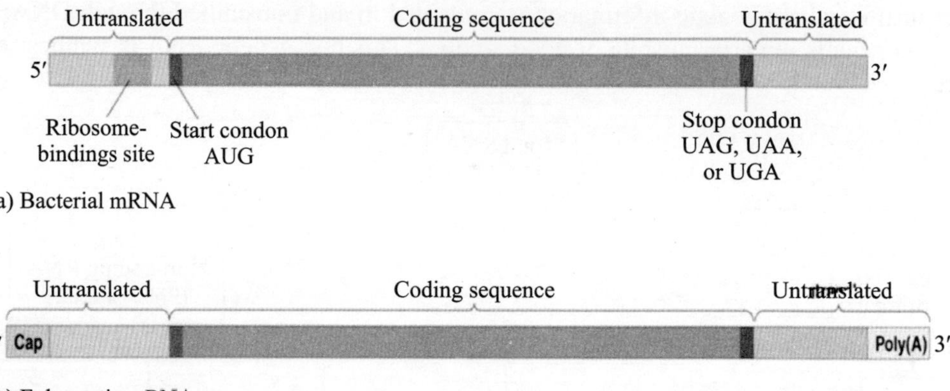

It consists of several components:

(a) **Cap**: It is blocked 7- methylated guanylate sequence (unusual base). It is found at 5′- end of mRNA. It helps in binding of m-RNA with ribosome very firmly and protects mRNA from degradative action of exonuclease enzyme at 5′-end. If it is removed then there is loss of translational process hence it controls protein synthesis.

(b) Non-coding region: It is rich in adenine and uracil residues and does not translate proteins hence called untranslated region (UTR).

(c) Initial codon: It is AUG/GUG.

(d) Coding region: It translates proteins. Its length depends on length of polypeptide chain.

(e) Terminating codon: It is UAA / UAG / UGA.

(f) Non-coding region: It does not translate proteins.

(g) Poly A tail: It is polyadenylate sequence found at 3′-end of mRNA. It is added in nucleus before m-RNA reaches the cytoplasm. It protects m-RNA from cellular enzyme and helps in detachment from ribosome.

Function: It conveys genetic information (message) from DNA to ribosome for the synthesis of particular type of polypeptides.

2. **Ribosomal RNA**: The term 'rRNA' was coined by Kurland.

(a) It is greatly coiled, heaviest and of the highest molecular weight (1.1×10^6).

(b) It is most stable.

(c) It constitutes 80% of total cellular RNA hence most abundant in cytoplasm.

(d) It is synthesised in nucleolus from nucleolar DNA (NOR of chromosome) in eukaryotes.

(e) It appears either in compact rod (coil) or in the form of extended region. Its length is variable.

(f) It is found in ribosome as its basic constituent and occurs in association with proteins.

(g) It is highly conserved among organisms and presumably present in the earliest form of life.

(h) Its sedimentation coefficient is variable.

(i) The sedimentation coefficient in chloroplast rRNA is 16 S and that of mitochondrial rRNA is 12 S.

Types	Size	Large subunit rRNA	Small subunit rRNA
Prokaryotic	70S	50S has 5 S rRNA (120 nucleotides) and 23 S (2906 nucleotides).	30 S has 16S rRNA (1542 nucleotides).
Eukaryotic	80S	60S has 5S rRNA (121 nucleotides), 5.8 S rRNA (156 nucleotides) and 28 S rRNA (5070 nucleotides)	40 S has 18S rRNA (1869 nucleotides)

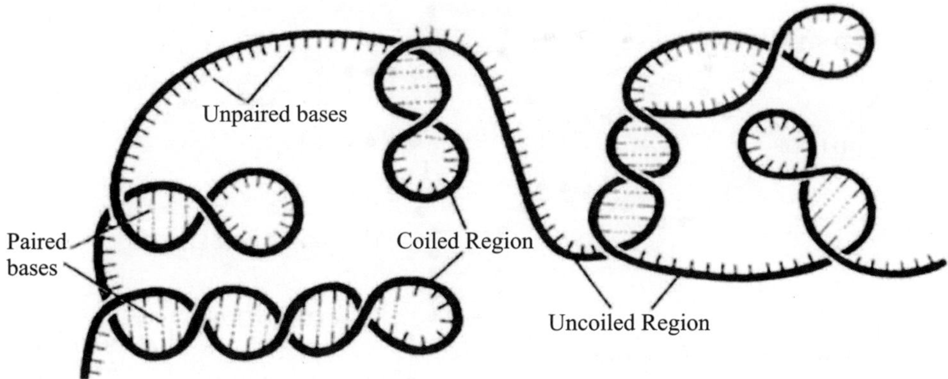

Fig. ***rRNA***

Functions:

(a) It provides stability to ribosomal subunits.

(b) It provides large surface area for spreading mRNA over ribosomes in translation.

3. **Transfer RNA**: The term 'tRNA' was coined by Hoagland.

(a) It is the smallest, shortest, lightest and lowest molecular weight.

(b) It is made up of 77 nucleotides (74-95).

(c) It is synthesised in nucleus on DNA template in the direction of $5' \rightarrow 3'$.

(d) It constitutes 15% of total cellular RNA in the cell.

(e) It is found in cytoplasm.

(f) Its sedimentation coefficient is 3.8 S.

(g) Its number is 20 in the cell carrying specific amino acids.

(h) It is quite stable in prokaryotes and less stable in eukaryotes.

(i) It is soluble in 10 M-HCl. Its life span is short.

(j) It contains several unusual bases like pseudouridine, dihydrouridine, methyl guanine, methyl cytosine, inosine, ribothymine etc which protect tRNA from degrading action of ribonuclease (RNase). These unusual bases are produced after transcription by chemical modifications such as deamination (e.g., inosine), methylation (e.g., methyl cytosine, methyl guanine etc.) and double bond saturation (e.g., dihydrouridine).

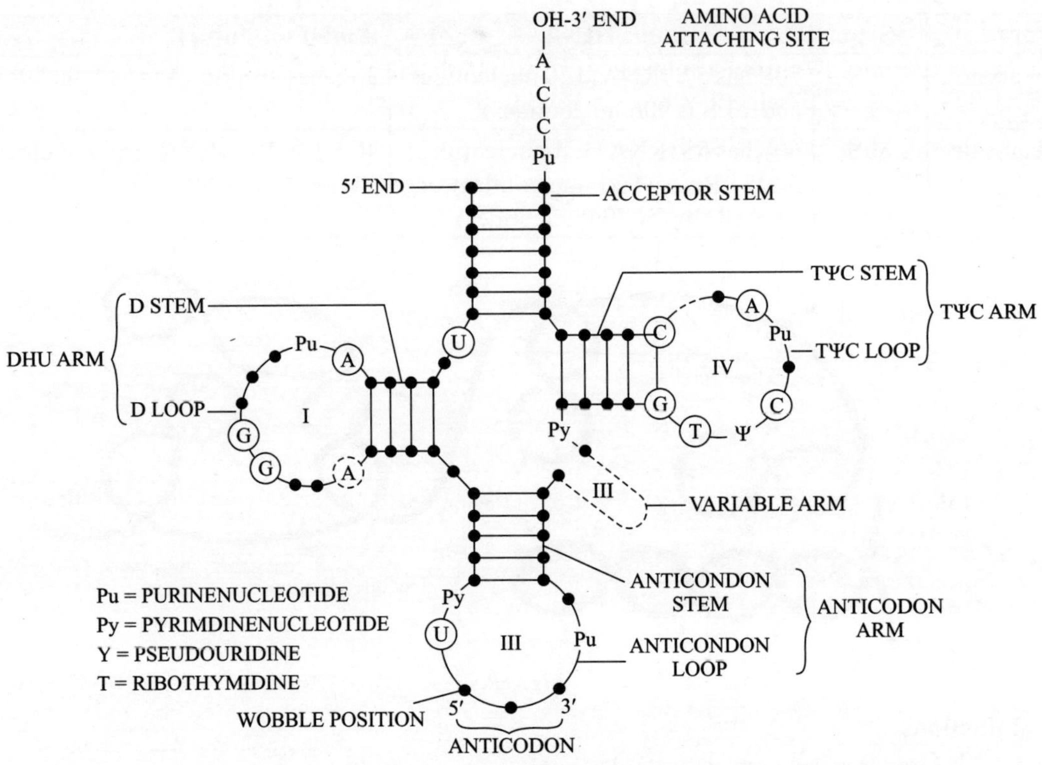

Fig. ***Clover leaf model of tRNA***

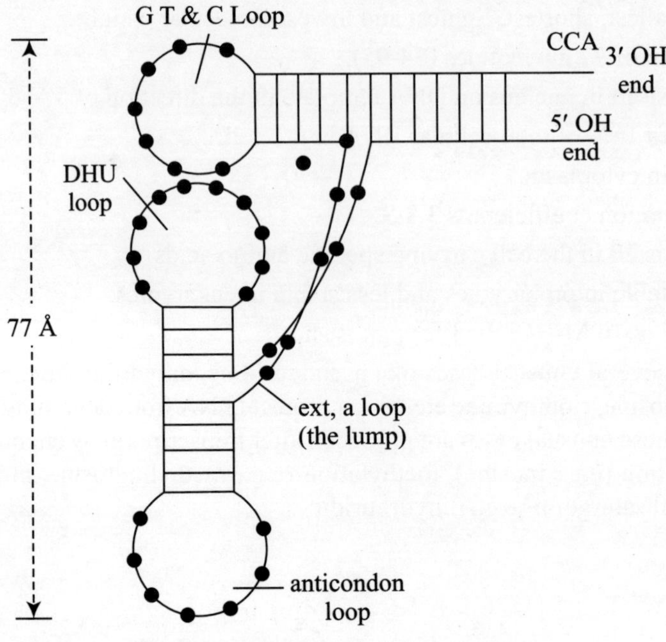

Fig. ***L-shaped model of tRNA***

Types:

(a) Monomeric: It arises from a single gene. It is found in prokaryotes. It is a single molecule with a leader sequence of nucleotides at 5′ end and trailer sequence at 3′-end.

(b) Multimeric: It arises from a cluster of genes.

Structure: Its tertiary structure (three dimensional) is L-shaped as proposed by Kim and Klug (1973). Its secondary structure (2-dimensional) of yeast alanyl t-RNA is clover leaf like as proposed by Robert Holley. It consists of several arms forming stem and loop.

(a) Acceptor arm: It contains 7 base pairs and 4 unpaired bases (nucleotides). Its 5′-end bears Guanine having terminal phosphate group and its 3′-end has CCA- OH group. It lacks loop. It is site of amino acid binding.

(b) Pseudouridine (TψC) arm: It contains 4 base pairs in stem and 7 unpaired bases (nucleotides) in loop. It has pseudouracil as unusual base. It is site for attachment with ribosome.

(c) Dihydrouridine (DHU) arm: It is the largest loop. It has 4 base pairs in stem and 8 unpaired bases (nucleotides) in loop. It has dihydrouracil as unusual bases. It is site for binding aminoacyl synthetase enzyme.

(d) Extra arm: It is variable arm present between pseudouridine arm and anticodon arm. It has variable number of nucleotides which cause difference in size of t-RNA. It forms two types of tRNA: Type I containing 4-5 nucleotides in extra arm and Type II having 10-24 nucleotides).

(e) Anticodon arm: It contains 5 base pairs in stem and 7 unpaired bases (nucleotides) in loop including three nucleotides called anticodons. It recognises and binds codons on mRNA by H-bonds.

Function: It collects amino acids from cytoplasm, transfers to ribosome and aligns to mRNA template by recognising the codon. Hence it is called adaptor molecule. It acts as physical link between the mRNA and the amino acid sequences of protein.

Other non-genetic RNA:

(a) **Small nuclear RNA** (snRNA): It is found in nucleus. There are U_1, U_2, U_4, U_5, U_6 used for formation of spliceosome leading to splicing of introns. It helps in processing of m-RNA and rRNA in nucleus.

(b) **Small cytoplasmic RNA** (scRNA): It is found in cytoplasm. It helps in binding of ribosome with ER in cytoplasm for producing secretory proteins.

(c) **Heterogeneous Nuclear RNA** (hnRNA): It is found in nucleus. It is precursor to mRNA. It is larger and higher molecular weight than mRNA. It is synthesised from split genes in eukaryotes.

(d) **Antisense RNA (**mRNA inhibiting complimentary (mic) RNA): It is synthesised sometimes on the strand complimentary to the one used for mRNA synthesis. It is used for regulation of DNA synthesis and gene expression.

Effect of temperature and pH on Nucleic acid: DNA undergoes denaturation at high temperature (82-90°C) and high pH in which H-bonds are broken between bases of both strands. A-T base pair can undergo easy denaturation (melting) since A-T base pairs have two H-bonds which denature comparatively at low temperature but G-C base pair can undergo denaturation (melting) at high temperature because G-C base pairs have three H-bonds hence G-C base pair is more stable and dense. The melting temperature (Tm) is a function of G+C content of DNA. Its range is 80-100°C. Denatured strand absorbs more light energy showing hyperchromatic effect.

Renaturation or annealing occurs when denatured DNA reassociates and forms duplex when it is cooled at 65°C or when pH is lowered. Rapid cooling allows only the formation of local regions of dsDNA formed by base pairing or joining of short regions of complimentary strands within or between strands but slow cooling allows time for the wholly complementary DNA strands to find each other and sample becomes fully double stranded.

Both DNA and RNA are hydrolysed completely to their constituents (bases, sugar and phosphate) when these are subjected to very high temperature (>100°C) or strong acid (perchloric acid) or urea.

RNA is more susceptible to hydrolysis than DNA due to the presence of 2-OH in RNA. It is perfectly positioned to participate in cleavage of RNA backbone by intramolecular attack on the phosphate of phosphodiester bond.

Chromicity of DNA: Both DNA and RNA absorb UV light due to conjugated aromatic nature of bases. The sugar phosphate backbone does not help in absorption. Both DNA and RNA absorb maximum UV light at wavelength of 260 nm which is quite distinct from protein which absorbs maximum UV light at wavelength of 280nm. The absorption at 260nm is maximum for isolated nucleotides, intermediate for ssDNA and RNA (hyperchromic) and minimum for dsDNA (hypochromic).

Concept of gene: Classical concept

1. **Mendel** used the term 'unit factor' for a specific, distinct and particulate unit of inheritance that takes part in the expression of trait from one generation to the next generation.
2. **Garrod** (1902) proposed the one mutant gene-one metabolic block (product) hypothesis.
3. **Bateson and Punnet** (1906) proposed presence or absence theory. The dominant character has determiner and recessive character lacks determiner.
4. **Johanssen** (1905) coined the term gene. Gene is an elementary unit of inheritance which can be assigned to a particular trait.
5. **Morgan** (1926) proposed 'Particulate gene theory'. Genes are arranged in linear order on chromosome and looked like beads on a string.
6. **Belling** (1928) proposed that chromosomes that appeared as granule can be gene.
7. **Beadle and Tatum** (1941) proposed one gene-one enzyme hypothesis. Gene is a unit of inheritance that specifies a single enzyme.
8. **Ingram** (1957) proposed one gene-one polypeptide hypothesis. Gene is a unit of hereditary material that specifies the synthesis of a single polypeptide.
9. **Yanofsky and Brenner** (1969) proposed colinearity between gene and polypeptide products. Gene is a linear sequence of nucleotides which determine the linear sequence of amino acids in a protein.

The classical concept defines gene is a unit if inheritance composed of a segment of DNA situated at specific locus which carries coded information associated with a specific physiological function and can undergo recombination and mutation.

Modern concept: The classical concept has changed after the discovery of DNA in 1953. Now it is confirmed that hereditary material is DNA and gene is a linear segment of DNA called cistron. Thus cistron (synonym with gene) is the largest element within a gene. It is unit of function.

It defines gene is a unit of genetic information and a segment of DNA that codes for a single polypeptide or one RNA molecule and includes the 5′ (upstream) and 3′ (downstream) regulatory sequence and noncoding sequence (intron).

Types of genes: Based on nucleotide sequence

1. **Normal genes:** These have continuous sequence of nucleotides which code for a single polypeptide chain.
2. **Pseudogenes:** These resemble other genes but do not contain useful genetic information. Hence these do not direct protein synthesis. These are useless to organisms and considered as defective copies of functional genes. These have been reported in alpha and beta globulin genes of mice; *Drosophila*, human being etc.
3. **Split genes:** These were discovered by Sharp and Roberts (1977). These are found in eukaryotes in which these contain non-essential introns (non-coding) interspersed with essential exon (coding region). These introns are required to be removed before entering into translation. Split gene is ancient feature of genome. The presence of introns represents the dominance of RNA world. These were firstly reported in ovalbumin gene of chicken.
4. **Jumping genes (Transposons):** These were discovered by Barbara Mc Clintock (1951) in maize. These move from one place to another in the chromosome and also between the chromosomes of same genome.
5. **Overlapping genes:** These can code for two or even more proteins. These are found in tumour producing viruses such as Φ x 174, SV 40.

Genetic code

It is the relationship or correspondence between sequences of bases in DNA / mRNA and sequence of amino acids in a polypeptide chain. It is set of rules by which information encoded in genetic material (DNA or RNA sequences) is translated into proteins by living cells. It is collection of codons. It allows DNA and RNA sequences to be 'decoded' into the amino acids of a protein.

Meaning of genetic code: The uniqueness of every cell, individual or species lies in the uniqueness of its proteins. The cells are enabled to synthesise their specific proteins by the information flowing from DNA.

Function of Genetic code: It determines the sequence of amino acids in polypeptide by the sequence of nitrogenous bases in DNA or mRNA.

Nature of genetic code: There are 20 types of amino acids and only four types of nucleotides.

If one base codes for one amino acid then only four amino acids would be specified, hence protein could contain only four amino acids. This singlet code is thus inadequate.

A
G
C
U

If a sequence of two bases codes for one amino acid, the four bases could specify 16 (4x4) amino acids. This is doublet code which is also insufficient.

AA	AG	AC	AU
GA	GG	GC	GU
CA	CG	CC	CU
UA	UG	UC	UU

If a sequence of three bases codes for one amino acid, then four bases could specify 64 (4x4x4) amino acids. This is triplet code which is sufficient for 20 amino acids.

George Gamow (1954) suggested triplet code. He postulated that sets of three bases must be employed to encode the 20 standard amino acids used by living cells to build protein.

Francis Crick (1961) demonstrated that codon consists of three DNA bases. by producing mutations involving addition or deletion of bases in T_4 bacteriophage.

Assignment of codons with unknown sequence (Polyuridylic acid-Poly UUU):

1. **Marshall Nirenberg** and **Mathaei** (1961) discovered and proved experimentally. They synthesised RNA (m-RNA) using only uracil, so that all along the length of mRNA. There was no other base. The mRNA consisted of same triplet UUU repeated many times. It was called polyuridylic acid. A series of 20 test tubes was prepared each containing cell free extracts of E. coli including ribosome, tRNA, ATP, enzymes and different radioactive labelled amino acids. Poly U was added to each test tube and left to allow in vitro synthesis of polypeptides to occur. Analysis of contents of test tube showed that a polypeptide had been formed only in the test tube containing the amino acid, Phenylalanine. Thus Nirenberg showed that base triplet of mRNA or codon UUU determines the position of Phenylalanine in a polypeptide chain.

   Later on Poly A produced polylysine: Poly C produced polyproline. It means UUU coded for Phenylalanine, AAA for lysine and CCC for proline. Using various polymers most of the remaining codons were then determined. Thus Nirenberg deciphered first codon, UUU which coded for phenylalanine.
2. **Khorana** synthesised first artificial (non-functional) gene, Alanine tRNA gene with 77 base pairs with defined combination of bases (homopolymers and copolymers).
3. **Khorana** synthesised second artificial (functional) gene, tyrosine tRNA gene with 207 base pairs with defined combination of bases (homopolymers and copolymers).

Properties of Genetic Code:

(a) **Triplet**: A genetic code is triplet because three adjacent bases called codon specifies one amino acid in polypeptide.

(b) **Degenerate**: It is degenerate or redundant because more than one codon (2-6) is required to specify an amino acid. The different codons for a given amino acid are said to be synonymous. The degeneracy or redundancy of code was discovered by Nirenberg. The degenerate code provides a protection to the organisms from many harmful mutations, because if one base of a codon is mutated, there are other codons which will code for the same amino acid and there will be no alteration in the polypeptide chain.

Partial degeneracy: It occurs when first two bases are identical and third base of degenerate codons differs. e.g., CUU and CUC code for Leucine.

Complete degeneracy: It occurs when any of four bases can take third position and still code for the same amino acid. e.g., UCU, UCC, UCA, UCG code for serine.

Exceptions: Single codon (AUG) codes for methionine and another single codon UGG codes for tryptophan.

Second base

First base	U	C	A	G	Third base
U	UUU, UUC – Phenylalanine F UUA, UUG – Leucine L	UCU, UCC, UCA, UCG – Serine S	UAU, UAC – Tyrosine Y UAA Stop codon UAG Stop codon	UGU, UGC – Cysteine C UGA Stop codon UGG Tryptophan W	U C A G
C	CUU, CUC, CUA, CUG – Leucine L	CCU, CCC, CCA, CCG – Proline P	CAU, CAC – Histidine H CAA, CAG – Glutamine Q	CGU, CGC, CGA, CGG – Arginine R	U C A G
A	AUU, AUC, AUA – Isoleucine I AUG M Methionine start codon	ACU, ACC, ACA, ACG – Threonine T	AAU, AAC – Asparagine N AAA, AAG – Lysine K	AGU, AGC – Serine S AGA, AGG – Arginine R	U C A G
G	GUU, GUC, GUA, GUG – Valine V	GCU, GCC, GCA, GCG – Alanine A	GAU, GAC – Aspartic acid D GAA, GAG – Glutamic acid E	GGU, GGC, GGA, GGG – Glycine G	U C A G

Fig. ***Genetic code***

(c) Non-ambiguous: It is non-ambiguous because one codon specifies only one amino acid hence same codon does not code for two different amino acids.

Exceptions: A few codons are ambiguous where one codon codes for more than one amino acid.

UUU codes for phenylalanine in streptomycin sensitive strain of *E.coli* but it also codes for leucine, isoleucine and serine when ribosome is treated with Streptomycin. This ambiguity is enhanced at high Mg^{+2} concentration, low temperature and in the presence of ethyl alcohol.

GUG codes for methionine initially but later codes for valine.

(d) Non-overlapping: It is non-overlapping because single letter (ribonucleotide) at a specific location in mRNA is not used for two different codons hence adjacent codons do not share any base. Thus each letter is read only once. e.g., AUGCCGUCA is read only as AUG and CCG and UCA. It is not as AUG GCC CGU UCA or AUG UGC GCC CCG

Exception: Overlapping codon is seen in bacteriophage ϕ x174.

(e) Commaless: It is commaless because there is no punctuation or gap between the codons. Once translation starts, the codons read one after another with no breaks or demarcating signals in between them.

(f) Universal: It is almost same for all organisms from bacteria to human beings because all living organisms contain some 20 common amino acids and same 5 bases. It is quite evident that all organisms on earth evolved from a common ancestor and code must have been established shortly after life originated.

Exceptions: There are variations in genetic code in mitochondrial genome of *Mycoplasma capricolum* (a ciliate protozoan), yeast and mammals (mice, cattle and human beings).

Universal / standard code	Non-universal code
There are 55 anticodons excluding terminal codons.	There are 22 anticodons.
AUG and GUG are initial codons.	(AUN=AUU, AUC, AUA and AUG) are initial codons for all.
UAA, UAG and UGA are terminating codons.	UGA codes for tryptophan amino acid.
AGA and AGG code for arginine.	These are terminating codons.
AUA codes for isoleuine.	AUA codes for methionine.
CUU, CUC, CUA and CUG code for leucine.	These code for threonine .

(g) Polarity: It has fixed polarity (direction) from left to right. It has fixed starting and termination ends. The messenger in mRNA reads in 5′→3′ direction and polypeptide chain is synthesised in N→C direction (from amino group to carboxyl group) hence polarity is from 5′-end to 3′-end.

(h) Initial codon: Genetic code has AUG which plays dual roles (acting as start signal for initiation of polypeptide chain and also coding formyl methionine in prokaryotes and only methionine in eukaryotes). Sometimes, GUG also acts as start signal in prokaryotes but it is less efficient since it has lesser affinity for f-met-RNA. It initiates protein synthesis when normal AUG is lost by addition.

(i) Terminating codons: Genetic code has three codons such as UAG (amber), UAA (ochre) and UGA (opal) which act as full stop to protein synthesis thus terminating the coding message. These codons are also known as nonsense codons because they do not code for any of the amino acid. UAG was first terminating codon to be discovered.

(j) Wobbling: Francis Crick proposed wobble hypothesis to explain the partial degeneracy (redundancy) of code which occurs at 3rd position of a codon. The degeneracy of code is not random. Mostly different codons for a particular amino acid have the same first two bases. The degeneracy of code accounts for unequal distribution of different amino acids with a protein. According to this, the specificity of codon is particularly determined by first two bases and third base of a codon is insignificant. For example CGU, CGC, CGA and CGG all code for arginine. It appears that CG specifies arginine and the third letter is no important so major degeneracy occurs at third position of a codon while first two bases of a codon does not change. This 3rd base is called wobble (abnormal) base which lacks specificity. This hypothesis states that the 3rd base in an mRNA can undergo non-Watson–Crick base pairing with the first base of a tRNA anticodon. These abnormal bases such as inosine (deaminated guanine), pseudouridine, tyrosine etc. can pair with more than one nitrogenous base at same position.

Thus organisms need not have the pressure of producing so many tRNAs hence wobbling reduces the number of tRNAs for protein synthesis thus overcoming the effect of degeneracy since several codons meant for same amino acid are recognised by same tRNA or it helps one t-RNA to read more than one codon. The tRNA can recognise either one, two or three codons depending upon wobble base (5′-anticodon base) hence at least 32 tRNAs are required to code 61 codons. This phenomenon gives rise to multiple codes for a given amino acid (degeneracy), possible suppression of point mutation in 3rd base of the codon, more easily removed deacylated tRNA during protein synthesis and fewer tRNA molecules than expected. There are not 61 tRNAs for 61 sense codons (each codon does not have a distinct tRNA). The 5′-nucleotide in the anticodon can vary. If the nucleotides in the second and third anticodon positions complement the first two bases in mRNA codon, an aminoacyl tRNA with proper amino acid usually binds to mRNA.

Loose base pairing will occur in first position (5′) of anticodon and third position of codon. Thus the pairing between unusual base of tRNA and wobble base of mRNA is called wobble pairing or wobbling. Three types of wobbling have been proposed. A wobble base pair is a pairing between two nucleotides in RNA that does not follow Watson-Crick base pair rule.

tRNA 5′-anticodon base	mRNA 3′-codon base	Pairing
Inosine pairs with	A, C or U	CCI anticodon can pair with GGU, GGA or GGC
Guanine pairs with	C or U	CCG anticodon can pair with GGC or GGU
Uracil	A or G	CCU can pair with GGA or GGG

Wobble hypothesis has the following attributes:

(a) The first two bases in the codon show the coding specificity. These bases form strong Watson-crick base pairs and bind strongly to the anticodon of the tRNA.

(b) While reading in the direction of 5′ to 3′ if first nucleotide in the anticodon (present on tRNA) pairs with the last nucleotide of the codon (present on mRNA) then it determines how many nucleotides the tRNA actually distinguishes. If the first nucleotide in the anticodon is a C or A then pairing is specific and acknowledges original Watson-crick pairing, that is only one specific codon can be paired to that tRNA. If the first nucleotide is U or G, the pairing is less specific and in fact two bases can be interchangeably recognised by the tRNA. Inosine displays the true qualities of wobble where it is first nucleotide in the anticodon then any of the bases in the original codon can be matched with the tRNA.

(c) Due to specificity inherent in the first two nucleotides of the codon, if one amino acid is coded for by multiple anticodons and those anticodons differ in either the second or third position (first or second position in the codon) then a different tRNA is required for that anticodon.

(d) The minimum requirement to satisfy all possible codons (61 excluding three stop codons) is 32 tRNAs (31 tRNAs for the amino acids and one initial codon).

Terms:

Open reading frame (ORF): It is a set of codons that run continuously between an initiation codon and a terminating codon. This is used to identify protein coding DNA sequences in genome sequencing project.

Second Genetic code: It refers to the interaction of specific tRNA with corresponding aminoacyl synthetase enzyme.

Second Half Genetic code: It refers to protein folding problem or to the rules governing the formation of a 3D structure (amino acid sequence) of a protein.

Protein Synthesis

Gene expression is the process by which information contained in the base sequences of genes is decoded to produce proteins that determine the phenotypic traits of organisms and responsible for cellular functions.

The proteins constitute the major part by dry weight of an actively growing cell. These proteins are built up from 20 amino acids which constitute the basic building blocks. In proteins, these amino acids are linked up by peptide bonds to form long chains called polypeptides.

The sequence of bases in particular segment of DNA will determine the sequence of amino acids in a particular polypeptide. This is central dogma. It explains how protein synthesis is controlled by nucleic acids. Protein synthesis occurs in G_1 and G_2 phase of cell cycle. It takes place in cytoplasm where it involves ribosomes. Therefore genetic information has to be carried from nucleus to cytoplasm. There are two modes of flow of information.

Unidirectional flow of information: It was proposed by Francis Crick (1953).

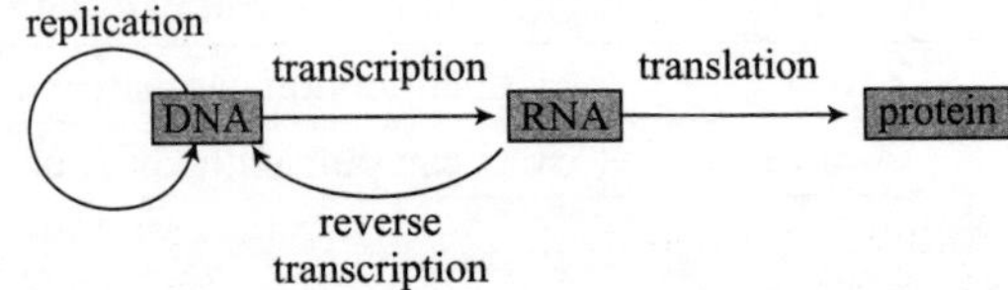

Reverse flow of information: It was proposed by Howard Temin and David Baltimore (1975) in retroviruses (HIV, Rous sarcoma virus). It occurs in the presence of reverse transcriptase enzyme. It is RNA dependent DNA polymerase (RNA directed DNA polymerase). It is found in both normal and tumour cell. It is used in the synthesis of complimentary or copy DNA (c-DNA) from mRNA leading to the preparation of cDNA library for study of genes expressed in specific tissues.

Protein synthesis occurs in following two steps:

Transcription

It is synthesis of mRNA from DNA in nucleus. It requires following components:

(a) Template strand of double stranded DNA or single stranded DNA.

(b) Four activated ribonucleotides (ATP,GTP, UTP,CTP)

(c) Mg^{+2}

(d) RNA polymerase: It was discovered by Severo Ochoa. It is DNA dependent or DNA directed enzyme. It does not require primer and exonuclease activity. A single RNA polymerase enzyme synthesises all types of RNA in prokaryotes but in eukaryotes three types of RNA polymerase enzymes are required to synthesis all types of RNA.

Features	RNA polymerase I	RNApolymerase II	RNA polymerase III
Location	Nucleolus	Nucleoplasm	Nucleoplasm
Sensitivity to α-aminitin	Insensitive	Strongly sensitive	Strongly sensitive
Promoters	Upstream of start point	Upstream of start point	Downstream of start point
Level of activity	50-70%	20-40%	1-10%
Carboxy terminal domain	Absent	Present	Absent
Function	Synthesises 18 S, 28S, 5.8 S rRNA	Synthesises heterogeneous nuclear RNA and mRNA	Synthesises 5S rRNA, tRNA and snRNA.

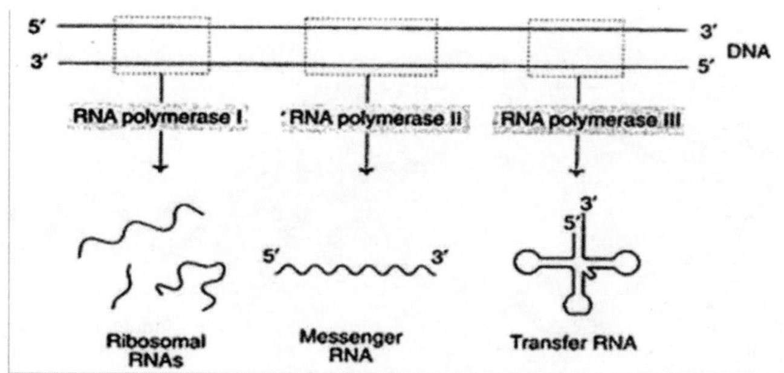

Each RNA polymerase enzyme (holozyme) consists of a core complex having four polypeptide chains and a sigma factor.

α-subunit	It recognises promoter.
α′-subunit	It recognises promoter.
β-subunit	It binds sigma factor.
β′-subunit	It binds DNA template.
Sigma factor	It recognises promoter signal on DNA template and directs binding of RNA polymerase to promoter region. It thus increases the specificity of core enzyme binding to DNA.

A typical transcription unit has

(a) Promoter at 5′-end of coding strand: It has 10 and 35 nucleotide pairs such as
- A-10 sequences Pribnow box (TATAAT) or Hogness box (eukaryotes): It is most conserved sequence.
- A-35 recognition sequences (TTGACA)

(b) Start point: It is mostly purine on template strand.

(c) Coding region:

(d) Terminator site:

35 sequences 10 sequences ←Upstream Downstream→

Recognition sequence (TTGACA)	17 bp (spacer)	Pribnow box (TATAAT)	7bp spacer	▌ Coding strand ▌ Start point Terminator site Template strand

Promoter site

Types of transcription: It is of two types:

(a) Asymmetrical Transcription: A given gene is transcribed from one strand of double stranded DNA.

(b) Symmetrical Transcription: A given gene is transcribed from both strand of double stranded DNA.

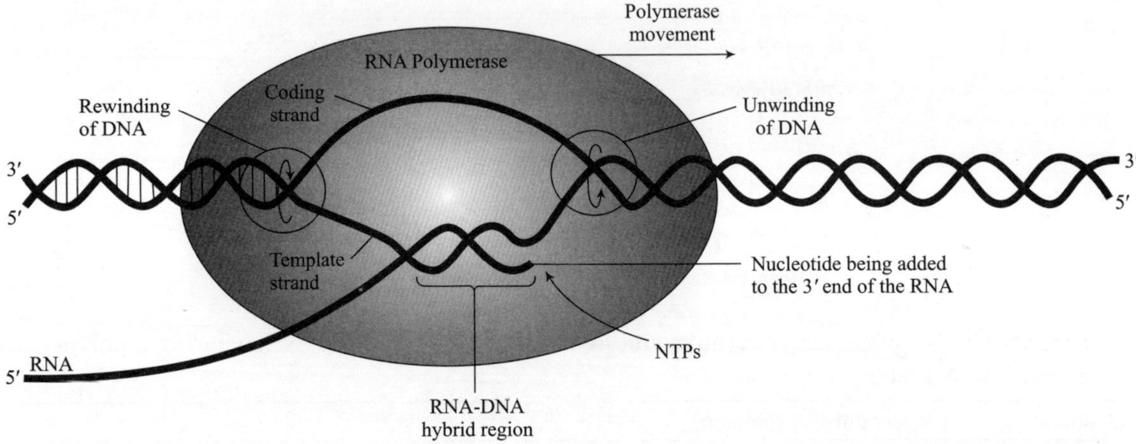

Fig. ***Mechanism of transcription***

Mechanism of transcription: Transcription cannot start randomly but starts specifically at the start of a gene. The signal for starting transcription occurs in promoter sequence. Promoter lies directly upstream of coding strand. Promoter contains specific DNA sequence which acts as pint of attachment for RNA polymerase.

(a) Sigma factor is attached with core enzyme by beta subunit forming holozyme. Sigma factor recognises the start point at -35 sequences on template strand.

(b) Transcription starts from left to right (5′→3′) on coding strand when holozyme binds to promoter region of DNA. Holozyme binds to about 70-80 bp extending from -58 to +20 to form closed binary complex on template strand.

(c) Holozyme now creates a bubble which separates the two strands of DNA at -10 box which is rich in weak A-T bonds forming open binary complex.

(d) mRNA starts forming on coding strand because correct ribonucleotides are aligned opposite to deoxyribonucleotides of template strand according to base pair rule. The first base is guanine transcribed for mRNA. When first two robonucleotides were aligned then holozyme forms phosphodiester bond between them.

(e) Holozyme with template strand containing two ribonucleotides forms ternary complex.

(f) Sigma factor now detaches from core enzymes of holozyme at promoter region for reuse because it is no longer required or its presence may disturb the process.

(g) The newly formed RNA chain has triphosphate group at 5′ end and OH at 3′-end.

(h) Core complex binds tightly to template strand and sequentially adds the correct nucleotides to 3′-end of RNA and copies the base sequence of coding strand into RNA copy.

(i) As RNA elongates, its 5′-end progressively separates from template strand.

(j) These single strands (coding and template) progressively reassociate to form duplex. Hence elongation involves progressive separation of duplex to form transient single strand.

(k) Core complex continues RNA synthesis at the speed of 40 nucleotides per sec till it reaches terminator site.

(l) Core complex continues RNA synthesis at the speed 40 nucleotides per sec till it reaches terminator site.

(m) Termination occurs non-randomly when core complex reaches terminator sire. In *E. coli*, it occurs at palindrome. It requires rho (ρ) factor which binds RNA transcript at specific site. It involves the following events:

(a) It stops adding ribonucleotides to RNA chain,

(b) It dissociates RNA chain from template strand by disrupting non-covalent bond that holds newly formed RNA to template DNA.

(c) It reassociates the separated strands to form DNA duplex,

(d) It causes disappearance of bubble and

(e) Finally it releases core enzyme from DNA.

Double strands of DNA

5′—GTCATGATCACGA—3′ : Non-template / coding / sense strand

3′—CAGTACTAGTGCT—5′ : Template / noncoding / antisense strand

↓

5′—GUCAUGAUCACGA—3′ : **mRNA** transcript

Prokaryotic transcription	Eukaryotic transcription
It occurs in cytoplasm.	It occurs in nucleus.
There is no definite phase for its occurrence.	It occurs in G_1 and G_2 phase.
It requires a single RNA polymerase for synthesising all RNAs.	It requires 3 RNA polymerase for synthesising all RNAs.
RNA is released and processed in cytoplasm.	RNA is released and processed in nucleus.
mRNA transcript has few surplus nucleotides.	mRNA transcript has large surplus nucleotides.
Transcription and translation are coupled.	Transcription and translation occur separately.

Post transcriptional processing: The released RNA is called primary transcript.

Prokaryotic mRNA undergoes little processing to become active because transcription and translation occur in same region. It results in beginning of translation even before mRNA is fully formed.

Eukaryotic mRNA usually undergoes chemical modification called RNA processing in the nucleus of eukaryotic cells to form functional RNA. In eukaryotes, gene is made up of several pieces of base sequence coding for amino acid called exon separated by stretches of noncoding sequence called intron. Thus the information on eukaryotic gene for assembling a protein is not continuous but split. During processing, elements that are no longer of any use are removed and functionally effective elements are added. In eukaryotes, transcribed mRNA is larger than processed or mature mRNA. The larger RNA containing pre-mRNA transcript is known as heterogenous nuclear RNA (hn-RNA). It contains both unwanted base sequence (transcribed from intron) alternated with useful base sequence (transcribed from exon). This primary transcript is converted into functional RNA after post-transcriptional processing.

Technique	Developer	Objective
Southern blotting	Edwin Southern	To identify DNA segments
Northern blotting	Alwine	To identify RNA
Western blotting	Burnette	To identify protein

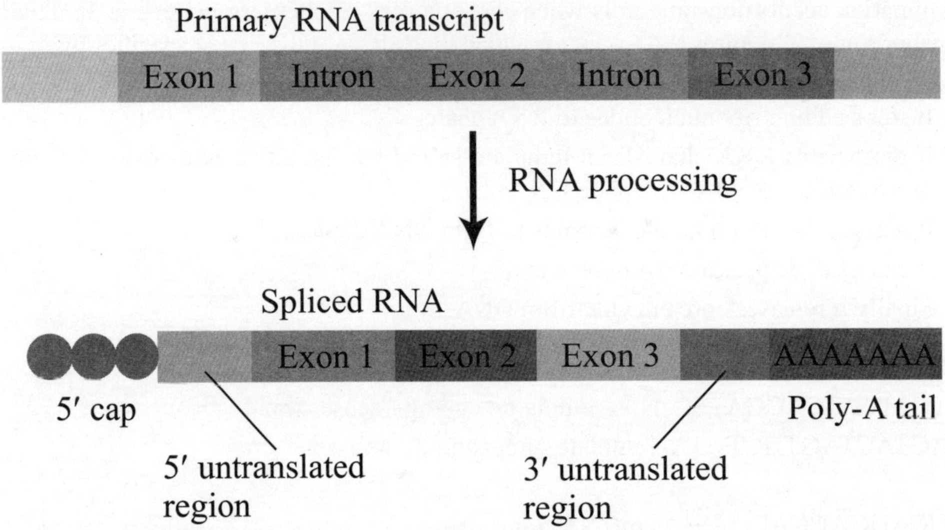

Primary RNA transcript
Exon 1 Intron Exon 2 Intron Exon 3
RNA processing
Spliced RNA
Exon 1 Exon 2 Exon 3 AAAAAAA
5′ cap
Poly-A tail
5′ untranslated region
3′ untranslated region

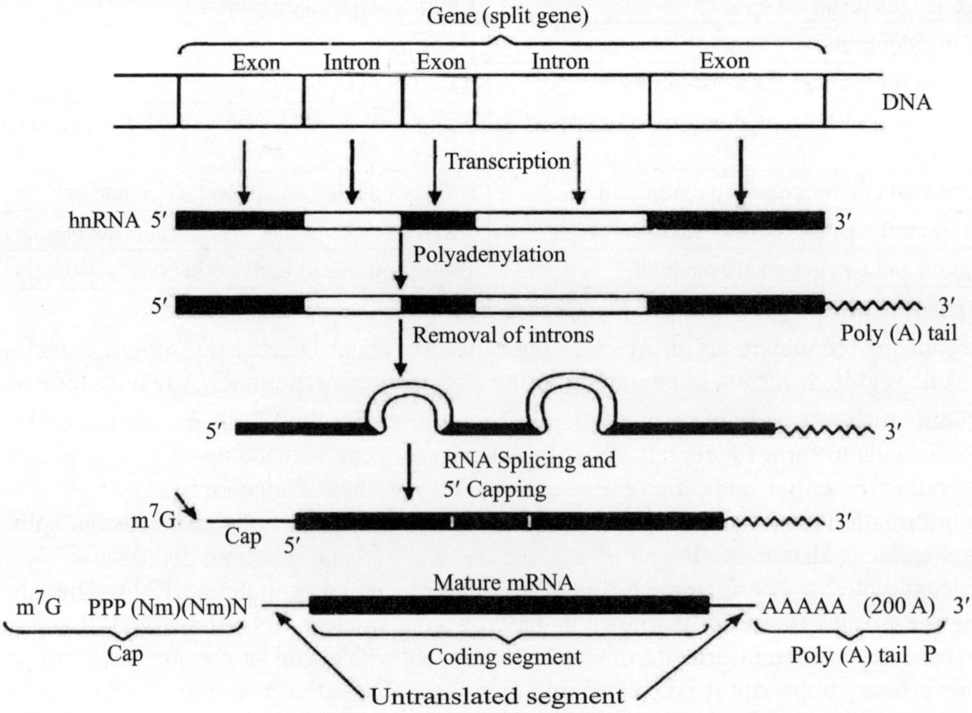

Gene (split gene)
Exon Intron Exon Intron Exon
DNA
Transcription
hnRNA 5′ 3′
Polyadenylation
5′ 3′
Removal of introns
Poly (A) tail
5′ 3′
RNA Splicing and 5′ Capping
m^7G
Cap 5′ 3′
Mature mRNA
m^7G PPP (Nm)(Nm)N AAAAA (200 A) 3′
Cap
Coding segment
Poly (A) tail P
Untranslated segment

Fig. ***Process of post-transcriptional processing***

Post-transcriptional processing: It involves several events which are as follows:

1. **Methylation (Capping)**: It is addition of $-CH_3$ group to guanylate sequence at 5′-end of nascent m-RNA by RNA polymerase II. It protects mRNA from degradation. Cap is essential for formation of mRNA-ribosome complex. Translation is not possible if cap is absent.

2. **Polyadenylation (Tailing):** It is terminal addition of polyadenylate sequence of 200-300 bases at 3′-end of mRNA (hnRNA) by polyadenylate polymerase enzyme. It occurs immediately after transcription. It stabilises mRNA and protects it from degradation of coding sequence at 3′-end by exonuclease enzyme. It transports mature mRNA from nucleus to cytoplasm.
3. **Splicing (Tailoring):** It is removal of introns from primary transcript by spliceosome. Each intron starts with dinucleotide GU and ends with dinucleotide AG (GU-AG rule). These are recognised by components of splicing apparatus of small nuclear ribonucleoprotein (snurp) complex such as U_1,U_2,U_4,U_5,U_6 called spliceosome which is formed between 5′-end (GU) and 3′-end (AG) of intron. Energy is obtained from ATP and it removes ATP. It occurs in two steps:
   (a) **Phosphodiester cleavage**: In this, 3′-hydroxyl group of Adenine and branch point sequence attacks phosphodiester bond 5′ to the G of GU (5′-splice site). The bond is broken releasing the 5 end of intron and attacking it to branch point sequence. The intron now forms a tailed loop structure called lariat. It does not require ATP.
   (b) **Ligation reaction**: In this, the 3′end of intron is cleaved after G of AG (3′- splice site). The intron is released and two exon sequences are joined together by RNA ligase enzyme. It requires ATP.

   Splicing is of two types:
   (a) **Constitutive splicing:** It is common and occurs in all cells without any variation. It produces only one mature mRNA.
   (b) **Alternative splicing:** It occurs rarely. It produces more than one mRNA.

Translation

It is a process in which RNA molecule (a particular nucleotide sequence present on mRNA) is translated into a particular amino acid sequence forming polypeptide chain with the help of ribosome. It is very complex process involving all three types of RNA. It occurs in following steps:

1. Activation of amino acids: It requires following components:

(a) Amino acids: There are 20 amino acids (monomers) which form the raw materials for protein synthesis. These are available in cytoplasm as amino acid pool. Polypeptide chain is brought about by an amino acid, methionine in eukaryotes and formyl methionine in prokaryotes.

Table: List of 20 amino acids which take part in protein synthesis

S.No.	Amino acid	Abberviation	S.No.	Amino Acid	Abbreviation
1.	Phenylalanine	Phe	11.	Histidine	His
2.	Leucine	Leu	12.	Glutamine	Gla
3.	Isoleucine	Ile	13.	Asparagine	Asn
4.	Methionine	Met	14	Lysine	Lys
5.	Valine	Val	15.	Aspartic acid	Asp
6.	Serine	Ser	16.	Glutamic acid	Glu
7.	Proline	Pro	17.	Cysteine	Cys
8.	Threonine	Thr	18.	Tryptophan	Tyr
9.	Alanine	Ala	19.	Arginine	Arg
10.	Tyrosine	Tyr	20.	Glycine	Gly

(b) tRNA:

(c) Aminoacyl tRNA: It recognises the specific codon of mRNA due to anticodon that base pairs with mRNA codon.

(d) ATP,

(e) Mg^{+2}

All amino acids except glycine show D and L forms (sterioisomerism or optical isomerism). Only L forms are involved in protein synthesis. Each of 20 amino acids occurs in cytoplasm in an inactive state. Hence each amino acid is activated by a charging enzyme, aminoacyl synthetase in the presence of ATP. This enzyme is substrate specific.

$$\text{Amino acid} + \text{ATP} \xrightarrow{\text{Aminoacyl synthetase, } Mg^{+2}} \text{Aminoacyl adenylate (AA-AMP)} + \text{PPi}$$

It is first energy requiring process in entire process. The hydrolysis of inorganic pyrophosphate with the help of pyrophosphatase enzyme provides energy for driving the initial reaction.

2. Charging of tRNA: Now the activated amino acid is transferred to 3′- end of tRNA. It occurs with the formation of high energy ester bond between –COOH group of amino acid and 3′-OH group of terminal adenosine of tRNA with the help of aminoacyl tRNA synthetase enzyme. It has two active sites: One for recognising anticodon loop and acceptor arm of tRNA and another for recognising specific amino acid.

$$\text{AA-AMP} + \text{tRNA} \xrightarrow{\text{Aminoacyl tRNA synthetase}} \text{Aminoacyl tRNA} + \text{AMP}$$

3. Chain initiation: It is slow process. It requires following components:

(a) Formylated methionyl tRNA (f Met tRNA):

(b) mRNA: It acts as template. It conveys genetic information from DNA to ribosome.

(c) Initial codon (AUG) of mRNA:

(d) Ribosome: It has small and larger subunits. These units are separate when not involved in protein synthesis. Several ribosomes line up on mRNA chain during protein synthesis called polysome or polyribosome. It has been established that proteins are synthesised on polyribosome not at single ribosome. This occurs in both prokaryotes and eukaryotes as well as for the cell organelles such as mitochondria and chloroplast.

Eukaryotic ribosome has a groove at the junction of two subunits (40S and 60S). A tunnel extends from the groove through large subunit and opens into the canal of ER. The polypeptides are synthesised in groove between these subunits and pass through the tunnel of large subunit into ER. The developing polypeptides are protected from the action of cellular enzymes. The small subunit forms cap over large subunit. The large subunit attaches to ER by two glycoproteins (ribophorin-I and ribophorin-II). Ribosome holds mRNA, tRNA and associated enzymes control until the formation of peptide bond between adjacent amino acids.

Prokaryotic ribosome consists of two subunits-30S and 50S in which 30S stabilises mRNA while 50S stabilises aminoacyl tRNA.

Thus ribosome has three sites: A site (aminoacyl tRNA site) receives all the incoming charged tRNA while P site (peptidyl tRNA site) holds tRNA carrying the growing polypeptide chain and E site (exit site) where discharged tRNA leaves the ribosome.

It is advantageous for DNA (information centre) and ribosome (manufacturing centre) to be present at different sites in the cell. If both were in nucleus, the ribosomes would be far away from energy sources and raw materials. Likewise, if both were in cytoplasm, the DNA would be exposed to respiratory breakdown.

(e) GTP: It is energy source.

(f) Mg^{+2}

(g) Initiation factors: There are 3 initiation factors in prokaryotes present in 30S subunit of ribosome.

IF_1	It binds over A site of 30S ribosome to prevent initiator tRNA from binding to A site and thus helps binding f-met tRNA with initial codon of mRNA at P site. It increases the affinity of 30S subunit for other factors. It also stimulates the activity of IF_2 and IF_3.
IF_2	It positions f-met-tRNA at P site in initiation complex. It binds with GTP hence it has ribosome dependent GTPase activity.
IF_3	It binds free 30S subunit of ribosome with mRNA.

Chain initiation occurs in following steps:

(a) Formation of formylated methionyl tRNA: Methionine carried by tRNA is formylated in the presence of transformylase enzyme and formyl tetrahydrofolate (formate) to form f-met tRNA.

$$\text{Met-tRNA} + \text{formate} \xrightarrow{\text{Transformylase}} \text{f-Met-tRNA}$$

(b) Binding of 30 S ribosome with mRNA: The 3′- end of 16 S rRNA in 30 S subunit of ribosome binds with 5′-end of Shine Delgarno sequence or purine rich leader sequence (5′-AGGAGGU-3′) of mRNA carrying initial codon (AUG) in the presence of IF_3. This attachment helps AUG to combine with initiator anticodon (UAC) on tRNA.

5′ ———— AGGAGGU ———— AUG ———— 3′
(Shine Delgarno sequence) (Initial codon)

$$30S + mRNA \xrightarrow{IF_3} 30S\text{-}mRNA\text{-}IF_3$$

(c) Binding of 30S-mRNA-IF_3 with f-met-tRNA: The anticodon (UAC) of f-met tRNA binds with codon (AUG) of mRNA with the help of IF_2 and GTP.

$$30S\text{-}mRNA\text{-}IF_3 + \text{f-met-tRNA} \xrightarrow{IF_2,\,GTP} IF_3\text{-}30S\text{-}mRNA\text{-f-met tRNA-}IF_2$$
(Initiation complex)

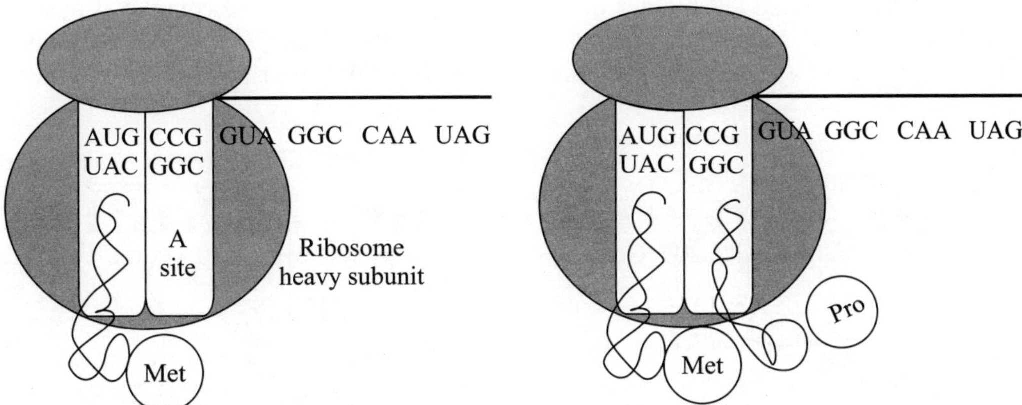

(d) Association of ribosomal unit: The initiation complex combines with large subunit (50S) of ribosome hydrolysing GTP into GDP and iP and releasing both initiation factors (IF_2 and IF_3). Initiation complex is now complete with active ribosome in which A site is vacant and P site is occupied by f-met tRNA.

$$IF_3\text{-}30S\text{-}mRNA\text{-f-met tRNA-}IF_2 + 50S \longrightarrow 70S\text{-}mRNA\text{-f-met-tRNA} + IF_3 + IF_2$$

4. Chain elongation: It requires following components:

(a) 70S ribosome,

(b) Codons of mRNA (other than initial codon and terminating codons),

(c) Aminoacyl tRNA,

(d) Elongation factors: There are three elongation factors required.

EF-Tu (Temperature unstable)	It directs next t-RNA to its correct position in ribosome.
EF-Ts (Temperature stable)	It is most abundant. It regenerates EF-Tu after hydrolysis of attached GTP.
EF-G (Translocase or GTPase)	It translocates mRNA

(e) GTP

(f) Mg^{+2}

Chain elongation is most rapid and cyclic process. It undergoes at the rate of 15 amino acids per sec at 37^0C in prokaryotes and 2 amino acids per sec in eukaryotes. It runs by regular addition of amino acids. It involves all the reaction from the formation of first peptide bond to that of last peptide bond of polypeptide chain. It occurs in following steps:

(a) Binding of aminoacyl tRNA at A site of ribosome (decoding): P site is occupied by f-met tRNA paired to AUG and A site is positioned over second codon.

Any charged aminoacyl tRNA except f-met tRNA enters A site of ribosome depending mainly on the codon of mRNA present at A site. Its anticodon locates (reads) and binds to complimentary codon of mRNA chain by H-bond at A site in the presence of GTP and two elongation factors (EFTu and EFTs). EFTu mediates the entry of aminoacyl tRNA at A site by firstly binding to GTP.

Aminoacyl t-RNA + EF-Tu +GTP $\longrightarrow$ Aminoacyl tRNA-EF-Tu-GDP

EF-Tu-GDP + EF-Ts $\longrightarrow$ EF-Tu-EF-Ts + GDP

EF-Tu-EF-Ts + GTP $\longrightarrow$ EF-Tu-GTP + EF-Ts

The activated EF-TU-GTP complex binds to tRNA. Thenafter, GTP of complex is hydrolysed to GDP and derives the binding of aminoacyl tRNA at A site. Then, EF-Tu is released leaving new tRNA at A site. EF-Ts mediates the release of EF-Tu-GDP from ribosome and regeneration of EF-Tu-GTP.

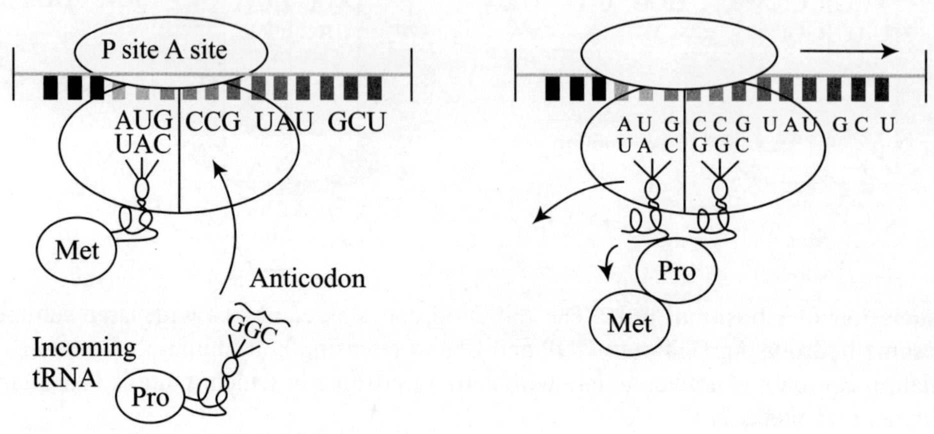

(b) Formation of Peptide bond (transpeptidation): A peptide bond is formed between free –COOH group of peptidyl tRNA at P site and free $-NH_2$ group of aminoacyl tRNA at A site which is just arrived at this site. Bond formation is catalysed by peptidyl transferase enzyme (a component of 23 S rRNA of large subunit of ribosome) and it is simultaneous with the transfer of amino acid from P site to A site. Peptide bond formation depends on release of EF-Tu from ribosome but not on the hydrolysis of GTP.

In peptide bond formation, the linkage between first amino acid and its tRNA is broken and the –COOH group now forms a peptide bond with free $-NH_2$ group of second amino acid. Thus, the second tRNA carries a dipeptide and first tRNA is without amino acid.

The reaction is accomplished by transfer of f-methione on f-met tRNA of P site to amino acid attached to tRNA bound to second codon at A site. After the formation of peptide bond, P site has a free tRNA (deacylated) while A site has peptidyl tRNA.

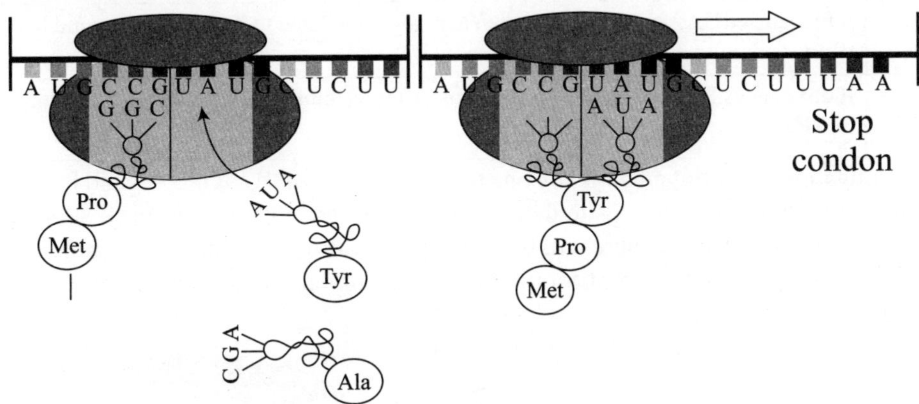

(c) Translocation of peptidyl tRNA from A site to P site: Ribosome now moves 3 bases along mRNA, this ejects free tRNA from P site and shifts it to Exit site and from here it leaves ribosome and returns to mixed pool of tRNA in cytoplasm.

Now dipeptidyl tRNA moves from A site to new free P site. It requires EF-G and GTP.

EF-G-GTP complex binds to EF-G site of 50S subunit. Then GTP is hydrolysed and EF-G is released that is essential for binding EF-Tu (present in ternary complex to A site).

In this manner, ribosome moves 3 bases at a time and one amino acid is added after every activity of translocation. The first tRNA , now uncharged, separates from mRNA chain at P site of the ribosome and returns to the mixed pool of tRNA in cytoplasm. Likewise the second tRNA leaves the mRNA chain vacating the Psite.

Thus the cycle repeats as each amino acid is added to polypeptide.

As the ribosome moves over mRNA, all the codons of mRNA arrive at A site one after another and peptide chain grows. Thus the amino acids are linked up in a polypeptide in a sequence communicated by DNA through mRNA. This is called nascent polypeptide.

The growing polypeptide chain always remains attached to its original ribosome and is not transferred from one ribosome to another. Only one polypeptide chain can be synthesised at a time on one ribosome.

Energy used for protein synthesis

- One ATP is hydrolysed to AMP during activation of amino acid.

- One GTP is hydrolysed to GDP as each successive amino acyl-tRNA complex attaches to the A site of ribosome.
- A second GTP is hydrolysed to GDP as ribosome moves to each new codon on mRNA.

Thus 2 GTP and 1ATP are required for the formation of each peptide chain.

$$1\text{ATP} + 2\text{GTP} \xrightarrow{\text{Formation of peptide bond}} \text{AMP} + 2\text{GDP} + 4\text{Pi}$$

5. Chain Termination: It requires following components:

(a) 70S ribosome:

(b) Terminating codons (UAA,UAG and UGA):

(c) Release factors (RF): There are three release factors in prokaryotes:

RF_1	It recognises UAG and UAA and then activates ribosome to hydrolyse the bond between tRNA and polypeptide.
RF_2	It recognises UAA and UGA and then activates ribosome to hydrolyse the bond between tRNA and polypeptide.
RF_3	It dissociates RF_1 and RF_2 from ribosome after termination and thus releases either RF_1 or RF_2.

The chain elongation is continued until any one terminating codon (UAG, UAA and UGA) is reached at A site of mRNA. When it reaches A site then it does not attract any aminoacyl tRNA hence it is not read and joined by anticodon. Thus there will be no further addition of amino acid to polypeptide chain. The terminating codon gives signal to ribosome for the attachment of a protein, release factors (RF) at A site.There are three release factors in prokaryotes:

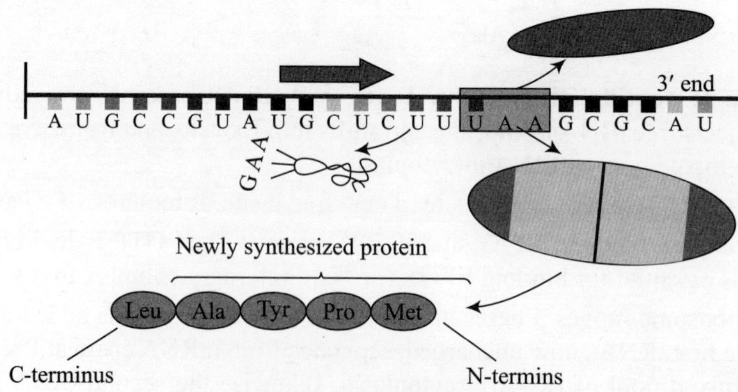

Fig. *Chain termination*

Thus release factor brings about termination of chain formation by following events:

(a) The polypeptide chain separates from tRNA at P site.

(b) tRNA is released from P site of ribosome.

(c) Ribosome is dissociated to subunits.

(d) mRNA is released.

Prokaryotic translation	Eukaryotic translation
It is carried out by 70S ribosome.	It is carried out by 80 S ribosome.
It starts while transcription is on.	It starts after completion of transcription.
It has polycistronic mRNA	It has monocistronic mRNA.
It starts with formyl methione (first amino acid).	It starts with methionine (first amino acid). No formylation is required.
It performs chain initiation when ribosome enters mRNA at AUG or at nearby Shine Delgarno sequence.	It performs chain initiation when ribosome enters at capped region of mRNA at 5′end then to AUG.
It requires 3 initiation factors (IF_1, IF_2 and IF_3).	It requires 10 initiation factors.
It requires 3 elongation factors (EFTs, EFTu and EFG).	It requires 2 elongation factors (EF_1 and EF_2).
It requires 3 release factors (RF_1, RF_2 and RF_3).	It requires 1 release factor.

Post translational processing: The released polypeptide chain is modified in bacteria in several ways:

(a) Deformylation**:** The formyl group of f-met is removed by deformylase enzyme by hydrolysing the entire f-methionine residues.

(b) Modification of released polypeptide: The released polypeptide being linear (primary structure) loses some amino acids from the end by activity of exopeptidase enzyme and then coil and fold on itself to acquire secondary or tertiary structure. It may combine with other polypeptides to assume quaternary structure.

(c) Cleavage of specific region: Proezyme is converted to enzyme by trimming of amino acids from the end or removal of internal peptides or removal of aminoterminal signal peptide sequence from secreted protein.

(d) Glycosylation: It is done by attachment of sugar.

(e) Phosphorylation;

(f) Hydroxylation:

(g) Acetylation

Role of antibiotics in protein synthesis: The antibiotics inhibit the process of protein synthesis at both level of transcription and translation.

Antibiotics	Action	Functions
Actinomycin D	Transcription	It blocks progression of RNA polymerase
Rifampicin	Transcription	It inhibits RNA polymerase by binding its beta subunit.
Streptomycin	Translation	It binds 30 S subunit and causes misreading.
Tetracycline	Translation	It inhibits binding of aminoacyl tRNA to 30S ribosomal subunit.
Neomycin	Translation	It inhibits interaction between mRNA and tRNA.
Chloramphenicol	Translation	It inhibits peptidyl transferase or peptide bond formation
Erythromycin	Translation	It inhibits translocation of mRNA
Puromycin	Translation	It causes premature release of growing polypeptides.

Protein sorting or protein trafficking: It is a process by which proteins synthesised in the cell are translocated to exact locations within the same cell or other cells.

A signal hypothesis has been proposed for this transport. According to this, protein sorting involves the interaction of cell membrane either with some parts of newly synthesised proteins or with a protein sequence added to this new protein. The transfer of proteins takes place by two modes:

(a) Cotranslational transfer: Transfer is coupled with translation process (proteins synthesised on ribosomes attached to ER)

(b) Post translational transfer: Transfer takes place after translation (proteins synthesised on ribosomes).

Ribosome synthesising proteins are attached to the membrane through leader sequences or signal sequences. These sequences are recognised by receptors located within membrane. Signal recognising particle (SRP) in the form of a ribonucleoprotein complex present in the protein and SRP receptors present in the membrane transfer ribosome (engaged in protein synthesis) to the membrane. The entire process of protein sorting requires ATP as a source of energy.

The proteins that are imported into mitochondria and chloroplast possess N-terminal leader sequences that target them to organellar membrane or to inside organelle. These leader sequences have regions which comprise envelope targeting signal or matrix targeting signal which position protein to particular location and when its function is over the leader sequence is cleaved by protease enzyme and degraded.

Regulation of gene expression

It is a process by which rate of gene expression is controlled in different tissues of organisms at different times because genes do not operate to produce enzymes all the time. Each cell of a living organism contains thousands of genes. But all genes do not function at a time..Genes function according to the requirements of cell. Genes control the phenotypic expression through the production of enzymes. The synthesis of enzymes is not constant hence sometimes high and sometimes low depending upon the requirements of cells. The organisms regulate the activity of their genes to conserve energy and resources so that only those products essential for functioning of cells are produced. It is essential for the organisms as it increases the versatility and adaptability by allowing the cell to express proteins when needed. The genes will be turned off or on as per the requirement.

There are two types of genes:

1. **Constitutive genes:** These are expressed continuously and remain active in a cell because their products are required for normal cellular activities. These are also called housekeeping genes. e.g., genes of glycolysis, Krebs cycle, ATPase etc.
2. **Non-constitutive genes**: These are not always expressed themselves in a cell. These are switched off or on according to the requirement of cellular activities. These are also called luxury genes. e.g., Structural genes in lactose system in *E. coli*. These are two types:

   (a) Inducible genes: These are switched on in the presence of inducer.

   (b) Repressible genes: These constitute to express themselves till a repressor stops their activity.

   $$\text{DNA} \xrightarrow{\text{Transcription}} \text{mRNA} \xrightarrow{\text{Translation}} \text{Protein} \xrightarrow{\text{Biosynthetic pathway}} \text{Phenotype}$$

Levels of regulation: The genes regulate at following levels during protein synthesis.

Process	Organisms	Examples
Transcription (during formation of transcript)	Prokaryotes and eukaryotes	*Lac* operon, *Trp* operon, *Gal* operon
Post transcription (during splicing)	Eukaryotes	
Translation (during transport of mRNA from nucleus to cytoplasm)	Eukaryotes	*Arg* operon
Post-translation	Eukaryotes	

Operon Model: Jacob, Monod and Lwoff (1961) gave the concept of operon model to explain the regulation of gene expression (action). Operon is transcriptional unit (unit of gene expression) in which several genes usually encoding proteins with related functions are regulated together. Thus operon consists of a group of structural genes whose transcription is regulated by a set of genes; regulator, promoter and operator genes. These genes encode enzymes for same biosynthetic or metabolic pathway. Operon is of two types:

1. **Inducible operon (*Lac* operon system)**: In prokaryotes, it contains genes that encode enzymes in the presence of inducer. These enzymes are used in catabolism of lactose hence this operon occurs in catabolic pathway (positively induced). It generally remains switched off but works only in the presence of inducer. The message produced by an operon is polycistronic. Lactose is apparent natural inducer but allolactose (isomer of lactose) is actual inducer. Isopropyl thiogalactoside (IPTG) is artificial inducer.

   Several genes involved in this system are as follows:

   (a) **Structural genes**: These are three genes in *E. coli* (5 in *Trp* operon of *E.coli* and 10 in *His* operon of *Salmonella typhimurium*) adjacent to one another and each gene is controlled separately depending upon substrate to be utilised. These genes are expressed together because in *E. coli* mRNA is polycistronic containing more than one coding region under same regulatory control. The enzymes produced by these genes are normally present at very low level but in the presence of lactose, their level rise rapidly. These genes are as follows:

   ***Lac* Z**: It is the largest gene having 3063 bp. It codes for β-galactosidase enzyme which hydrolyses lactose into glucose and galactose. It also converts lactose into allolactose.

   ***Lac* Y**: It contains 800bp. It codes for permease enzyme which transports lactose into cell through cell wall.

   ***Lac* A**: It contains 800bp. It codes for trans-acetylase enzyme which transfers acetyl group from acetyl CoA to β-galactosidase to hydrolyse lactose. It is not essential.

   (b) **Operator gene (*Lac* O)**: It is small gene having 35 bp. It is situated between base -5 and +21 at 5'end of *Lac* promoter region. It serves as binding site for repressor protein. It directs and regulates structural genes to transcribe. If mutation occurs in operator gene, then repressor cannot bind.

   (c) **Promoter gene (*Lac* P)**: It is the smallest gene containing 30 bp. It serves as binding site for RNA polymerase at recognition point. It has three sites: a recognition site, a binding site and mRNA initiaton site.

   (d) **Regulator gene (*Lac* I)**: It is moderately large gene having 1290 bp. It lies upstream of operon. It codes a repressor protein which is synthesised all the time constitutively. Repressor protein is allosteric and exists in two forms; in one form it is paratactic to operator and binds with it and in another form it is paratactic to inducer (lactose).

Lac **operon** system shows two types of control:

1. **Negative control**: It produces **active repressor** protein whose binding will stop transcription. It operates in a biosynthetic pathway.
   (a) If lactose is absent: Regulator gene produces active repressor protein. Repressor binds with operator gene. It stops binding of RNA polymerase on recognition point of promoter gene. Hence transcription of structural genes will not occur **(Switch off).**
   (b) If lactose is present: Regulator gene produces active repressor protein. Cell encounters lactose, a few molecules of *Lac* enzymes present in the cell allows lactose to be taken up and metabolised. Lactose (inducer) binds to active repressor which becomes inactive repressor by changing its conformation. Now inactive repressor does not bind to operator gene. Thus RNA polymerase binds to promoter gene. Transcription of structural genes will occur (**Switch on**).

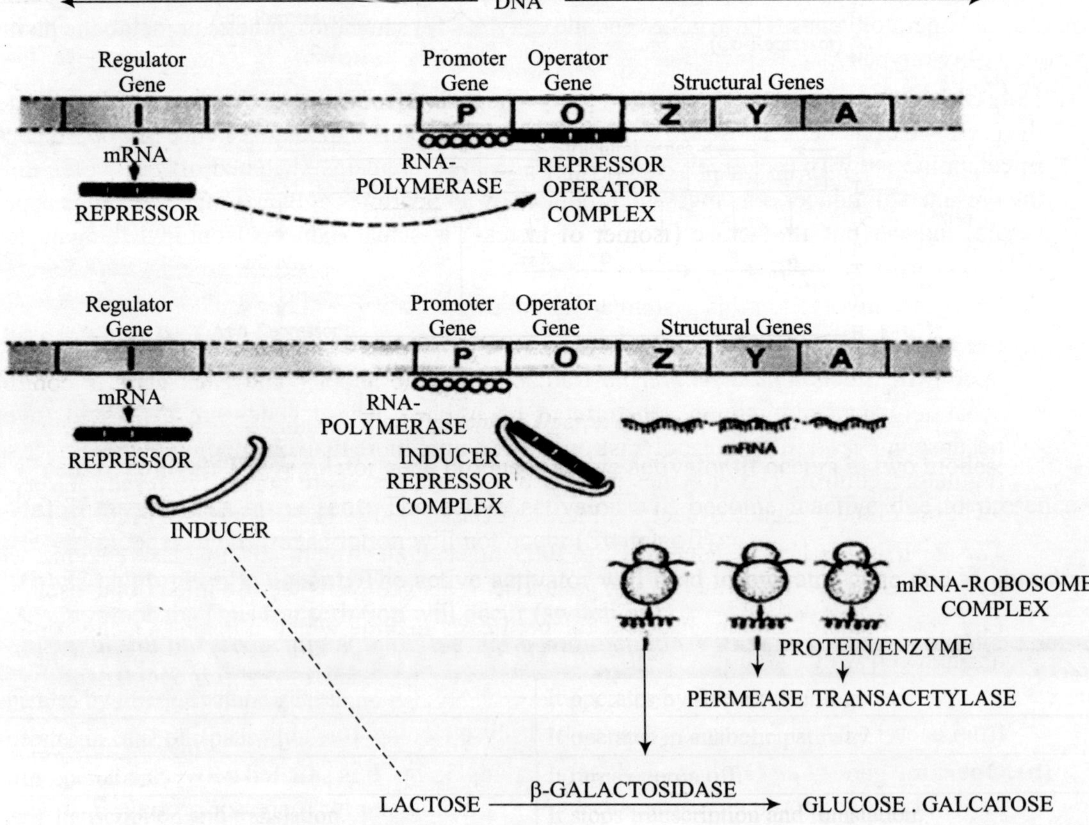

Fig. ***Negative control (Switch off and Switch on)***

2. **Positive control**: It produces **inactive activator** (Catabolite activator protein or CAP) whose binding will start transcription.
   (a) If lactose is absent: Regulator gene produces inactive activator protein. It cannot bind RNA polymerase on recognition point of promoter gene. Hence transcription of structural genes will not occur (**Switch off).**

Regulator gene | Promoter Operator | Structural genes

Lac I	Lac P	Lac O	Lac Z	Lac Y	Lac A

Activator (inactive) — No transcription ≠ ≠ ≠ **(Switch off)**

(b) If lactose is present: Regulator gene produces inactive activator protein. Cell encounters lactose, a few molecules of *Lac* enzymes present in the cell allows lactose to be taken up and metabolised. Lactose (inducer) binds to inactive activator which becomes active by changing its conformation. Now active activator binds to operator gene. Thus RNA polymerase binds to promoter gene. Transcription of structural genes will occur (**Switch on**).

Regulator gene Promoter gene Operator gene Structural genes

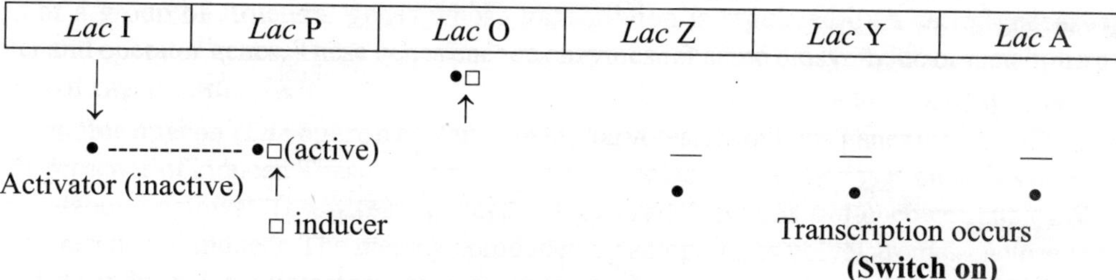

Case:

1. **If glucose is present**: *Lac* operon will not operate indefinitely in the presence of lactose in the medium. If glucose accumulates in medium then *lac* operon is switched off and *E.coli* uses glucose instead of lactose. Glucose inhibits operon by decreasing the level of cAMP because glucose inhibits adenylate cyclase enzyme. Under this condition, catabolite activator protein (CAP) does not bind upstream of promoter and *lac* operon is transcribed at very low level. To switch on *Lac* operon, the high level of cAMP and CAP are required to bind to promoter region of DNA.
2. **If glucose is low or absent**: The level of cAMP will increase and thus cAMP binds CAP forming cAMP-CAP complex which binds promoter gene of *Lac* operon. This enhances RNA polymerase binding to promoter thus transcription is increased by 50 fold.
3. **If both glucose and lactose are present together**: In this case cell will use the glucose first hence transcription of *Lac* operon will occur at low level. However, when glucose is fully used up then catabolite repression will end and transcription of *Lac* operon increases allowing available lactose to be utilised.

Lactose is low	Glucose is low	The *Lac* operon is not expressed.
Lactose is low	Glucose is high	The *Lac* operon is not expressed.
Lactose is high	Glucose is low	The *Lac* operon is expressed.
Lactose is high	Glucose is high	The *Lac* operon is expressed.

2. Repressible operon/Tryptophan operon system: It contains genes that encode enzymes involved in anabolic pathway. *Trp* operon works as corepressor and inhibits its own synthesis through feedback inhibition (end product) or by alternative mechanism called attenuation. It contains five structural genes. There are two controls:

1. **Negative control**: The regulator gene produces aporepressor (inactive repressor). It occurs in two modes:

(a) **If tryptophan is present:** The aporepressor becomes active repressor in the presence of corepressor, tryptophan. The active repressor binds to operator gene. Transcription will not occur

(b) **If tryptophan is absent**: The aporepressor will not be active due to absence of corepressor, tryptophan. Thus aporepressor cannot bind operator gene. Hence transcription will occur.

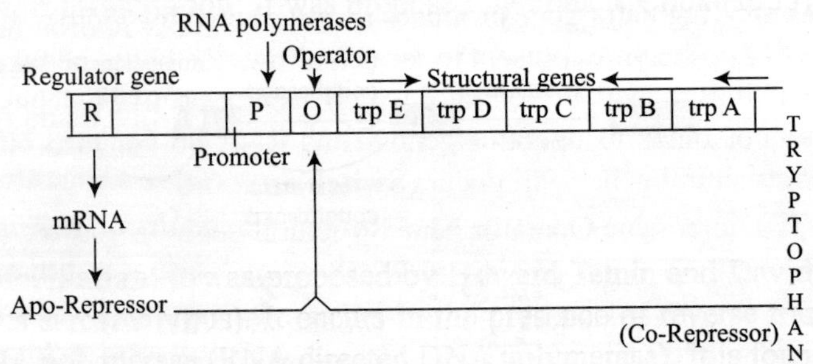

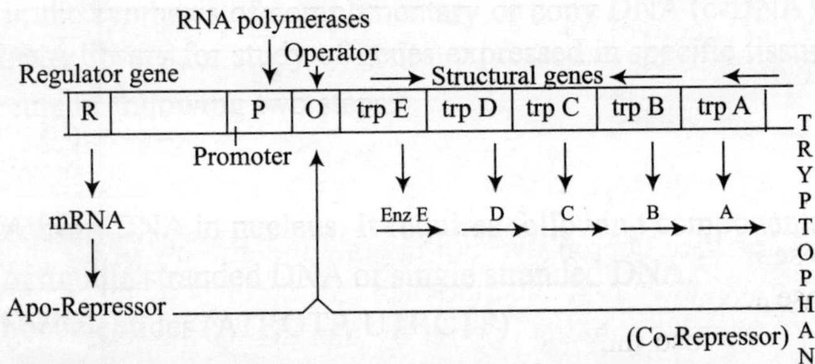

Fig. ***Tryptophan Operon of E. coil***

2. **Positive control**: The regulator gene produces active activator. It occurs in two modes:
   (a) **If tryptophan is present**: The active activator will become inactive due to presence of tryptophan. Thus transcription will not occur (Switch off).
   (b) **If tryptophan is absent**: The active activator will bind to operator gene due to absence of tryptophan. Thus transcription will occur (switch on).

Inducible operon	Repressible operon
It operates by three structural genes.	It operates by five structural genes.
It operates in catabolic pathway.	It operates in anabolic pathway.
It turns operon on.	It turns operon off.
It starts transcription and translation.	It stops transcription and translation.
Its regulator gene produces active repressor.	Its regulator gene produces inactive aporepressor.
It uses lactose which changes active repressor to inactive to bring about transcription.	It uses tryptophan which changes inactive aporepressor to active corepressor to stop transcription.

Human Genome Project

The term 'Genome' was coined by Winkler. Genome is total DNA (complete set of genes) in a single chromosome (haploid set) and inherits as a unit from one parent. Thus single genome is found in haploid cell, double genome in diploid cell and more than two genomes in polyploid cell. Genome includes both nuclear DNA and extranuclear DNA (of mitochondria and chloroplast. Genome of prokaryotes consists mostly of unique DNA sequences but genome of eukaryotes contain repetitive, interspersed and cluster DNA sequences.

The term 'Genomic' was coined by Roderick (1986). Genomics is the study of genome through analysis, sequencing and mapping of genes and non-cistronic areas.

Human genome project is an international collaborative research programme to complete mapping and understanding of genome of human beings at the level of nucleotide sequence. The genome sequenced by the project is a composite one based on the individual genomes of men and women from a variety of races and ethnic groups.

It was launched on 1st October, 1990. It was launched by National Human Genome Research Institute (NHGRI) and coordinated by US Department of Energy and National Institute of Health involving 14 countries (Australia, Brazil, Canada, Denmark, France, Germany, Israel, Italy, Japan, Netherland, Russia, Sweden, United Kingdom and US). It is called a mega project because it involved lots of money, most advanced technologies, many scientists, numerous computers and a long span of time. If data is to be stored in books, with each book having 1000 pages and each page having 1000 letters then some 3300 books will be required.

It was scheduled to be completed in 2005 but completed on April 14, 2003 (only sequence of chromosome I was completed in 2006). Its total cost was US 9 billion dollars.

First genome to be sequenced	1999	*Haemophilus influenzae* (a bacterium)
First eukaryotic genome to be sequenced	1999	Yeast, *Coenorhabditis elegans* (a nematode)

HGP also includes mapping of genome of other species

Organisms	Number of genes	Number of base pairs (Genome size)
Bacteriophage	100	10,000 bp
E.coli	4600	4.6 million bp
Yeast	6000	12 million bp
Arabidopsis	25,000	130 million bp
Rice	50,000	407 million bp
Coenorhabditis	19,000	97 million bp
Drosophila	16,000	180 million bp
Human being	30,000	3164.7 million bp or 3.2 billion bp

Classification of human genome

1. **Extragenic DNA** (70-80%): It contains
   (a) Low copy number (70-80%):
   (b) Highly repetitive (20-30%): It contains Clustered repeats (60%) which have microsatellite DNA, minisatellite DNA and satellite DNA; and Dispersed repeats (40%) which have short interspersed elements (SINES) and Long interspersed elements (LINES).
2. **Genes and related sequences** (20-30%): These contain
   (a) Coding sequences (<10%)
   (b) Non-coding sequences (>90%): These contain gene fragments, pseudogenes, Introns, promoters, leaders and trailers.

Objectives of HGP:

1. To identify all genes in human DNA.
2. To determine the structure and function of genes.
3. To determine the sequences of genes of all 23 pairs of chromosomes.
4. To construct detail genetic and physical map of all base pairs.
5. To point out specific genes that cause genetic disorders.
6. To develop tools for data analysis.
7. To store the information from DNA sequences in database.
8. To address ethical, legal and social issues arising from this project.
9. To establish bioinformatics

Methods:

1. **Sequence tagged site (STS)**: It is a short DNA segment that occurs only once in a genome and whose exact location and order of bases is known. STS serves as landmark on physical map of a genome. It is also known as Expressed sequence tag (EST)-the genes which are expressed as RNA.
2. **Sequence annotation (SA):** It is sequencing of whole genome (coding and noncoding region) and later assigning functions to different regions in the sequence. This method is followed to sequence human genome.

Process:

(a) The total DNA is isolated from a cell.

(b) DNA is digested into small fragments. Each piece of DNA fragment is amplified.

(c) The small fragment of DNA is ligated into a vector (Yeast artificial chromosome or Bacterial artificial chromosome) to form recombinant DNA.

(d) The r-DNA is introduced into yeast or bacteria.

(e) The fragment is sequenced from two ends by automated sequencer (developed by F. Sanger).

(f) The sequences are aligned based on overlapping segments by specialised computer based programme.

(g) The sequences are annotated and assigned to each chromosome.

The Celera Genomics, USA used a strategy of sequencing from both ends of human DNA fragment.

Findings:

1. Human body is made up of 100 trillion cells grouped into over 260 types.
2. Every cell contains the same number of chromosomes and same level of genome.
3. The size of genome is not associated with complexity of body organisation.
4. Human genome contains 3164.7 million base pairs (or almost 3.2 billion base pairs) in the 23 pairs of chromosomes within the nucleus of our cells.
5. The 99.9% of total base pairs are exactly same in all humans.

6. The total number of genes is 30,000.
7. Chromosome I has the highest (2968) number of genes.
8. Y chromosome has the lowest (231) number of genes.
9. Among somatic chromosomes, 22nd chromosome has the lowest (272) number of genes.
10. X chromosome is larger than Y chromosome in size and has more (2000) genes than Y (231) chromosome.
11. Gene has an average size of 3000 base pairs.
12. The largest gene is dystrophin containing 2.4 million base pairs long situated on X chromosome.
13. The smallest gene is testis determining factor (TDF), a holandric gene containing 14 base pairs long situated on Y chromosome.
14. Genes occupy only 1.1% of chromosomal space and rest is junk DNA or spacer.
15. Genes work in networks rather in isolation.
16. Less than 2% (around 1.5%) of genome is exon which codes for protein and the remaining is intron (either pseudogene or repeat sequence) which does not code for protein.
17. Repeat sequences or junk DNA repeat 100 to 1000 times.
18. Three chromosomes such as 1st, 6th and 10th are killer chromosomes which are responsible for genetic disorders. The I^{st} chromosome is responsible for 350 illness.
19. Single nucleotide polymorphism (SNP) has been identified at 1.4 million locations.

Single nucleotide polymorphism (SNP): It is a variation in a single nucleotide that occurs at a specific position in the genome where each variation is present to some appreciable degree within a population. On an average, SNP occurs once in 300 nucleotides. Hence, out of three billion (3×10^9) base pairs, there will be 10^7 SNP. Thus, its number will be 1 % or more.For example, at a specific base position in human genome, the base Cytosine may appear in most individuals but in some individuals the position is occupied by base Adenine so there is a SNP at this specific base position and the two possible nucleotide variations-Cytosine and Adenine are said to be alleles for this base position. SNP is single base pair position in genomic DNA at which different nucleotides occur in different individuals of a population. Each nucleotide at such a position denotes an allele of SNP. The SNP acts as biological marker. It is caused by mutation.

Significance:

(a) It is used to study interaction between various genes, protein formed in tissues and organs.
(b) It helps to diagnose the disease by revelation of genes present on a particular chromosome.
(c) It is useful to treat incurable genetic disorders through gene therapy.
(d) It is helpful in understanding human genetics.
(e) It has developed forensic science.
(f) It is used to develop designer drugs and genetically modified diet.
(g) It is useful to achieve healthier living by studying changes in different developmental stages.

DNA finger printing

It was developed by Alec Jeffrey (1985) while working on the sequence of myoglobin gene. It facilitates the identification at gene level by matching DNA fragments using DNA probes.

This technology was refined by developing a more sensitive DNA profiling technique focusing mainly on highly variable minisatellites in human genome.

Genome (DNA of single set of chromosome) of a person contains about 3 billion base pairs in which 99.9% of base pairs are same in all human beings and only 0.1% of total base pairs vary and provide individuality to each human beings. The chemical structure of everyone's DNA is same. The only difference between people is the sequence of base pairs. Everyone has different sequence hence every person would be identified solely by the sequence of base pairs.

DNA molecule can be divided into coding sequences (exons) which carry information required to synthesise proteins and non-coding sequences (introns) which do not supply any relevant information. The changes usually occur and accumulate within these non-coding sequences. These (about 10-100 base pairs) are responsible for variation among individuals and can be employed as tool to create a genetic fingerprint of an individual. Some of these non-coding sequences contain highly repetitive sequences of base pairs. These are tandemly repeated many times and are found at many places (centromere, telomere and heterochromatic area) throughout the length of DNA. The number of repeats is very specific in each individual and it is inherited. It is unlikely that the repeats of one individual are found in DNA are considered whether two samples of DNA are from same person, related persons or unrelated persons.

DNA fingerprint is the loci (region) where test DNA hybridises with probe. When DNA is subjected to density gradient centrifugation, these repeat sequences are called satellite because these float over rest of DNA.

Types of satellite DNA: On the basis of length of repeats and size of clusters.

1. **Microsatellite DNA** (Simple sequence repeat): It is short repeat sequence of 4-8 (less than 10) base pairs which repeat 10 to 100 times. It is random or dispersed and frequently distributed throughout all chromosomes hence it is abundant. It shows large stable polymorphism due to variation in the number of repeat units. It is ideal molecular marker for gene mapping in Human Genome Project.
2. **Minisatellite DNA**: It is small repeat sequence of 10-60 base pairs which repeat 1000 times. It varies in number (hypervariable) from person to person hence called variable number of tandem repeats (VNTR). It is very specific in each individual and varies in number from person to person. It shows high degree of polymorphism and is inherited from parents. It is found at many places but frequently at telomere and centromere. It is used as genetic marker and acts as probe that recognises complimentary sequences in DNA. Every individual receives 50% of DNA from father and 50% from mother thus the number of VNTRs in a particular area of the two homologous chromosome or DNA molecules will be different. Hence, an individual will have a distinct combination of VNTR.

Genetic fingerprinting exploits highly variable repeating sequences called minisatellites. Two unrelated persons will be unlikely to have the same number of satellites at a given locus. It is used to distinguish one person from another with the exception of monozygotic twin. The probability of two unrelated persons having same pattern of location and repeat number of minisatellite is one in ten million. For instance, one person has the unit ATG repeated 60 times at one gene locus, 120 times at second locus and so on whereas another person is likely to have different number of repeats at these loci. Such polymorphic gene loci are called simple tandem repeats.

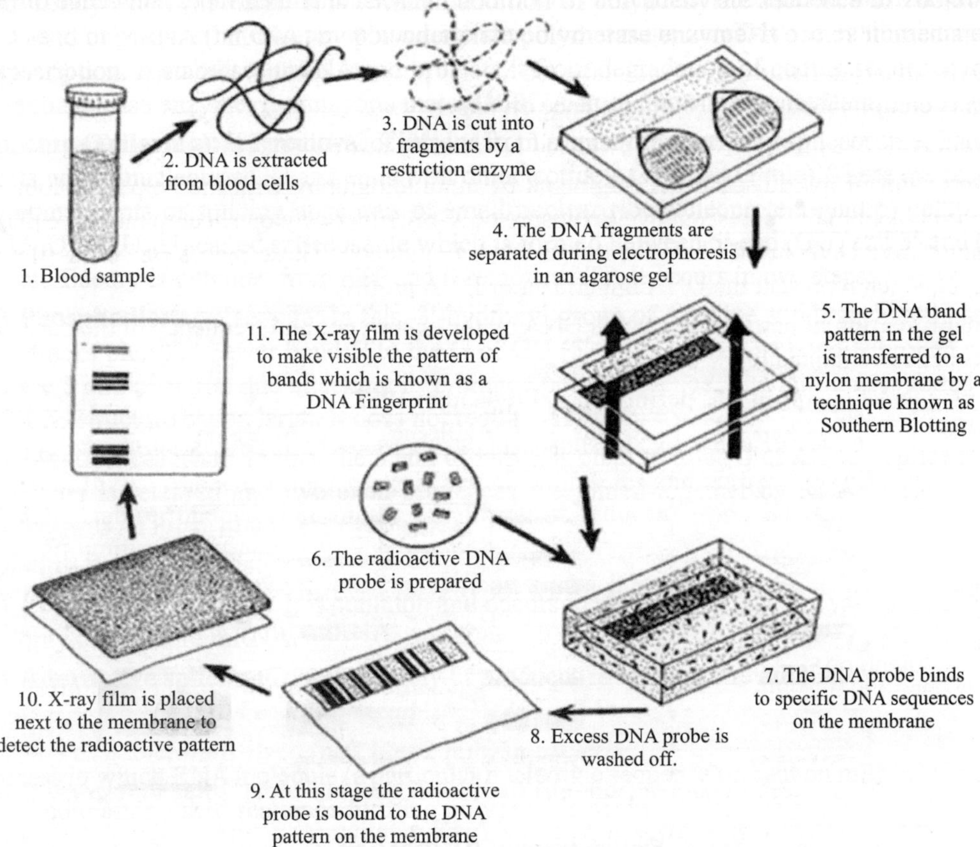

Fig. ***Steps involved in DNA fingerprinting***

Procedures:

1. **Collection** of any sample (WBC of blood stain, sperm of semen, egg of vaginal fluid, cell of hair root follicle, buccal cell of saliva, cell of skin, cell of bone marrow, cell present in urine and cell of any body tissue).
2. **Isolation (Extraction)** of DNA from any samples. It is done chemically using detergent to wash extra material from DNA. It is also done mechanically applying a large amount of pressure to squeeze DNA. If DNA is in low amount, then it can be amplified by PCR. One microgram is sufficient for DNA finger printing.
3. **Digestion (Fragmentation)** of DNA into several fragments containing VNTR of different size by a restriction endonuclease enzyme (e.g., *Alu* I or *Hae* III).
4. **Separation** of DNA fragments (VNTR) on the basis of size by agarose gel electrophoresis or polyacrylamide gel electrophoresis (PAGE). The VNTRs are denatured into single stranded DNA in the presence of alkaline chemical (NaOH).
5. **Transfer** of separated ssDNA to synthetic nylon membrane or nitrocellulose paper on the gel by Southern blotting.

6. **Preparation** of radioactive single stranded DNA probe containing repeat sequence of base pairs complimentary to VNTR to analyse Southern blot.
7. **Hybridisation (Binding)** of radiolabelled VNTR probe to specific DNA repeat sequence of base pairs complimentary to VNTR sequence on paper.
8. **Removal (Washing)** of the unhybridised DNA using buffer.
9. **Detection** of hybridised DNA fragments by autoradiography with the help of colour inducing (chromogenic) or fluorescent dye (exposure of X-ray film on nylon sheet to mark the place where radioactive DNA probes have bound to DNA fragments). These places on X-ray film are dark band which represent the DNA fingerprint or DNA profile.
10. **Identification** of particular person's DNA by the occurrence and frequency of the particular gene sequence contained in the probe.

    First DNA fingerprint was performed in England in 1987.

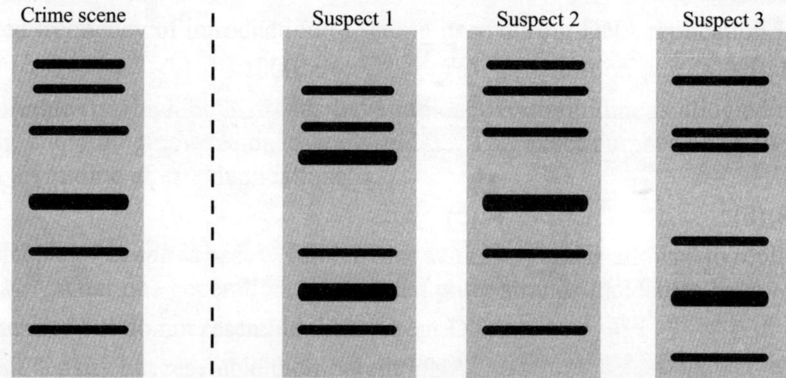

Fig. ***Matching of DNA finger prints***

Applications:

(a) DNA finger printing identifies the culprit or criminal or real victim by matching DNA of suspect with DNA collected from site.

(b) It solves case of paternity or maternity because a person inherits his or her VNTR from his or her parents (every son or daughter receives half of chromosome from father and half from mother).

(c) It identifies the victims of disasters like plane crash, building collapse, fire breakout, flood or earthquake.

(d) It determines the evolutionary relationship or link among organisms or races.

(e) It identifies human remains and matches organ donors.

(f) It solves the immigration dispute by establishing whether a hopeful immigrant is close relative of already an established resident.

SHORT ANSWER TYPE QUESTIONS

1. What are the criteria of a genetic material?
2. Distinguish between R strain and S strain.
3. Distinguish between nucleoside and nucleotide.

4. Explain Edwin Chargaff's rule.
5. Explain the base ratio.
6. Distinguish between B form of DNA and Z form of DNA
7. Distinguish between prokaryotic DNA and eukaryotic DNA.
8. Write notes on nucleosome.
9. Distinguish between conservative replication and semiconservative replication of DNA.
10. Mention the evidences in favour of RNA as primitive genetic material.
11. Distinguish between monocistronic mRNA and polycistronic mRNA.
12. Draw awell labelled diagram of tRNA.
13. Mention the functions of mRNA, rRNA and tRNA..
14. Write notes on split genes and overlapping genes.
15. Define the classical concept of gene.
16. Define the modern concept of gene.
17. Explain the linear flow of information.
18. Write notes on RNA polymerase.
19. Discuss the post transcriptional processing.
20. Distinguish between prokaryotic transcription and eukaryotic transcription.
21. What are the reactions in chain initiation process of translation process?
22. Distinguish between prokaryotic translation and eukaryotic translation.
23. Explain the degeneracy of genetic code.
24. Discuss the events taking place during chain termination.
25. Mention the functions of structural genes.
26. Distinguish between inducible operon and repressible operon.
27. What are the objectives of Human genome project?
28. What are the important findings of human genome project?
29. What are the significance of human genome project?
30. Distinguish between microsatellite DNA and minisatellite DNA.

LONG ANSWER TYPE QUESTIONS

1. Describe the transformation experiment top prove DNA a genetic material.
2. Describe the blender experiment top prove DNA a genetic material.
3. Describe the double helical structure of DNA and show how it differs from RNA?
4. Describe the mechanism of DNA replication.
5. Describe the different types of RNA.
6. Describe the mechanism of transcription in protein synthesis.
7. Describe the mechanism of translation in protein synthesis in brief.
8. Discuss the properties of genetic code.
9. Describe the regulation of gene expression with respect to the concept of Lac operon.
10. Describe the procedure of DNA finger printing.

MULTIPLE CHOICE QUESTIONS

1. The base ratio is less than one in
   (a) E. coli (b) Plants (c) Fungi (d) Animals
2. The length of DNA in prokaryotes is
   (a) 1.56 mm (b) 1.56 cm (c) 1.56 m (d) 15.6 cm
3. The core of nucleoplasm contains
   (a) Dimer of histones (b) Tetramer of histones
   (c) Hexamer of histones (d) Octamer of histones
4. The nature of DNA replication is
   (a) Semi-conservative and continuous (b) Semi-conservative and semi-discontinuous
   (c) Conservative and continuous (d) Conservative and discontinuous
5. The expected frequency of introduction of wrong base during DNA replication Is
   (a) 1 : 100 (b) 1 : 1000 (c) 1 : 1 million (d) 1 : 1 billion
6. A DNA molecule in which both strands have radioactive thymidine is allowed to duplicate in an environment containing non-radioactive thymidine. The exact number of DNA containing some radioactive thymidine after 3 duplications is
   (a) 1 (b) 2 (c) 4 (d) 8
7. A DNA molecule of *E. coli* is heavy (full labelled with N^{15}) and is allowed to replicate in a medium containing N^{14}. After one generation, the two daughter strands molecules have
   (a) Same density but do not resemble their parent DNA
   (b) Different density but resemble their parent DNA
   (c) Different density but do not resemble their parent DNA
   (d) Both the strands differ in density but resemble their parent DNA
8. RNA and DNA were artificially synthesised *in vitro* by, respectively
   (a) Ochoa and Nirenberg (b) Ochoa and Korenberg
   (c) Kornberg and Nirenberg (d) Nirenberg and Khorana
9. Nirenberg synthesised a mRNA containing 34 polyadenine (A-A-A-A…) and found a polypeptide formed of 11 polylysine. It proved that genetic code for lysine is
   (a) A singlet (b) A-A doublet (c) A-A-A triplet (d) Many adenines
10. The 5′-end of tRNA always ends in the base
    (a) Adenine (b) Guanine (c) Thymine (d) Uracil
11. The minimum length of cistron in a base pair which synthesises a polypeptide of 50 amino acid codon is
    (a) 50 bp (b) 100 bp (c) 150 bp (d) 200 bp
12. Considering that we have four nucleotides A,G,T and C, the number of base substitutions that can occur in the amino acid codons are
    (a) 261 (b) 264 (c) 549 (d) 535
13. In a codon, wobbling or degeneracy of genetic code is generally restricted to
    (a) First N base (b) Second N base
    (c) Third N base (d) Aromatic amino acid

14. The formation of polyribosome from ribosomes is done in the presence of
(a) Ca^{+2} (b) Mg^{+2} (c) Na^{+} (d) Cu^{+2}
15. Nitrogenous bases are linked to sugar by
(a) Hydrogen bond (b) Glycosidic bond (c) Phosphodiester bond (d) All of above
16. Positively charged basic proteins that are found in eukaryotes are called
(a) Histones (b) Non-histones (c) Polyamines (d) All of above
17. Which of the following enzyme catalyses the transcription of all types of RNAs in eukaryotes?
(a) DNA dependent RNA polymerase (b) DNA dependent DNA polymerase
(c) RNA dependent DNA polymerase (d) RNA dependent RNA polymerase
18. Cell free system for protein synthesis was used by
(a) Khorana (b) Nirenberg (c) Francis Crick (d) Richard Gamow
19. In protein synthesis, tRNA carrying the amino acid enters
(a) A site of ribosome (b) P site of ribosome (c) E site of ribosome (d) Anywhere
20. Transfer RNA is attached to mRNA by its
(a) I loop (b) II loop (c) III loop (d) Variable arm
21. Amino acid binding site in tRNA is
(a) Anticodon loop (b) DHU loop (c) CCA end (d) 5′end
22. Which of the following rRNA acts as structural RNA as well as ribozyme in bacteria?
(a) 23 S rRNA (b) 16 S rRNA (c) 5 S rRNA (d) 5.8 S rRNA
23. Positively regulatory proteins are called
(a) Repressor (b) Activator (c) Promoter (d) Enhancer
24. Southern blotting technique involves the transfer of DNA from
(a) Membrane to gel (b) Gel to membrane (c) Solution to gel (d) Gel to solution
25. The sensitivity of DNA finger printing can be increased by using
(a) Intron sequence (b) Exon sequence (c) PAGE (d) PCR
26. If base sequence of a codon in mRNA is 5′-AUG-3′, the sequence of tRNA pairing with it must be
(a) 3′-CAU-5′ (b) 5′-CAU-3′ (c) 5′-UAC-3′ (d) 3′-UAC-5′
27. The number of nucleosomes in human cell with 6 billion base pairs is
(a) 3×10^5 (b) 3×10^7 (c) 2×10^5 (d) 2×10^7
28. RNA polymerase was discovered by
(a) Kornberg (b) Ochoa (c) Khorana (d) Nirenberg
29. The total number of nitrogenous bases in one helical turn of DNA is
(a) 5 (b) 10 (c) 12 (d) 20
30. The enormous diversity of protein molecules is mainly due to diversity of
(a) Peptide bonds (b) Amino group on the amino acids
(c) R groups on the amino acids (d) Amino acid sequences
31. In double stranded DNA molecule, the percentage of cytosine is 18. The percentage of adenine will be
(a) 18% (b) 32% (c) 36% (d) 50%

32. The number of genome types present in a typical green plant cell is
(a) One (b) Two (c) Three (d) Four
33. The step of translation which does not consume a high energy phosphate bond is
(a) Activation of amino acid (b) Formation of peptide bond
(c) Binding of aminoacyl tRNA to ribosome (d) Translocation of aminoacyl tRNA
34. A polypeptide just released by a ribosome from mRNA has 75 amino acids. What would be the number of nucleotides in coding region of mRNA including the initiation codon and terminating codon ?
(a) 225 (b) 228 (c) 220 (d) 230
35. How many H-bonds are present in segment of DNA if one strand contains A-T-G-C nucleotide sequence?
(a) 4 (b) 8 (c) 5 (d) 10
36. If DNA exclusively consists of only 6 base pairs then what would be minimum coding ratio to code only 20 amino acids?
(a) 1 (b) 2 (c) 3 (d) 4
37. If a polynucleotide contains 5 bases then what will be the number of different sequences?
(a) 125 (b) 625 (c) 512 (d) 1024
38. How many signal codons are there?
(a) 1 (b) 2 (c) 3 (d) 4
39. Considering the average molecular mass of a base to be 500 Da, what is the molecular mass of a ds DNA of 10 base pairs?
(a) 5 KDa (b) 10 KDa (c) 100 KDa (d) 500 KDa
40. If molecular weight of a polypeptide is 15.3 KDa, what would be the minimum number of nucleotides in the m RNA that codes for this polypeptide if molecular weight of each amino acid is 90 KDa?
(a) 123 (b) 170 (c) 340 (d) 510
41. Which of the following sets of bacterial strains was selected by Griffith for his transformation experiment?
(a) S II and R II (b) S II and R III (c) S III and R II (d) S III and R III
42. Which of the following sequences in blender experiment occurred?
(a) Infection- Blending- Centrifugation (b) Blending-Infection-Centrifugation
(c) Centrifugation-Infection-Blending (d) Blending –Centrifugation-Infection
43. In case of pyrimidines, N atoms are present on
(a) 1st and 7th positions (b) 1st and 5th positions
(c) 1st and 3rd positions (d) 1st and 9th positions
44. The number of nucleosomes in solenoid is
(a) 2 (b) 4 (c) 6 (d) 8
45. The core of nucleosome contains
(a) 60 bp (b) 100 bp (c) 140 bp (d) 200 bp

46. Taylor proved the semi-conservative mode of replication in the roots of
(a) *Pisum sativum* (b) *Vicia faba*
(c) *Lathyrus odoratus* (d) *Oenothera lamarckiana*

47. DNA polymerase III lacks
(a) 5′→3′ exonuclease activity (b) 5′→3′ endonuclease activity
(c) 3′→5′ exonuclease activity (d) 3′→5′ endonuclease activity

48. Which of the following enzymes relieves tension during DNA replication?
(a) Helicase (b) Gyrase
(c) DNA polymerase I (d) DNA polymerase III

49. The gap left by removal of primer is filled by
(a) 5′→3′ polymerase activity of DNA polymerase I
(b) 3′→5′ polymerase activity of DNA polymerase I
(c) 5′→3′ polymerase activity of DNA polymerase III
(d) 3′→5′ polymerase activity of DNA polymerase III

50. The removal of mismatched bases (proof reading) is done by
(a) 3′→5′ exonuclease activity of DNA polymerase I
(b) 5′→3′ exonuclease activity of DNA polymerase I
(c) 3′→5′ exonuclease activity of DNA polymerase II
(d) 3′→5′ exonuclease activity of DNA polymerase III

51. Which of the following enzymes catalyses the formation of RNA from DNA template?
(a) RNA polymerase (b) DNA polymerase
(c) Reverse transcriptase (d) *Taq* polymerase

52. The first codon to be discovered by Nirenberg and Mathaei was
(a) AUG (b) UUU (c) UAG (d) GUG

53. Which is not true according to Chargaff's rule?
(a) A = T (b) G = C (c) A + G = T + C (d) A + T = G + C

54. In the sequence of the processed mRNA ready for translation
5′-AUG CUA UAC CUC CUU UAU CUG UGA-3′. The number of different tRNA molecules required to translate this mRNA is
(a) 4 (b) 6 (c) 7 (d) 8

55. Telomere of eukaryotic chromosome contains short segment of
(a) A rich (b) G rich (c) T rich (d) C rich

56. A gene encoding polypeptides of 50 amino acids get mutated at 25th codon, UAU becoming UAG. The result would be
(a) Polypeptide of 24 amino acids (b) Polypeptide of 50 amino acids
(c) Polypeptide of 25 amino acids (d) Polypeptide of 49 amino acids

57. In DNA segment having 6 coils, there are 22 base pairs linked by two H-bonds. How many cytosine bases are found in that segment?
(a) 22 (b) 38 (c) 44 (d) 76

58. The removal of RNA polymerase III from nucleoplasm will affect the synthesis of
(a) mRNA (b) rRNA (c) tRNA (d) All of the above
59. The DNA fingerprinting has proved to be useful in forensic science. It involves the use of
(a) Microsatellite DNA (b) Minisatellite DNA (c) cDNA (d) rRNA
60. The length of DNA molecule consisting of 5000 nucleotide pairs will be
(a) 0.17 μm (b) 1.7 μm (c) 17 μm (d) 170 μm
61. A DNA molecule having A+G/T+C =0.71 shows that the molecule is
(a) Linear (b) Circular (c) Single stranded (d) Double stranded
62. Fuelgen staining for DNA is based on the reaction of Schiff's reagent with
(a) Free aldehyde group (b) Purines at N-7 positions
(c) Deoxyribose at C-2 positions (d) Pyrimidines at N-1 position
63. RNA polymerase II which synthesises mRNA in eukaryotes is generally
(a) Resistant to alpha amanitin (b) Sensitive to alpha aminitin
(c) Resistant to actinomycin-D (d) Sensitive to rifampicin
64. Genomes of prokaryotes consist mostly of
(a) Repetitive DNA sequences (b) Unique DNA sequences
(c) Interspersed DNA sequences (d) Clustered repeated DNA sequences
65. RFLP (restriction fragment length polymorphism) study is a technique for
(a) Transferring genes from unrelated species (b) Isolating single genes
(c) Isolating single gene products (d) Identifying genetic (DNA) homologies
66. Which one of the following is known to act as a biological mutagen?
(a) Acridine orange (b) Nitrosoguanidine (c) Streptomycin (d) Transposon
67. Which of the following is incorrectly matched?
(a) Southern blot- DNA hybridisation (b) Northern blot- RNA sequencing
(c) Western blot- Protein sequencing (d) Eastern blot- Amino acid sequencing
68. Which one of the following combinations possesses the integrated protein synthesising machinery?
(a) Ribosome, tRNA, amino acid (b) mRNA, tRNA, amino acid
(c) Ribosome, t RNA, mRNA, amino acid (d) tRNA, ribosome, nucleus, mRNA
69. Which one of the following is not a PCR based molecular marker?
(a) RAPD (b) RFLP (c) AP-PCR (d) DNF
70. Which one of the following is not correct with respect to tRNAs?
(a) The 5′- end is phosphorylated (b) They have unusual bases
(c) The anticodon loop is identical (d) The 3′- end base sequence is CCA
71. An increased melting temperature for duplex DNA results from a high content of
(a) Adenine + Guanine (b) Cytosine + Thymine
(c) Guanine + Cytosine (d) Thymine + Cytosine
72. What will happen to transcription in the absence of sigma factor of RNA polymerase?
(a) Transcription will not take place
(b) Transcription will initiate from the initial codon
(c) Transcription will initiate from many random sites
(d) Transcription will initiate from the other strand

73. The number of ribosome binding sites used by tRNA during protein synthesis is
    (a) 2 (b) 3 (c) 4 (d) 6
74. The replication of single stranded viral RNA is brought about by
    (a) Reverse transcriptase (b) RNA polymerase
    (c) DNA primer (d) DNA ligase
75. Which one of the following sequences is located upstream of transcription initiation sites in E. coli?
    (a) Hogness box (b) Pribnow box (c) CAAT box (d) MADS box
76. During initiation in the process of translation, a GTP molecule is required in
    (a) Association of 30S-mRNA with formyl-met tRNA
    (b) Association of 50 S subunit of ribosome with initiation complex
    (c) Formation of formyl-met- tRNA
    (d) Binding of 30S subunit of ribosome with mRNA
77. Which of the following is incorrect with respect to *lac* operon concept?
    (a) It is inducible operon? (b) Absence of glucose is must for transcription
    (c) It is applicable for all prokaryotes (d) Repressor is a part of *lac* operon
78. How much energy is required for the formation of each peptide chain?
    (a) 1 GTP and 1 ATP (b) 2 GTP and 1ATP (c) 1GTP and 2 ATP (d) 3GTP and 2 ATP
79. Single nucleotide polymorphisms (SNP) has been identified at
    (a) 1.4 million locations (b) 2.4 million locations
    (c) 1.4 billion locations (d) 2.4 billion locations
80. Minisatellite DNA is a small repeat sequence of
    (a) 4-8 bp (b) 10-60 bp (c) 80-100 bp (d) 100-120
81. Which of the following is correct about operons?
    (a) These are uniform in size
    (b) These are found in all eukaryotic genes
    (c) These do not bind protein
    (d) These are short in lower eukaryotes than higher forms
82. Penicillin inhibits bacterial multiplication because
    (a) It checks spindle formation (b) It destroys chromatin
    (c) It inhibits RNA formation (d) It inhibits cell wall formation
83. All of the following are potential control mechanisms for regulation of gene expression in eukaryotes except
    (a) Gene amplification (b) Degradation of mRNA
    (c) Lactose operon (d) Transcription
84. DNA fragments are transferred from a gel to a membrane via a process called southern blotting. Its purpose is to
    (a) Analyse the RFLPs in the DNA
    (b) Separate out the PCR
    (c) Attach the DNA fragments permanently to a substrate
    (d) Separate the two complimentary DNA strands

85. DNA and RNA synthesis takes place in
   (a) $5' \rightarrow 3'$ direction
   (b) $3' \rightarrow 5'$ direction
   (c) Both a and b
   (d) DNA in $5' \rightarrow 3'$ and RNA in $3' \rightarrow 5'$ direction
86. Which of the following amino acids is entirely incorporated in the nucleotide bases?
   (a) Glutamic acid (b) Alanine (c) Glycine (d) Serine
87. Modern molecular taxonomy involves comparison of
   (a) Chromosome morphology
   (b) r DNA
   (c) C-DNA
   (d) tRNA
88. Which of the following techniques is used for biodiversity characterisation?
   (a) RAPD (b) RFLP (c) SSR (d) SNP
89. Human genome contains 3×10^9 base pairs having all four base pair in equal properties. A restriction endonuclease can cut a segment of 4 bp long, then how many fragments will be obtained after digestion?
   (a) 1.2×10^7 (b) $1,2 \times 10^6$ (c) 1.2×10^5 (d) 1.2×10^4
90. Which of the following is not an example of genomic control of gene?
   (a) Alternative DNA splicing
   (b) DNA rearrangement
   (c) Gene amplification
   (d) Gene deletion
91. Which of the following is an allosteric protein that is inactive until it binds to cAMP to activate transcription?
   (a) RNA polymerase
   (b) Catabolite activator protein
   (c) ATP
   (d) All of the above
92. Histone structure can be altered by
   (a) DNA rearrangement (b) Acetylation (c) Sigma factor (d) Methylation
93. In eukaryotes, DNA packing affects gene expression by
   (a) Controlling access to DNA
   (b) Positioning related structural genes near each other
   (c) Protecting DNA from mutation
   (d) Enhancing recombination of genes
94. Which of the following is arranged in correct order by size (from largest to smallest)?
   (a) Codon-chromosome-gene-nucleotide
   (b) Gene- chromosome-codon-nucleotide
   (c) Nucleotide-chromosome-gene-codon
   (d) Chromosome-gene-codon-nucleotide
95. Multigene families arise as a result of
   (a) Nucleosome (b) Introns (c) Heterochromatin (d) Euchromatin
96. DNA is destabilised by
   (a) Polyglycine (b) PEG (c) Sucrose (d) Formamide
97. Which of the following is most conserved amongst organisms?
   (a) Mitochondrial DNA
   (b) Chloroplast DNA
   (c) Recombinant DNA
   (d) DNA that codes for tRNA
98. The melting temperature would be maximum for
   (a) *Sarcina* (b) *E. coli* (d) *Drosophila* (d) Human

99. Which of the following conditions is not essential for nucleic acid hybridisation?
(a) High salt concentration (b) Primer
(c) High temperature (d) Complimentary strands

100. Which of the following groups is present at the extreme 3′ end of single polynucleotide strand?
(a) Phosphate group (b) Hydroxyl group (c) Oxo group (d) Alcoholic group

101. The genome size of *E. coli* is 4.6 x 10^6 bp. If a 6 bp cutter is utilised to obtain restriction fragments, then what will be total number of fragments obtained?
(a) 1.12×10^3 (b) 7.66×10^5 (c) 25×10^4 (d) 3.83×10^3

102. The first step in the initiation of protein synthesis requires
(a) 30S ribosomal subunit (b) ATP
(c) Ribosomal A site (d) Translational factor Tu

103. If the molecular mass of a amino acid is 150 Dalton, then the molecular weight of its tripeptide will be
(a) 450 (b) 486 (c) 414 (d) 504

104. An open reading frame is one that has
(a) No start and stop codon (b) A start and stop codon
(c) No start but stop codon (d) A start but stop codon

105. The consensus sequence of 5′ and 3′ splice junctions in eukaryotic m RNA contains
(a) GU- GA (b) GU- AG (c) AG- GU (d) CG- AG

106. The uncoiled DNA in differentiating cells is generally involved in synthesis of
(a) DNA (b) mRNA (c) Protein (d) Glucose

107. The functioning of structural genes is controlled by
(a) Operator (b) Regulator (c) Promoter (d) Enhancer

108. Which of the following will be produced in a cell if there is a nonsense mutation in the *Lac* Y gene?
(a) Beta galactosidase (b) Permease (c) Transacetylase (d) All of the above

109. *E.coli* has 4.6×10^6 base pairs and completes the process of replication within 38 minutes. What will be the average rate of replication?
(a) 100 bp/sec (b) 200 bp/sec (c) 100 bp/sec (d) 2000 bp/sec

110. If a polynucleotide contains 6 bases then what will be the number of different sequences?
(a) 2064 (b) 2460 (c) 3456 (d) 4096

111. How many H-bonds are present in a segment of DNA if one strand contains A-T-G-C nucleotide sequence?
(a) 5 (b) 10 (c) 12 (d) 16

112. The coding ratio in all organisms is 3 (three nucleotides code one amino acid). If DNA exclusively consists of only 6 base pairs, then what would be the minimum coding ratio be assuming that these are only 20 amino acids to be coded?
(a) 1 (b) 2 (c) 3 (d) 4

113. The total number of thymine bases present in ds DNA if it transcribes mRNA which reads as follows: 5′ AUG CAU CAU GCA-3′
(a) 5 (b) 6 (c) 7 (d) 8

114. If there is double bond between A and T and triple bond between G and C, then which sequence will be most stable at higher temperature?
(a) ATTGTACCAAA (b) AATTATATATA
(c) AGCAGAGAGTT (d) AGGCCGGCCCT

115. The relative amount (mole %) of nitrogenous bases in a genomic DNA sample is as such: A = 25, G = 25, T = 32, C = 18. The most likely source of DNA is
(a) Virus (b) Yeast (c) Bacteria (d) Salmon sperm

116. Hershey and Chase performed blender experiment to prove DNA as genetic material on
(a) T_2 bacteriophage (b) T_3 bacteripophage (c) T_4 bacteriophage (d) T_5 bacteriophage

117. Which of the following enzymes is used in replication of DNA?
(a) DNA polymerase I (b) DNA polymerase II (c) DNA polymerase III (d) All of the above

118. The diameter of solenoid is
(a) 10 nm (b) 20 nm (c) 30 nm (d) 300 nm

119. Which of the following types of RNA is the smallest and shortest?
(a) mRNA (b) rRNA
(c) tRNA (d) None of the above

120. Testis determining factor (TDF) is the smallest gene containing
(a) 10 base pairs (b) 14 base pairs (c) 20 base pairs (d) 24 base pairs

121. Codon-anticodon pairing occurs by
(a) Hydrophobic interaction (b) Electrostatic interaction
(c) Hydrogen bond (d) Covalent bond

122. *Trp* operon is
(a) Negatively repressed (b) Negatively induced
(c) Positively repressed (d) Positively induced

123. Griffith performed transformation experiment to prove DNA as genetic material on
(a) Mycobacterium tuberculosis (b) Salmonella typhimurium
(c) E. Coli (d) Streptococcus pneumonia

124. Which of the sets of isotopes of elements was used by Hershey and Chase to prove DNA as genetic material?
(a) S^{35} and P^{32} (b) S^{35} and C^{14} (c) P^{32} and C^{14} (d) C^{14} and N^{15}

125. The nitrogen atoms in purines are present on
(a) Two positions (b) Three positions (c) Four positions (d) Six positions

126. The IUPAC name of adenine is
(a) 6-aminopurine (b) 5-aminopurine (c) 2-aminopurine (d) 4-aminopurine

127. Which of the following positions of carbon atom of sugar is occupied by base?
(a) 1st (b) 2nd (c) 3rd (d) 5th

128. The phosphate group present in DNA is
(a) PO_4^{-3} (b) PO_4^{-2} (c) PO_4^{-} (d) PO_3

129. How many H-bonds (on average) are present within each complete turn of DNA double helix?
(a) 18 (b) 20 (c) 25 (d) 30

130. The distance of phosphate from axis is

(a) 8A° (b) 10A° (c) 12A° (d) 20A°

131. RNA is less stable due to

(a) Presence of uracil
(b) Presence of 2-OH group
(c) Both of the above
(d) None of the above

132. Which of the following has circular DNA?

(a) Prokaryotic DNA (b) Mitochondrial DnA (c) Chloroplast DNA (d) All of the above

133. The length of DNA in eukaryotes is

(a) 2.2 mm (b) 2.2 cm (c) 2.2 m (d) 2 m

134. Which of the histone proteins does not contain in core of nucleosome?

(a) H_1 (b) H_2a (c) H_2b (d) H_3

135. The average number of base pairs in linker DNA of nucleosome is

(a) 40 (b) 60 (c) 100 (d) 160

136. RNA as genetic material was proved by Fraenkel-Conrat and Singer on

(a) Bacteriophage
(b) Tobacco mosaic virus
(c) Pox virus
(d) HIV

137. Ribozyme is actually RNA catalysing as enzyme. It was discovered in

(a) Mycoplasma (b) Tetrahymena (c) Oenthera (d) E. coli

138. Which of the following RNA is most abundant?

(a) mRNA
(b) rRNA
(c) tRNA
(d) None of the above

139. Which of the following RNA is soluble?

(a) mRNA
(b) rRNA
(c) tRNA
(d) None of the above

140. The cap of eukaryotic mRNA contains methylated residue of

(a) Adenine (b) Guanine (c) Thymine (d) Cytosine

141. The 3-D structure of tRNA is

(a) Clover leaf shaped (b) L shaped (c) J shaped (d) V shaped

142. How many anticodons are present in anticodon loop of tRNA?

(a) Two (b) Three (c) Five (d) Seven

143. Renaturation or annealing of DNA occurs at

(a) 40°C (b) 65°C (c) 80°C (d) 94°C

144. DNA absorbs maximum UV light at

(a) 200 nm (b) 260 nm (c) 280 nm (d) 300 nm

145. Which of the following sets of codons is considered as exception to degenerate codons?

(a) AUG and UGG (b) UGG and UAG (c) UGG and UGA (d) AUG and UGA

146. The first terminating codon discovered was

(a) UAA
(b) UGA
(c) UAG
(d) All were discovered simultaneously

147. Reverse flow of information for protein synthesis occurs in
(a) Reo viruses (b) Adeno viruses (c) Toga viruses (d) Retro viruses

148. tRNA in eukaryotes is synthesised by
(a) RNA polymerase I (b) RNA polymerase II (c) RNA polymerase III (d) None

149. Which of the following subunits of RNA polymerase binds with sigma factor in transcription?
(a) α'subunit (b) α subunit (c) β' subunit (d) β subunit

150. Which of the Initiation factors binds free 30S subunit of ribosome with mRNA?
(a) IF_1 (b) IF_2 (c) IF_3 (d) IF_4

151. How many release factors are in prokaryotes?
(a) One (b) Two (c) Three (d) Four

152. Which of the following is first event when chain termination occurs at the end of translation process?
(a) Separation of polypeptide (b) Separation of tRNA
(c) Dissociation of ribosome (d) Release of mRNA

153. Which of the following antibiotics inhibits protein synthesis at transcription stage?
(a) Streptomycin (b) Erythromycin (c) Tetracycline (d) Rifampicin

154. Which of the following genes is the longest in *Lac* operon?
(a) *Lac* Z (b) *Lac* Y (c) *Lac* A (d) Regulator gene

155. The *Lac* Y of structural genes of operon codes for
(a) β-galactosidase enzyme (b) Permease
(c) Transacetylase (b) Lactase

156. Human genome project was launched on
(a) 1st October, 1988 (b) 1st October, 1990 (c) 1st October, 1995 (d) Ist October, 2000

157. Dystrophin is the largest gene containing
(a) 1 million base pairs (b) 1.4 million base pairs
(c) 2 million base pairs (d) 2.4 million base pairs

158. The smallest gene in *Lac* operon is
(a) Regulator gene (b) Operator gene
(c) Lac A of structural gene (d) Promoter gene

159. The first step in chain elongation of protein synthesis is
(a) Binding of aminoacyl tRNA at A site of ribosome
(b) Binding of aminoacyl tRNA at P site of ribosome
(c) Translocation of peptidyl tRNA
(d) Formation of peptide bond

160. The Shine Delgarno sequence in Transcription process of protein synthesis is
(a) AUGAGGU (b) GUGAUGG (c) AGGAUGG (d) AGGAGGU

161. Which of the following properties does not belong to genetic code?
(a) It is degenerate (b) It is comma less
(c) It is overlapping (d) It is non-ambiguous

162. The inducible operon follows
(a) Anabolic pathway (b) Catabolic pathway
(c) Inhibitory pathway (d) Stimulatory pathway

163. Which of the following countries did not take part in Human genome project?
(a) Japan (b) Italy (c) China (d) Germany

164. Which of the following samples is used while doing DNA fingerprinting?
(a) RBC (b) WBC (c) Plasma (d) Anyone

165. Which of the following processes is not involved in DNA fingerprinting?
(a) Agarose gel electrophoresis (b) Southern blotting
(c) Autoradiography (d) Homogenisation

166. Which of the following condons does not justify degeneracy rule of genetic code?
(a) UGG (b) CUG (c) GGG (d) GAG

167. The enzyme required for post-transcriptional event and represents the dominance of RNA world is
(a) SNRNPs (b) Guanyl transferase (c) Peptidyl transferase (d) Methylase

168. *Lac* operon is regulated by
(a) Lac repressor alone (b) CAP-cAMP complex
(c) Initiation complex (d) Both a and b

169. If there are 81 bases in RNA that codes for protein with 27 amino acids and the base at 61th position is deleted such that the length of RNA becomes 80 bases, how many condons will be altered
(a) 7 (b) 9 (c) 11 (d) 15

170. A sequential expression of a set of human genes occur when a steriod molecule binds to the
(a) Ribosome (b) m-RNA (c) t-RNA (d) DNA sequence

171. Which of the following is used to know the phylogeny?
(a) DNA (b) mRNA (c) rRNA (d) tRNA

172. Which of the following is an ambiguous codon?
(a) GUG (b) CUG (c) GGG (d) GAG

173. Which of the following rRNAs acts as structural RNA as well as ribozyme in bacteria?
(a) 23 S rRNA (b) 5 S rRNA (c) 18 S rRNA (d) 5.8 S rRNA

174. Which of the following countries is not involved is human Genome Project?
(a) Israel (b) Japan (c) China (d) Brazil

175. Telomeric repetitve DNA sequence controls the function of eukaryotic chromosomes because they
(a) prevent chromosome loss (b) act as replicons
(c) are RNA transcription factor (d) help chromosome pairing

176. The angle of helical twist in B form of DNA is
(a) 6.3° (b) 36° (c) 60° (d) 90°

177. Ribozyme, a catalytic RNA was discovered in
(a) *Saccharomyces cerevisiae* (b) *Tetrahymena thermophila*
(c) *Coenorhabditis elegans* (d) *Drosophila melanogaster*

178. The DHU loop in transfer RNA is site of
   (a) Amino acid binding
   (b) Aminoacyl synthetase enzyme binding
   (c) Condon recognising
   (d) Codon binding on mRNA

179. The β-subunit of RNA polymerase
   (a) recognises promoter
   (b) binds sigma factor
   (c) binds DNA template
   (d) terminates transcription

180. The structural genes such as Lac Z, Lac Y and Lac A of operon model donot involve in synthesis of
   (a) β-galactosidase enzyme
   (b) permease enzyme
   (c) trans-acetylase enzyme
   (d) Glucose isomerase enzyme

ANSWERS

1.a	2.a	3.d	4.b	5.d	6.b	7.a	8.b	9.c	10.b
11.c	12.c	13.c	14.b	15.b	16.a	17.a	18.b	19.a	20.b
21.c	22.a	23.b	24.b	25.d	26.d	27.b	28.b	29.d	30.d
31.b	32.c	33.b	34.b	35.d	36.b	37.d	38.d	39.b	40.d
41.c	42.a	43.c	44.c	45.c	46.b	47.a	48.b	49.a	50.a
51.c	52.b	53.d	54.c	55.b	56.a	57.b	58.c	59.b	60.b
61.c	62.a	63.b	64.b	65.d	66.d	67.d	68.c	69.d	70.c
71.c	72.c	73.a	74.a	75.b	76.a	77.d	78.b	79.a	80.b
81.a	82.d	83.c	84.a	85.a	86.c	87.b	88.a	89.a	90.a
91.b	92.b	93.a	94.d	95.c	96.d	97.a	98.b	99.b	100.b
101.d	102.a	103.c	104.b	105.b	106.b	107.b	108.a	109.d	110.d
111.b	112.b	113.c	114.d	115.d	116.a	117.c	118.c	119.c	120.b
121.c	122.a	123.d	124.a	125.c	126.a	127.a	128.a	129.c	130.b
131.c	132.d	133.c	134.a	135.b	136.b	137.b	138.b	139.c	140.b
141.b	142.b	143.b	144.b	145.a	146.c	147.d	148.c	149.d	150.c
151.c	152.a	153.d	154.a	155.b	156.b	157.d	158.d	159.a	160.d
161.c	162.b	163.c	164.c	165.d	166.a	167.a	168.d	169.a	170.d
171.c	172.a	173.a	174.c	175.a	176.b	177.b	178.b	179.b	180.d

17

Biotechnology: Principles and Processes

Biotechnology is scientific and wilful manipulation of living organisms at molecular and gene level for the benefits of human beings

The term Biotechnology was coined by Karl Erky, a Hungarian Engineer in 1917 to describe a process for large scale production of useful substances.

Branches of Biotechnology:

1. Traditional or Old Biotechnology: Biotechnology has been practiced since the beginning of history. It has been used to bake bread, produce dairy products, brew alcoholic beverages, and breed food crops and domestic animals.
2. Modern Biotechnology: It is concerned with manipulation of genes and based on genetic engineering. It includes in vitro fertilisation, development of antibiotics, vaccines, recombinant proteins, hormones, gene therapy etc.
3. Non-gene Biotechnology: It is concerned with whole cells, tissues, or even individual organisms.
4. Gene Biotechnology: It is concerned with transfer of genes from one organism to another or genetic engineering.

Initiatives by Govt. of India: National Biotechnology Board (NBTB) was setup in 1982 to encourage and coordinate various research activities in biotechnology. Now it was replaced by Department of Biotechnology (DBT) within ministry of science and technology in 1986.

Genetic Engineering

It is technique of altering the genetic constitution of cells or individuals by the selective removal, insertion or modification of individual gene or gene sets. It is also called genetic modification. Or it is technique to alter the nature of genetic material to introduce it into host organism and then change the phenotype of host organism. It is done by linking the foreign genes to bacterial control sequences then it is possible to achieve expression of these genes and there by produce bacterial strains capable of producing new commercial products.

The term 'genetic engineering' was coined by Jack Williamson.

It is also called DNA cloning where there is production of a large number of identical DNA molecules from a single ancestral DNA molecule.

Precursor: Pasteur took the opportunity to unravel the mystery why wine turned sour. He utilised microbes for fermentation.

Facilitator: Watson and crick elucidated the double helical model of DNA in 1953.

The first company to focus on genetic engineering, Genetech, was founded in 1976 and started the production of human proteins.

Revolutions in Biotechnology:

1. Fermentation or Gelatinisation or solubilisation of starch into butanol and acetone by *Clostridium acetobutyricum* in 1920 as reported by Chain Weisman.
2. Discovery of Penicillin antibiotic from *Penicillium notatum* in 1929 by Alexander Fleming
3. Establishment of Recombinant DNA technology.

Achievements:

1. Paul Berg (1972) is regarded as father of genetic engineering. He developed first recombinant DNA by combining DNA (gene) of a monkey or Simian Virus (SV-40) with lambda phage (virus).
2. Stanley Cohen and Herbert Boyer (1973) constructed first recombinant DNA and created the first transgenic organism with the help of restriction endonuclease and ligase enzyme by inserting antibiotic resistance gene of *Salmonella typhimurium* into plasmid of *E. coli*.

Tools:

1. Enzymes: Restriction endonuclease enzyme: It was discovered by Arber, Nathans and Smith (1970). It comes under the category of nuclease enzyme which is of three types such as exonuclease (cutting only one strand of ds DNA at external position either 5' or 3' end), endonuclease (cutting only one strand of dsDNA only at any internal position) and **restriction endonuclease.** It is an enzyme which cuts both strands of DNA duplex at specific site (recognition sequence or palindromic sequence) by breaking phosphodiester bond between sugar and phosphate component.

It is also known as molecular scalpel or scissor. It generally occurs in bacteria as a chemical weapon against viruses. Its number is more than 3000 which have been isolated from 230 strains of bacteria and were fully characterised. All enzymes require Mg^{+2}. It does not act on DNA of own cell because of presence of parallel enzyme produced by the cell which keeps on adding $-CH_3$ group to specific Adenine and cytosine residues within recognition sequences.

The first restriction enzyme discovered was Hind II from *Haemophilus influenzae* strain Rd. Later Eco R1 was isolated from *Escherichia coli* strain RY13.

Nomenclature of Restriction endonuclease (as in *Eco*RI)

1. The first letter of Restriction enzyme denotes the name of genus (**E**: *Escherichia*)
2. The second and third letters denotes the epithet of species (**co**: coli)
3. The fourth letter (in Roman) indicates the strain of species (**R**: strain)
4. The fifth letter (in Roman numeral) indicates the order of identification of different enzymes from the same bacterium it produces (**I**: first identified).

Types of restriction endonuclease:

Type I: It has three different subunits**.** It is not used in RDT because it cuts DNA at site other than recognition site forming random cleavage patterns. It requires ATP.

Type II: It has two identical subunits**.** It is used in RDT because it cuts DNA at specific site (palindromic site) and produces restriction fragments. It does not require ATP. It requires Mg^{+2}. It is most stable**.** Most of restriction endonuclease enzymes belong to this type.

It forms palindromes by acting on recognition sequence (4-6 base pairs which are predominantly G-C rich) with rotational symmetry. The recognition sequences are different and specific for the different restriction endonuclease enzymes. Thus in palindrome, the base sequence in both strands of

DNA duplex reads the same when read from the same end (5′- or 3′- end of both strands). The enzyme which recognise 4 bases will cut DNA more frequently (called Tetra cutter) than those which recognise 6 bases (called Hexa cutter).

Cleavage pattern: Restriction endonuclease enzymes produce two types of cuts:

(a) **Staggered**: The enzymes cut both strands of DNA at different locations generating protruding ends or sticky ends or cohesive ends (one strand of duplex extends some bases beyond the other but these end readily pair with each other). This is required in RDT

(b) **Even**: The enzymes cut both strands of DNA at same locations generating blunt or flush ends.

Type III: It has two different subunits. It is not used in RDT because it cuts DNA at site other than recognition site forming random cleavage patterns. It also requires ATP

List of restriction enzymes with their source, cleavage site and products

Enzymes	Source (Organism and strain)	Cleavage site	Products
Hind II	*Haemophilus influenzae* Rd	GTC↓GAG CAG↑CTC	Blunt ends
Hin d III	*Haemophilus influenzae* Rd	A↓AGCTT T TCGA↑A	Sticky ends
Hae III	*Haemophilus aegypticus*	GG↓CC CC↑GG	Blunt ends
Hpa II	*Haemophilus parainfluenzae*	C↓CGG GGC↑C	Sticky ends
Eco R I	*Escherichia coli* Ry13	G↓AATTC C TTAA↑G	Sticky ends
Eco R II	*Escherichia coli* R24	C↓CTGG GGAC↑C	Sticky ends
Eco R V	*Escherichia coli*	GAT↓ATC CTA↑TAG	Blunt ends
Alu I	*Arthrobacter luteus*	AG↓CT TC↑GA	Blunt ends
Bam H I	*Bacillus amyloliquefaciens* H	G↓GATCC CCTAG↑G	Sticky ends
Sal I	*Streptomyces albus* G	G↓TCGAC CAGCT↑G	Sticky ends

Taq I	*Thermus aquaticus* YT1	T↓CGA AGC↑T	Sticky ends
Pst I	*Providencia stuartii*	CTGCA↓G G↑ACGTA	Sticky ends
Pvu II	*Proteus vulgaris*	CAG↓CTG GTC↑GAC	Blunt ends
Sca I	*Streptomyces caespitosus*	AGT↓ACT TCA↑TGA	Blunt ends
Stu I	*Streptomyces tubercidicus*	AGG↓CCT TCC↑GGA	Blunt ends
Sma I	*Serratia marcescens*	CCC↓GGG GGG↑CCC	Blunt ends
Bgl II	*Bacillus globigli*	A↓GATCT TCTAG↑A	Sticky ends
Kpn I	*Klebsiella pneumoniae*	GGTAC↓C C↑CATGG	Sticky ends
Xba I	*Xanthomonas badrii*	T↓CTAGA AGATC↑T	Sticky ends
Sau 3AI	*Staphylococcus aureus* 3A	G↓ATC CTA↑G	Sticky ends
Nla III	*Neisseria lactamica*	↓CATG GTAC↑	Sticky ends
Ava II	*Anabaena variabilis*	G↓GACC CCTG↑G	Sticky ends
Not I	*Nocardia otitidis*	GC↓GGCCGC CGCCGG↑CG	Sticky ends

DNA Ligase: It was discovered by Mertz and Davis (1972). It seals the nicks between adjacent nucleotides in DNA duplex by forming new sugar-phosphate bond with 5′-phosphate and a 3′-OH. It is also known as molecular glue. It is present in both prokaryotes and eukaryotes. There are two types:

(i) T_4 DNA ligase (Viral ligase): It ligates both sticky and blunt ends. It requires ATP as cofactor.

(ii) *E.coli* DNA ligase (Bacterial ligase): It ligates sticky ends only. It requires NAD^+ as cofactor.

Both enzymes perform annealing when cofactor undergoes splitting and forms an Enzyme-AMP complex. The complex binds to the nick and makes a phosphodiester bond which seals nick. Then the enzyme and AMP are finally released.

Alkaline phosphatase: It removes phosphate group from the 5′-ends of DNA/RNA. It is used to prevent sticky ends from being ligated from one another (recircularisation).

2. Vector: It is DNA which carries foreign inserted DNA fragment into host for cloning and expression. It is also known as vehicle DNA.

Features of good vector:

(a) It should be easy to isolate and purify from the cells.

(b) It should replicate autonomously independently of host chromosome and should have origin of replication (*Ori*). The *Ori* is a specific sequence in DNA from where replication starts. It is single in prokaryote but several in eukaryotes. It also controls the copy number of linked DNA as per the requirement.

(c) It should be easily introduced into host cell.

(d) It should show normal activity in host cell.

(e) It should have unique recognition or cloning sites which are required to insert foreign DNA into vector for ligation.

(f) It should have one or two marker gene which help in identification and selection of transformed from non transformed cell; and recombinant containing transformed cell from non recombinant containing transformed cell. The marker gene may be selectable which encodes a product that allows the transformed cell to survive and grow under the condition that kill or restrict the growth of non-transformed cells or screenable which gives a product that can be detected easily

Types of vector:

(a) Cloning vector: It is used for cloning

(b) Expression vector: It is production of protein by translation and transcription by specified DNA insert.

Examples:

(a) Plasmid: It was discovered by Lederberg and William Hayes. It is extrachromosomal, small (1000 times smaller than nucleoid), circular, double stranded, closed and supercoiled DNA. It can be easily isolated

It occurs naturally in bacteria but it can be produced artificially. Its size is 2 kb to 200 kb and it contains 2-250 genes. It can undergo autonomous replication independently of nucleoid. It may exist independently or becomes integrated into bacterial chromosome forming episome. It contains about 2% of genetic information which has sufficient for its own replication and additional information for host properties such as antibiotic resistance. It frequently carries the genes which are useful to cell and is not essential for normal growth and division.

It may be either **Stringent** type when it has low copy number (1 or 2 copies) or **Relaxed** type when it has high copy numbers (10 or more)

It may be conjugative when it mediates DNA transfer through conjugation (e.g., F-plasmid, R-plasmid) or Non-conjugative (e.g., Col plasmid).

It can clone DNA insert up to the size of 10 Kb.

An ideal plasmid must have the following features:

(a) It has small size (2-4 Kb) which makes easy to purify from bacterial culture.

(b) It has only one recognition site for at least one restriction endonuclease.

(c) It has at least two selectable marker genes.

(d) It undergoes more than one replication for each replication of host genome.

Artificial plasmids:

pBR322: It is plasmid developed by Bolivier and Rodriguez (1977). It is most popular, widely studied and widely used. It is constructed from plasmids of *E.coli* (pBR318 and pBR320). Its size is 4,363 bp. It undergoes much faster replication than bacterial DNA. It possesses two genes which confer resistance to antibiotics- *amp*^{r} and *tet*^{r}. It has unique recognition sites for 20 restriction endonucleases. There are two sites such as Pst I and Pvu I are located within amp^r gene and *Bam* H I and *Sal* I are within tet^r gene. Some other recognition sites such as Eco R I, *Cla* I, *Hind* III are also exist. The rop codes for the proteins involved in the replication of plasmid.

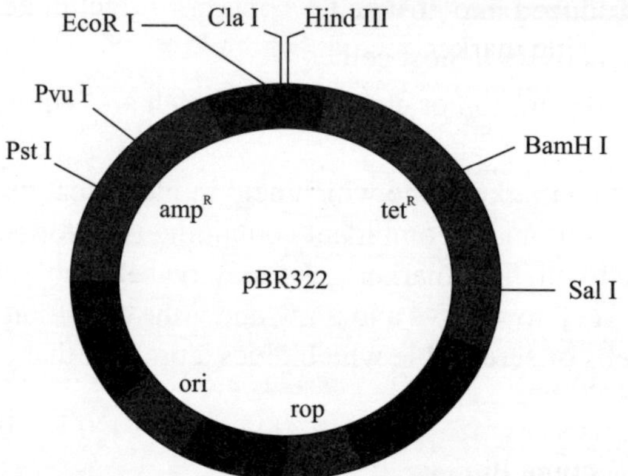

pUC 18: It is a plasmid developed in University of California by Messing and Vieria. It is much smaller than pBR322. Its size is 2,688 bp. It is derivative of pBR322 and possesses all essential parts of pBR322. It has following advantages over pBR322.

(i) Each bacterial cell produces up to 700 copies without any treatment.

(ii) Cells containing r-DNA are selected in a single step.

(iii) It also allows sequencing of DNA insert.

pUC 8: It is the smallest plasmid. Its size is 2,100 bp

pSC 101: It is a plasmid developed by Stanley Cohen and Boyer (1973). It was recombinant plasmid capable of being replicated in host (*E. coli*).

(b) Lambda phage: It is the best studied and genetically modified strain of bacteriophage. Its size is 48,502 bp. It contains an origin of replication, genes for Head and Tail protein, enzymes for DNA replication, lysis and lysogeny. It also contains single stranded protruding cohesive ends of 12 bases. It has double stranded and linear DNA in Head region but in *E. coli*, the two cohesive ends join to form a circular molecule that is necessary for replication. The sealed cohesive ends are called ***cos*** site which has 12 unpaired and complimentary bases. This *cos* site is site of cleavage and essential for packaging of mature phage DNA into phage Head.

It can clone DNA insert up to the size of 23 kb

(c) M-13: It is commonly called Phagemid. Its size is 6,407 bp. It has single stranded and circular DNA. When it infects E. coli, its single stranded DNA replicates to produce a double stranded DNA known as Replication Form (RF). Hence RF molecule can be regarded as being similar to plasmid. It can clone DNA insert up to size of 1-4 kb.

(d) Cosmid: It was developed by Collins and Hons (1978). It is hybrid and artificial vector constructed by *cos* site of lambda phage and plasmid. Its size is 5400 bp. It contains a minimum of 250 bp of λ-DNA which includes a *cos* site of 12 bases of lambda phage and sequence needed for binding of and cleavage by terminase. It can clone DNA insert up to size of 45 kb which makes it more efficient. It is also used for gene mapping.

(e) **Ti-Plasmid**: It is the largest plasmid. Its size is 213 kb. It is produced from *Agrobacterium tumefaciens* which infects mostly in dicots. This bacterium invades plant tissue and causes a tumour called crown gall. During infection part of Ti plasmid called T-DNA integrates into plant chromosomal DNA.

(f) **Bacterial Artificial chromosome (BAC):** It contains F-factor gene for replication, unique restriction site and a selectable marker gene. It is better than Yeast Artificial Chromosome (YAC) because it is more stable, simpler and performs easier transformation. It can clone large DNA insert up to size of 300 kb. It is also used analysis of large genome (genome sequencing project).

(g) **Yeast Artificial chromosome (YAC):** It is linear vector that behaves like a yeast chromosome. It is capable of replicating in two different hosts (prokaryotes and eukaryotes). It contains ARS (*ori*) sequence for replication; CEN-Y sequence for centromeric function; TEL sequence (telomeric) present at two ends for protection from exonuclease; two selectable marker genes (TRP-1 and URA-3); one selectable marker gene (SUP-4) into which DNA insert is integrated and necessary sequence from *E. coli* plasmid for selection and propagation in *E.coli*.

It can clone large DNA insert of size between 200 kb and 1000 kb. It is also used in analysis of large genome (Human Genome Project).

(h) **Shuttle Vector:** It is constructed for use in both prokaryotes and eukaryotes. e.g., Yeast Episomal Plasmid (YEP).

(i) **Plant Viruses:** Several plant viruses such as Cauliflower mosaic virus, Tobacco mosaic Virus, Gemini virus etc are also used as vector in eukaryotes esp. in plant cells

(j) **Animal Viruses:** Several animal viruses such as Adenovirus, Bovine Papilloma virus, Retro virus, Simian Virus-40 etc have also been used as vector in eukaryotes esp. animal cells.

3. **Host organism:** The features of a good host (competent cell) are as follows:

(a) It should grow fast and in large quantity under ordinary culture method

(b) It should be non-pathogenic

(c) It should support the replication of recombinant DNA.

(d) It should be easy to transform.

(e) It should not any undesirable elements which interfere the replication of recombinant DNA.

(f) It should not contain active restriction enzymes.

(g) It should not have methylase enzyme because it could methylate the r-DNA hence it can become resistant to restriction enzyme.

(h) It should have well developed genome.

(i) It should secrete recombinant protein from expressed foreign gene.

Usually *E. coli*, yeast, plant cell and mammalian cell are used as host cell depending upon the aim of cloning experiment. But *E. coli* is mostly preferred for eukaryotic genes because

(i) It has genome size of 4.6 million base pairs.

(ii) It has 2923 genes.

(iii) It is easy to grow in ordinary culture medium at 37^0C.

(iv) It shows rapid rate of multiplication.

(v) It is generally non-pathogenic.

(vi) It is usually sensitive to antibiotics.

4. Bioreactor: It is also called fermentor. It is a large vessel made up of glass, stainless steel or polypropylene used for culturing the host (plant/animal) on large scale to extract the desired biotechnological product under optimum condition of temperature, pH, substrate, salt, vitamins, O_2 etc. Its capacity is 100 to1000 litres. Stirred tank reactor is simple and commonly used. It is a large vessel in which cell culture is stirred mechanically. It is suitable for bacterial and animal cell cultures. It is well suited for the production of primary metabolites (ethanol, acetone, butanol, amino acids) and secondary metabolites (antibiotics, plant hormones).

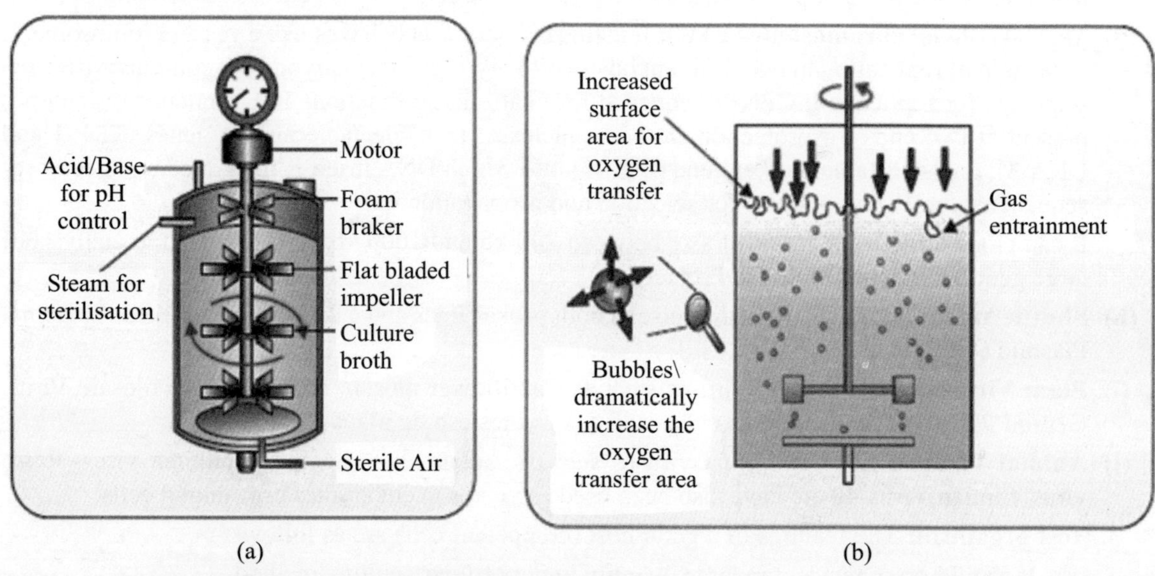

(a) Simple stirred-tank bioreactor; (b) Sparged stirred-tank bioreactor through which sterile air bubbles are sparged

Procedures of Genetic Engineering:

1. **Isolation of DNA**: It can be done by the following methods.

   (a) Conventional method: If the source of DNA is bacterial cell then it is treated with lysosome to dissolve cell wall. If source of DNA is fungal cell then it is treated with chitinase to dissolve cell wall. If source of DNA is plant cell then it is treated with cellulase to dissolve cell wall. If source of DNA is animal cell then it is treated with Sodium Dodecyl Sulphate (SDS) to lyse cell.

   The naked protoplast so obtained is purified. In case of eukaryotes, DNA is enclosed within membrane, the cell is broken open to release DNA along with RNA and protein. Then RNA is removed by RNase and protein is removed by protease. Now pure DNA is precipitated by adding chilled alcohol. The precipitate appears as fine threads in the suspension which can be removed by spooling. Hence the gene of interest is now isolated. Now centrifugation is done and followed by density gradient centrifugation. The localisation of DNA in the gradient can be done by instrumental method (absorption at 260 nm or visually by fluorescence). Finally DNA is isolated by inserting a syringe into the rod of soft plastic centrifuge tube and withdrawing the bands visible in UV radiation.

(b) **cDNA method**: A copy or complimentary DNA is used when eukaryotic gene is to be expressed in prokaryotes. In this process, m-RNA of desirable gene is isolated from the cell and then the complimentary DNA strand of RNA is produced by reverse transcriptase enzyme. Later DNA: RNA hybrid molecule is formed. Now RNA strand is digested by RNase H or alkaline hydrolysis. The 3′-end of DNA strand serves as template for synthesis of its complimentary strand by reverse transcriptase enzyme. Hairpin is opened by Si nuclease. Finally C-DNA duplex so produced is used for cloning.

2. **Fragmentation of DNA**: The purified DNA is cut into fragments by restriction endonuclease enzyme under optimum conditions. The different restriction endonuclease enzymes produce different number of fragments and one of these contains the gene of interest.
3. **Isolation of desired fragment of DNA**: The desired fragment of DNA is separated by Gel Electrophoresis.

Gel Electrophoresis: It was developed by Tiselius (1937). It is a technique used to separate a mixture of molecules (charged particles) by their differential migration through stationary phase (gel) in an electric field. It is efficient tool to separate, identify and purify the DNA and other biomolecules.

Procedure: It is done in electrophoresis box which contains anode (positive electrode), cathode (negative electrode), a shelf designed to hold gel, a comb used to form well within gel and a power supply. It involves the following steps:

(a) **Preparation of agarose gel of appropriate concentration:** Agarose is linear polymer of D-galactose and anhydro L-galactose. It is purified powder of isolated from agar. It has large pore size which acts as molecular sieve. This allows the larger molecules to move slowly than smaller molecules. It forms gel when dissolved in water and boiled. Agarose gel is cast by boiling agarose in the presence of buffer. A desired concentration of agarose gel (0.8%) is added in electrophoresis buffer TAE (Tris Acetate EDTA). Gel comb is inserted to make well.

(b) **Loading of DNA sample into well to achieve separation**: The appropriate quantity of DNA is loaded in well and electric field (150V DC) is applied across gel to start electrophoresis. DNA, a negatively charged particle at neutral pH, migrates towards anode at rates determined by their molecular size and conformation. The smaller DNA fragments will travel faster. The speed of migration (electrophoretic mobility) of a molecule depends on its charge and molecular mass. All DNA fragments migrate nearly the same distance into gel and form bands. Gel electrophoresis is carried out in buffers ranging from pH 3-10.

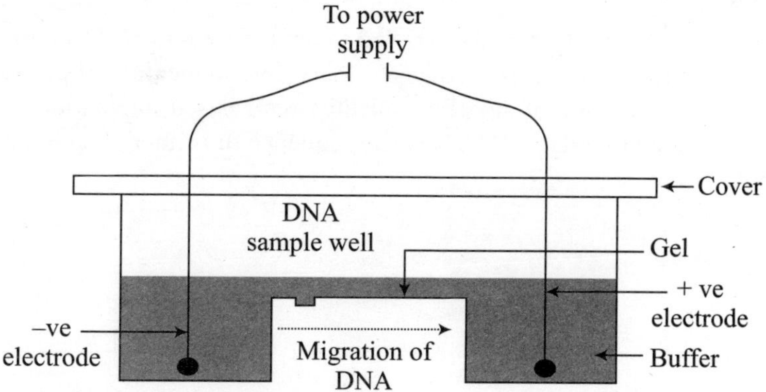

(c) **Staining the gel**: The gel is now removed from chamber and immersed in a solution of sucrose and loading dye (ethidium bromide / methylene blue) for 45 minutes for staining. Sucrose increases the density of DNA preparation and dye increases the visibility of preparation. The tracking dye such as bromophenol is used. The orange bands appear due to ethidium bromide and blue bands due to methylene blue at room temperature. The separated bands of DNA were cut from agarose gel and extracted from gel piece. This is called elution. Hence DNA fragments are purified for constructing recombinant DNA by joining with vector.

Polymerase Chain Reaction (PCR): It was developed by Kary Mullis (1985). It is method of formation of multiple copies of desired DNA fragment in vitro with the help of RNA primer and DNA polymerase. It results in selective amplification of a specific region (target sequence) of DNA. It generates billion copies of desired DNA in a few hours.

Requirements:

(a) DNA template: It is taken from any source which contains one or more target DNA. The RNA can also be used by first making a DNA copy using reverse transcriptase enzyme.

(b) Two nucleotide primers: Primers are short and single stranded DNA (20 bp long) that hybridise to target DNA region, one to each strand of duplex is required. These primers should be complimentary and are oriented with their ends facing each other allowing synthesis of DNA towards one another.

(c) DNA polymerase enzyme: *Taq* polymerase is required. It is isolated from *Thermus aquaticus*, a bacterium. It is DNA dependent DNA polymerase enzyme. It is stable at 92^0C. Other thermostable enzymes such as *Pfu* polymerase (isolated from *Pyrococcus furiosus*) and Vent polymerase (isolated from *Thermococcus litoralis*) are considered more efficient than *Taq* polymerase because they have proof reading function which produces high fidelity amplification.

(d) Four deoxyribonucleotides (dATP, dGTP, dTTP and dCTP).

(e) Mg^{+2}

Procedure: It is a chain reaction because newly synthesised DNA strands will act as templates for subsequent DNA synthesis. The reaction mixture was prepared containing all four components and Mg^{+2}. It involves the following steps:

(a) **Denaturation:** The reaction mixture containing target DNA is heated at high temperature (94^0C) resulting into the separation of two strands. Each single strand of target DNA acts as template for DNA synthesis.

(b) **Primer annealing**: The mixture is now cooled to temperature from 50 to 70^0C (depending upon the melting temperature of the expected DNA) that allows annealing of primer to complimentary sequences in DNA. Two sets of small chemically synthesised nucleotides. Primer hybridise to each of single stranded template DNA because sequence of primer is complimentary to 3- end of template DNA.

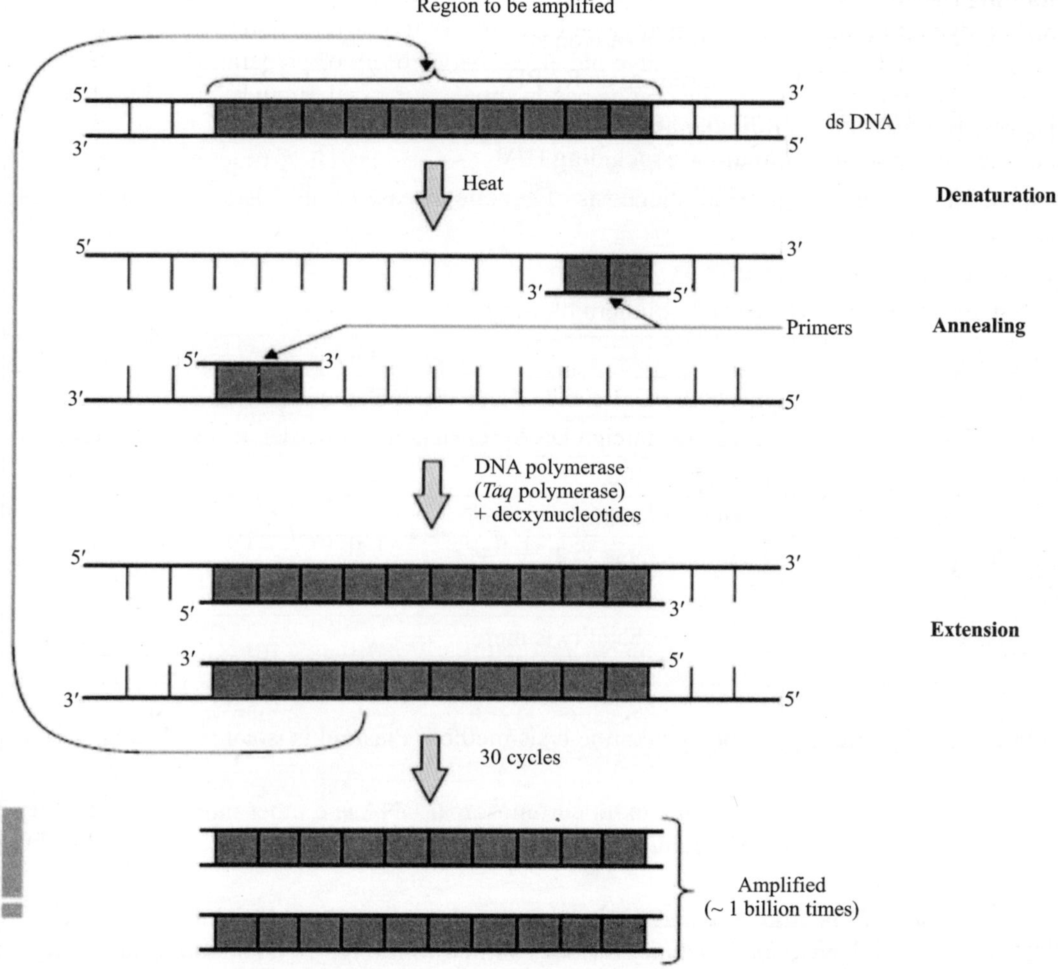

(c) Polymerisation (Primer extension): *Taq* polymerase synthesise DNA region between the primers using deoxynucleotide triphosphate (dNTP) and Mg^{+2}at 74^0C. Hence primer is extended towards each other so that DNA segment lying between two primers is copied.

In first cycle, DNA are synthesised which extend beyond target sequence. This is because there is nothing to prevent DNA polymerase from continuing to copy the template beyond the end of target sequence.

However in subsequent cycles, newly synthesised DNA which ends with the primer sequence acts as template and limits synthesis of target sequence so that amplified DNA contains only the target sequence.

The cycle is repeated many times until desired level of amplification is achieved. After each cycle the number of templates is doubled. This number is calculated by this formula: $M_f = M_i \times 2^n$ where M_f is the final number of DNA molecules produced by PCR., M_i is the initial number of molecules or template and n is the number of cycles performed.

PCR products have much higher mutations (base substitution) than DNA obtained from in vivo replication because *Taq* polymerase does not have proof reading ability.

Applications of PCR:

(a) It is used for gene cloning (generation of clones).

(b) It helps in study of DNA polymorphism as RAPD.

(c) It is useful in DNA fingerprinting, gene therapy, site directed mutagenesis.

(d) It detects bacterial and viral disease including HIV.

(e) It is useful in antenatal (prenatal) diagnosis of genetic disease or disorders and determination of sex embryos.

(f) It is used in DNA sequencing

(g) It prepares probe in Northern and Southern blotting.

PCR	Gene cloning
It is *in vitro* manipulation.	It is both *in vitro* and *in vivo* manipulation.
It requires DNA.	It requires foreign DNA, restriction endonuclease, vector and host cell.
Its cost is less.	Its cost is more.
It is more efficient.	It is less efficient.
It requires less skill.	It requires more skill.
It takes a few hours.	It takes 2-4 days.
Its error probability is less.	Its error probability is more.
It has more application.	It has less application.

4. Isolation of plasmid: It is done by alkaline lysis method. Plasmid is isolated when bacterial cell is treated with detergents.

Cell membrane is solubilised. Plasmid, extra chromosomal DNA and other molecules are released. This mixture is called lysate (supernatant). The lysate is subjected to several chemicals and finally plasmid is obtained.

5. Fragmentation of Plasmid: The plasmid is treated with same restriction endonuclease to generate sticky ends which must have complimentary residues similar to foreign DNA but opposite in direction to prevent recircularisation of digested plasmid.

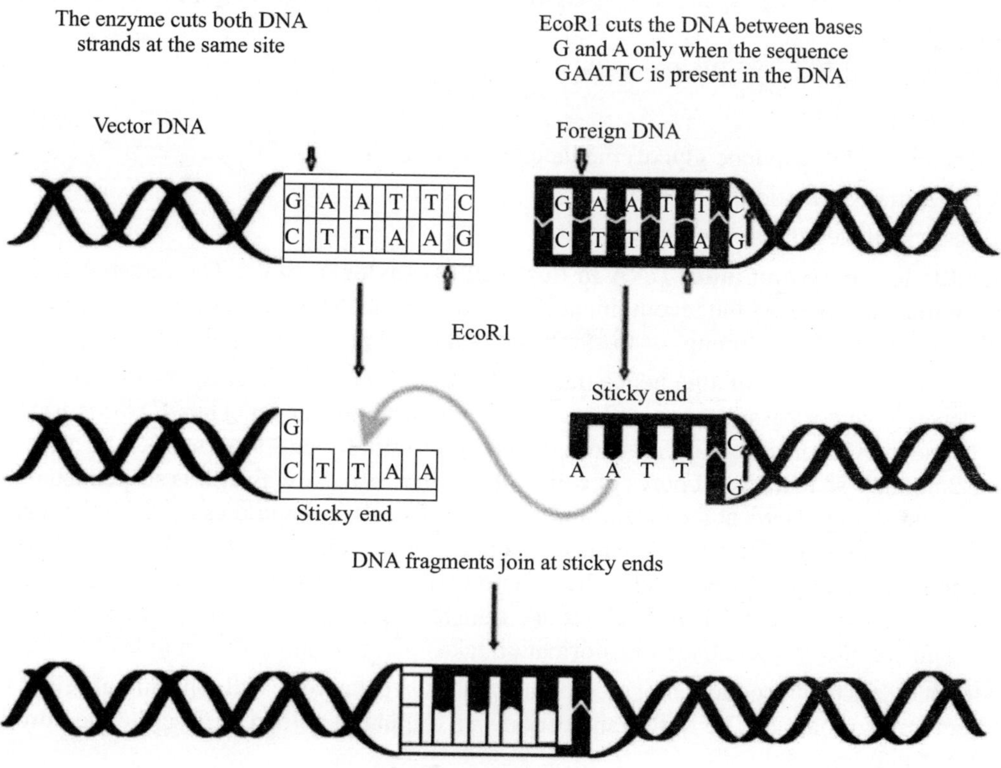

6. Ligation of desired DNA fragment into vector to form r-DNA: The ends of both plasmid and foreign DNA are cohesive (sticky) and are mixed under annealing conditions which permits pairing between them. The nicks (the broken covalent bonds between 5-phosphate and 3-OH of two neighbouring nucleotides within DNA strand) are joined and ligated by DNA ligase enzyme. The recombinant DNA is formed that combines the genetic information of two different organisms.

Direct or Vector-less Gene transfer:

(a) **Electroporation:** It is most common and efficient method of delivering genes in bacteria, plant, yeast and animal cell. In this both DNA and host cell are mixed in a suspension and subjected to electric field of an optimum voltage (350 V) for a very short time. Thus small pore is developed into cell wall and plasma membrane of host cell. Then r-DNA enters into host cell through pores.

(b) **Microinjection:** It is technical, time consuming and more efficient method. It is easier in animal cell (egg, embryo) due to lack of cell wall which interferes the process of transformation. It is most successful in animal cell (non-vacuolated embryonic). In this, foreign DNA is directly injected inside the nucleus or cytoplasm of plant or animal cell with the help of micropipette or microneedle under powerful microscope.

(c) **Short Gun/Gene gun/Microprojectile/Biolistic/Ballistic method:** It is costly method. It is used in leaf cell of plant cell. It does not require removal of cell wall. It is also used to transform shoot apex, embryos, mature pollen. It is much easier to transform in meristematic cells than in non-dividing cells. In this, DNA is coated with very fine gold or tungsten (1-2μm) particles and shot at target cell at very high speed. It is used more effectively in producing transgenic plants.

Other methods:

(a) Macroinjection
(b) Lipofection
(c) Chemical (Polyethylene glycol) mediated gene transfer
(d) Pollen transformation
(e) Laser mediated transformation

7. Introduction of recombinant DNA in host cell (Transformation): The recombinant DNA is introduced into *E.coli* to select the recombinant DNA from the unchanged vector or obtain many copies of recombinant DNA (Gene cloning) or to synthesise a desired protein(recombinant protein).

E. coli is not easily transformable hence poorly accessible to r-DNA because r-DNA is hydrophilic molecule hence it cannot pass through cell membrane due to lipid bilayer. Thus *E.coli* is treated with cold $CaCl_2$ under hypotonic condition at low temperature initially and then temperature is raised to 42^0C for 1-2 minutes so that size of pores of cell wall will increase. Now E. coli is suspended in 50 mM $CaCl_2$ to take up foreign DNA at a concentration of 10^8-10^{10} cells/ml and these cells are incubated for 12-24 hours to increase the efficiency of transformation. The efficiency of transformation depends on type of vector used and type of host cell. The transformation efficiency of 10^5 μg would be adequate for simple cloning experiment. There is about one transformation per 10,000 plasmids. The r-DNA is added and it enters into *E. coli* hence transformation takes place within a few minutes.

Transfection: It is the introduction of recombinant DNA in eukaryotic cell esp. animal cell. It is more efficient than transformation. The term transfection was originally coined to describe the introduction of phage DNA into bacterial cells.

8. Selection and screening of Transformed cell: The transformed cells are suitably diluted and spread thinly or plated on a suitable medium to increase the numbers so that each cell is well separated and produce a separate colony. Now transformed cells are isolated and identified from among these that contain unaltered vector. It is done by Colony Hybridisation (sequence dependent Screening) method. It is most efficient method. Firstly bacterial colonies are replica plated. Cells are lysed and their DNA is denatured and probe (either DNA or RNA having small molecule of as few as 15 bases) is prepared. Now filter is incubated with specific radioactive (P^{32} labelled) probe under annealing condition. Now probe is washed out leaving only those probes that have hybridised with denatured DNA from bacterial cells. Colonies with whose DNA the probe has hybridised are identified by autoradiography. These colonies are isolated from master plate. Now the transformed clones are then replica plated on ampicillin containing medium. The clones containing r-DNA will be sensitive to ampicillin due to inactivation of amp^r gene by DNA insert. Such clones are identified and isolated from master plate.

Following types of bacterial cells are obtained after transformation.

(a) Majority are non-transformed cells without vector.
(b) A proportion of transformed cells are non-recombinant cells with unaltered vector (without r-DNA).
(c) Rest transformed cells are recombinant cells with recombinant vector (with r-DNA).

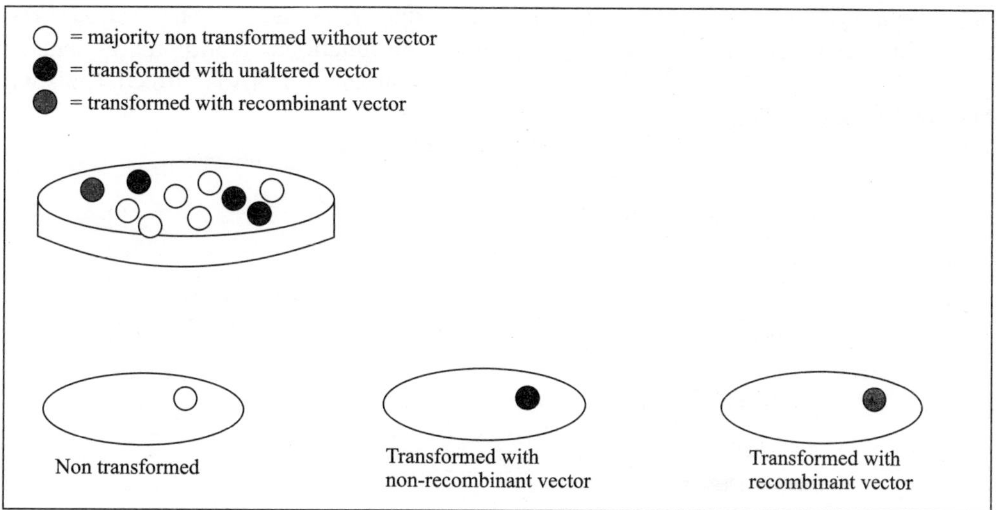

Fig. ***Master plate showing three types of cells***

Identification of transformed bacterial cell requires the presence of suitable marker genes on the vector molecule. Marker genes may be either selectable markers or screenable markers.

The selectable markers are antibiotic resistance genes. These are used to differentiate transformed from untransformed cells. A good selectable marker gene is one that provides resistance to the antibiotic which arrests the growth of particular cell. A host cell strain is selected that is sensitive to a particular antibiotic, ampicillin, tetracycline, kanamycin. The non-transformed cells are eliminated by integrating a suitable marker gene (reporter gene) into vector. A marker gene produces a phenotype which helps in quick selection of transformed cells. When a population of bacterial cell is placed on a medium containing ampicillin, then those cells survive and form colonies which have ampicillin resistance gene. The reporter gene such as *npt* II (neomycin phosphotransferase II) is marker gene whose phenotypic expression can be monitored and thus report about the activity or change in advance of the effect of modification..

But in case of vector (pBR322) there are two selectable marker genes in which the DNA insert may be placed within one of the marker amp^r gene. The other marker tet^r gene is used for the elimination of non-transformed cells.

The recombinants are screened from non-recombinants on the basis of their ability to produce colour in the presence of chromogenic substrate. A screenable marker gene produces enzyme which gives a specific colour in the presence of particular substrate. The *lac* Z gene encodes β-galactosidase, an enzyme that splits lactose into glucose and galactose. If r-DNA is inserted within the coding sequence of β-galactosidase enzyme then enzyme gets inactivated. The presence of a chromogenic substrate gives blue coloured colonies if the plasmid does not have foreign DNA insert (non-recombinants). If there is no blue colour, then the vector has foreign DNA insert and such colonies are called recombinants.

Insertional Inactivation: It is used to detect the presence of foreign DNA within plasmid in transformed cell. It occurs by following two methods:

(a) Restriction enzyme (*Pst* 1 or *Pvu* 1) is used. If foreign DNA insert is placed at *Pvu* 1 site of **ampicillin resistant** gene. The amp^r gene becomes non functional. Antibiotic resistance by amp^r gene is lost. Cells do not grow on ampicillin. But recombinant DNA is formed. Cells will grow in the presence of tetracycline antibiotic.

(b) Restriction enzyme (*Bam* H1 or *Sal* 1) is used. If foreign DNA insert is placed at *Bam* H1 site of **tetracycline resistant** gene. The tet^r gene becomes non functional. Antibiotic resistance by tet^r gene is lost. Cells do not grow on tetracycline. But recombinant DNA is formed. Cells will grow in the presence of ampicillin antibiotic.

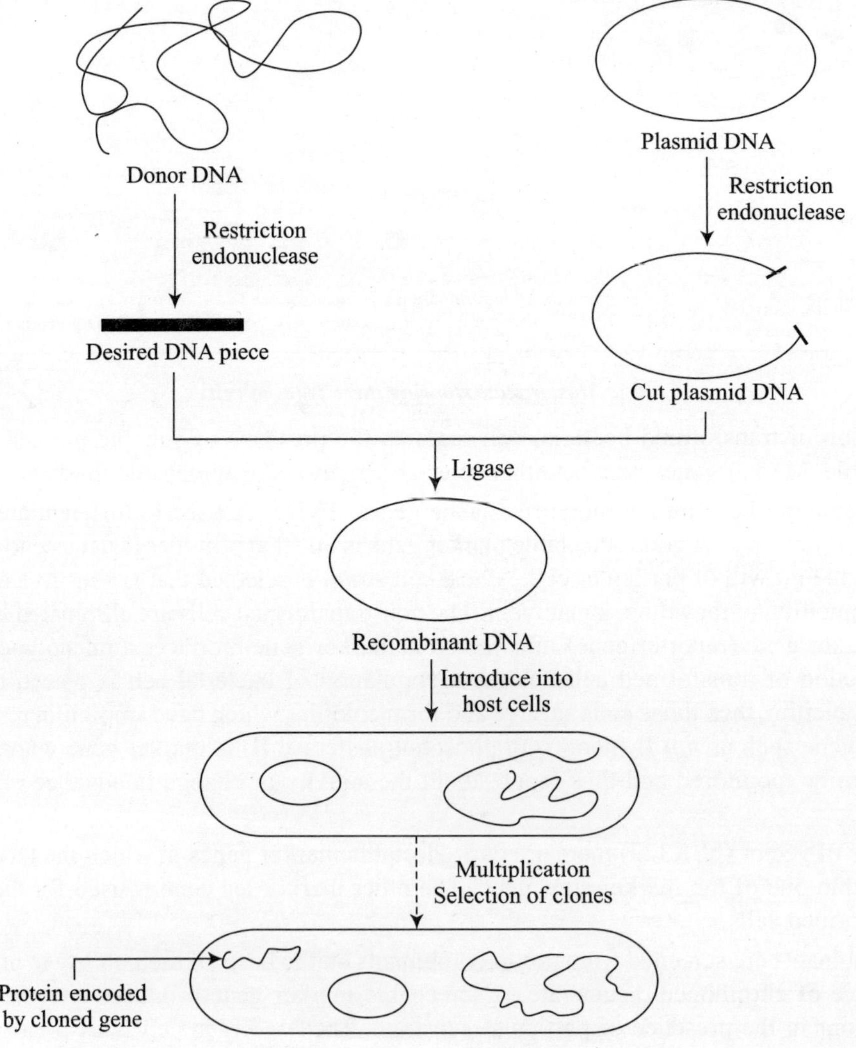

Fig. ***Process of genetic engineering***

9. Multiplication of Transformed cell: The transformed cell is multiplied in continuous culture in a bioreactor containing suitable medium. The cells reach in physiologically most active phase-log phase where multiplication is rapid. This may be introduced into bacterium for the production of protein encoded by insert. If any protein encoding gene is expressed in heterologous host, it is called recombinant protein. This produces a larger biomass to get higher yield of desired proteins.

10. Downstream processing: After the formation of product in bioreactor, it undergoes some processes before ready for marketing. It involves

(a) Separation or removal of particles (insolubles): It is the separation of cells, cell debris and other particles from fermentation broth. It can be done by filtration, centrifugation, sedimentation, flocculation, electro-precipitation or gravity settling.

(b) Isolation of product: It can be done by solvent extraction, adsorption, ultrafiltration, precipitation etc.

(c) Purification of product: It separates contaminants which resemble the product in physical and chemical properties. It is done by chromatography, crystallisation and fractional precipitation.

(d) Packaging of Products: It is done by crystallisation, desiccation, lyophilisation, spray drying.

The product is subjected to quality control testing and kept in suitable preservative. If drug is to be manufactured then it has to undergo clinical trials. The downstream processing and quality control testing are different from product to product.

SHORT ANSWER TYPE QUESTIONS

1. What are the contributions of Paul Berg and Stanley Cohen in genetic engineering?
2. Write notes on pBR322.
3. Distinguish between endonuclease and restriction endonuclease enzymes.
4. Mention five examples of restriction endonuclease enzymes.
5. Write notes on DNA ligase.
6. Distinguish between sticky ends and blunt ends.
7. What are the features of a good vector?
8. Write notes on cosmid.
9. Distinguish between BAC and YAC.
10. What are the features of a good host for transformation process?
11. Write notes on bioreactor.
12. Write notes on downstream processing.
13. Distinguish between PCR and gene cloning.
14. Write notes on gene gun.
15. Explain insertional inactivation.

LONG ANSWER TYPE QUESTIONS

1. What are the features of plasmid? Discuss its role in genetic engineering.
2. Describe the different steps in recombinant DNA technology.
3. Discuss the procedure of PCR and mention its significance.
4. Discuss the procedure of gel electrophoresis.
5. Describe the types of restriction endonuclease enzymes and discuss their role in genetic engineering.

MULTIPLE CHOICE QUESTIONS

1. Lysozyme is used to remove cell wall in case of
(a) Fungus (b) Plant (c) Bacterium (d) All of the above
2. The most widely used plasmid is
(a) pTi (b) pBR 322 (c) pRi (d) pUC
3. A vector specially designed for cloning 45 kb DNA fragment is
(a) Plasmid (b) Cosmid (c) Phagemid (d) pBR322
4. The vector accommodating the largest size of DNA is
(a) Plasmid (b) YAC (c) Cosmid (d) BAC
5. Which of the following is a palindrome?
(a) CCGTA/GGCAT (b) GAATTC/CTTAAG (c) CCCC/GGGG (d) CATTG/GTAAC
6. Which one of the following is not a PCR based molecular marker?
(a) RAPD (b) RFLP (c) AP-PCR (d) DNF
7. Restriction enzymes cut DNA at specific sites known as
(a) Telomeric sequences (b) Palindromic sequences
(c) Terminator sequences (d) Attenuator sequences
8. The enzyme involved in self splicing of introns in *Tetrahymena thermophila* was
(a) Primase (b) Reverse transcriptase
(c) Ribozyme (d) RNA polymerase
9. The most commonly used method of choice of gene transfer using immature embryos as explants is
(a) Microprojectile (b) Electroporation
(c) Liposome mediated (d) Chemically stimulated
10. The frequency of *Agrobacterium* mediated genetic transformation is increased by
(a) Ascorbic acid (b) Aflatoxin (c) Acetosyringone (d) Abscissic acid
11. A shuttle vector is defined as
(a) A plasmid capable of replicating in one organism
(b) A plasmid capable of accepting only large DNA insert
(c) A plasmid incapable of replicating in cells
(d) A plasmid capable of replicating in two different organisms.
12. The selectable marker that provides resistance to the antibiotic Kanamycin is
(a) Gentamicin acetyltransferase (gat) (b) Hygromycin phosphotransferase (hpt)
(c) Streptomycin phosphotransferase (spt) (d) Neomycin phosphotransferase (npt II)
13. The pBR322, a frequently used vector for cloning gene in *E. coli* is
(a) An original bacterial plasmid (b) A modified bacterial plasmid
(c) A viral genome (d) A transposon
14. The gene promoter sequence commonly used in gene cloning experiments are derived from
(a) TMV (Tobacco Mosaic Virus) (b) TSV (Tobacco Satellite Virus)
(c) CMV (Cucumber Mosaic Virus) (d) CaMV (Cauliflower Mosaic Virus)

15. Ti plasmid used in genetic engineering is obtained from
    (a) *Bacillus thuringiensis*
    (b) *Agrobacterium rhizogenes*
    (c) *Agrobacterium tumefaciens*
    (d) *Pseudomonas putida*
16. Genetically engineered male sterile crop plants have been produced by inserting
    (a) Opaque 2 gene
    (b) Virus coat-protein gene
    (c) Barnase gene
    (d) Chitinase gene
17. Transgenic plants can act as living bioreactors for the production of chemicals and pharmaceutical. This is known as
    (a) Organic farming
    (b) Molecular farming
    (c) Pharmacotechnology
    (d) Hybridoma technology
18. DNA coated with tungsten or gold particles is used in direct or vector less transformation in
    (a) Biolistic method (b) Electroporation (c) Microinjection (d) Lipofection
19. The enzyme used in PCR technique is
    (a) RNA polymerase
    (b) DNA polymerase
    (c) Reverse transcriptase
    (d) *Taq* polymerase
20. Which one of the following is not a thermostable enzyme?
    (a) *Taq* DNA polymerase
    (b) RNA polymerase
    (c) *Pfu* DNA polymerase
    (d) Vent polymerase
21. Crown gall disease was caused by
    (a) Bacteria (b) Pi plasmid (c) Ti plasmid (d) Virus
22. Gel electrophoresis is used for
    (a) Isolation of DNA
    (b) Fragmentation of DNA
    (c) Separation of DNA
    (d) Ligation of DNA
23. Restriction enzyme *Eco*R1 cleaves DNA at the sequence
    (a) AAGCTT (b) GAATTC (c) TATAGC (d) CGTACG
24. The blunt ends are formed by
    (a) *Alu* I (b) *Taq* I (c) *Hind* II (d) All
25. Plasmid can clone DNA insert up to the size of
    (a) 10 kb (b) 25 kb (c) 45 kb (d) 200 kb
26. Which of the following vector is hybrid?
    (a) Plasmid (b) Cosmid (c) BAC (d) YAC
27. Who first used a gene of SV-40 in the experiment of genetic engineering?
    (a) Arber, Nathans and Smith
    (b) Stanley cohen
    (c) Boliver and Rodriguez
    (d) Paul Berg
28. The vector used for ligation of foreign DNA in eukaryotes is
    (a) Cosmid (b) BAC (c) Plant virus (d) pBR^{322}
29. The first restriction endonuclease enzyme to be discovered was
    (a) *Eco* RI (b) *Hind* II (c) *Hind* III (d) *Alu* I

30. The loading dye used in the agarose gel electrophoresis is
(a) Acridine dye (b) Safranin
(c) Acetocarmine (d) Ethidium bromide
31. Which of the following process is not associated with PCR?
(a) Denaturation (b) Autoradiography (c) Annealing (d) Extension
32. During transformation, *E. coli* was treated with
(a) $CuSO_4$ (b) $HgCl_2$ (c) $NaNO_3$ (d) $CaCl_2$
33. Molecular glue used in the experiment of genetic engineering is
(a) Restriction endonuclease (b) DNA polymerase
(c) Helicase (d) DNA ligase
34. The process in which separated bands of DNA are cut out from agarose gel and extracted from the gel piece is called
(a) Eluviation (b) Alluviation (c) Illuviation (d) Elution
35. Genomic DNA is digested with *Alu* I, a restriction enzyme which is a four base cutter. What is the frequency with which it will cut the DNA assuming a random distribution of bases in the genome
(a) 1/16 (b) 1/32 (c) 1/64 (d) 1/256
36. When the DNA fragments are observed under UV light, they are seen as
(a) Orange coloured bands (b) Yellow coloured bands
(c) Blue coloured bands (d) All of the above
37. In gel electrophoresis, the restriction enzyme digested DNA is loaded in a well near
(a) Anode (b) Cathode (c) Centre of gel (d) Anywhere in gel
38. In insertional inactivation, the recombinant DNA is inserted with coding sequence of
(a) Restriction enzyme (b) Tetracycline resistant gene
(c) Ampicillin resistant gene (d) Beta galactosidase enzyme
39. Recombinant colonies in inserted inactivation are differentiated on the basis of showing
(a) Blue colour (b) Red colour (c) Orange colour (d) No colour
40. A selectable marker gene is incorporated in a plasmid to
(a) Increase its copy number
(b) Increase its transformation efficiency
(c) Eliminate the non-transformants
(d) Increase the expression of the gene of interest
41. Foreign DNA cannot pass through cell membrane because it is
(a) Hydrophilic (b) Hydrophobic (c) Rich in protein (d) Heavy
42. A single strand of nucleic acid tagged with a radioactive molecule is called
(a) Plasmid (b) Vector (c) Prion (d) Probe
43. PCR involves use of *Taq* polymerase enzyme which is
(a) Thermostable (b) Thermolabile (c) Psychrophilic (d) All of the above
44. A method used to distinguish DNA of one individual from another is
(a) DNA sequencing (b) Hybridisation (c) RFLP (d) RAPD

45. T4 polynucleotide kinase is used for
    (a) Labelling 3′ ends of DNA
    (b) Labelling 5′ ends of DNA
    (c) Creating blunt ends of DNA
    (d) Dephosphorylation of DNA
46. A plasmid cloning vector shoul contain all of the following except
    (a) Origin of replication
    (b) Inducible promoter
    (c) Selectable marker gene
    (d) Multiple cloning sites
47. Which one of the following sequences is most likely to be a restriction enzyme recognition site?
    (a) CGGC
    (b) CGC
    (c) GTAATG
    (d) GTCGAC
48. If bacterial genome and plasmid are allowed to replicate in same manner then
    (a) Plasmid genome will replicate first
    (b) Bacterial genome will replicate first
    (c) Both will replicate at the same time
    (d) Depends on GC content
49. Which of the following types of restriction endonuclease enzymes is used in genetic engineering?
    (a) Type I
    (b) Type II
    (c) Type III
    (d) All of the above
50. The polymerase chain reaction was discovered by
    (a) Stanley Cohen
    (b) Paul Berg
    (c) Arber, Nathans and smith
    (d) Kary Mullis

ANSWERS

1.c	2.b	3.b	4.b	5.b	6.d	7.b	8.c	9.b	10.c
11.d	12.d	13.b	14.d	15.c	16.c	17.b	18.a	19.d	20.b
21.c	22.c	23.b	24.d	25.a	26.b	27.d	28.c	29.b	30.d
31.b	32.d	33.d	34.d	35.d	36.c	37.c	38.d	39.d	40.c
41.a	42.d	43.a	44.c	45.b	46.b	47.d	48.a	49.b	50.d

18

Biotechnology and its Applications

Applications of Biotechnology

1. Role in agriculture: Production of transgenic plants: Biotechnology has produced more transgenic crops in a shorter time period. These plants contain foreign gene of trans species inserted artificially at level of protoplast, gamete or zygote by recombinant DNA technology. These have been produced in more than 60 plants of angiosperms esp. dicots. The transgenic plant has been produced only one gymnosperm (e.g., *Picea*). Tobacco was the first transgenic plant produced. A genetically modified or engineered organism is one in which its genetic material was altered by any method.

The first genetically modified organism was bacteria in 1973.

Objectives of transgenic plants:

(a) To introduce a new trait to the plant (not naturally occurring).

(b) To bring together in one plant useful genes from a wide range of living species.

(c) To increase yield or productivity and

(d) To develop features like insect resistance, disease(virus) resistance, herbicide resistance, nematode resistance and tolerance to heat, cold and drought.

Transgenic plants are useful in several ways:

(a) Insect resistant: The crop plants have been engineered to express the insecticidal toxin gene of *Bacillus thuringiensis* so that insects attempting to eat these plants get killed. *Bacillus thuringiensis* is Gram positive, aerobic, motile, peritrichous, endospore forming soil bacterium. It is species specific and does not harm other forms. It contains about 40 *cry* genes in its large plasmid. These genes have been cloned from bacteria and introduced in plants such as cotton, corn, rice, wheat, soya bean. It forms a solid, intracellular, insoluble and toxic crystal protein (α, β, γ-exotoxins and δ-endotoxin in crystalline form) in the vegetative cell during endospore formation. The transgenic plants having *cry* genes act as bioinsecticides. The insecticide protein is secreted as inactive protoxin. The insects ingest inactive protoxin and reach their midgut where alkaline pH due to protease enzyme solubilise it and convert to active toxin. The toxic function occurs in N-terminal of protein and C-terminal is involved in forming crystal protein. Active toxin binds to epithelial cell of midgut of insect. Then lysis of cell occurs due to osmotic imbalance. The insects die within 30 minutes to 3 days after ingestion. Bt- toxin does not harm fish, birds, humans and other mammals. This Bt toxin does not kill *Bacillus* because toxin is present in the form of inactive protoxin.

Types of gene	Target
Cry I A(a)	Butterfly (lepidopteran)
Cry I A(b)	Corn borer (lepidopteran)
Cry I A(c)	Boll worm (lepidopteran)
Cry I B	Moth and butterfly (lepidopteran)
Cry I C	Moth and butterfly (lepidopteran)
Cry I D	Moth and butterfly (lepidopteran)
Cry I E	Moth and butterfly (lepidopteran)
Cry II A	Bollworm (lepidopteran)
Cry II B	Flies and mosquitoes (Dipteran
Cry III A, B, C	Beetle and boll weevil (Coleopteran)
Cry IV A, B, C, D	Black flies and mosquitoes

Crop plants	Product
Cotton	Bollgard
Maize	Yield guard, Maximiser
potato	New leaf

(b) Nematode resistant: It was discovered by Fire and Melo. Transgenic plants such as Tomato, Tobacco and Brinjal etc. have been developed which are resistant to a nematode (*Meloidogyne incognita*) which causes root knot disease in which plants cannot absorb water and mineral and thus root develops crack causing secondary infection. Nematode resistance in tobacco is based on RNA interference (RNAi) technique. All eukaryotic organisms use RNAi as a method of cellular defense.

Procedure: It involves the following steps:

(a) Nematode specific gene is isolated.

(b) It is ligated to *Agrobacterium* vector.

(c) It is introduced into host Tobacco.

(d) Nematode specific gene is triggered to produce single stranded RNA which is complimentary to RNA produced by nematode in order to infect host tobacco plant.

(e) This complimentary RNA binds to m-RNA of nematode to form double stranded RNA. The source of ds-RNA can be infection of ribovirus or transposon.

(f) The ds-RNA is fragmented into short interference RNA (siRNA).

(g) The siRNA develops silencing complex which prevents specific m-RNA of nematode to perform translation.

(h) The m-RNA becomes toxic. The nematode cannot survive in transgenic Tobacco host expressing specific interfering RNA (RNAi). Thus nematode is unable to infect the transgenic tobacco plant and it gets protected from parasite (nematode).

(c) Virus resistant: The first transgenic virus resistant plant was Tobacco produced in1986.

Transgenic plants such as Tobacco, Tomato, Potato resistant to Tobacco mosaic virus were developed due to introduction of gene for viral coat protein into plants. When these plants were inoculated with

TMV then symptoms fail to develop. The resistance occurs due to blocking of the process of uncoating of virus particles which is essential for replication and expression of viral genome.

Transgenic plants such as Tobacco and Tomato resistant to Tobacco Rattle Virus were developed due to introduction of gene for nucleoprotein into plants.

Transgenic plants such as Tobacco and Tomato resistant to Tobacco Satellite Virus were developed due to introduction of gene for viral satellite RNA into plants.

Crop plant	Product
Squash	Freedom II

(d) Herbicide resistant: These plants tolerate herbicides which kill the surrounding flora but leave cultivated crop plants intact.

Transgenic *Petunia* resistant to herbicide, **Glyphosate** was developed due to insertion of gene for Enol pyruvyl shikimate Phosphate synthetase (EPSPS) of *Salmonella typhimurium* into vector, *Agrobacterium*. Its herbicide action is broad spectrum. It inhibits EPSPS and kills plants by interfering with shikimate pathway in plants.

Transgenic Tobacco resistant to **atrazine** was developed due to insertion of gene for QB protein of *Arabidopsis* into a vector, A*grobacterium*. Its herbicide action is broad spectrum.

Transgenic tobacco, corn, wheat and soyabean resistant to **phosphinothricin** were developed due to insertion of *bar* gene of *Streptomyces* into a vector, *Agrobactertium*. Its herbicide action is broad spectrum.

Transgenic tobacco resistant to **Salphonyl urea** was developed due to insertion of acetolacetate synthase gene of *Arabidopsis*. Its herbicide action is selective.

Crop plants	Product
Cotton	Roundup Ready
Maize	Roundup Ready
Soyabean	Roundup Ready
Oilseed rape (Canola)	Roundup Ready

(e) Nutrient rich (Golden rice): It was developed by Ingo Potrykus (1990). It is yellow or orange coloured due to presence of beta carotene. It is genetically enriched with β-carotene to combat vitamin A deficiency. It was developed due to insertion of three genes (*psy* gene coding for the enzyme, phytoene synthase; *lyc* gene for lycopene cyclase from Daffodil (*Narcissus pseudonarcissus*) and *crt1* gene coding for carotene desaturase from bacterium (*Erwinia uredovora*) in Ti plasmid of *Agrobacterium* to form r-DNA.

Procedure: Now the r-DNA is introduced into *Agrobacterium*. Transformed bacteria are cultured and transferred to Petridish containing embryo. These bacteria infect rice embryo resulting in the transfer of desirable gene necessary for the synthesis of β-carotene. Thus golden rice shows golden coloured grains. It involves the following processes:

Isoprenoid unit→Geranyl diphosphate (Carotenoid) → Phytoene → Lycopene→β-carotene → Retinal → Retinol (vitamin A)

(f) Delayed ripening: Actually ripening occurs due to slow breakdown of polygalacturonic acid, component of cell wall of pericarp of fruit. It is based on antisense transcription (normally m-RNA is synthesised on sense strand of DNA. If it occurs on complimentary strand of sense strand, then it is called antisense transcription).

Procedure: Transgenic tomato (**Flavr Savr**) is produced as such:

(a) Firstly antisense copy of gene encoding polygalacturonase enzyme of tomato is isolated.

(b) It is inserted into *Agrobacterium tumefaciens* to form r-DNA.

(c) Now r-DNA is introduced in fresh tomato.

(d) Antisense copy of gene forms antisense RNA (opposite to sense RNA).

(e) Complimentary sense and antisense RNA are formed in cell forming ds-RNA. It blocks translation. Then usual amount of polygalacturonase enzyme is not formed in tomato. Thus it will delay ripening or softening of fruit hence it will increase shelf life and nutrients that add flavour. It is the first commercially grown genetically modified food crop and it was made more resistant to rotting by Californian company, Calgene.

(g) Biodegradable plastic: Transgenic *Arabidopsis* is developed due to insertion of Polyhydroxybutyrate (PHB) gene of *Alcaligenes eutrophus* for the production of biodegradable plastic.

(h) Molecular farming: Transgenic potato, tomato and tobacco plants are used as bioreactor or factories for manufacturing pharmaceuticals such as sugar, fatty acid, starch, cellulose, wax etc.

(i) Pest resistant: Transgenic tobacco plant is developed due to insertion of Cow Pea Trypsin inhibitor (CpTI) gene.

(j) Hybrid seed: Transgenic *Brassica napus* was produced for the production of hybrid seed. Male sterile plant was produced due to insertion of *barnase* gene (encoding a ribonuclease) derived from *Bacillus amyloliquefaciens*. Likewise fertility restorer plant was produced due to insertion of *barstar* gene (encoding ribonuclease inhibitor) derived from *Bacillus amyloliquefaciens*.

(k) Abiotic stress resistant: Transgenic tobacco plant tolerates adverse environmental conditions like drought, frost, salinity, heavy metal toxicity etc.

2. Roles in medicine:

(a) Production of Therapeutic drugs: The recombinant DNA technology is used for the production of around 30 recombinant drugs (therapeutic drugs) for human use. These drugs do not induce immunological reactions as some other drugs obtained from non-human sources. These are produced from genetically engineered microbes esp. *E.coli* and yeast). The genes are expressed in animal cell culture.

Recombinant proteins	Application
Humulin	Teatment of diabetes mellitus
Blood clotting factor-VIII	Treatment of haemophilia-A
Blood clotting factor-IX	Treatment of haemophilia-B
Reo pro	Prevention of blood clot
Tissue plasminogen activator (tPA)	Treatment of heart disease by thrombolysis (it dissolves blood clot)
Urokinase	Treatment of heart disease by thrombolysis (it dissolves blood clot)
OKT-3	Reversal of acute kidney transplantation
Asparaginase	Treatment of cancer
Tumour necrosis factor	Treatment of cancer
Interleukin	Treatment of cancer
DNase	Treatment of cystic fibrosis

Monaclonal antibodies	Diagnosis and treatment of diseases
Interferon	Treatment of Hepatitis-C, cancer, AIDS, viral infection
Hepatitis-B surface antigen (HBSAg)	Treatment of hepatitis-B
Hirudin	Anticoagulant
Erythropoietin	Treatment of anaemia
Human growth hormone	Treatment of dwarfism
Calcitonin	Treatment of rickets
Chorionic gonadotrophin	Treatment of infertility
Epidermal growth factor	Treatment in wound and burn healing
Platelet derived growth factor	Treatment of Diabetic ulcers
Antitrypsin	Treatment of emphysema

Production of Humulin (Human insulin): Insulin is produced by β-cells of islet of Langerhans of pancreas which controls blood sugar level. In human body, two genes A and B genes produce polypeptides independently which undergo association in cytoplasm.

Humulin was synthesised by Tsan in 1978.

Procedure:

(a) m-RNA synthesising pro-insulin is extracted from Pancreatic cell.

(b) c-DNA (gene) is synthesised by reverse transcriptase enzyme using m-RNA.

(c) c-DNA is ligated to pBR 322 (vector) to form r-DNA.

(d) Now recombinant DNA is introduced into E. coli (transformation).

(e) Transformed bacterial cell divides rapidly every 20 minutes. Hence numerous transformed cells containing r-DNA are produced within short span of time.

(f) Transformed bacterial cells for the presence of r-DNA are selected using marker gene (ampicillin).

(g) Transformed cells are grown in large bioreactor. Lactose is used as an inducer to switch on the *Lac* Operon linked to the gene of interest in *E.coli*.

(h) The hybrid protein containing β-galactosidase and proinsulin is isolated from this culture and purified. The hybrid protein is then treated with cyanogen bromide to separate β-galactosidase and pro-insulin. Finally pure human insulin (humulin) is formed.

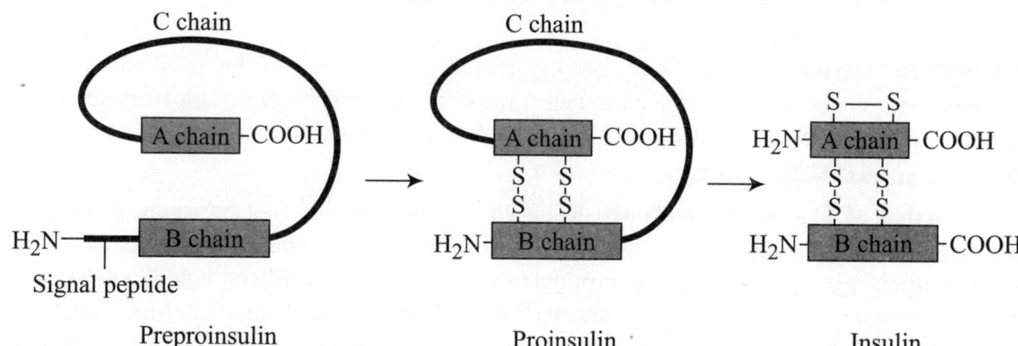

The freshly synthesised polypeptide is pre-proinsulin. It is composed of 109 amino acids in a chain. In first step, pre-proinsulin undergoes processing and loses about 23 amino acids and becomes proinsulin.

In second step, it undergoes denaturation using trypsin and carboxypeptidase hence losing 35 amino acids to become functional insulin. The functional insulin is made up of 51 amino acids arranged in two polypeptide chains (A chain contains 21 amino acids and B chain contains 30 amino acids) held by three disulphide bonds.

N-terminal ——— B-chain ——— C-chain ——— A-chain ——— C-terminal

Preproinsulin (109 amino acids)
↓ -23 amino acids
Proinsulin (86 amino acids)
↓ -35 amino acids
Insulin (51 amino acids)

(b) Gene therapy: It is treatment of disease by inserting a normal and functional extra gene into a cell or tissue of an individual with the intention that gene product will play therapeutic role. Michacle Blease and French Anderson performed first clinical gene therapy successfully in 1989 on a human patient(4 year old girl) suffering from Severe Combined Immuno Deficiency (SCID). This disease is caused by the absence of adenosine deaminase (ADA) an enzyme involved in purine metabolism and is essential for immune system to function. Its deficiency destroys selectively T- and B-cells of the immune systems leading to SCID (Bubble boy disease) in which patients die off ordinary disease due lack of immunity. It is caused due to defect in gene for expressing the enzyme, Adenosine deaminase.

Gene replacement therapy: It is cure of disease by replacing the abnormal gene with normal gene.

Procedure:

(a) Defective gene is identified.
(b) Normal healthy gene encoding Adenosine Deaminase (ADA) is either isolated or synthesised.
(c) It is ligated to vector (retrovirus).
(d) Lymphocyte (cell) is isolated from the blood of bone marrow of patient.
(e) A functional recombinant ADA gene is introduced in lymphocyte at embryonic stage to function. The difficulties of gene therapy include the choice of vector being used.

Use:

Patient therapy: Cells with healthy gene are introduced in affected tissue so that healthy gene overcomes the effect without affecting the inheritance of patient.

Embryo therapy: The genetic constitution of embryo at post zygotic level is altered so that inheritance will be altered.

3. Roles in industries: Microbes have been genetically engineered for the production of enzymes, hormones, organic acids, antibiotics, Single cell protein, alcoholic beverages, biofuels such as ethanol and butanol. These microbes are also useful in extraction of minerals by bacterial leaching.

4. Roles in animal biotechnology:

(a) Production of Transgenic animals: Rudolph Jaenisch created first transgenic animal (mouse) in 1980 by introducing foreign DNA encoding human growth hormone into its embryo. Most of the transgenic animals were produced by microinjection of DNA into fertilised egg. Transgenic animals have also been developed in cow (Rossie), sheep (Polly and Molly), goat, and fish. Maximum transgenic animal has been developed in mice due to short estrous cycle, short gestation period, short generation time, production of several offspring, convenient *in vitro* fertilisation and successful culture of embryo.

Procedure:

(a) The desirable gene is isolated.

(b) The vector is selected.

(c) The desired gene is fused with vector to form recombinant DNA.

(d) The adult female is injected with hormone for superovulation (collection of eggs).

(e) The eggs are mated to form fertilised egg.

(f) The recombinant DNA is injected into fertilised egg.

(g) The transformed fertilised egg is allowed to grow for some time.

(h) The selected transformed egg is implanted into uterus of surrogate mother.

Importance: The transgenic animals are used

(a) to study the expression of normal gene on normal body function and development.

(b) to test the toxicity of drugs

(c) to test the vaccines before use on human beings.

(d) to serve as model to study the role of gene for various human diseases such as cystic fibrosis, cancer, Alzheimer disease, rheumatoid arthritis.

(e) to serve as model to exhibit the disease symptoms to treat the diseases.

(f) to produce a desired substance (recombinant protein) in large quantity (bioreactors) and

(g) to produce organs for organ transplantation.

(b) Cloning: Clone (identical organism) is produced by fusion of nucleus of adult somatic cell from one sheep with unfertilised enucleated egg from other followed by implantation into surrogate mother. Dolly (clone of sheep) was successfully created by Ian Wilmut in Feb, 1996 after failures of 277 times. This lamb was exactly similar to her mother. Other examples are George and Charlie (cloned calves) and ANDI (World's first genetically altered primate, monkey) produced by inserting a jelly fish gene into the embryo of rhesus monkey.

Procedure:

(a) Mammary cell was isolated from donor sheep (6 year old) and cultured *in vitro.*

(b) Egg cell was isolated from another donor sheep and its nucleus was removed.

(c) Both cells were fused and this fused cell contains nucleus of mammary cell.

(d) This fused cell was grown (cultured *in vitro*) into early embryo up to blastula or morula stage.

(e) The embryo was implanted into the uterus of a third sheep (surrogate mother) who has been treated hormonally to stimulate pregnancy. Finally surrogate mother gave birth to a normal lamb which grew into normal adult sheep (Dolly) that is genetically identical to the mammary cell donor.

Importance: The cloned animals are used in

(a) Multiplication of desirable genotype without the risk of variation

(b) Understanding of curing genetic disease

(c) Genetic research.

5. **Role in environment:** Genetically engineered microbes are used for

(a) Pollution control: The Oil spill in water body is cleared by digesting hydrocarbons of crude oil. It is done by developing super bug or oil eating bug (as developed by Anand Mohan

Chakrabarty) due to introduction of plasmid from different strains into a single cell of *Pseudomonas putida*. The super bug degrades oil (hydrocarbon) into non-polluting and harmless chemicals.

(b) Detoxification of waste and effluents (waste treatment).

(c) Removal of xenobiotics (non-biological wastes).

(d) Restoration of degraded land.

Hazards of Genetic engineering:

(a) The genetically modified organisms for herbicide and pest resistance can transfer the genes into sexually compatible wild relatives or weeds that acquire resistance to become super weeds and cause more harm.

(b) The genetically modified organisms may cause allergic reaction because mostly allergens are protein.

(c) The genetically modified organisms may disturb local ecosystem by inhibiting or promoting growth of indigenous plants.

(d) The transgenic plants for pest resistance may be toxic to human beings because these produce toxic substance.

(e) The transgenic plants may produce harmful pathogens in environment.

Safeguards of Genetic engineering and Bioethics:

(a) Care should be taken to avoid accidental escape of dangerous genetically modified microbes.

(b) Strict laboratory procedures should be followed to protect people from infection by genetically modified microbes

(c) Extreme precaution should be taken while cloning harmful genes.

(d) Prior approval should be required from the concerned Govt. authority while doing research work.

Regulatory agencies of India

(a) Recombinant DNA Advisory Committee (**RDAC**): To evolve guidelines for research involving RDT.

(b) Review Committee on Genetic Manipulation (**RCGM**): To recommend special suitable conditions for experiment and trials.

(c) Genetic Engineering Approval Committee (**GEAC**): To monitor large scale manufacturing and release of genetically modified organisms into environment; import and export of genetically modified organisms.

Biosafety issues: Modern Biotechnology has resulted in a legal characterisation and treatment of trade related biotechnological process and products commonly called Intellectual Property. Simply it is product of mind. The Intellectual property right (IPR) is the right to protect this property. The IPR includes right related scientific works, inventions in all fields, industrial design, trademarks, patent, copyright, trademarks, trade secrets, RDT.

Biopatent: It is the protection or right given by the Government to inventor of a biological material to prevent others from commercial use of his invention. It prevents others from making, using or selling the invention.

A patent is a personal property which can be licensed or sold like any other property. Different countries have different patent laws and these can be modified from time to time. Infringement of patents is decided by court.

Biopatent is the patent of life forms. The biopatents are granted by the Government for

(a) Discovery of new strain of microbes
(b) Transgenic or genetically modified plants and animals
(c) New biological or biotechnological method/process
(d) Discovery of DNA sequence
(e) Proteins encoded by DNA sequences
(f) Production process
(g) Product application
(h) New varieties of plant developed through breeding technique.

Biopiracy: It is the unlawful use of wild plants, biological compounds or genetic sequences by advanced countries without obtaining consent. Several examples of biopiracy are as follows:

(a) Government of India challenged the patent of Indian seed oil of **Neem** granted to US Deptt. of Agriculture in 1992. This patent was get cancelled.
(b) Government of India challenged the patent of Indian **basmati rice** granted to Rice Tech of USA in 2001. This patent was get cancelled.
(c) Government of India challenged the patent of Indian **turmeric** powder granted to University of Mississippi medical centre of USA in 1995. This patent was get cancelled.

SHORT ANSWER TYPE QUESTIONS

1. What are the objectives of transgenic plants.
2. Write notes on golden rice.
3. Discuss the procedure of nematode resistant plants.
4. Explain antisense technology in producing Flavr Savr.
5. Mention five examples of therapeutic proteins.
6. Discuss the role of genetically engineered microbes in environment.
7. Discuss the procedure of producing clone.
8. Write notes on gene therapy.
9. Write notes on biopatent.
10. Write notes on super bug.

LONG ANSWER TYPE QUESTIONS

1. How insect resistant transgenic plant is produced?
2. Discuss the role of recombinant DNA technology in medicines with reference to humulin.
3. Discuss the procedure of producing transgenic animals.

MULTIPLE CHOICE QUESTIONS

1. The microbe provides a natural system for gene transfer to produce transgenic plants is
   (a) E.coli (b) Pseudomonas putida
   (c) Bacillus thuringiensis (d) Agrobacterium tumefaciens
2. Transgenic plants are produced by introducing new genes by using
   (a) Mutation (b) Crossing over (c) Radiation (d) Gene gun
3. Silencing of mRNA has been used in producing transgenic plants resistant to
   (a) Boll worms (b) Nematodes (c) Viruses (d) Bacteria
4. Human insulin (Humulin) is being commercially produced from transgenic species of
   (a) Rhizobium (b) Yeast (c) E.coli (d) Agrobacterium
5. The first clinical gene therapy was given to a SCID patient who suffered from the deficiency of
   (a) Restriction endonuclease (b) Polygalacturonase
   (c) Adenosine deaminase (d) Neuroamine deaminase
6. Insect resistant transgenic cotton has been produced by inserting a piece of DNA from
   (a) an insect (b) a bacterium (c) a virus (d) a yeast
7. In transgenics, expression of transgene in target tissue is determined by
   (a) Enhancer (b) Promoter (c) Reporter (d) Attenuator
8. The number of amino acids contained in Proinsulin is
   (a) 106 (b) 86 (c) 51 (d) 21
9. Cry 1 A (b) gene is resistant to
   (a) Boll worm (b) Corn borer (c) Mosquitoes (d) Honey bees
10. A therapeutic drug, Tissue plasminogen activator (tPA) is used in the treatment of heart disease because
    (a) It does not form clot in blood vessel (b) It dissolves blood clot in blood vessel
    (c) It decreases blood pressure (d) It provides oxygen.
11. The RDAC stands for
    (a) Recombinant DNA Approval Committee
    (b) Recombinant DNA Advisory Committee
    (c) Recommendation of DNA in Advisory Committee
    (d) Report of DNA on Approval Committee
12. The first introduced transgenic crop in India (2002) is
    (a) Bt-cotton (b) Bt-brinjal (c) Bt-rice (d) Bt-turmeric
13. A transgenic plant Golden rice was discovered by
    (a) Tsan (b) Ingo Potrykus (c) Paul Berg (d) Bayliss
14. Root knot disease was caused by a
    (a) Fungus (b) Virus (c) Bacterium (d) Nematode
15. FLAVR SAVR is a transgenic plant of tomato based on
    (a) Same sense RNA technology (b) Antisense RNA technology
    (c) Terminator technology (d) Nonsense technology

16. Which country has decided to 'label' those Genetically modified (GM) foods that are found to be safe?
   (a) US (b) England (b) Japan (d) Mexico
17. The most common trait that was improved in commercialised crops was
   (a) Nutritional quality (b) Disease resistance (c) Drought resistance (d) Insect resistance
18. The crystals of Bt-toxin produced by some bacteria do not kill the bacteria themselves because
   (a) Bacteria are resistant to the toxin (b) Toxin is inactive
   (c) Toxin is immature (d) Bacteria enclose toxin in a special sac
19. In RNAi, genes are silenced using
   (a) ss RNA (b) ds RNA (c) ss DNA (d) ds DNA
20. In which of the following plants, resistance against a nematode was introduced by implying RNAi?
   (a) Tomato (b) Tobacco (c) Golden rice (d) Bt-cotton
21. A functional ADA cDNA can be introduced into the stem cell of the patients receiving gene therapy by using vector
   (a) Retrovirus (b) *E. coli*
   (c) *Bacillus thuringiensis* (d) *Agrobacterium*
22. Which of the following genes was introduced in first transgenic cow?
   (a) *Cry* I Ac (b) α-1-antitrypsin
   (b) β- 1-antitrypsin (d) Human α-lactalbumin
23. Which of the following has been covered under broad patent category?
   (a) *Brassica* (b) *Triticum* (c) *Pisum* (d) *Oryza*
24. The new strain of bacteria produced by biotechnology in alcohol industry is
   (a) *E. coli* (b) *Saccharomyces cerevisiae*
   (c) *Pseudomonas putida* (d) *Bacillus subtilis*
25. Which of the following microbes was used to clear oil spill in water body?
   (a) *Bacillus megatherium* (b) *Anabaena azollae*
   (c) Pseudomonas denitrificans (d) *Pseudomonas putida*
26. Dolly is a clone of sheep was successfully created by
   (a) Ingo Potrykus (b) Ian Wilmut (c) Paul Berg (d) Stanley Cohen
27. In transgenics, the expression of trans gene in target tissue is determined by
   (a) Reporter (b) Promoter (c) Enhancer (d) All of the above
28. The first transgenic plant produced was
   (a) Tomato (b) Tobacco (c) Potato (d) Cotton
29. Bollgard is transgenic plant of
   (a) Tomato (b) Tobacco (c) Potato (d) Cotton
30. Transgenic plant was found to be resistant to a nematode known as
   (a) *Coenorhabditis elegans* (b) *Meloidogyne incognita*
   (c) *Ascaris lumbricoides* (d) *Wuchereia bancrofti*

31. The first transgenic virus resistant plant produced was
    (a) Tomato (b) Tobacco (c) Potato (d) Cotton
32. Transgenic *Petunia* was found to be resistant to a herbicide known as
    (a) Glyphosate (b) Atrazine (c) Phosphinothricin (d) Sulphonyl urea
33. Which of the following therapeutic recombinant proteins is used in the treatment of heart disease?
    (a) Humulin (b) Urokinase
    (c) Tissue plasminogen factor (d) Hirudin
34. Humulin contains
    (a) 21 amino acids (b) 51 amino acids (c) 86 amino acids (d) 109 amino acids
35. Which of the following diseases is treated by Gene therapy?
    (a) Tuberculosis (b) Huntingto chorea (c) SCID (d) AIDS
36. Rossie is transgenic animal of
    (a) Mice (b) Cow (c) Sheep (d) Monkey
37. Maximum transgenic animal has been developed in
    (a) Mice (b) Cow (c) Sheep (d) Monkey
38. ANDI was world's first genetically altered
    (a) Mice (b0 Cow (c) Sheep (d) Monkey
39 Which of the following regulatory agencies monitors import and export of genetically modified organisms?
    (a) GEAC (b) RDAC (c) RCGM (d) CPCB
40. The Government of India in 2001 challenged the patent a certain hybrid of
    (a) Cotton (b) Rice (c) Turmeric (d) Brinjal

ANSWERS

1.d	2.d	3.b	4.c	5.c	6.b	7.c	8.b	9.b	10.b
11.b	12.b	13.b	14.d	15.b	16.c	17.d	18.b	19.b	20.b
21.a	22.d	23.a	24.d	25.b	26.b	27.a	28.a	29.d	30.b
31.b	32.a	33.c	34.b	35.b	36.b	37.a	38.d	39.a	40.b

19

Strategies for Enhancement in Food Production

Plant breeding: It is the development of improved varieties of plants possessing desirable characters from the existing one or production of new crop varieties superior to earlier ones. It is purposeful manipulation of plant species to develop desired genotypes and phenotypes. It is the need of hour to meet the food requirement for the population. In India agriculture sector employs 62% of population and contributes to 33% of its GDP. The plants have been improved in the following ways:

(a) Enhanced potential for more vigorous growth and yield

(b) Increased resistance to all types of pests

(c) Production of hybrids

(d) Selection of genetic variants with desirable qualities

Objectives of Plant Breeding:

1. To get higher yield of seeds, grain, fruits, fibres, oils etc.
2. To improve the quality of seeds and fruits in size, shape, colour, flavour, taste, nutrients and storability.
3. To improve the resistance against disease, insect pests, herbicides, drought, frost, salinity, and lodging.
4. To prevent premature falling of buds, fruits etc.
5. To change the duration of crops (developing early maturing or late maturing varieties).
6. To change the habit of growth (producing dwarf or tall varieties)
7. To develop photo-insensitive and thermo-insensitive varieties.
8. To increase the responsiveness to fertilisers.
9. To develop the varieties which can grow under different climatic and soil condition.
10. To make harvesting easier.

Conventional technique of Plant Breeding: It involves the following methods to produce plants with desirable traits.

1. Introduction: It is the easiest and most common method of crop improvement. It is taking up of a plant or a group of plants or germplasm (cuttings or seeds) with desirable characters from growing locality to a new locality. If plants were introduced from foreign country to a new area then these are called exotic. If plants were collected from same country then these are called indigenous.

It may be **primary introduction** in which the introduced variety is well adapted to new environmental condition and released as such for cultivation without any alteration.

It may also be **secondary introduction** in which the introduced variety will undergo acclimatisation for several generations to adapt in the new or changed climatic condition, otherwise plants will perish.

All introductions are subjected to quarantine in which these are carefully examined for the presence of weds, insects and pathogens (disease causing organisms).

Some of the introduced varieties of crops

Crop	Variety	Introduced from
Wheat	Sonora 64, Lerma Rojo	Mexico
	Ridley	Australia
Rice	IR-8	Philippines
	Taichung native 1	Taiwan
Maize	Dixie-11, Texas-21	USA
Oat	Kent	Australia
Pea	Rimpus	Germany

2. Selection: It is the oldest method and the basis of all crop improvement. It may undergo Natural selection in which it takes a longer duration to bring about improvement. It may also undergo artificial selection in which it takes a few years.

It is done by following three methods:

(a) Mass Selection: It is the simplest and easiest method. It is done to improve mainly cross-pollinated crops. It is based on phenotype or external character of crops such as size of fruit and seed. It involves marking out plants of desired quality and then selected. The selected plants are harvested and seeds are mixed together and later mixed seeds are sown to get a new crop. The plants are selected in next generation and grown for subsequent generation till the best result is obtained. The selected varieties contain better qualities.

Advantages:

(i) It eliminates the inferior type

(ii) It does not require much scientific knowledge

(iii) It needs to control pollination.

Limitations:

(i) It does not increase yield.

(ii) The improvement of crop plants is short-lived.

(b) Pure line selection (Head to Row selection/Inbred selection): Pure line is progeny of single, homozygous, self-pollinated and self fertilised plant. The concept of pure line was given by Johanssen (1903) who worked on princess variety of French bean (*Phaseolus vulgaris*). It is done to improve the local varieties of self-pollinated crops. It involves 50-100 plants or heads are selected from mixed population of the field before harvest. Their seeds are collected separately. The progeny of each such plant is grown in a separate line. Plants with desirable characters are selected from each line again. Around 25-50 seeds from each such plant are grown in separate rows in plot. Same process is followed for 7 to 8 generations until a new variety is obtained.

Advantages:

(i) It does not control pollination.

(ii) It produces varieties which are uniform in appearance.

(iii) It selects lesser number of plants.

Limitations:

(i) It is laborious process.

(ii) It takes longer time up to 10 years than mass selection.

(iii) Its yield and adaptability are low due to extreme homozygosity.

(c) Clonal selection: Clone is progeny of single plant obtained by vegetative propagation. Each member plant of clone is called ramet. It is the selection of desirable clone from the mixed population of vegetatively propagated plants. It is done to improve the vegetatively propagated plants such as banana, sugarcane, potato. It involves:

- Several clones are selected.
- Each member of clone is evaluated on the basis of yield and quality.
- Disease and pests are artificially inoculated.
- Disease free and high yielding clones are selected.
- Finally multiplication of clone is done.

Advantages:

(i) It produces stable and uniform varieties of crop because variation does not occur.

(ii) It does not lose desirable characters in plants even after many years of cultivation.

(iii) It utilises and preserves hybrid vigour of particular plant for many generations.

3. Hybridisation: It is the production of new variety of crop with desirable characters by crossing two genetically different individuals of same species or different species. It is the production of hybrid as a result of genetic recombination. It does not change the genetic content of organisms because new characters are not formed but of course, new combination of gene is created. It is practised when no further improvement can be achieved in a local variety by other methods. It is done to improve both self-pollinated and cross pollinated crops.

Historical background:

1. Cotton Mather (1716) developed first natural hybrid in corn.
2. Thomas Fairchild (1717) produced first natural interspecific hybrid (Fairchild's mule) by crossing Sweet William and Carnation.
3. Joseph Kolreuter (1760) developed the technique of hybridisation for plant improvement in tobacco. He produced successful hybrids through artificial crosses.

Objectives:

1. To combine all the desirable characters of two plants in a single variety.
2. To increase the range of genetic variation
3. To utilise the hybrid varieties

Procedure:

(a) Collection of germplasm: The germplasm is the sum total of all the alleles of genes present in a crop. The superior germplasm such as seeds (common) ,seedlings, buds, rooted cuttings underground modified stem are collected from cultivated improved varieties, old local varieties, pure lines produced by plant breeders and wild crop species.

(b) Selection of parent: The germplasm is evaluated to select plants with desirable characters of plants. It is better to select local healthy plants as parent because these are already adapted to particular climate.

(c) Selfing of parent plant: The selected plants are self-pollinated to increase homozygosity or to minimise heterozygosity. It helps in removing the undesirable characters. It does not require in naturally self-pollinated crops. The plants obtained after self pollination are called inbred lines.

(d) Emasculation: It is the removal of stamens or anthers (male reproductive organ) from a bisexual flower without affecting female reproductive part at bud stage just before anthesis or maturity to prevent self fertilization. It is done by removing anther by forceps in plants of large flowers or by dipping the flower buds in hot water at 45-50^{0}C for about one to ten minutes in plants of small flowers or by treatment of chemicals (e.g.,2,4-D, Maleic hydrazide). It is not needed in dioecious (unisexual) plants but essential in monoecious (bisexual) plants of self as well as cross pollinated crops. The male sterile lines can also be used if emasculation is not possible.

(e) Bagging: The emasculated flower of female parent plant is covered by butter paper (most commonly used) or cellophane or polythene bag and closed at their mouth to prevent undesirable and foreign pollens to fall on stigma of female flower to check cross pollination. Female parent plant is kept bagged by butter paper till seed formation and opened only once in between for artificial pollination.

(f) Tagging: The emasculated and bagged flower is properly labelled or tagged with the following information:

(i) Number of plant

(ii) Date of emasculation

(iii) Date of pollination (crossing)

(iv) Name of female parent (written first) followed by a cross sign (x) and then male parent. e.g., A × B

(g) Crossing: It is artificial crossing or cross pollination between genetically different (closely or distantly) plants. In this, firstly pollens from the flower of selected male parent are selected and then bag of emasculated flower is opened for awhile. Finally pollens are transferred to stigma by brush and then flower is again bagged immediately. There are several types of crossing.

Hybridisation methods in Self-pollinated crops

(i) Back cross method: It is rapid method in which there is back crossing of F_1 and subsequent generations to new parent repeatedly.

(ii) Bulk method: It is simple method in which F_2 and subsequent generations are grown in bulk. Thus purelines are obtained through individual selection of plant

(iii) Pedigree method: It is complex method in which individual plants are selected in F_2 and subsequent generations. Thus pedigree record is obtained. It is the basis of selection in later generations.

Hybridisation methods in cross-pollinated crops

(i) Single cross: It is the cross between two inbred lines or varieties.

$$X = A \times B$$

(ii) Double cross: It is most common. It is the cross between two single crosses or between hybrids of two single crosses.

$X = (A \times B) \times (C x D)$

(iii) Three way cross: It is the cross between a single cross and inbred line or it is intermediate between single crosses and double cross

$X = A \times B \times C$

(iv) Multiple cross: It is a cross involving more than one inbred line.

$X = (A \times B) \times (C \times D) \times (E \times F)$

(h) Production of F_1 generation plants: Seeds are harvested after maturing and later dried, cleaned and stored properly. In next season seeds are sown to raise F_1 generation hybrid plants.

(i) Production of F_2 plants: The F_1 hybrids are selfed to produce F_2 plants. The desirable characters of F_2 plants are carefully identified and developed further by different selection methods. The process is repeated several times.

(j) Evaluation, Release and Commercialisation of new cultivar: The newly selected lines or cultivar are tested at various climatic conditions for atleast three growing seasons. Then research field is ideally with respect to irrigation, fertilisers, elimination of weeds. The selected hybrids are superior to both parents and local cultivar and released in the market providing information to cultivators regarding climatic conditions, water requirement and dose of fertilisers. The high quality seeds (part of plant which is used for growing the crop, e.g; stem of sugarcane, tuber of potato, grain of maize) are certified by testing laboratories for sale.

Types of hybridisation:

(a) Intravarietal cross: It is a cross between two plants of same variety but having different genotype. It is useful for the improvement of self pollinated crops.

Cabbage × Cabbage

(b) Intervarietal or Intraspecific cross: It is most common. it is across of plants of different varieties of same species. It is useful for improvement of self pollinated and cross-pollinated crops.

Wheat	NP-52 × NP -165	NP 710
Rice	TN-1 × TN-141	Jaya

(c) Interspecific cross or Intrageneric cross: It is across between plants of different species of same genus. it is useful for producing disease, pest and drought resistant plants.

Cotton	*Gossypium arboretum* × *Gossypium hirsutum*	Devraj
Cotton	*Gossypium herbaceum* × *Gossypium hirsutum*	Devitej
Rice	*Oryza indica* x *Oryza japonica*	ADT-137

(d) Intergeneric cross: It is most difficult cross. It is a cross between plants of different genera of same family. It is also called wide cross or distant cross hybridisation because parent plants are taxonomically distant plants. It produces agriculturally important plants.

Raphanus sativus x *Brassica oleracea* var. *capitata*	Raphanobrassica
Triticum dicoccum x *Secale cereale*	Triticale
Potato x Tomato	Pomato
Brinjal x Tomato	Bromato

Advantages of Hybridisation:

(a) It produces hybrid vigour in plants

(b) It brings all useful characters which are scattered over several chromosomes due to interaction of many genes, into a single variety.

(c) It produces variations which are heritable and source of evolution in plants.

(d) It produces several new varieties such as

Rice	Jaya, Padma, Sabarmati, Krishna
Wheat	Kalyan sona, Sonalika
Sugarcane	*Saccharum barberi, Saccharum officinarum*

Disadvantages of Hybridisation:

(a) It is complicated and highly technical.

(b) It is time consuming and expensive method.

Hybrid vigour: It is the superiority of hybrid over either of the parents on one or more traits. The term hybrid vigour was coined by George Harrison Shull (1914). It was first reported in maize. It is also known as Heterosis. If it is obtained by crossing over it is called balanced heterosis. If it is obtained by mutation it is called mutational heterosis.

It is most profitably utilised in vegetatively propagated crops because they do not involve sexual reproduction hence there is no loss of hybrid superiority. Hybrids have greater yield, quicker growth, more viable and more resistant to disease, drought and pests.

Hybrids show greater yield due to larger size of root in carrot, radish and sugar beet and boll in cotton; greater number and size of tubers in potato; greater number and size in cucumber and brinjal, larger size of grains in cereals; larger size of leaf in cabbage; high sugar, high yield, thick steam of sugarcane and its ability to grow in North India.

Hybrid vigour is best maintained by (a) continuous breeding, (b) vegetative reproduction and (c) maintaining parental lines.

Hybrid vigour in animals is concerned with greater production of milk, greater amount of meat and greater number of eggs.

Inbreeding: It is mating between close relatives or self fertilisation by monoecious condition. It is quite common in plants.

It increases the homozygosity in recessive alleles that express their harmful effects. If self fertilization occurs then the proportion of heterozygotes will become halved at each generation because half of combination of gametes will form heterozygote from homozygous zygotes. The problem is that harmful recessive alleles present in low frequency will become homozygous in inbred offspring. Thus continuous inbreeding (self fertilisation) resulted in gradual loss of vigour, fertility and productivity and this phenomenon is called **Inbreeding depression**. It also results in appearance of defective traits. This is not a serious problem for those plants (Pea, wheat) which normally inbreed because selection eliminates harmful alleles. It is a serious problem for those plants which normally outbreed. The inbreeding depression is lost inF_2 generation in self pollinated plants and slowly declines in cross-pollinated plants.

Modern Plant Breeding: It involves genetical principles.

1. **Polyploid breeding**: It is done by method of polyploidy. It gives rise to new varieties of grains, vegetables and fruits which are better in quality, taste, yield, and nutrients. The polyploids are obtained by treatment of colchicine (aqueous solution=0.1-0.5%). Colchicine arrests and breaks spindle fibres so that a cell division occurs without cytokinesis leading to doubling of chromosome number. This is possible in three ways: Soaking by seeds, soaking by axillary buds or injecting at seedling stage in cortical cell.

   Varieties developed: Raphanobrassica, Triticale, Modern wheat (*Triticum aestivum*), Cotton (*Gossypium hirsutum*), Seedless water melon, Seedless banana.

2. **Mutational breeding:** It occurs by causing induced mutation due to physical treatment in plant parts. Seed is the best material to be irradiated but bud is the best part to be irradiated in case of vegetatively propagated plants. It involves following steps:

   (a) Inducing mutation on seed by several means under controlled condition.

   (b) Germination of propagule in Petriplate.

   (c) Transfer of propagule in the field.

   (d) Screening of plants for resistance.

   (e) Growing the selected plants in the field.

   Varieties developed: Most of the mutant varieties were developed in rice, followed by barley and wheat.

Wheat	Sharbati Sonora (amber grained variety) was developed by γ-ray treatment of Sonora-64 (red grained variety) by M.S. Swaminathan.
Wheat	Pusa Lerma was developed by γ-ray treatment of Lerma Rojo-64.
Wheat	NP 836
Rice	Reimei
Rice	Taichung Native 1(Taiwanese variety)
Rice	IR-8 (Phillippinese variety)-Miracle rice
Rice	Atomita was developed from Pelita-1(Indonesian variety),
Rice	Jagannath, Padmini, IIT 60
Maize	Protina, Rattan
Barley	Erectoides, Erectiform
Peppermint	Todd's Mitcham
Tomato	Pusa Lal Meeruti
Cotton	Mu 7, Indore-2

Disadvantages:

(a) Most of the mutations are lethal and invaluable

(b) The rate of mutation is very low

(c) Mostly mutations are recessive

(d) The mutants have the tendency to revert back to original type.

Green revolution: It is breakthrough in production of food grains due to introduction of new semi-dwarf varieties of Wheat and rice in India.

India is an agricultural country contributing 33% of GDP and providing employment to 62% of population.

It was launched in1960s which increased food production not only to meet the national requirement but also for export. It occurred due to introduction of high yielding varieties (HVYs), increased irrigation facilities, distribution of hybridised seeds, fertiliser application, weed, insect and pathogen control, multiple cropping and better agricultural management.

It was the need of hour and launched due to poor adaptability of traditional and conventional varieties, geometrical growth of human population, arithmetical growth of food grain, occurrence of famine in 1961 etc.

Father of Green revolution is Norman Borlaug who is credited with saving over a billion people from starvation. Norman Borlaug was invited to India by Indian Government to start a programme. Punjab was selected by the Indian Government to be the first state to try the new crops because of its reliable water supply.

Father of Indian Green revolution of is M.S. Swaminathan (Father of radiation genetics).

Benefits:

(a) Production of high yielding, quick yielding, semi-dwarf and disease resistant varieties in wheat, rice and maize as a result of mutational breeding.
(b) Expansion in area under cultivation of cereal crops
(c) Jump in productivity of cereals.
(d) Import of two superior genes such as Norin-10 (a dwarfing gene of wheat from Japan to develop Sonora-64 and Lerma Rojo) and Dee-Geo-Woo-Gen (a dwarfing gene of rice from Taiwan to develop IR-8).

Disadvantages:

(a) It needs more water and fertilisers
(b) It needs frequent weedicides and pesticides.
(c) It produces less fodder.

Plant breeding for Insect Resistance

Insect resistance in host plants is due to morphological, biochemical or physiological characters.

Solid stem in wheat is resistant to stem saw fly.

Smooth leaved and nectar free cotton are resistant to Boll worm.

Low content of Nitrogen & sugar and high content of aspartic acid in maize are resistant to Stem borer.

Methods for producing Insect resistant plants

(a) Conventional breeding
(b) Mutational breeding
(c) Genetic engineering

Insect resistant varieties developed:

Variety	Crop	Resistant to insects
Pusa Gaurav	Mustard	Aphids
Pusa Sem-2,-3	Flat bean	Aphids, Fruit borer, Jassids
Pusa Swani, Pusa A-4	Lady's finger	Fruit and shoot borer

Plant Breeding for Disease resistance

Disease resistance is the ability of plants to withstand, overcome and resist the attacks of pathogens such as bacteria, viruses, and fungi. The development of disease depends on three factors such as (a) host genotype (b) pathogen genotype and (c) environment.

List of common viral diseases of plants

Disease	Causal organism
Tobacco mosaic	Tobacco mosaic virus
Potato leaf curl	Potato leaf curl virus
Tomato leaf curl	Tobacco leaf curl virus
Little leaf of Brinjal	Brinjal little virus
Bunchy top of Banana	Banana bunchy top virus

List of common bacterial diseases of plants

Disease	Host	Causal organism
Canker	Citrus plant	*Xanthomonas citri*
Crown gall	Apple	*Agrobactertium tumefaciens*
Black rot	Cabbage	*Xanthomonas campestris*
Scab	Potato	*Streptomyces scabies*
Fire blight	Apple, Pear	*Erwinia amylovora*
Gummosis	Sugarcane	*Xanthomonas vasculorum*
Tundu	Wheat	*Corynebacterium tritici*
Angular leaf spot	Cotton	*Xanthomonas malvacearum*

List of common fungal diseases

Disease	Causal organism
Stem rust of wheat	*Puccinia graminis tritici*
Loose smut of wheat	*Ustilago tritici*
Red rot of sugarcane	*Colletotrichum falcatum*
Wilt of cotton	*Fusarium vasinfectum*
Brown spot of rice	*Helminthosporium oryzae*
Downy mildew of cereals	*Sclerospora graminicola*
Powdery mildew	*Erysiphe graminicola*

Early blight of potato	*Alternaria solani*
Late blight of potato	*Phytophthora infestans*

There are two types of disease resistance

(a) Horizontal resistance: It is a type of resistance in which host shows resistance against all pathogens.

(b) Vertical resistance: It is a type in which host shows resistance against a few pathogens.

Methods for breeding disease resistant plants:

(a) Conventional breeding

(b) Mutational breeding

(c) Genetic engineering

Some disease resistant varieties developed

Variety	Crop	Resistant to disease
Himgiri	Wheat	Leaf or stripe rust, Hill bunt (Fungal)
Pusa swarnim	Mustard	White rust (Fungal)
Pusa shubhra, Pusa snow ball K-1	Cauliflower	Black rot (Bacterial)
Pusa sadabahar	Chilly	Chilly mosaic, Leaf curl (Viral)
Pusa komal	Cow pea	Blight (Bacterial)

Plant Breeding for improved food quality (Biofortification)

It is method of breeding crops to increase their nutritional value to improve health in human beings. The biofortification focuses on making plant foods more nutritive during growth of plants but ordinary fortification focuses on adding nutrients to the foods when they are being processed. It is undertaken to improve the following objectives:

(a) Protein content and quality with respect to essential amino acids (consumed from diet) such as Isoleucine, Leucine, lysine, Methionine, Phenylalanine, Threonine, Tryptophan, Valine. Cereals are deficient in Lysine and Tryptophan and Pulses are deficient in Methionine, Cysteine (Sulphur containing amino acids).

(b) Oil content and quality with respect to unsaturated fatty acids.

(c) Vitamin content.

(d) Minerals and micronutrients content.

Methods: This can be achieved by

1. **Selective breeding**: It is the simplest, cheapest and quickest method. In this a particular variety of crop (naturally nutrient rich) is crossbred with another variety of crop (High yielding) to obtain high yielding and nutrient rich seed.
2. **Genetic engineering**: It is complicated and technical process. It is the production of transgenic plant such as Golden rice to supplement Vitamin A. It contains genes from a soil bacterium, *Erwinia uredovora* and Daffodil (*Narcissus pseudonarcissus*) which increase the level of β-carotene and finally it is converted into Vitamin A in the body.

List of Fortified foods:

Crops	Nutrients rich in
Carrot, Spinach, Pumpkin, Maize, Cassava, Sweet potato,	Vitamin-A
Maize (protina, shakti, rattan varieties)	Lysine, Tryptophan
Spinach, Bathua	Iron and Calcium
Wheat (Atlas-66), Broad bean, French Bean	Protein
Rice, Pearl millet	Iron
Rice, wheat, Bean	Zinc
Mustard, Tomato, Bitter gourd	Vitamin-C

Antinutritional Factors: These are biochemicals present in food items which produce harmful effect on the health of human beings.

Examples:

1. Lectin is a toxic protein produced from Red kidney bean or French bean (*Phaseolus vulgaricus*). It causes agglutination of RBCs.
2. Cyanoalanine is a neurotoxin produced from Sweet pea (*Lathyrus sativus*). It causes a type of paralysis called Favism (muscular weakness and tremor).
3. Erucic acid and Glucosinolates are chemicals produced from Rape seed (*Brassica napus*). These affect the functioning of thyroid gland.
4. Some other chemicals such as tannins, oxalates, betaglucans present in the leaves of vascular plants are harmful for animals.

List of Institutes engaged in Plant breeding

International Rice Research Institute (IRRI)	Philippines
International Centre for maize and wheat improvement	Mexico
International Crops Research Institute for the Semi-arid Tropics (ICRISAT)	Hyderabad
International Potato Centre	Peru
Central Potato Research Institute	Shimla
International Plant Genetic Resource Institute (IBPGR)	Delhi
Indian Agricultural Research Institute (IARI)	Delhi
Sugarcane breeding Institute	Coimbatore

Single Cell Protein

It is the protein derived from microbial cells for use as food or feed supplement. Thus it is edible unicellular microorganisms. It is the alternative source of mixed protein extracted from either pure or mixed culture of bacteria, cyanobacteria and fungi. It is rich in high quality protein (60-80%), vitamins and minerals but poor in fats.

The term 'Single Cell Protein' was coined by Wilson. Microbes produce SCP better than conventional crop because

(a) These show rapid rate of multiplication.

(b) These are non-pathogenic to plants, animals and humans.

(c) These have short generation time (20 minutes to 2 hours for bacteria; 1-3 hours for fungi; 2-6 hours for algae) hence these are easily produced.

(d) These are easily modified genetically.

(e) These utilise a large number of different carbon sources (raw materials)

(f) These are independent of seasonal and climatic variations.

(g) These have minimum land requirement.

Group	Microorganism	Substrate
Bacteria	*Methylophilus methylotrophus*	Methyl alcohol
	Cellulomonas	Cellulose
Cyanobacteria	*Spirulina maxima*	CO_2 and sunlight
Algae	*Chlorella pyrenoidisa, Scenedesmus acutus*	CO_2 and sunlight
Fungi (unicellular)	*Saccharomyces cerevisiae*	Molasses (Sugar)
	Candida utilis	Potato starch waste
Fungi(filamentous)	*Chaetomium cellulolyticum*	Cellulosic waste
	Aspergillus niger	Molasses (Sugar)
	Trichoderma viridae	Straw, Starch

Procedures to produce SCP:

(a) Cultivation of suitable strains of microbes

(b) Preparation of suitable medium with specific carbon source

(c) Prevention of contamination

(d) Separation of microbial biomass.

Advantages:

(a) It reduces pressure on agricultural production for the supply of protein

(b) It reduces environmental pollution because it is based on industrial effluents.

(c) It can be obtained in laboratory or in a small area of land throughout the year.

(d) It is not dependent on climate.

Disadvantages:

(a) Microbes (Bacteria and yeast) contain high nucleic acid content which can be hazardous to health.

(b) Microbes (bacteria) may cause contamination.

(c) Microbes may be toxic and cause allergic reactions.

(d) Microbes (algae) contain high chlorophyll content which is not suitable for human use.

Tissue Culture: It is a latest method of growing cells, tissues, or organs and their differentiation on artificial culture medium under aseptic condition and controlled physical and chemical conditions such as light, temperature, pH and nutrients.

It involves all types of plant cultures such as cell, callus, protoplast, anther, meristem, embryo, organ etc.

It is based on cellular totipotency which is the ability of a somatic cell to divide to produce complete organism. Plant cells are thus totipotent.

$$\text{Mature cell} \xrightarrow{\text{Dedifferentiation}} \text{Callus} \xrightarrow{\text{Redifferentiation}} \text{Whole plant}$$

Auxenic Culture: It is pure culture without any contamination.

Discoveries:

1. Gottlieb Haberlandt (1902) first developed the *in vitro* technique to demonstrate the cellular totipotency. He is regarded as father of Tissue culture.
2. Hanning (1904) isolated embryo of crucifer in mineral salts and sugar solution.
3. Gautheret (1939) cultured the cambium tissue of carrot on Knop solution supplemented with other chemicals.
4. White (1939) cultured the tumour tissue of tobacco.
5. Skoog (1944) reported shoot formation on *in vitro* culture of tobacco pith tissue.
6. Skoog and Miller (1957) proposed root-shoot differentiation *in vitro* regulated by auxin-cytokinin ratio.
7. Stewart (1958) demonstrated the regeneration ability on callus tissue culture of carrot.
8. Guha and Maheshwari (1967) first produced hybrid plant from pollen grain by culturing anther of *Datura*.

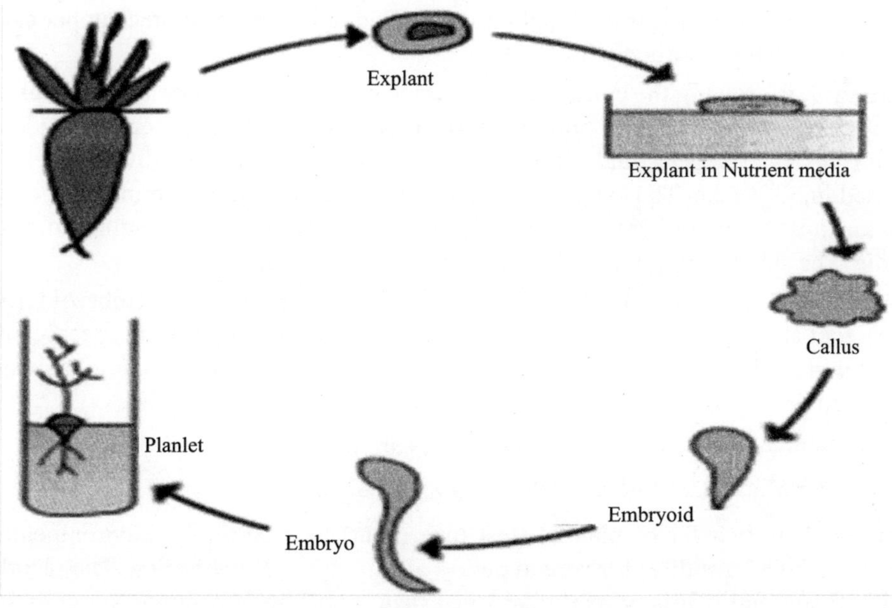

Fig. ***Different stages in Tissue culture***

Procedure:

1. **Preparation of Nutrient medium**: It is nutrient solution for the desired growth and development of explant. In tissue culture, MS medium is usually used that was developed by Murashige and Skoog. It contains
   (a) Carbon source: Sucrose is preferred because it is relatively stable to autoclaving and readily assimilated by plant cell in the form of glucose and fructose after hydrolysis.
   (b) Inorganic salts: Micronutrients and meganutrients
   (c) Organic supplements: Vitamins (Thiamine, Niacin, Pyridoxine), Amino acid (Arginine), Sugar alcohol (Myoinositol).

(d) Natural extracts: Coconut milk, Yeast extract, Malt extract, Hydrolysed casein etc.

(e) Growth regulators: Auxins (IAA, IBA, 2,4-D,NAA), Cytokinins (Benzyl amino purine, kinetin) are used. Synthetic auxin is commonly used because it is stable.

(f) Solidifying agent: Agar-agar is a polysaccharide which does not react with other media constituents, not digested by plant enzymes and stable at incubation temperature.

(g) pH = 5.6-5.8

2. **Selection of explant**: The explant is the excised (cut) meristematic plant part such as stem tip, root tip, axillary bud tip. These regions are free from viruses because vascular supply is absent so no chance to enter into this meristematic tissue and further rate of cell division in meristematic zone is much higher than multiplication of virus.
3. **Sterilisation of Medium and explants**: Medium and culture vessels are sterilised by autoclaving at121°C for 15 minutes. The explant is surface sterilised by disinfectants such as sodium hypochlorite (NaOCl) and the washing with distilled water to avoid microbial contamination. The process of sterilisation kills microbes completely.
4. **Inoculation:** The explant is transferred into slant medium in inoculation chamber called laminar air flow cabinet which is filled with UV-tube.
5. **Incubation**: It is growing the culture in plant tissue culture room under artificial light (16 hours), temperature (-26°C) and relative humidity (60%). The explant produces callus which is a mass of irregular (loosely arranged), undifferentiated and unorganised actively dividing cells. The callus is obtained in2-3 weeks. The callus is subjected to sub-culturing. It is transfer of cells/tissues in new culture vessel containing fresh media at regular interval. If it is present in same position then it will die. The subculturing will promote the formation of callus.
6. **Regeneration/Morphogenesis/Organogenesis**: The callus produces embryoid (non zygotic embryo), root and stem hence plantlet is formed. It is controlled by certain specific ratio of auxin and cytokinin. If the concentration of auxin and cytokinin is equal then completely disorganised callus is formed.

   $$\text{Callus} \xrightarrow{\text{High auxin (NAA) + low cytokinin (BAP)}} \text{Roots are formed}$$

   $$\text{Callus} \xrightarrow{\text{Low auxin (NAA) + high cytokinin (BAP)}} \text{Shoots are formed}$$
7. **Hardening**: It is transfer of plantlet (test tube plant) to a particular environmental condition (reduced and high humidity) for certain period of time for acclimatisation. The plantlet is placed in sterilised soil and if it is properly hardened then it will be transferred to pot or soil or green house or field condition.

Importance:

(a) It produces a large number of plantlets vegetatively (micropropagation) used in forestry and floriculture.

(b) It produces virus free plants (e.g., Apple, Banana, Potato etc.)

(c) It produces plants in a small space within short span of time throughout the year.

(d) It conserves rare and endangered plants.

(e) It produces genetically identical plants (somaclones).

(f) It produces somatic hybrids.

(g) It shows somaclonal variations present among plant cells. These variations have produced wheat tolerant to rust and high temperature; rice to leaf hopper and tungro virus; potato to late blight.

(h) It produces transgenic plants for valuable compounds such as recombinant proteins, antibiotics, essential oils, pigments, flavouring agents, sweeteners etc.

Protoplast Fusion/Somatic hybridisation: It is the production of somatic hybrid by fusion of somatic cell of two different species. It is also called parasexual hybridisation because hybrids are not produced through conventional method of hybridisation because of sexual incompatibility.

Carlson (1972) developed first somatic hybrid between two species of Tobacco (*Nicotiana glauca* and *Nicotiana langsdorfii*).

Procedure:

(a) Firstly mature leaf is collected from two different healthy plants.

(b) Both leaves are sterilised.

(c) The lower epidermis of sterilised leaves is peeled.

(d) Mesophyll cell (preferable) is isolated.

(e) Cell wall of mesophyll cell is removed by cellulase and pectinase (macerozyme) to form protoplast.

(f) The protoplasts of two plants are fused under pH10 and Ca^{+2} by chemicals such as Polyethylene glycol (commonly used) / Sodium nitrate (chemofusion) or with the supply of high voltage electric current (electrofusion). The electrofusion is simpler, quicker and more efficient.

(g) The fused protoplast having heterokaryon (two unfused nuclei) is grown (plated) on culture medium forming somatic hybrid cell having synkaryon (single fused nuclei) which develops own wall.

(h) Somatic hybrid cell is selected through visual marker such as pigmentation of protoplast.

(i) Somatic hybrid cell is now cultured on a suitable medium forming hybrid callus.

(j) The hybrid callus undergoes differentiation or organogenesis or regeneration forming somatic hybrid plant.

(k) Finally somatic hybrid is characterised and evaluated.

Types of Somatic Hybrid:

1. Symmetric hybrids: These retain full somatic chromosome complements of both parental types. For example,
   (a) Interspecific somatic hybrid: Radish x Cabbage = Raphanobrassica
   (b) Intergeneric somatic hybrid: Potato x Tomato = Pomato
   (c) Intertribal somatic hybrid: *Arabidopsis* x Mustard = Arabidobrassica
2. Asymmetric hybrid: These contain full somatic chromosome of one parent. e.g., Interfamilial somatic Hybrid: Soya bean x Tobacco

Importance:

(a) It produces new varieties of crop plants.

(b) It produces hybrid in asexually reproducing plants.

(c) It is easy to manipulate genetically for introducing new genes such as cytoplasmic male sterility, herbicide resistance etc.

SHORT ANSWER TYPE QUESTIONS

1. What are the objectives of plant breeding?
2. Mention the introduced varieties of plants in India.
3. How mass selection differs from pure line selection?
4. Write notes on heterosis.
5. How emasculation is done and what is its importance in hybridisation?
6. Write notes on inbreeding depression.
7. Write notes on green revolution.
8. Mention the disease resistant varieties of plants.
9. Mention the insect resistant varieties of plants.
10. Distinguish between mutational breeding and polyploidy breeding.
11. Mention some of the varieties developed as a result of mutation.
12. Write notes on biofortification.
13. Explain morphogenesis in tissue culture.
14. Write notes on single cell protein.
15. Discuss the importance of biofertilisers.

LONG ANSWER TYPE QUESTIONS

1. Describe the different steps in the process of hybridisation and mention its significance.
2. Describe the different steps in the process of tissue culture and mention its significance.
3. Discuss the procedure of protoplast fusion.

MULTIPLE CHOICE QUESTIONS

1. Which of the following plants is gift of new world to old world?
   (a) Maize (b) Wheat (c) Rice (d) Sugarcane
2. Sonalika and Kalyan sona are the varieties of
   (a) Maize (b) Wheat (c) Rice (d) Millet
3. Which method of crop improvement can be practised by a farmer if he is inexperienced?
   (a) Mass selection (b) Pure line selection (c) Clonal selection (d) Hybridisation
4. Colchicine is an alkaloid involved in production of
   (a) Haploids (b) Polyploids (c) Hybrids (d) Somatic hybrids
5. Triploids are said to be more useful than other polyploids because these are
   (a) Fertile and healthier
   (b) Fruitless and are always propagated by stem cuttings
   (c) Produced when diploid nucleus fuses with male gamete
   (d) Sterile and show increased vigour
6. Sharbati Sonora variety of wheat was obtained by
   (a) Crossing with varieties of wheat
   (b) Hybridisation between wild varieties of wheat
   (c) Irradiation of Sonora-64 with gamma rays
   (d) X-ray treatment
7. The most convenient way to maintain plant germplasm is
   (a) Tissue culture (b) By developing new varieties
   (c) By storing seeds (d) By treating plants at low temperature
8. Dwarf wheat varieties brought to India from Mexico were
   (a) Sonora-64 and Sonalika (b) Sharbati sonora and Lerma rojo
   (c) Sharbati sonora and Sonalika (d) Sonora-64 and Lerma rojo
9. Hybrid vigour can be best maintained for long periods by
   (a) Inbreeding (b) Cross pollination (c) Clonal selection (d) Mutation
10. Which of the following is pseudocereal?
    (a) *Oryza sativa* (b) *Fagopyrum esculentum*
    (c) *Triticum turgidum* (d) *Triticum spelta*
11. Genetic diversity in agricultural crops is threatened by
    (a) Introduction of high yielding varieties (b) Extensive intercropping
    (c) Intensive use of fertilisers (d) Intensive use of biopesticides
12. Natural resistance to late blight of potato is present in
    (a) *Solanum tuberosum* (b) *Solanum demissum*
    (c) *Solanum stoloniferum* (d) *Solanum acaule*
13. Which of the following amino acids lack in cereals and millets?
    (a) Lysine and glycine (b) Methionine and tryptophan
    (c) Tryptophan and glycine (d) Lysine and tryptophan
14. Which of the following varieties of maize is lysine rich?
    (a) Protina (b) Rattan (c) Shakti (d) All of the above
15. Dee-geo-woo-gen is dwarfing gene of
    (a) Rice (b) Wheat (c) Maize (d) Barley

16. Which of the following amino acids lacks in pulses?
(a) Cysteine and methionine (b) Cysteine and lysine
(c) Glycine and tryptophan (d) Methionine and tryptophan
17. Evaluation of newly evolved varieties is carried out by
(a) IARI (b) ICAR (c) CSIR (d) NBPGR
18. Which of the following bacteria is good source of SCP?
(a) *Bacillus megatherium* (b) *Methylophilus methylotrophus*
(c) *Streptomyces griseus* (d) *Saccharomyces cerevisiae*
19. Protoplast fusion occurs with the help of
(a) PEG and NaCl (b) PEG and KCl (c) PEG and $NaNO_3$ (d) PEG and $MgSO_4$
20. Pusa sem-2 is an insect resistant variety of
(a) Mustard (b) Cow pea (c) Wheat (d) Flat bean
21. In tissue culture, whole plant is developed from mature cell which undergoes
(a) Differentiation and dedifferentiation (b) Differentiation and redifferentiation
(c) Dedifferentiation and redifferentiation (c) All of the above
22. Callus will form shoot if treated with
(a) Low auxin and high gibberellin (b) Low auxin and high cytokinin
(c) High auxin and low cytokinin (d) Low gibberellin and high cytokinin
23. In hybridisation, emasculation is done to
(a) Prevent cross pollination (b) Prevent self-pollination
(c) Prevent fertilisation (d) All of the above
24. In crop improvement, haploids are useful because
(a) They grow better even under adverse condition
(b) They are useful in studies of meiosis
(c) They have less requirement for energy inputs
(d) They give homozygous lines following diploidisation
25. Embryoids arise from
(a) Zygote in culture (b) Anther in culture
(c) Synergids in culture (d) Somatic cells in culture
26. Embryo rescue technique has been used for hybridisation between
(a) Wheat and rye (b) Common bean and wild bean
(c) Potato and tomato (d) None of the above
27. Somaclonal variations are produced in
(a) Protoplast fusion (b) Tissue culture
(c) Somatic hybridisation (d) Sexual reproduction
28. Haploid plants can be produced by
(a) Pollen culture (b) Embryo culture
(c) Meristem culture (d) Endosperm culture
29. Genetic erosion is caused by
(a) Deforestation
(b) Shifting cultivation
(c) Adoption of genetically uniform varieties of crops
(d) All of the above

30. Which of the following is most important source of food?
(a) Algae (b) Fungi (c) Lichen (d) Angiosperms
31. Desired improved varieties of economically useful crops are raised by
(a) Natural selection (b) Hybridisation (c) Mutation (d) Biofertilisers
32. Emasculation in maize plant is done
(a) Before anthesis (b) After anthesis (c) Before pollination (d) After pollination
33. A cross in which parents are distantly related is called
(a) Intravarietal cross (b) Intervarietal cross
(c) Interspecific cross (d) Intergeneric cross
34. Minor variations appear in the progeny of a pureline due to
(a) Inbreeding (b) Cross pollination (c) Mutation (d) Environment
35. Newly introduced plants from other place will be tested in
(a) Greenhouse (b) Quarantine (c) Autoclave (d) Incubator
36. Dwarf varieties of wheat were developed by
(a) Swaminathan (b) Vavilov (c) Borlaugh (d) Mendel
37. Genetic variability can be created by
(a) Mass selection (b) Inbreeding (c) Hybridisation (d) Mutation
38. The dividing and undifferentiated cells are known as
(a) Primordium (b) Promeristem (c) Callus (d) Embryo
39. Cellular totipotency is observed in
(a) Only germ cells (b) All plant cells
(c) All eukaryotic cells (d) Only bacterial cells
40. Micropropagation is commercially used for growing
(a) Mango (b) Cotton (c) Rice (d) Banana
41. A cybrid is a cytoplasmic hybrid carrying
(a) Genomes and cytoplasm of two different plants
(b) Cytoplasms of several plants but genomes of two plants
(c) Cytoplasms of two different plants
(d) Cytoplasms of two different plants but genome of one plant
42. The technique that was employed to produce haploids of *Datura* was
(a) Embryo culture (b) Meristem culture (c) Anther culture (d) Protoplast culture
43. The loss of hybrid vigour in successive generation is due to
(a) Inbreeding (b) Induced mutation
(c) Cross breeding (d) Self-incompatibility
44. Heterotic effect in a hybrid is due to
(a) Over dominance at many loci (b) Nonallelic interaction
(c) Interallelic interaction (d) Additive gene effects
45. Virus free clonal plants can be produced by
(a) Cuttings (b) Anther culture
(c) Shoot tip culture (d) Micropropagation
46. Post zygotic sexual incompatibility can be overcome by
(a) Ovule culture (b) Protoplast culture (c) *In vitro* pollination (d) Bud pollination

47. Plant breeders are interested in apomixes because
    (a) It enlarges genetic variability (b) It improves seed set
    (c) It yields clonal progeny (d) It shortens the life cycle
48. In view of suicides committed by the poor farmers towards raising the commercial F_1 hybrid seeds, a plant breeder wishes to supply them the hybrid seeds at a cheaper cost. Which one of the following types of genetic stocks can be done?
    (a) Inbred lines (b) Back cross progeny (c) Male sterile lines (d) Pedigree clones
49. Which of the following is incorrectly matched
    (a) IR-8 variety –Rice (b) Protina variety-Maize
    (c) Pusa lerma variety- Wheat (d) Pusa gaurav- Millet
50. The inbreeding depression results in
    (a) Loss of vigour (b) Appearance of defective traits
    (c) Loss of productivity (d) All of the above
51. International Rice Research Institute is situated in
    (a) Indonesia (b) Malaysia (c) Philippines (d) Mexico
52. Favism occurs due to consumptionof sweet pea which contains
    (a) Cyanoalanine (b) Lectin (c) Erucic acid (d) Glucosinolates
53. Golden rice will supplement
    (a) Vitamin-A (b) Vitamin-B (c) Vitamin-C (d) Vitamin-D
54. Norin-10 is a dwarfing gene of
    (a) Rice (b) Wheat (c) Maize (d) Mustard
55. Which of the following is not a variety of rice?
    (a) Reimei (b) IR-8 (c) Atomita (d) Sharbati Sonora
56. Which of the following is not produced from polyploid breeding?
    (a) Raphanobrassica (b) Triticale (c) Modern wheat (d) Modern maize
57. Which of the following is not required in the process of hybridisation?
    (a) Emasculation (b) Bagging (c) Tagging (d) Inoculation
58. Maize variety shows resistance towards maize stem borer due to
    (a) High aspartic acid (b) Low nitrogen content
    (c) Low sugar content (d) All of the above
59. Which of the following is not used for crop improvement?
    (a) Selection (b) Introduction (c) Inbreeding (d) Mutation
60. Which of the following is generally used for introducing mutagenesis in crop plants?
    (a) X-rays (b) Gamma rays (c) UV rays (d) IR rays

ANSWERS

1.a	2.b	3.a	4.b	5.d	6.c	7.d	8.d	9.c	10.b
11.a	12.b	13.d	14.d	15.a	16.a	17.b	18.b	19.c	20.d
21.b	22.a	23.b	24.d	25.a	26.a	27.b	28.a	29.d	30.d
31.b	32.a	33.d	34.d	35.a	36.b	37.c	38.c	39.b	40.d
41.d	42.c	43.a	44.a	45.c	46.c	47.c	48.a	49.d	50.d
51.c	52.a	53.a	54.b	55.d	56.d	57.d	58.d	59.c	60.b

20

Microbes in Human Welfare

Microbes are microscopic organisms which are too small to be seen with the unaided eye. Microbes include bacteria, viruses and subviral agents (viroids and prions), some fungi, algae and protists. Louis Pasteur was the first person who demonstrated the role of microbes in food. He reported that buttermilk and Beer are the products of fermentation carried out by yeast. Microbes play an important role in the processing of many foods, improving their nature, digestibility, nutritive value, flavour and appearance of raw material and the modified product is called a fermented food.

Roles of microbes in Household products: Microbes are helpful in the preparation of household products such as

1. **Dairy products**: Milk is an aqueous solution (pH=7.0) of proteins (casein and lactalbumin), fats and carbohydrates (lactose) with numerous vitamins and minerals. *Lactobacillus* form of bacteria grows in milk and converts lactose sugar of milk into lactic acid. It causes coagulation and partial digestion of casein into calcium paracaseinate.

   (a) **Curd**: It is prepared by inoculating milk with *Lactobacillus acidophilus* at 30-40^{0}C. Curd is more nutritious than milkas it contains a number of organic acids and vitamins including vitamin B_{12}.

   (b) **Yoghurt**: It is prepared by inoculating milk with *Lactobacillus bulgaricus* and *Streptococcus thermophilus* at 45^{0}C.

   (c) **Buttermilk**: It is prepared by inoculating skimmed milk with combination of *Lactobacillus acidophilus*, *Streptococcus cremoris* and *Streptococcus lactis* at 22^{0}C for 18 hours. Buttermilk (milk remains after cream) is acidulated liquid after churning of butter from curd.

   (d) **Sour cream**: It is prepared by inoculating churned milk with *Streptococcus lactis* for producing lactic acid.

   (e) **Cheese**: It is the oldest food item. It is prepared by separating curd from liquid part or whey by *Propionibacterium shermanii*. The large holes in cheese occur due to production of large amount of CO_2 released by bacteria. Cheese may be soft (>50% water), semisolid (40-50%) and hard (<40% water).

   Roquefort cheese is ripened by growing *Penicillium roqueforti* for flavour.

   Camembert cheese is ripened by growing *Penicillium camemberti* for flavour.

2. **Bakery products:**

   (a) **Bread**: Bread is a staple food which is prepared by baking, steaming or frying dough. It mainly consists of wheat flour, water, salt and a leavening agent, Baker's yeast (*Saccharomyces cerevisiae*). Firstly selected strains of yeast are grown on molasses and these are harvested after sufficient growth. These are converted into powder or cakes. A small quantity of yeast is added to wheat flour and is kneaded. The kneaded flour is kept at warm temperature for a few hours. Finally it swells up called leavening which is caused by secretion of three enzymes by yeast.

Firstly amylase breaks down starch into maltose; maltase converts maltose into glucose and zymase brings about fermentation of glucose forming ethyl alcohol and CO_2. These two compounds cause swelling or leavening of the dough. The leavened dough is baked and finally ethyl alcohol and CO_2 evaporate making the bread porous and spongy.

Roles of Microbes in Industrial products: Microbes are suitable for industrial fermentation processes because

(a) They grow very rapidly after inoculation.
(b) They grow very easily on simple nutrient requirements.
(c) Their growth conditions such as temperature, pH can be controlled very precisely.
(d) They require temperature for the processes lower than conventional processes.
(e) They can be easily manipulated through genetic engineering.
(f) They produce wide variety of chemicals (products).

Procedure: It involves three stages which are collectively called scaling up.

1. **Laboratory scale:** It involves several steps such as search of large number of strains soon after discovery of microorganisms; selection of most suitable strain; multiplication of suitable strain and establishment of glass fermentor and other parameters such as nutrients, pH, temperature, aeration, by products, time of optimum production, separation and purification of products.
2. **Pilot plant scale:** It involves several steps such as testing of laboratory scale processes; evaluation of quality of products; replacement of glass fermentor by metallic container and optimisation of pH, temperature, aeration etc.
3. **Manufacturing Unit**: Microbes are added in bioreactor in several ways such as support growth system or surface of nutrient medium; suspended growth system or surface of nutrient medium or immobilised or column growth system where required microorganisms are placed in calcium alginate beads kept in calcium alginate.

1. Brewery: It is the production of alcohol from sugar with the help of microbial enzymes as a result of fermentation. It is anaerobic process carried out by yeast. It was demonstrated for the first time by Louis Pasteur. A variety of alcoholic beverages are made by using the yeast-*Saccharomyces* acting on different substrates. There are several species of yeast which are involved in brewery.

Baker's yeast or Brewer's yeast	*Saccharomyces cerevisiae*
Wine yeast	*Saccharomyces ellipsoideus*
Sake yeast	*Saccharomyces sake*
Food yeast	*Torulopsis utilis*

Process:

(a) Selection of strain: The microbes used by fermentation industries are mutant strains of fungi or bacteria that selectively synthesise large amounts of various metabolic intermediates or metabolites.
(b) Preparation of medium: The alcoholic beverages require specific substrate which provides flavour, colour, aroma etc. These differ with respect to the production process but microbe is the same in all cases.

Fermentation of barley malt	Beer and Whisky
Fermentation of grape juice	Wine, Champagne and Brandy
Fermentation of sugarcane molasses	Rum
Fermentation of rice starch	Sake
Fermentation of rye malt	Gin
Fermentation of potato starch	Vodka

(c) Sterilisation of medium and bioreactor (fermentor).

(d) Inoculation of selected strains of yeast in bioreactor: Fermentation is carried out in three ways:

(i) Batch process: It is discontinuous process in which entire culture medium with organism is removed from bioreactor after every batch.

(ii) Continuous process: A fraction of medium with fermented product is drawn out at regular intervals and fresh sterilised medium and inoculums are fed into bioreactor continuously.

(iii) Fed batch process: A small amount of medium is regularly fed in bioreactor to optimise the working of fermenting microorganisms without inhibition.

(e) Filtration: The culture fluid is filtered when growth of microorganisms is over. Later yeast cells (biomass) are separated from the supernatant.

(f) Recovery of products: The products are recovered from supernatant. It depends on culture medium, size of fermentor, culture condition, potential microorganism and design for separation of solid and liquid. There are two types of beverages which are as follows:

(i) Fermented beverages such as beer (5% alcohol), wine (10% alcohol) and sake (15% alcohol) are produced without distillation.

(ii) Distilled beverages such as gin (40% alcohol), rum (40% alcohol), vodka (40% alcohol), whisky (50% alcohol) and brandy (70% alcohol) are produced by distillation of fermented medium. The process of distillation is required for hard or strong liquors containing high concentration of alcohol. Actually when the fermenting mixture reaches an alcoholic concentration between 14% and 18% then yeast cells die. As a result, the fermenting mixture must be distilled heating it above boiling temperature of ethyl alcohol (78%) and below boiling temperature of water (100%).

2. Production of enzymes: Microbial enzymes may be intracellular (produced inside the cell and start functioning in the same cell) and extracellular (produced inside the cell, later secreted out the cell and start functioning outside the cell). Microbial enzymes have advantages over the plant and animal enzymes because these are produced on large scale within limited space and time and further these are easily purified. Microbes produce wide variety of enzymes.

Enzymes	Sources	Application	Industry
Amylase	*Bacillus subtilis*	Starch coating	Paper
	B.amyloliquefaciens	Removal of stains	Detergent
	Aspergillus niger	Digestive aid	Pharma
	Aspergillus oryzae, Rhizopus oryzae	Softening of bread	Baking
Protease	*Bacillus subtilis*	Removal of stain	Detergent
	B. licheniformis	Tenderisation of meat	Meat
	Aspergillus oryzae	Bread	Baking

Lipase	*Aspergillus niger*	Removal of stain	Detergent
	Candida lipolytica	Hydrolysis of butter fat	Dairy
Cellulase	*Streptomyces* spp.	Clarification of wine	Brewery
	Aspergillus niger	Clarification of wine	Brewery
	Trichoderma reesei	Softening of fruit or hydrolysis of fruit juices	Fruit processing
Pectinase	*Bacillus subtilis*	Pectin hydrolysis in fruit juices	Fruit processing
	Aspergillus niger	Clarification of wine	Brewery
	Rhizopus oryzae	Retting of fibres	Textile
Lactase	*Saccharomyces fragilis*	Coagulation of milk protein	Dairy
	Torula cremoris	Hydrolysis of lactose in milk and whey	Dairy
Invertase (Sucrase)	*Saccharomyces cerevisiae*	Hydrolysis of sucrose	Confectionary
Glucanase	*Bacillus subtilis* *Aspergillus niger*	Hydrolysis of fruit juices	Fruit processing
Glucose isomerise	*Bacillus coagulans*	Conversion of corn starch into high fructose corn syrup	Soft drink
Glucose oxidase	*Aspergillus niger*	Removal of glucose or oxygen	Food
Streptokinase (Tissue Plasminogen Activator)	*Streptococcus*	Dissolution of blood clots. Plasminogen→plasmin	Medicine
Cyclosporin-A	*Trichoderma polysporum*	Inhibition of T-cells and prevents rejection reaction in organ transplantation as immunosuppressant	Medicine
Statin	*Monoascus purpureus*	Lowering of blood cholesterol	Medicine

3. Production of organic acids:

(a) Acetic acid: It is the earliest microbial product used by human being. It is produced in two steps through anaerobic fermentation of sugarcane juice or molasses by yeast and *Acetobacter aceti*. It is used in vinegar, sauce, pickle, pharma industry and also used as preservative of meat and vegetables.

(b) Lactic Acid: It is first acid to be produced through fermentation of corn starch by *Lactobacillus delbrueckii*. It is used in confectionary, soft drinks, fruit juices, leather and textile industry. It is also used as preservative of fruits and vegetables.

(c) Citric acid: It is produced through fermentation of beet molasse (sucrose), glucose by *Aspergillus niger*. It is used in dye, pharma industry and also used as flavouring agent or preservative in jam, jelly, and confectionary.

(d) Propionic acid: It is produced through fermentation of sugar by *propionibacterium*.

(e) **Butyric acid**: It is produced through fermentation of sugar by *Clostridium acetobutyricum*. It is used in preparation of solvents.

(f) **Fumaric acid**: It is produced through fermentation of sugar by *Rhizopus nigricans*, *Rhizopus stolonifer*. It is used as wetting agent and in manufacture of resins.

(g) **Gluconic acid**: It is produced through fermentation of glucose by *Aspergillus niger, Penicillium purpurogenum*. It is used in pharma industry as it forms calcium gluconate and ferrous gluconate to provide calcium and iron in human body.

(h) **Oxalic acid**: It is produced through fermentation of sugar by *Aspergillus niger*. It is used as organic solvents.

(i) **Gallic acid**: It is produced through fermentation of sugar by *Aspergillus niger*. It is used in manufacturing ink and dye.

(j) **Itaconic acid**: It is produced through fermentation of sugar by *Aspergillus terreus*. It is used in manufacturing plastics.

(k) **Kojic acid**: It is produced through fermentation of sugar by *Aspergillus oryzae*. It is used in manufacturing insecticides and fungicides.

4. Production of vitamins: Vitamins are small organic compounds provided to body in small amount. These usually make up all or part of the enzyme cofactors. These help in normal maintenance and optimal growth of an organism.

Vitamin	Discoverer
A (Retinol)	McCollum
B_1 (Thiamine)	Casimir Funk
C (Ascorbic acid)	Szent Gyorgy
D (Calciferol)	Mellanby

(a) **Vitamin C** (Ascorbic acid): It was the first vitamin to be produced from fermentation of sugar by *Acetobacte*r, *Corynebacterium* etc.

(b) **Vitamin B_2** (Riboflavin): It is produced through fermentation of maize steep liquor by *Ashbya gossypii, Clostridium acetobutyricum, Mycobacterium smegmatis* etc.

(c) **Vitamin B_6** (Pyridoxine): It is produced from *E.coli*.

(d) **Vitamin B_{12}** (Cyanocobalamin): It is produced through fermentation of sugar by *Propionibacterium shermanii, Pseudomonas denitrificans* and *Bacillus megatherium*.

(e) **Vitamin K**: It is produced from fermentation by *E. coli*.

5. Production of amino acids:

(a) **Lysine**: It is produced through fermentation of acetate by *Corynebacterium glutamicum* and *E.coli*.

(b) **Glutamic acid**: It is produced through fermentation of glucose by mutant strains of *Corynebacterium glutamicum* and *Brevibacterium*

6. Production of Dextran: It is polysaccharide produced from *Leuconostoc mesenteroides*. It is plasma expander used in blood transfusion.

7. Production of steroids: These are synthesised by microbial bioconversion or biotransformation.

8. Production of Antibiotics: These are chemical substances produced by certain microorganisms at very low concentration that inhibit or kill other microorganisms without affecting host. The term 'antibiotic' was coined by Paul Vuillemin. Babes (1855) demonstrated the phenomenon of antibiosis.

Alexander Fleming (1928) discovered first antibiotic, Penicillin (wonder drug) from *Penicillium notatum* which inhibits the growth of *Staphylococcus aureus*. This antibiotic was widely used to treat wounded American soldiers in World War II. It is first natural compound showing antimicrobial property. More than 7000 antibiotics have been isolated so far, only 100 or so are being used to treat human, animal and plant diseases. The antibiotic was however, commercially extracted by Chain and Florey.

Sources:

(a) Bacteria: *Streptomyces* (mostly) *Bacillus, Pseudomonas, Noccardia, Micromonospora, Actinoplanes* etc.

(b) Fungi: *Penicillium, Cephalosporium* etc.

Features of a good antibiotic:

(a) It is nontoxic to host cells (humans and animals).

(b) It does not affect normal microflora of alimentary canal of host.

(c) It does not interfere with the normal host defence mechanisms.

(d) It shows quick action.

(e) It does not cause allergic to host.

(f) It remains stable and active in body fluids.

Procedures for Commercial production of penicillin:

(a) Selection of high yielding microbial strain.

(b) Preparation of inoculum in the form of suspension.

(c) Preparation of the suitable medium (broth).

(d) Inoculation of the medium in the fermentor.

(e) Incubation with forced aeration during fermentation.

(f) Separation of fungal biomass.

(g) Extraction, purification and recovery of the penicillin

Chemical nature:

(a) Aminoglycosides: Streptomycin, Neomycin, Kanamycin, Gentamicin

(b) Polypeptides: Polymixin-B, Bacitracin.

(c) Beta-lactams : Penicillin, Ampicillin, Amoxycillin, Cephalosporin.

(d) Tetracyclines: Aureomycin (Chlorotetracycline), Terramycin (Oxytetracycline).

(e) Macrolides: Erythromycin, Azithromycin, Carbomycin, Novobiocin.

Range of effects:

(a) Narrow spectrum: Bacitracin, Polymixin, Penicillin

(b) Broad spectrum: Streptomycin, Cyclohexamide, Tetracycline, Terramycin, Gentamicin, Rifampicin

Mode of action: Antibiotics function as bactericides (killing bacteria) or bacteriostatic (inhibiting growth of bacteria). These perform their action by following ways:

(a) Disruption of cell wall synthesis: Penicillin, Bacitracin, Cephalosporin

(b) Destruction of plasma membrane synthesis: Amphotericin, Polymixin, Nystatin etc.

(c) Inhibition of 30S ribosome function in protein synthesis: Streptomycin, Tetracycline, Kanamycin etc.

(d) Inhibition of 50S ribosome function in protein synthesis: Erythromycin, Chloramphenicol.

(e) Inhibition of 30 S ribosome function in protein synthesis: Streptomycin

(f) Inhibition of chain initiation in protein synthesis: Tertracycline

(g) Inhibition of DNA synthesis: Rifampicin

Reasons for resistance to antibiotics: Bacteria show resistance to antibiotics by undergoing the following processes:

(a) Bacteria modify the structure of their cell wall to reduce the ability of antibiotic to penetrate.

(b) Bacteria modify the antibiotics enzymatically. e.g., Penicillinase or β- lactamase deactivates Penicillin, Ampicillin, Cephalosporin etc.

(c) Bacteria methylate their ribosomes obscuring the target of antibiotics.

(d) Bacteria undergo mutation to alter the target of antibiotics.

List of antibiotics and their sources

Antibiotics	Source
Subtilin	*Bacillus subtilis*
Bacitracin	*Bacillus licheniformis*
Polymixin-B	*Bacillus polymyxa*
Polymixin-E	*Bacillus colistinus*
Tyrothricin	*Bacillus brevis*
Nisin	*Streptococcus cremoris*
Streptomycin	*Streptomyces griseus*
Grisin	*-do*
Cyclohexamide	*-do*
Aureomycin (Chlorotetracycline)	*Streptomyces aureofaciens*
Tetracycline	*-do*
Rimocidin	*Streptomyces rimosus*
Terramycin (Oxytetracycline)	-do
Chloramphenicol (Chloromycetin)	*Streptomyces venezuelae*
Erythromycin	*Steptomyces erythreus*
Kanamycin	*Streptomyces kanamyceticus*
Neomycin	*Streptomyces fradiae*
Streptothricin	*Streptomyces lavendulae*
Carbomycin	*Streptomyces halstedii*

Cycloserine	*Streptomyces orchidaceous*
Novobiocin	*Streptomyces niveus*
Nystatin	*Streptomyces noursae*
Vancomycin	*Streptomyces orientalis*
Viomycin	*Streptomyces puniceus*
Oleandomycin	*Streptomyces antibioticus*
Lincomycin	*Streptomyces lincolensis*
Gentamicin	*Micromonospora purpurea*
Rifampicin (Rifamycin)	*Noccardia mediterranei*
Thermorubin	*Thermoactinomyces antibioticus*

Role of microbes in sewage treatment:

Sewage is waste water of domestic, commercial and industrial units. It is rich in organic wastes and microbes many of which are pathogens. Industrial sewage is called effluent and it carries huge amount of organic residues. The effluents of sugar industries (factories), distilleries, paper and cloth factories are heavily loaded with organic wastes and thus require proper treatment before draining them to any freshwater bodies. When these wastes are drained in untreated state to river they are unable to recycle them and then water becomes heavily polluted.

Sewage treatment is the process of removing contaminants from waste water or sewage through physical, biological and chemical methods to produce environmentally safe water.

Sewage treatment is also known as waste water treatment.

Pollution level of water is measured by level of Biochemical oxygen Demand (BOD). It is the amount of oxygen consumed by various microbes in aerobic decomposition of organic wastes present in one litre of sewage water. If organic wastes are more in water, then the BOD of water will be increased hence the Dissolved oxygen (DO) will be decreased due to more turbidity of water.

The sewage water needs to be treated to decrease the BOD and increase the DO before draining in water bodies.

Treatment of water occurs through the following **three** stages once the level of pollutants in waste water is determined. These stages are as follows:

1. **Primary treatment (Physical or mechanical method):** It removes 50% of suspended solids. It involves the following steps:
   (a) The large, suspended and floating solid matters are physically removed by sequential filtration.
   (b) The filtrate is now kept in large, open and circular settling tank where grit containing soil, sand, silt and pebbles settle down.
   (c) Grit is now removed by sedimentation.
   (d) All sediments or solids settle down to form primary sludge and supernatant forms primary effluent which is industrial sewage containing huge amount of organic residues.
   (e) Effluent (settled sewage) is now moved to secondary treatment.
2. **Secondary treatment (Biological method):** It involves microbial degradation of organic matter. It removes 85% of impurities. It employs following three methods: Activated sludge system, Oxidation tank and Trickling filter system.

Activated sludge system: It involves the following steps

(a) The primary effluent is taken to large and rectangular aeration tank.

(b) Liquid is regularly agitated mechanically.

(c) Air is pumped into water (primary effluent) to encourage the growth of large number of aerobic and heterotrophic bacteria (*Bacillus, Arthrobacter, Beijerinckia, Mycobacterium, Pseudomonas, Enterobacter, Corynebacterium* and Actinomycetes), protozoans and yeast.

(d) These saprotrophs form flocs which are bacterial colonies entangled with sewage fungal mycelia floating over surface of effluent.

(e) These organisms decompose (digest) a lot of organic matter in effluent, convert it into microbial biomass and release a lot of minerals. This action reduces BOD.

(f) Effluent is now passed into settling tank.

(g) The flocs (bacterial colonies) are allowed to sedimentation and sediment of settling tank is called activation sludge. The effluent or supernatant is generally passed into river or it is treated chemically to purify further.

(h) A large amount of activation sludge is passed into a large tank called anaerobic sludge digester and remaining small part is used as inoculum in aeration tank.

(i) The aerobic microbes get killed in it.

(j) The anaerobic heterotrophic bacteria (Methanogens) digest the organic mass as well as aerobic microbes of activated sludge to produce biogas (CH_4, CO_2, H_2S etc).

(k) Now effluents from secondary treatment are generally released into natural water bodies such as river and streams. If the waste water is to be used for domestic purpose then it will be subjected to tertiary treatment.

Primary sludge	Secondary sludge
It is produced from primary treatment.	It is produced from secondary treatment.
It does not form flocs.	It forms flocs.
It does not require aeration.	It requires aeration.
It requires little decomposition.	It requires large decomposition.

Tertiary treatment (Chemical method): It removes 99% of all impurities from sewage. It involves removal of inorganic nutrients and killing of pathogenic microbes by the following processes:

(a) UV filtration: It kills pathogenic microbes (viruses and bacteria)

(b) Precipitation: It removes phosphates

(c) Chlorination: It removes NH_3 and kills pathogenic microbes.

(d) Carbon filter: It removes residual toxins.

(e) Reverse Osmosis: It removes Total dissolved solids (TDS) in the form of salts.

Role of Microbes in Biogas Production

Biogas is a fuel gas produced by anaerobic breakdown of biomass with the help of methanogens. It is also called as gobar gas, sewage gas, landfill gas, biofuel. It is colourless and odourless gas. It contains average 60% CH_4, 40% CO_2, 4% N_2, 1% H_2S and traces of H_2O, H_2, and O_2.The energy released from biogas depends on the proportion of methane present in it. It is renewable and alternative source of energy. China is the largest producer of biogas. The biogas production was first started in India by the joint efforts of Indian Agricultural Research Institute (IARI) and Khadi and Village Industries Commission (KVIC).

It is mainly produced from agriculture waste, plant materials, manure, municipal waste, animal wastes esp. dung because these contain cellulosic materials and methanogens. Cattle dung is most important substrate for biogas production as it contains cellulose as well as methanogens.

Production Unit:

1. Biogas plant consists of 10-15 feet deep and air tight cylindrical concrete tank (digester) in which slurry of dung is fed
2. A floating cover is placed over slurry which keeps on rising as the gas is produced in the tank by microbial activity.
3. It has an outlet which is connected to pipe for removal of biogas.
4. It has another outlet to remove spent slurry which can be used as fertiliser.

Process: Plant and animal wastes (cattle dung and manure from poultry) containing polymers (cellulose, hemi-cellulose, starch, lignin, protein, fats) are the raw materials for biogas. Biogas is produced in three stages:

1. **Solubilisation of polymers** (enzymatic breakdown) by decomposer microbes (facultative anaerobes) into simpler and soluble compounds (monomers).
2. **Acidogenesis of monomers** by fermenting microbes (facultative anaerobes esp. Bacteroides and *Clostridium*) into organic acid (acetic acid).
3. **Methanogenesis of organic acid** by methanogens (obligate anaerobes esp. *Methanobacterium* and *Methanococcus*) into **biogas** (Methane and Carbon dioxide).

Factors useful for biogas production

(a) Type of waste: Cow dung is preferable. If human excreta are added then biogas production will increase because it is rich in nitrogen content.

(b) Temperature: Optimum temperature should be 35 to 38^{0}C. Low temperature will reduce its production.

(c) Optimum pH should be 7.

(d) Slurry composition where water and solid should be in the ratio of 1:1

(e) Anaerobic condition

(f) Presence of inhibitors: There should not be ammonia, sulphates and antibiotics. Agricultural waste should be avoided because it is rich in ammonia. Likewise paper residues should be avoided because it is rich in sulphates.

Advantages:

(a) It provides fertilisers or manure by its left over

(b) It is cheaper than fuel for cooking and generating electricity.

(c) It does not cause pollution hence it can also be compressed like CNG.

(d) It does not release offensive odour

(e) It is ideal for small scale production

(f) It improves health condition and sanitation by checking spread of faecal pathogens.

(g) It can be purified to yield good grade of methane,

Disadvantages:

(a) It can be ecological problem as methane absorbs infrared radiation and thus is a greenhouse gas. Its overproduction may cause global warming.

(b) Its production is very low.

(c) It has low heat value

(d) It is expensive for large scale production

(e) It contains some gases as impurities.

(f) It is quite difficult to store, transport and distribute methane

Biofertilisers:

These are microorganisms which bring about nutrient enrichment of soil by enhancing the availability of nutrients to crops. These are also known as microbial inoculants. These microorganisms act as biofertilisers in three ways:

1. **Nitrogen fixing biofertilisers:**
   (a) Free living asymbiotic bacteria: *Azotobacter, Bacillus, Clostridium*
   (b) Free living loose symbiotic bacteria: *Azospirillum*
   (c) Free living asymbiotic cyanobacteria: *Anabaena, Nostoc, Aulosira* (in rice field), *Cylindrospermum* (sugarcane and maize field), *Plectonema, Tolypothrix*.
   (d) Symbiotic nodule forming bacteria in leguminous plants: *Rhizobium*
   (e) Symbiotic nodule forming bacteria in non leguminous plants: *Frankia*
   (f) Symbiotic cyanobacteria: *Anabaena azollae*
2. **Phosphate solubilising biofertilisers:** A fungus, *Glomus* develops symbiotic association with roots of higher plants forming endomycorrhiza which converts insoluble phosphorus into soluble phosphorus for uptake of crop plants.
3. **Phosphate mobilising biofertilisers:** These facilitate unavailable and insoluble inorganic phosphorus into soluble and usable form of phosphorus (phosphates) to crop plants. These may be
   (a) Bacteria: *Bacillus megatherium* var. *phosphaticum, Bacillus circulans, Pseudomonas striata*
   (b) Fungi: *Penicillium, Aspergillus awamori*

Advantages:

(a) It is cheaper and can be used by small and marginal farmers.

(b) It increases the yield of crops by 25%.

(c) It improves soil texture, water holding capacity and organic matter.

(d) It does not cause water pollution.

(e) It does not allow pathogens to flourish.

(f) It produces vitamins and growth promoting chemicals

Green manuring: It is a farming practice where a leguminous plant which has derived enough benefits from its association with species of *Rhizobium*, is ploughed into soil and then a nonlegume is grown and allowed to take the benefits of already fixed nitrogen. Some of the important leguminous plants used as green manure are as follows:

(a) Sunhemp (*Crotolaria juncea*)
(b) Dhaincha (*Sesbania aculeata*)
(c) Cluster bean (*Cyamopsis tetragonoloba*)
(d) Cow pea (*Vigna sinensis*)
(e) Sweet clover/Senji (*Melilotus pamniflora)*
(f) Egyptian clover/ Berseem *(Trifolium alexandrium*)
(g) Lentil (*Lens esculenta*)
(h) Horse gram (*Macrotyloma uniflorum*)

Importance:

(a) It adds organic matter and nitrogen to soil.
(b) It controls the growth weed.
(c) It helps in reclamation of alkaline soil.

Biocontrol

It is the natural method of control of pests (weeds, insects and pathogens) with the help of living organisms (viruses, bacteria, fungi, plants and animals). These organisms are called biopesticides.

Types of biopesticides:

1. **Bioherbicides**: These are living organisms or their products to either control or kill weeds or unwanted plants without harming useful plants. These are as such
   (a) Transgenic plants such as tobacco, tomato show tolerance to specific herbicide.
   (b) Smoother crops such as sweet clover, soya bean, rye, alfalfa, sunflower do not allow weeds to grow in the vicinity through chemicals. If crop rotation is done with smoother crops then it will reduce the chance of growth of weeds.
   (c) Predatory herbivore such as Cochineal insect (*Cactoblastis cactorum*) checks the growth of cacti (*Opuntia*).
   (d) Predatory herbivore such as beetle (*Zygogramma bicolorata*) checks the growth of carrot grass or congress grass (*Parthenium hysterophorus.*
   (e) A fungus, *Cercospora rodmanii* controls the growth of water hyacinth (*Eicchornia*).
   (f) Mycoherbicide such as devine (a liquid suspension of chlamydospores of *Phytophthora palmivora* is sprayed to control Milk weed vine (*Morrenia odorata*).
   (g) Mycoherbicide such as collego (a dry powder formulation of a fungus, *Colletotrichum gloespirodes* is sprayed to control a leguminous weed, Northern joint Vetch (*Aeschyynomena*)
2. **Bioinsecticide**: These are living organisms which kill harmful insects. These are as such
   (a) Transgenic plants such as tomato show resistance to hornworm larvae.
   (b) Predators such as Dragon fly kills mosquitoes.
   (c) Predators such as Lady bird (Ladybug beetle) and Praying mantis kill aphid, scale insects of crop plants, respectively.
   (d) Fishes such as Gambusia and Guppy kill mosquito larvae.
   (e) A free living fungus, *Trichoderma* kills several plant pathogens.

(f) Baculoviruses (under genus, Nucleopolyhedral virus or NPV) control many lepidopteran insects such as boll worm, gypsy moth, butterfly, beetle, ants, wasps of cotton, rice and vegetables. These are species specific and have narrow spectrum application. These do not cause harm to plants, fish, birds, mammals or even non-target insects.

(g) **Natural insecticides** are formed which are as follows:

- Rotenone: It is prepared from root of *Derris elliptica*
- Azadirachtin: It is prepared from leaf of *Azadirachta indica.*
- Nicotine: It is prepared from leaf of *Nicotiana tabacci.*
- Pyrethrum and Cinerin: These are prepared from flowers of *Chrysanthemum.*

(h) A **commercial insecticide** such as sporeine is developed from *Bacillus thuringiensis*. It was first such insecticide developed in France. It is species specific and active against lepidopteran and dipteran larvae, cockroaches, mosquitoes etc.

SHORT ANSWER TYPE QUESTIONS

1. Why microbes are suitable for industrial processes?
2. Distinguish between distilled beverage and fermented beverage.
3. Mention the sources of five important enzymes.
4. What are the applications of streptokinase, cyclosporine and statin?
5. What are the features of a good antibiotic?
6. Distinguish between narrow spectrum antibiotic and broad spectrum antibiotic.
7. Discuss the mode of action of antibiotics.
8. What are the reasons for resistance against antibiotics?
9. Mention the sources of five important antibiotics.
10. Distinguish between primary sludge and secondary sludge.
11. Distinguish between phosphate solubilising biofertilisers and phosphate mobilising biofertilisers.
12. Mention five important nitrogen fixing microbes.
13. Discuss the importance of biofertilisers.
14. Write notes on green manuring.
15. Write notes on natural insecticides.

LONG ANSWER TYPE QUESTIONS

1. Describe the role of microbes in producing household products.
2. Describe the role of microbes in brewery.
3. Describe the role of microbes in producing organic acids.
4. Describe the role of microbes in sewage treatment
5. Describe the role of microbes in producing biogas.
6. Define biocontrol. Discuss the role of different organisms in control of pests.

MULTIPLE CHOICE QUESTIONS

1. During formation of bread it becomes porous due to release of CO_2 by the action of
   (a) Bacteria (b) Yeast (c) Protozoans (d) Archaebacteria
2. In the manufacture of wine, as the alcohol concentration increases, yeast cells die. The cells can tolerate alcohol up to
   (a) 5% (b) 10% (c) 15% (d) 30%
3. When huge amount of sewage is dumped into a river, the BOD will
   (a) Decrease (b) Increase
   (c) Slightly decrease (d) Remain unchanged
4. The group of bacteria used in biogas production is
   (a) Methanotrophs (b) Methanogens (c) Thermoacidophiles (d)Organotrophs
5. A common biocontrol agent for the control of plant disease is
   (a) *Fusarium* (b) *Bacillus thuringiensis*
   (c) *Glomus* (d) *Trichoderma*
6. Which of the following organisms helps in absorption of phosphorus from soil by plants?
   (a) *Glomus* (b) *Frankia* (c) *Anabaena* (d) *Rhizobium*
7. The nitrogen fixing bacteria associated with maize roots belong to the genus
   (a) *Azotobacter* (b) *Aulosira* (c) *Azospirillum* (d) *Azorhiza*
8. Vesicular arbuscular mycorrhizae (VAM) are important in
   (a) Nitrate nutrition of plants (b) Phosphate nutrition of plants
   (c) Carbonate nutrition of plants (d) Sulphate nutrition of plants
9. Green manuring increases crop yield by
   (a) 5-10% (b) 20-25% (c) 30-50% (d) 80-90%
10. Sporeine is based on
    (a) Spores of algae (b) Spores of fungi (c) Bacteria (d) Viruses
11. Which of the following N_2 fixing bacteria forms loose association with roots of grasses?
    (a) *Azospirillum* (b) *Azotobacter* (c) *Rhizobium* (d) *Bacillus*
12. Which of the following plants are used as green manure in crop fields and in sandy soils?
    (a) *Saccharum munja* and *Lantana camara* (b) *Dichanthium annulatum* and *Azolla nilotica*
    (c) *Crotolaria juncea* and *Alhagi camelorum* (d) *Calotropis procera* and *Phyllanthus niruri*
13. Chloramphenicol antibiotic is isolated from
    (a) *Streptomyces fradiae* (b) *Streptomyces venezuelae*
    (c) *Streptomyces rimosus* (d) *Streptomyces aureus*
14. Genes for antibiotic resistance are located in
    (a) Nucleus (b) Plasmid (c) Chromosome (d) Cell wall
15. Antibiotics destroy
    (a) Bacteria and viruses (b) Bacteria and fungi
    (c) Fungi and viruses (d) Bacteria, viruses and fungi

16. Fermentation of sugar to yield alcohol is carried out by
    (a) High temperature (b) Microorganisms (c) Zymase (d) Cellulose
17. Distillation of wine is necessary because
    (a) It decreases toxicity
    (b) It has no action on the nature of wine
    (c) It increases turbidity
    (d) It prevents further fermentation
18. Cyclosporin produced from *Trichoderma* is useful because
    (a) It lowers blood cholesterol
    (b) It lowers blood glucose level
    (c) It removes toxic materials in body
    (d) It prevents rejection reaction in transplantation
19. Mycorhiza improves the soil structure by
    (a) Secreting carbonic acid
    (b) Releasing glucosides
    (c) Absorbing heavy metals
    (d) Adding glycoprotein
20. Primary treatment of sewage involves
    (a) Filtration and flocculation
    (b) Flocculation and sedimentation
    (c) Sedimentation and Homogenisation
    (d) Filtration and sedimentation
21. Which of the following alcoholic drinks is produced without distillation?
    (a) Brandy (b) Rum (c) Whisky (d) Wine
22. In sewage treatment, the bacterial flocs are allowed to sediment in a settling tank. This sediment is called
    (a) Primary sludge (b) Secondary sludge (c) Activated sludge (d) Anaerobic sludge
23. BOD of waste water is estimated by measuring the amount of
    (a) Oxygen consumption
    (b) Total organic matter
    (c) Oxygen evolution
    (d) Total dissolved solids
24. Methanogens are not found in
    (a) Rumen of cattle
    (b) Gobar gas plant
    (c) Activated sludge
    (d) Bottom of waterlogged paddy field
25. Biogas production technology was developed by
    (a) IARI and CSIR (b) IPRI and IARI (c) CSIR and KVIC (d) IARI and KVIC
26. Cheese is prepared by separating curd from liquid part by
    (a) *Bacillus megatherium*
    (b) *Penicillium camemberti*
    (c) *Lactobacillus acidophilus*
    (d) *Propionibacterium shermanii*
27. In preparation of bread , the sequence of enzymes involved is
    (a) Amylase, maltase and zymase
    (b) Maltase, amylase and zymase
    (c) Zymase, maltase and amylase
    (d) Amylase, zymase and maltase
28. The scientific name of wine yeast is
    (a) *Saccharomyces cerevisiae*
    (b) *Saccharomyces ellipsoideus*
    (c) *Saccharomyces sake*
    (d) *Torulopsis utilis*

29. Statin is produced from
    (a) *Trichoderma polysporus* (b) *Trichoderma viridiae*
    (c) *Monoascus purpureus* (d) *Saccharomyces cerevisiae*
30. Which of the following acids was first to be produced through fermentation?
    (a) Acetic acid (b) Lactic acid (c) Citric acid (d) Oxalic acid
31. The function of streptokinase is to
    (a) Lower blood cholesterol (b) Suppress the immune system
    (c) Lower blood glucose level (d) Dissolve blood clots
32. Which of the following vitamins was first to be produced through fermentation?
    (a) Vitamin-A (b) Vitamin-B (c) Vitamin-C (d) Vitamin-D
33. Which of the following antibiotics is broad spectrum?
    (a) Penicillin (b) Bacitracin (c) Polymixin (d) Chloramphenicol
34. The main composition of biogas is
    (a) CH_4, CO_2 and O_3 (b) CH_4, O_3 and SO_2 (c) CH_4, NO and O_3 (d) CH_4,CO_2 and N_2
35. The production of biogas involves
    (a) Solubilisation (b) Acidogenesis (c) Methanogenesis (d) All of the above
36. The example of phosphate mobilising biofertiliser is
    (a) *Bacillus megatherium* (b) *Lactobacillus vulgaris*
    (c) *Propionibacterium shermanii* (d) *Clostridum butyricum*
37. Dextran is plasma expander produced from
    (a) *Bacillus megatherium* (b) *Clostridium glutamicum*
    (c) *Trichoderma polyporus* (d) *Leuconostoc mesenteroides*
38. Which of the following fungi forms endomycorhiza?
    (a) *Agaricus* (b) *Glomus* (c) *Polyporus* (d) *Puccinia*
39. Which of the following is incorrect regarding baculovirus?
    (a) It is species specific (b) It has narrow spectrum applications
    (c) It controls lepidopteran insects (d) All of the above
40. Which of the following parts of plant produces a natural insecticide, Pyrethrum?
    (a) Leaf (b) Flower (c) Fruit (d) Seed

ANSWERS

1.a	2.c	3.b	4.b	5.d	6.a	7.c	8.b	9.b	10.c
11.a	12.c	13.b	14.b	15.b	16.c	17.c	18.d	19.d	20.d
21.d	22.c	23.a	24.d	25.d	26.d	27.a	28.b	29.c	30.b
31.d	32.c	33.d	34.d	35.d	36.a	37.d	38.b	39.d	40.b

21

Organism and Population

Ecology is study of interrelationship between organisms and their environment.

The term 'Ecology' (Gr. *Oikos*=home, *logos*-study) was coined by Reiter (1868).

Ernst Haeckel (1886) first correctly defined ecology as 'the science which deals with reciprocal relationship of organisms and the external world".

Warming is regarded as father of Ecology. He wrote a book" Oecology of Plants". Famous ecologists are Clarke, Warming, Clements, Odum, Schimper. Ramdeo Misra is father of Indian Ecology. Other famous Indian ecologists are G.S.Puri, S.C.Pandeya etc.

Branches of ecology:

1. Autecology: It was coined by Schroeter. It is study of a single individual organism (individual species) with respect to its environment.
2. Synecology: It was coined by Kirchner. It is study of a group of organisms (population) with respect to their environment.
3. Applied ecology: It deals with application of ecological concepts for human welfare.
4. System Ecology: It deals with interpretation of ecological concepts in terms of mathematical principles.
5. Genecology: It is study of ecological adaptations with respect to genetic variability.

Levels of organisation of living and nonliving matter:

1. Physical: Proton,neutron,electron→Atom→Compound→Organelle→Cell→Tissues
2. Biological: Organs→Organ system→Organism→Population
3. Ecological: Biotic community→Ecosystem→Biome→Biosphere

Terms:

1. Species: It is organism (individual) of same kind. The organism is considered as basic unit of study in ecology.
2. Race: It is group of organisms within a species.
3. Population: It group of organisms belonging to same species living in a particular geographical area in a particular time. e.g., Bacteria in a culture plate, Fish in a pond, Birds in a grove, Elephant in a forest etc.
4. Community: It is group of organisms of different species (such as bacteria, fungi, plants, animals) in a particular geographical area interacting with one another through competition, predation, mutualism etc. e.g., Pond community

5. Ecosystem: It is an interacting system of biotic community with physical world (environment). e.g., Terrestrial ecosystem (Grassland and forest) and aquatic ecosystem (Pond and ocean).
6. Biome: It is a group of ecosystems stretching over a large physical area characterised by distinctive life forms or in other words, it is a community of plants and animals existing ina large area showing common features of environment.e.g., Tropical rain forest, Temperate deciduous forest, Temperate grassland, Desert, Tundra, Taiga etc.
7. Biosphere: It is entire inhabitable portion of earth in which ecosystem operates. It is the largest biological system. It receives solar energy for maintenance of life on earth and releases heat energy to environment by cellular respiration and decomposition of their dead organic remains. It includes
   (a) Lithosphere: Solid component (soils, rocks, minerals of earth crust)
   (b) Hydrosphere: Liquid component (water of ocean, sea, river, lake, pond)
   (c) Atmosphere: Gaseous cover around lithosphere and hydrosphere.

Terms:

1. **Habitat**: It is a place where organism lives. It is controlled by temperature, water, light, soil and other biotic components such as parasites, predators, competitors and pathogens.
2. **Niche**: It was coined by Grinnel (1917). It is specific part of habitat occupied by individual limited by its range of tolerance, range of movement, microclimate, type and availability of food, shelter and type of predator. Every niche has its own specific environment. It is specific position and total role of the individual (species) in a habitat. It shows the profession of species. Each species has distinct niche. Thus no two species occupy exactly the same niche for long span of time. For example, tadpole and frog occupy different ecological niche because tadpole is aquatic and herbivorous whereas frog is terrestrial and carnivorous.
3. **Biota**:It is total collection of organisms of a geographical region.
4. **Environment:** It is set of things and condition around organism.
5. **Weather**: It is condition of atmosphere at any place at a particular time.
6. **Climate:** It is long term weather condition and factors specific to an area.

Major climatic regions:

Zone	Latitude	Mean annual temperature	Vegetation
Tropical	0 to 20	Above 24°C	Tropical forest
Subtropical	20 to 40	17 to 24°C	Subtropical deciduous forest
Temperate	40 to 60	7 to 17°C	Coniferous forest
Arctic and Antarctic	60 to 80	Below 7°C	Arctic forest (alpine forest)

Ecological Factors: These are environmental factors which affect living organisms. These factors do not act in isolation but interact with one another to influence organisms. The life of plants and animals is very intimately related to environmental factors of place and gets fitted to particular set of environmental condition. They face difficulties in survival if there is any alteration in these factors. These factors may be abiotic and biotic: Abiotic factors may also be **direct factors** which influence growth and distribution of organism directly, e.g., temperature, light, water, humidity of air, soil air and soil water and may also be **indirect factors** which influence organisms through direct factors by causing changes in these factors, e.g., wind, rainfall, soil structure, soil organisms and all topographic factors. These factors include:

Temperature: It is most relevant and direct factor. Its source is sun. It is maximum at equator and plain and minimum at pole and mountain top. It is maximum in summer season and minimum in winter. It is measured by thermometer and temperature under water at desired depths is measured by thermistor. It exceeds 100^0C in unique habitat such as thermal springs and deep seas hydrothermal vents. It varies due to change in latitude, altitude, topography slope, vegetation, season and humidity. It shows lapse rate (temperature is decreased by 1° when there is 1° change in latitude and 90 m rise in altitude.

Temperature effects morphology, geographical distribution of plants and animals (Mango trees do not grow in temperate condition and snow leopard is not found in tropical regions), transpiration, photosynthesis, seed germination, bud sprouting, growth and metabolism (low temperature inactivates enzymes and high temperature denatures enzymes).

Low and high temp cause injury to plants but some plants and animals tolerate low temperature of arctic region and high temperature of desert depending on functioning of enzymes.

Thermophilic bacteria and fungi tolerate 80-90°C. *Pinus* tolerates 85°C and praying mantis tolerates 62°C.

Temperature produces four climatic zones: (1) Tropical (2) Subtropical (3) Temperate and (4) Arctic /Alpine.

Types of vegetation on basis of temperature requirement:

1. **Megatherms**: Plants require constant high temperature throughout the year. e.g., Tropical dry deciduous forest
2. **Mesotherms**: Plants require alternate low and high temperature. e.g., Subtropical deciduous forest.
3. **Microtherms**: Plants require very low temperature but they cannot tolerate high temperature during summer. e.g., Coniferous forest
4. **Hekistotherms**: Plants require very low temperature and these plants tolerate it without any injury. These occur in Arctic region. The plants are of very short vegetative phase. Condensed short rosette appearance, small leaves. The animals show thick coat of hairs, scales, feathers and subcutaneous fats. e.g., Alpine forest (*Rhododendron, Juniperus*) above 12000 ft in temperate zone and above 16000 ft in tropical zone.

Types of animals on the ability to tolerate variation in temperature:

1. **Eurythermal**: These organisms tolerate and survive in a wide range of temperature e.g., Cyclops, Toads, Lizards, Birds and most mammals
2. **Stenothermals**: These organisms live in a narrow range of temperature because of the requirement of nearly constant temperature throughout the year. e.g., Corals, Fishes, Amphibians, Snake, Polar bear etc.

Types of organisms on the ability to regulate body temperature:

1. **Homoiothermal** (Endothermal / Warm-blooded): These animals maintain constant body temperature in spite of changes in temperature of environment. These have high assimilation efficiency and low production efficiency e.g., Birds and most mammals.

2. **Poikilothermal** (Ectothermal /cold-blooded): These animals change body temperature with the change in temperature of environment. These have low assimilation efficiency and high production efficiency e.g., All invertebrates, fish, amphibians, reptiles etc.

Water: It is second important direct factor. Its source is precipitation (rainfall, snowfall, hail storm). Water from rainfall is present in sea, river, pond, lake, ground and soil. Out of total rainfall, 45% is runoff or runaway water which flows to water bodies; 35% is evaporated and rest 20% is absorbed by ground as a result of percolation. Earth contains 71% of water (97.5% in ocean and 2.5% in fresh water bodies such as rivers, lakes etc.

It determines the growth and distribution of plants and animals on the basis of clarity of water, salt content, depth, water waves and pH. The salt content is measured as salinity per 1000. It is less than 5 (<0.5%) in inland water, between 30 and 35 (3-3.5%) in sea and more than 100 (>10%) in superhaline (hypersaline) lagoon. It is essential for turgidity of cells, transport of minerals and organic solutes, photosynthesis and growth.

It determines the nature of soil formation. In humid climates, the laterisation and podzolisation occur due to thorough eluviations of top soil layer by good rainfall but in arid climates, the calcification occurs due to very poor eluviations of top soil layer by scanty and occasional rainfall.

Types of vegetation on the basis of amount and timing of rainfall:

1. Desert: Scanty rainfall (0-10″)
2. Tropical deciduous forest: Less rainfall (30″) in summer and more (50″) in winter
3. Grassland: More rainfall (30″) in summer and less (10″) in winter
4. Tropical rain forest: High rainfall (>50″) throughout the year.

Types of animals on the basis of ability to tolerate salinity:

1. **Euryhaline**: These animals tolerate wide range of salinity. These are fewer in numbers. These cannot live in freshwater. e.g., Lamprey, Herring, Sea urchin; Stingray, Salmon, Eels etc.
2. **Stenohaline**: These animals tolerate narrow range of salinity. These are more in number. These die in seawater e.g., Most freshwater fish such as goldfish, Shark.

Types of plants on the basis of availability of water:

1. Hydrophytes
2. Xerophytes
3. Mesophytes

Light: Its source is sun. Its visible range is 390-760 nm (photosynthetically active region). The visible light has seven colours of different wavelengths. It is maximum at equator and latitude and minimum at poles and altitude. It is maximum in summer and minimum in winter season. The intensity of light is measured by Photometer or Luxmeter. Out of total amount of sunlight incident on leaf, 79% is lost as heat in environment, 12% is reflected back in environment, 5% is transmitted through leaf surface and 4% is utilised by chlorophyll in the process of photosynthesis.

Its quality, duration and intensity influence distribution of plants, transpiration, pigmentation (formation of chlorophyll and orientation of chloroplast, photosynthesis, photoperiodism, growth, germination of seeds, phenology (timing of seasonal activities of organisms), daily rhythm in animals-diurnal, nocturnal, auroral (active at early morning), vesperal (active at sunset) and crepuscular (active at both dawn and dusk), bird migration, animal breeding etc.

Types of plants on the basis of requirement of light:

1. Heliophytes
2. Sciophytes

Soil: It is surface layer of earth's crust which is rich in humus and supports the terrestrial life. Study of soil is called Paedoloy or Edaphology.

Warming was the first ecologist who recognised the importance of soil. Soil provides anchorage, water and minerals to plants. It also determines the growth and patterns of vegetation.

Paedogenesis (formation of soil) is formed by the following processes

(a) Weathering: It is a process in which large rocks are broken into small pieces by physical means by water, wind, gravity, glaciers; chemical means by oxidation, reduction, hydrolysis, carbonation and chelation; and biological means by activities of microbes, lichens, fungi.

(b) Humification: It is a process by which organic matter (humus) is formed by the death of bacteria, fungi, lichen, arthropods etc.

(c) Eluviation: It is leaching or washing down of materials from upper strata to inside soil.

(d) Iluviation: It is deposition of washed down materials in lower strata.

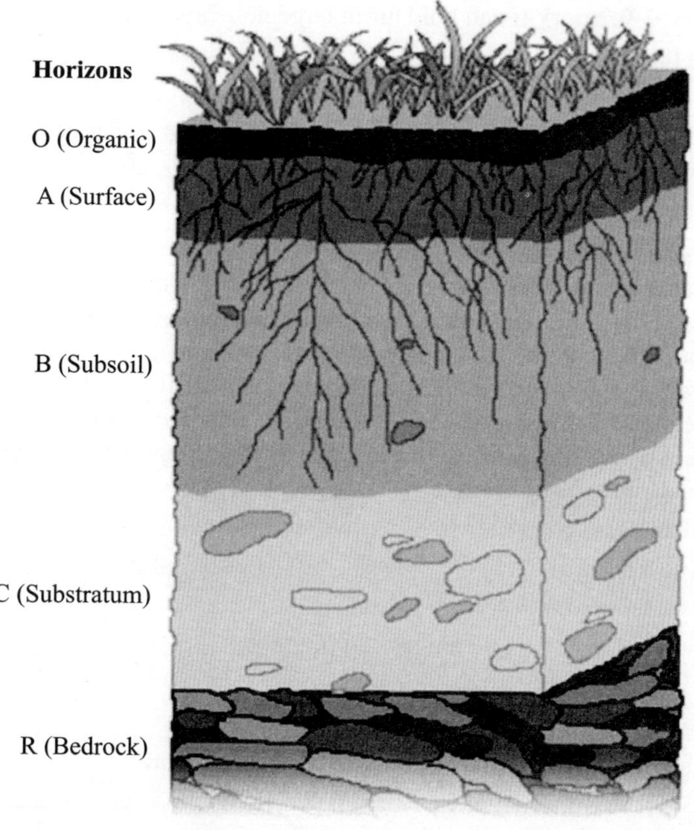

Soil Profile: It is a vertical cross-section of layers of soil found in a given area. It has several layers or horizons which are as follows:

1. **O-horizon (Organic)**: It is surface layer of organic matter which lies above top or true soil. It contains fresh or partially decomposed organic matter.
2. **A-horizon (Top soil** or surface): It is uppermost horizon of soil. It contains minerals and humus. It is rich in microbes. It contains embedded roots. It forms feeding zone (zone of eluviation).
3. **B-horizon (subsoil)**: It is dark in colour due to accumulation of leached substances like clay, iron and aluminium. It contains minerals but less humus. It is poor in aeration and biological activity. It contains poorly developed roots. It is also called zone of accumulation or illuviation.
4. **C-horizon (Subsoil** or substratum): It contains irregular and partially weathered (fragmented) rocks. It does not show biological activity. It is saturated with water and represents water table.
5. **D-horizon (Bedrock)**: It contains unweathered bed rocks.

Composition of soil:

(a) **Mineral matter** (40%): It is obtained by weathering of rocks. It consists of particles of different size:

(i) Clay (< 0.002mm): It can hold water in fine capillary space.

(ii) Silt (0.002-0.02 mm): It can hold water and air.

(iii)Fine sand (0.02-0.2mm): It can hold air in large non-capillary pore.

(iv) Coarse sand (0.2-2 mm)

(v) Gravel/stone (>2mm)

(b) **Organic matter (10%):** It is formed as a result of death and decay of plants and animals. It is present in the following forms:

(i) Litter: It is freshly and recently fallen leaves and twigs.

(ii) Duff: It is partially decomposed organic matter.

(iii) Humus: It is completely decomposed or broken down organic matter as a result of microbial action (decomposer microbes). It is amorphous, dark brown and gummy substance present in top soil. It is slightly soluble in water and rich in nutrients (humic complex, phenolic complex, soluble sugar, amino acids, protein and organic acids). It may be Mor humus (raw humus or coarse textured humus) and Mul humus (fine textured humus). It divides soil into two types: (1) Eutrophic soil (soil containing optimum nutrients) and (2) Oligotrophic soil (soil containing suboptimum nutrients).

Functions of Humus:

(a) It increases soil fertility.

(b) It increases water holding capacity of soil

(c) It helps in soil aggregation and makes soil granular.

(d) It provides inorganic nutrients by mineralisation

(e) It secretes growth promoting chemicals.

(f) It contains acid for solubilisation of minerals present in soil

(c) **Soil Water (25%):** It is measured by tensiometer. It contains various dissolved solutes such as carbonates, sulphates, nitrates, chlorides of calcium, magnesium, potassium etc. It is of several types:

1. **Capillary water:** It is only available water present in narrow interparticles space or micropores of soil held by capillary forces which work against gravitational forces.

2. **Gravitational water:** It is unavailable water present in deep strata of soil beyond the reach of root hairs. It is drained downwardly after good rain or irrigation due to the influence of gravitational forces. It is also called run away or runoff water which reaches water body
3. **Hygroscopic water:** It is unavailable water present in thin film around soil particles held by very strong attractive forces.
4. **Combined water**: It is unavailable water present attached with chemicals such as hydrated oxides of silicon, iron, aluminium in soil.
5. **Water vapour**: It is unavailable water present in the form of vapour (gas).

Terms:

(i) Holard: It is total amount of water. It may be
   (a) Chresard: It is available water such as capillary water.
   (b) Echard: It is unavailable water such as gravitational water, hygroscopic water, combined water and water vapour.

(ii) Field capacity: It is amount of all types of water present in soil after complete drainage of gravitational water. It is total water without gravitational water.

(iii) Water holding capacity: Field capacity without hygroscopic water.

Types of soil:

1. Physically dry soil: Water is unavailable in soil.
2. Physiologically dry soil: Water is present in soil but it is not absorbed. e.g., acidic soil, salty soil etc.

(d) Soil air (25%): It depends on a number of macropores connected with soil surface. It is rich in CO_2 (richer by 1% from normal) and poor in O_2 (poorer by 1% from normal). It is essential for absorption of water. If soil contains less O_2 the absorption of water is slow due to less development of root hairs and if it contains more O_2 then absorption of water will be fast due to excess development of root hairs. Waterlogged soil is highly deficient in oxygen.

(e) Soil organisms: These include bacteria and cyanobacteria which increase soil fertility by fixing N_2 and fungi, algae, protozoa, rotifers, earthworms, insects, molluscs which improve the condition of soil.

Types of soil:

1. Residual soil: It is soil which is found in same area where parent rocks are present earlier.
2. Transported soil: It is soil which is transported to another place from the original place of weathering. It is of several types:
   (a) Alluvial soil: It is transported by running water (river).
   (b) Colluvial soil: It is transported by force of gravity of hill.
   (c) Eolian soil: It is transported by wind.
   (d) Glacial soil: It is transported by melting ice (glacier).

Texture of Soil: It is physical structure of soil. It is divided into

(a) Clay soil: (50% clay and 50% silt): It is compact (solid) hence penetration of root. It is fertile soil containing inorganic nutrients. It has maximum water holding capacity hence water logging occurs. It has low temperature. It has poor aeration hence development of root and root hair is poor.

(b) Loamy soil (40% silt, 40% sand and 20% clay): It is highly fertile and best for growth and development of plant. It has moderate water holding capacity. It has moderate temperature. It has good mineral content, good aeration and hydration.

(c) Sandy soil (80% sand and 20% clay): It is porous hence water percolates very easily and becomes nutrient deficient. It is not fertile. It has minimum water holding capacity. It has very high temperature. It has maximum aeration.

Types of vegetation on the basis of soil condition:

(a) Halophytes: Plants growing in saline soil
(b) Oxylophytes: Plants growing in acidic soil
(c) Psammophytes: Plants growing in sandy soil
(d) Psychrophytes: Plants growing in cold soil
(e) Chasmophytes: Plants growing on rock crevices (cracks)
(f) Lithophytes: Plants growing on rock surface
(g) Helophytes: Plants growing on mud
(h) Chersophytes: Plants growing on wasteland
(i) Eremophytes: Plants growing on desert.
(j) Hygrophytes: Plants growing on wet soil.
(k) Cryophytes: Plants growing on ice.
(l) Cryptophytes/Geophytes: Plants growing below soil with underground bud.

Humidity: It is moisture in the form of invisible vapours in atmosphere. It is measured by Psychrometer or Paper Hydrometer. It is expressed in terms of Relative Humidity (RH) which is amount of vapour in atmosphere as percentage of total amount which the air or atmosphere can hold at the existing temperature.

$$RH = \frac{\text{Absolute humidity}}{\text{Saturation Point}} \times 100$$

It increases if atmosphere will become cooler and it decreases if atmosphere will become warmer. Hence RH is inversely proportional to temperature. It determines precipitation (formation of cloud, dew and fog). It is influenced by solar radiation, temperature, wind and water. It regulates the rate of evaporation of water in land plants as a result of transpiration. The rate of evaporation will be more at low RH. The plants such as lichens, mosses, orchids which grow in high RH are called hygrophyllous.

Wind: It is air in motion. It is measured by Anemometer. It commonly occurs at sea coast and mountain. It may be gale (60-90 km speed per hour), storm (90-120km per hour) and hurricane (more than120 km). It helps in transpiration, pollination, dispersal of fruits and seeds. It causes lodging of herbs, uprooting of plants, distortion in shape, abrasion (or injury) and desiccation (loss of turgidity of cells due to reduction in cell size).

pH: The optimum pH is required for most of organisms. The fertile soil has pH 6-7. It determines the distribution of land plants and soil organisms. It determines two types of plants: (1) Calcifuge (plants growing in low pH or at less than 3.5 and (2) Calcicole (plants growing in high pH or greater than 6.5.

Atmosphere: It is gaseous cover over earth surface. The atmosphere of earth is up to height of 300 km. in which 90% of gases are present up to height of 20 km and rest of gases are found from 20 km to 300 km. It contains nitrogen (78.08% by volume), oxygen (20.84%), argon (0.93%), CO_2 (0.03% = 300 ppm). The atmosphere is divided into

(a) Troposphere: It is lowermost zone which is first 20 km from earth. It contains 90% of all mass of gases and 99% of water vapours. It shows formation of cloud and occurrence of lightning and thundering. In this temperature goes on decreasing up to -65°Cas we move higher and decrease is 6.5°C in temperature with every km. The extreme limit is tropopause which merges with next zone.

(b) Stratosphere: It is about 30 km from troposphere (50 km from earth). In this temperature increases from -65°C to -2°C due to ozone formation. The extreme limit is stratopause which merges with next zone.

(c) Mesosphere: It is about 40 km stratosphere (90 km from earth). In this temperature further decreases from -2°C to -80°C. The extreme limit is mesopause.

(d) Ionosphere: It is uppermost zone (thermosphere) beyond Mesosphere. In this all gases are in ionised forms and air is thin. It shows the presence of satellites.

Topographic factors (Physiographic factors): These are indirect factors and connected with physical geography of earth. These influence other abiotic factors such as temperature, water, wind, pressure etc. These include:

(a) Altitude of place: If altitude increases the temperature will decrease.

(b) Slope of land: The slope of land cannot absorb sufficient water because rainwater runs away rapidly on slope hence it remains dry and shows few plants to grow. The well exposed slope shows different vegetation.

(c) Direction of valley and mountain: These determine the isolation, speciation (formation of species) and geographical distribution of organisms. If slope of mountain receives more light and wet air then luxuriant growth of plants occurs (dense vegetation). If slope of mountain receives less light and dry wind then xerophytes will grow.

Response to abiotic factors: These factors influence the functioning of organisms. As time passes the abiotic factors vary and in order to maintain the efficiency of all biochemical reactions and physiological functions, these organisms develop homeostasis (constancy in internal environment) to maintain equilibrium between organisms and environment. The homeostasis protects organisms from harmful changes. The living organisms perform in following ways:

(a) Regulate: It is a condition in which some (1%) organisms such as birds, mammals, tuna fish, sword fish maintain their constant body temperature and osmotic concentration in spite of changes in external environment. Hence these are called regulators. They perform homeostasis mostly through osmoregulation and thermoregulation by physiological adjustments. They are more active and higher metabolic rate (about five times of conformers). This regulation requires energy, so regulators tend to have high metabolic rates. Mammals got success in dominating the earth because they have developed their ability to maintain constant body temperature in spite of living in hot and cold condition.

In summer season (the external temperature is more than body temperature) the superficial blood vessels of mammals dilate and bring more blood to skin. The excess sweating occurs because sweat glands are more active hence body temperature is lowered.

In winter (the external temperature is less than body temperature) the superficial blood vessels of mammals constrict to check loss of heat. The erection of hairs occurs and provides insulation. The body starts shivering to raise the temperature of body. Camel raises its body temperature during daytime and lower at night to conserve energy.

Most of vertebrates and a number of aquatic invertebrates are osmoregulators. They eliminate excess water in freshwater and drink sea water in sea to compensate loss of water due to presence of salt excreting glands. Marine fishes cannot survive in freshwater but migratory fishes such as Hilsa, Salmon have hormone controlled mechanism either to absorb salt in freshwater or excrete salt in sea. If marine fish is placed in fresh water aquarium then it will die because of loss of excess mineral ions from body and entry of excess water inside body because of lack of mechanism to control entry of excess water and loss of mineral ions from body.

Rules:

Allen's rule	The size of extremities (ears, snout, tail, legs) in warm blooded animals is smaller in cold areas than in warm areas.
Bergman's rule	The general size of warm blooded animals is larger in colder areas than in warm area.
Gloger's rule	The pigmentation in warm blooded animals is less in cold and dry areas than in warm and humid areas.
Rensche's rule	The size of wing in birds is smaller in cold areas than warm areas.
Jordon's rule	The number of vertebrae in fish is more in cold water than hot water.

(b) Conform: It is a condition in which 99% of organisms (99% animals and all plants) cannot maintain constant body temperature and osmotic concentration. These are also called poikilothermal or ectothermal or cold blooded. They have not evolved the mechanism of homeostasis (thermoregulation and osmoregulation) because it is energetically very expensive and requires five times higher metabolic rate. The smaller animals such as shrew, humming bird have larger surface area relative to their volume. They tend to lose body heat quickly if it is cold outside so these animals have to expend much energy to produce body heat through metabolic processes. That is why smaller animals cannot regulate hence these are rarely found in polar and alpine region.

(c) Partial regulate or partial conform: It is a condition in which some aquatic vertebrates perform homeostasis up to a certain limit beyond that they behave as conformers. Such organisms change their body temperature to certain level of ambient temperature but if the temperature goes on increasing they can maintain a steady temperature. These are intermediate between regulators and conformers. e.g., Cray fishes

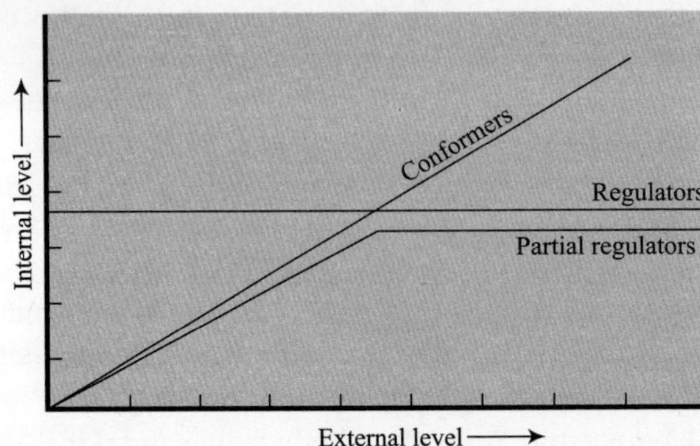

(d) Migrate: It is two way movement of whole population of organisms from stressful habitat to hospitable area for food, shelter, favourable climate. It occurs when stressful conditions are localised or remain only for a short duration and such organisms return back when unfavourable period is over. It includes

1. Migration of fishes: It is of two types:
   (a) Anadromous fish such as Salmon, Shad, Sturgeon which move from sea to fresh water and
   (b) Catadromous fish such as *Anguilla* which move from freshwater to sea.
2. Migration of Locust
3. Migration of birds

(e) Suspend: It is the condition in which organisms stop their activity during stressful period. They resume their activity with the start of favourable condition. It includes:

1. Aestivation (summer sleep): It occurs in summer. e.g., Snail, Fish
2. Hibernation (winter sleep): It occurs in winter. e.g., Frogs, Snakes
3. Diapause: It occurs in pond and lake. e.g., Zooplanktons (Pink cotton bollworm).
4. Dormancy: It is seen in seeds of higher plants where seeds do not germinate.
5. Thick walled spores: These are found in bacteria, fungi and lower plants (algae).

Biotic factors: These are categorised into three parts:

1. Interrelationship among plants:
   (a) Epiphytes (plants growing on another plants) such as orchids
   (b) Lianas(woody climbers)
   (c) Parasites such as *Rafflessia*, Dodder, *Cuscuta*
   (d) Saprophytes
   (e) Symbionts such as lichens
2. Interrelationship between plants and soil microbes: These organisms live in soil and affect yield of crop
   (a) Rhizobium
   (b) Protozoans
   (c) Algae
   (d) Fungi
2. Interrelationship between plants and animals:
   (a) Insectivorous plants (Carnivorous plants) such as *Drosera, Dionea, Utricularia, Nepenthes* kill insects.
   (b) Herbivorous animals (Herbivores) such as cattle graze plants
   (c) Human beings use certain plants or plant parts.
   (d) Ficus, Yucca and Salvia show relationship with animals for pollination

Adaptations: It is the feature which enables organisms to survive and reproduce in its particular habitat. It makes the organisms better fitted to obtain food and space.

Types of ecological adaptations:

(a) Phenotypic adaptations: These are favourable morphological and physiological changes which develop in response to changes in environmental condition. These adaptations disappear when environmental conditions become normal.

Phenotypic Plasticity: It is the ability of genitally similar population to undergo phenotypical changes in response to change in environment.

Ecad (Ecophene): It is genetically similar population but show environmentally induced variation.

(b) Genotypic adaptations: These are favourable genetical changes which enable a organism to adapt a particular habitat and environmental condition.

Ecotype: It is genetically adapted local population which develop from a species.

Adaptations to water scarcity:

(a) Kangaroo rat (*Dipodomys*) of arid areas manages scarcity of water by reducing loss of water. Water loss is prevented by living in burrows during day by nasal counter current mechanism and excreting concentrated urine and solid faeces.

(b) Camel of desert manages scarcity of water and dehydration up to 40% due to counter current mechanism by excreting concentrated urine and solid faeces.

(c) Several desert plants (xerophytes) minimise loss of water by developing thick cuticle, sunken stomata, spines, phylloclades and showing CAM metabolism.

Adaptations to Environmental stress: Archaebacteria living in hot springs tolerate high temperature due to high temperature.

(a) Aestivation: Lower animals undergo summer sleep. e.g., Cold-blooded animals (Snails, fish)

(b) Hibernation: Lower animals undergo winter sleep. e.g., Cold-blooded animals (Frogs, snakes)

(c) Altitude sickness: Human beings suffer from altitude sickness (nausea, fatigue and palpitation) while moving upwards (>3500 m). This occurs due to low atmospheric pressure of high altitude where body does not get enough oxygen. But if a person stays there for sometime then person gets acclimatised and does not suffer from altitude sickness. The body compensates low availability of O_2 by increasing production of RBCs, decreasing the binding capacity of haemoglobin and by increasing the rate of breathing.

(d) Migration: Fish and birds migrate to better place from original habitat.

(e) Diapause: Zooplanktons and insect larvae suspend their activities.

Adaptations for protection from predators:

(a) Camouflage (Cryptic colouration): It is the ability of animals to mix with surroundings (resembling leaves) to remain unnoticed for (a) aggression (attack) in case of praying mantis, reptiles and mammals and (b) protection in case of grasshopper, leaf insects, stick insects.

(b) Mimicry: It is the resemblance of one species with another in order to obtain advantage esp. against predators. It has mimic (animal which imitates) and distasteful and ferocious model (animal which is imitated). It is of two types:

(i) Batesian Mimicry: It was reported by Walter Bates. In this, harmless species (viceroy Butterfly) imitates the warning signal of harmful or distasteful or unpalatable species (monarch butterfly) which contains a harmful substance, cardiac glycosides in body due to eating upon *Calotropis* hence predator does not eat upon it.

(ii) Mullerian mimicry: It was reported by Friedrich Muller. In this, two or more closely related harmful or unpalatable species (monarch butterfly and queen butterfly) look very similar in order to ward off common predator by developing warning colouration on body for mutual benefit. It increases the representation of lack of palatability among potential prey. It is also observed in bees, hornet, wasp etc.

Plant adaptations to light requirement:

Heliophytes (Sun plants)	Sciophytes (Shade plants)
These are profusely branched.	These are less branched.
Stem is thick with short internode.	Stem is thin with long internode.
Leaves are vertically oriented.	Leaves are horizontally oriented.
Leaf colour is pale green with shinning surface.	Leaf colour is bright green with dull surface.
Cuticle is thick.	Cuticle is thin.
Stomata is thick, sunken and present on lower surface.	Cuticle is thin, on level and present on both surface.
Palisade cells are well-developed and spongy cells are less developed.	Palisade cells are less developed and spongy cells are well developed.
Leaf cells have lesser and smaller chloroplast.	Leaf cells have more and larger chloroplast.
These are more photosynthetic.	These are less photosynthetic.
Mechanical tissues are well developed.	Mechanical tissues are less developed.
Vegetative growth is less but more flowering and fruiting tendency.	Vegetative growth is more but less flowering and fruiting tendency.
These are more resistant to heat and cold.	These are less resistant to heat and cold.
These have more metabolic and respiratory activities.	These have less metabolic and respiratory activities.
These have high osmotic pressure.	These have less osmotic pressure.

Plant adaptations to water: The plants grow in water are known as hydrophytes. Warming and Schimper classified these plants into several groups on the basis of requirement of water.

1. **Hydrophytes:** These grow in water. These are of three types.
   - **(a) Submerged**: These remain under water. These may be

     Rooted: These are fixed in the bottom of water body. e.g., *Hydrilla, Vallisneria, Potamogeton.*

     Suspended: These are not fixed in mud. e.g., *Ceratophyllum, Najas, Utricularia*
   - **(b) Floating**: These float on water. These may be

     Fixed: These are fixed in the bottom of water bodies by roots but leaves float on surface. e.g., Lotus (*Nelumbo*), *Nymphaea, Trapa, Victoria* etc.

     Free: These float freely on water surface but not connected to soil. e.g., *Azolla, Eichhornia, Pistia, Jussiaea, Wolffia, Lemna* (Duckweed), *Salvinia* etc.
   - **(c) Amphibuous** (Emergent): These have submerged and fixed roots but shoot is aerial. e.g., *Ranunculus, Sagittaria, Marsilea, Typha, Limnophila, Cyperus, Carex* etc.

Morphological adaptations:

(a) Roots are absent as in *Utricularia* and *Wolffia*

(b) Root hairs and root cap are absent as in Lemna and Pistia.

(c) Root cap is replaced by root pocket for balance as in *Eicchornia* and *Pistia.*

(d) Stem is narrow and slender as in submerged plants

(e) Stem is short as in floating plants

(f) Stem is well developed as in amphibious plants

(g) Leaves are thin and membranous as in *Hydrilla.*

(h) Leaves are dissected in *Vallisneria.*

(i) Leaves are long ribbon like as in *Potamogeton.*

(j) Leaves show heterophylly (two types of leaves where submerged leaves are dissected and aerial leaves are not dissected (with entire margin) as in *Limnophyla, Azolla, Salvinia, Ranunculus, Sagittaria.*

(k) Flowers and seeds are poorly developed.

Anatomical adaptations:

(a) Aerenchyma is formed in cortex for buoyancy and gaseous exchange.

(b) Mechanical tissues (collenchyma and sclerenchyma) are absent.

(c) Conducting tissues (xylem and phloem) are either poorly developed or absent.

(d) Cuticle is absent as in leaves of submerged plants.

(e) Mesophyll cells are not differentiated into palisade and spongy cells.

(f) Stomata is absent but present only on upper surface of leaves as in floating plants.

(g) Secondary growth is absent.

Physiological adaptations:

(a) Absorption of water and minerals occurs through general body surface by diffusion.

(b) Osmotic pressure is low

(c) Vegetative reproduction occurs

2. **Xerophytes**: These occur in dry habitat. These are of four types on the basis of life cycle and water storage.

(a) Ephemerals: These complete their life cycle only in two months during rainy season hence these are short lived and called **drought escapers**. These are not true xerophytes because they do not face summer season and do not show any morphological and anatomical modifications.
e.g., *Argemone, Artimisia, Astragalus, Cassia, Euphorbia* etc.

(b) Annuals: These complete their life cycle in several months even after stoppage of rain. These are not true xerophytes because they show some modifications to reduce transpiration by formed smaller size of leaf, leathery leaves and thick cuticle hence these are called **drought evaders.**

e.g., *Chenopodium rubrum, Echinops, Solanum surattense* etc.

(c) Succulents: These resist drought by forming succulent (water storage) organs due to shortage of water and mucilage. Hence these are called **drought resistants**. These are true xerophytes.

e.g., *Asparagus* and *Ceiba* have root as succulent organ

Opuntia , *Cocoloba* and *Euphorbia* have stem as succulent organ

Agave, Aloe, Begonia and *Bryophyllum* have leaves as succulent organ.

(d) Nonsucculents: These tolerate long drought externally and internally hence called **drought endurer**. These are xerophytes.

e.g., *Acacia, Calotropis, Casuarina, Capparis, Nerium, Prosopis, Salvadora, Ruscus,* and *Zizyphus* etc.

Morphological adaptations:

(a) Root system is well developed and extensive to increase absorption of water.

(b) Stem is stunted, woody and hard with thick bark

(c) Stem is modified into phylloclade as in *Opuntia* and *Cocoloba*

(d) Stem is modified into cladode as in *Asparagus*

(e) Stem is covered with hairy or waxy coating as in *Calotropis*

(f) Leaf is thick and leathery as in *Calotropis* and *Salvadora*.

(g) Leaf is needle like as in *Pinus*

(h) Leaf shows rolling due to presence of bulliform cells in upper epidermis as in *Ammophila, Agropyron, Poa* (xerphytic grasses)

(i) Leaf is modified into spine as in *Opuntia*

(i) Leaf is modified into scale as in *Asparagus, Ruscus* and *Casuarina* etc.

Anatomical modifications:

(a) Cuticle is thick on upper surface of leaves.

(b) Epidermis is thick walled

(c) Multiple epidermis is found on both upper and lower epidermis as in *Nerium*.

(d) Stomata is sunken and generally present

(e) Hypodermis is thick walled and sclerenchymatous as in *Pinus* needle.

(f) Water storing parenchyma is well developed.

(g) Mechanical tissues (collenchyma and sclerenchyma) are well developed.

(h) Conducting tissues (xylem and phloem) are well developed.

Physiological adaptations:

(a) Osmotic pressure is high.

(b) Transpiration is reduced

(c) Mucilage is present to hold water.

(d) Resistance to desiccation occurs.

(e) CAM pathway occurs in some plants (e.g., *Begonia,*and *Bryophylum*) in which stomata is closed during day and open during night.

Physical xerophytes: These grow in physically dry soil in which water is unavailable. e.g., *Casuarina, Cocoloba, Opuntia, Ruscus*.

Physiological xerophytes: These grow in soil which contains water but unavailable due to salinity or frosting. e.g., Mangrove plants.

3. **Halophytes**: These occur in saline habitat such as salt lake, marine condition.
   e.g; *Atriplex, Salsola, Salicornia, Tamarix*, Mangrove plants.

Morphological adaptations:

(a) Pneumatophores are present

(b) Stem and leaves are succulent

(c) Leaves are small, leathery and spinous

(d) Vivipary germination of seed occurs

Anatomical adaptations:

(a) Cuticle is thick

(b) Stomata is sunken

(c) Salt gland is found for secreting salts as in *Avicennia*

Physiological adaptations:

(a) Osmotic pressure is high

(b) Ability to tolerate salt stress.

Population

It is a group of organisms of same species living in a particular geographical area in a particular time. These organisms interbreed freely. These do not establish relationship of eating and being eaten. The local population of a species is called deme. Local population of a species of different areas is called sister population. The local population adapted genetically to its particular environment is called Ecotype. The intermediate form between two ecotypes is called ecoline. The entire group is called metapopulation. e.g., bacteria in a culture plate, fish in a pond, birds in a grove, elephants in a forest.

Features:

(a) Birth rate: It is the number of individuals added in initial size of population.

$$B = \frac{\text{Number of individuals added}}{\text{Initial size of population}}$$

(b) Death rate: It is the number of individuals removed in initial size of population. It often increases with age.

$$D = \frac{\text{Number of individuals removed}}{\text{Initial size of population}}$$

(c) Sex ratio: It is the number of females in a population per 1000 males.

$$\text{Sex ratio} = \frac{\text{Number of females}}{\text{1000 males}}$$

(d) Age distribution: It is the distribution of individuals of different age groups. There are three ecological groups in a population such as pre-reproductive, reproductive and post-reproductive individuals. These determine the three types of growth.

(i) Positive growth: A population has greater number of pre-reproductive individuals.

(ii) Zero growth: A population has equal number of pre-reproductive and post-reproductive individuals.

(iii) Negative growth: A population has greater number of post-reproductive individuals.

(e) Age pyramid: It is graphic representation of different age groups found in a population. It shows following three different status of growth.

(i) Triangular: It has more pre-reproductive (younger) individuals, moderate reproductive individuals and less post-reproductive individuals. It is indicator of growing population (positive growth) because birth rate is high.

(ii) Bell shaped: It has same number of pre-reproductive and reproductive individuals but the number of post reproductive individuals is less. It is indicator of stable population (neither growing nor declining) because birth rate is equal to death rate hence there is zero population growth.

(iii) Urn shaped: It has lesser number of pre-reproductive individuals than both reproductive and post-reproductive individuals. It is indicator of declining population (negative growth) because birth rate is low.

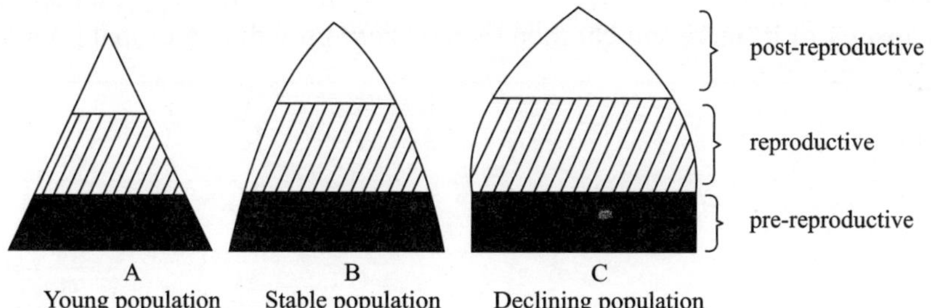

(f) Population density (size): It is the number of individuals of a species in a unit area (space) at a given time.

$$D = \frac{\text{Number of individuals}}{\text{Space}}$$

Space is expressed in two dimensions (m^2) for terrestrial organisms and in three dimensions (m^3) for aquatic organisms and for the organisms suspended in space. Population density depends on

(i) Natality: It is the number of births (or hatching / germination) per unit population per unit time. It adds new number to the population hence increases the density (size) of population.

(ii) Mortality: it is the number of deaths pet unit population. It moves the individuals from population hence decrease the density (size) of population.

(iii) Immigration: It is the permanent inward movement of same individuals into a local population. It increases the size of local population.

(iv) Emigration: It is the permanent outward movement of same individuals from a local population. It decreases the size of local population.

Under normal conditions, birth and death are important factors for influencing population density. Under special conditions, immigration and emigration are important factors for influencing population density.

$$\text{Density} = (\text{Natality} + \text{Immigration}) - (\text{Mortality} + \text{Emigration})$$

Population density is determined by the following factors:

(a) Counting all the individuals of an area

(b) Measuring biomass

(c) Determining relative density

(d) Counting indirectly such as pug marks and faecal pellets in case of tigers in national parks and reserves.

Growth of population is controlled by the interaction between three factors which are as follows:

(a) Biotic potential (Reproductive potential): It is the natural capacity of a population to increase at its maximum rate when all the environmental conditions are favourable. It is very high for all animals. If it is unchecked then the numbers of any species will quickly overrun the world.

(b) Environmental resistance: It is the sum of inhibitory factors such as unavailability of food, and shelter, drought, flood, fires, temperature fluctuations, accidents, pathogens and parasites and predators to check the biotic potential.

(c) Carrying capacity: It is the highest limit of population size or it is capacity beyond which growth is not possible. It is the number of individuals that a particular favourable environment can support over long period of time (indefinitely). It is symbolised as K. It is property of environment. It varies over space and time with the abundance of limiting resources. If a population reaches the carrying capacity of its environment then there is zero growth rate of that population.

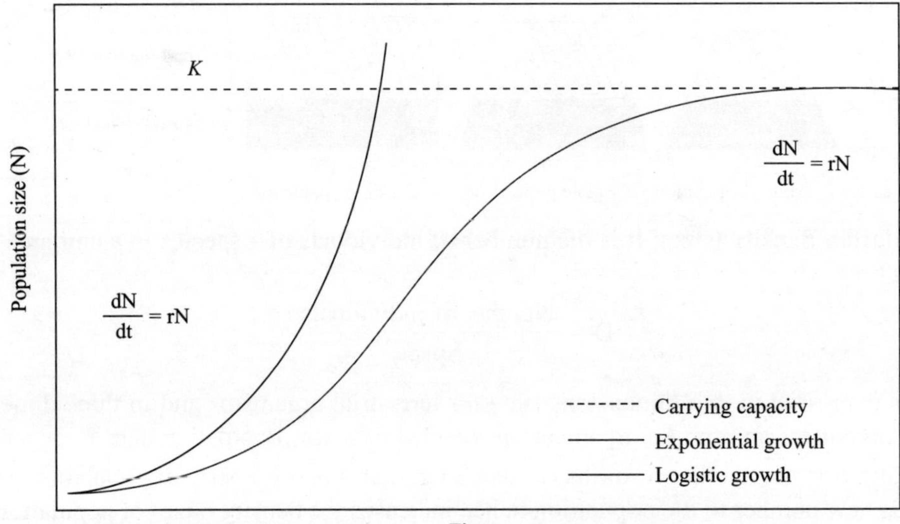

Population growth models: There are two models which are as such

(a) Exponential growth: It occurs when resource (food and space) is unlimited. It undergoes **lag phase** in which growth occurs slowly because individuals adapt themselves to the new environment and **log phase** in which growth occurs very rapidly and continues till enough food is available. With the excessive increase in population, the food is exhausted (deficiency of food) and environmental resistance increases hence there is mass mortality by which the population density (size) suddenly declines.

It shows J-shaped curve.

Biotic potential is realised because birth rate (natality) is higher than death rate (mortality).

It is shown by algal blooms, *Paramecium*, Insects, Lemming of Tundra and annual plants. The population of insects shows explosive increase in numbers during rainy season and disappear at the end of season. It is expressed as

$$dN/dT = rN = (B-D)N$$

where dN /dt = change in population size per unit time,

r = intrinsic rate of natural increase

N = Population size, B = Birth rate, D=death rate

The integral form of exponential growth is represented as

$N_t = N_o e^{rt}$ where Nt=population size after time t; No= population size in time t and e = base of natural logarithm.

(b) Logistic growth: It occurs when resource (food and space) is limited hence it is more realistic than exponential growth. It is also called Verhulst-Pearl logistic growth. It undergoes the following phases:

Lag phase: Growth occurs slowly because individuals adapt themselves to the new environment.

Log phase: Growth occurs rapidly.

Inflexion point: Growth slows down as environmental resistance increases and finally

Stationary phase or plateau phase: Growth stops and population stabilises around carrying capacity because the rate of mortality and natality become equal to each other.

It shows sigmoid or S-shaped curve.

The equilibrium number of individuals or growth of organisms in population is determined by carrying capacity (K) only.

It is shown by bacteria, yeast cells in a culture medium and most of organisms.

It is represented as dN/dt = $rN\frac{(1-N)}{K}$

where dN/dt = Rate of change in population size

r = intrinsic rate of natural increase (natality-mortality)

N = size of population, K = carrying capacity

K-N/ K = environmental resistance.

Exponential growth	Logistic growth
It is J-shaped.	It is S-shaped
It occurs when resources are abundant or unlimited.	It occurs when resources are limited or finite.
Population grows beyond carrying capacity.	Population cannot grow beyond carrying capacity.
Declining phase is not seen.	Declining phase is seen before reaching the equilibrium.
Exponential phase or log phase is of longer duration.	Exponential or log phase is of shorter duration.
Population crash occurs after growing to maximum population.	Population crash is generally not seen.

Population interaction

It is essential for the survival of different species and the community. It is categorised into neutral, positive and negative interactions depending upon the nature of effect. There are three types of interaction:

Neutral interaction: It is neither beneficial nor harmful interaction.

1. **Neutralism**: It is the interaction between two individuals of two different species in which both organisms are neither benefited nor harmed. e.g., Rat, shrew and rabbit live in the same area but none will be affected.

Positive interaction: It is the interaction in which at least one partner is benefitted and neither is harmed. It includes:

1. **Symbiosis** (obligate mutualism): It is the interaction between two individuals of two different species in which both organisms are benefited from each other and this association is obligatory (essential for the existence of both organisms). The term 'symbiosis' was coined by De Bary. It shows coevolution (both partners develop certain features to be successful).

   Examples:

   (a) Algae and fungi form lichen.

   (b) Algae and coelenterates form coral reef.

(c) Fungi and roots of higher plants form mycorrhiza.
(d) *Rhizobium* and leguminous plants form root nodule.
(e) Cyanobacteria and coralloid root of *Cycas* perform nitrogen fixation.
(f) *Salvia* and bees perform pollination.
(g) Fig and gall wasp cause pollination.
(h) *Yucca* and pronuba moth cause pollination.
(i) Mango and ants cause pollination.
(j) *Chlorella vulgaris* (green alga) and *Hydra.* The alga provides O_2 and food to *Hydra* which provides shelter and nitrogenous components to alga.
(k) *Trichonymph* (Protozoa) and Termite. *Trichonymph* secretes cellulase enzyme which helps in digestion of cellulose from wood by termitc which provides shelter, suitable environment and food to *Trichonymph.*
(l) *E. coli* and human beings. *E. coli* feeds on undigested food inside human's intestine and in return it releases vitamins and checks the growth of putrefying bacteria for humans.

2. **Proto-cooperation** (Facultative or Partial mutualism): It is the interaction of two individuals (organisms) of different species in which both organisms are mutually benefited but this association is non-obligatory (they can live independently).

Examples:

(a) Egret (bird) eats lice and tick of the body of cattle and thus cattle gets relief.
(b) Egyptian plover (bird) eats leeches and decaying meat stuck in the teeth of crocodile when it opens mouth and thus crocodile gets relief.
(c) Ox pecker (bird) eats lice and ticks of the body of rhinoceros and buffalo and thus these mammals gets relief.
(d) Hermit crab (*Eupagurus*) and sea anemone (*Adamsia*). Hermit crab takes sea anemone on its back to other feeding zone and in return sea anemone provides protection to crab from enemies.

3. **Commensalism**: It is the interaction of two individuals of different species in which one organism is benefited and other is neither benefited nor harmed.

Examples:

(a) Orchid (epiphyte) grows on mango plant for space
(b) Algae grows animals (epizoic algae)
(c) Spider builds web on a tree.
(d) Lianas (vascular plants rooted in soil) get support from other plant for their erectness.
(e) Clown fish is protected from enemies by carrying sea anemone.
(f) Barnacle (an arthropod) is attached to whale for space or shelter.
(g) Remora or *Echenesis* (sucker fish) are attached by dorsal fin to shark or whale for dispersal, protection from predators and obtaining nutrition from leftover of shark or whale.
(h) Vulture eats the leftovers of lion.
(i) Jackal eats the leftovers of lion.
(j) Bacteria grow on the surface of root (Rhizoplanic bacteria).
(k) Bacteria grow on the surface of leaf (Phylloplanic bacteria).

Negative interaction: It is an interaction in which at least one partner is harmed.

1. **Predation:** It is the interaction between two organisms (individuals) of different species in which one organism (predator) captures, kills and eats the other (prey) to obtain food. The predators are mostly animals and have small population whereas preys are also animals which have large population due to their high reproductive (biotic) potential. Prey will exponentially grow in the absence of predators.

   Significance:

   (a) Predators act as conduits between trophic levels for transfer of energy

   (b) Predators maintain species diversity by reducing the competition among competing prey species (It was observed in American pacific coast where all the starfish (*Pisater*) were removed from particular area then several species of invertebrates became extinct within a year due to intraspecific competition).

   (c) Predators keep the population of prey in check to maintain equilibrium.

   (d) Predators help in biological control of pests.

   (e) Predators help in growth of vegetation by checking the population of herbivore.

   Examples:

   (a) Predacious fungi such as *Dactylella, Dactylaria, Arthrobotrys* and *Zoophagus* kill annelids. These fungi have rings in mycelium which acts as noose and after that haustoria are released.

   (b) Insectivorous plants such as *Nepenthes, Dionea, Utricularia* etc kill insects

   (c) Cochineal insect (*Cactoblastis canctorum*) kills *Opuntia.*

   (d) Gambusia fish kills mosquito larvae.

   (e) Lady bird beetle kills the aphids.

   (f) Mynah kills red locust.

   (g) Herbivores eat plants.

   (h) Carnivores eat herbivores.

2. **Parasitism**: It is the interaction between two organisms (individuals) of different species in which one organism (parasite) is benefitted by obtaining nutrition from other organism (host) which is harmed.

   Parasitic adaptations:

   (a) Loss of digestive system as in *Taenia*

   (b) Occurrence of anaerobic respiration as in *Ascaris, Taenia* and *Fasciola.*

   (c) Presence of adhesive organs(suckers) as in leech and *Taenia*

   (d) Loss of sense organs

   (e) Well developed reproductive system

   (f) Excessive multiplication

   Types of parasite:

   1. **Ectoparasite**: These live on external surface of host. e.g., Lice, Tick, Flea, Aphid
   2. **Endoparasite**: These live inside body of host. It may be intracellular e.g., *Plasmodium*; body fluid e.g., *Trypanosoma*; Intestinal worms such as *Ascaris, Taenia.*
   3. **Temporary**: These live only for brief period. e.g., Female mosquito, bedbug, Leech

4. **Permanent**: These live for life time. e.g., *Ascaris, Entamoeba, Taenia*, Lice.
5. **Partial or semi parasite**: These are green and prepare their own food but they depend on host only for water and minerals.

   Root parasite: These obtain nutrition from root. e.g., *Santalum album* (largest parasite) found on *Albizzia, Dalbergia*, Morinda etc. *Striga* found on sorghum, sugarcane etc.

   Stem parasite: These obtain nutrition from stem e.g., *Loranthus* (found on *Acacia, Albizzia, Dalbergia*, mango), *Viscum*.
6. **Total or Holoparasite or obligate parasite**: These are non-green and fully dependent on host for food and water. They obtain nutrition by haustoria.

   Root parasite: These obtain nutrition from root. e.g., *Rafflessia* (found on *Ficus*), *Orobanche* (found on crucifers, tomato, potato), *Epifagus*

   Stem parasite: These obtain nutrition from stem. e.g., *Cuscuta* (found on Citrus, *Zizyphus*); *Arceuthobium* (smallest parasite) found on *Pinus, Juniperus* etc.
7. **Pathogenic:** These cause disease. e.g., Fungi (*Puccinia, Ustilag)* and Bacteria (*Mycobacterium, Vibrio, Corynebacterium)*.
8. **Non pathogenic:** They do not cause disease but only they obtain small amount of nutrition from host. e.g., *Entamoeba coli*, Aphid eating sap from plant.
9. **Hyperparasite or Epiparasite:** They live on another parasite e.g., Bacteriophage on bacteria (parasite), *Pasteurella pestis* on rat flea (*Xenopsella cheopsis*), a parasite.
10. **Brood parasitism:** In fact it is case of co-evolution. In this cuckoo (parasitic bird) lays eggs in nest of crow (host) and let the host incubate them. The eggs of cuckoo develop to resemble the host's egg in size and colour to reduce the chance of cuckoo to locate the foreign eggs.

    Significance:

    (a) Parasite shows host specificity.

    (b) Parasite has high reproductive or biotic potential.

    (c) Parasite is smaller than host.

    (d) Parasite reduces survival, growth and reproduction of host.

    (e) Parasite makes host weak so it is prone to be attacked by predators.

    (f) It is case of co-evolution (if host develops certain mechanism to resist or avoid parasite then the parasite will also develop certain features or mechanism to counteract host to be successful with the same host species.

3. **Amensalism**: It is the interaction of two individuals of different species in which one organism is harmed and other is neither harmed nor benefited.

   Examples:

   (a) *Penicillium notatum* inhibits the growth of a bacterium, *Staphylococcus aureus* by secreting allelochemical (Penicillin).

   (b) *Chlorella vulgaris* produces chlorellin which is toxic to other species.

   (c) *Trichoderma* inhibits the growth of *Aspergillus.*

   (d) *Convolvulus arvensis* checks the growth and germination of wheat.

   (e) Roots of *Parthenium argentatum* check the growth of other plants by secreting trans-cinnamic acid.

(f) Roots of *Juglans regia* (Walnut) check the growth of apple, alfalfa and tomato etc by secreting juglone.

(g) Roots of smoother crops (barley, sunflower and *Sorgum*) inhibit the growth of weeds by secreting allelochemicals.

4. **Competition:** It is the interaction (rivalry) between two or more individuals for obtaining same resource (food and shelter etc) in which one organism will die. It is of two types:
   (a) Intraspecific: It is the competition between individuals of same species. It controls the size of population. It is more acute and intense because all the individuals have similar requirements of food, light, water, space, shelter and mate.
   (b) Interspecific: It is competition between individuals of different species for the same resources. It is essential for organic evolution.

Different terms related to competition:

1. **Competitive exclusion**: It was studied by Gause (1984). In this two closely related compatible species competing for the same limited resource cannot coexist in the same place hence one of them gets eliminated after sometime due to strong competition. Thus one species uses the resource more efficiently and reproduce more rapidly than others. It was shown by *Paramecium*. (*Paramecium aurelia* increases its number gradually and there is decrease in number of *Paramecium caudata*).
2. **Competition release**: Two closely related compatible species competing for same limited resource cannot coexist in the same place hence one inferior species is restricted to a small geographical area due to the presence of competitively superior species. It was shown by Barnacle.
3. **Competition coexistence (Resource partitioning):** Two closely related compatible species competing for the same source coexist and avoid competition by choosing different feeding behaviour. It was shown by Warbler birds.
4. **Cannibalism**: It is type of predation in which larger individual of a species eats the smaller one of same species. The large fish eats small fish.

SHORT ANSWER TYPE QUESTIONS

1. Explain ecological niche.
2. Define biosphere.
3. Mention the major climatic regions.
4. Distinguish between eurythermal and stenothermal animals.
5. Distinguish between homiothermal and poikilothermal animals.
6. Distinguish between euryhaline and stenohaline organisms.
7. Mention the types of vegetations on the basis of rainfall.
8. Discuss the steps in formation of soil.
9. Write notes on humus.
10. Explain field capacity and water holding capacity.
11. Distinguish between c;ay soil and sandy soil.
12. Discuss the several parts of atmosphere.
13. Explain the Allen's rule.
14. Write notes on biotic factors.

15. Explain altitude sickness.
16. Write notes on Batesian and Mullerian mimimcry.
17. Distinguish between sun plants and shade plants.
18. Distinguish between physical xerophytes nad physiological xerophytes.
19. Explain biotic potential and carrying capacity.
20. Distinguish between exponential growth and logistic growth.
21. Distinguish between partial parasite and total parasite.
22. Write notes on predation.
23. Write notes on amensalism.
24. Distinguish between intraspecific competition and interspecific competition.
25. Explain competitive exclusion.

LONG ANSWER TYPE QUESTIONS

1. Describe the role of temperature, light and water in ecology.
2. Discuss the several responses of organisms to abiotic factors
3. Describe the morphological and anatomical adaptations in hydrophytes.
4. Describe the morphological and anatomical adaptations in xerophytes.
5. Discuss the population growth models.
6. Discuss the different features of population.
7. Explain positive population interaction with response to symbiosis, proto-cooperation and commensalism.
8. Describe the types of parasite nad show their adaptations and significance.

MULTIPLE CHOICE QUESTIONS

1. The number of level of organisation of living and non-living matter is
(a) One (b) Two (c) Three (d) Four
2. Race is defined as
(a) Group of organisms of same species (b) Group of organisms within a species
(c) Group of organisms of different species (d) Group of complex biotic community
3. The complex biotic community stretching over a large geographical area is called
(a) Ecosystem (b) Biome (c) Ecotone (d) Biosphere
4. The temperate climatic region is situated at latitude
(a) 0-20° (b) 20-40° (c) 40-60° (d) 60-80°
5. Which of the following comes under indirect factor?
(a) Temperature (b) Light (c) Water (d) Rainfall
6. Which of the following plant comes under hekistotherm?
(a) Teak (b) Sal (c) Pinus (d) Rhododendron
7. The size of extremeties such as ear, tail, snout, legs etc is considered in
(a) Allen's rule (b) Bergman's rule (c) Gloger's rule (d) Jordon's rule
8. Salt content in sea is
(a) Less than 2% (b) Between 3-3.5% (c) Between 5-10% (d) More than 10%
9. Which of the following processes is not concerned with paedogenesis?
(a) Humification (b) Eluviation (c) Elution (d) Illuviation
10. The most ecologically relevant environmental factor is
(a) Temperature (b) Light (c) Water (d) soil
11. In soil, the principal cation in alkaline soil is
(a) Na^+ (b) Ca^{+2} (c) K^+ (d) Mg^{+2}
12. Which of the following is incorrect?
(a) A-horizon of soil contains minerals and humus
(b) Amount of mineral matter is 40%
(c) Clay particle is the smallest in size
(d) Humus is partially decomposed organic matter
13. Which of the following water is available for plants?
(a) Hygroscopic water (b) Gravitational water (c) Capillary water (d) All of the above
14. If birth rate is 50 and death rate is 40 and population size is 80 then dN/dT is
(a) 200 (b) 400 (d) 800 (d) 1600
15. A person studied the population of cat in which the average natality was 100, average mortality 90, the average immigration 10 and average emigration 15 then the net increase in population is
(a) 5 (b) 10 (c) 20 (d) 40
16. If the birth rate is 100 and death rate is 10 in the population of 1000 individuals. The percentage natural growth rate is
(a) 5% (b) 9% (c) 10% (d) 20%
17. In a pond there are 20 lotus plants. Due to reproduction 10 new plants were added last year taking the correct population of 30. The birth rate is
(a) 0.5 (b) 1 (c) 1.5 (d) 2

18. A person caught, marked and released 80 fish in pond. Later 100 fish were caught randomly in which 40 were found to be marked. The number of fish in pond is
    (a) 120 (b) 150 (c) 180 (d) 200
19. The growth rate of a population according to logistic model is zero when
    (a) N/K is equal to zero (b) N/K is exactly one
    (c) N is close to carrying capacity (d) Death rate is higher than birth rate
20. The ratio between mortality and natality is called
    (a) Population ratio (b) Vital index (c) Density coefficient (d) Carrying ratio
21. Which one of the following correctly represents an organism and its ecological niche?
    (a) *Hydrilla* and pond (b) Desert locust and desert forest
    (c) Aphid and leaf (d) Vulture and dense forest
22. Which of the following is maximum contributor of oxygen?
    (a) Grassland (b) Herbs and shrubs (c) Dense forest (d) Phytoplankton
23. Xeric environment is characterised by
    (a) Precipitation (b) Low atmospheric humidity
    (c) Extreme temperature (d) High rate of vaporisation
24. Which of the following is incorrect?
    (a) About 1% of organisms are regulators
    (b) About 99% of animals and all plants are conformers
    (c) Conformers have not evolved the mechanism of homeostasis
    (d) The smaller animals tend to lose heat very slowly
25. The condition of diapauses is noticed in
    (a) Fish (b) Snail (c) Zooplanktons (d) Frog
26. Monarch butterfly is distasteful or unpalatable because it contains harmful substance, cardiac glycosides due to eating upon
    (a) *Parthenium* (b) *Nerium* (c) *Calotropis* (d) *Crocus*
27. Heterophylly condition is seen in
    (a) *Hydrilla* (b) *Wolffia* (c) *Limnophylla* (d) *Vallisneria*
28. Bell shaped pyramid is indicator of
    (a) Stable population (b) Growing population
    (c) Declining population (d) None of the above
29. In logistic growth, growth of organisms in population is determined by
    (a) Carrying capacity (K) only (b) Size of population (N) only
    (c) Biotic potential (d) Both carrying capacity and Biotic potential
30. Which of the following is incorrect?
    (a) Exponential growth occurs when resource is unlimited
    (b) It shows J-shaped growth
    (c) Population crash occurs
    (d) Biotic potential is not realised
31. Population density of terrestrial organisms is measured in terms of individual per
    (a) Meter (b) Meter^2 (c) Meter^3 (d) Meter^4
32. Carrying capacity of a population is determined by
    (a) Natality rate (b) Mortality rate
    (c) Predation (d) Limiting resources

33. Population termed r-strategists have
    (a) J-shaped growth curve
    (b) S-shaped growth curve
    (c) Pioneer species
    (d) All of the above
34. Population growth curve in most animals except human beings, is
    (a) J-shaped
    (b) S-shaped
    (c) J-shaped with tail
    (d) S-shaped with tail
35. A cuckoo laying eggs in the nest of other species of birds shows
    (a) Hyperparasitism (b) Ectoparasitism (c) Brood parasitism (d) Commensalism
36. Which of the following interaction acts as conduit for energy transfer across trophic level?
    (a) Parasitism (b) Amensalism (c) Protocooperation (d) Predation
37. Trichonymph and termite exhibit
    (a) Parasitism (b) Symbiosis (c) Protocooperation (d) Commensalism
38. Remora and Shark exhibit
    (a) Protocooperation (b) Commensalism (c) Amensalism (d) Symbiosis
39. Which of the following is incorrect about parasitism?
    (a) It is a case of co-evolution
    (b) Plasmodium is intracellular endoparasite
    (c) Parasite shows host specificity
    (d) Parasitic adaptations do not occur in *Ascaris*, *Taenia* and *Hirudo*.
40. Which of the following is incorrect about competition?
    (a) Intraspecific competition is more acute than interspecific competition.
    (b) Competition exclusion in paramecium was explained by Gause
    (c) Competition release was shown by Barnacle.
    (d) Resource partitioning was not shown by Warbler birds.
41. The effective population size for completely monogamous species having 40 males and 10 females would be
    (a) 10 (b) 20 (c) 30 (d) 40
42. Which of the following ecological factors is density independent?
    (a) Food
    (b) Predation
    (c) Disease
    (d) Extreme temperature
43. Which of the following shows most different from others?
    (a) Algae embedded in coral tissue
    (b) *Salmonella* in human gastric tract
    (c) Cellulolytic bacteria in termite gut
    (d) Pollen collecting bees visiting flowers
44. A term 'biotype' means
    (a) All individuals having same phenotype
    (b) All individuals having same genotype
    (c) All individuals having different phenotype
    (d) All individuals having different genotype
45. A term ' ecotype' means
    (a) Genetically different but phenotypically similar individual
    (b) Genetically similar but ecologically different individuals
    (c) Genetically adapted ecological race
    (d) Genetically and phentypically dissimilar individuals.
46. The characteristic population growth curve of bacteria is
    (a) Straight (b) S-shaped (c) Random (d) J-shaped

47. Soil in naturally growing vegetation is
    (a) Acidic due to bicarbonates in soil
    (b) Basic due to bicarbonates in soil
    (c) Neutral due to bicarbonates in soil
    (d) Acidic due to sulphuric acid in soil
48. When a marine fish is placed in fresh water, it dies because
    (a) Entry of water a nd it floats
    (b) Exit of water and it shrinks
    (c) Deficiency of salts
    (d) Deficiency of nutrients
49. The birds in tropics are generally smaller in size due to
    (a) Increase in surface area to volume ratio
    (b) Decrease in surface area to volume ratio
    (c) Easy flight
    (d) Aestivation
50. Which of the following is not a limiting factor to the growth of population?
    (a) Availability of food
    (b) Increasing population of predators
    (c) Introduction of disease
    (d) Long term abundance of space
51. Which of the following zones shows the occurrence of coniferous forests?
    (a) Temperate zone (b) Tropical zone (c) Subtropical zone (d) Arctic zone
52. Which of the following particles of soil is the smallest in size?
    (a) Silt (b) Clay (c) Fine sand (d) Gravel
53. Which of the following types of water is present in the soil in thin film?
    (a) Gravitational water (b) Hygroscopic water (c) Capillary water (d) Combined water
54. Field capacity of water does not include
    (a) Capillary water
    (b) Hygroscopic water
    (c) Combined water
    (d) Gravitational water
55. Which of the following animals shows partial regulate response to abiotic factor?
    (a) Reptiles (b) Birds (c) Mammals (d) Cray fish
56. Logistic growth is represented as
    (a) $dN/dt = rN$ (b) $dN/dt = rN(1-N/K)$ (d) $dN/dt = rN(1-K/N)$ (d) $dN/dt = r(1-N/K)$
57. Competitive exclusion is best observed in
    (a) Warbler birds (b) Bonellia (c) Pramecium (d) Euglena
58. Which of the following acts as conduit between trophic levels for transfer of energy?
    (a) Parasite (b) Symbiont (c) Commensal (d) Predator
59. Hermit crab and sea anemone show best example of
    (a) Symbiosis (b) Protocooperation (c) Commensalism (d) Amensalism
60. Monarch butterfly is distasteful to predators because it contains cardiac glycosides in its body due to consumption of
    (a) *Nerium* (b) *Euphorbia* (c) *Parthenium* (d) *Calotropis*

ANSWERS

1.c	2.b	3.b	4.c	5.d	6.d	7.a	8.b	9.a	10.a
11.b	12.d	13.c	14.c	15.a	16.b	17.a	18.d	19.b	20.b
21.c	22.d	23.b	24.d	25.c	26.c	27.c	28.a	29.a	30.d
31.b	32.d	33.a	34.a	35.b	36.d	37.b	38.b	39.d	40.d
41.b	42.d	43.b	44.a	45.c	46.d	47.b	48.c	49.d	50.d
51.a	52.b	53.b	54.d	55.d	56.b	57.c	58.d	59.b	60.d

22

Ecosystem

Ecosystem is an interactive system between living and nonliving components of environment where both are interdependent.The term 'ecosystem' was coined by Tansley (1935).

It is an open system because there is continuous and variable entry and loss of energy and materials. It is structural and functional unit of environment.

Terms of Ecosystem:

1. Biocoenosis proposed by Carl Mobius
2. Microcosm proposed by Forbes
3. Biogeocoenosis proposed by Sukhachev
4. Holocoen proposed by Friederich

Types of ecosystem:

1. **Natural ecosystems:** These are naturally developed and operate without human intervention. These are of following types
   (a) Terrestrial ecosystem: Grassland, forest, desert, garden
   (b) Aquatic ecosystem:
      (i) Freshwater (Limnetic): lake, pond (lentic or stagnant); river, stream (lotic or running) and
      (ii) Marine: Ocean, sea, estuary
2. **Artificial ecosystems**: These are developed by human beings hence called man-made.. e.g., agriculture (largest), lake, dam, aquarium, garden, spacecraft etc.

Types of ecosystems on the basis of size:

1. Megaecosystem: It is of large size. e.g., Ocean
2. Mesoecosystem: It is of medium size. e.g., Desert, forest, grassland
3. Microecosystem: It is of small size. e.g., Lake, pond
4. Nanoecosystem: It is of very small size. e.g., Aquarium, garden, log of wood etc.

Components of ecosystem: The ecosystem has two functional components which are as follows:

1. **Abiotic components**: These include kinds, quantity, quality and distribution of physical and chemical factors. These determine the maintenance of biotic components. These are temperature, light, water, soil, wind, humidity, topography, inorganic substances such as N, P, S, Ca etc. and organic substances such as carbohydrate, protein, lipid etc.
2. **Biotic components**: These include kinds, numbers and distribution of living organisms. These obtain inorganic nutrients and energy from abiotic components.

(a) **Producers (Autotrophs)**: These are organisms which are capable of synthesising food in the presence of sunlight and chlorophyll by using inorganic compounds. These are source of food for all other organisms which are incapable of synthesising their own food.

These are also called **converters or transducers** because they convert light energy to chemical energy. They take part in building of their bodies by synthesising organic compounds as well as release energy for various physiological activities. They maintain CO_2/O_2 balance in nature. They pick up CO_2 from atmosphere and release O_2 as a result of photosynthesis.

In aquatic ecosystem, major producers are phytoplanktons or rooted or large floating plants in freshwater ecosystem and algae in marine ecosystem but in terrestrial ecosystem, rooted plants (trees) are major producers.

Other producers are photoautotrophic and chemoautotrophic bacteria, cyanobacteria, photosynthetic protists such as diatoms, dinoflagellates, and euglenoids.

(b) **Consumers** (Heterotrophs): These are organisms (animals) which are incapable of synthesising their own food and depend on other organisms for food hence these are known as phagotrophs.

(i) **Primary consumers**: (Consumers of first order): These are organisms which directly feed on plants only. These are also called herbivores. These are also known as key industry animals because they convert plants into animals. e.g., grasshopper, deer, cattle (terrestrial ecosystem); molluscs, tadpole and mosquito larvae (aquatic ecosystem).

(ii) **Secondary consumers** (Consumers of second order): These are organisms which feed upon herbivores. These are also called primary carnivores. e.g., Frog, spider, dog, cat, tiger and lion (in terrestrial ecosystem); small fish, *Hydra* and frog (in aquatic ecosystem).

(iii) **Tertiary consumers** (consumers of third order): These are organisms which upon secondary consumers. These are also called secondary carnivores. e.g., snake, hawk (terrestrial ecosystem); shark, whales (aquatic ecosystem).

(iv) **Quaternary consumers** (consumers of fourth order): These cannot be preyed upon by other animals. These are also called top carnivores. e.g., lion, vulture, falcon, peacock.

(c) **Decomposers** (Saprotrophs): These are organisms which feed upon dead bodies of organisms and organic wastes of living organisms. These are also called **osmotrophs** because they secrete extracellular enzymes to digest the organic matter hence the digested form of organic matter is partly absorbed by microbes for their own assimilation. The remaining adds new materials and minerals back into substratum. These are called **microconsumers** because of their small size. These are also called **reducers** because they remove or degrade the dead bodies of organisms. These may be called **mineralisers** because they make the nutrients available to the producers thus they maintain a flow of nutrients in an ecosystem through biogeochemical cycle.

(d) **Parasites**: They obtain food directly from other organisms. e.g., bacteria, fungi, some worms, some insects etc.

(e) **Detrivores**: They breakdown organic matter as they feed upon detritus. e.g., Earthworm, Termite.

(f) **Scavengers**: They feed on dead bodies of organisms. e.g., Crow, vulture, Carrion beetles.

Different components of ecosystem:

1. **Aquatic Ecosystem (Pond)**:

(a) Abiotic components: Water, temperature, light, pH of water, inorganic and organic substances

(b) Biotic components:

Producers: Plants

Consumers: Primary-zooplanktons (insect larvae, rotifers, molluscs)

Secondary-Insects, medium fish, frog, snake

Tertiary- Large fish (Shark)

Decomposers: Bacteria, Fungi

2. **Terrestrial Ecosystem (Forest):**

(a) Abiotic components: Temperature, light, inorganic and organic substances

(b) Biotic components:

Producers: Plants (herbs, shrubs, trees)

Consumers: Primary- Ants, Bugs, Birds, Herbivores

Secondary- Snake, lizard, large bird

Tertiary- Tiger, lion

Decomposers: Bacteria, Fungi

Features of Ecosystem:

(a) Species composition: It is the variety of species present in an ecosystem. It depends upon geography, topography and location. It varies in different ecosystem. It is minimum in desert and arctic region. It is maximum in tropical rain forest (in terrestrial region) and coral reef (in aquatic region).

(b) Stratification: It is presence of vertical layers which depends on density of vegetation. It is minimum (3 layers such as herbs, shrubs and trees) in desert and arctic region and maximum (7 layers such as Tall trees, canopy trees, under storey trees, tree layer, shrub layer, herb layer, grass layer) in tropical rain forest.

(c) Trophic structure: It is presence of specific food chain and food web in ecosystem. It is usually grazing food chain in grassland.

(d) Standing crop: It is the amount of living matter or biomass (number of living organisms present in unit area) in an ecosystem present at any time. It is usually measured in dry weight.

(d) Standing state: It is the amount of inorganic nutrients (materials present in soil and water) in an ecosystem per unit area at anytime. It varies in different season and different ecosystems.

Functional Components of Ecosystem:

1. Productivity

It is rate of production of organic matter (biomass) at any trophic level per unit area in a given period of time. It is measured as weight ($g/m^2/yr$) or energy ($Kcal/m^2/yr$).

(a) Primary productivity: It is amount of organic matter produced per unit area per unit time.

(i) Gross primary productivity (GPP): It is total amount of organic matter synthesised by producers as a result of photosynthesis. It depends on photosynthetic efficiency of plants, availability of solar energy and availability of inorganic nutrients.

(ii) Net primary productivity (NPP): It is net amount of organic matter stored by producers after loss in respiration (R) in plants. It is available biomass for the consumption of heterotrophs.

$$NPP = GPP - R$$

(b) Secondary productivity: It is rate of accumulation of organic matter by consumers level per unit area per unit time. It depends on loss while transferring energy from previous trophic level and consumption due to respiration and predation.

(i) Gross Secondary productivity (GSP): It is amount of food energy absorbed.

(ii) Net secondary productivity (NSP): It is amount of food energy stored.

$$NSP = GSP - R$$

(c) Net community productivity: It is rate of storage of organic matter not used by heterotrophs (consumers) or it is rate of net synthesis of organic matter by a community per unit area per unit time.

Features:

(a) The bulk of earth's living mantle consists of plants and only small fraction constitutes animals.

(b) Primary productivity depends on plant species.

(c) It varies in different ecosystems.

(d) It is minimum in desert, arid land and deep sea.

(e) It is moderate in temperate (coniferous) forest, agricultural crops, grassland.

(f) It is maximum in tropical rain forest, ocean, coral reef, estuary, sugarcane field. Coral reef grows in areas having sufficient light, enough warm water and abundant nutrients.

Measurement of productivity: It is measured in following ways:

(a) Ecological efficiency**:** It is the percentage of energy transferred from one trophic level to another. It differs in different species, population or communities.

$$\text{Ecological efficiency} = \frac{\text{Energy converted into biomass at one trophic level}}{\text{Energy present in biomass at previous torphic level}} \times 100\%$$

(b) Ecological Efficiency = Exploitation efficiency × Assimilation efficiency × Net production efficiency

(c) Exploitation efficiency: It is prey biomass which is eaten by predator.

(d) Assimilation efficiency: It is proportion of consumed energy that is assimilated.

(e) Net production efficiency: It is proportion which assimilated energy is incorporated into growth, storage and reproduction.

(f) Net production efficiency $= \dfrac{NPP}{GPP} \times 100$

(g) Photosynthetic efficiency $= \dfrac{\text{Gross primary productivity}}{\text{Incident total solar radiation}} \times 100$

It is 0.8% in forest, 1.15% in grassland, 5% in cropland and 10% in sugarcane.

Factors affecting primary productivity:

(a) Temperature: Primary productivity is less in temperate plants when temperature is minimum and it is more in tropical plants when temperature is maximum

(b) Light: It is less in polar region because light is not available and more in tropical region because light is abundant.

(c) Water: It is less in desert because water content is low and more in tropical rain forest because water content is high.

(d) Nutrients: It is less in desert and sea because nutrient content is low and more in estuary and coral reef because nutrient content is high.

(e) Photosynthetic efficiency: It is less in forest and more in crop lands. It is less in C_3 plants and more in C_4 plants. It is maximum (10%) in sugarcane.

2. Decomposition

It is the breakdown of complex organic compounds by bacteria and fungi. The organic remains (dead plant parts, animal remains and excreta) are called detritus. The upper layer of soil is main site of decomposition in terrestrial ecosystem. It occurs as below ground detritus (dead roots, underground dead animals) and above ground detritus (dried plant parts, leaf litter, remains and excreta of animals).

Process:

(a) **Fragmentation of detritus**: A large amount of detritus was fed upon by detrivores such as earthworm, termites and carrion flies and leave small fragment. They cause fragmentation or pulverisation of detritus in digestive tract of herbivore and a part comes out. The part digested by detrivores is immobilised. Due to fragmentation, leftover detritus comes to have large surface area.

(b) Catabolism: The complex and insoluble organic substances are converted into simple and soluble organic and inorganic substances by extracellular enzymes released by decomposers such as bacteria and fungi. A part of broken down food is taken up by decomposers and immobilised.

(c) Leaching: Soluble substances (sugars and inorganic nutrients) drain or enter in deeper layer of soil by percolating water.

Detrivores	Decomposers
They eat detritus (dead organic remains).	They obtain nutrition from detritus.
They do not decompose organic matter but ingest it	They decompose organic matter.
They do not release enzymes	They release enzymes.
They cause pulverisation of detritus	They cause humification and mineralisation.
e.g., earthworm, Termites, Carrion flies	e.g., Bacteria, Fungi

(d) Humification: It is formation of humus by partial decomposition of detritus. It is partially decomposed, amorphous, dark coloured, slightly acidic and colloidal in nature. Humus is rich in cellulose and lignin. It is quite resistant to microbial action. It is reservoir of NH^{+4}, Ca^{+2}, Mg^{+2} and K^+ nutrient. These nutrients are released slowly as humus is slowly decomposed. It maintains the soil moisture and aeration.

(e) Mineralisation: It is release of inorganic substances such as minerals (NH^{+4}, Ca^{+2}, Mg^{+2}, K^+) and non-minerals (CO_2 and H_2O) from organic matter. This process is slow due to immobilisation and trapping in humus.

Detritus

↓Fragmentation, Catabolism and Leaching

Products

↓Humification

Humus

↓Mineralisation

NH^{+4}, Ca^{+2}, Mg^{+2} (minerals) and CO_2, H_2O (non minerals)

Factors affecting decomposition:

(a) **Temperature**: Optimum temperature for decomposition is 25°C. The rate of decomposition of detritus decreases if temperature is less than 10°C and increases if temperature is more than 25°C.

(b) **Moisture**: Optimum moisture is required. The rate of decomposition of detritus decreases if there is low moisture and also decreases if there is excess moisture.

(c) **pH of soil**: The rate of decomposition of detritus decreases in acidic soil and increases in neutral or slightly alkaline soil.

(d) **Oxygen content (aeration)**: The rate of decomposition of detritus decreases in anaerobic condition and increases in aerobic condition.

(e) **Composition of detritus**: The rate of decomposition of detritus is very slow if detritus is rich in cellulose, chitin, lignin etc and very rapid if detritus is rich in proteins, fatty acids and water soluble sugars.

3. Energy flow

It is continuous flow of energy starting from sun to biotic components and then to physical environment. It is governed by first and second laws of thermodynamics.

Features:

(a) It is the basis of life. Living organisms use energy in the form of radiant energy (sunlight) and fixed energy (potential chemical energy bound in various organic substances).

(b) Its source is sunlight. Less than 50% of solar energy incident over earth is photosynthetically active region (PAR) and rest more than 50% is lost (20% reflected by cloud +10% absorbed by cloud + 10% absorbed by gases/vapours + 10% scattered by dust). Out of 50% PAR, 45-49% is lost from the plants by reflection, radiation or evaporation and the remaining 1-5% of incoming radiation is utilised by chlorophyll and used in photosynthesis (gross primary productivity). Out of 1-5%, 0.8-4% is utilised in net primary productivity and rest 0.2-1% is lost in respiration.

(c) It is unidirectional. Sunlight ⟶ Producers ⟶ Consumers). It does not pass back to autotrophs.

(d) It starts with the fall of sunlight on the green surface of plants, a part of it is transformed into chemical energy which is stored in various organic products in the plants. When the herbivores consume plants as food and convert chemical energy accumulated in plants products into kinetic energy, degradation of energy will occur through its conversion into heat. When carnivores consume herbivores then further degradation will occur likewise.

(e) It decreases with rise in trophic level (Energy of Producers > energy of primary consumer > energy of secondary consumer > top consumer).

(f) It determines the diversity and density of organisms.

(g) It determines the developmental and functional status of ecosystem.

(h) It is either passed on to the higher trophic level or becomes available to detrivores or decomposers after death of organisms.

Ten percent law: It was proposed by Raymond Lindeman (1942). During the transfer of energy from one trophic level to next, about 10 % of energy is stored in flesh for body building and remaining 90 % of energy is lost in ingestion, respiration, incomplete digestion and maintenance of body heat. The respiratory loss gradually increases in successive trophic levels. It is 20 % at producer level, 30 % at consumer level and 60 % at top carnivore level.

Producer biomass (1000Kcal)
10% transferred ↓ 90 % lost
Primary consumer biomass (100 Kcal)
10% transferred ↓ 90 % lost
Secondary consumer biomass (10 kcal)
10% transferred ↓ 90 % lost
Top consumer biomass (1 Kcal)

Single channel energy flow model	Lindemann
Y-shaped energy flow model	E.P.Odum

Food chain: It is the sequence of organisms of an ecosystem through which food and its energy passes with each member in a continuous manner. It can be seen in any ecosystem. It was firstly proposed by Charles Elton (1927).

Features:

(a) It is generally straight and simple food chain rarely occurs in nature. (Producers → Herbivores → Carnivores → Decomposers).
(b) It shows gradual decreases in biomass, number and energy of individual with rise in trophic level.
(c) It is sustained by producers and decomposers.
(d) The shorter the food chain, the greater is the energy available.
(e) It does not include decomposers because they operate at all level.

Types of food chain:

1. **Grazing food chain** (Predatory food chain): It is most common. It directly depends on sunlight. It gets energy from sunlight. It starts from green plants (small organism), goes to grazing herbivores and passes to carnivores (large animals). It supports detritus food chain by providing organic matter. It is of two types:
   (a) Terrestrial ecosystem: It shows the following examples:
   - Grass→Grasshopper→Bird→Hawk
   - Grass→Grasshopper→Lizard→Hawk
   - Grass→Grasshopper→Frog→snake→Peacock→Hawk
   - Grass→Rabbit→fox→Wolf→Lion
   - Grass→Deer→lion
   - Grass→Insects→Insectivorous insect→Bird→Hawk

   (b) Aquatic ecosystem: It shows the following examples:
   - Phytoplankton→Zooplankton→Small fish→Large fish→Crocodile/Shark
   - Phytoplankton→Zooplankton→Frog→Snake
   - Phytoplankton→Zooplankton→Crustacean→Predatory insect→Small fish→King fisher
2. **Detritus food chain** (Saprophytic food chain): It is less common. It does not depend on sunlight hence producers are absent. It starts from detritus (organic remains of plants and animals). It gets energy from detritus. It supports grazing food chain by providing inorganic matter.
   - Detritus→Earthworm (detrivore)→Sparrow→Falcon (large bird)
   - Detritus→Earthworm (detrivore)→Frog→Snake→Hawk

3. **Parasitic food chain:** It is also known as accessory or subsidiary food chain. It starts with host (producer or large organism) and ends in parasite (small organism) without outright killing. It is often connected with grazing food chain
   - Grass→Cattle→Bacteria
   - Tree→Fruit eating bird→Tick/lice→Bacteria/Fungi

Food web: It is network of several food chains which become interconnected at different trophic levels forming a number of feeding connections amongst different organisms. One particular animal may feed upon organisms of different trophic levels. e.g., snake may feed upon mice (herbivore) and frog (carnivore). Likewise Jackal is both carnivore and scavenger. Sparrow is a primary consumer when it eats seeds, fruits etc. as well as secondary consumer when it eats insects and worms. Human consumes vegetables (producer), animals (primary consumer or herbivore) and Salmon (secondary consumer or carnivore).

Features:

(a) It is more real than food chain.
(b) It consists of a number of food chains.
(c) It is not straight as food chain.
(d) It provides a number of alternative foods to consumers which become polyphagous.
(e) It maintains balance and stability in ecosystem.
(f) It increases adaptability and competitiveness of the organ.

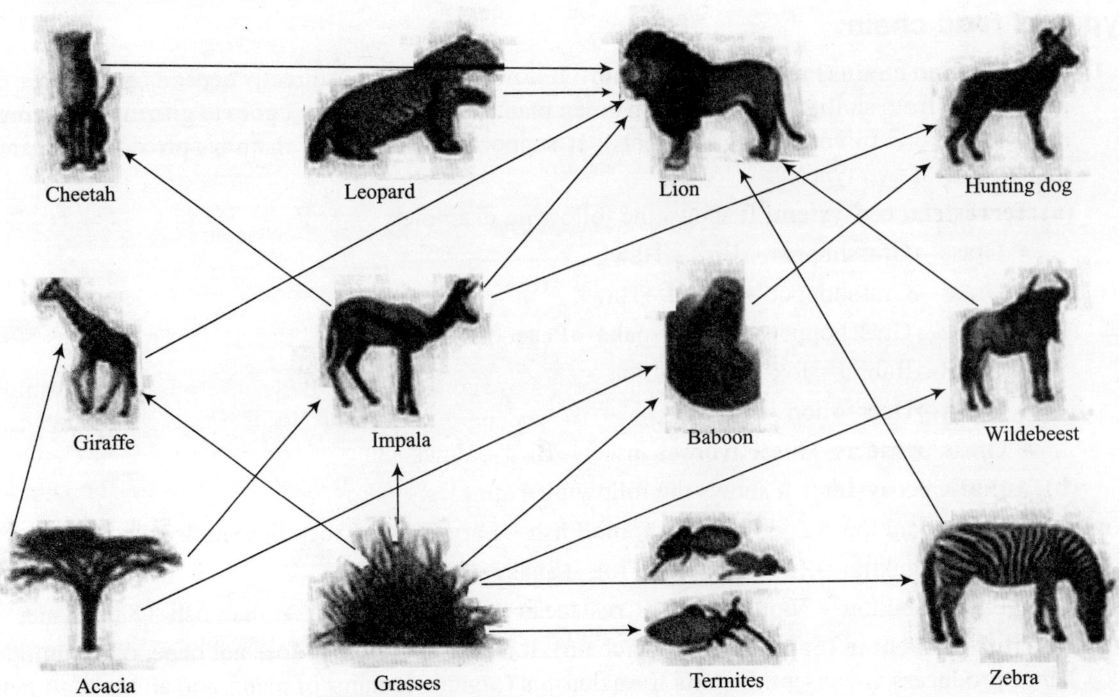

Fig. ***Terrestrial food web***

Trophic level: It is level in food chain which is characterised by the method of obtaining its food. The parasites have no fixed trophic level since they feed on producers, herbivores as well as carnivores of various levels. Sparrow comes under more than one trophic level. Its number is 3 to 6. The number of trophic level is less (3 in tropical forest, 4 in grassland) in terrestrial ecosystem. The number of trophic level is more (5 in coastal ocean ecosystem and 6 in open ocean ecosystem. It is shown as such:

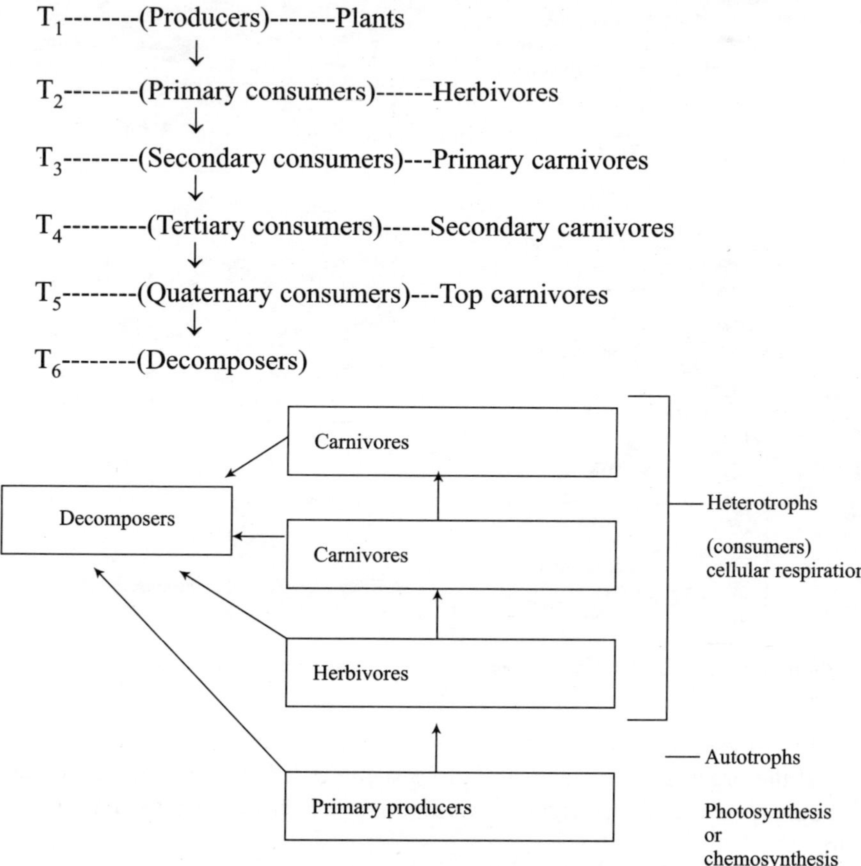

4. Ecological pyramid

It was proposed by Charles Elton (1927). It is diagrammatic representation of the nutritional relationship of organisms of different trophic level and the flow of energy from one trophic level to another in ecosystem. It gives an idea about the structure of ecosystem consisting of interaction of food chains. The ecological pyramid is not suitable for detrivores and decomposers. It does not represent food web. It does not include omnivore. It is of three types:

(a) Pyramid of Number: It actually shows the number of individuals involved in each step of food chain.

Upright	Grassland ecosystem (terrestrial), Pond ecosystem (aquatic)
Inverted	Parasitic food chain
Spindle shaped	Forest or Tree ecosystem (terrestrial)

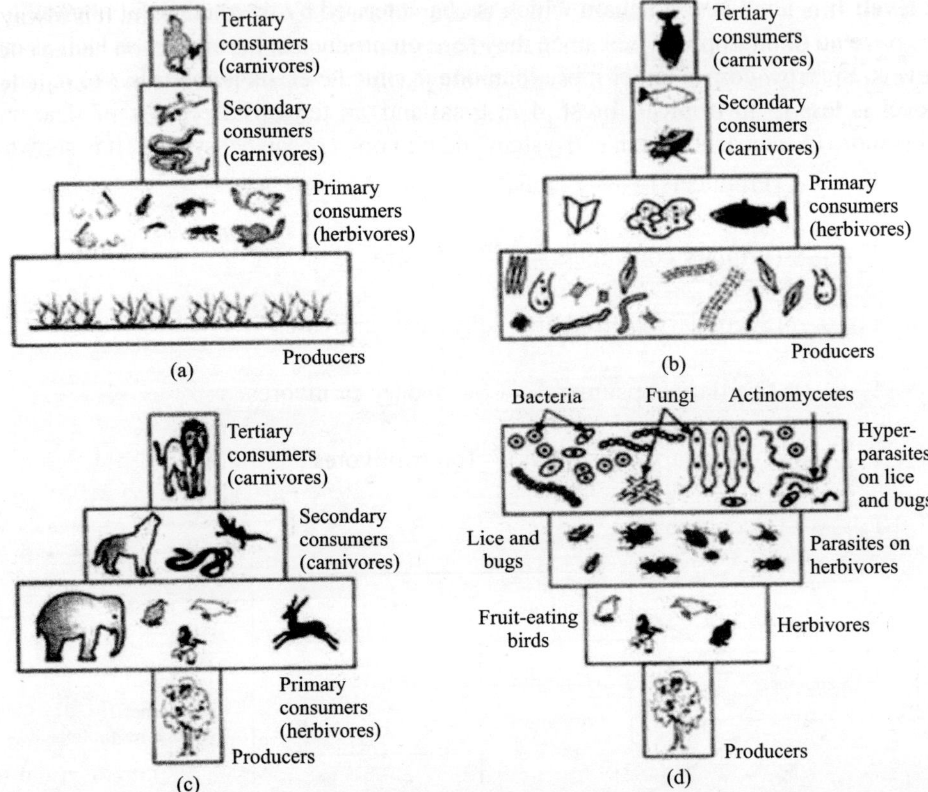

Fig. Pyramid of number: (a) Grassland ecosystem, (b) Pond ecosystem, Forest ecosystem and (d) Parasitic food chain

(b) Pyramid of biomass: It shows the relative amount of biomass (total organic matter or total body weight of all the organisms) in each step of food chain. It is more real than pyramid of number because it considers the size (weight) of individual. It is easier to estimate the dry weight in an ecosystem.

Upright	Grassland and forest ecosystem (terrestrial)
Inverted	Pond ecosystem (aquatic)

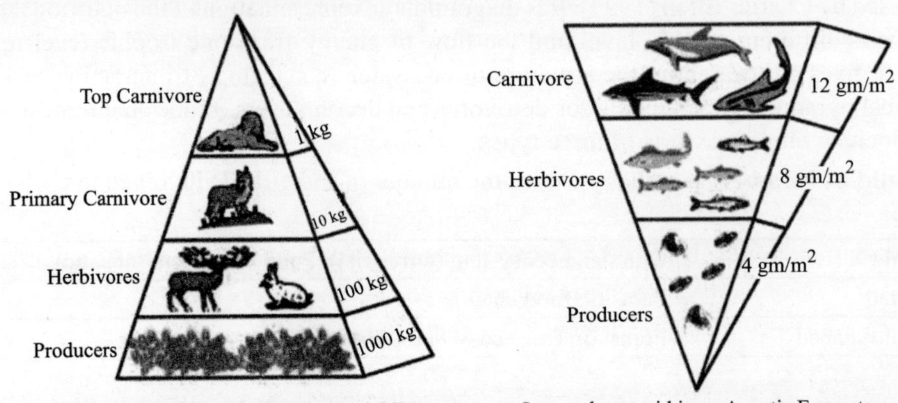

(c) Pyramid of energy: It shows the amount of energy in each step of food chain. It is always upright because energy of producers is always more than primary consumer and so on hence energy goes on decreasing. It is clear that only a part of energy (10%) is transferred to next trophic level from the previous trophic level and a major part (90%) is lost as heat.

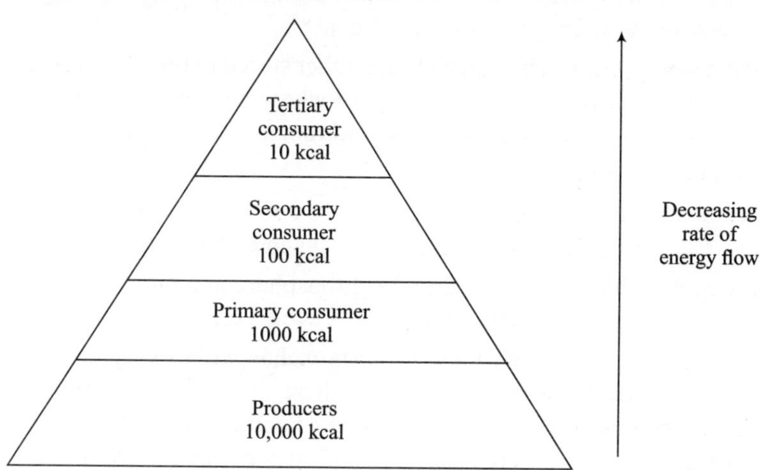

Ecological pyramid energy

Terms related to ecosystem:

(i) **Ecotone:** It is transitional area between two communities. It is rich in diversity called edge effect because it has large number of species at edge. Estuary is ecotone of marine and freshwater habitat.

(ii) **Ecological equivalents:** These are organisms which occupy the same ecological niche in different geographical condition.

(iii) **Keystone species (Ecological dominants):** It is species which influences more on its community or ecosystem. It controls relative abundance of other species. It is small in number but sufficient and it dominates other species in biomass. It has huge impact on community's organisation and survival. Its loss can lead to sharp population drop and extinction of other species that depend on it for certain services and causes serious disruption in functioning of community. There are several examples which are as follows:

(a) **Fig** (Ficus): It bears fruits during season when no other fruits are available. If Fig is removed then many animals die of starvation during period of food scarcity.

(b) **Starfish** (Pisaster): It is predator in intertidal zone of Pacific coast of North America. If it is removed then there will be overproduction of mussel (*Mytilus*) in entire zone.

(c) **Kangaroo rat**: It is herbivorous and lives in desert. It eats large seeds. If it is removed then there will be increase of large number of herbs and shrubs thus whole desert will become green desert grassland.

(d) **African elephant**: It destroys shrubs and small trees by its feeding habit and converts entire woodland habitat to open grassland.

5. Biogeochemical cycle

It is cycling and recycling of the non mineral elements (C, H, O and N) and mineral elements (P, S, Ca, Mg, K, Fe) through biological (biotic) and physical (abiotic) world. It is the interrelationship between living organisms and soil and chemicals. The nonliving materials are made available to the living organisms in an ecosystem though biogeochemical cycle.

It is also called nutrient cycling. The nutrients are either stored in reservoir pool (phosphates in rocks) where materials are slowly transferred to cycling pool where nutrients are emptied and replenished by exchange between biotic and abiotic components (soil moisture, pH temperature).

It is categorised into two groups:

(a) **Gaseous Cycle**: The reservoir pool is atmosphere and biogenetic material is gaseous. It occurs very fast. e.g., Carbon cycle, Hydrogen cycle, Oxygen cycle, Nitrogen cycle etc.

(b) **Sedimentary cycle**: The reservoir material is lithosphere and biogenetic material is nongaseous. It occurs slowly. e.g., Sulphur cycle, Phosphorus cycle etc.

Carbon cycle: It is gaseous cycle by which carbon is exchanged between the lithosphere, hydrosphere and atmosphere. It is considered as perfect because there is rapid and continuous cycling of carbon between abiotic and biotic system and there is no loss of carbon. The natural flow of carbon between atmosphere, hydrosphere and lithosphere is fairly balanced and stable. Carbon is most abundant element and forms 49% of dry weight of living organism.

Sources: There are four major reservoirs of carbon which are as follows:

(a) Atmosphere: Carbon is present in the form of a gaseous CO_2.

(b) Hydrosphere (ocean): Carbon is in the form of carbonic acid and bicarbonates. Ocean contains the largest pool (71%).

(c) Lithosphere (soil and rocks): Carbon is present in the form of carbonates and graphite.

(d) Fossil fuel

Steps:

(a) Carbon moves from atmosphere to plants which make their own food as a result of photosynthesis.

(b) It moves from plants to animals through food chain.

(c) It moves from plants and animals to soil after death and their bodies are buried and bring carbon into the ground where these form fossil fuels.

(d) Carbon now moves from plants and animals as a result of respiration to atmosphere in the form of CO_2.

(e) It also moves from fossil fuel to atmosphere as a result of combustion (burning) by human beings.

(f) It now moves from atmosphere to ocean and other water bodies where ocean absorbs some carbon from the atmosphere. The carbon is dissolved in water.

(g) Carbon moves from atmosphere to lithosphere (rocks) through rain and forms carbonic acid after combining with water. This acid dissolves rocks due to chemical weathering.

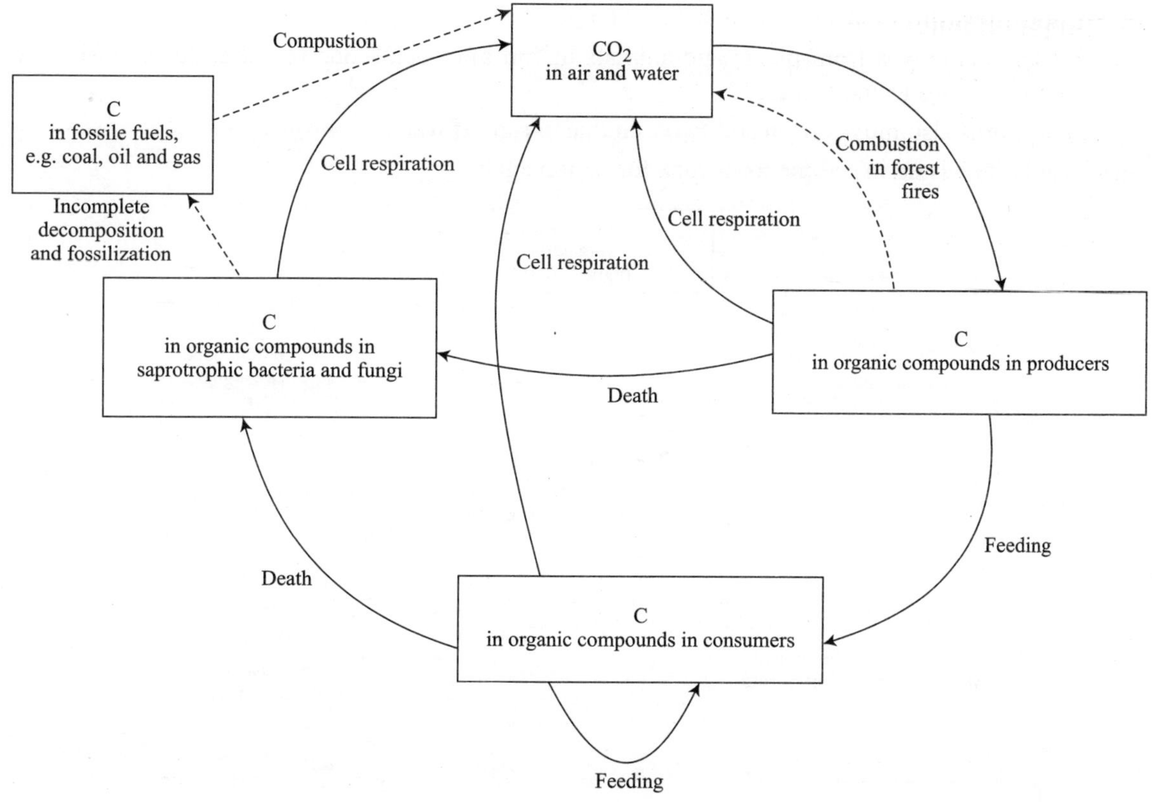

Fig. ***Carbon cycle***

Phosphorus cycle: Phosphorus is second most critical element and often the limiting factor for plant growth since its quantity in soil is small. Phosphates are exchanged between lithosphere and hydrosphere. It describes the movement of phosphorus through lithosphere and hydrosphere. Atmosphere does not play any role in movement of phosphorus because phosphorus and its compounds are mostly solids and these compounds are not present in the air as gases. The natural reservoir is rock.

It is considered as open cycle because of the general tendency for phosphorus to be carried from land to ocean. It is lost from ecosystem by erosion. It is one of the press enter slowest biogeochemical cycles because movement of phosphorus through the soil or ocean is slow. **Sources:** There are two major reservoirs of phosphorus which are as follows:

(a) Lithosphere (rock): Phosphorus is present in the form of orthophosphate (PO_4^{-3}) salt. Its natural reservoir is rock.

(b) Hydrosphere (ocean): Phosphorus is present in the form of orthophosphoric acid in water.

Steps:

(a) Phosphorus brings ocean sediments to land.

(b) It moves from rocks and minerals to soil due to weathering where it is dissolved in soil water and later it is absorbed by plants.

(c) It is transformed into organic compounds in plants.

(d) It now moves from plants to animals through food chain.

(e) Phosphorus moves from plants and animals to soil after death due to either decomposition or excretion of urine and faeces.

(f) Phosphorus also moves from soil to ocean due to run off water.

(g) Finally it ends up in sediments or rock formation again.

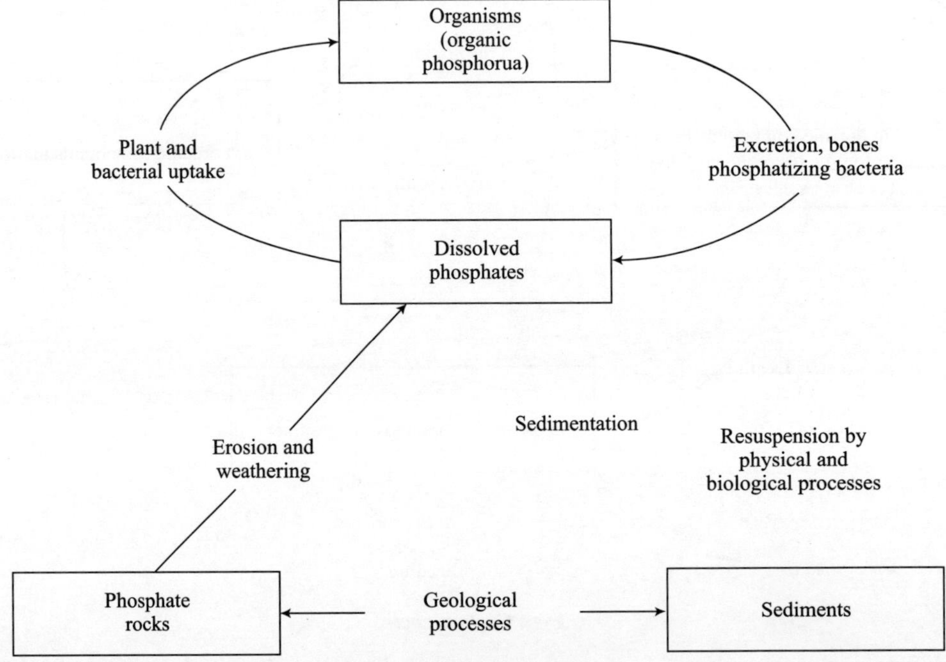

Fig. *Phosphorus cycle*

Sulphur cycle: It is basically an oxidation-reduction cycle with organic, inorganic and gaseous sulphur pools. Sulphur is an essential element being constituent of many proteins and vitamins and hormones. The sulphur containing proteins are first degraded into their constituent amino acids (cysteine and methionine) by proteolytic enzymes secreted by a large number of soil organisms.

Sources: There are four major sources of sulphur which are as follows:

(a) Lithosphere (sedimentary rocks): Sulphur is maximum in earth's crust in the form of minerals such as gypsum and pyrite.

(b) Hydrosphere (ocean): It is present in the form of sulphate (SO_4^{-2}) salt.

(c) Atmosphere: It is present in the form of gas (SO_2).

(d) Fossil fuel: It is released in the form of gas (SO_2) due to combustion of fossil fuel.

Steps:

(a) Sulphur is released due to weathering of sedimentary rocks. It is also emitted directly into the atmosphere due to volcanic eruptions and combustion (burning) of fossil fuel.

(b) Sulphur comes into contact with air where it is converted into sulphate.

(c) Sulphur in the form of sulphate is abundant in soil and is absorbed by plants and microbes and finally it is converted into organic form.

(d) The organic form of sulphur is mineralised into inorganic form (H_2S, S) which are oxidised to SO_4^{-2}.

(e) The organic form of sulphur moves from plants to animals through food chain.

(f) Sulphur is now converted to sulphate due to death and decomposition of plants and animals and is present in soil.

(g) Sulphur comes to earth with rainfall and moves to ocean via river and lake due to runoff.

(h) Sulphur moves through marine communities by food chain and a portion of sulphur is emitted back to atmosphere.

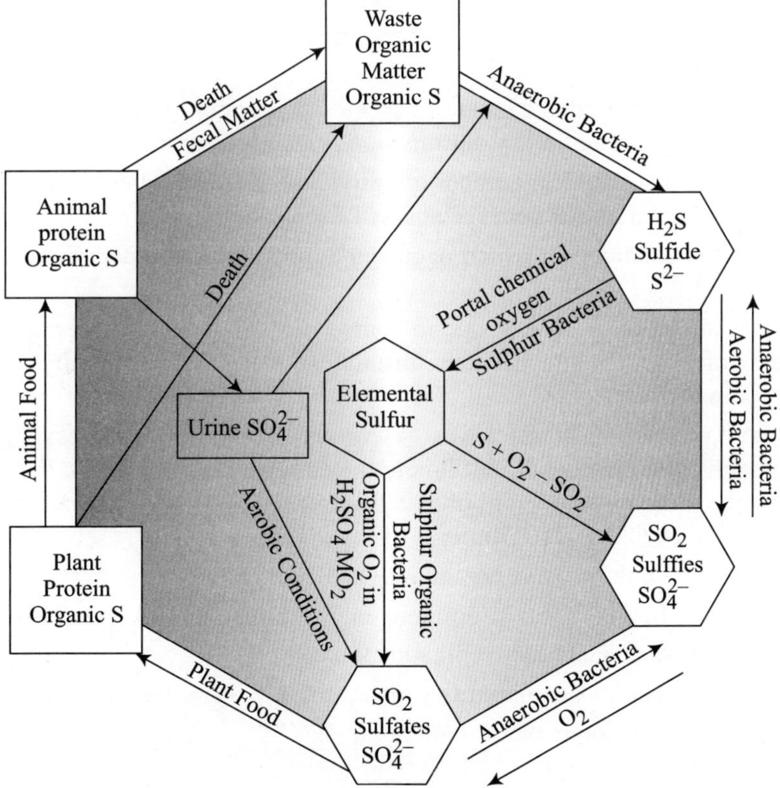

Fig. ***Sulphur cycle***

6. Plant Succession

It is successive colonisation of different plant communities on the same area during the course of development of vegetation. The term 'succession' was coined by Hult (1885). It is long term, universal, unidirectional and cumulative change in vegetation during ecological time.

During succession some species colonise an area and their populations become more numerous whereas populations of other species start declining and even disappear. The succession may be **autogenic** in which changes or replacement in communities take place due to existing organisms and their interactions and **allogenic** in which changes or replacement in communities take place due to external conditions. The succession may be **autotrophic** which is characterised by early and continued dominance by autotrophic organisms and **heterotrophic** which is characterised by early dominance of heterotrophs.

Process of succession

1. **Nudation:** It is formation of bare areas without any form of life. The secondary bare area may develop due to
   (a) Biotic causes: by cutting, cultivation, grazing, deforestation (destruction of forest).
   (b) Climatic causes: by fire, frost, hail, storm, glaciations and long dry period.
   (c) Topographic causes: by wind, water current, rainfall.
2. **Invasion**: It is establishment of a species in bare area. It takes place by
   (a) Migration: It is just arrival of seeds, spores, vegetative propagules to bare area as a matter of chance by air, water or animal.
   (b) Ecesis: It is successful establishment of species (immigrant) in a new area due to adjustment with the favourable condition by germination or multiplication.
   (c) Aggregation: It is increase in number of individual of species due to reproduction in a given area. The population of different species will form community.
3. **Competition**: It is condition in which aggregated individuals compete for water, nutrients and solar energy.
4. **Reaction**: It is alteration in habitat due to changes in soil, water, temperature, light etc hence particular environment of habitat becomes unsuitable for existing community and is soon replaced by another community.
5. **Stabilisation**: It is condition in which final or terminal community becomes stabilised for a longer period of time in prevailing condition. It is in the complete harmony with the environment. The final community is not replaced until drastic disturbance occurs so it is called climax. The term climax was coined by Clements. **Climax** is of several types:
   (a) Climatic climax: In this climax community is determined by general climate.
   (b) Edaphic climax: In this climax community is determined by soil water, soil nutrients etc.
   (c) Subclimax: It is stage just before climatic climax.
   (d) Disclimax: In this climax community is determined by disturbance caused by human being.
   (e) Preclimax: It is condition in which community has life forms lower than those in the expected climatic climax.
   (f) Postclimax: It is condition in which community has life forms higher than those in the expected climatic climax.

Theories of Climax

(a) Monoclimax theory: It was proposed by Clements. It considers only one climax is determined solely by climate.
(b) Polyclimax theory: It was proposed by Tansley. It considers several climaxes are determined by several factors such as soil moisture, soil nutrients, topography, slope exposure, fire etc.

Types: On the basis of occurrence

Primary Succession	Secondary Succession
It occurs in bare areas where earlier community did not exist.	It occurs in those areas where earlier community existed but destroyed by flood, fire or deforestation.
It occurs on bare rock, newly cooled lava, newly created pond or reservoir.	It occurs on burnt forest, abundant farm land, flooded plain.
It is less common.	It is more common.
It occurs very slowly hence it can observed only after a few decade.	It occurs very fast hence it can be observed in a few years.
It develops when pioneer community comes from outside.	It develops partly from previous occupants and partly from migrants.
It takes longer time for completion (reaching climax stage).	It takes shorter time for completion (reaching climax stage).
It does not show the presence of reproductive structure of previous community.	It shows the presence of reproductive structures of earlier community.
It does not show the presence of soil in the beginning.	It shows the presence of soil from the very beginning.
It has several seral or transitional communities.	It has a few seral or transitional communities.

Stages in succession: There are three types of communities with respect to time and stability. These communities are as follows:

1. **Pioneer community**:
   (a) It is first biotic community which develops on bare area.
   (b) It takes long time to change the environment for invasion of next community.
   (c) It develops soil.
   (d) It grows very fast hence its life span is short.
   (e) It has very little diversity.
   (f) It has a few small sized organisms.
   (g) It is unstable and soon replaced by next seral community.
2. **Seral communities** (Transitional communities):
   (a) These communities exist between pioneer community and climax community.
   (b) These are developed from pioneer community.
   (c) These are also unstable and replaced by next community.
3. **Climax Community**:
   (a) It is final community which develops in the last.
   (b) It is established over an area previously occupied by transitional communities.
   (c) It grows on already formed soil.
   (d) It grows very slowly hence its life span is long.
   (e) It has maximum diversity.
   (f) It has numerous large and small sized organisms.
   (g) It is stable and not replaced by any other community.

Trends or changes in early and late succession stages:

Early trends	Late trends
Environment is variable and unpredictable.	Environment is stable and predictable
It shows aquatic or dry condition.	It shows mesic or mesophytic condition.
Successional stage is early and starts with pioneer community.	Successional stage is late and ends in climax community.
Stability is poor hence entropy is high	Stability is good hence entropy is low.
Stratification is not well organised.	Stratification is well organised.
Organic matter and energy content are small.	Organic matter and energy content are large.
Species diversity (plants and animals) is low.	Species diversity is high.
Growth form shows r-selection in which species are of small body size, large in number, early and rapidly reproduced and simple &short life span (<1 year).	Growth form shows k-selection in which species are of large body size, small in number, delayed and slowly reproduction and complex& long life span (>1 year).
Gross productivity is high.	Gross productivity is low.
Food chain is simple and grazing.	Food chain is web-like and detritus.

Examples of Ecological succession:

1. **Hydrosere**: It is ecological succession on aquatic habitat (freshwater). Succession progresses from hydric to mesic (moderate) condition. The various stages of a hydrosere are as follows:
   - **(a) Phytoplanktons stage**: It is **pioneer** community. These organisms are bacteria, cyanobacteria, diatoms, dinoflagellates, green algae etc. These are small autotrophs which divide rapidly. Later zooplanktons are developed that eat upon autotrophs. When they die then cause formation of organic matter. .
   - **(b) Submerged stage**: These are *Hydrilla, Vallisneria, Potamogeton* etc. which are rooted and form dense growth. Later sand, silt and stigma are deposited. Later suspended plants such as *Ceratophyllum, Najas* grow. These are replaced by floating stage.
   - **(c) Floating stage**: These are fixed floating plants such as *Nelumbo, Nymphaea, Trapa* which are replaced by free floating plants such as *Eichhornia, Pistia, Lemna, Wolffia, Azolla*.
   - **(d) Amphibuous/Emergent/Reed Swamp stage**: These are plants such as *Sagittaria, Typha, Ranunculus* etc. which grow where water body becomes shallow. These form more organic matter and silt.
   - **(e) Marsh meadow/sedge stage**: These plants such as *Carex, Cyperus, Juncus, Mentha* etc. grow on silt and form more organic matter and soil.
   - **(f) Woodland stage:** These plants such as *Cassia, Cornus, Populous, Salix, Terminalia* etc. tolerate more sunlight and transpire. These build up more soil.
   - **(g) Tree stage**: It is **climax** stage in which coniferous forest develops in temperate region and rain forest occurs in moist tropical area. Finally water body is converted into land.

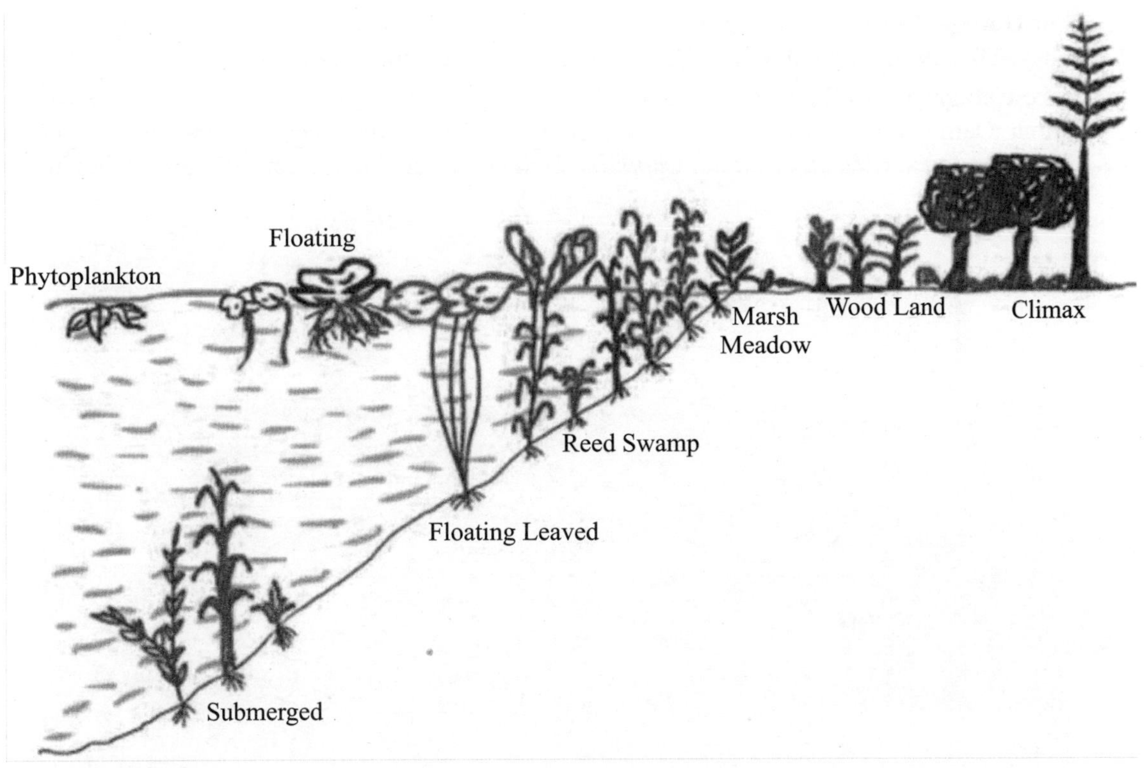

Fig. ***Hydrosere***

2. **Xerosere**: It is ecological succession on dry habitat. Succession progresses from xeric to mesic (moderate) condition. It is divided into
   (a) Lithosere: Succession starts on bare rocks.
   (b) Halosere: Succession starts on saline condition.
   (c) Psammosere: Succession starts on sandy area like sand dune.

The various stages of **xerosere** are as follows:

(a) Crustose lichen stage: It is **pioneer** community. These lichens are *Graphis, Lecanora, Rhizocarpon* etc. which tolerate desiccation and cause weathering of rocks by releasing lichenic acid and carbonic acid. Later these lichens die and mix with particles of rocks. This is replaced by foliose lichens.

(b) Foliose lichen stage: These lichens are *Dermatocarpon, Parmelia, Peltigera, Physcia* etc. which contain large leaf. These cause further cracks on rocks accumulate more soil and organic matter. This is replaced by mosses.

(c) Moss stage: These are *Grimmia, Polytrichum, Pogonatum, Tortula* etc. which accumulate more soil particles and organic matter. This is replaced by annual grasses.

(d) Annual grass stage: These are *Aristida, Eleusine, Festuca, Poa* etc. which grow by germination of their seeds on mat created by mosses. Their roots penetrate more in the rocks.

(e) Perennial grass stage: These are *Cymbopogon, Heteropogon* etc. which grow on the substratum formed by earlier stage.

(f) **Shrub stage**: These are *Capparis, Phytocarpus, Rhus, Zizyphus* which cause further cracks in rocky substratum. These develop shade in area and invite hardy trees.

(g) **Forest stage**: It is **climax** stage in which mesophytic trees are developed. Evergreen forest (Birch, Oak) and coniferous forest (*Abies, Cedrus, Pinus, Taxus*) occur in temperate region and deciduous forest (*Acacia, Albizzia, Capparis, Dalbergia, Shorea* develops in tropical region.

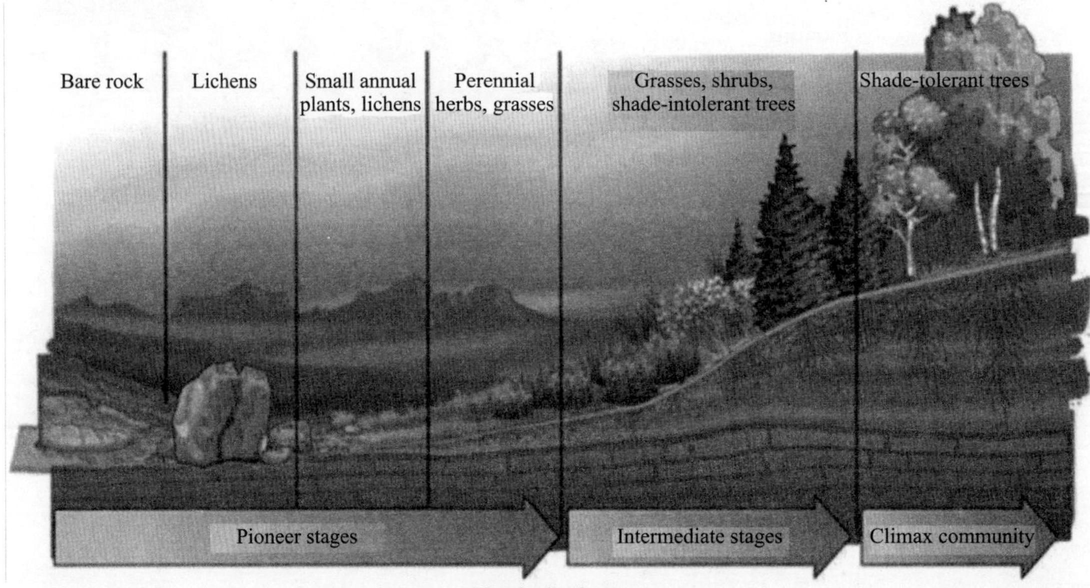

Fig. ***Xerosere***

7. Ecosystem services: The ecosystem performs several functions which are as follows:

(a) Producers of ecosystem perform photosynthesis by which they take up CO_2 from atmosphere and convert into organic compound.

(b) Producers of ecosystem liberate a lot of O_2 as a result of photosynthesis. Amazon rain forest (lung of our planet) produces nearly 20% of O_2.

(c) Several plants in ecosystem get pollinated by particular animals by which both partners are mutually benefited.

(d) Several plants need animals such as bird, squirrel for dispersal of seeds.

(e) The plants prevent soil from erosion.

SHORT ANSWER TYPE QUESTIONS

1. Mention the several types of ecosystem on the basis of size.
2. Distinguish between decomposers and detrivores.
3. Explain the terms standing crop and standing state.
4. Explain the ecological efficiency.
5. Write notes on food web.
6. Explain ten percent law.

7. Write notes on trophic level.
8. Explain the ecotone and ecological equivalents.
9. Explain keystone species.
10. Distinguish between gaseous cycle nad sedimentary cycle.
11. Draw well labelled diagram of phosphorus cycle.
12. Discuss the process of plant succession.
13. Distinguish between pioneer community and climax community.
14. Distinguish between primary succession and secondary succession.
15. What are the benefits of ecosystem?

LONG ANSWER TYPE QUESTIONS

1. Describe the biotic components of ecosystem.
2. Describe the functional component of ecosystem with reference to decomposition.
3. Describe the functional component of ecosystem with reference to productivity.
4. Discuss the energy flow in ecosystem and comment upon the ten percent law.
5. Describe the features and types of food chain in ecosystem.
6. Discuss the three types of ecological pyramids.
7. Define biogeochemical cycle. Describe the carbon cycle.
8. Discuss the early and late trends in succession and comment upon the climax community.
9. Describe the ecological succession in water.
10. Describe the ecological succession in dry habitat.

MULTIPLE CHOICE QUESTIONS

1. Cropland ecosystem is characterised by
   (a) Less diversity and less resilience (b) Less diversity and high resilience
   (c) High diversity and less resilience (d) High diversity and high resilience
2. Which occupy more than one trophic level in pond ecosystem?
   (a) Phytoplankton (b) Fish (c) Zooplankton (d) Frog
3. Maximum productivity is seen in
   (a) Grass land ecosystem (b) Forest ecosystem
   (c) Desert ecosystem (d) Aquatic ecosystem
4. Pitcher plant (*Nepenthes*) is an example of
   (a) Primary producer (b) Primary consumer (c) Secondary consumer (d) Both a and b
5. How much solar energy that fall on the leaves of a plant is converted into chemical energy by photosynthesis?
   (a) Less than 1% (b) 2-10% (c) 10-20% (d) More than 20%
6. Which of the following must be preserved in an ecosystem if the system is to be maintained?
   (a) Producer (b) Producer and Consumer
   (c) Decomposer (d) Producer and decompose
7. The first plants to reappear in a badly burnt forest area will be
   (a) Grasses (b) Mosses (c) Ferns (d) Liverworts
8. Energy flow and energy transformation in living system follow
   (a) Law of minimum (b) Law of limiting factors
   (c) Law of tolerance (d) Law of thermodynamics
9. Which one is most essential for recycling of materials?
   (a) Producers (b) Consumers (c) Decomposers (d) All of above
10. Which of the following is imperfect cycle?
    (a) Carbon cycle (b) Nitrogen cycle (c) Oxygen cycle (d) Phosphorus cycle
11. About 70% of total global carbon is found in
    (a) Grassland (b) Forest (c) Ocean (d) Agrosystem
12. Energy for detritus food chain comes from
    (a) Air (b) Water (c) Sunlight (d) Organism
13. Atmosphere is not involved in
    (a) Carbon cycle (b) Nitrogen cycle (c) Phosphorus cycle (d) Sulphur cycle
14. The most stable ecosystem is
    (a) Forest (b) Ocean (c) Desert (d) Mountain
15. If phytoplanktons are destroyed from sea, then
    (a) Algae will grow heavily (b) Primary consumers will grow luxuriantly
    (c) Food chain will be affected (d) There will be no effect

16. How much of NPP of a terrestrial ecosystem is eaten and digested by herbivores?
    (a) 1% (b) 10% (c) 5% (d) 20%
17. Which of the following ecological pyramids may be upright and inverted?
    (a) Pyramid of number only (b) Pyramid of energy only
    (c) Pyramid of number and energy (d) Pyramid of number and biomass
18. The reservoir of sedimentary cycle exisrts in
    (a) Organic sediment (b) Limestone
    (c) Earth's crust (d) Calcareous sediment
19. Secondary succession occurs in
    (a) Newly cooled lava (b) Newly created pond (c) Flooded land (d) Bare rock
20. Cropland ecosystem is the largest manmade ecosystem characterised by
    (a) Less diversity and high productivity (b) High diversity and high productivity
    (c) Less diversity and less productivity (d) High diversity and low productivity
21. Extinction of a species in a food chain is compensated in
    (a) Food web (b) Food chain (c) Ecological pyramid (d) All of the above
22. Detritus food chain begins with
    (a) Viruses (b) Bacteria (c) Algae (d) Protozoa
23. Highest annual net productivity occurs in
    (a) Alpine forest (b) Coniferous forest
    (c) Subtropical deciduous forest (d) Tropical rain forest
24. The Great barrier reef of Australia is categorised as
    (a) Population (b) Community (c) Ecosystem (d) Biome
25. The biomass available for consumption by the herbivores and the decomposers is called
    (a) Gross primary productivity (b) Net Primary productivity
    (c) Secondary productivity (d) Community productivity
26. Which of the following is characteristic of a biological community?
    (a) Stratification (b) Mortality (c) Natality (d) Sex ratio
27. Pyramid of number in parasitic food chain is
    (a) Upright (b) Inverted
    (c) Spindle shaped (d) Dumb bell shaped
28. The largest population in a food chain found in
    (a) Producers (b) Primary consumers (c) Top consumers (d) Decomposers
29. What fraction of the assimilated energy is used in respiration by herbivores?
    (a) 10% (b) 20% (c) 30% (d) 50%
30. In a food chain such as grass--dear—lion, the energy cost of respiration as proportion of total assimilated energy at each level would be
    (a) 10-20-30% (b) 20-30-60% (c) 10-20-40% (d) 20-40-80%

31. The process of decomposition does not involve
    (a) Catabolism (b) Leaching (c) Humification (d) Stratification
32. Which of the following is incorrect regarding keystone species?
    (a) It is small in number
    (b) It influences more on its community
    (c) Its loss can cause extinction of dependent species
    (d) It has no impact on community's organisation and survival
33. The successful establishment of species in a new area is known as
    (a) Nudation (b) Stabilisation (c) Ecesis (d) Climax
34. The sequence of succession follows
    (a) Invasion-nudation-competition-reaction-stabilisation
    (b) Invasion-nudation-reaction-competition-stabilisation
    (c) Nudation-invasion-competition-reaction-stabilisation
    (d) Nudation-invasion--reaction-competition-stabilisation
35. Which of the following is incorrect regarding climax community?
    (a) It is final community (b) It has minimum diversity
    (c) It is not replaced by any other community (d) It grows very slowly
36. The r-selection refers to the growth of species which shows
    (a) Small in number (b) Late reproduction (c) Long life span (d) Small body size
37. The pioneer community in xerosere is
    (a) Foliose lichen (b) Crustose lichen (c) Moss (d) Phytoplankton
38. Secondary succession occurs in
    (a) Burnt forest (b) Flooded plain (b) Abundant farm land (d) All of the above
39. Which of the following is the largest pool of carbon?
    (a) Atmosphere (b) Ocean (c) Rocks (d) Fossil fuel
40. The organisms which occupy the same ecological niche in different geographical condition are known as
    (a) Ecological dominants (b) Ecotone
    (c) Keystone species (d) Ecological equivalents
41. In recycling of mineral elements within an ecosystem, the responsible direct acting organisms are called
    (a) Producers (b) Decomposers
    (c) Primary consumers (d) Secondary consumers
42. Primary productivity at the climax stage of succession is
    (a) Higher than consumption (d) Lower than consumption
    (c) Equal to consumption (d) Not related to consumption

43. The pyramid of number of a parasitic food chain in forest ecosystem is
    (a) Always inverted (b) Always upright
    (c) Mixture of upright and inverted (d) Sometimes upright and sometimes inverted
44. The most stable ecosystem is
    (a) Pond (b) Ocean (c) Desert (d) Forest
45. The rate of storage of organic matter not used by heterotrophs is called
    (a) Gross primary productivity (b) Net primary productivity
    (c) Net productivity (d) Secondary productivity
46. Rapid nutritional cycling takes place in
    (a) Coral (b) Desert (c) Grassland (d) Ocean
47. Which of the following has large reservoir of nitrogen that is unusable by most organisms?
    (a) Soil (b) Ocean (c) Space (d) Atmosphere
48. Early successional stages are tolerant of the harsh and and abiotic condition is known as
    (a) Weedy a-selected (b) Weedy r-selected (c) Weedy g-selected (d) Weedy k-selected
49. The lowest productivity per sq. metre is seen in
    (a) An open sea (b) A coral reef (c) A small marsh (d) A grassland
50. Which of the following has the largest amount of free energy?
    (a) Producer (b) Primary consumer (c) Top consumer (d) Decomposer
51. The highest value in a grassland ecosystem is found in
    (a) GPP (b) NPP (c) SP (d) TP
52. All of the organisms in a particular area make up a food chain are called
    (a) Population (b) Community (c) Ecosystem (d) Niche
53. Which of the following is not a characteristic of r-selected species?
    (a) Short maturation time (b) Short life span
    (c) Small size of offspring (d) More than one time offspring production
54. In grassland ecosystem, trees donot replace the grasses as part of ecological succession because of
    (a) Insects and fungi (b) Limited sunlight and nutrients
    (c) Cool temperature (d) Water limit or overgrazing
55. The net productivity of closed ecosystem is
    (a) Zero (b) Less than zero (c) More than zero (d) +1/–1
56. Which of the following is not an example of terrestrial ecosystem?
    (a) Garden pot (b) Human skin (c) Volcano site (d) Rotten log
57. In a pond ecosystem, net productivity by zooplanktons is P and biomass consumed by small fish is C, then the ratio of C/P is termed as
    (a) Consumption efficiency (b) NPP
    (c) Assimilation efficiency (d) Conversion efficiency

58. Which of the following has the largest number of trophic level?
    (a) Grassland (b) Tropical forest (c) Coastal ocean (d) Open ocean
59. Single channel energy flow model was proposed by
    (a) Lindemann (b) O.P Odum (c) Clements (d) Charles Elton
60. Which of the following animals come under the category of detrivores?
    (a) Earthworm (b) Termites (c) Carrion flies (d) All of the above

ANSWERS

1.b	2.b	3.d	4.d	5.b	6.d	7.b	8.d	9.c	10.d
11.c	12.b	13.c	14.b	15.c	16.a	17.d	18.c	19.c	20.a
21.a	22.b	23.d	24.c	25.b	26.a	27.c	28.d	29.b	30.b
31.d	32.d	33.c	34.c	35.b	36.d	37.b	38.d	39.b	40.d
41.b	42.c	43.a	44.b	45.a	46.a	47.d	48.b	49.a	50.a
51.a	52.d	53.d	54.d	55.a	56.a	57.a	58.d	59.a	60.d

23

Biodiversity and Conservation

Biodiversity is number and variety of plants, animals and other organisms. The term biodiversity was coined by Walter Rosen but popularised by Edward Wilson.

All species of plants taken together are known as flora and about 0.3 million of plants are known to date. All species of animals taken together are known as fauna and about 1.2million of animals are known to date. Among animals 70 percent are insects. Animals constitute more than 70% of all species and share of plants is about 22% on earth.

India is the 7th largest country of the world with a geographical area of 32,87,591 sq.km, 15200 km long land frontiers and about 7,520 km of coastline. It accounts for only 2.4% of total land area of earth's surface. It contributes 8.1% of global diversity. India ranks the 6th among 12 mega diversity regions of the world. In India 45,000 plants (7% of world flora) and 81,000 animals (8% of world fauna) are found. Among invertebrates, arthropods are maximum and among vertebrates, fishes are maximum. In case of plants, angiosperms are maximum (dicots are more than monocots).

Approximate Number of different species (animals) on earth

Animal groups	Number of species in world	Number of species in India
Mammals	5,416	390
Birds	9,956	1,232
Reptiles	8,240	456
Amphibians	6,199	209
Fish	30,000	2,546
Molluscs	81,000	5,000
Arthropods	990,000	68,389
Other invertebrates	1,30,000	8,400

Approximate Number of different species (plants) on earth

Plant Groups	Numbers in world	Numbers in India	Percentage in India
Angiosperms	2,60,000	17,500	7%
Gymnosperms	980	64	10%
Pteridophytes	13,000	1100	12%
Bryophytes	15,000	2850	17%
Algae	12,000	6500	18%
Lichens	10,000	2000	16%
Fungi	70,000	14500	20%

Levels of biodiversity: On the basis of hierarchy

1. **Genetic diversity**: It is variation of genes within an individual (species) and among different species of population. It occurs in alleles, number and type of genes or in structure of chromosome by recombination during gamete formation, mutation and random mating (sexual fertilisation).

   Features:

   (a) It is a trait of species.

   (b) It helps in speciation (evolution of species).

   (c) It helps in producing varieties of plants and animals. Rice has more than 50,000 genetically different strains. Likewise, mangoes have thousands of varieties.

   (d) It provides resistance to disease in species.

   (e) It helps in adaptation to environmental conditions.

   (f) It maintains diversity at the species and community levels.

2. **Species diversity**: It is variety (number) of different species within an area.

   Features:

   (a) It is a trait of community.

   (b) It influences biotic interaction and stability of the community.

   (c) It is more in western ghat than eastern ghat.

   (d) It is measured by Shannon-Weiner index.

   (e) Its importance was described by **Rivet-Popper hypothesis**.

   Rivet-Popper hypothesis: It was proposed by Paul Ehlrich. He suggested in an aeroplane (ecosystem) all parts are joined together using thousands of rivets (species). If every passenger starts popping a rivet (causing a species to become extinct), it may not affect flight safety (functioning of ecosystem) initially but as more and more rivets are removed, the plane become dangerously weak. It shows significant loss of a certain species affects the function and diversity of an ecological community in ecosystem.

3. **Ecosystem diversity**: It is the variety of ecosystem showing diversity in the number of niches, trophic levels, food webs, and nutrient cycles. It is high in India due to the presence of a number of ecosystems like rain forest, desert, mangroves, coral reefs, estuary, wet land.

Types of Ecological diversity: Whittaker (1972) divided it into three types:

1. α-diversity: It is diversity of species in a given community or habitat (within habitat diversity). It shows limited variation. There is lots of competition, adjustment and interrelationship among members of same community.
2. β-diversity: It is diversity between two communities at two habitats (between habitat diversity) due to replacement of species with the change in community. It arises due to the presence of microhabitat and niches, differences in environmental condition and change in habitat.
3. γ-**diversity**: It is diversity of habitat (regions) over large geographical area. It is also known as ecosystem diversity (regional diversity) due to occurrence of large and different types of ecosystem such as rain forest, desert, deciduous forest, coral reefs, mangroves, wetlands, estuary, alpine. It describes the number of niches, trophic levels and various other ecological processes.

Patterns of biodiversity:

1. Latitudinal gradient:

(a) Species diversity is maximum at equator (tropical region) and minimum at pole (arctic region). It decreases from equator to pole or from low latitude to high latitude.

(b) It is maximum in tropical rain forest (Amazon rain forest (lung of planet) in South America) due to lack of glaciation (a type of disturbance), presence of uniform and stable climatic conditions (less seasonal variation), stability in environment and greater surface area.

(c) It is maximum at plain and minimum at high altitude. It decreases as we move from plain region to top region due to drop in temperature and greater seasonal variation.

(d) Insects are more numerous among animals.

(e) It shows surface area relationship.

Alexander Van Humboldt observed the relationship between species richness and area (without considering taxa or region) is considered as straight line (linear) on log scale which is represented as

$$\text{Log } S = \log C + Z \log A$$

where, S= species richness; C= Y-intercept;

Z = slope of the line (regression coefficient); A= Area.

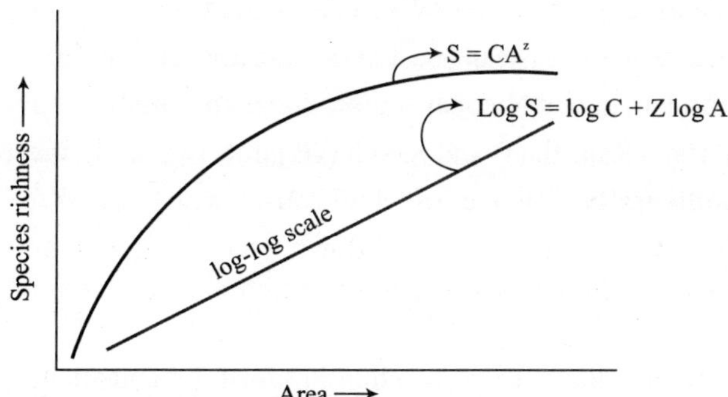

The relationship between species richness and area is rectangular hyperbola for a wide variety of organisms (taxa) where the slope of "Z" is much steeper.

$$S = CA^2$$

The value of Z lies in the range of 0.1-0.2 irrespective of taxonomic group or the region. The value of "Z" is 0.6-1.2 for large geographical area such as continent. In growth curve, the slope of regression shows species richness with area. As area increases, the species richness will also increase but only upto a limit. It shows more number of species in less area.

Importance of biodiversity:

(a) It stabilises ecosystem by maintaining gaseous composition of atmosphere.

(b) It protects land from soil erosion and loss of nutrients.

(c) It controls climate by forest.

(d) It checks flood by absorbing excess water.

(e) It produces wide variety of useful products such as foods and fibres.

(f) It promotes ecotourism.

Mega biodiversity in world: There are 12 mega biodiversity regions which constitute 65% of all biodiversity. These are Australia, Brazil, China, Columbia Ecuador, India, Indonesia, Madagascar, Malaysia, Mexico, Peru and Zaire. The rank of India is sixth.

Biodiversity (Biogeographical region) in India: India is one of the 17th megadiversity richest regions in the world. It has 8.1% of species of world and occupies 2.4% of land of world. India has 10 biogeographical regions which are as follows:

1. Western ghat: It is the richest in diversity. It is located in parallel to western coast of Indian peninsula in Maharashtra, Karnataka, Kerala, Tamil Nadu. It is mostly evergreen such as silent valley of Kerala (containing tropical rain forest) and Agasthyamalai hill.
2. North east: It is located in region of Brahmaputra valley and North east hill. It is second richest in diversity. It is called cradle of civilisation
3. Deccan plateau: It is central highland, Chhotanagpur plateau. It has the largest coverage area (42%).
4. Semiarid: It is seen in Punjab and Gujarat region. It includes Aravalli hill range. It has second largest coverage area (16.6%).
5. Himalaya: It is found in north, west, east and central Himalayan region. It covers 7.2 % of area.
6. Trans Himalaya: It is seen in Ladakh region. It covers 5.7% of area.
7. Gangetic plain: It is northern region. It covers 11% of area.
8. Coast: It is seen in western and eastern coast and lakshadweep region.
9. Island: It is seen in Andaman and Nicobar region. It has the smallest coverage area (0.3%).
10. Desert: It is seen in Thar (Rajasthan) and Kutch (Gujarat) region. It covers 6.9 % of area.

Causes for the loss of biodiversity: There are several causes which are as follows:

1. Destruction of habitat: It occurs due to cutting down trees, ploughing grassland, overgrazing, soil erosion, burning forest and human settlement by which plants are eliminated and animals move out in alien land.
2. Overexploitation: It occurs due to excessive hunting for food and shelter.
3. Invasion of alien or exotic species: The introduction of non-native (alien) species leads to loss of native species because alien species are generally bigger, stronger, more aggressive and faster growing than native species. These alien species are as follows
   (a) Water hyacinth (*Eichhornia crassipes*) causes death of several aquatic plants and animals in water bodies. Initially it was introduced in India to reduce water pollution. This plant is also known as terror of Bengal.
   (b) Carrot grass or Congress grass (*Parthenium hysterophorus*) kills several herbs and shrubs.
   (c) *Eupatorium odoratum* reduces the population of *Tectona grandis.*
   (d) *Lantana* (a shrub) replaces several species in forest of India.
   (e) *Periplaneta americana* replaces *Blatta orientalis.*
   (f) Nile perch (predatory fish) kills cichlid fish into lake Victoria in East central Africa.
   (g) African catfish (*Clarias gariepinus*) threatens native catfish (*Clarias bacterachus*).
4. Co-extinction: It occurs when the extinction of one species leads to the extinction of the other which was mutually associated with the former species.

   Note: The above four causes are called **'evil quartet'**.

5. Disturbances: It may be natural such as forest fire, tree fall, pest infestation, locust attack and manmade activities such as felling of trees and use of fire for clearing.
6. Mass extinction: It occurs due to volcanic eruption, drought, flooding, excessive heating and cooling.
7. Water Pollution: It occurs due to use of chemical fertilisers, acid rain etc.
8. Eruption of disease: It occurs due to breakout of viral, bacterial and fungal diseases from time to time.
9. Population attributes: It occurs due to large body size (e.g., Rhinoceros); small population and low reproductive potential (e.g., Blue whale, Giant panda), high status of trophic level in food chain (e.g., Bengal tiger, Bald eagle), fixed habitat, fixed migratory route and inability to switch over alternative foods.

Impact of loss of biodiversity:

(a) Plant productivity will be reduced.
(b) Resistance to drought will be reduced.
(c) Food will be less available.
(d) Genetic and evolutionary variation will be reduced.

IUCN (International Union of Conservation of Nature &natural resources): It is now called World Conservation Union (WCU). Its head quarter is Morges (Switzerland). It works in the field of nature conservation and sustainable use of natural resources. It prepares red data book or red list which provides information regarding taxa facing risk of extinction. It was initiated in 1963.

Objectives of Red Data book:

(a) It provides information regarding degree of threat.
(b) It prepares conservation strategy.
(c) It provides global index about decline of diversity.
(d) It identifies the species at high risk of extinction.

Categories of species: Red data book has mentioned 8 categories of species based on rate of decline, population size, and area of geographical distribution. These include:

1. **Extinct**: Species does not exist and it has been completely eliminated. e.g., Dodo (Mauritius), Passenger pigeon, Pink headed duck, Mountain quail, Lesser 1- horned rhinoceros, Quagga=Zebra type (Africa), Thylacine =Tasmanian wolf (Australia), Steller's Sea Cow (Russia).
2. **Extinct in wild**: Species survives only in cultivation (plants) or in captivity (animals). It is not found in natural habitat. e.g., *Ginkgo biloba* (plant), Alagoas curassow (bird).
3. **Critically endangered**: Species is under high risk of extinction and can become extinct any time in future. e.g., Pigmy hog, Jerdons courser, Great Indian bustard.
4. **Endangered**: Species can become extinct if causal factors continue to operate. e.g., Red Panda, Snow leopard, Black buck, Bengal tiger, One horned rhinoceros, wild Asiatic ass, Asiatic lion, Indian rock python, Blue whale, Lion tailed macaque.

Endangered ornamental plant	*Rhododendron*
Endangered medicinal plants	*Atropa acuminata, Rauwolffia serpentina, Podophyllum hexandrium*
Endangered scientific plants	*Nepenthes, Balanophora, Dischidia, Sapria himalayensis*
Endangered economical plants	*Pinus gerardiana, Santalum album, Dioscorea*

5. **Vulnerable**: Species likely to move into endangered category in near future if causal factors continue to operate. e.g., Black buck

   (**Threatened category** includes three categories such as critically endangered, endangered and vulnerable which contains 272 species (maximum in mammals) in India).
6. **Lower risk**: Species is least concerned and does not come to any threatened categories.
7. **Data deficient**: There is insufficient information to assess risk of extinction.
8. **Not evaluated**: Species has not been assessed for risk of extinction.

   Two more categories were added:
9. **Rare**: These are species with small and localised population. These are not at present endangered and vulnerable but are at risk. e.g., Clouded leopard.
10. **Indeterminate**: Species is in danger but reason is unknown. e.g., Mexican prairie dog, 3-banded Armadillo of Brazil

Green Data book: It is catalogue of rare plants growing in protected areas and botanical garden.

Conservation Approaches:

(a) **Narrow utilitarian**: Conservation is done because we get benefit from food (cereals, pulses, vegetables and fruits), fibres, fire woods, industrial and medicinal products.

(b) **Broadly utilitarian**: Conservation is done because we get supply of oxygen from plants as a result of photosynthesis and supply of fruits and seeds as a result of pollination by pollinators (insects, birds, bats etc).

(c) **Ethical**: Conservation of microbes, plants and animals is done because we share the planet and morally bounded together.

Conservation strategy: It is for the protection, restoration or preservation of wildlife (plants and animals) and natural resources such as forest and water. The world wildlife week is celebrated on first week of October every year.

(a) All threatened species should be protected on priority basis.

(b) The natural habitat should be protected and restored.

(c) The reserves and protected areas should be developed

(d) Alien species should not be introduced

(e) Pollution should be controlled.

(f) National wildlife protection law (1972) should be enacted.

(g) Public awareness should be increased.

Types of conservation strategy:

***In situ* conservation**: It is onsite conservation and protection of plants and animals in their native ecosystem (natural habitat). It includes

1. **Hot spot**: It is a small area which contains the richest and most threatened species of plants and animals. It shows high level of species diversity in which endemic species (present in a particular area not found anywhere else) are present. Maximum endemic species has been noticed in amphibians. There are 34 hot spots in the world. In India there are 3 hot spots such as Himalaya, Western ghat, and Indo-Burma region which cover less than 2% of total area and contain 44% of species.

2. **National Park**: It is a protected area meant for protection of both flora and fauna. In this, cultivation of land, private ownership and grazing of animals are not allowed. It occupies 1.1% of geographical area in India. There are more than 100 national parks in India. The first national park established in India was Jim Corbett National park (Uttaranchal). UNESCO has declared several National Parks (Kaziranga of Assam, Keoladeo of Rajasthan, Manas of Assam, Nanda Devi of Uttaranchal and Sunderbans of West Bengal) as world heritage sites.
3. **Wild life Sanctuary**: It is a protected area meant for protection of animals (fauna) only. In this, cultivation of land, private ownership and grazing of animals are allowed. It occupies 3.6% of geographical area in India. There are around 450 sanctuaries in India.
4. **Biosphere Reserve**: It is a multi-purpose protected area to preserve biodiversity. It was launched in 1971 under Man and Biosphere programme (MAB) of UNESCO. The first biosphere reserve was established in 1979 but around 600 biosphere reserves were formed in more than 105 countries. The Indian Government has established 18 biosphere reserves so far in India. Each biosphere reserve consists of following 3 zones:
   (a) **Core/Natural zone**: It is innermost undisturbed area where there is no human activity.
   (b) **Buffer zone:** It is central area which is surrounded by core zone in where limited human activity (education, research and training) is allowed.
   (c) **Transition/Manipulation zone**: It is outermost zone in which human settlement, cropping, recreation and forestry are permitted.

Biosphere reserves	Year	Regions
Nilgiri (First in India)	1986	Kerala, Karnataka,Tamil Nadu
Nanda Devi	1988	Uttaranchal
Nokrek	1988	Meghalaya
Manas	1989	Assam
Sunderbans	1989	West Bengal
Gulf of Mannar	1989	Tamil Nadu
Great Nicobar	1989	Andaman &Nicobar
Simlipal	1994	Orissa
Dibru-Saikhowa	1997	Assam
DihangDibang	1998	Arunachal Pradesh
Panchmarhi	1999	Madhya Pradesh
Khangchendzong	2000	Sikkim
Agasthyamalai hill	2001	Kerala
Achanakmar-Amarkantak	2005	Madhya Pradesh, Chattisgarh
Great Rann of Kutch	2008	Gujarat
Cold desert	2009	Himachal Pradesh
Seshachalam hill	2010	Andhra Pradesh
Panna	2011	Madhya Pradesh

Objectives of Biosphere reserves:
(a) To conserve representative samples of ecosystem of the world.
(b) To promote and facilitate ecological and environmental research.
(c) To promote appropriate sustainable managements of the living resources.

5. Sacred groves and lakes: These are most undisturbed forest without any human impact around place of worship. These are protected by tribal communities due to religious sanctity. e.g., Jaintia & khasi hills of Meghalaya; Aravalli hills of Rajasthan; Deodar groves in Kumaon; Pushkar lake in Ajmer (Rajasthan); Areas of Bastar in Madhya Pradesh.
6. Cold desert: Ladakh (Jammu and Kashmir)
7. Hot desert: Thar (Rajasthan)
8. Wet land: Assam and N-E states
9. Forest: Tropical moist evergreen forest in Assam
10. Saline swampy area:
11. Coral reef: Palk bay, Gulf of Mannar, Gulf of Kutch, Andaman and Nicobar, Lakshadweep
12. Mangroves: Sunderbans in West Bengal

***Ex situ* conservation**: It is off site conservation of rare or endangered plants and animals outside their natural habitat in artificial environment.

1. Botanical garden: It is an area of land where plants of different species and areas are grown.
2. Zoological park / Zoo: It is protected area of land where wild animals are kept under human care.
3. Wildlife safari Park: It is large area where animals are kept in semi-captivity under semi-natural condition.
4. Aquaria: It is a tank, pool or bowl filled with water for keeping live fish and aquatic animals.
5. Arboreta: It is garden with shrubs and trees.
6. Orchard: It is home garden.
7. Seed bank: Seeds are stored in cold storage for many years. It has recalcitrant and orthodox seeds.
8. Tissue culture method: It is technique of growing cells/tissues/organs under aseptic condition.
9. Cryopreservation: It is preservation of tissues, embryos, spores, seeds, gametes at -196°C under liquid nitrogen.

Convention on Biological Diversity (CBD): The United Nations Convention on Biological Diversity is intended to ensure an international action to control the destruction of biological species, habitats and ecosystems. It was opened for signature at the UN conference on Environment and Development- the **Earth Summit** in Rio de Janeiro, Brazil on 5 June, 1992. The CBD sets three goals, i.e. (i) the conservation of biodiversity, (ii) sustainable use of biological resources, and (iii) equitable sharing of benefits which would accrue from such use. It was signed by 152 countries. The convention subsequently came into force on December 29, 1993. India became a party to this convention in May, 1994.

Legislations by Government of India:

National park Act	1955
Wildlife (protection) Act	1972 (amended in 1991)
Project Tiger	1972 but launched on 1st april,1973
Biodiversity Act	2002

Days to remember

21st March	World forestry day
22nd March	World water day
22nd April	World earth day
5th June	World environment day
11th July	World population day
3rd October	World animal day
2nd December	National pollution control day
3rd December	World conservation day
29th December	World biological diversity day

Symbolic animals:

India	Tiger
Australia	Kangaroo
New Zealand	Kiwi
USA	Bald eagle
China	Giant Panda
WWF	Red Panda

SHORT ANSWER TYPE QUESTIONS

1. Write notes on Rivet-popper hypothesis.
2. Draw a graph between species richness and area.
3. Mention the biogeographical regions of India.
4. What are the objectives of red data book.
5. Write notes on threatened category of species.
6. Mention some of the endangered plants.
7. Write notes on biosphere reserve.
8. Mention any five biosphere reserves of India.
9. Write notes on Earth summit of Brazil.
10. What are the legislations by Government of India to conserve biodiversity?

LONG ANSWER TYPE QUESTIONS

1. Describe the levels, types and patterns of biodiversity.
2. Discuss the causes for the loss of biodiversity.
3. Describe the *in situ* and *ex situ* conservation strategies.

MULTIPLE CHOICE QUESTIONS

1. Greatest biodiversity is found in
   (a) India (b) South America (c) England (d) frica
2. Maximum number of species diversity occur in
   (a) Algae (b) Fungi (c) Mosses (d) Lichens
3. The relationship between species richness and area for a wide variety of taxa shows
   (a) Straight line (b) Sigmoid
   (c) J-shaped (d) Rectangular hyperbola
4. The relationship between the species richness and the area for a wide variety of taxa appears as
   (a) Straight line (b) Sigmoid
   (c) Parabola (d) Rectangular hyperbola
5. In Rivet-popper hypothesis, rivet and popper are concerned with, respectively
   (a) Ecosystem and species (b) Species and community
   (c) Community and species (d) Species and ecosystem
6. Rivet-Popper hypothesis explains
   (a) Role of species in an ecosystem (b) Role of biological diversity in an ecosystem
   (c) Role of disturbances in ecosystem (d) Role of pollutants in environment
7. Which of the following groups of vertebrates comprises the highest number of endangered species?
   (a) Fish (b) Reptiles (c) Birds (d) Mammals
8. A threatened category does not include
   (a) Critically endangered (b) Endangered
   (c) Vulnerable (d) Rare
9. If at high altitude, birds become rare, the plants likely to disappear are
   (a) Orchids (b) Pines (c) Oaks (d) Rhododendron
10. Western ghat comes under category of hot spot because of
    (a) High elevation (b) Tropical climate (c) Good location (d) High endemism
11. The exotic species introduced in India are
    (a) Lantana and Rhododendron (b) Lantana and Eichhornia
    (c) Eichhornia and Vallisneria (d) Apistia and Eichhornia
12. Which of the following is not observed in hot spots?
    (a) Species richness (b) Lesser interspecific competition
    (c) Endemism (d) Accelerated species loss
13. Maximum genetic diversity in India has been found in
    (a) Mango (b) Wheat (c) Ground nut (d) Rice
14. The contribution of India in global diversity is
    (a) 5% (b) 7.1% (c) 8.1% (d) 10%
15. Which of the following groups of plants comprises the least number of species?
    (a) Algae (b) Bryophyta (c) Pteridophyta (d) Gymnosperms

16. Silent valley of Kerala has
    (a) Temperate forest
    (b) Alpine forest
    (c) Tropical deciduous forest
    (d) Tropical rain forest
17. Great Indian bustard comes under category of
    (a) Extinct
    (b) Extinct in wild
    (c) Critically endangered
    (d) Endangered
18. World biodiversity day is celebrated on
    (a) 5th June (b) 29th December (c) 11th July (d) 21st March
19. Which of the following does not come under strategy of *In situ* conservation?
    (a) Hot spot (b) Sacred groves (c) Biosphere reserve (d) Botanical garden
20. The Earth summit took place in Rio de Janeiro of Brazil on
    (a) 5th June, 1990 (b) 5th June, 1991 (c) 5th June 1992 (d) 5th June 1993
21. Which of the following has permanent status and governed by central legislation?
    (a) National park (b) Forest (c) Sport garden (d) Wild sanctuary
22. The most effective way to conserve the plant diversity of an area is
    (a) Biosphere reserve
    (b) Botanical garden
    (c) Seed bank
    (d) Tissue culture method
23. Rivet popper hypothesis was proposed by
    (a) Ernst Mayr (b) Ernst Haeckel (c) O.P.Odum (d) Paul Ehlrich
24. Dodo is an extinct animal belonged to
    (a) Australia (b) Maldives (c) Mauritius (d) Malaysia
25. Mangroves are found in
    (a) Silent valley of Kerala
    (b) Sunderbans of West Bengal
    (c) Thar of Rajasthan
    (d) Ladakh of Jammu and Kashmir
26. Which of the following is known as between habitat diversity?
    (a) Alpha diversity (b) Beta diversity (c) Gamma diversity (d) Delta diversity
27. Amazon rain forest is found in
    (a) North America (b) South America (c) Africa (d) Europe
28. Human activity is not allowed in
    (a) Core zone
    (b) Buffer zone
    (c) Transition zone
    (d) Manipulation zone
29. Project tiger was launched on
    (a) Ist April,1970 (b) Ist April,1971 (c) Ist April,1972 (d) Ist April,1973
30. Which of the following is known as cradle of civilisation?
    (a) Western ghat (g) Desert (c) Island (d) North-east

ANSWERS

1.b	2.b	3.d	4.d	5.a	6.a	7.d	8.d	9.d	10.d
11.b	12.b	13.d	14.c	15.d	16.d	17.c	18.b	19.d	20.c
21.a	22.c	23.d	24.c	25.b	26.b	27.b	28.a	29.d	30.d

24

Environmental Issues

Pollution is undesirable change in physical, chemical or biological characteristics of air, water and soil that causes harm in the living organisms. The basic causes are industrialisation, human population and urbanisation.

Pollutant: It is chemical, geochemical or biological substance which deteriorates natural environment of living organisms or it is a substance which causes pollution.

Types of Pollutants:

1. **Primary pollutants**: These are released directly from source. e.g., CO_2, SO_2, gaseous hydrocarbon, fly ash, DDT.
2. **Secondary pollutants**: These are released by reaction of primary pollutants in the atmosphere. These are more harmful than primary pollutants. e.g., O_3, PAN, benzopyrene.
3. **Natural pollutants**: CO, CO_2, SO_3, O_3, Pb, Hg
4. **Synthetic pollutants**: Detergents, Fertilisers, Pesticides, Industrial effluents, Drugs
5. **Biodegradable pollutants**: These are degraded or decomposed very rapidly either automatically due to heat or biologically due to bacterial action in natural environment. e.g., Sewage, sludge (solid part of sewage), garbage of living organisms
6. **Non-biodegradable Pollutants**: These are either not degraded or degraded very slowly in natural environment. e.g., Pesticide (e.g., DDT), Heavy metals (Pb, Hg, Cr, Ni), plastics, Polyethene bag, Aluminium cans.

Air pollution: It is undesirable change in physical, chemical or biological characteristic of air.

Sources: Each source produces pollutants in the form of solid, liquid or gas. These are of several types:

(a) Mobile source: Pollutants are released from moving things. e.g., Automobile, train, aircraft.
(b) Stationary source (Point source): Pollutants are released from single source. e.g., Industries, factories, power plants, oil refineries, metal refinery (smelter).
(c) Area source: Pollutants are released from whole area. e.g., Wood burning, fire places.
(d) Natural source: Pollutants are released naturally. e.g., dust storm, wild fires, volcanoes, pollens etc.

Pollutants and their effects:

1. **CO**: It is colourless, odourless, tasteless and non-irritating gas. It is the largest single air pollutant (50% of total air pollutants). It is produced mostly from combustion of fuel from vehicles and engines; and some amount from cigarette smoke. It combines with haemoglobin (due to its

greater affinity for Hb than O_2) to form carboxyhaemoglobin which reduces the transport of O_2 to different tissues and organs. It causes heart diseases.

2. **SO_2**: It is produced from metal refineries (smelter), fuel combustion, industries, electric utilities, thermal plants and volcanoes. It combines with water in air to form H_2SO_4 in rainy season by which acid rain occurs. It has mutagenic properties. It causes discolouration and deterioration of buildings, sculpture, fabric, paper, leather etc. It causes corrosion of metals. It causes chlorosis (loss of chlorophyll) and necrosis (death of tissue) in plants. It causes membrane damage, metabolic inhibition and yield reduction. It causes asthma, eye trouble in human beings.
3. **Nitrogen oxides (NO_x):** These are produced from combustion of fuel of vehicles and engines. These react with unsaturated hydrocarbons to form Peroxy Acetyl Nitrate (PAN).These form photochemical smog. These are carcinogen and mutagen. These cause necrosis, lesion, dieback and defoliation in plants. These cause eye irritation, respiratory trouble and even death by combining with Hb if their concentration is high in inhaled air.
4. **Smog**: It is dark fog having condensed water vapours, smoke, dust and gases (SO_2, H_2S, NO_x). It causes silvering or glazing and necrosis in plants. It causes asthma, chronic bronchitis in human beings.

   It is of two types:

Classical smog (London smog)	Photochemical smog (Los Angeles smog)
It occurred in London in 1952.	It occurred in Los Angeles in 1940.
It is dark brown, reducing and less toxic gas.	It is light brown, oxidising and more toxic gas.
It is formed by condensation of several gases.	It is formed by photochemical reaction of nitrogen oxides and hydrocarbons.
It contains condensed water vapours, smoke, dust, and excess SO_2, H_2S.	It contains condensed water vapours, smoke, dust, NO_x, O_3, PAN.
It occurs at low temperature.	It occurs at high temperature.
It does not produce secondary pollutants.	It produces secondary pollutants.

5. **Particulate matters**: These are produced from combustion of coal and petroleum; industrial chemical reactions, farming etc. These include dust, soot, fur, fly ash, hair, spores, pollen grains etc.

   These may be **settleable** (their size is greater than 10μm and remain in air for less than one day) and **suspended** (their size is less than 10μm and remain in air for more than one day such as aerosol, dust and mist). The particles of 2.5 μm cause great harm according to Central Pollution Control Board (CPCB). The particles less than 10 μm in size can enter into human nasal tract, and particles less than 2.5μm in size can reach further inside the lungs causing serious lung disease. These worsen lung or heart disease if their exposure is for short term. These cause lung or heart disease if their exposure is for long term. Cotton dust causes byssinosis (brown lung disease), pneumoconiosis (lung fibrosis), emphysema, siderosis etc.
6. **Ozone**: It is produced from photochemical reaction of NO_x and volatile organic compounds (VOC) such as aldehyde, ketone etc. It is strong oxidant and secondary pollutant. It causes necrosis. It causes irritation of eyes and respiratory trouble.
7. **Peroxy Acetyl Nitrate (PAN):** It is produced from photochemical reaction of Nitrogen oxides and Volatile organic compounds (VOC) such as aldehyde and ketone. It is secondary pollutant. It causes irritation of eyes.

8. **CO_2**: It is least harmful gas. It is produced from combustion of fuel and deforestation. It causes head ache and nausea.
9. **Acid rain**: It was coined by Robert Angus Smith (1879). It is formed when oxide of nitrogen and sulphur produced from combustion of fuel combine with moisture in the atmosphere to form sulphuric acid (60%) and nitric acid (40%).

   $2NO_2 + O \longrightarrow N_2O_5$ $\quad$ $2SO_2 + O_2 \longrightarrow 2SO_3$

   $N_2O_5 + H_2O \longrightarrow 2HNO_3$ $\quad$ $SO_3 + H_2O \longrightarrow H_2SO_4$

   It is rainfall due to deposition of acidic components in rain, snow or fog. Its pH is 4 but normal rain water has pH 5.6. It causes acidification of soil. It affects soil fertility by damaging soil microbes. It affects the productivity of forest, grassland and crop by causing senescence of leaves. It also causes damage to buildings, built up marbles, lime stones and sand stones, ceramics, textiles, painted surface.
10. **Other inorganic substances:**
    (a) **Hg:** It is produced from burning or combustion of coal, and smelting (metal refinery). It enters into water body through air. Microbes in water convert Hg into highly toxic form, methyl mercury which is consumed by fish. Humans suffer from minamata disease due to consumption of contaminated fish.

    (b) **Pb:** It is produced from metal refineries, combustion of leaded gasoline of engines and motor vehicles; battery etc. It is present as antiknock additive in the form of $Pb(CH_3)_4$ or $Pb(CH_3CH_2)_4$ in petroleum. It causes plumbism or lead poisoning (loss in IQ, memory and learning; anaemia and renal problem).

    (c) **NH_3**: It is produced from fertilisers, dye, lacquer and explosive industries. It causes irritation to upper respiratory tract.

    (d) **Phosgene**: It is produced from pesticide and dye industries. It forms methyl isocyanate by reacting with methyl amine.

    (e) **Methyl Isocyanate (MIC):** It is produced from pesticide industry. It is a chemical used as raw material for the synthesis of carbaryl, a chemical pesticide. Its accidental leakage from union carbide factory caused Bhopal gas tragedy on 3rd December, 1984.

    (f) **Benzopyrene:** It is produced from tobacco smoke, automobile exhaust and industries. It is carcinogen.
11. **Noise:** It is loud disturbing sound dumped into the ambient atmosphere. Its main source includes loud music, transportation noise electricity generator etc. It is expressed in decibel (dB). It is now considered as air pollutant. It is considered as air pollutant if the value of noise will be more than 80 dB. It is unbearable at 140 dB. It causes partial or permanent loss hearing, anxiety and stress. Green muffler or green belt vegetation is row of trees grown to absorb noise.

Air pollution indicators:

(a) Lichens (*Cladonia, Parmelia, Usnea*) and Moss (*Sphagnum*) are indicators of SO_2 pollution.

(b) Tobacco is an indicator of O_3 pollution.

Air pollution metabolisers:

(a) *Daucus, Ficus* and *Phaseolus* metabolise CO.

(b) *Pinus* and *Vitis* metabolise $(NO)_x$

(c) Neem, Peepal, *Albizzia* etc. metabolise SO_2.

Control measures:

(a) Compressed Natural gas (CNG) should be used in vehicles.
(b) Unleaded petrol should be used.
(c) Old vehicles should be phased out gradually.
(d) Tall chimney should be installed.
(e) Biogas should be produced as an alternative source.
(f) Solar energy should be used as an alternative source.
(g) Burning of excessive vegetation should be stopped
(h) Afforestation should be done on mass scale.

Treatment of Pollutants:

(a) Electrostatic precipitator removes particulate matter (dust and smoke).
(b) Catalytic converter (fitted in vehicles) converts CO into CO_2, NO_x into N_2 and unburnt hydrocarbon into CO_2 and H_2O.
(c) Scrubber removes dust and gas (SO_2) to make air clean.
(d) Cyclone separator or collector settles particulate matter (coarse dust).
(e) Trajectory separator removes dirty air.

Euro norms: It is emission standard instituted by Govt. of India to regulate the output of air pollutants from motor vehicles. Now Bharat stage norms are based on European regulation. Thus Bharat stage is equivalent to Euro norms. The standard is set by Central Pollution Control Board (CPCB).

Higher the stage, the more will be stringent norms. Euro II norms mentions that amount of sulphur has to be 350 ppm in diesel and 150 ppm in petrol. The objective is to reduce sulphur to 50 ppm in petrol and diesel. There are several norms which are as follows:

(a) Bharat stage II norms **(Euro II)** were applicable nationwide from 1st April, 2005.
(b) Bharat stage III norms (Euro III) were applicable nationwide from 1st April, 2010. (In 13 major cities, Bharat stage IV norms have been effective from April, 2010).
(c) Bharat stage 1V norms (Euro IV) were supposed to be applicable nationwide from 1st April, 2017.
(d) Bharat stage V norms (Euro V) were supposed to be applicable nationwide from 1st April, 2019.
(e) Bharat stage VI norms (Euro V1) were supposed to be applicable nationwide from 1st April, 2023. (In 2016, Indian government announced that country would stop Bharat stage v norms altogether and adopt Bharat stage VI norms from 1st April, 2020 in which sulphur must be 10 ppm).

Water Pollution:

It is undesirable change in physical, chemical or biological characteristics of water which harmfully affects the living organisms. It is contamination of water bodies.

Types of water pollutants:

1. Physical pollutants: These are hot water, oil spills and soil carried in rainwater.

2. Chemical pollutants: These are organic wastes, organic insecticides such as dichlorodiphenyl trichloroethane (DDT), benzene hexachloride (BHC), polychlorinated biphenyl (PCB), and inorganic chemicals such as nitrates, phosphates, Pb, Hg, Ni, Cd, As, fluorides.
3. Biological pollutants: These are pathogenic microorganisms such as viruses, bacteria, protozoans, algae, helminths etc.

Sources:

1. Point source: The pollutants are discharged from single source or localised location. e.g., factories, tanker etc.
2. Nonpoint source: The pollutants are discharged from multiple sources. e.g., contaminated water after rain, leaching out nitrogenous compounds from fertilisers

Pollutants and their effects:

1. **Sewage:** It is municipal waste water which contains 0.1% impurities. These impurities contain suspended solids (sand, silt, clay), colloidal materials (bacteria, faecal matter) and dissolved materials (nitrates, phosphates, NH_3, Ca).

   Sewage causes **eutrophication** in which water body becomes nutrient enriched. These stimulate the luxuriant growth of algae or algal bloom (e.g., *Microcystis*) and weed (*Eichhornia*). This causes death and decomposition of algae by which amount of dissolved oxygen is reduced in water body. The whole process is called eutrophication by which water body receives excess inorganic nutrients mainly nitrates and phosphates. Domestic sewage is rich in biodegradable organic matter. It stimulates the activity of several decomposers increasing the biochemical oxygen demand (BOD). Increase in BOD will decrease the dissolved oxygen (DO) in polluted water. BOD is amount of oxygen required for 5 days in one litre of water at 20°C for the microbes to metabolise organic waste. It is directly proportional to amount of organic waste present in water. It measures the degree of impurities in water. It uses oxygen hence amount of dissolved oxygen is decreased leading to death of aquatic organisms. The aquatic organisms will subsequently die.

   The BOD is maximum in distillery effluent, followed by sugar mill effluent, domestic sewage and minimum in paper mill effluent.

   Types of lakes on the basis of richness in nutrients:

   (i) Oligotrophic lakes: These lakes are poor in nutrients and biologically unproductive.
   (ii) Mesotrophic: These lakes have moderate amount of nutrients and biologically productive.
   (iii) Eutrophic: These are very rich in nutrients and biologically very productive. Algal bloom is abundant.

2. **Fertilisers:** The excessive use of fertilisers are sprayed over crops and passed into water bodies due to surface run off or rain water. These will cause enrichment of water bodies called **eutrophication** due to presence of nitrates and phosphates. The biochemical oxygen demand (**BOD**) is increased and hence amount of dissolved oxygen is decreased. The aquatic animals will die due to asphyxiation. The excess use of fertiliser esp. urea has increased the level of nitrates in soil. These nitrates go to leaves and fruits and then to human beings. In alimentary canal the bacteria convert nitrates to toxic nitrites which combine with haemoglobin of blood to form methaemoglobin due to which O_2 transport is reduced. It causes methaemoglobinaemia or cyanosis (skin becomes blue coloured and leads to respiratory problems).

Types of water	BOD (mg/l or ppm)
Clean water	Less than 1
Low polluted water	1-2
Moderately polluted water	2-8
Heavily polluted water	More than 8

3. **Organic pesticides**: These are sprayed over crops and passed into water bodies due to surface run off.

DDT is a persistent, broad spectrum, fat soluble and non biodegradable pesticide. It persists for a long time in environment.

It causes **biomagnification** (accumulation of chemicals in organisms in increasingly higher concentration at successive trophic levels). The residual DDT enters the food chains and is transferred to different organisms. Since it is non-degradable hence it is stored in tissues of body of animals and it is increased day by day (cumulative effect). The concentration builds up in at each step of food chain and maximum concentration is found in top consumer. It becomes lethal. In water body, higher amount of pesticides disturbs calcium metabolism of bird (Sea gull or bald eagle) resulting in thinning of shell of egg. When bird tries to incubate the egg, the weight of bird breaks the shell of egg hence embryo gets killed. The population of such birds has declined.

Silent spring is a novel written by Rachel Carlson (1962) mentioning the effect of DDT on birds.

Water (DDT= 0.003 parts per billion)

↓

Phytoplanktons (DDT= 0.003 parts per million)

↓

Zooplankton (DDT= 0.04 ppm)

↓

Small fish (DDT= 0.5 ppm)

↓

Large fish (DDT= 2 ppm)

↓

Fish eating birds such as Sea gull, Bald Eagle (DDT=25 ppm)

4. **Industrial wastes**: The industrial wastes produced from chemical and fertiliser factories, petroleum refineries, paper mills, tanneries etc. are discharged into water bodies due to lack of proper waste management. These are non-usable toxic chemicals and heavy metals. These change the colour of water, cause biomagnifications and pose serious hazards to human beings. These include:

   (a) Hg: It is transformed in methylated mercury in the bottom sediment of water and present in the form of $(CH_3)_2Hg$. It is non biodegradable, highly persistent and causes biomagnification. It is absorbed by aquatic plants once it is dissolved and enters food chain through herbivorous fish. Finally it reaches human body where it inhibits chromosomal disjunction causing minamata disease (numbness of lip, tongue, limb, haemolysis, convulsion, impairment of hearing and vision) as reported in Japan in 1952.

   (b) Cu: It causes hypertension.

   (c) Pb: It damages liver, kidney and brain. It inhibits enzymes.

(d) **Cd**: It causes itai-itai or ouch-ouch disease (liver and lung damage; and growth retardation) as reported in Japan in 1947.

(e) **Co**: It causes diarrhoea and bone defects.

(f) **Cr**: It causes gastro-intestinal cancer.

(g) **As** (Arsenic): It causes black foot disease (arsenicosis).

(h) **Asbestos**: It causes liver and lung cancer and mesothelioma (respiratory trouble).

(i) **Fluoride**: It causes skeletal fluorosis (knock-knee syndrome in which pain in bones and joints and there is outward bending of legs from knees).

5. **Thermal pollution**: Its main sources are industries, thermal power plants, oil refineries etc. Water is used as coolant in thermal power plants and hot water flowing out of this unit is released to water body. The sudden rise in temperature causes reduction in dissolved oxygen hence killing fish and other aquatic organisms. It also increases the respiration of algae.
6. **Siltation**: It is sediment deposited by running water .It increases turbidity of water hence aquatic animals esp. fish will not survive.
7. **Microbial growth**: The luxuriant growth of aerobes and anaerobes occurs when hazardous substances enter into surface water and later reach water bodies. These use more O_2 in water. If O_2 is depleted, aerobes will die but anaerobes will grow more to produce more harmful toxins such as NH_3 and sulphides. These microbes include *Giardia, Salmonella, Burkholderia, Cryptosporidium* etc.

Water pollution indicators:

(a) *E. coli*, *Tubifex* (an annelidian worm) and Chironomous larvae: Their number will increase due to pollution.

(b) *Daphnia* (a crustacean): Its number will decrease due to pollution.

General harmful effects:

(a) The aquatic organisms will die.

(b) The food chain is disrupted.

(c) The water borne diseases (jaundice, dysentry, cholera, diarrhoea, typhoid) may erupt.

(d) The ecosystem is destructed.

Control measures:

(a) The fertilisers and pesticides should be used in controlled manner.

(b) Sewage should be treated by undergoing primary, secondary and tertiary treatment before discharging into water bodies.

(c) Cooling of hot waste plant should be done.

(d) Precipitation of colloidal particles and heavy metals should be done.

(e) Neutralisation of acids and alkalis must be undertaken.

(f) Ganga action plan was launched in 1985 and it covers 27 cities to prevent untreated sewage and industrial effluents into river Ganga.

Radioactive pollution: It occurs due to release of radioactive substances or high energy particles into air, water or soil. It occurs by accident or design. It is also known as radioactive contamination when improperly disposed nuclear wastes enter into water bodies and also carried away by wind and spread to wide area and later settle on soil. It is related to all major life supporting systems-air, water and soil.

Sources:

(a) **Natural**: It comprises cosmic rays which reach surface of earth from space and terrestrial radiations from radio-nuclides present in earth crust such as $Radium^{224}$, $Uranium^{235}$, $Thorium^{232}$ etc.

(b) **Artificial**: It comprises mining and refining of Thorium, Plutonium; preparation of radioisotopes and production and explosion of nuclear weapons, nuclear power plants and fuel. The nuclear test and explosion occurring in Hiroshima and Nagasaki of Japan; spent fuel of atomic reactor and accidental leakages such as Three Mile Island, Pennsylvania (USA) on 28th March, 1979 and Chernobyl atomic reactor (Ukraine) on 26th April, 1986 produced very large neutron flux conditions which make other materials in the surrounding radioactive.

Degree of pollution:

1. Low level: Small amount of radioactivity occurs when radioactive wastes are produced by testing lab, irradiation centre, radiotherapy, inducing mutation and studying metabolic pathways. It is not harmful.
2. Intermediate level: Moderate amount of radioactivity with liberation of heat occurs. These can cause hair loss, subcutaneous bleeding.
3. High level: Large amount of radioactivity with liberation of large amount of heat occurs. These wastes kill plants, animals and humans.

Effects:

(a) Sr^{90} is the most dangerous radioactive pollutant. It causes bone cancer, leukaemia and degeneration of tissue.

(b) $Iodine^{131}$ damages thyroid gland, bone marrow, WBC, lung cancer, lymph node and skin cancer

(c) Cs^{137} accumulates inside body muscles. It replaces sodium and potassium in many metabolic reactions.

(d) It destroys cells in living organisms.

(e) It has burning effect.

(f) It can leave soil infertile.

Control measures:

(a) Protective garments should be used by workers.

(b) Proper and careful disposal of radioactive wastes should be done.

(c) Nuclear waste should be disposed in 500 m deep below earth's surface.

(d) Alternative source of energy should be developed.

Soil pollution: It is alteration in soil caused by addition or removal of substances which decrease the productivity. It is also called land pollution. It is of three types:

1. Positive soil pollution: It is due to the addition of undesirable substances (agrochemicals, industrial wastes, pesticides, discarded food, carcasses) hence quality of soil is decreased.

2. Negative soil pollution: It is due to the erosion or overuse of soil hence quality of soil is decreased.
3. Third pollution (landscape pollution): It is due to the direct dumping of wastes hence fertile land is converted to barren area.

Control measures:

(a) The pneumatic pipes should be installed for collection and disposal of solid wastes.

(b) The wastes should be incinerated.

Green House Effect

It is warming of earth's surface due to increase in green house gases esp. CO_2. It is natural phenomenon. It was first described by Fourier (1827).

Mechanism: The earth receives energy from the sun in the form of solar radiation (UV, visible and IR radiations) and most of which passes through atmosphere without being absorbed and reflected. Out of total amount of solar energy available at the top of atmosphere, about 20% is reflected back to space by the clouds, 19% of incoming solar radiation (< 0.3μm) is absorbed by atmosphere (stratosphere by O_2 and O_3) and cloud; 6% is scattered from atmosphere and rest 4% is reflected by earth's surface. The large fraction (51%) of the remaining energy is absorbed at the earth's surface (colder). Because it is warm, the earth's surface reradiates far infra red (IR) thermal radiations that have wavelength much longer (>4μm) than the wavelength absorbed. Most of this thermal radiation is absorbed by the gases of atmosphere such as water vapour (H_2O), CO_2, N_2O and CH_4. This absorption heats the atmosphere which in return radiates energy back to earth. The green house gases act as thermal blanket around earth raising the earth's temperature. This effect is called Green house effect. The Greenhouse gases has started rising resulting in enhanced Greenhouse effect that is increasing the mean global temperature (15°C). It is Global warming.

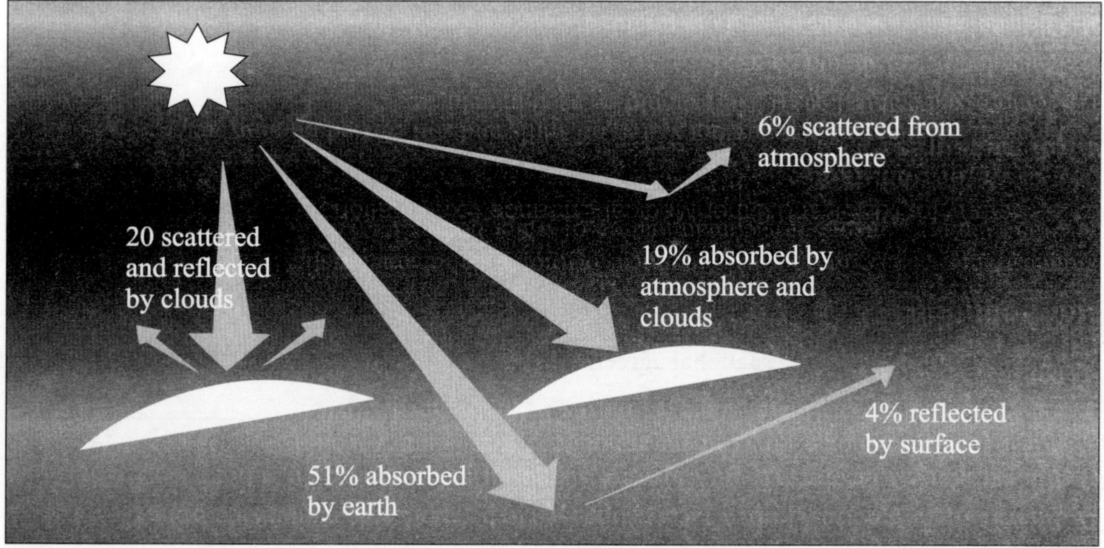

Intergovernmental Panel on Climatic Change (IPCC) assesses abundance of Greenhouse gases and their impact on global environment.

Contribution of Gases to Green house effect:

(a) Water vapour:

(b) CO_2 (60%): It is produced due to large scale deforestation, change in land use and large scale combustion of fossil fuels.

(c) CH_4 (20%): It is produced due to incomplete combustion of biomass, incomplete decomposition by methanogenesis, freshwater wetland and flooded farm land.

(d) Chlorofluorocarbons (14%): These are colourless, nontoxic, non-inflammable, man-made and chemically inert propellant. These are synthetic gas of carbon and halogen. It is used in aerosol cans, jet fuel, refrigerator, air-conditioner and fire extinguisher. It is leaked from the above sources. These are Chlorofluorocarbon (CFC-11), Hydrochlorofluorocarbon (HCFC) and Hydrofluorocarbon (HFC-23) etc. The potency of CFCs as green house gas is much higher than CO_2 because of their high life time and showing chain reaction.

(e) N_2O (6%): It is produced from combustion of fuel, breakdown of nitrogenous fertilisers in soil and burning of biomass.

Effects:

(a) Green house effect increases the mean air temperature near the earth's surface by 0.8 to 1.02°C to normal temperature (global warming).

(b) It increases the sea level due to melting of ice caps (mountain/polar) and thermal expansion of water.

(c) It will change the pattern of rainfall due to variation in convectional current.

(d) It will change the season and seasonal characters by causing drought in some areas and flood in other areas (El Nino effect).

(e) It will reduce the agricultural productivity due to increase in respiration and decrease in photosynthesis by which there is shortage of food.

Control measures:

(a) Forest cover should be increased for photosynthetic utilisation of CO_2.

(b) Use of nitrogenous fertilisers should be reduced to reduce emission of N_2O.

(c) Alternate source of energy such as solar energy and wind energy should be used.

(d) Vehicular traffic should be checked to minimise the use of fossil fuel hence emission of CO_2 is reduced.

(e) CFC free refrigerator and air-conditioner should be used.

(f) Population growth should be checked.

Kyoto Protocol: It is an international agreement linked to UN Framework Convention on Climatic Change (UNFCC).

Held in	Kyoto (Japan)
Adopted on	11th December,1997
Effective from	16th February, 2005
Signatories	83 nations
Partner	192 nations
Expiry of first commitment	31st December, 2012
Objective:	To fight global warming by reducing greenhouse gases by 5.2 % by 2010.

Recent Conference of the Parties (COP) held to negotiate the Kyoto protocol

COP-21	Paris (France)	From 30 November to 12 December, 2015
COP-22	Marrakech (Morocco)	From 7-18 November, 2016
COP-23	Bonn (Germany)	From 6-17 November, 2017
COP-24	Katowice (Poland)	From 3-14 December, 2018
COP-25	—	From 11-22 November, 2019

Ozone layer and depletion

Ozone layer was discovered by French scientist, Charles Fabry. About 90% of ozone is found in stratosphere where its amount is 2-8 ppm and 0.3 ppm in troposphere. Its amount is measured by Dobson meter from the ground in Dobson unit. Its average amount throughout world is 300 DU. Its highest level is found in spring and lowest in autumn in northern hemisphere. It is maximum in tropics and minimum in poles. Its thickness varies seasonally and geographically.

Ozone is created by UV light which strikes oxygen molecule (O_2) splitting into individual oxygen atom. The atomic oxygen then combines with unbroken O_2 to create O_3.

$$O_2 \text{——} \text{UV (242 nm)} \longrightarrow O + O$$

$$O + O_2 \longrightarrow O_3$$

Ozone layer absorbs 97-99% of harmful UV radiations of sun.

Types of UV radiations: These are of three types based on their wavelength.

1. UV-A (315-400nm): It is least harmful. It is transparent to ozone hence so most of it reaches the earth surface.
2. UV-B (315-280nm): It is harmful and causes sunburn, skin cancer, cataract, snow blindness (inflammation of cornea), suppression of immune system and mutation. Mostly it is absorbed by ozone.
3. UV-C (280-100 nm): It is very harmful to all living organisms hence it is absorbed by ozone in atmosphere.

Ozone depletion widely occurs in stratosphere. It can be depleted by free radical catalysts such as NO, N_2O, atomic Cl, atomic Br, hydroflurocarbon (HFC), hydrochlorofluorocarbon (HCFC), CCl_4, methyl chloroform. Each Chlorine and bromine radical breaks down 100,000 ozone molecules into oxygen. Bromine atom is believed to be 40 times destructive than chlorine atom.

The thinning of ozone layer is called ozone hole. It was discovered over Antarctica in 1985 and over Arctica in 1990.

September 16 is an international day for the preservation of ozone layer.

Control measures:

(a) The number of vehicles should be less.

(b) The use of nitrous oxide should be limited.

Montreal protocol

Held in	Montreal (Canada)
Signed on	16th September,1987
Effective from	1st Jan,1989
Initial signatories	175 nations
Objective	To phase out ozone depleting substances by year 2000.

Helsinki declaration: Around 82 countries modified Montreal protocol on May, 1989 to phase out production and consumption of CFCs by 2000.

Recent Meeting of Parties (MOP) to negotiate the Montreal protocol

MOP-27	Dubai (UAE)	From 1-5 November, 2015
MOP-28	Kigali (Rwanda)	From 10-14 October, 2016
MOP-29	Montreal (Canada)	From 20-24 November, 2017
MOP-30	Ecuador	From 5-9 November, 2018

India Government's initiative to control pollution:

Water (Prevention &control of Pollution) Act	1974
Air (Prevention & control of Pollution) Act. It was amended in 1987 to include noise as air pollutant.	1981
Environment (Protection) Act	1986
Motor vehicles Act	1988

Agrochemicals

These are substances which help manage an agricultural system or community of organisms in a farming area. These are chemical products used in agricultural practices to control crop harming organisms. These are usually referred as pesticides.

Types of agrochemicals

Insecticides	These kill harmful insect pests that damage field crops.
Fungicides	These kill harmful fungi that damage field crops.
Herbicides	These kill weeds.
Rodenticides	These kill rats.
Nematicides	These kill nematodes.

Advantages:

(a) These are used to increase food production.
(b) These improve soil condition.
(c) These protect crops from pests.
(d) These decrease the cost of food.
(e) These allow consumers to consume high quality produce.

Disadvantages:

(a) Most of the agrochemicals are highly toxic. To humans and animals.
(b) The repeated use can cause an imbalance in the pH of the soil.
(c) The excessive use of fertilisers will cause contamination of water bodies (eutrophication).
(d) The use of insecticides such as DDT can lead to biomagnifications thus affecting the aquatic animals.

Solid waste management

Solid waste is useless, unwanted and abandoned garbage produced from animals and human beings. These wastes are in the form of solid, liquid or gas and each have different methods of disposal. Its main objective is to minimise the harmful effects of wastes on human health and environment. The waste management differs among countries (developed and developing), regions (urban or rural) and sectors (residential or industrial).

Types of wastes:

Garbage	These are decomposable wastes from food
Rubbish	These are non-decomposable wastes or dry materials (either combustible such as paper, wood, cloth or non-combustible such as metal, glass, ceramics).
Ashes	These are residue that remains after burning.
Sewage	These are waste matter carried in drains.
Dead animals	These are decomposed bodies of animals.
Industrial wastes	These are chemicals, paints and sand
Agricultural wastes	These are animal manures and crop residues
Large wastes	These are construction debris and demolished trees

Methods of solid waste disposal: It is collection, transport, treatment and disposal of solid wastes. Its strategy includes

(a) Reduction of the generation of wastes
(b) Reuse of wastes
(c) Recycling of wastes includes composting (organic part of reuse is allowed to decompose)
(d) Recovery of materials in the form of heat, fuel or electricity
(e) Disposal in sanitary land-fills or through incineration (burning in an enclosed container).

Classification of wastes:

1. Hazardous wastes: These cause immediate harm to exposed individuals or environment. These are
   (a) Toxic: Pesticides, heavy metals, arsenic, asbestos
   (b) Ignitable: Organic solvents, oils and paint wastes
   (c) Corrosive: Acids, alkali, cleaning agents, discarded battery
   (d) Reactive: Fire crackers
   (e) Infectious: wastes of hospital containing harmful microbes
2. Non-hazardous wastes: These are called refuse. The refuse includes garbage and rubbish.

Advantages:

(a) It produces several useful products.
(b) It reduces the adverse effects on health of human beings.
(c) It increases the quality of air and water
(d) It reduces the emission of green house gases.
(e) It reduces the odour released by rotten garbage.

Control measures:

(a) Indiscriminate dumping of wastes should be stopped.
(b) Burning of large amount of wastes in open should be stopped.
(c) Wastes should be poured in pit.

Afforestation

Forests occupy central position in nature. They restore ecological balance of all ecosystems, maintain biological diversity, act as catchments for soil and water conservation and prevent floods and safeguard future of tribal people.

Forest conservation: It is the practice of planning and maintaining forest areas for the benefit and sustainability of future generations. The forest cover of geographical area includes all lands having tree density of 10% and more and has a minimum area of one hectare. India is tenth most forest rich country in the world. Forests have been reported to cover 21.34% of land which are distributed as such

1. Peninsular hill and plateau region=57%
2. Himalayan region=18%
3. Western ghat and coastal region=10%
4. Eastern ghat=10%
5. Northern plain=5%.

Constitution of Indian forests: There are six types of forests present in India.

1. Moist Tropical forests
2. Dry Tropical forests
3. Montane sub-tropical forest
4. Montane temperate forest
5. Sub-alpine forest
6. Alpine forest

The forest cover is 33% for urban areas and 67% for rural areas as envisaged by India's national forest policy.

Legislation by Govt. of India:

(a) The Forest (conservation) act came into force in 1980.
(b) Ministry of environment and forests was created in 1985.
(c) National forest policy was made in 1988.
(d) National forest commission was set up in 2003.
(e) Forest Rights bill became a law in 2007.

Reasons of deforestation

(a) Trees were cut down for timber, firewood etc.

(b) Forest land was converted to agricultural area.

(c) Jhumming (Shifting cultivation) was practised in which trees are cut down and set on fire and finally crops are grown on resulting ashes.

Impact of deforestation:

(a) Soil erosion and loss of soil productivity

(b) Increase in frequency of floods and drought

(c) Imbalance in ecosystem

(d) Loss of wild life

(e) Shortage of forest products

The Govt. of India has launched the following programmes:

Conservation strategies by Govt. of India:

1. **Social forestry programme:** It was started in 1976 to produce firewood, fodder, and small timber for the use of rural community to relieve pressure on existing forests. The programme includes raising, planting and protecting trees with multiple uses. The species to be planted should have the following features: (i) fast growth, (ii) stress resistance, (iii) less palatable to cattle and other animals, (iv) early propagable, (v) high caloric value, (vi) high yield of biomass, and (vii) disease/ pest resistant.

   Some of important plants are: *Acacia nilotica, Albizzia lebbek, Anthocephalus chinensis, Azadirachta indica, Bauhinia variegata, Butea monosperma, Cassia fistula, Dalbergia sissoo, Eucalyptus globules, Ficus glomerata, Madhuca indica, Morus alba, Populus ciliata, Terminalia arjuna, Toona ciliata, Salix alba.*
2. **Joint forestry programme**: It was started in 1980 to protect and manage the forests with the participation of local communities.
3. **Agroforestry programme**: It aims at reviving ancient land use practice where same land is used for farming, forestry and animal husbandry.
4. **Urban forestry programme**: It aims at planting trees for aesthetic purpose in urban areas.

Movements against deforestation:

1. **Bishnoi movement:** It was started 400 years ago by a sage known as Sombaji. Later it was Amrita Devi who propagated the movement in 1731. Government of India recently instituted the Amrita Devi Bishnoi Wildlife protection Award in her honour.
2. **Chipko movement**: It was started in 1972 in Chamoli district of Tehri Garhwal areas of Uttaranchal by Sundarlal Bahuguna and Chandi Prasad Bhatt.
3. **Appiko movement**: It was started in 1983 in Uttar Kannada areas of Karnataka by Pandurag Hegde.

SHORT ANSWER TYPE QUESTIONS

1. Distinguish between primary pollutants and secondary pollutants.
2. Distinguish between biodegradable pollutants and non-biodegradable pollutants.
3. Distinguish between classical smog and photochemical smog.
4. Write notes on acid rain.
5. Mention some of the air pollution indicators.
6. Write notes on Euro norms.
7. What are the types of water pollutants?
8. Distinguish between eutrophication and biomagnification.
9. Mention some of the water pollution indicators.
10. What are the harmful effects of radioactive pollution?
11. Mention the different types of soil pollution.
12. Write notes on Kyoto protocol.
13. What are the types of UV radiations?
14. Write notes on Montreal protocol.
15. Discuss the initiatives of Government of India to control pollution.
16. Write notes on agrochemicals.
17. Discuss the types of solid wastes.
18. Discuss the various programmes for afforestation.
19. What are the impacts of deforestation?
20. Write notes on Bishnoi movement, Chipko movement and Appiko movement.

LONG ANSWER TYPE QUESTIONS

1. Describe the sources, effects and control measures of air pollution.
2. Describe the sources, effects and control measures of water pollution.
3. Discuss the mechanism, effects and control measures of Green house effect.

MULTIPLE CHOICE QUESTIONS

1. Which method is used to remove particulate matters present in exhaust of thermal power plants?
   (a) Electrostatic precipitator (b) Gravitational method
   (c) Scrubber (d) Absorber
2. Which of the following metal is used as catalyst in catalytic converter?
   (a) Platinum (b) Palladium (c) Rhodium (d) All of above
3. Euro norms II stipulate to control
   (a) Sulphur at 350 ppm in diesel and 150 ppm in petrol
   (b) Sulphur at 150 ppm in diesel and 350 ppm in petrol
   (c) Sulphur at 250 ppm in diesel and 150 ppm in petrol
   (d) Sulphur at 150 ppm in diesel and 250 ppm in petrol
4. Sound becomes a hazardous noise pollutant if its level exceeds
   (a) 80dB (b) 100dB (c) 120dB (d) 150dB
5. Maximum permissible noise as per noise pollution rule is
   (a) 65dB (b) 75dB (c) 90dB (d) 100dB
6. Green muffler is related to pollution of
   (a) Air (b) Noise (c) Water (d) Radioactive
7. High level of radioactive waste can be managed by
   (a) Open dumping (b) Composting
   (c) Incineration (d) Dumping in sealed container
8. Pollutant emitted by paddy field is
   (a) CO (b) CO_2 (c) CH_4 (d) $(NO)_x$
9. The exhaust of motor vehicle does not produce
   (a) CO (b) SO_2 (c) Hydrocarbons (d) Fly ash
10. Taj Mahal is affected by
    (a) $(NO)_x$ (b) SO_2 (c) CO (d) O_3
11. Sewage water can be purified for recycling with the action of
    (a) Aquatic plants (b) Microbes (c) Antibiotics (d) Fish
12. Spraying DDT on crops causes pollution of
    (a) Air (b) Water (c) Soil (d) All of above
13. Methyl Isocyanate of Bhopal gas tragedy was
    (a) Carbamate (b) Organochlorine (c) Organophosphate (d) Drug
14. The blue baby syndrome results from
    (a) Excess of TDS (b) Excess of chloride (d) Excess of DO (d) Methemoglobin
15. Which of the following is absent in polluted water?
    (a) Blue green algae (b) Hydrilla (c) Water hyacinth (d) Larva of stone fly
16. The most common indicator organism of polluted water is
    (a) *E.coli* (b) *Entamoeba* (c) *Vibrio* (d) *Clostridium*
17. Relative biological effectiveness is concerned with
    (a) Pollution (b) Radiation (c) Low temperature (d) High temperature

18. Phosphate pollution is mainly caused by
    (a) Sewage and phosphate rocks (b) Sewage and fertilisers
    (c) Sewage and insecticides (d) Fertilisers and insecticides
19. A sewage treatment process in which a part of decomposer bacteria present in the waste is recycled into the beginning of the process, is called
    (a) Primary treatment (b) Activated sludge treatment
    (c) Physical treatment (d) Tertiary treatment
20. The effect of pollution is observed firstly on
    (a) Microbes (b) Vegetation (c) Herbivores (d) Food crops
21. Photochemical smog does not contain
    (a) CO_2 (b) $(NO)_x$ (c) O3 (d) PAN
22. According to CPCB, thc size of particulate matter most harmful is
    (a) <1 μm (b) 2.5 μm
    (c) Between 2.5 and 5μm (d) >5μm
23. The Bhopal gas tragedy took place because MIC reacted with
    (a) CO_2 (b) Water (c) NH_3 (d) DDT
24. The two gases making highest relative contribution to the green house gases are
    (a) CO_2 and N_2O (b) CO_2 and CH_4 (c) CH_4 and N_2O (d) CO_2 and SO_2
25. Chlorofluorocarbons (CFCs) are believed to be associated with cancer because
    (a) They cause mutation (b) They inhibit DNA replication
    (c) They stop protein synthesis (d) They destroy ozone layer
26. Which one of the following is not non-biodegradable pollutant?
    (a) DDT (b) Heavy metals (c) Plastics (d) Sewage
27. Which one of the following pollutants is carcinogenic as well as mutagenic?
    (a) CO (b) SO_2 (c) (NO)x (d) Smog
28. Which of the following pollutants is secondary pollutant?
    (a) PAN and O_3 (b) PAN and NO_2
    (c) O_3 and NO_2 (d) Particulate matters
29. Which of the following is incorrectly matched?
    (a) Hg- Minamata disease (b) Pb- Plumbism
    (c) MIC- Bhopal gas tragedy (d) NH_3- Emphysema
30. Lichens and mosses are indicators of
    (a) CO pollution (b) (NO)x pollution
    (c) SO_2 pollution (d) Particulate matter pollution
31. Catalytic converter is fitted in vehicles because
    (a) It converts CO into CO_2
    (b) It converts $(NO)_x$ into N_2
    (c) It converts unburnt hydrocarbon to CO_2 and H_2O
    (d) All of the above
32. Which of the emission norms will be effective from 1st April, 2020?
    (a) Bharat stage III (b) Bharat stage IV (c) Bharat stage V (d) Bharat stage VI

33. Eutrophication is caused by
    (a) Sewage and pesticides (b) Pesticides and fertilisers
    (c) Sewage and fertilisers (d) Sewage and industrial wastes
34. Which of the following water pollutants cause biomagnification?
    (a) Sewage (b) Fertilisers (c) Organic pesticides (d) Industrial wastes
35. Which of the following is incorrectly matched?
    (a) Cadmium- Itai-itai disease (b) Arsenic- Black foot disease
    (c) Mercury- Minamata disease (d) Asbestos- Knock knee syndrome
36. Strontium^{90} causes
    (a) Bone cancer (b) Congenital defects (c) Lung acncer (d) Skin cancer
37. Which of the following is correct?
    (a) Kyoto protocol- To reduce greenhouse gases
    (b) Montreal protocol- To phase out ozone depleting substances
    (c) Helsinki declaration- To phase out production and consumption of CFCs
    (d) All of the above
38. Noise was included as air pollutant in the year
    (a) 1981 (b) 1984 (c) 1986 (d) 1987
39. Which of the following UV radiations is most harmful?
    (a) UV-A (b) UV-B (c) UV-C (d) UV-D
40. Chipko movement was started against deforestation by
    (a) Pandurag Hegde (b) Hemvati Nandan Bahuguna
    (c) Sundarlal Bahuguna (d) Sombaji
41. Pollution rate increases in winter due to
    (a) SPM (b) SO_2 and NO_2 (c) Burning of fuel (d) Low temperature
42. The concentration of which gas can be reduced by preventing forest depletion.
    (a) O_2 (b) CH_4 (c) CO_2 (d) N_2O
43. What is approximate increase in global surface temperature in the past century?
    (a) 1^0F (0.5°C) (b) 1.5°F(0.8^{0}C) (c) 2^0F(1.1^{0}C) (d) 5^0F(2.8^{0}C)
44. What has been the approximate rise in sea level over the past century?
    (a) 2-4 inches (b) 6-8 inches (c) 10-12 inches (d) 14-16 inches
45. Which of the following is most capable of slowing global warming?
    (a) Decomposition (b) Resp[iration (c) Photosynthesis (d) Chemosynthesis
46. The ozone layer absorbs
    (a) UV-A (b) UV-B (c) UV-A and UV-B (d) UV-B and UV-C
47. Global warming is due to
    (a) Absorption of UV by ozone (b) Absorption of UV by CO_2
    (c) Absorption of IR by ozone (d) Absorption of IR by CO_2
48. The maximum water pollution is caused by
    (a) Detergents (b) Pesticides (c) Industrial wastes (d) Ammonia
49. Bhopal gas tragedy took place in 1984 because MIC reacted with
    (a) Water (b) Ammonia (c) Carbon dioxide (d) DDT

50. The biggest impact is made on the environment by
(a) Human interference (b) Predation (c) Competition (d) Migration
51. CPCB is abbreviated form of
(a) Central Pollution Control Bureau (b) Central Pollution Control Board
(c) Central Pollutants Control Board (d) Control of Pollutants and Chemicals Board
52. The major contributor of acid rain is
(a) Sulphuric acid (b) Nitric acid (c) Hydrochloric acid (d) Nitrous acid
53. Black foot disease is caused by
(a) Mercury (b) Strontium (c) Lead (d) Arsenic
54. Plumbism is caused by
(a) Mercury (b) Strontium (c) Lead (d) Arsenic
55. El-Nino effect is concerned with
(a) Increase in Ozone depleting substance (b) Increase in green house gases
(c) Melting of polar ice caps (d) Changes in seasonal pattern
56. Montreal protocol was signed on
(a) 5th June (b) 11th July (c) 16th September (d) 29th December
57. The international day for the preservation of ozone layer is observed on
(a) 5th june (b) 11th july (c) 16th September (d) 29th December
58. The agrochemicals include
(a) Fertilisers (b) Biofertilisers (c) Humus (d) Insecticides
59. In India, maximum forest is located in
(a) Western ghat (b) Eastern ghat
(c) Himalayan region (d) Peninsular and plateau region
60. Joint forestry programme was started in
(a) 1976 (b) 1980 (c) 1986 (d) 1990

ANSWERS

1.a	2.d	3.a	4.d	5.b	6.b	7.d	8.c	9.d	10.b
11.b	12.d	13.a	14.d	15.d	16.a	17.b	18.b	19.b	20.b
21.a	22.b	23.b	24.b	25.d	26.d	27.c	28.a	29.d	30.c
31.d	32.d	33.c	34.c	35.d	36.a	37.d	38.d	39.b	40.c
41.a	42.c	43.d	44.a	45.c	46.d	47.d	48.a	49.b	50.a
51.b	52.a	53.d	54.d	55.d	56.c	57.c	58.d	59.d	60.b